Advances in Biocompatible and Biodegradable Polymers-Volume II

Advances in Biocompatible and Biodegradable Polymers-Volume II

Editors

José Miguel Ferri
Vicent Fombuena Borràs
Miguel Fernando Aldás Carrasco

MDPI • Basel • Beijing • Wuhan • Barcelona • Belgrade • Manchester • Tokyo • Cluj • Tianjin

Editors
José Miguel Ferri
Department of Nuclear and Chemical Engineering
Universitat Politècnica de València
Alcoy
Spain

Vicent Fombuena Borràs
Department of Nuclear and Chemical Engineering
Universitat Politècnica de València
Alcoy
Spain

Miguel Fernando Aldás Carrasco
Center of Polymers Applied Research (CIAP)
Escuela Politécnica Nacional
Quito
Ecuador

Editorial Office
MDPI
St. Alban-Anlage 66
4052 Basel, Switzerland

This is a reprint of articles from the Special Issue published online in the open access journal *Polymers* (ISSN 2073-4360) (available at: https://www.mdpi.com/journal/polymers/special_issues/biocompat_biodegrad_polym).

For citation purposes, cite each article independently as indicated on the article page online and as indicated below:

LastName, A.A.; LastName, B.B.; LastName, C.C. Article Title. *Journal Name* **Year**, *Volume Number*, Page Range.

Volume 2
ISBN 978-3-0365-7406-6 (Hbk)
ISBN 978-3-0365-7407-3 (PDF)

Volume 1-2
ISBN 978-3-0365-7402-8 (Hbk)
ISBN 978-3-0365-7403-5 (PDF)

Contents

About the Editors

José Miguel Ferri

Dr. José Miguel Ferri (h-index 14—Scopus) works at the Department of Nuclear and Chemical Engineering of the Escuela Politécnica Superior de Alcoy (EPSA) of the Universitat Politècnica de València, (UPV) Alcoy, Spain. He holds degrees in Technical Engineering, Industrial Mechanical Engineering, and Materials Engineering. He joined the Technological Institute of Materials (ITM) in 2014 as a research technician and obtained an international PhD in engineering and industrial production in 2017. His field of research concerns the development of new metallic materials (Mg foams for the medical sector), polymeric materials (the improvement of the thermal, mechanical, and processing properties of biodegradable polymers), and composite materials (metallic matrices for the electronics and aeronautics sectors and polymeric matrices for the aeronautics and construction sectors).

Vicent Fombuena Borràs

Dr. Vicent Fombuena Borràs (h-index 18—Scopus) works at the Department of Nuclear and Chemical Engineering of the Escuela Politécnica Superior de Alcoy (EPSA) of the Universitat Politècnica de València, (UPV) Alcoy, Spain. He holds degrees in Technical Engineering, Industrial Chemical Engineering, and Materials Engineering. He joined the Technological Institute of Materials (ITM) in 2008 as a research technician and obtained an international PhD in engineering and industrial production in 2012. His research work concerns the implementation of a circular economic model in the polymer industry. He has focused his efforts on obtaining multiple active compounds from seeds and agroforestry residues for application in thermoplastic and thermosetting polymers. He has extensive experience in developing techniques for the chemical, thermal, and mechanical characterization of materials.

Miguel Fernando Aldás Carrasco

Miguel Fernando Aldás Carrasco (h-index 10—Scopus) has been a lecturer and researcher at the Escuela Politécnica Nacional (Quito, Ecuador) since 2008. He actively collaborates with the Centro de Investigaciones Aplicadas a Polímeros (Center of Polymers Applied Research) (CIAP) and the Faculty of Chemical Engineering and Agroindustry. He holds degrees in Chemical Engineering from Escuela Politécnica Nacional (Quito, Ecuador), a Master in Innovative Materials degree from Université Claude Bernard Lyon 1 (Lyon, France), and a Doctorate in engineering from Universitat Politècnica de València (Valencia, Spain). He has worked under the direction of Prof. Juan López Martínez and Dr. Marina P. Arrieta, constituting a portion of the team of researchers at the Materials Technology Institute (ITM-UPV) at the Alcoy campus within the Development of Materials for Sustainable Structures (DEMES) group. His field of research concerns the use of natural additives for biopolymers and biodegradable materials, the development of thermoplastic starch-based materials, the study of the degradation of synthetic and biodegradable plastic materials, and the technology behind processing techniques.

Preface to "Advances in Biocompatible and Biodegradable Polymers-Volume II"

Among the strategies for reducing the negative effects on the environment effected by the uncontrolled consumption and low potential for the recovery of conventional plastics, the synthesis of new biodegradable and recyclable plastics represents one of the most promising methods for minimizing the negative effects of conventional non-biodegradable plastics. The spectrum of existing biodegradable materials is still very narrow; thus, to achieve greater applicability, research is being carried out on biodegradable polymer mixtures, the synthesis of new polymers, and the incorporation of new stabilizers for thermal degradation, alongside the use of other additives such as antibacterials or new and more sustainable plasticizers. Some studies analyze direct applications, such as shape memory foams, new cartilage implants, drug release, etc.

The reader can find several studies on the degradation of biodegradable polymers under composting conditions; however, novel bacteria that degrade polymers considered non-biodegradable in other, unusual conditions (such as conditions of high salinity) are also presented.

José Miguel Ferri, Vicent Fombuena Borràs, and Miguel Fernando Aldás Carrasco
Editors

polymers

MDPI

Article

Major Factors Influencing the Size Distribution Analysis of Cellulose Nanocrystals Imaged in Transmission Electron Microscopy

Hui Qian

Nanotechnology Research Center, National Research Council of Canada, 11421 Saskatchewan Drive, Edmonton, AB T6G 2M9, Canada; Hui.Qian@nrc-cnrc.gc.ca

Abstract: Size distributions of cellulose nanocrystals (CNCs), extracted from softwood pulp via strong sulfuric acid hydrolysis, exhibit large variability when analyzed from transmission electron microscopy (TEM) images. In this article, the causes of this variability are studied and discussed. In order to obtain results comparable with those reported, a reference material of CNCs (CNCD-1) was used to evaluate size distribution. CNC TEM specimens were prepared as-stained and dried with a rapid-flushing staining method or hydrated and embedded in vitreous ice with the plunge-freezing method. Several sets of bright-field TEM (BF-TEM), annular dark-field scanning TEM (ADF-STEM) and cryogenic-TEM (cryo-TEM) images were acquired for size distribution analysis to study the contributing factors. The rapid-flushing staining method was found to be the most effective for contrast enhancement of CNCs, not only revealing the helical structure of single CNCs but also resolving the laterally jointed CNCs. During TEM specimen preparation, CNCs were fractionated on TEM grids driven by the coffee-ring effect, as observed from contrast variation of CNCs with a stain-depth gradient. From the edge to the center of the TEM grids, the width of CNCs increases, while the aspect ratio (length to width) decreases. This fractionated dispersion of CNCs suggests that images taken near the center of a droplet would give a larger mean width. In addition to particle fractionation driven by the coffee-ring effect, the arrangement and orientation of CNC particles on the substrate significantly affect the size measurement when CNC aggregation cannot be resolved in images. The coexistence of asymmetric cross-section CNC particles introduces a large variation in size measurement, as TEM images of CNCs are mixed projections of the width and height of particles. As a demonstration of how this contributes to inflated size measurement, twisted CNC particles, rectangular cross-section particles and end-to-end jointed CNCs were revealed in reconstructed three-dimensional (3D) micrographs by electron tomography (ET).

Keywords: nanoparticle size distribution; cellulose nanocrystals; negative staining; cryo-TEM ADF-STEM; electron tomography; coffee-ring effect; fractionation

Citation: Qian, H. Major Factors Influencing the Size Distribution Analysis of Cellulose Nanocrystals Imaged in Transmission Electron Microscopy. *Polymers* **2021**, *13*, 3318. https://doi.org/10.3390/polym13193318

Academic Editors: José Miguel Ferri, Vicent Fombuena Borràs and Miguel Fernando Aldás Carrasco

Received: 3 September 2021
Accepted: 26 September 2021
Published: 28 September 2021

1. Introduction

Cellulose nanocrystals (CNCs) are crystalline particles derived from naturally abundant plant or animal cellulose sources (wood, cotton, tunicate and bacteria, etc.) via strong acid hydrolysis. Depending on the origin of bulk cellulose and acid hydrolysis conditions, the crystallinity, shape and geometric dimensions of extracted CNCs exhibit great variety. In general, CNCs come in rod, ribbon or whisker-like shapes, with lengths ranging from tens of nanometers to several microns and widths ranging from 3 to 50 nm [1–4]. Compared with bulk cellulose with greater amorphous fractions, CNCs exhibit a higher aspect ratio (length-to-width) with the reactive surface of hydroxyl side groups, a greater axial elastic modulus and unique liquid crystalline properties. These properties, along with their natural abundance and biocompatibility, make CNCs attractive for many industrial applications, such as sustainable energy and electronics, biomedical engineering, water treatment, etc. [5–8].

CNCs are reminiscent of crystalline regions within elementary fibrils, and the cross-sections of terminating surfaces are either square, rectangle or hexagon [9]. However, CNCs derived following acid hydrolysis display considerable variability in crystallinity and morphology despite being from the same cellulosic source. The processing conditions during the exaction and post-drying of CNCs may result in the variability of CNC products [3,10]. On the other hand, the characterization of products, as a bridging procedure between production and utilization, may also provide inconsistent results due to varying characterization techniques. This variability makes further functionalization and application of CNCs inconsistent. Therefore, the development of consistent, reliable and accurate measurement protocols is critical to understand the various processes required for optimizing CNC production and utilization.

The structure and size distribution of CNCs, as significant physicochemical properties, have been characterized using light scattering and electron microscopy for several decades. Since the early 1950s, transmission electron microscopy (TEM) has been used to reveal the morphology of CNCs extracted from cotton, ramie and bacterial cellulose [1,11]. Samira et al. performed comprehensive characterization using TEM/cryogenic-TEM, atomic force microscopy (AFM) and small- and wide-angle X-ray scattering (SAXS and WAXS) to determine the shape and size distribution of CNCs from several cellulose sources by [4]. However, challenges in CNC size distribution analysis still remain, and a standard analysis protocol has yet to be established, although numerous efforts have been made in several aspects, including analysis methods and material treatment [12–14]. Johnston's team recently reported interlaboratory comparisons of CNC size distributions measured with both TEM and AFM. The results show great variability between the participating laboratories, and a skew-normal distribution method was proposed to accommodate the variability from different datasets [15,16]. Interestingly, the mean width (7.5 nm) measured in TEM is still approximately twice the mean height (3.4 nm) measured in AFM. This may indicate that not all CNC particles are composed of single crystallites, CNC particles are composed of single crystallites in different sizes or CNCs are laterally jointed bundles. More recently, both the width and height of CNCs were measured using AFM images with an internal calibration standard to evaluate the broadening effect of the AFM tip on width measurement. The results show 28% of CNCs with an approximately symmetric transverse cross-section (square) and the remainder with an asymmetric cross-section [17]. However, a validation procedure is required for width measurement in AFM, which may introduce another source of variability from gold nanoparticles used as the internal calibration standard.

Unlike AFM topography images, TEM images are the projection of three-dimensional (3D) objects along the electron-beam direction on a camera or detector as a two-dimensional (2D) image. Therefore, the arrangement and orientation of CNCs on substrate/TEM grids will affect the projected image, which is used for size measurement. As illustrated in the schematic diagram (Figure S1) in the Supplementary Materials, the width measured from a projected image is equal to the height when a rod-shaped particle has a symmetric cross-section. However, with an asymmetric cross-section, they may be different, depending on which side lies on the substrate. Therefore, the measured width distribution from TEM images is a mix of the width and height of 3D particles. In addition to the orientation on the substrate, CNC particle arrangement may also introduce variation in width and length measurements when individual particles cannot be resolved. Furthermore, sessile droplets on TEM grids may cause the CNC particles to be distributed fractionally during drying, leading to an inaccurate size distribution due to the missing representation of particles from certain regions of the TEM grids. In this article, several image sets with different imaging conditions were processed to evaluate the size distribution of CNCs. The major factors influencing the size distribution of CNCs will be discussed in depth based on observations and analysis.

2. Materials and Methods

2.1. TEM Specimen Preparation

The CNC material used is a National Research Council Canada-certified reference (NRCC), CNCD-1 [12,15–18]. The CNCs were extracted with sulfuric acid hydrolysis of softwood pulp, followed by neutralization and sodium exchange, purification and spray drying. The CNC aqueous solution was prepared by dispersing dry CNC powders in deionized water with a concentration of 0.02% (*w*/*v*). Vortex or sonication of the CNCs in aqueous solution was performed before preparing TEM specimens.

Continuous and perforated carbon-film-supported TEM grids were treated with glow discharge in air for 15 s at 15 mA current (PELCO easiGlow™, Ted Pella, Inc., Redding, CA, USA) before applying samples. CNC TEM specimens were prepared under four different conditions as described below. The first specimen (Sp1) was CNCs negatively stained with 2% uranyl acetate (UA) on TEM grids. Briefly, a 3 µL droplet of CNC aqueous solution was deposited on an ultra-thin continuous carbon-film TEM grid. Excess solution was blotted off with filter paper from the edge of the grid or the top of the droplet after 10 s incubation. The TEM grid was then tilted 30 to 45 degrees, rapidly washed with two running UA droplets and placed with a 2 µL droplet. After about 30 s of incubation, UA was blotted away with filter paper from the edge of the grids, and a thin layer of stain was left to air-dry on the TEM grids. The second specimen (Sp2) was pristine CNCs deposited on carbon TEM grids with the same preparation procedure as Sp1, except that the droplet incubation time was 60 s without UA staining. The third specimen (Sp3) was negatively stained CNCs on carbon TEM grids with gold nanoparticles as fiducial markers for electron tomography (ET). Colloid gold nanoparticles with a 5 nm diameter (0.1 mg/mL) were deposited on carbon-coated TEM grids and air-dried prior to CNC deposition. The same procedure for Sp1 was then followed to obtain negatively stained CNCs.

The fourth specimen (Sp4) was CNCs embedded in vitreous ice, and the cryo-TEM specimen was prepared using the plunge-freezing method [19]. Briefly, the plunge freezer Dewar/chamber (EMS-002 Rapid Immersion Freezer, Electron Microscopy Sciences, Hatfield, PA, USA) was precooled with liquid nitrogen for 30 min. Compressed ethane gas was liquefied in a Falcon tube surrounded with liquid nitrogen. The liquid ethane was then poured into a cryogen container immersed in a liquid nitrogen chamber with a temperature of −184 °C. The container was ready for use when the ethane was lightly frozen at the bottom and the walls of the container. One 4 µL droplet of CNC aqueous solution was placed on the perforated carbon side of a TEM grid facing up for 2 min. The tweezer securing the TEM grid was then mounted to the releasing anvil. Excess solution was blotted using filter paper from the back side of the TEM grid for 4 s, and the TEM grid was rapidly plunged into the liquid ethane. The frozen CNCs on the TEM grid were kept in liquid nitrogen and transferred for cryogenic-TEM (cryo-TEM) imaging.

2.2. Image Acquisition

All bright-field TEM (BF-TEM) and annular dark-field scanning TEM (ADF-STEM) images of CNCs were obtained using a JEOL 2200FS TEM/STEM at a 200 kV accelerating voltage with an in-column Omega energy filter. BF-TEM images with a pixel size of 2048 × 2048 were acquired on a Gatan slow-scan camera with pixel resolution at selected magnifications calibrated with a standard specimen (MAG*I*CAL, Electron Microscopy Science, USA). For imaging CNCs embedded in vitreous ice, the frozen specimen was transferred to a cryo-TEM holder (Gatan Inc., Pleasanton, CA, USA) while being kept below −170 °C. During TEM imaging, the specimen was kept at −180 °C, and the electron beam was blanked between image acquisitions to minimize beam damage on CNCs. The beam dose for each image was 2 e^-/A^2. Ten to fifteen TEM images of CNCs from different regions on the TEM grid were acquired at each magnification.

In STEM mode, ADF-STEM images of CNCs were acquired using a 0.5 nm probe with a pixel dwelling time of 10 µs and an image size of 512 × 512 pixels. The pixel resolution was 0.5 nm, which was the same as the scanning step to avoid oversampling between pixels. Approximately 15 ADF-STEM images of CNCs taken from different areas were used for size measurement.

The TEMography® software package (Frontier Inc., Tokyo, Japan) was used for the acquisition of tilt-serial BF-TEM images, fine alignment of 2D images, reconstruction of 3D data, rendering and visualization of CNCs in 3D. The tilt-serial dataset was collected by tilting the holder along the x-axis from −70° to +70° with a 2° increment and a defocus value of 1 μm. Gold nanoparticles were used as fiducial markers and focusing objects during image acquisition and fine alignment.

2.3. Size Measurement and Distribution Analysis

ImageJ (Version 1.53c), an open-source software, was used for image processing [20]. Single CNC particles or particles with clear separation outlines were manually selected for size measurement. The size of CNC particles was measured directly from TEM/STEM images. As shown in Figure 1, the length (L) is defined as the distance in a straight line between the two furthest points along a single CNC particle (major axis), and the width (W) (minor axis) is measured at the midpoint of its length unless a particle is clearly asymmetric. An ImageJ macro was used to open sequential images automatically after manual measurement of the length and width of selected CNCs on an active image. Measurement results were exported in a data format suitable for further statistical analysis in OriginPro 2017 or Excel. Approximately 200 particles were measured to give adequate statistical confidence. Histograms of length and width distributions for each dataset were plotted, and Gaussian fitting was applied for descriptive statistics. Box plots labeled with median, mean and one standard deviation (SD) above and below the mean were also used for comparing size distributions in datasets of stained CNCs and CNCs in vitreous ice.

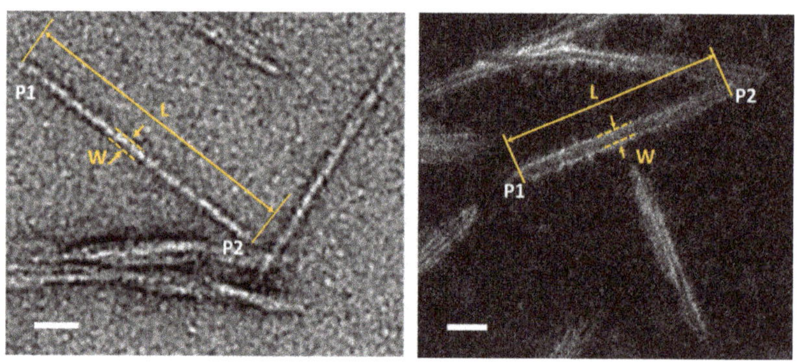

Figure 1. The length (L) is defined as the straight line between the furthest two points (P1 & P2) along the individual CNC particle; the width (W) is defined as the distance between two edges of CNC particle, as indicated in BF-TEM image (**left**) and ADF-STEM image of CNCs (**right**). Scale bars are 20 nm.

3. Results and Discussion

3.1. Contrast of CNC Particles in a TEM Image

CNCs are low-density (1.6 g/cm^3), high-aspect-ratio and electron-beam-sensitive biopolymers. The mass contrast of CNC nanoparticles in a normal BF-TEM image is very weak (Figure 2a). To enhance the contrast, conventional negative staining is widely used, and generally, 1–2% aqueous uranyl acetate is used as the staining reagent. However, due to the different staining methods/protocols, inconsistent quality and properties of the supporting films on TEM grids, as well as individual operator skills, staining results vary even for CNCs deposited on the same TEM grids [21]. As shown in Figure S2, a typical region of CNCs stained with 2% uranyl acetate has a gradient stain depth, resulting in CNCs with different contrasts. The very shallow stain cannot cover the overall contour of

CNCs, resulting in a weak contrast (Figure 2b). The CNCs in the deep stain appear bright with a fairly uniform dark background (Figure 2d), whereas those in the shallow stain are outlined by the stain (dark outlines), and the helical structure of the CNC particles is revealed clearly (Figure 2c). Despite this stain-depth variation, the rapid-flushing staining method described in the TEM specimen preparation was found to be the most suitable method among all those tested. To ensure that particles selected for image processing and size measurement were as homogenous as possible, regions of CNCs with similar stain depths were chosen as one dataset.

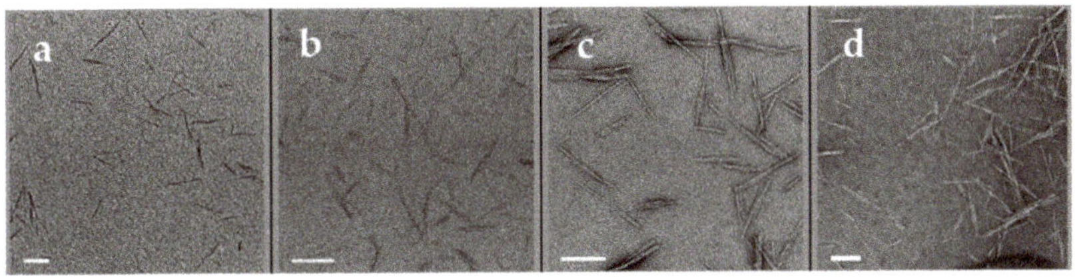

Figure 2. Variation in stain-depth of CNCs shown in BF-TEM images. (**a**) unstained; (**b**) very shallow; (**c**) shallow and (**d**) deep stained CNCs, respectively. Scale bars: 100 nm.

In addition to heavy-metal staining such as UA, the contrast of CNCs can also be enhanced with other approaches such as electron energy filtering, objective lens defocusing and phase-plate insertion in the BF-TEM mode. As shown in Figure 3a, a zero-loss BF-TEM image of unstained CNCs was taken with a 10 eV energy filter and a defocus of −3 µm, in which the helical-like structure of CNCs was revealed, which indicates that the helical structure in stained CNCs is not an artifact. In the last decade, phase-plate imaging has also been explored in contrast enhancement of soft materials. However, considering the various phase-plate types, phase-shift interpretation and access to instruments with a phase plate, it might be challenging to standardize the size distribution with this approach. In addition to the above approaches of contrast enhancement, STEM imaging is an important method of contrast tuning for polymers and biomolecules [22,23]. The contrast of an ADF-STEM image is related to mass thickness and atomic numbers in the specimen. The pristine CNCs in the ADF-STEM images give high contrast, the individual CNC particles are distinguishable and some CNC particles (pointed by arrows) with nonuniform width along their long axis can be revealed (Figure 3b). On the other hand, electron-beam-induced contaminations can easily build up during scanning and blur the image at high resolution. For CNCs stained with uranyl acetate, edges of CNCs exhibit sharp contrast from the stain with a shallow depth in the ADF-STEM image, as shown in Figure 3d. It is easier to identify the individual CNCs than those in the BF-TEM images (Figure 3c) taken from an adjacent area, even though the helical or twist structure of CNCs is not revealed in the ADF-STEM images, which may not be important for size measurement.

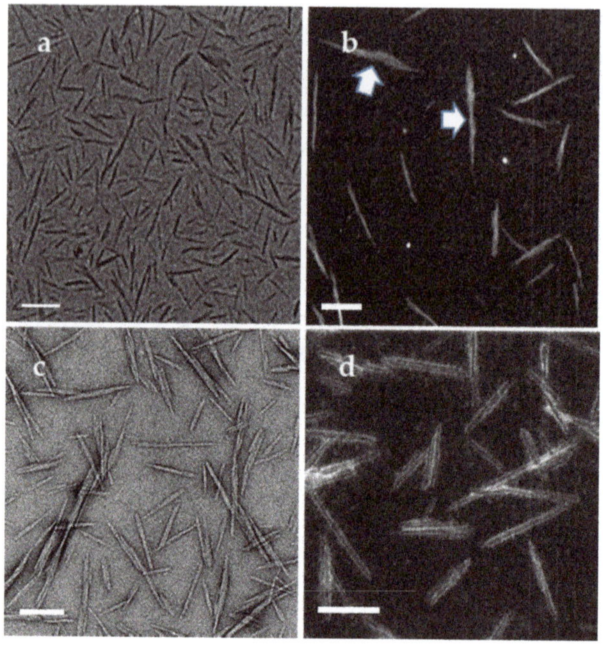

Figure 3. (**a**) under-focused (−3 μm) BF-TEM and (**b**) ADF-STEM images of unstained CNCs; (**c**) BF-TEM and (**d**) ADF-STEM image of stained CNCs. Scale bars: 100 nm.

With the contrast enhancement methods discussed above, sets of BF-TEM and ADF-STEM images of stained and unstained CNCs deposited on carbon TEM grids (Sp1 and Sp2 specimens) were acquired. The length and width of selected CNCs imaged at different conditions were measured, and the descriptive statistics of their distributions are summarized in Table S1. For curiosity, a set of BF-TEM images of CNCs was taken for each of the areas with a gradient stain depth, indicated as Zones A1, A2 and A3 in Figure S2, and the image resolution was 0.41 nm/pixel, 0.41 nm/pixel and 0.29 nm/pixel, respectively. The mean length of CNCs in Zones A1, A2 and A3 is 100.5 nm, 88.8 nm and 79.6 nm, and the mean width is 5.8 nm, 6.4 nm and 6.7 nm, respectively. Moving toward inward areas with a shallow stain (a similar stain depth as in Zone A3), an image set of ADF-STEM for Zone A4 was taken with an image resolution of 0.5 nm/pixel. The mean length and width of CNCs in Zone A4 are 76.3 nm and 7.0 nm, respectively. The histograms of length and width distributions are all included in Figure S3. Box plots of length, width and aspect ratio (length to width) for Zones A1 to A4 are shown in Figure 4. From Zones A1 to A4 with decreasing stain depths, the CNC width increases, while the CNC length and aspect ratio (AR) decrease, which suggests a correlation between size distribution and stain depth or imaging zone. Thus, the question to be answered is: does the stain depth affect the size of CNCs, or are CNC particles distributed differentially or fractionally on the substrate by size during droplet drying?

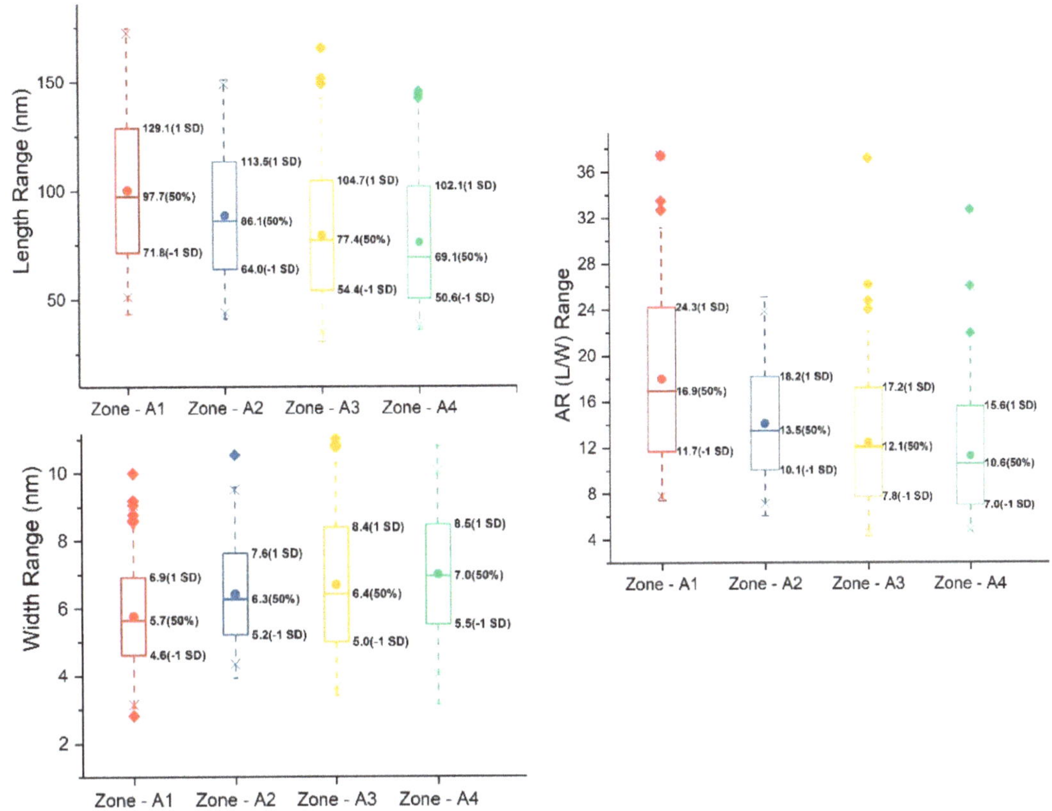

Figure 4. Box plots of length, width and aspect ratio (L/W) distribution of stained.CNCs at zones A1 to A4 with deceasing stain-depth. The line and dot in the box are the median and mean, respectively. The values of the median and one standard deviation (SD) above and below the mean are labelled to the left of boxes.

3.2. Dispersion of CNCs Adsorbed on Carbon-Film-Supported TEM Grids

As known, when a sessile droplet containing dispersed solid dries on a solid surface, a characteristic "coffee ring" is formed due to capillary flow toward the droplet edge [24]. In general, during the evaporation of a sessile droplet, the distribution of solids can be affected by capillary flow direction [24], Marangoni flows [25], particle−substrate interactions, particle–particle interaction and particle shape [26]. During TEM specimen preparation, the coffee-ring effect starts once the initial droplet is placed on TEM grids, and the degree of this effect depends on CNC concentration, incubation time, droplet volume and film surface properties. In the initial droplet, rod-shaped particles can be fractionated by particle size along the capillary flow direction when the contact lines of droplets recede on the substrate [27,28]. Particles with a high AR can be orderly arranged near the ring (droplet) edge with their major axis parallel to the contact lines [29]. After a certain incubation time, the initial capillary flow direction is interrupted by shear force from blotting, and the initial droplet is divided into many smaller droplets scattered on the carbon film surface, which might cause redistribution of the CNC particles on the substrate. In subsequent staining, the staining reagent may change the CNC distribution and orientation, but the lateral movement of CNCs should be confined within small areas. After drying, coffee rings might be formed, resulting in gradient thickness of stain. In Specimen Sp1 of stained CNCs, 3 μL of CNC solution and a 10 s incubation time for the initial droplet were used to minimize the coffee-ring effect so that CNCs of all sizes are included for quantitative analysis. However, it is difficult to void the coffee-ring effect completely, as seen in the gradient depth of stain in Figure S2. The results from Zone A1 to Zone A4 with decreasing stain depth, showing that the mean width increases while the mean length and the mean AR decrease, are due to the coffee-ring effect fractionating CNC dispersion on the substrate, instead of different stain thicknesses.

To avoid the influence of staining reagent on CNC distribution, unstained CNCs dried on TEM grids (Specimen Sp2) were prepared with the same volume of the initial droplet (3 μL) as Sp1 but with a longer incubation time (60 s). As shown in Figure 5a, well-dispersed CNCs are surrounded by more concentrated CNCs lying along their major axis in a coffee-ring pattern (only part of the ring is shown). The major axis of CNCs was redirected to follow the blotting direction rather than being parallel to the pinned line. The size distribution of CNCs in the center area is shown as Dataset A5 in Table S1. The mean length and width are 85.6 nm and 8.9 nm, respectively. The mean length is comparable to those in Datasets A1 to A4, while the mean width is much larger. Such a large mean width was also reported by Johnston's team with a consensus distribution of 7.7 ± 2.2 nm [15]. In their case, a 10 μL volume and a 4 min incubation time were used for the initial droplet of CNCs, and an extra washing step with water was added before the subsequent staining of CNCs. The larger droplet volume and longer incubation time made the fractionation and ordered arrangement of CNCs on the substrate driven by the coffee-ring effect more predominant. The washing step may have removed or redistributed some loose CNCs, but the subsequent staining should not affect the dispersion of CNCs already adsorbed on the carbon film surface. Due to the polydispersity of CNCD-1 materials, the fractionated dispersion of CNCs caused by the coffee-ring effect may introduce variability of size distribution when CNCs are imaged and selected from different regions on TEM grids, especially in the radial direction (TEM grid center to edge). The measured mean width of CNCs adsorbed on the center of the TEM grids is larger than that near the edge. Unfortunately, the location of images taken for each laboratory was not specified. In general, areas near the center of the TEM grids would be used for imaging as the default location, likely resulting in biased size distributions.

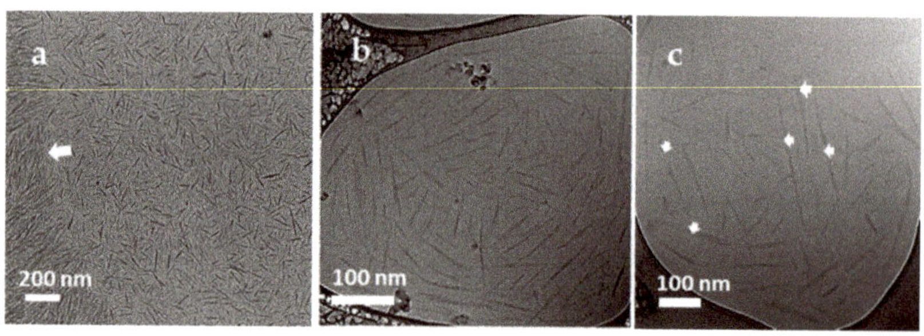

Figure 5. (**a**) coffee-ring formed while unstained CNCs dried on carbon film and more CNCs are concentrated on the edge (white arrow) with well-dispersed CNCs in the center; (**b**) cryo-TEM images of CNCs embedded in vitreous ice taken from the areas near the rim of TEM grids; CNCs lined up parallel to the hole edge and filled across the hole; and (**c**) less ordered CNCs embedded in the ice near the hole edge taken from near the center of TEM grids; possible end-to-end or slightly side-by-side arranged CNCs are pointed by arrows.

To further prove this coffee-ring effect or the dispersion of CNCs within a droplet, Specimen Sp4 was prepared using the plunge-freezing method to vitrify the CNC aqueous solution without a further drying process. Compared with Specimens Sp1 and Sp2, the incubation time of Sp4 was longer so that most CNCs could settle down within the droplet. Despite blotting from the backside of the droplet, the majority of CNCs near the edge of the droplet were arranged with their major axis parallel to each other (Figure 5b) and dispersed across the entire carbon film hole. CNCs near the center of the TEM grids were less ordered and dispersed than those near the edge of the carbon film hole (Figure 5c). Cryo-TEM images were acquired from regions near the center of the TEM grids as Datasets A6 and near the edge as Dataset A7, respectively. The size distribution in each dataset is summarized in Table S1. The histograms of length and width distributions are shown in Figure S4. The mean width and length with standard deviation for A6 and A7 are 7.0 ± 1.1 nm and 112.4 ± 40.6 nm, 6.2 ± 1.2 nm and 96.4 ± 30.0 nm, respectively. The population of CNCs near the center (A6) has a larger length and width than that near the edge (A7), as shown in the box plots in Figure 6. The ARs in both regions are similar, in contradiction to the dried CNC specimen, Sp1. The blotting direction and TEM grid surface used for the cryo-TEM specimen and the dried CNC specimen were different, which may interfere with the CNC distribution differently during blotting. As illustrated in Figure S5, shear force is induced once the filter paper touches the edge of the continuous carbon-film-coated TEM grid, causing CNCs near the air–water interface to drain along the shear force direction. However, when blotting from the backside of perforated film, the shear force induced flow is through the holes so that CNCs near the air–water interface are drawn within thinned aqueous film rather than being removed [30,31]. CNCs within frozen aqueous film can be dispersed at different heights along the beam direction, which is different from CNCs drying on continuous film after blotting. As shown in Figure 5c, the larger particles marked by arrows might be composed of multiple end-to-end or slightly side-by-side CNC particles or overlapped projections from two particles embedded in different vertical positions, contributing to size inflation for both length and width.

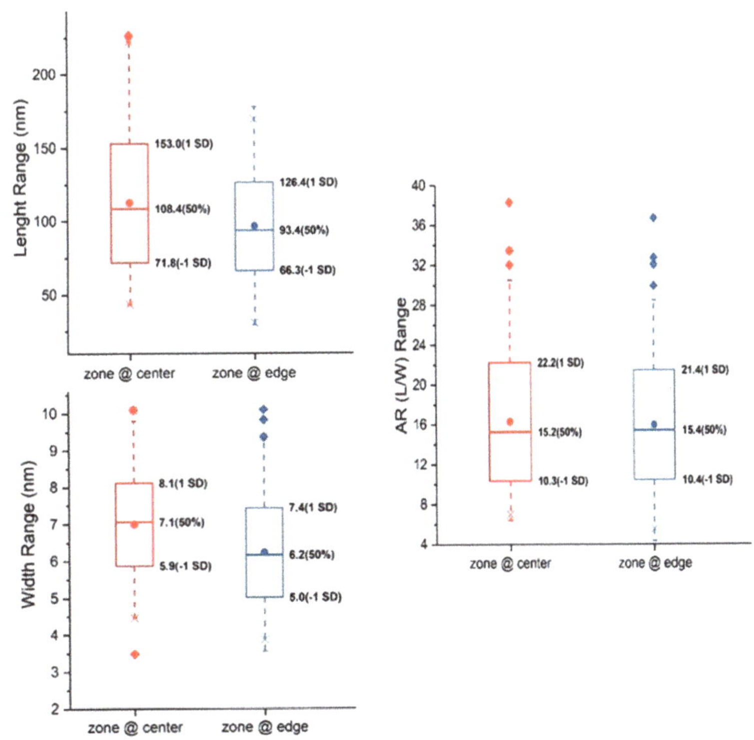

Figure 6. Box plots of length, width and aspect ratio (L/W) distribution of CNCs embedded in vitreous ice near the center and edge of TEM grids. The line and dot in the box is the median and mean, respectively. The values of the median and one standard deviation (SD) above and below the mean are labelled to the left of boxes.

3.3. Orientation of CNCs on Carbon-Film-Supported TEM Grids

As discussed above, CNCs on TEM grids could be fractionated during TEM specimen preparation, resulting in variability of size distribution when CNCs images are taken from different regions in the radial direction. In addition to the fractionated dispersion, the orientation of CNCs on the substrate also has a significant influence on size measurement from TEM images, as illustrated in Figure S1. In the CNCD-1 materials, about 28% of CNCs are particles with a square or a symmetric cross-section, while the remainder are asymmetric with one axis 2–3 times longer than the other, as analyzed in AFM [17]. When these asymmetric particles adsorb on the substrate surface with random orientations, the width distribution measured from TEM images may have large heterogeneity, as the TEM images are mixed projections of width and height from the minor-axial side of CNCs. Due to the strong hydrogen bond, CNCs tend to be laterally or twistedly jointed, and the widest side is normally the preferred orientation on the substrate. Figure 7 shows a typical example of aggregated CNCs in one region, which are either in a lateral (NP1, NP2), vertical and lateral (NP3) or twisted arrangement (NP4). NP3 and NP4 may be composed of more than two CNCs, and the width of NP4 along the long axis is not uniform. All these arrangements and orientations of CNCs on the substrate cause the variability of size measurement both in TEM and AFM. With proper staining of CNCs, the separation line between laterally jointed CNCs can be resolved in TEM. As shown in the line profile of NP1 in Figure 7, two particles with widths of 5.34 nm and 5.64 nm are well revealed and resolved. However, for unstained CNCs dried or embedded in vitreous ice or overstained CNCs, the separation between two CNC particles may not be revealed, resulting in inflated width measurement in TEM. In AFM, due to its limited resolving power in the lateral direction, it is difficult to resolve or separate laterally jointed CNCs. Therefore, the mean width measured with AFM may be overestimated when the orientation of CNCs is on the wider side. When CNC particles are arranged end-to-end or slightly side-by-side as illustrated in Figure S1, the length measurement will be affected if the separation cannot be resolved in either TEM or AFM.

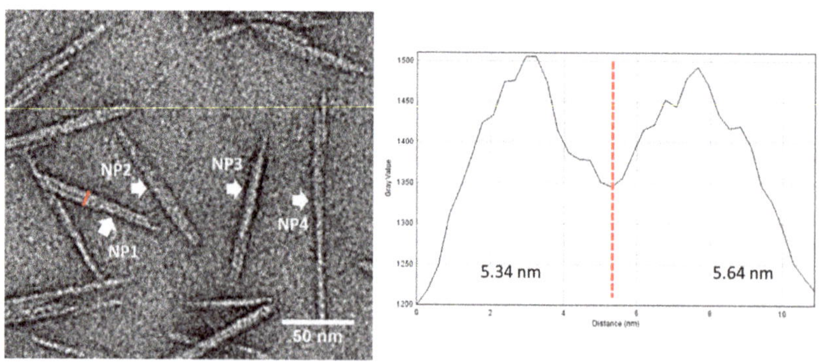

Figure 7. Right: Aggregated CNCs with different orientations marked by arrows (NP1, NP2, NP3 & NP4) on carbon film supported TEM grids. **Left**: The line profile of NP1 composed of two laterally jointed CNCs with width of 5.34 nm and 5.64 nm, respectively.

To reveal particle orientation and arrangement on the substrate, typical regions of CNC particles on Specimen Sp3 were used for ET. Figure 8a shows a reconstructed 3D micrograph of two geometrically twisted CNCs, which are identified as one individual CNC particle (marked with an orange arrow) in the BF-TEM image (Figure 8b). These two CNCs are screw-like or helical along the long axis, as revealed in the reconstructed volume views. The length and width of these two CNCs measured from the reconstructed 3D data are 57 nm and 3.5 nm and 60 nm and 3.8 nm, respectively. Moreover, in the 2D TEM image (Figure 8b), the width measured for this twisted particle is 5.7 nm, which is used for width distribution analysis. In the same field of view, there is another particle marked with a white arrow. Tilting the particle along the long axis by about 20 degrees showed the particle is actually composed of three end-to-end-connected single CNC particles (Figure 9). Thus, the length measurement for size distribution will give a larger value.

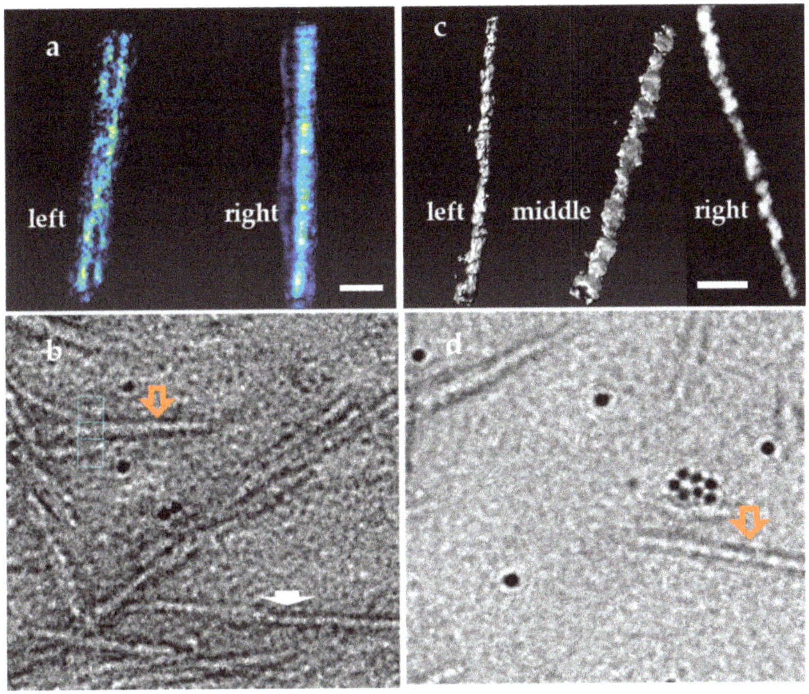

Figure 8. Geometrically twisted CNCs bundles and ribbon-liked CNCs coexists with rod-like CNCs (Sp3 specimen). Scale bars 10 nm. (**a**) Reconstructed 3D volume view of two helical CNCs twisted with an angle (left) and overlapped view (right); (**b**) 2D BF-TEM image of the twisted CNCs (marked as orange arrow) used for 3D reconstruction; (**c**) Reconstructed 3D isosurface of a ribbon-like CNC viewed at an angle with the narrowest width (left), widest width (middle) and the slice view (right); (**d**) BF-TEM image of the CNC used for 3D reconstruction (marked as orange arrow).

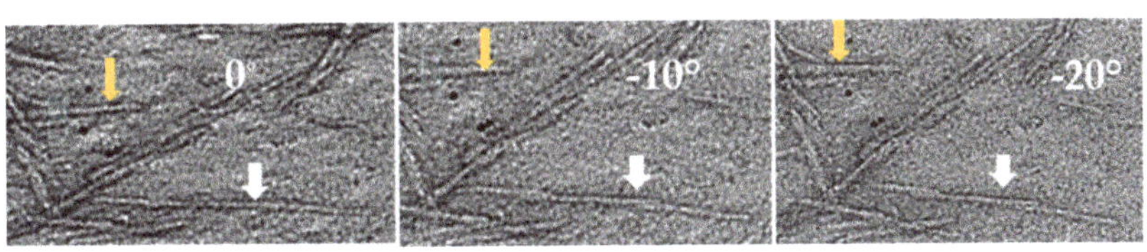

Figure 9. The end-to-end arranged CNC particles is shown as one particle in BF-TEM (**Left**); as two particles (**Middle**) tilted at 10° along long-axis; and as three particles tilted at 20° (**Right**).

In addition to the particle–particle interactions and arrangement on TEM specimens, the intrinsic shape of CNC particles also contributes to size measurement variability. Ribbon-like CNCs marked with an orange arrow in Figure 8d coexist in the same materials, as shown in Figure 8c of a typical reconstructed 3D isosurface view. The reconstructed 3D data reveal that the CNC particle is helical-like, and its width is not uniform along the long axis. The length, height (thickness) and width of this particle are 60 nm, 3.6 nm (left in Figure 8c) and 5.7 nm (right in Figure 8c), respectively. Thus, this asymmetric or rectangular cross-section suggests that CNC particles with a large width may be composed of more than one single crystallite.

Therefore, the shape, geometrical variation and orientation of CNCs on the substrate all contribute to the size measurement. The size measured from 2D images may not be the true parameter of CNC particles; rather, it is the average size of the projected images of all possible shapes and geometrical arrangements of CNCs deposited on the substrate, resulting in large variability of size distribution.

4. Conclusions

High-quality CNC images with good contrast and resolution can mitigate human bias in particle selection and size measurement. The traditional method of staining with heavy metal to enhance contrast for soft materials or biopolymers is still a very effective approach for CNCs. The rapid-flushing staining method has demonstrated that it can reveal the helical structure of single CNCs and separate laterally jointed CNC particles (Figure 2c,d, Figures 3c and 7). It can also reduce the preferential orientation of CNCs on the substrate by running droplets. AFD-STEM imaging is also a good approach for contrast enhancement, as shown in Figure 3, revealing the pristine structure of CNCs. To overcome beam-induced contamination and damage in future work, cooling the TEM holder at liquid nitrogen temperature can be performed when imaging unstained CNCs in ADF-STEM. ADF-STEM images of CNCs with a shallow stain depth give a sharp contrast of CNC particle edges, which may be helpful in particle identification and measurement.

Ideal statistical analysis of size distribution should include CNC particles of all sizes with equal representation. Due to the polydispersity of CNCs and their propensity to agglomerate, it is very challenging to obtain well-dispersed single CNC particles across TEM grids. During TEM specimen preparation, CNC particles could be fractionated across the substrate due to the coffee-ring effect, as seen in dried and hydrated cryo-TEM specimens. Droplet volume, drop-casting incubation time and substrate surface properties will affect the degree of fractionation. A smaller droplet volume and a shorter incubation time are recommended to minimize the coffee-ring effect while preparing CNCs dried on TEM grids. CNCs from different regions of TEM grids, especially in the radial direction, are recommended to be imaged and analyzed for size distribution. Specimen Sp1 is a typical example of fractionated CNCs on a substrate. The size distribution was obtained using the combined datasets from Zones A1 to A3, giving a mean width and length of 6.3 ± 1.3 nm and 89.9 ± 26.2 nm, respectively. Histograms of length and width and descriptive statistics are shown in Figure S6. Compared with the reference values in the certificate (mean width of 7.3 ± 1.8 nm and length of 87.0 ± 35 nm), the length distribution agrees well, but not for the width distribution. The higher mean width of the reference is likely due to lower CNC representation from the edge of droplets with a lower mean width in the measurement sampling, as the center of the droplet is typically the default starting measurement location when TEM grids are loaded inside the instrument. In addition to fractionation, the arrangement and orientation of CNC particles on the substrate significantly affect the size measurement when CNC aggregates cannot be resolved in images. The large mean width in Dataset A5 of unstained CNCs indicates that a large population of laterally jointed or twisted CNCs lying with the widest side on the substrate was included in the analysis. Furthermore, when a large CNC population with an asymmetric cross-section coexists with a symmetric population, the random CNC orientations on the substrate contribute to the variability of the width distribution because the measured width in the TEM images contains the width and height of 3D CNC particles.

In short, to obtain an accurate size measurement of CNCs from TEM images, high-quality CNC TEM specimens and images are essential. To avoid unintended bias of size distribution, imaging and analyzing CNCs from all regions across the TEM grids is recommended. For future work, correlated AFM and TEM imaging can be explored for size distribution analysis of CNCs deposited on the same substrate, which may provide more insight into cross-section shape distribution.

Supplementary Materials: The following are available online at https://www.mdpi.com/article/10.3390/polym13193318/s1, Figure S1: Schematics of the transverse cross-section, orientation on substrate and projected image on camera of rod-shaped nanoparticles and the possible particle–particle arrangements, Figure S2: Stain-depth variation of CNCs deposited on TEM grid, Table S1: Length and width distributions measured from BF-TEM, ADF-STEM and cryo-TEM images of CNCs, Figure S3: Histograms of length, width and aspect ratio distributions of CNCs measured from images taken in Zones A1, A2 and A3 of stained CNCs on TEM grids (Specimen Sp1), Figure S4: Histograms of length and width distributions for unstained CNCs from TEM specimen Sp2 and cryo-TEM specimen Sp4, Figure S5: Initial droplet of CNC aqueous solution on continuous carbon-film-coated TEM grids and perforated carbon-film-coated TEM grids, Figure S6: Size distribution of CNCs by combining Datasets A1, A2 and A3 measured from Specimen Sp1.

Funding: This research received no external funding.

Institutional Review Board Statement: Not applicable.

Informed Consent Statement: Not applicable.

Data Availability Statement: The data presented in this study are available on request from the corresponding author.

Acknowledgments: The author would like to thank Shan Zou with the Metrology Research Center, NRC, for providing the CNCD-1 material and many insightful discussions on AFM imaging.

Conflicts of Interest: The author declares no conflict of interest.

References

1. Ranby, B.G. Physico-chemical investigations on bacterial cellulose. *Ark. Kemi* **1952**, *4*, 249.
2. De Rodriguez, N.L.G.; Thielemans, W.; Dufresne, A. Sisal cellulose whiskers reinforced polyvinyl acetate nanocomposites. *Cellul.* **2006**, *13*, 261–270. [CrossRef]
3. Beck-Candanedo, S.; Roman, M.; Gray, D.G. Effect of Reaction Conditions on the Properties and Behavior of Wood Cellulose Nanocrystal Suspensions. *Biomacromolecules* **2005**, *6*, 1048–1054. [CrossRef] [PubMed]
4. Elazzouzi-Hafraoui, S.; Nishiyama, Y.; Putaux, J.-L.; Heux, L.; Dubreuil, F.; Rochas, C. The Shape and Size Distribution of Crystalline Nanoparticles Prepared by Acid Hydrolysis of Native Cellulose. *Biomacromolecules* **2008**, *9*, 57–65. [CrossRef] [PubMed]
5. Zhu, H.; Luo, W.; Ciesielski, P.N.; Fang, Z.; Zhu, J.Y.; Henriksson, G.; Himmel, M.E.; Hu, L. Wood-Derived Materials for Green Electronics, Biological Devices, and Energy Applications. *Chem. Rev.* **2016**, *116*, 9305–9374. [CrossRef]
6. Jorfi, M.; Foster, E.J. Recent advances in nanocellulose for biomedical applications. *J. Appl. Polym. Sci.* **2015**, *132*, 41719. [CrossRef]
7. Sunasee, R.; Hemraz, U.D.; Ckless, K. Cellulose nanocrystals: A versatile nanoplatform for emerging biomedical applications. *Expert Opin. Drug Deliv.* **2016**, *13*, 1243–1256. [CrossRef]
8. Karim, Z.; Claudpierre, S.; Grahn, M.; Oksman, K.; Mathew, A.P. Nanocellulose based functional membranes for water cleaning: Tailoring of mechanical properties, porosity and metal ion capture. *J. Membr. Sci.* **2016**, *514*, 418–428. [CrossRef]
9. Moon, R.J.; Martini, A.; Nairn, J.; Simonsen, J.; Youngblood, J. Cellulose nanomaterials review: Structure, properties and nanocomposites. *Chem. Soc. Rev.* **2011**, *40*, 3941–3994. [CrossRef]
10. Hamad, W.Y.; Hu, T.Q. Structure-process-yield interrelations in nanocrystalline cellulose extraction. *Can. J. Chem. Eng.* **2010**, *88*, 392–402. [CrossRef]
11. Mukherjee, S.; Woods, H. X-ray and electron microscope studies of the degradation of cellulose by sulphuric acid. *Biochim. Biophys. Acta (BBA)-Bioenerg.* **1953**, *10*, 499–511. [CrossRef]
12. Jakubek, Z.; Chen, M.; Couillard, M.; Leng, T.; Liu, L.; Zou, S.; Baxa, U.; Clogston, J.D.; Hamad, W.Y.; Johnston, L.J. Characterization challenges for a cellulose nanocrystal reference material: Dispersion and particle size distributions. *J. Nanoparticle Res.* **2018**, *20*, 98. [CrossRef]
13. Bai, W.; Holbery, J.; Li, K. A technique for production of nanocrystalline cellulose with a narrow size distribution. *Cellul.* **2009**, *16*, 455–465. [CrossRef]
14. Beck, S.; Bouchard, J.; Berry, R. Dispersibility in Water of Dried Nanocrystalline Cellulose. *Biomacromolecules* **2012**, *13*, 1486–1494. [CrossRef]
15. Meija, J.; Bushell, M.; Couillard, M.; Beck, S.; Bonevich, J.; Cui, K.; Foster, J.; Will, J.; Fox, D.; Cho, W.; et al. Particle Size Distributions for Cellulose Nanocrystals Measured by Transmission Electron Microscopy: An Interlaboratory Comparison. *Anal. Chem.* **2020**, *92*, 13434–13442. [CrossRef] [PubMed]
16. Bushell, M.; Meija, J.; Chen, M.; Batchelor, W.; Browne, C.; Cho, J.-Y.; Clifford, C.A.; Al-Rekabi, Z.; Vanderfleet, O.M.; Cranston, E.D.; et al. Particle size distributions for cellulose nanocrystals measured by atomic force microscopy: An interlaboratory comparison. *Cellulose* **2021**, *28*, 1387–1403. [CrossRef]

17. Chen, M.; Parot, J.; Hackley, V.A.; Zou, S.; Johnston, L.J. AFM characterization of cellulose nanocrystal height and width using internal calibration standards. *Cellulose* **2021**, *28*, 1933–1946. [CrossRef]
18. Available online: http://www.nrc.ca/crm (accessed on 18 June 2021).
19. Qian, H.; Jia, Y.; McCluskie, M.J. Application of cryogenic transmission electron microscopy for evaluation of vaccine delivery carriers. In *Vaccine Delivery Technology: Methods and Protocols*; Methods in Molecular Biology; Pfeifer, B., Hill, A., Eds.; Humana: New York, NY, USA, 2020; Volume 2183, ISBN 978-1-0716-0794-7.
20. Rasband, W.S. U.S. National Institute of Health, Bethesda, Maryland, USA. Available online: https://imagej.nih.gov/ij/ (accessed on 18 June 2021).
21. Scarff, C.A.; Fuller, M.J.G.; Thompson, R.F.; Iadaza, M.G. Variations on Negative Stain Electron Microscopy Methods: Tools for Tackling Challenging Systems. *J. Vis. Exp.* **2018**, *10*, e57199. [CrossRef] [PubMed]
22. Wall, J.S.; Hainfeld, J.F. Mass Mapping with the Scanning Transmission Electron Microscope. *Annu. Rev. Biophys. Biophys. Chem.* **1986**, *15*, 355–376. [CrossRef]
23. Williams, D.B.; Cater, C.B. *Transmission Electron Microscopy: A Text Book for Material Science*; Springer: New York, NY, USA, 2009; Chapter 22.
24. Deegan, R.D.; Bakajin, O.; Dupont, T.F.; Huber, G.; Nagel, S.R.; Witten, T.A. Capillary flow as the cause of ring stains from dried liquid drops. *Nature* **1997**, *389*, 827–829. [CrossRef]
25. Hu, A.H.; Larson, R.G. Marangoni Effect Reverses Coffee-Ring Depositions. *J. Phys. Chem. B* **2006**, *110*, 7090–7094. [CrossRef] [PubMed]
26. Yunker, P.J.; Still, T.; Lohr, M.A.; Yodh, A.G. Suppression of the coffee-ring effect by shape-dependent capillary interactions. *Nat. Cell Biol.* **2011**, *476*, 308–311. [CrossRef] [PubMed]
27. Fan, F.; Stebe, K.J. Size-Selective Deposition and Sorting of Lyophilic Colloidal Particles on Surfaces of Patterned Wettability. *Langmuir* **2005**, *21*, 1149–1152. [CrossRef] [PubMed]
28. Ryu, S.-A.; Kim, J.Y.; Kim, S.Y.; Weon, B.M. Drying-mediated patterns in colloid-polymer suspensions. *Sci. Rep.* **2017**, *7*, 1079. [CrossRef] [PubMed]
29. Dugyala, V.R.; Basavaraj, M.G. Evaporation of Sessile Drops Containing Colloidal Rods: Coffee-Ring and Order–Disorder Transition. *J. Phys. Chem. B* **2015**, *119*, 3860–3867. [CrossRef]
30. Zheng, Y.; Lin, Z.; Zakin, J.L.; Talmon, Y.; Davis, H.T.; Scriven, L.E. Cryo-TEM Imaging the Flow-Induced Transition from Vesicles to Threadlike Micelles. *J. Phys. Chem. B* **2000**, *104*, 5263–5271. [CrossRef]
31. Glaeser, R.M. Proteins, interfaces, and cryo-EM grids. *Curr. Opin. Colloid Interface Sci.* **2018**, *34*, 1–8. [CrossRef] [PubMed]

polymers

MDPI

Review

A Review on Synthesis, Properties, and Applications of Polylactic Acid/Silica Composites

Mosab Kaseem [1,*], Zeeshan Ur Rehman [2], Shakhawat Hossain [3], Ashish Kumar Singh [4] and Burak Dikici [5,*]

1 Department of Nanotechnology and Advanced Materials Engineering, Sejong University, Seoul 05006, Korea
2 School of Materials Science & Engineering, Changwon National University, Changwon 641-773, Korea; Zeeshan.physics@gmail.com
3 Department of Industrial and Production Engineering, Jashore University of Science and Technology, Jashore 7408, Bangladesh; shakhawat.ipe@just.edu.bd
4 Department of Applied Sciences, Bharati Vidyapeeth's College of Engineering, New Delhi 110063, India; ashish.singh.rs.apc@itbhu.ac.in
5 Department of Metallurgical and Materials Engineering, Ataturk University, Erzurum 25240, Turkey
* Correspondence: mosabkaseem@sejong.ac.kr (M.K.); burakdikici@atauni.edu.tr (B.D.)

Abstract: Polylactic acid (PLA)/silica composites as multifunctional high-performance materials have been extensively examined in the past few years by virtue of their outstanding properties relative to neat PLA. The fabrication methods, such as melt-mixing, sol–gel, and in situ polymerization, as well as the surface functionalization of silica, used to improve the dispersion of silica in the polymer matrix are outlined. The rheological, thermal, mechanical, and biodegradation properties of PLA/silica nanocomposites are highlighted. The potential applications arising from the addition of silica nanoparticles into the PLA matrix are also described. Finally, we believe that a better understanding of the role of silica additive with current improvement strategies in the dispersion of this additive in the polymer matrix is the key for successful utilization of PLA/silica nanocomposites and to maximize their fit with industrial applications needs.

Keywords: polylactic acid; silica; composites; thermal stability; toughness; biodegradability; 3D printing applications

Citation: Kaseem, M.; Ur Rehman, Z.; Hossain, S.; Singh, A.K.; Dikici, B. A Review on Synthesis, Properties, and Applications of Polylactic Acid/Silica Composites. *Polymers* **2021**, *13*, 3036. https://doi.org/10.3390/polym13183036

Academic Editors: José Miguel Ferri, Vicent Fombuena Borràs and Miguel Fernando Aldás Carrasco

Received: 19 August 2021
Accepted: 7 September 2021
Published: 8 September 2021

1. Introduction

Owing to its versatility and good performance, polylactic acid (PLA) as a biodegradable polymer is regarded as one of the most favorable polymers to substitute petroleum-based plastics [1–3]. The low toughness and poor thermal stability of PLA, however, restrict the usage of PLA in applications, such as food packaging and medical implants [3]. To improve the performance, therefore, several approaches, such as blending with other polymers, mixing with inorganic fillers, and copolymerization, have been suggested by several research groups [4–10]. In particular, the mixing with inorganic additives, such as TiO_2, carbon nanotube, ZnO, Al_2O_3, MgO, and SiO_2, was reported to be a useful method to boost the performance of neat PLA [5,7].

Silicon dioxide (SiO_2), commonly referred to as silica, which may exist in the amorphous and crystalline structure, was found to be a useful filler for improving the mechanical performance of polymeric materials [11,12]. The silica, as an additive, is used in coatings, food, and biomedical applications. For example, silica as a bio-safe additive can be incorporated along with silver nanoparticles to improve the antibacterial properties and corrosion properties of biomaterials [13–23]. The amorphous silica nanoparticles extracted from natural sources, such as rice husk and fly ash, can be used as a nucleating agent when added to thermoplastic polymers in very low amounts, owing to their large surface area [24]. The improving dispersibility of silica within the PLA matrix during fabrication of PLA/silica composites would lead to enhancing the mechanical, thermal, and rheological properties of neat PLA [21]. The presence of silanol groups, siloxane bridges, and hydroxyl groups on the silica would facilitate the surface functionalization of silica for biomedical, catalysis, and sensor applications [12].

Therefore, by taking into consideration the attractive properties of silica, the fabrication of PLA/silica composites would be a beneficial procedure to enhance the performance of PLA. To the best of our knowledge, this is the first systematic article deliberating the recent progress in PLA/silica. Thus, this work aims to discuss the recent developments in the field of PLA/silica composites in terms of fabrication, properties, and applications.

2. Synthesis of PLA/Silica Composites

The direct melt-mixing, solution mixing, sol–gel process, and in situ polymerization methods are considered as the main methods utilized for the synthesis of PLA/silica composites. The melt-mixing method involves the direct mixing of PLA with silica nanoparticles, while the sol–gel process can be performed either in the existence of PLA or simultaneously during the polymerization of lactic acid monomers [25]. The solution mixing method starts from the dissolution of polymers in a suitable solvent with nanoparticles together with the evaporation of the solvent, or precipitation [25]. As for the in situ polymerization method, the silica nanoparticles should be distributed in the monomers before polymerization [25]. In addition, it is worth mentioning that the freezing-drying process would also be a promising method to fabricate PLA composites. In the freezing-drying process, the colloidal dispersion of the aerogel precursors is frozen, with the liquid component freezing into different morphologies depending on a variety of factors, such as the precursor concentration, type of liquid, temperature of freezing, and freezing container [25]. However, a lack of works was found on the utilization of this method to fabricate PLA/silica composites.

The surface modification by physical or chemical methods is a common procedure to increase the compatibility between PLA and silica nanoparticles [26–30]. The functionalization of hydrophilic silica nanoparticles, which is usually conducted on the reactive silanol end-groups, would improve the hydrophobicity of silica nanoparticles, improving their dispersion in the PLA matrix. This process, i.e., the formation of hydrophobic-fumed silica, can be obtained by chemical treatment of hydrophilic silica with silanes or siloxanes. As a result of the improved dispersion of silica particles in the PLA matrix, the rheological properties of the composites are improved. In general, the large surface area and the smooth surface of silica nanoparticles could increase the interactions not only between silica nanoparticles but also between PLA and silica nanoparticles. Thus, good physical interactions between the silica and PLA matrix can be achieved, leading to significant enhancements in composites properties. Several functionalization (coupling) agents, such as tetraethyloxysilane (TEOS) and γ-glycidoxypropyltrimethoxysilanes (GOPTMS) were used to improve the dispersibility of silica nanoparticles in the PLA matrix. In the study of Hakim et al. [26], a reactive extrusion method was utilized for the melt-mixing of PLA with 2.5 wt.% of silica nanoparticles. The schematic illustrations of SiO_2 before and after surface modification by organic chains are displayed in Figure 1. As shown in Figure 1b, the surface -OH groups of both surface-modified silica types were mostly substituted by functionalized organic chains. Thus, many PLA chains could be grafted on one silica nanoparticle, which would have risen the local shear field applied to the agglomerated nanoparticles during mixing, thereby improving the dispersibility of silica nanoparticles in the polymer matrix. However, transmission electron microscopy (TEM) observations implied that the uniformity and dispersibility of the nanoparticles under the experimental processing conditions were found not to be affected by the enhanced interfacial interaction of both surface-treated silica types (Figure 1c–e).

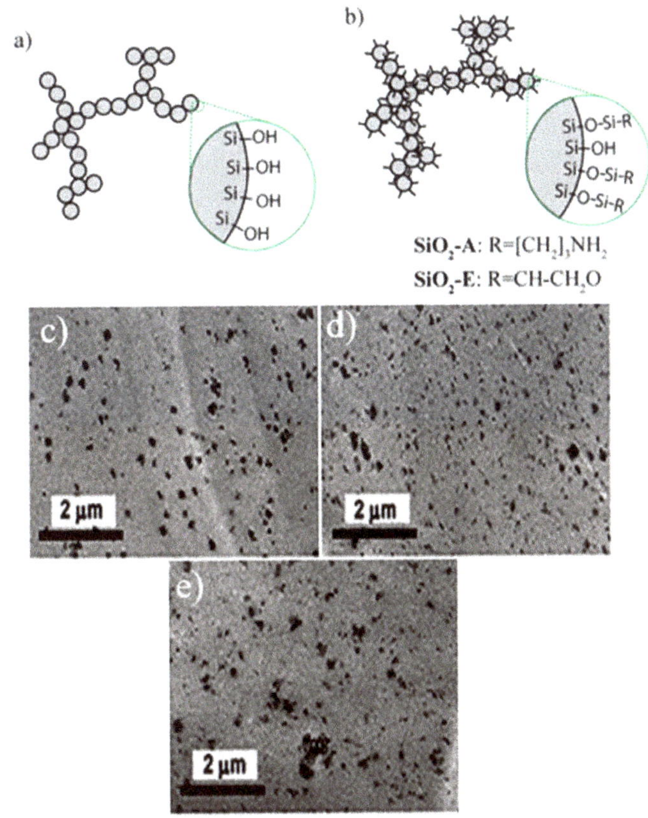

Figure 1. Schematic representation of the (**a**) surface-unmodified silica and (**b**) surface-modified silica aggregate. (**c**–**e**) TEM images of the PLA composites containing (**c**) un-modified SiO_2, (**d**) SiO_2-A, and (**e**) SiO_2-E [26].

The fabrication of PLA/silica composites via the sol–gel method was first reported by Yan et al. [31], who aimed to obtain plasticized composites. The fabrication process of the composites involved an in situ synthesis of silica nanoparticles via condensation reactions of TEOS and GOPTMS in the presence of PLA and polyethylene glycol as a plasticizer in tetrahydrofuran. Hydrochloric acid was utilized as a catalyst during the synthesis method. The infrared results proved the formation of a silica network structure within the PLA matrix, while the results of mechanical tests indicated that the incorporation of 4 wt.% of silica nanoparticles would increase the tensile strength of PLA from 15 to 18 MPa. In another work, Yan et al. [32] polymerized L-lactic acid (PLLA) in the presence of silica without the use of catalysts, but in solution. In this case, the silica surface was not organically treated. The polycondensation was carried out in toluene to remove the water formed by azeotropic dehydration. The result is silica grafted with PLLA oligomers. The authors were able to witness the grafting by infrared spectroscopy characterization (unfortunately, the molar masses are not specified). The grafted silica was then dispersed in PLLA and its good dispersion helped to improve mechanical properties, as compared to the PLLA composites with non-grafted silica.

Wu et al. [32] also polycondensed L-lactic acid in the presence of silica nanoparticles, but this time in bulk. The authors first mixed an aqueous solution of L-Lactic acid (LA) with an acidic silica sol containing silica particles of 12 nm. The mixture was dehydrated under vacuum with sonication treatment to well disperse the particles. After complete drying of the mixture, the polycondensation was performed under vacuum conditions to remove the water formed. The grafting occurred as above, with the polycondensation using the SiOH groups on the surface of the silica nanoparticles. The results showed that the molar mass of the grafted PLLA was about 31,100 g·mol^{-1}. Liu et al. [33] fabricated PLA/silica composites through ring-opening polymerization of lactide initiated by modified silica nanoparticles in the presence of stannous octoate as a catalyst. The experimental conditions, such as the weight ratio of silica to the lactide monomer, reaction temperature, and reaction time were optimized to be 1:20, 140 °C, and 72 h, respectively. The morphological observations revealed that silica nanoparticles tended to be distributed uniformly within the PLA matrix, improving compatibility between PLA and silica nanoparticles. Accordingly, the thermal and mechanical properties of the composites were improved significantly in comparison to that of pure PLA. In other works [24,34,35], the melt-mixing method to fabricate PLA/silica composites was carried out at 175 °C. In comparison to neat PLA, the PLA/silica composites containing various contents of silica (from 1 to 10 wt.%) exhibited significant improvements in the thermal stability and the barrier properties against nitrogen and oxygen gases, which were connected with the establishment of the silica network structure as approved by Fourier-transform infrared (FTIR) and rheological results. In the work of Liu and co-workers [36], amino-functionalized nano-SiO_2 (m@g-SiO_2) was first prepared through a coupling reaction on the surface of silica nanoparticles before its melt-mixing with PLA. Molecular dynamics simulation was carried out to discover the correlation between PLA and silica nanoparticles before and after the organic functionalization of silica. The crystallization behavior and the mechanical properties of PLA exhibited significant improvements after the melt-mixing with functionalized silica nanoparticles which were attributed to the fact the organic modification of silica nanoparticles would lead to enhancements in the interaction energy and mobility of PLA chains.

To improve the dispersion of silica in the PLA matrix, an in situ melt condensation of l-lactic acid was carried out in the presence of silica nanoparticles [37]. However, when the content of silica exceeded 10 wt.%, some agglomerated nanoparticles appeared in the TEM images of the prepared composites. The agglomeration of silica nanoparticles would be attributed to the strong van der Waals forces, which tended to reduce the physical properties of the obtained composites. Thus, it is believed that the physical mixtures of PLA and organo-modified SiO_2 resulted in the separation in discrete phases, leading to inferior mechanical properties [38,39]. Zou and coworkers tried to mix silica nanoparticles with a copolymer made of PLA and epoxidized soybean oil [40]. The FTIR and thermogravimetric analysis (TGA) results confirmed the reactions between silica and epoxidized soybean oil, which, in turn, led to improve crystallization behavior and mechanical properties of PLA, while Sepulveda et al. [41] demonstrated that the direct grafting of L-lactic acid oligomer onto the silica surface through its silanol groups was a good strategy to enhance the physical, thermal, and mechanical properties of PLA/silica composites. As such, Zhu et al. [42] found that the surface functionalization of fumed silica nanoparticles oleic acid by oleic acid would help to improve the rheological, thermal, and mechanical properties of PLA/silica composites prepared via the melt-mixing method, which was linked to the good interfacial adhesion in the composites containing functionalized-silica nanoparticles.

3. Rheological Properties

Understanding the rheological properties is very important due to their considerable effects on molecular weight, morphology, chain structure, and chain motions [43–49]. Basilissi et al. [50] demonstrated that the melt viscosity of PLA/silica composites fabricated by bulk ring-opening polymerization can be improved by the silane-based modification of silica nanoparticles. Li et al. [51] reported that both the storage modulus and biodegradation rate of PLA tended to be improved by the addition of silica nanoparticles into the PLA matrix. The formation of a silica network structure was responsible for storage modulus improvements, while the enhancement in the biodegradation rate was ascribed to the easy release of silica aggregates from the PLA matrix. Nerantzaki et al. [52] prepared a series of poly(DL-lactide) (PDLLA)/SiO_2 composites by a novel two-step technique (ring-opening polymerization (ROP)—polycondensation). The concentration of SiO_2 was varied from 2.5 to 5, 10, and 20 wt.%. The results of this work are presented in Figure 2. Based on the intrinsic viscosity results, it was demonstrated that the average molecular weight of PDLLA (Mn $\approx$ 38,097 g mol^{-1}) tended to reduce with an increase in the silica content. Moreover, the average Mn values of PLA were found to decrease when the content of silica increased from 2.5 to 20 wt.%. This finding was elucidated based on the fact that the addition of silica to the PDLLA matrix would hinder the increment in the molecular weight of the polymer. Besides, silica nanoparticles can interact with DL-lactide through their silanol groups. In another study [53], the silica was functionalized with TEOS and GOPTMS through graft-condensation reaction. Afterward, the functionalized silica was melt-mixed with PLA by reactive extrusion technique. In comparison to neat PLA, the addition of functionalized silica to the PLA matrix caused considerable improvements in values of the complex viscosity and the storage modulus of the PLA/silica composites. The inclusion of silica was expected to enhance the hydrolysis resistance by the formation of a stable silica network. Although the molecular weight of PLA was reduced by 12 wt.% during the processing, a slighter reduction of less than 10% was reported upon the addition of TEOS, which also resulted in increasing the value of zero-shear rate viscosity (η_0) obtained via the Cross model.

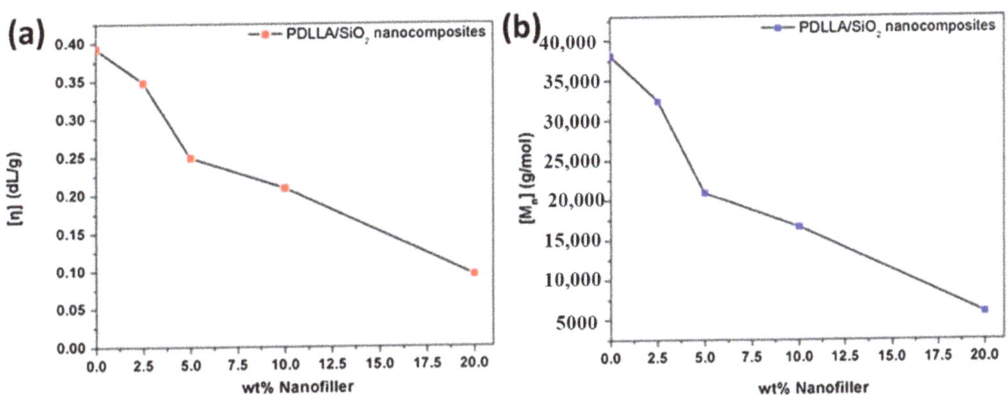

Figure 2. Effect of silica content on the (**a**) intrinsic viscosity and (**b**) average molecular weight of PDLLA [52].

As reported by Hao's group [54], the inclusion of 1.1, 2.8, 5.8, and 9.0 vol.% of silica particles into the PLA matrix led to the improvement in the viscoelastic properties of PLA. Here, silica with different sizes (same density), namely, silica 300 (7 nm), silica OX50 (40 nm), and silica 63 (9000 nm), were used. As shown in Figure 3, the low-frequency G' increased with the addition of silica and reached an approximately frequency-independent plateau at the high content (above 2.8 vol.%). Moreover, it was observed that the effect induced by silica 63 on the storage modulus (G0) and complex viscosity (η^*) of PLA/silica 63 composites was extremely weak. With the increased silica loading, G0 varied negligibly in the high-frequency region and just a slight increase was found at the low-frequency range. Moreover, the times obtained from the plots of storage modulus ($G'(t)/G'$ onset versus time) at 180 °C in the case of neat PLA were found to be 5000 s, while longer times of 8000 s were reported for PLA/silica composites. The results obtained in this work clearly show that the rheological properties of PLA/silica composites are strongly affected not only by the particle size but also by the particle content.

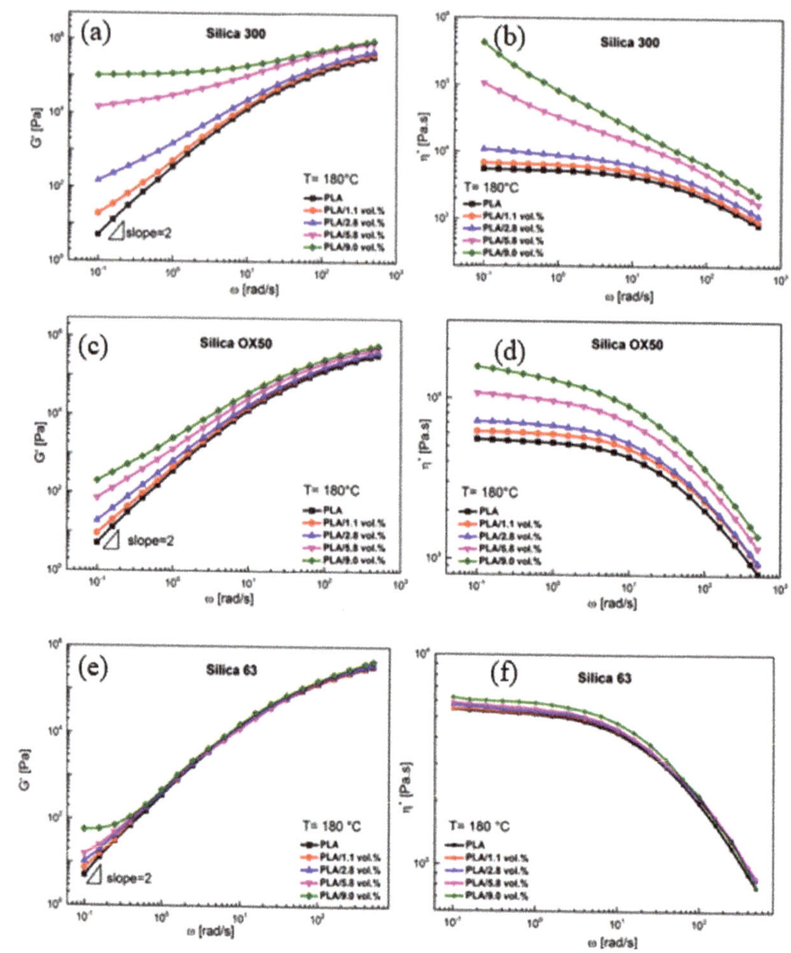

Figure 3. The influence of silica type on the rheological properties of PLA/silica composites: (**a**,**b**) PLA/silica 300; (**c**,**d**) PLA/silica OX50; (**e**,**f**) PLA/silica 63 [54].

4. Thermal Properties

The thermal stability of PLA can be improved by the incorporation of silica nanoparticles. In the works of Wen et al. [24,55], PLLA/silica composites were fabricated via a melt-mixing method. The thermal stability of PLA showed significant improvement with the addition of silica which was linked to the barrier effect of the silica network structure, which was also responsible for the improvements in the rheological properties of PLA discussed in the previous chapter. Zhang et al. [56] reported experimentally and theoretically, using a molecular dynamics simulation, that the thermal properties of PLA tended to be enhanced with the addition of silica into the polymer matrix during the processing in a twin-screw extruder. The results indicate that the glass transition temperature (T_g) of PLA was increased by 1.34 °C, while the thermal stability was increased by 12 °C when 2 wt.% of fumed silica was added into the PLA matrix. Klonos and Pissis [57] examined the thermal behavior of PLA/composites by taking the role of H-bonds, formed due to the reaction between the carbonyl groups in PLLA with hydroxyl groups on the silica surface, into account. Examinations of findings involved a combination of assessments on initially amorphous and on semicrystalline (annealed) samples. No change in the T_g by the silica was noted by differential scanning calorimetry (DSC), whereas the heat capacity step decreased in the PNCs. The segmental relaxation in the broadband dielectric relaxation spectroscopy (DRS) became, however, quicker and weaker in the PLA/silica composites. They reported that T_g and the temperature difference between onset (T_{onset}) and end (T_{end}) of the event ($DT_{GT} = T_{end} + T_{onset}$) tended to increase after annealing of crystallization, by 6–8 K and by a factor of ~2, respectively, as compared to the amorphous samples, due to constraints imposed by crystallites and heterogeneities. As for semicrystalline composites, T_g was increased by ~2 K, as compared to neat PLA. Simultaneously, DT_{GT} decreased slightly with increasing silica content. The authors attributed these results to the changes in semicrystalline morphology (Figure 4), polymer diffusion, and porosity/dispersion of silica particles.

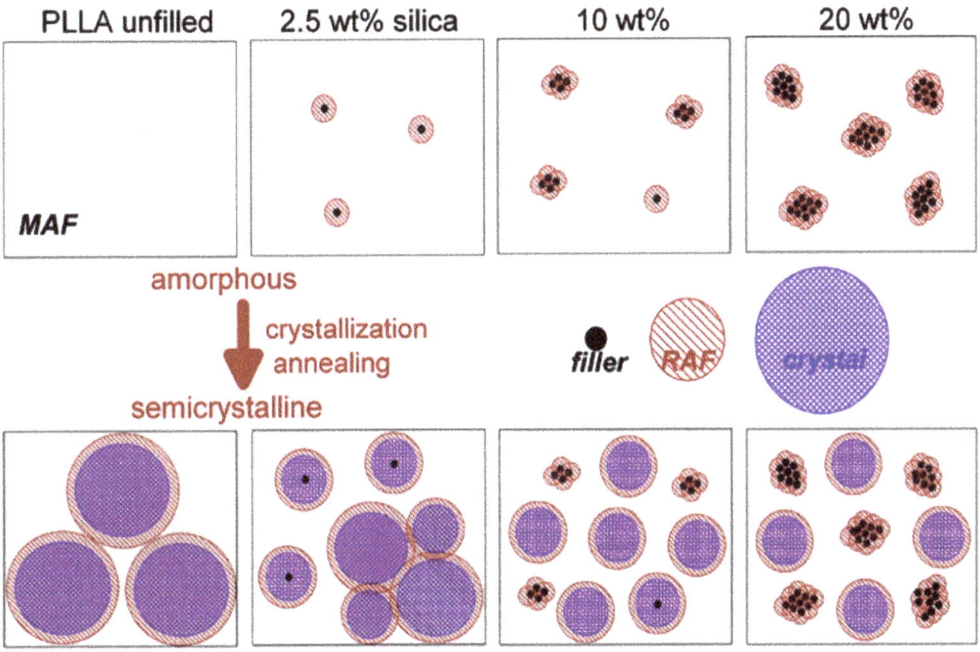

Figure 4. The proposed distribution of the silica in the PLA matrix before and after annealing of crystallization [57].

In the work of Bouamer et al. [58], a casting method was used to fabricate hybrid PLA composites with silica and AlO particles. In this regard, 10 wt.% of either silica or AlO particles was incorporated into the PLA matrix to fabricate PLA/silica and PLA/AlO composites, respectively, while 5 wt.% of SiO_2 and 5 wt.% of AlO particles were added to the PLA matrix to formulate PLA/silica/AlO composites. Based on the XRD patterns, it was reported that the crystallinity of the PLA acid film could be increased with the addition of silica/AlO particles into the PLA matrix. This result was attributed to the nucleating role of silica and AlO particles, which would facilitate the crystallization of PLA. However, Santos et al. [59] reported that the crystallinity of PLA was not affected by the inclusion of silica into the PLA matrix, while the dual addition of silica and cellulose into the PLA matrix resulted in significant improvements in the crystallization behavior of PLA. This finding was ascribed to the synergism between the two types of nanoparticles in which the agglomeration of cellulose nanoparticles could be prevented in the presence of silica. Thus, the nucleation was related not only to the chemical nature of the particles but also to the increased contact surface between the silica and cellulose nanoparticles. Zou et al. [60] demonstrated that the inclusion of a proper amount of silica nanoparticles (nucleating agent) would cause a reduction in the nucleation barrier shortening the nucleation period of PLA. The higher crystallinity was obtained in the composites containing 1.5 wt.% of silica nanoparticles. Such an amount of silica was expected to increase the interfacial compatibility and crystallinity of the composites thus enhancing the thermal stability. According to Prapruddivongs et al. [61], the crystallinity of the PLA film would have been affected by the type of silica added to the PLA matrix. Here, two types of silica, such as commercial silica ($CSiO_2$) and silica extracted from rice husks ($RSiO_2$), were added. The DSC thermograms of PLA composites (containing Triallyl isocyanurate and dicumyl peroxide) indicated that the silica nanoparticles affected the crystallinity and the melting behavior of PLA by impeding the chemical crosslinking reactions which were reflected by the change in the FTIR functional band of silica/CPLA composites at 1685 cm^{-1}.

Prapruddivongs and coworkers [62] studied the properties of PLA/silica and chemically crosslinked PLA (CrPLA)/silica composites prepared via the melt-mixing method in the presence of triallyl isocyanate and dicumyl peroxide as crosslinking agents. Here, two types of silica, such as $CSiO_2$ and $RSiO_2$, were used. Irrespective of the silica type, the thermal properties of the PLA/silica CrPLA/silica composites were improved. The addition of 1 wt.% of $CSiO_2$ and $RSiO_2$ led to an increase in the T50 value (the temperature associated with loss of 50 wt.%) of PLA from 321 °C to 339 °C and 342 °C, respectively. It was found also that the degradation temperature of PLA composites tended to increase in the presence of triallyl isocyanate and dicumyl peroxide. Moreover, the thermal stability of CrPLA/$RSiO_2$ composites was better than the thermal stability of CrPLA/$CSiO_2$, implying that $RSiO_2$ was an efficient additive for enhancing the thermal stability of PLA and CrPLA. Vidakis and coworkers [63] prepared PLA/silica composites via the melt-mixing of PLA with different contents of silica, such as 0.5, 1, 2, and 4 wt.%. Although the mass loss for the PLA/4 wt.% silica composite was lower than for PLA/1 wt.% silica, the overall thermal stability of the PLA seemed not to be influenced by the existence of the silica filler, which would be related to the strong H-bond interactions between Si–OH in silica nanoparticles with PLA chains. This is, in turn, restricted their release into the environment.

Lv et al. [64] examined the thermal properties of the PLA/silica composites prepared via the melt-mixing method. As displayed in Figure 5a, the increase in the silica content increased the degradation temperature of PLA/silica composites. For example, the inclusion of 10 wt.% of silica led to increasing the degradation temperature by 15 °C, indicating a significant increase in the thermal stability of PLA was achieved. From Figure 5b, it was found that the cold crystallization temperature (T_{cc}) of neat PLA acid tended to be shifted to lower temperatures upon the inclusion of silica nanoparticles, suggesting that plying the silica nanoparticle worked as nucleating agents or it could hinder crystallization from the melt. While the low-temperature melting peaks became weak, the high-temperature melting peaks showed a gradual increase with the increase in the silica content in the composites. This result would suggest that the inclusion of silica causes a reduction in the defective crystals, increasing the perfect crystals for the PLA phase. According to Ge et al. [65], the dispersibility of silica nanoparticles in the PLA matrix could be improved when the content of silica was less than 3 wt.%. Accordingly, the crystallinity of PLA was improved. The best crystallization behavior was obtained when 1 wt.% of silica was added to the PLA matrix. However, although some agglomerated nanoparticles were observed, the thermal stability of PLA was enhanced by the addition of silica nanoparticles.

The effect of silica content on the T_g of PLA/silica composites was explored by Pilić and coworkers [66]. It was reported that the addition of low amounts of silica, such as 0.2 and 0.5 wt.%, into the PLA matrix led to an increase in the T_g of neat PLA (47.6 °C) to 48.9 and 50.6 °C, respectively. This behavior was ascribed to the fact that the chain mobility throughout the PLA matrix volume tended to be decreased in the presence of silica nanoparticles, while the high loadings of silica, such as 1, 2, 3, and 5 wt.%, resulted in a decrease in the values of T_g in comparison to that in neat PLA, which was assigned to the agglomeration of silica nanoparticles within the PLA matrix which affected the chain mobility of the polymer. However, Wen et al. [56] reported that the addition of various loadings of silica, such as 1, 3, 5, 7, and 10 w.%, did not affect the value of T_g of PLA, suggesting that the impacts of silica on the chain mobility of PLA were insignificant.

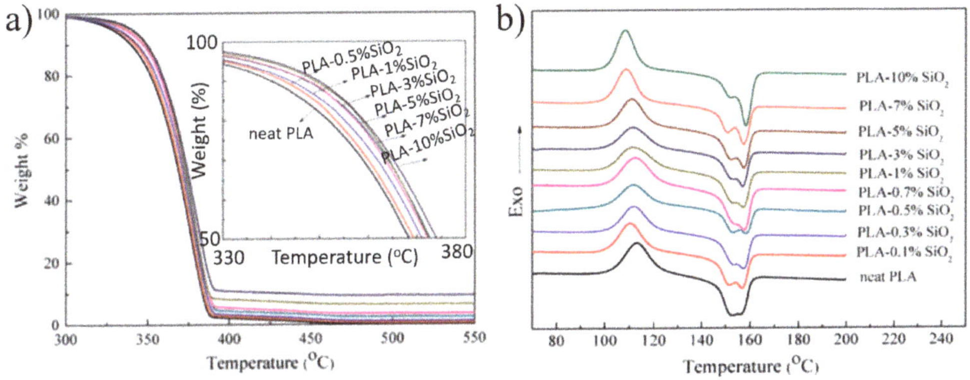

Figure 5. The effects of SiO_2 content on the thermal behavior of neat PLA and PLA/silica composites: (**a**) thermogravimetric analysis; (**b**) crystallization and melting behavior [64].

Techawinyutham et al. [67] examined the thermal stability of PLA and porous silica-containing capsicum oleoresin (SiCO)-modified PLA composites before and after the accelerating weathering test. The results imply that the accelerated weathering test induced the photolysis and partial hydrolysis, which resulted in improvements in the crystallization behavior of PLA. In addition, the accelerated weathering reduced the storage and loss moduli due to the increased chain mobility caused by the chain scission of the polymer chains.

Wu et al. [68] first grafted PLA on the surface of silica nanoparticles, then examined the thermal properties of PLA/PLA-grafted silica composites. Based on the DSC results, it was found that the addition of PLA-grafted silica could accelerate the crystallization rate of PLA. Besides, PLA/PLA-grafted silica composites exhibited typical homopolymer-like behavior in the final structure, regardless of the PLA-grafted silica content. The PLLA/silica composites containing 2.5 wt.% of silica exhibited a shielding effect to the evolution of gases that were released during the decomposition, improving mostly the initial stages of thermal degradation [69]. The thermal stability of PLA could be increased by 20 °C when low amounts of stearic acid-modified silica nanoparticles (0.1–1.5 wt.%) were added to the PLA matrix [40]. Similarly, Khankrua et al. [70] demonstrated that the presence of 5 wt.% of silica nanoparticles in the polymer matrix could cause significant improvements in the thermal stability of PLA. The effect of untreated silica nanoparticles on the thermal stability of PLA was examined by Basilissi et al. [51]. They found that the thermal stability of the composites tended to improve with the increase in the silica content. For instance, the temperature corresponding to 5 wt.% weight loss in the case of PLA composites containing 2 wt.% silica was higher by 70 °C than the counterpart corresponded to neat PLA, which was connected to the formation of silica network structure in the composites. The formation of a silica network structure hindered the diffusion of volatile decomposition products out of PLA and the diffusion of oxygen into the matrix. Similar results were reported by Wen et al. [24], who fabricated PLA/silica composites via the melt-mixing method and found that the thermal stability of PLA could be increased by the addition of $\leq$5 wt.% silica. However, the higher loading of silica ($\geq$5 wt.%) would cause an agglomeration of the nanoparticles in the polymer matrix due to the weak dispersion, which led to a reduction in thermal stability of PLA. In contrast, Lv et al. [64] found that the thermal stability of PLA/silica composites made by melt-mixing method tended to increase with the increase in the silica content—even when high loadings of silica, such as 7 and 10 wt.%, were added to the PLA matrix. Similarly, Mustapa et al. [71] postulated that the thermal stability of PLA could be increased with the addition of 2.5 and 7.5 wt.% of silica nanoparticles. In another works by Mustapa and coworkers [72–74], it was reported that the melting temperature and crystallization behavior were affected by the addition of silica nanoparticles, which acted as nucleation sites that induced the crystallization phenomenon of PLA. Lai and Li [75] functionalized the silica by the melt-mixing with polyurethane and found that the functionalized silica could greatly trigger the nucleation and crystallization behavior of PLA, when compared to the counterpart composites with non-functionalized silica. The influence of silica particles on the thermal properties of PLA composites is summarized in Table 1.

Table 1. The influence of silica incorporation on the thermal properties of PLA composites.

Silica (wt.%)	T_g (°C)	T_{cc} (°C)	T_{onset} (°C)	T_{max} (°C)	ΔH_m (J/g)	Ref.
10	-	-	342.3	370.9	-	[24]
5.0	59	135	-	-	8	[35]
2.0	64.5	0	-	-	31.32	[36]
0.5	52.3	102.8	113.5	-	54.3	[49]
3.0	61.22	107.87	162.60	168.70	34.57	[55]
2.0	52.23	-	-	-	-	[56]
2.5	60	-	-	-	37	[57]
1.5	-	-	-	300	-	[60]
4.0	60.8	108.3	146.6	153.6	24.9	[61]
4.0	-	-	273	374	-	[63]
10	-	-	-	370	-	[64]
5.0	-	120	-	-	70	[65]
0.5	50.6	-	-	-	-	[66]
3.08	57.66	112.33	147.5	153.41	22.73	[67]
0.5	61.02	124.26	-	-	-	[68]
0.2	-	-	345.54	362.56	-	[70]
7.5	59.0	-	355.5	412.3	-	[71]
7.5	59.0	124.94	-	-	6.22	[72]
7.5	60.0	96.3	355.5	412.3	9.9	[73]
7.5	57.4	117	-	-	-	[74]

5. Mechanical Properties

Several research teams have aimed to improve the mechanical properties of PLA with the addition of silica to the polymer matrix. For example, He et al. [76] reported that the inclusion of silica nanoparticles is a good strategy to improve the mechanical properties of PLA. Indeed, the tensile strength (TS), Young's modulus (YM), and impact strength of the PLA/silica composites made by the melt-mixing method were greatly increased with the addition of silica nanoparticles, which was linked to the good dispersibility of the particles in the polymer matrix, as well as PLA-silica interactions [77]. The low loading of silica (below 5 wt.%) was found to be distributed homogenously within the PLA matrix, while some agglomerates were observed at the high loading (higher than 5 wt.%) [36]. Ahmed et al. [78] studied the impact of silica additive into PLA matrix-3D printing. When the silica addition amount was 10 wt.%, its TS, flexibility, YM, and other properties were greatly enhanced. The application of this kind of composite material is feasible, but once the silica content exceeds 15%, the performance is reduced. The improved mechanical properties promote the recycling of PLA while retaining its biodegradability. As such, YM and ultimate tensile stress (UTS) of PDLLA were increased by 106% and 63.7%, respectively, upon the addition of 3 wt.% silica into the polymer matrix [79].

Yan et al. [32] prepared PLA/silica composites via two steps: grafting of L-lactic acid oligomer onto the silica surface, followed by melt-mixing with PLA. This procedure led to fabricating composites with improved mechanical properties. In another work by Yan and co-workers [31], the TS of the plasticized PLA/silica composite could be increased to large values even in the presence of low loadings of silica. Although the addition of the silica nanoparticles accelerated the biodegradation rate of PLA, the resultant composites had improved mechanical properties, as reported by Georgiopoulos et al. [80].

To obtain better dispersion in the PLA matrix, silica nanoparticles with 24 nm in average diameter were functionalized via stearic acid [38]. Different contents of silica, such as 0, 0.1, 0.3, 0.5, 0.8, 1, and 1.5 wt.%, were mixed with PLA via an extrusion technique. The mechanical tests revealed that the addition of 0.8 wt.% of silica nanoparticles into the PLA matrix led to increasing the Izod impact strength from 1.57 to 5.13 kJ/m^2. The impact strength values of the PLA/silica composites increased with increasing silica content, then decreased when the silica content exceeded 0.8 wt.%. Moreover, the presence of nano-silica in the PLA matrix caused improvements in the EB and TS of PLA (Figure 6a). As shown in Figure 6b, dimple morphologies were observed on the fractural surfaces of PLA/silica composites, suggesting that neat PLA was less ductile than PLA/silica composites. According to the lubrication sliding and the ductile behavior, it was confirmed the toughness of PLA could be largely improved by the incorporation of silica nanoparticles. The effects of silica content (1, 3, and 5 wt.%) on the mechanical performance of PLA film were studied Zirak and Tabari [81]. It was found that the values of the TS and YM of PLA tended to be increased upon the incorporation of silica nanoparticles into the PLA matrix, regardless of the silica content. For example, the addition of 5 wt.% of silica into PLA led to an increase in the values of TS and YM for PLA from 29 MPa and 2.3 GPa to 43 MPa and 3.1 GPa, respectively, which was attributed to the reinforcing effect of SiO_2 nanoparticles. In contrast, the value of EB of PLA (17.5%) was reduced with the inclusion of silica nanoparticles, where a value of 10.5% was found in the composite containing 5 wt.% SiO_2.

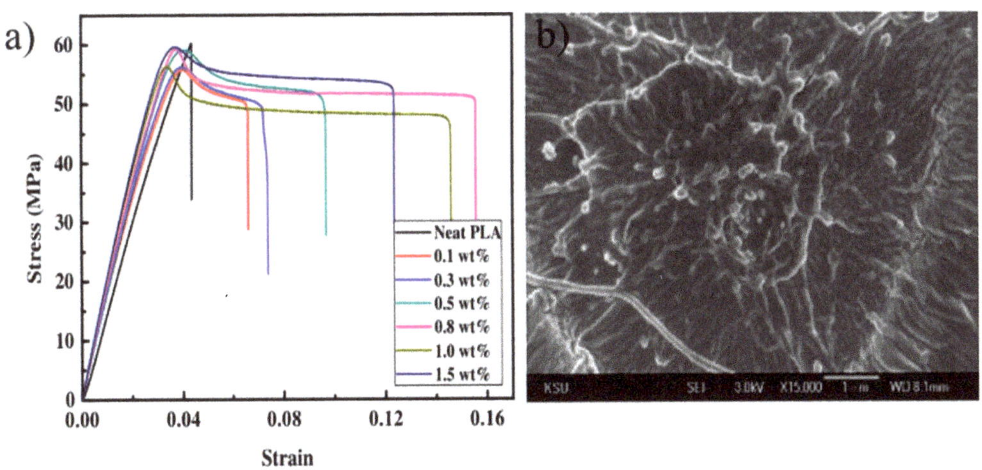

Figure 6. (**a**) The stress–strain curve of neat PLA and PLA composites. (**b**) The fracture morphology after the tensile test of PLA composite [38].

In Yan's research [32], it was approved that the grafting of L-lactic acid oligomer on the silica nanoparticles by condensation reaction without catalyst would result in the formation of a good filler, helping to increase the toughness of PLA. Chrissafis et al. [82] reported that the inclusion of 2.5 wt.% of fumed silica and montmorillonite into the PLA matrix would result in the formation of composites having higher values of TS and YM than neat

PLA. However, the addition of fumed silica and montmorillonite to PLA led to a reduction in the value of EB, implying that those two additives acted as reinforcing agents. Lai and Hsieh [83] examined the influence of the surface functionalization of silica on the mechanical properties of PLA. To form the modified silica, polyethylene glycol methyl ether was grafted on the silica in the presence of aminosilane. The interfacial interaction between silica and PLA was improved in the presence of modified silica; thereby, the value of TS of the composites with modified silica was higher than the counterpart composites containing unmodified silica.

Battegazzore et al. [84] used different contents, such as 5, 10, 20, and 30 wt.%, of silica powder, obtained by the conventional extraction method from rice husk, to fabricate PLA/silica nanocomposites via the melt-mixing method. Based on Archimedes law, the silica density was calculated to be 1.82 and 1.88 $g{\cdot}cm^{-3}$ for the extracted silica and commercial silica, respectively, while the density of PLA was 1.25 $g{\cdot}cm^{-3}$. The density values were used to calculate the volumetric fractions. When the extracted silica nanoparticles were melted with PLA, the YM of neat PLA was increased, while the oxygen permeability of PLA was slightly reduced. As compared to the composites including 10 and 30 wt.% of commercial silica, the counterpart composites with extracted silica exhibited better mechanical properties. Considering the economic analysis of the whole process and by reusing the energy recovered from burning rice husk, the authors considered the composites containing 20 wt.% of extracted silica as economically sustainable materials.

It is worth mentioning that silica can improve the mechanical properties of other PLA composites. For example, the addition of silica can improve the jute/PLA interfacial adhesion, which results in the enhancement of the mechanical performances of the resultant composites [85].

On the other hand, the influence of other silica-based materials on the properties of PLA was explored by many research groups. For instance, the addition of glass fibers to the PLA matrix improved the mass flow rate and flexural modulus, while their effects on the impact and flexural strength, as well as thermal stability of PLA, were insignificant [86]. Besides, the TS and EB were reduced with the addition of glass fibers.

6. Biodegradbility and Other Properties

In general, significant research still needs to be performed to achieve the final target of ideal biodegradable PLA/silica composites that exhibit high performance and easy biodegradability when their roles are completed. The hydrophilic nature of silica nanoparticles is expected to affect the degradation of PLA. The hydroxyl groups in silica are bound together by hydrogen bonds and can assist the hydrogen bonding interaction with the functional groups in PLA or a covalent bonding with a macromolecular chain [87]. Thus, silica nanoparticles are expected to facilitate the hydrolysis or enzymatic attacks of ester groups of PLA, leading to a fast biodegradation rate. For instance, Li et al. [50] reported that the weight loss during the biodegradation process of PLA/silica composites, fabricated by melt-mixing method, was larger than that in neat PLA. Indeed, the incorporation of 9 wt.% of silica nanoparticles into the PLA matrix led to a biodegradation rate of 0.36 $mg\ cm^{-2}\ h^{-1}$, which was 6.5 times higher than that of neat PLA. Figure 7a shows the fast biodegradation of the PLA composites compared to neat PLA. This result was confirmed by the DSC curves shown in Figure 7b. As shown in Figure 7b, all samples were amorphous at a T_g of 60 °C. The improvements in the degradation rate were attributed to the easy release of silica particles from the PLA matrix. The hydrophilic silica facilitated the hydrolysis and enzymatic attack of ester groups of PLA chains. The biodegradation was more pronounced in the composite membrane after 2 months of in vitro tests [88]. On the other hand, it was reported that the flame retardant properties of PLA could be improved greatly by adding treated silica nanoparticles into the PLA matrix [64]. Indeed, the weight loss ratios of PLA/silica composites at different burning time intervals tended to decrease with the increase in the silica content, which was linked to the excellent dispersion of particles in the PLA matrix. So, the larger specific surface area in the composites provided a better effect of thermal insulation [64]. SiO_2-fluorinated PLA composites can be used as reversible and highly hydrophobic coatings to protect the exterior of buildings [89]. Similarly, it was reported that the flame retardancy of PLA could be improved by the addition of fumed silica and Ni_2O_3 into the polymer matrix [90].

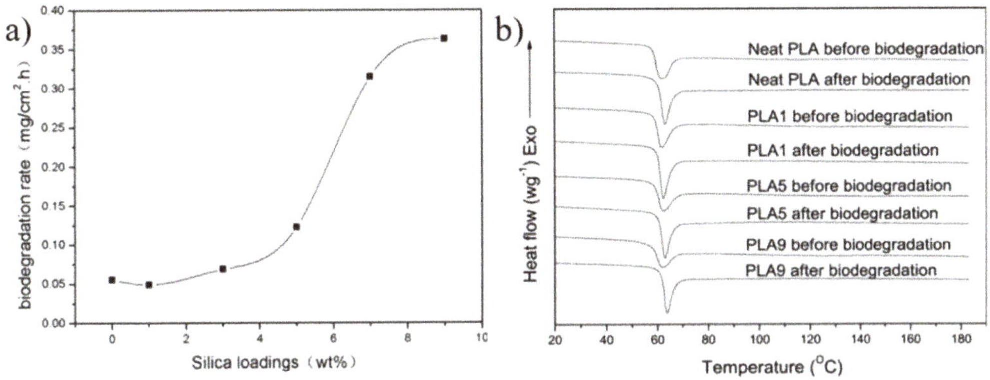

Figure 7. (**a**) The effect of silica content on the biodegradation rate of PLA/silica composites. (**b**) The DSC curves of the PLA/silica composites before and after degradation of 24 h [50].

A study has been reported by Zhao et al. [91], where *N*-halamine precursor with epoxy and hydantoin structures, 3-(4′-epoxyethyl-benzyl)-5, 5-dimethylhydantoin (EBDMH) was utilized for the *N*-halamine-modified silica nanoparticles. EBDMH was fabricated and immobilized onto aminofunctionalized silica nanoparticles to form EBDMH–SiO_2 nanoparticles (Figure 8a). PLA was mixed with EBDMH–SiO_2 nanoparticles via the melt-mixing method. The efficiency of the PLA/EBDMH–SiO_2 composite as antimicrobial material was evaluated against *S. aureus* and *E. coli*, respectively, and the obtained results are shown in Figure 8b. The PLA/EBDMH–SiO_2 composite exhibited excellent bactericidal efficiency. Indeed, the PLA/EBDMH–SiO_2 composite with a contact time of 10 min neutralized about 90.2% of *S. aureus* and 89.4% of *E. coli*. When the contact time was increased to 180 min, a kill efficiency of 99.97% and 99.91% against *S. aureus* and *E. coli*, respectively, was obtained. Due to improved biocompatibility between PLA and silica, as well as the excellent antibacterial efficiency, the PLA/EBDMH–SiO_2 composite could be utilized for hygienic product packaging and filters, as well as medical textiles.

The impact of silica nanoparticles on the interfacial tension between PLA and supercritical CO_2 at high temperature and high pressures was examined by Sarikhani and coworkers [92]. The addition of a low loading of silica (less than 2 wt.%) led to reducing the interfacial tension, while the interfacial tension tended to increase when the silica content was larger than 2 wt.%, which was linked to the fact that higher levels of silica originated attractive lateral capillary forces due to the perturbation of the PLA–CO_2 interface by particles. The interfacial interactions between PLA and silica nanoparticles were found to be decreased with an increase in the CO_2 content, which facilitated the adsorption behavior at higher pressures. Based on the experiments carried out by Seng and co-workers [93], it was reported that the addition of 1 wt.% of silanol treated-silica into the PLA matrix led to a decrease of 40% in the hygroscopicity of PLA, while the other loadings of silica caused improvements between 3 and 19% in the reduction in hygroscopicity. Chen et al. [94] reported that the hydrophilicity of PLA/silica composites tended to be improved with increasing the silica content, where the inclusion of 5 wt.% of silica into the PLA matrix led to a decrease in the water contact angle of PLA from 82° to 68°. In other words, the less silica in the composite, the larger the contact angle is, thereby, the higher the hydrophobicity of the composite surface is. Thus, the hydrolytic degradation ability of PLLA/silica composites was accelerated in the presence of higher loadings of silica in the polymer matrix.

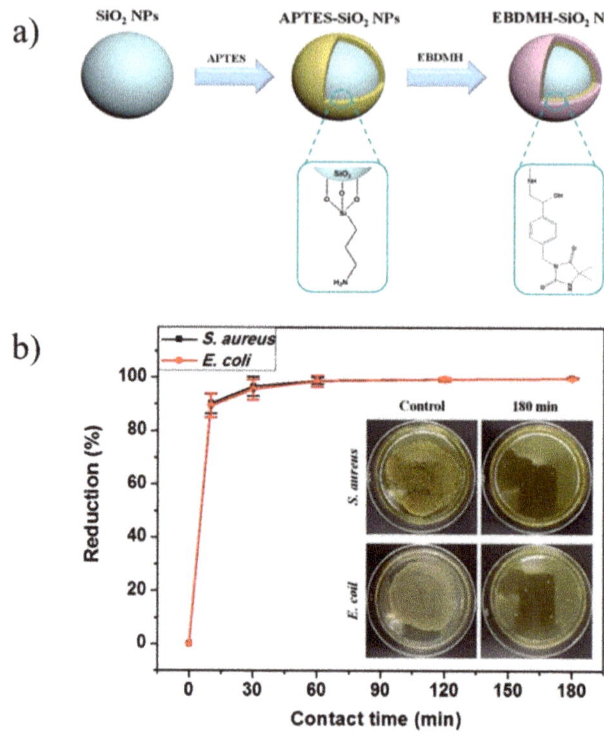

Figure 8. (**a**) Synthesis of *N*-halamine precursor-modified silica nanoparticles (EBDMH–SiO_2 NPs). (**b**) Antibacterial tests of PLA/EBDMH–SiO_2-9 plastic sheets after chlorination against *S. aureus* and *E. coli*. In the inset are photographs showing the bacterial culture plates of *S. aureus* and *E. coli* upon 180 min contact with the control and PLA/EBDMH–SiO_2-9 plastic sheets [91].

Kosowska and Szatkowski [95] examined the impacts of silica addition on the ultraviolet aging of PLA nonwovens obtained by the electrospinning technique. The inclusion of silica nanoparticles greatly increased the percentages of crystalline and amorphous phases in the fabricated films and altered the photodegradation mechanism. In the work of Jin-Bo [96], a novel porous membrane composed of PLA with the addition of SiO_2–CaO by the sol–gel method was designed. The surface potentials of the composite membranes became more negative with higher SiO_2–CaO contents. Apatite with an orderly ring structure nucleated and grew on the surface of the composite membranes after immersion in SBF for 7 days, implying that the incorporation of SiO_2–CaO significantly improves the bioactivity of PLA. Based on the dielectric spectroscopy study [97], it was reported that the lifetime and thermal stability of the electric state in PLA can be increased with the addition of SiO_2, where the optimal concentration of silica for negative corona electrets was found to be 4 wt.%.

7. Potential Applications of PLA/Silica Composites

Owing to its high resistance to heat and good thermal stability, silica has received considerable attention in 3D-printing applications. The enhanced melt viscosity of PLA/silica composites could facilitate the molding and processing of these materials which could lead to a variety of applications [54]. Since the usage of PLA used in the bone tissue applications would not be able to withstand the high-load resistance associated with such applications, the incorporation of silica would make PLA a suitable candidate in these applications, as the flexural modulus and the tensile strength of PLA tended to be improved significantly with the incorporation of a silica filler. For this reason, it was reported that the biorenewable PLA/SiO_2 composites can be used as promising materials in the bone repair processes in animal models [98]. As indicated by cell-culture measurements conducted by Abe et al. [99], PLA/silica composite could exhibit excellent cytocompatibility. Interestingly, the ROS-responsive LDLR peptides-conjugated PLA-coated mesoporous silica nanoparticles were reported to have an important opportunity for oxidative stress therapy in the central nervous system [100]. Moreover, the good thermal stability and mechanical properties of PLA/silica composites would make these materials suitable for structural applications where high thermal stability and high mechanical strength are needed [71]. Thanks to their physical, thermal, rheological, and mechanical properties, the PLA/silica composites can be used in 3D-printing applications, as indicated by Thongsang and coworkers [101]. Moreover, the PLA/silica composites have a great opportunity to be used as smart and active packaging. According to Jaikaew [102], the CO_2/O_2 permeability ratio of PLA/silica composite films can be tuned by varying the types of silica particles and their compositions. Light transmission reduction in both the UV and visible regions is achieved in PLA/modified-silica bio-composite films. As approved by Opaprakasit et al. [103], the films obtained from PLA/modified silica composites have strong potential to be used as biodegradable packaging materials with tunable gas permeability.

8. Future Outlooks of PLA/Silica Composites

The state of knowledge discussed above clearly shows that the fabrication, properties, and applications of PLA/silica composites will keep growing in the future. We thought that future trends of the research on PLA/silica composites should be focused on several relatively new directions, as follows:

- Although PLA/silica composites have been utilized in 3D-printing applications, more research would be needed in this area, as the main challenge in 3D-printing applications is enhancing the flexibility, as well as controlling the viscosity, of PLA.
- The dispersion of silica nanoparticles in the PLA matrix should be improved, as various undiscovered applications in various fields, such as the aerospace, energy, and chemical industries, are expected to be available shortly.
- The utilization of PLA/silica composites to produce hierarchically porous PLA materials for biomedical applications, such as tissue scaffolding, would be a promising research area shortly. The processing conditions, however, should be selected carefully to successfully fabricate such materials.
- Even PLA/silica composites exhibited some levels of biodegradability; they are still far from being considered a solution for plastic waste accumulation.

9. Conclusions

Based on the brief discussion reported in this work, the inclusion of silica nanoparticles into the PLA matrix is one of the promising methods to enhance the performance of PLA while retaining its biodegradability. The fabrication methods, namely, melt blending and in situ polymerization, as well as the organo-modification of silica nanoparticles to increase their affinity to hydrophobic PLA, would play a pivotal role in the final performance of PLA/silica composites. Indeed, the surface functionalization of silica nanoparticles had critical impacts on the rheological performance of the silica within the polymer matrix. The thermal stability, biodegradability, YM, and TS of PLA were greatly improved by the inclusion of silica nanoparticles into the polymer matrix. Such improved properties qualified the PLA/silica composites to be suitable candidates in food packaging, 3D printing, and biomedical applications.

Author Contributions: M.K.: Conceptualization, writing—original draft, writing-review and editing; funding acquisition; Z.U.R., S.H., A.K.S. and B.D.: writing—review and editing. All authors have read and agreed to the published version of the manuscript.

Funding: This work was supported by the National Research Foundation of Korea (NRF) grant funded by the Korea government (MSIT) (No. NRF-2019R1G1A1099335).

Institutional Review Board Statement: Not applicable.

Informed Consent Statement: Informed consent was obtained from all subjects involved in the study. The data presented in this study are available on request from the corresponding authors.

Data Availability Statement: The data presented in this study are available on request from the corresponding author.

Conflicts of Interest: The authors declare no conflict of interest.

References

1. Chandra, R.; Rustgi, R. Biodegradable polymers. *Prog. Polym. Sci.* **1998**, *23*, 1273–1335. [CrossRef]
2. Briassoulis, D. An overview on the mechanical behaviour of biodegradable agricultural films. *J. Polym. Environ.* **2004**, *12*, 65–81. [CrossRef]
3. Hamad, K.; Kaseem, M.; Ayyoob, M.; Joo, J.; Deri, F. Polylactic acid blends: The future of green, light and tough. *Prog. Polym. Sci.* **2018**, *85*, 83–127. [CrossRef]
4. Hamad, K.; Kaseem, M.; Yang, H.W.; Deri, F.; Ko, Y.G. Properties and medical applications of polylactic acid: A review. *Express Polym. Lett.* **2015**, *9*, 435–455. [CrossRef]
5. Kaseem, M.; Hamad, K.; Deri, F.; Ko, Y.G. A review on recent researches on polylactic acid/carbon nanotube composites. *Polym. Bull.* **2017**, *74*, 2921–2937. [CrossRef]
6. Kaseem, M.; Ko, Y.G. Melt flow behavior and processability of polylactic acid/polystyrene (PLA/PS) polymer blends. *J. Polym. Environ.* **2017**, *25*, 994–998. [CrossRef]
7. Kaseem, M.; Hamad, K.; Rehman, Z.U. Review of recent advances in polylactic acid/TiO_2 composites. *Materials* **2019**, *12*, 3659. [CrossRef] [PubMed]
8. Hamad, K.; Kaseem, M.; Deri, F. Preparation and characterization of binary and ternary blends with poly(lactic acid), polystyrene, and acrylonitrile-butadiene-styrene. *J. Biomater. Nanobiotechnol.* **2012**, *3*, 405–412. [CrossRef]
9. Kaseem, M. Poly(lactic acid) composites. *Materials* **2019**, *12*, 3586. [CrossRef]
10. Hamad, K.; Ko, Y.G.; Kaseem, M.; Deri, F. Effect of acrylonitrile–butadiene–styrene on flow behavior and mechanical properties of polylactic acid/low density polyethylene blend. *Asia-Pac. J. Chem. Eng.* **2014**, *9*, 349–353. [CrossRef]
11. Kadhum, S.A. The Effect of two types of nano-particles (ZnO and SiO_2) on different types of bacterial growth. *Biomed. Pharmacol. J.* **2017**, *10*, 1701–1708. [CrossRef]
12. Li, H.; Liebscher, M.; Curosu, I.; Choudhury, S.; Hempel, S.; Davoodabadi, M.; Dinh, T.T.; Yang, J.; Mechtcherine, V. Electrophoretic deposition of nano-silica onto carbon fiber surfaces for an improved bond strength with cementitious matrices. *Cem. Concr. Compos.* **2020**, *114*, 103777. [CrossRef]
13. Singh, L.; Karade, S.; Bhattacharyya, S.; Yousuf, M.; Ahalawat, S. Beneficial role of nanosilica in cement based materials A review. *Constr. Build. Mater.* **2013**, *47*, 1069–1077. [CrossRef]
14. Sikora, P.; Chougan, M.; Cuevas, K.; Liebscher, M.; Mechtcherine, V.; Ghaffar, S.H.; Stephan, D. The Effects of Nano-and Micro-Sized Additives on 3D Printable Cementitious and Alkali-Activated Composites: A Review. *Appl. Nanosci.* **2021**, 1–19. Available online: https://link.springer.com/article/10.1007/s13204-021-01738-2 (accessed on 16 August 2021).
15. Ren, Z.; Zhou, W.; Qing, Y.; Duan, S.; Pan, H.; Zhou, Y.; Li, N. Microwave absorption and mechanical properties of SiCf/SiOC composites with SiO2 fillers. *Ceram. Int.* **2021**, *47*, 8478–8485. [CrossRef]
16. Kaseem, M.; Choi, K.; Ko, Y.G. A highly compact coating responsible for enhancing corrosion properties of Al-Mg-Si alloy. *Mater. Lett.* **2017**, *196*, 316–319. [CrossRef]
17. Liu, W.; Wu, X.; Zhan, H.; Yan, F. Synthesis of bioactive poly(ethylene glycol)/SiO_2-CaO-P_2O_5 hybrids for bone regeneration. *Mater. Sci. Eng. C* **2012**, *32*, 707–711. [CrossRef]
18. Jia, H.; Hou, W.; Wei, L.; Xu, B.; Liu, X. The structures and antibacterial properties of nano-SiO_2 supported silver/zinc–silver materials. *Dent. Mater.* **2008**, *24*, 244–249. [CrossRef]
19. Petousis, M.; Tzounis, L.; Papageorgiou, D.; Vidakis, N. Decoration of SiO_2 and Fe_3O_4 nanoparticles onto the surface of MWCNT-grafted glass fibers: A simple approach for the creation of binary nanoparticle hierarchical and multifunctional composite interphases. *Nanomaterials* **2020**, *10*, 2500. [CrossRef]
20. Fatimah, S.; Kamil, M.P.; Kwon, J.H.; Kaseem, M.; Ko, Y.G. Dual incorporation of SiO_2 and ZrO_2 nanoparticles into the oxide layer on 6061 Al alloy via plasma electrolytic oxidation: Coating structure and corrosion properties. *J. Alloys Compd.* **2017**, *707*, 358–364. [CrossRef]
21. Tsai, P.-A.; Chiu, W.-M.; Lin, C.-E.; Wu, J.-H. Fabrication and characterization of PLA/SiO_2/Al_2O_3 composites prepared by Sol-Gel process. *Polym. Plast. Technol. Eng.* **2013**, *52*, 1488–1495. [CrossRef]
22. Kim, G.H.; Hwang, S.W.; Jung, B.N.; Kang, D.; Shim, J.K.; Seo, K.H.E. Effect of PMMA/Silica hybrid particles on interfacial adhesion and crystallization properties of Poly(Lactic Acid)/Block acrylic elastomer composites. *Polymers* **2020**, *12*, 2231. [CrossRef]

23. Zehra, T.; Kaseem, M.; Hossain, S.; Ko, Y.G. Fabrication of a protective hybrid coating composed of TiO_2, MoO_2, and SiO_2 by plasma electrolytic oxidation of titanium. *Metals* **2021**, *8*, 1182. [CrossRef]
24. Wen, X.; Zhang, K.; Wang, Y.; Han, L.; Han, C.; Zhang, H.; Chen, S.; Dong, L. Study of the thermal stabilization mechanism of biodegradable poly(L-lactide)/silica nanocomposites. *Polym. Int.* **2011**, *60*, 202–210. [CrossRef]
25. Basu, A.; Nazarkovsky, M.; Ghadi, R.; Khan, W.; Domb, A.J. Poly(lactic acid)-based nanocomposites. *Polym. Adv. Technol.* **2017**, *28*, 919–930. [CrossRef]
26. Hakim, R.H.; Cailloux, J.; Santana, O.O.; Bou, J.; S_anchez-Soto, M.; Odent, J.; Raquez, J.M.; Dubois, P.; Carrasco, F.; Maspoch, M.L. PLA/SiO_2 composites: Influence of the filler modifications on the morphology, crystallization behavior, and mechanical properties. *J. Appl. Polym. Sci.* **2017**, *134*, 45367. [CrossRef]
27. Yan, S.; Yin, J.; Yang, J.; Chen, X. Structural characteristics and thermal properties of plasticized poly(L-lactide)-silica nanocomposites synthesized by sol-gel method. *Mater. Lett.* **2007**, *61*, 2683–2686. [CrossRef]
28. Dash, S.; Mishra, S.; Patel, S.; Mishra, B.K. Organically modified silica: Synthesis and applications due to its surface interaction with organic molecules. *Adv. Colloid Interface Sci.* **2008**, *140*, 77–94. [CrossRef]
29. Parida, S.K.; Dash, S.; Patel, S.; Mishra, B.K. Adsorption of organic molecules on silica surface. *Adv. Colloid Interface Sci.* **2006**, *121*, 77–110. [CrossRef] [PubMed]
30. Bagwe, R.P.; Hilliard, L.R.; Tan, W. Surface modification of silica nanoparticles to reduce aggregation and nonspecific binding. *Langmuir* **2006**, *22*, 4357–4362. [CrossRef] [PubMed]
31. Mai, Y.-W.; Yu, Z.-Z. *Polymer Nanocomposites*; Woodhead Publishing Limited: Cambridge, UK, 2006.
32. Yan, S.; Yin, J.; Yang, Y.; Dai, Z.; Ma, J.; Chen, X. Surface-grafted silica linked with L-Lactic acid oligomer: A novel nanofiller to improve the performance of biodegredable poly(L-Lactide). *Polymer* **2007**, *48*, 1688–1694. [CrossRef]
33. Wu, L.; Cao, D.; Huang, Y.; Li, B.G. Poly(L-lactic acid)/SiO_2 nanocomposites via in situ melt polycondensation of L-lactic acid in the presence of acidic silica sol: Preparation and characterization. *Polymer* **2008**, *49*, 742–748. [CrossRef]
34. Liu, L.Z.; Ma, H.J.; Zhu, X.S. Preparation and properties of polylactide/nano-silica in situ composites. *Pigment. Resin. Technol.* **2010**, *39*, 27–31. [CrossRef]
35. Chang, C.C.; Tsai, Y.L.; Chiu, J.J.; Chen, H. Preparation of phase change materials microcapsules by using PMMA network-silica hybrid shell via sol-gel process. *J. Appl. Polym. Sci.* **2009**, *112*, 1688–1694. [CrossRef]
36. Fukushima, K.; Tabuani, D.; Abbate, C.; Arena, M.; Rizzarelli, P. Preparation, characterization and biodegradation of biopolymer nanocomposites based on fumed silica. *Euro. Polym. J.* **2011**, *47*, 139–152. [CrossRef]
37. Luo, D.; Zhen, W.; Dong, C.; Zhao, L. Performance and multi-scale investigation on the phase miscibility of poly(lactic acid)/amided silica nanocomposites. *Inter. J. Biol. Macromol.* **2021**, *177*, 271–283. [CrossRef] [PubMed]
38. Ma, Z.; Wu, L.; Peng, B. Poly(L-lactic acid)/silicone dioxide nanocomposites prepared via in situ melt polycondensation of L-lactic acid in the presence of acidic silica sol: Dispersion stability of nanoparticles during dehydration/oligomerization. *J. Appl. Polym. Sci.* **2012**, *124*, 3980–3987. [CrossRef]
39. Wu, J.H.; Yen, M.S.; Kuo, M.C.; Chen, B.H. Physical properties and crystallization behavior of silica particulates reinforced poly(lactic acid) composites. *Mater. Chem. Phys.* **2013**, *142*, 726–733. [CrossRef]
40. Ray, S.S.; Okamoto, M. Polymer/layered silicate nanocomposites: A review from preparation to processing. *Prog. Polym. Sci.* **2003**, *28*, 1539–1641.
41. Zou, J.; Ma, T.; Zhang, J.; He, W.; Huang, F. Preparation and characterization of PLLA–ESO/surface-grafted silica nanocomposites. *Polym. Bull.* **2011**, *67*, 1261–1271. [CrossRef]
42. Sepulveda, J.; Villegas, C.; Torres, A.; Vargas, E.; Rodriguez, F.; Baltazar, S.; Prada, A.; Rojas, A.; Romero, J.; Faba, S.; et al. Effect of functionalized silica nanoparticles on the mass transfer process in active PLA nanocomposite films obtained by supercritical impregnation for sustainable food packaging. *J. Supercrit. Fluids* **2020**, *161*, 104844. [CrossRef]
43. Zhu, A.; Diao, H.; Rong, Q.; Cai, A. Preparation and properties of polylactide–silica nanocomposites. *J. Appl. Polym. Sci.* **2010**, *116*, 2866–2873. [CrossRef]
44. Kaseem, M.; Hamad, K.; Deri, F. Slit die rheology of thermoplastic starch during extrusion process. *Inter. J. Polym. Mater.* **2013**, *17*, 51–60. [CrossRef]
45. Kaseem, M.; Hamad, K.; Deri, F. On-line rheological and mechanical measurement of ABS/starch composites. *Inter. J. Polym. Mater.* **2013**, *62*, 260–264. [CrossRef]
46. Kaseem, M.; Hamad, K.; Park, J.H.; Ko, Y.G. Rheological properties of ABS/wood composites. *Euro. J. Wood. Wood. Prod.* **2015**, *73*, 701–703. [CrossRef]
47. Kaseem, M.; Hamad, K.; Yang, H.W.; Deri, F.; Ko, Y.G. Melt rheology of poly(vinylidene fluoride) (PVDF)/low density polyethylene (LDPE) blends. *Poly. Sci. Part. A* **2015**, *57*, 233–238. [CrossRef]
48. Kaseem, M.; Hamad, K.; Deri, F.; Ko, Y.G. Effect of wood fibers on the rheological and mechanical properties of polystyrene/wood composites. *J. Wood. Chem. Technol.* **2017**, *37*, 251–260. [CrossRef]
49. Kaseem, M.; Hamad, K.; Deri, F. Preparation and studying properties of thermoplastic starch/acrylonitrile–butadiene–styrene blend. *Int. J. Plast. Technol.* **2012**, *16*, 39–49. [CrossRef]
50. Basilissi, L.; di Silvestro, G.; Farina, H.; Ortenzi, M.A. Synthesis and characterization of PLA nanocomposites containing nanosilica modified with different organosilanes II: Effect of the organosilanes on the properties of nanocomposites: Thermal characterization. *J. Appl. Polym. Sci.* **2013**, *128*, 3057–3063. [CrossRef]

51. Li, Y.; Han, C.; Bian, J.; Han, L.; Dong, L.; Gao, G. Rheology and biodegradation of polylactide/silica nanocomposites. *Polym. Compos.* **2012**, *33*, 1719–1727. [CrossRef]
52. Nerantzaki, M.; Prokopiou, L.; Bikiaris, D.N.; Patsiaoura, D.; Chrissafis, K.; Klonos, P.; Kyritsis, A.; Pissis, P. In situ prepared poly(DL-lactic acid)/silica nanocomposites: Study of molecular composition, thermal stability, glass transition and molecular dynamics. *Thermochim. Acta* **2018**, *669*, 16–29. [CrossRef]
53. Pandis, C.; Trujillo, S.; Matos, J.; Madeira, S.; Ródenas-Rochina, J.; Kripotou, S.; Kyritsis, A.; Mano, J.F.; Ribelles, J.L.G. Porous polylactic acid–silica hybrids: Preparation, characterization, and study of mesenchymal stem cell osteogenic differentiation. *Macromol. Biosci.* **2015**, *15*, 262. [CrossRef]
54. Hao, X.; Kaschta, J.; Schubert, D.W. Viscous and elastic properties of polylactide melts filled with silica particles: Effect of particle size and concentration. *Compos. Part. B* **2016**, *89*, 44–53. [CrossRef]
55. Wen, X.; Lin, Y.; Han, C.; Zhang, K.; Ran, X.; Li, Y.; Dong, L. Thermomechanical and optical properties of biodegradable poly(L-lactide)/silica nanocomposites by melt compounding. *J. Appl. Polym. Sci.* **2009**, *114*, 3379–3388. [CrossRef]
56. Zhang, J.; Lou, J.; Ilias, S.; Krishnmachari, P.; Yan, J. Thermal properties of poly(lactic acid) fumed silica nanocomposites: Experiments and molecular dynamics simulations. *Polymer* **2008**, *49*, 2381–2386. [CrossRef]
57. Klonos, P.; Pissis, P. Effects of interfacial interactions and of crystallization on rigid amorphous fraction and molecular dynamics in polylactide/silica nanocomposites: A methodological approach. *Polymer* **2017**, *112*, 228–243. [CrossRef]
58. Bouamer, A.; Benrekaa, N.; Younes, A. Characterization of polylactic acid ceramic composites synthesized by casting method. *Mater. Today Proc.* **2021**, *42*, 2959–2962. [CrossRef]
59. Santos, F.A.D.; Tavares, M.I.B. Development and characterization of hybrid materials based on biodegradable PLA matrix, microcrystalline cellulose and organophilic silica. *Polímeros* **2014**, *24*, 561–566. [CrossRef]
60. Zuo, Y.; Chen, K.; Li, P.; He, X.; Li, W.; Wu, Y. Effect of nano-SiO2on the compatibility interface and properties of polylactic acid-grafted-bamboofiber/polylactic acid composite. *Inter. J. Biol. Macromol.* **2020**, *157*, 177–186. [CrossRef]
61. Prapruddivongs, C.; Wongpreede, M.A.T. Effect of silica resources on the biodegradation behavior of poly (lactic acid)and chemical crosslinked poly (lactic acid) composites. *Polym. Test.* **2018**, *71*, 87–94. [CrossRef]
62. Prapruddivongs, C.; Rukrabiab, J.; Kulwongwit, N.; Wonpreedee, T. Effect of surface-modified silica on the thermal and mechanical behaviors of poly(lactic acid) and chemically crosslinked poly(lactic acid) composites. *J. Thermoplast. Compos. Mater.* **2020**, *33*, 1692–1706. [CrossRef]
63. Vidakis, N.; Petousis, M.; Velidakis, E.; Mountakis, N.; Tzounis, L.; Liebscher, M.; Grammatikos, S.A. Enhanced mechanical, thermal and antimicrobial properties of additively manufactured polylactic Acid with optimized nano silica content. *Nanomaterials* **2021**, *11*, 1012. [CrossRef] [PubMed]
64. Lv, H.; Song, S.; Sun, S.; Ren, L.; Zhang, H. Enhanced properties of poly(lactic acid) with silica nanoparticles. *Polym. Adv. Technol.* **2016**, *27*, 1156–1163. [CrossRef]
65. Ge, H.; Zhu, Z.; Yin, H.; Zhang, X.; Wang, R. Fabrication and properties of sc-PLA/SiO_2, Fabrication and Properties of sc-PLA/SiO_2 Composites. *IOP Conf. Ser. Mater. Sci. Eng.* **2014**, *62*, 012031. [CrossRef]
66. Pilic, B.M.; Radusin, T.I.; Ristic, I.S.; Silvestre, C.; Lazic, V.L.; Balos, S.S.; Duraccio, D. Hydrophobic silica nanoparticles as reinforcing filler for poly (lactic acid) polymer matrix. *Hem. Ind.* **2016**, *70*, 73–80. [CrossRef]
67. Techawinyutham, L.; Siengchin, S.; Dangtungee, R.; Parameswaranpillai, J. Influence of accelerated weathering on the thermo-mechanical, antibacterial, and rheological properties of polylactic acid incorporated with porous silica-containing varying amount of capsicum oleoresin. *Compos. Part. B* **2019**, *175*, 107108. [CrossRef]
68. Wu, F.; Lan, X.; Ji, D.; Liu, Z.; Yang, W.; Yang, W.; Yang, M. Grafting polymerization of polylactic acid on the surface of nano-SiO2 and properties of PLA/PLA-grafted-SiO2 nanocomposites. *J. Appl. Polym. Sci.* **2013**, *129*, 3019–3027. [CrossRef]
69. Chrissafis, K.; Bikiaris, D. Can nanoparticles really enhance thermal stability of polymers? Part I: An overview on thermal decomposition of addition polymers. *Thermochim. Acta* **2011**, *523*, 1–24. [CrossRef]
70. Khankrua, R.; Pivsa-Art, S.; Hiroyuki, H.; Suttiruengwong, S. Thermal and mechanical properties of biodegradable polyester/silica nanocomposites. *Energy Procedia* **2013**, *34*, 705–713. [CrossRef]
71. Mustapa, I.R.; Ansari, M.N.M.; Shanks, R.A. Poly(Lactic acid) Hemp composites combined with nano-silica. *TechConnect Briefs* **2011**, *3*, 7136–7138.
72. Mustapa, I.R.; Shanks, R.A.; Kong, I. Poly(lactic acid)-hemp-nanosilica hybrid composites: Thermomechanical, thermal behavior and morphological properties. *Inter. J. Adv. Sci. Eng. Technol.* **2013**, *3*, 192–199.
73. Mustapa, I.R.; Shanks, R.A.; Kong, I. Multiple melting behavior of poly(lactic acid)-hemp-silica composites using modulated-temperature differential scanning calorimetry. *J. Polym. Eng.* **2014**, *34*, 895–903. [CrossRef]
74. Mustapa, I.R.; Shanks, R.A.; Kong, I. Melting behaviour and dynamic mechanical properties of poly(lactic acid)-hemp-nanosilica composites. *Asian Trans. Basic Appl. Sci.* **2013**, *3*, 29–37.
75. Lai, S.M.; Li, P.W. Effect of thermoplastic polyurethane-modified silica on melt- blended poly(Lactic Acid) (PLA) nanocomposites. *Poly Poly Compos.* **2017**, *25*, 583–592. [CrossRef]
76. He, H.; Pang, Y.; Duan, Z.; Luo, N.; Wang, Z. The Strengthening and toughening of biodegradable poly (lactic acid) using the SiO2-PBA core—Shell nanoparticle. *Materials* **2019**, *12*, 2510. [CrossRef]
77. Lei, W.C.; Zhang, M.Q.; Rong, M.Z.; Friedrich, K. Silica nanoparticles filled polypropylene: Effects of particle surface treatment, matrix ductility and particle species on mechanical performance of the composites. *Comp. Sci. Tech.* **2005**, *65*, 635–645.

78. Ahmed, W.; Siraj, S.; Al-Marzouqi, A.H. 3D Printing PLA waste to produce ceramic based particulate reinforced composite using abundant silica-sand: Mechanical properties characterization. *Polymers* **2020**, *12*, 2579. [CrossRef]
79. Régibeau, N.; Tilkin, R.G.; Grandfils, C.; Heinrichs, B. Preparation of poly-d,l-lactide based nanocomposites with polymer-grafted silica by melt blending: Study of molecular, morphological, and mechanical properties. *Polym. Compos.* **2021**, *42*, 955–972. [CrossRef]
80. Georgiopoulos, P.; Kontou, E.; Meristoudi, A.; Pispas, S.; Chatzinikolaidou, M. The effect of silica nanoparticles on the thermomechanical properties and degradation behavior of polylactic acid. *J. Biomater. Appl.* **2004**, *29*, 662–674. [CrossRef]
81. Zirak, M.F.; Tabari, M. PLA-SiO2 nanocomposite films: Morphological and mechanical properties and specific end-use characteristics. *Nanomed Res. J.* **2018**, *3*, 140–145.
82. Chrissafis, K.; Pavlidou, E.; Paraskevopoulos, K.; Beslikas, T.; Nianias, N.; Bikiaris, D. Enhancing mechanical and thermal properties of PLLA ligaments with fumed silica nanoparticles and montmorillonite. *J. Therm. Anal. Calorim.* **2011**, *105*, 313–323. [CrossRef]
83. Lai, S.M.; Hsieh, T.T. Preparation and properties of polylactic acid (PLA)/silica nanocomposites. *J. Macromol. Sci. Part. B* **2016**, *3*, 211–228. [CrossRef]
84. Battegazzore, D.; Bocchini, S.; Alongi, J.; Frache, A. Rice husk as bio-source of silica: Preparation and characterization of PLA–silica bio-composites. *RSC Adv.* **2014**, *4*, 54703–54712. [CrossRef]
85. Song, X.; Fang, C.; Li, Y.; Zhang, Y.; Wang, P. The characterization of mechanical properties of Jute/PLA composites with modified nano Sio2 by coupling agent. *Res. Sq.* **2021**. [CrossRef]
86. Malinowski, R.; Janczak, K.; Rytlewski, P.; Kaczor, A.R.; Moraczewski, K.; Żuk, T. Influence of glass microspheres on selected properties of polylactide composites. *Compos. Part B* **2015**, *76*, 13–19. [CrossRef]
87. Bikiaris, D.N.; Vassiliou, A.A. *Nanocomposite Coatings and Nanocomposites Materials*; Trans Tech Publications Ltd.: Freienbach, Switzerland, 2009.
88. Stodolak-Zych, E.; Szumera, M.; Blazewicz, M. Osteoconductive nanocomposite materials for bone regeneration. *Mater. Sci. Forum* **2013**, *730*, 38–43. [CrossRef]
89. Pedna, A.; Pinho, L.; Frediani, P.; Mosquera, M.J. Obtaining SiO2–fluorinated PLA bionanocomposites with application as reversible and highly-hydrophobic coatings of buildings. *Prog. Org. Coat.* **2016**, *90*, 91–100. [CrossRef]
90. Gong, J.; Tian, N.; Wen, X.; Chen, X.; Liu, J.; Jiang, Z.; Mijowska, E.; Tang, T. Synergistic effect of fumed silica with Ni2O3 on improving flame retardancy of poly(lactic acid). *Polym. Deg. Stab.* **2014**, *104*, 18–27. [CrossRef]
91. Zhao, Y.; Wei, B.; Wu, M.; Zhang, H.; Yao, J.; Chen, X.; Shao, Z. Preparation and characterization of antibacterial poly(lactic acid) nanocomposites with N-halamine modified silica. *Inter. J. Biol. Macromol.* **2020**, *155*, 1468–1477. [CrossRef]
92. Sarikhani, K.; Jeddi, K.; Thompson, R.B.; Park, C.B.; Chen, P. Adsorption of surface-modified silica nanoparticles to the interface of melt poly (lactic acid) and supercritical carbon dioxide. *Langmuir* **2015**, *31*, 5571–5579. [CrossRef]
93. Seng, C.T.; Noum, S.Y.A.E.; Sivanesan, S.K.A.; Yu, L.J. Reduction of hygroscopicity of PLA filament for 3D printing by introducing nano silica as filler. *AIP Conf. Proceed.* **2020**, *2233*, 020024. [CrossRef]
94. Chen, H.M.; Wang, Y.P.; Chen, J.; Yang, J.H.; Zhang, N.; Huang, T.; Wang, Y. Hydrolytic degradation behavior of poly(L-lactide)/SiO2 composites. *Polym. Degrad. Stab.* **2013**, *98*, 2672–2679. [CrossRef]
95. Kosowska, K.; Szatkowsk, P. Infuence of ZnO, SiO2 and TiO2 on the aging process of PLA fbers produced by electrospinning method. *J. Therm. Anal. Calorim.* **2020**, *140*, 1769–1778. [CrossRef] [PubMed]
96. Bo, L.J.; Yong, L.X.; Feng, L.W.; Hao, Z.J. Preparation and characterization of bioactive poly (Lactic Acid)/SiO_2-CaO composite membranes. *J. Inorg. Mater.* **2011**, *26*, 998–1002.
97. Galikhanov, E.; Lounev, I.; Guzhova, A.; Gusev, Y.; Galikhanov, M.; Vasilyeva, M. Study of electret state in polylactic acid with nanosized filler by dielectric spectroscopy. *AIP Conf. Proc.* **2016**, *1748*, 020007. [CrossRef]
98. Majola, A.; Vainionpää, S.; Vihtonen, K.; Vasenius, J.; Törmälä, P.; Rokkanen, P. Intramedullary Fixation of Cortical Bone Osteotomies with Self-Reinforced Polylactic Rods in Rabbits. *Int. Orthop.* **1992**, *16*, 101–108. [CrossRef]
99. Abe, S.; Sasaki, A.; Iwadera, N.; Akasaka, T.; Uo, M.; Watari, F. Biodistribution and biocompatibility of poly(lactic acid)-coated SiO_2 particle. *Nano Biomed.* **2011**, *3*, 300–305.
100. Shen, Y.; Cao, B.; Snyder, N.R.; Woeppel, K.M.; Eles, J.R.; Cui, X.T. ROS responsive resveratrol delivery from LDLR peptide conjugated PLA-coated mesoporous silica nanoparticles across the blood-brain barrier. *J. Nanobiotechnol.* **2018**, *16*, 13. [CrossRef]
101. Thongsang, S.; Sontikaew, S.; Kachapol, K.; Kaewkate, T.; Aunchanlung, W. Comparision of filler types in polylactic acide composites for 3D printing applications. *Matter Int. J. Sci. Technol.* **2020**, *5*, 98–109. [CrossRef]
102. Jaikaew, N.; Petchsuk, A.; Opaprakasit, P. Preparation and properties of polylactide bio-composites with surface-modified silica particles. *Chiang Mai J. Sci.* **2018**, *45*, 2059–2068.
103. Opaprakasit, P.; Boonpa, S.; Jaikaew, N.; Petchsuk, A.; Tangboriboonrat, P. Preparation of surface-modified silica particles from rice husk ash and its composites with degradable polylactic acid. *Macromol. Symp.* **2015**, *354*, 48–54. [CrossRef]

polymers

MDPI

Article

Safety and Efficacy of Kartigen® in Treating Cartilage Defects: A Randomized, Controlled, Phase I Trial

Yen-Liang Liu [1,2], Chun-Che Yen [3], Tzu-Shang Thomas Liu [4], Chih-Hung Chang [5,6], Tiffany Ting-Fang Shih [7], Jyh-Horng Wang [8], Ming-Chia Yang [9], Feng-Huei Lin [10] and Hwa-Chang Liu [8,11,*]

1 Master Program for Biomedical Engineering, College of Biomedical Engineering, China Medical University, Taichung 406040, Taiwan; allen.liu@cmu.edu.tw
2 Graduate Institute of Biomedical Sciences, College of Medicine, China Medical University, Taichung 406040, Taiwan
3 Kartigen Biomedical Inc., Taipei 100047, Taiwan; cyenslk@gmail.com
4 Southern California Bone and Joint Clinic, Apple Valley, CA 92307, USA; redcometsports@gmail.com
5 Department of Orthopaedic Surgery, Far Eastern Memorial Hospital, New Taipei 220216, Taiwan; orthocch@mail.femh.org.tw
6 Graduate School of Biotechnology and Bioengineering, Yuan Ze University, Taoyuan 320315, Taiwan
7 Department of Medical Imaging and Radiology, National Taiwan University Hospital, Taipei 100225, Taiwan; ttfshih@ntu.edu.tw
8 Department of Orthopaedic Surgery, National Taiwan University Hospital, Taipei 100225, Taiwan; jhwang@ntuh.gov.tw
9 Biomedical Technology and Device Research Laboratories, Industrial Technology Research Institute, Hsinchu 310401, Taiwan; s1979329@gmail.com
10 Department of Biomedical Engineering, College of Engineering, National Taiwan University, Taipei 106319, Taiwan; double@ntu.edu.tw
11 Department of Orthopaedic Surgery, Taiwan Adventist Hospital, Taipei 105404, Taiwan
* Correspondence: hcliu@ntuh.gov.tw

Citation: Liu, Y.-L.; Yen, C.-C.; Liu, T.-S.T.; Chang, C.-H.; Shih, T.T.-F.; Wang, J.-H.; Yang, M.-C.; Lin, F.-H.; Liu, H.-C. Safety and Efficacy of Kartigen® in Treating Cartilage Defects: A Randomized, Controlled, Phase I Trial. *Polymers* **2021**, *13*, 3029. https://doi.org/10.3390/polym13183029

Academic Editors: José Miguel Ferri, Vicent Fombuena Borràs and Miguel Fernando Aldás Carrasco

Received: 30 July 2021
Accepted: 31 August 2021
Published: 7 September 2021

Abstract: Here, we aimed to investigate the safety and preliminary efficacy of Kartigen®, a matrix with autologous bone marrow mesenchymal stem cell-derived chondrocyte precursors embedded in atelocollagen. As a surgical graft, Kartigen® was implanted onto the cartilage defects at the weight-bearing site of the medial femoral condyle of the knee. Fifteen patients were enrolled and stratified into two groups, undergoing either Kartigen® implantation (n = 10) or microfracture (control group, n = 5). The primary endpoint was to evaluate the safety of Kartigen® by monitoring the occurrence of adverse events through physician queries, physical examinations, laboratory tests, and radiological analyses for 2 years. There were no infections, inflammations, adhesions, loose body, or tumor formations in the Kartigen®-implanted knees. The preliminary efficacy was assessed using the International Knee Documentation Committee (IKDC) score, visual analog scale, and second-look arthroscopy. The postoperative IKDC scores of the Kartigen® group significantly improved in the 16th week (IKDC = 62.1 ± 12.8, p = 0.025), kept increasing in the first year (IKDC = 78.2 ± 15.4, p < 0.005), and remained satisfactory in the second year (IKDC = 73.6 ± 13.8, p < 0.005), compared to the preoperative condition (IKDC = 47.1 ± 17.0), while the postoperative IKDC scores of the control group also achieved significant improvement in the 28th week (IKDC = 68.5 ± 6.1, p = 0.032) versus preoperative state (IKDC = 54.0 ± 9.1). However, the IKDC scores decreased in the first year (IKDC = 63.5 ± 11.6) as well as in the second year (IKDC = 52.6 ± 16.4). Thirteen patients underwent second-look arthroscopy and biopsy one year after the operation. The Kartigen® group exhibited integration between Kartigen® and host tissue with a smooth appearance at the recipient site, whereas the microfracture group showed fibrillated surfaces. The histological and immunohistochemical analyses of biopsy specimens demonstrated the columnar structure of articular cartilage and existence of collagen type II and glycosaminoglycan mimic hyaline cartilage. This study indicates that Kartigen® is safe and effective in treating cartilage defects.

Keywords: cartilage defect; knee; Kartigen®; chondrocyte precursors; stem cell therapy

1. Introduction

Cartilage defects are highly prevalent joint disorders and leading causes of disability and chronic pain in the world [1]. Articular cartilage defects are mainly attributed to trauma, osteoarthritis, osteonecrosis, and osteochondritis dissecans. Chondral lesions were found in around 60% of the patients undergoing knee arthroscopy [2–4].

Back in 1743, Hunter described the challenge of cartilage repair, stating that "once the cartilage is destroyed, it never recovers [5,6]." His observation still holds today. The avascular characteristics of cartilage constrain its self-regeneration from injury. If left untreated, the damaged cartilage gradually progresses into severe osteoarthritis. Through decades of effort, multiple surgical treatments have been developed to promote cartilage healing, such as abrasion arthroplasty [7], microfracture [8,9], and mosaicplasty [10,11]. However, these surgical approaches are usually associated with fibrocartilage formation [12,13], limited tissue sources [14], and donor-site morbidity [15], and its long-term efficacy remains controversial [16].

Cell therapy for cartilage repair was proposed in the 1980s by Robert Langer and Charles Vacanti using the approaches of tissue engineering. They clearly defined that a fine reconstruction of cartilage defect must include selected cells for transplantation, excipients, which are seeded with selected cells, and functional restoration of defect areas as before [17]. The cellular therapeutic innovation was realized in 1994. Autologous chondrocyte implantation (ACI) was introduced to treat cartilage defects in the knee [18]. In the past two decades, ACI has demonstrated its efficacy for knee osteoarthritis. Two systematic reviews concluded that, relative to microfracture and mosaicplasty, ACI may be the best option for large defects in active young patients who have had the symptoms for a short period and have not undergone a chondral surgery before [13,19]. In addition, a scaffold-based ACI, matrix-induced autologous chondrocyte implantation (MACI), outperformed microfracture in a 2-year randomized study [20]. Still, only a few products are available on the market: Carticel (FDA-approved in 1997), Chondron (South Korea MFDS-approved in 2001), and MACI (FDA-approved in 2016). The scarcity implies that some limitations remain, such as the limited source of chondrocytes, donor site morbidity, uncertain hyaline cartilage formation, low recovery of the recipient site, and questionable longevity of these implants or their derivative tissues [21].

The discovery of adult stem cells aroused a paradigm shift in regenerative medicine. The features of self-renewal and multipotency of stem cells make them ideal cell sources for cellular therapy. A variety of stem cell-based therapeutic innovations have been developed using mesenchymal stem cells (MSCs) derived from bone marrow [22], adipose tissue [23], synovium [24], peripheral blood [25], or periosteum [26]. Several clinical trials have demonstrated the safety and therapeutic efficacy of autologous MSC implantation for cartilage repair [27–30]. However, the implantation of undifferentiated MSCs cannot guarantee certain chondrogenesis in vivo, which might lead to heterogeneity of regenerated tissues.

The chondrogenesis of MSCs can be guided using growth factors [31] or biophysical/biomechanical stimuli [32] to improve the functional properties of the derived neo-cartilage tissues, including mature matrix formation [29,33]. Our previous study identified a unique population of chondrocyte precursors (CPs) derived from bone marrow mesenchymal stem cells (BMSCs) during chondrogenic induction [34]. These atelocollagen-embedded CPs (Kartigen®) can secrete glycosaminoglycan (GAG) and collagen type II but without lacunae formation.

A variety of biomaterials have been used for cartilage tissue engineering, including collagen [35], alginate [36], poly-lactic-glycolic acid [37], and tri-copolymer [38]. In this study, we selected atelocollagen because it is a low-immunogenic derivative of collagen obtained by the removal of N– and C–terminal telopeptide components [39]. Atelocollagen has been broadly applied in the regeneration of cartilage [34,40,41], intervertebral disc [42], cornea [43], periodontal tissues [44], and skin [45] to serve as a carrier for cell delivery and to provide an appropriate microenvironment for tissue regeneration.

A 9-year follow-up trial demonstrated that Kartigen® integrated with the host tissue and resulted in the formation of hyaline-like cartilage, thereby improving the impaired knee functions [34]. Due to the lack of a randomized control group in the previous study, we initiated this controlled and randomized trial to evaluate the safety and efficacy of Kartigen for repairing cartilage defects in the weight-bearing site of medial femoral condyles through the comparison with the microfracture treatment.

2. Materials and Methods

2.1. Ethical Approval

This study was an open-label, controlled, randomized, single-center, phase I clinical trial to evaluate the clinical safety of Kartigen® and its clinical improvements versus microfracture. This study was approved by the Taiwan Food and Drug Administration (TFDA, study number 1076026300) and by the Institutional Review Board of Taiwan Adventist Hospital (IRB number: 105-B-09). According to the guidelines of the Declaration of Helsinki, informed consent was obtained from each subject.

2.2. Study Population

This study enrolled 15 patients with cartilage defects at the weight-bearing site of the medial femoral condyle. The size of cartilage defects ranged from 0.6 to 4 cm^2. Patients were enrolled in this study between September 2018 and June 2020. Patients were selected according to inclusion and exclusion criteria in Table S1. They were randomly allocated into groups of Kartigen® implantation ($n = 10$) or microfracture treatment ($n = 5$). Based on our previous study [34], the seeding density of CPs was 1.6–3.3 $\times 10^6$ cells/cm^2, and the total cell number was less than 1.32 $\times 10^7$ cells. The schedule of study procedures and assessments is shown in Table S2.

2.3. Manufacture of Kartigen®

The manufacture of Kartigen® was described previously [34]. In brief, cell culture was performed under standard operative procedures at the cell processing unit (CPU) of Kartigen Biomedical Inc. (Taipei, Taiwan), following the Good Tissue Practice regulations. Fifteen mL of heparinized blood was aspirated from the iliac crest, collected in sterile 50-mL tubes. The MSCs were isolated using the density-gradient medium Ficoll (Cat. No. 17-5446-52, GE Healthcare, Little Chalfont, UK). The nucleated cells were collected from the interface, washed twice in PBS, and then suspended in Dulbecco's Modified Eagle's Medium (DMEM-LG, Cat. No. 31600, Gibco, Carlsbad, CA, USA) supplemented with the patient's serum. BMSC were expanded to a sufficient number, enough to fill the cartilage defects of patients. The required cell number was estimated according to the size of the defects. Before the BMSCs were induced into CPs, a small portion of BMSCs were used to assess the number, viability, and the immunophenotype of BMSCs, and to undergo sterility testing. The cell number and viability were quantified using trypan blue staining and an automated cell counter. BMSC's immunophenotype must be fulfilled by the minimal ISCT criteria: More than 95% of BMSCs are CD90-positive, and less than 2% of BMSCs are CD34-negative. Sterility testing must be negative for aerobic and anaerobic bacteria, mycoplasma, and <0.5 EU/mL endotoxin. The expanded cells were then seeded in excipient (Kartigen®) and cultured in CPs' induction medium.

2.4. Surgical Operation

The knee lesions were assessed arthroscopically. The defect size was measured through the arthroscope. For the Kartigen® implantation group, the required cell number of CPs was estimated before the collection of bone marrows. We aimed to implant 2 $\times 10^6$ cells/cm^2 in the defect. The cartilage defect at the medial femoral condyle was approached either by mini-medial arthrotomy of 2 to 3 cm in length or by the arthroscope only. After debridement, the cartilage defect was filled with Kartigen® and then sealed with fibrin glue (TISSEEL, Baxter AG). For microfracture, we followed a standard arthroscopic procedure as reported by Steadman et al. [46]. The multiple holes (microfractures) were about 3–4 mm apart, to preserve the structure and function of the subchondral plate.

2.5. Rehabilitation

Both groups of patients underwent the same rehabilitation regimen. The operated knee was kept in 20- to 30-degree flexion for 72 h, and then the knee was allowed to move freely. Partial weight-bearing was started 24 h after the operation, and full weight-bearing was allowed 4 weeks after the operation.

2.6. Safety Evaluation

A series of evaluations were assessed to assure the safety of the study for 2 years after treatments. These evaluations included physicians' queries, physical examinations, laboratory tests, and radiological studies. The Common Terminology Criteria for Adverse Events was used to elicit and report toxic effects [47]. We specifically monitored the possible severe adverse events, including the severity of pain, infection of the operation site, joint adhesion, the abnormal gross appearance of the knee, active range of motion, loose body formation, and tumorigenesis.

2.7. Efficacy Evaluation

The preliminary efficacy was assessed using the International Knee Documentation Committee (IKDC) scoring system [48], visual analog scale (VAS), and arthroscopy. Arthroscopy was performed 1 year after the operation to observe the changes of defects. At the same time, a biopsy specimen (2 mm in diameter) was taken for histological analysis.

2.8. Arthroscopic and Histological Analysis

Second-look arthroscopy was done with the patients' written consent. Alcian blue staining and immunohistology were applied to evaluate the existence of GAG and collagen type II. The International Cartilage Repair Society (ICRS) arthroscopic assessment scale was used to evaluate the degree of cartilage repair [49,50]. Alcian blue staining (Cat. No. B8438, Sigma-Aldrich, Saint Louis, MO, USA) was used to assess the amount of GAG accumulation in the implanted tissues. Nuclear Fast Red solution (Cat. No. H-3403, Vector Laboratories, Burlingame, CA, USA) was used for counterstaining. The immunohistochemistry was applied to quantify the protein expression level of collagen type II. An anti-collagen type II antibody (Cat. No. Ab34712, Abcam, Cambridge, UK) and UltraVision Quanto Detection System (Cat. No. TL-060-QHL, Thermo Scientific, Fremont, CA, USA) were used for the immunohistochemistry, according to the manufacturer's instructions. All samples were processed and stained using the same procedure.

2.9. Statistical Methods

The statistical analyses of IKDC scores and VAS scores were performed using t-test and Wilcoxon signed-rank test, respectively. A two-tailed paired t-test was used to test the difference of IKDC scores at different time points (before and after the operation) of the same treatment group. To compare the efficacy of treatments, we used a one-tailed t-test to test the null hypothesis: IKDC scores of microfracture treatment were higher than that of Kartigen® implantation. The statistical significance of the t-tests was set as $p < 0.05$. All statistical analyses were performed using Microsoft Excel.

3. Results

3.1. Demography

Fifteen patients were enrolled and randomized into two groups. The control group underwent microfracture, and the study group underwent Kartigen® implantation. No patients were withdrawn from this trial. The flow chart of the study is shown in Figure S1. The demographic information of the patients is shown in Table 1. There was no significant difference in age between the two groups, but the defect sizes were significantly different.

Table 1. Demography of the patients.

Group	Kartigen® Group (n = 10)	Microfracture (n = 5)
Gender (F:M)	5:5	2:3
Age (year)	22–77 (54.8 ± 18.0)	55–77 (67.8 ± 8.5)
Defect size (cm^2)	1.3–4.0 (2.9 ± 0.8) *	0.6–1.5 (1.0 ± 0.4)
Defect treatment	Kartigen®	Microfracture

* Note: The defect size of the Kartigen® group was significantly larger than that of microfracture group ($p < 0.005$).

3.2. Safety Assessment

In the Kartigen® group, three patients exhibited three treatment-emergent adverse events (TEAEs): renal stone with hematuria, cervical spondylosis, and upper respiratory tract infection, respectively. In the control group, two patients demonstrated four TEAEs, including urinary tract infection, cataract, prostate hypertrophy, and coronary artery stenosis. All seven TEAEs were not treatment-related. These TEAEs are summarized in Table S3. By X-ray examination and MRI study, there was no inflammation, joint adhesion, loose body, or tumorigenesis in the Kartigen®-implanted knee and microfracture-treated knee. Physical examination and laboratory tests revealed no infection among the 15 patients. Moreover, neither mortality nor complications were noted after operations in this study. Some laboratory values were abnormal, such as blood sugar, GOT, GPT, and urinary red blood cell in both groups. However, they were felt to be reflective of each patient's underlying medical condition and not as results of participation in the study.

3.3. Efficacy Assessment

We applied IKDC subjective knee evaluation criteria to evaluate the knee function before and after the treatments. Figure 1 showed a tendency that the IKDC scores continuously improved in the Kartigen® group, but the control group achieved the highest IKDC score in the 28th week and gradually decreased. In the study group, the IKDC score before Kartigen® implantation was 47.1 ± 17.0 (mean ± standard deviation) and then significantly improved to 62.1 ± 12.8 (p-value = 0.025) 16 weeks after implantation. The IKDC scores continuously increased to 78.2 ± 15.4 with a statistical significance (p-value < 0.005) one year after the operation and still maintained at 73.6 ± 13.8 2 years after the operation. While IKDC scores slightly decreased in the second year, the scores were still significantly higher than that of preoperation (p-value < 0.005). Meanwhile, although the control group demonstrated significantly improved IKDC scores from 54.0 ± 9.1 (preoperation) to 68.5 ± 6.1 (28 weeks after the operation) with a p-value of 0.032, the score decreased to 63.5 ± 11.6 1 year after the operation and to 52.6 ± 16.4 2 years after the operation. The comparison of IKDC scores between the Kartigen® group and the microfracture group in the second year exhibited statistical significance with the p-value of 0.029. Still, a longer follow-up will be needed to evaluate the long-term efficacy and longevity of the regenerated cartilage.

The VAS was used to measure the pain before and after treatments. Both the Kartigen® implantation and microfracture treatment effectively achieved pain relief at the 10th week after the operation (Figure 2). However, the Kartigen® implantation exhibited consistent pain relief through the 1-year follow-up, which was validated statistically using Wilcoxon signed-rank test with p-values < 0.005.

3.4. Arthroscopy and Histological Analysis

The regeneration of cartilage defect was grossly observed by knee arthroscopy, and a specimen of 2 mm in diameter was taken 1 year after surgery. Nine out of 10 patients in the Kartigen® group and 4 out of 5 patients in the control group consented to second-look arthroscopy. The gross appearance of regenerated cartilage tissues was examined under arthroscope and compared with their corresponding arthroscopic images taken before treatment (Figure 3). Eight out of 9 of the Kartigen® group exhibited smooth and elastic surfaces at the recipient sites (Figure 3A), but none of the control group showed normal or near-normal appearance. Only fibrillated surface was noted (Figure 3B). The ICRS scale is shown in Table 2 with the classification of grade I (normal tissue), grade II (nearly normal tissue), grade III (abnormal), grade IV (severely abnormal).

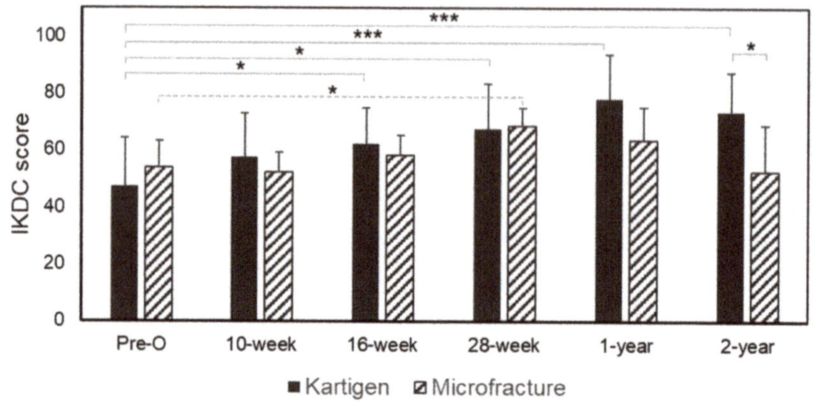

Figure 1. Assessment of knee functions of daily activities by IKDC scores. The knee functions of the patients were significantly improved 16 weeks after the Kartigen® implantation and 28 weeks after microfracture, respectively. The paired *t*-test was conducted to determine the difference between the means of the IKDC scores before and after the treatments at different time points among the same groups. Two years after operation, the IKDC score of the Kartigen® group was higher than the control group. The error bars stand for standard deviations. Asterisks represent the statistical significance: * $p < 0.05$, *** $p < 0.005$.

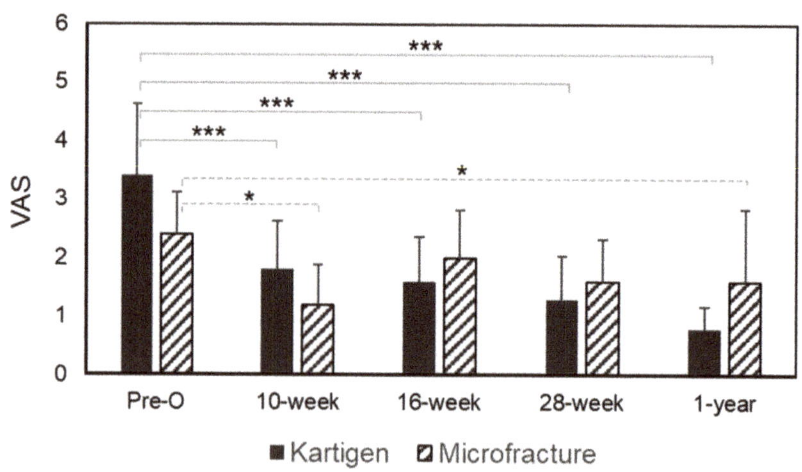

Figure 2. VAS assessment. The pain relief was effective and significant in the Kartigen®-treated group, as VAS scores decreased since the 10th week after operation, but the VAS scores flatulated in the control group. The error bars stand for standard deviations. Asterisks represent the statistical significance: * $p < 0.05$, *** $p < 0.005$.

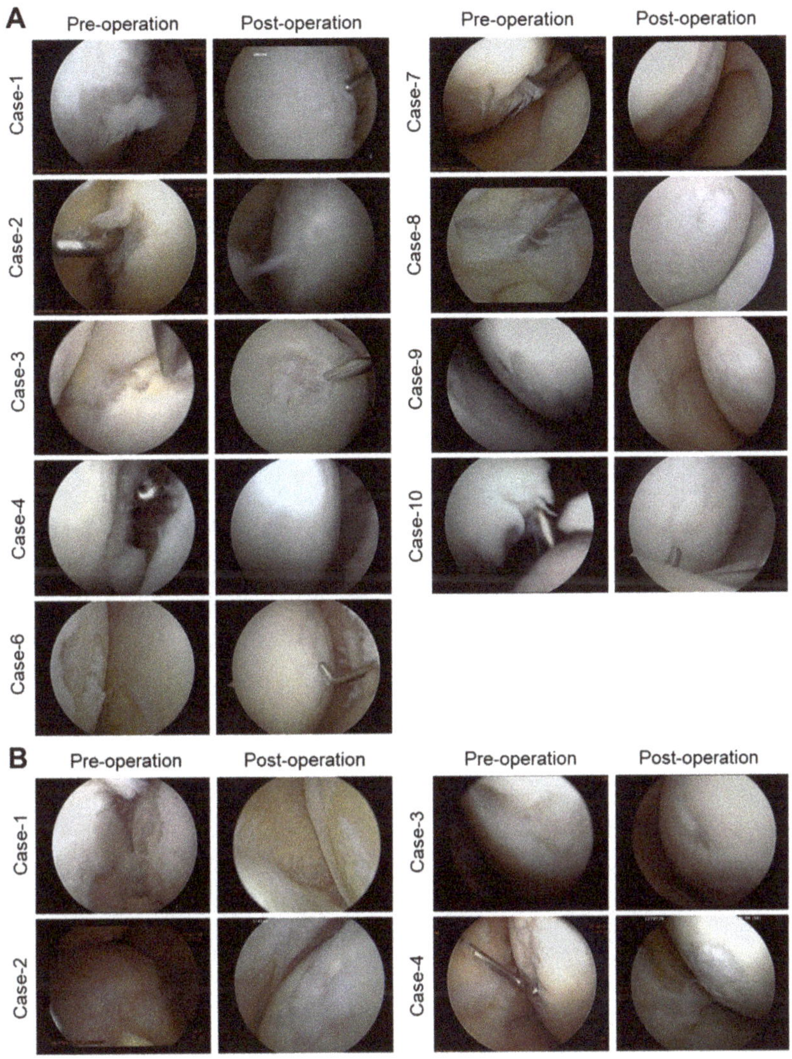

Figure 3. Arthroscopic images before and 1 year after operation. Images of Kartigen® group (**A**) and microfracture group (**B**).

Table 2. ICRS arthroscopic assessment.

ICRS Cartilage Repair Assessment			
Kartigen® Group (n = 10)		**Microfracture Group (n = 5)**	
Grade I	3	Grade I	0
Grade II	5	Grade II	0
Grade III	1	Grade III	3
Grade IV	0	Grade IV	1
Total	9	Total	4

The Kartigen® group demonstrated better results in both arthroscopic observations and histological analysis (Figure 4). In the control group, the regenerated cartilage did not exhibit columnar chondrocyte distribution but only fibrocartilage (Figure 4A). GAG and collagen type II were observed in the biopsy specimens, and a total of 12 specimens (Kartigen®, n = 8; control, n = 4) are presented in Figure 4B,C. New chondrocytes at defect had the same appearance as the original ones (Figure 4B). However, the chondrocytes in the repaired defect in the Kartigen® group were smaller and denser than those in the original cartilage.

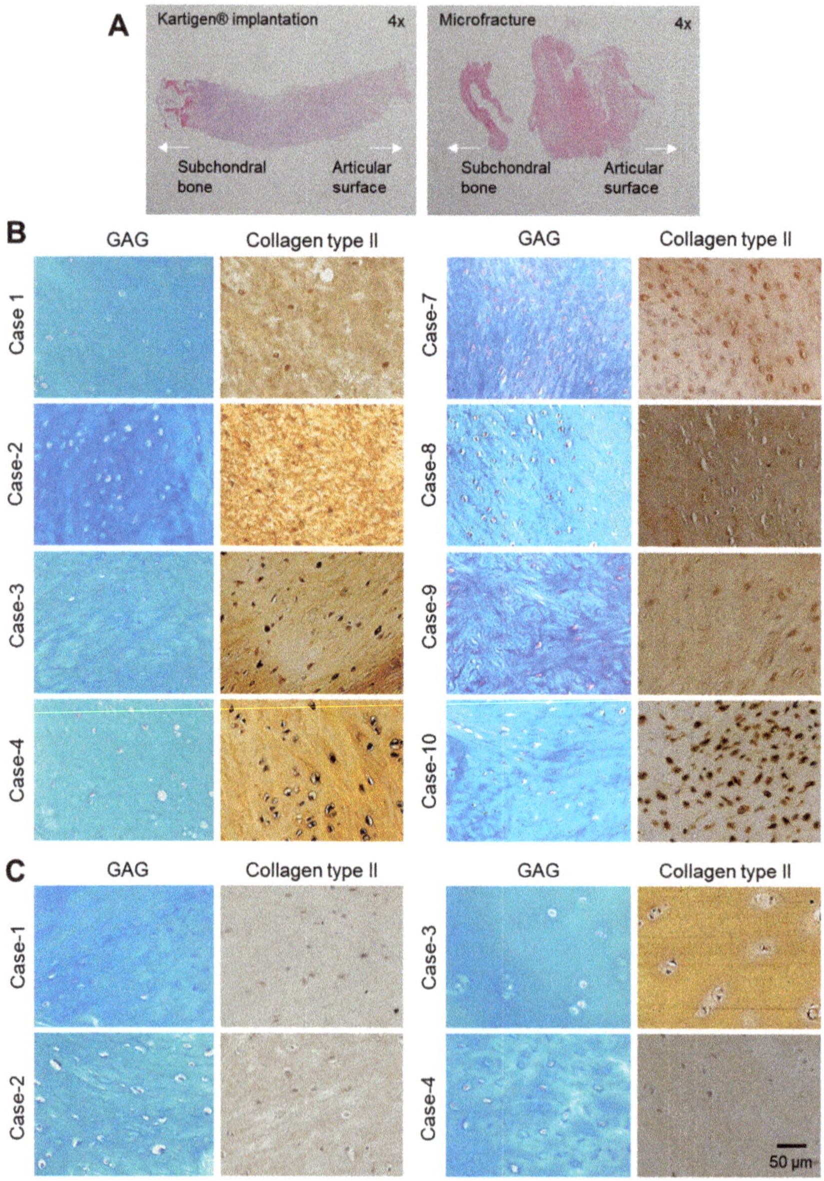

Figure 4. Immunohistochemical analyses of biopsy specimens. (**A**) Whole-slide histological images of regenerated cartilage under 4× objective lens. Alcian blue staining and immunochemical study show expression levels of GAG/Collagen type II, respectively, in Kartigen group (**B**) and microfracture group (**C**).

4. Discussion

The development of Kartigen® was started from our previous study using atelocollagen-embedded chondrogenic BMSCs to repair the full thickness of cartilage defect in miniature pigs [51]. The promising results encouraged us to initiate a case series study to test the safety and efficacy of Kartigen® for treating cartilage defects. In the 9 years of follow-up study, improvement of knee functions was satisfactory and sustainable [34]. Most importantly, infection, inflammation, joint adhesion, loose body, and tumor formation were not found in the Kartigen®-implanted knees. However, in the case series study, the contralateral knees of the same patients were used as a control group. To further validate the therapeutic efficacy of Kartigen® in comparison with standard surgical treatments, such as microfracture, a new controlled and randomized phase I clinical trial was conducted. As expected, no severe adverse events were found, and all the TEAEs were not treatment-related. Although clinical outcome was similar for Kartigen implantation and microfracture in 1-year follow-up, a sustained improvement of knee functions was found in the Kartigen® group in the 2 years follow-up.

Microfracture was introduced for focal articular cartilage repair in the 1980s and soon became a standard treatment for cartilage defects [9,46,52]. However, its results are usually associated with fibrocartilage production [12,13], and its long-term efficacy remains poor [16,53]. Mosaicplasty provides a better improvement than microfracture, but there are still several disadvantages, such as the donor-site morbidity for autografts [54], limited sources of grafts, a technical challenge in leveling the graft during operation [55], and poor integration of the implant into host tissue [56,57]. On the contrary, the procedure of Kartigen® implantation is technically simple and can be carried out with a small arthrotomy or using arthroscope. Furthermore, there is no donor site morbidity, and sources of cells are not limited.

The ability of engrafted cells to integrate into the recipient site and participate in the repair process is crucial for successful clinical outcomes. It has been reported that intra-articular injection of autologous MSCs reduced the degeneration of cartilage defects and provided pain relief [58], but integration of injected MSCs into damaged cartilage defects is unclear and doubtful. On the contrary, using atelocollagen as a cell carrier, Kartigen® prevents cell loss, is easy for implantation for cartilage defects of any shape, and achieves uniform cell distribution at the recipient site. These advantages contribute to the efficacy and durability of Kartigen® implantation. In addition, atelocollagen has been proven to enable a gradual proliferation and matrix synthesis of chondrocytes, which allow chondrocytes to maintain their phenotype for up to 4 weeks in vitro [40]. Atelocollagen gel can also support cell proliferation, matrix synthesis, and chondrogenic differentiation of MSCs [42]. Recently, the Adachi group reported the efficacy of repairing osteochondral defects with minced cartilage embedded in atelocollagen gel [41]. Our previous clinical trial with 9 years of follow-up also demonstrated the safety and efficacy of atelocollagen in repairing cartilage defects together with CPs [34].

Unlike previous studies using mature chondrocytes for cartilage repair in ACI [59], our study showed CPs in Kartigen® exhibiting sufficient integration capacity. Under arthroscopic examination, the integration between the graft and the recipient site was complete (Figure 3). The histological analyses of the biopsy specimens also demonstrated the integration of the implanted tissue into the surrounding articular cartilage (Figure 4). The accumulation of GAG and collagen type II was confirmed in the biopsy specimens (Figure 4). However, there are several limitations to this study. Because this is a phase I study, the sample size number is small. The 2 years of follow-up are not long enough to reach a final conclusion for cartilage defect repair.

Recent advances in 3D bioprinting [60–63] have enabled reconstructions of functional living cartilage to recapitulate the complexity and architecture [64,65] of an articular surface. We are looking into partnerships to integrate that technology and potentially further improve the outcome with Kartigen® implantation. This may result in more favorable biomechanical properties at the recipient sites and allow earlier weight-bearing and range of motion without concern of graft detachment. Since MSCs are multipotent, additional biochemical and biomechanical stimulations can delicately manipulate the chondrogenic differentiation and maturation of seeded cells [66,67] to improve the functional properties of the derived neo-cartilage tissues [29]. The integration of 3D bioprinting and biomimetic in vitro chondrogenesis will drive advanced therapeutic innovations for cartilage repairs.

5. Conclusions

In this study, Kartigen® containing CPs was proven safe and free of adverse events, such as infection, inflammation, joint adhesion, loose body, or tumor formation.

Supplementary Materials: The following are available online at https://www.mdpi.com/article/10.3390/polym13183029/s1. Table S1: Inclusion and exclusion criteria, Table S2: Scheduled visits of the study design, Table S3: Adverse events, Figure S1: Flowchart of this clinical trial.

Author Contributions: H.-C.L. and F.-H.L. conceived the project. H.-C.L., Y.-L.L., C.-C.Y. and T.-S.T.L. wrote the article. C.-C.Y. and M.-C.Y. supervised the preparation of Kartigen® and conducted immunohistochemical staining. H.-C.L. performed the surgery and collected clinical information. C.-H.C. and J.-H.W. interpreted the arthroscopic and histology images. T.T.-F.S. assessed the radiological study (MRI). H.-C.L. performed arthroscopy. H.-C.L. supervised the project. All authors have read and agreed to the published version of the manuscript.

Funding: The study was funded by Kartigen Biomedical Inc.

Institutional Review Board Statement: The study was conducted according to the guidelines of the Declaration of Helsinki, and approved by the the Taiwan Food and Drug Administration (TFDA, study number 1076026300) and by the Institutional Review Board of Taiwan Adventist Hospital (IRB number: 105-B-09).

Informed Consent Statement: Informed consent was obtained from all patients involved in the study. Written informed consent was obtained from the patients.

Acknowledgments: We would like to convey our sincere appreciation to the data and safety monitoring board consisting of Yi-Sheng Chan at Chang Gung Memorial Hospital, Chun-Hsiung Huang at Mackay Memorial Hospital, and Chih-Hao Chang at National Taiwan University Hospital. We thank Ying-Tai Jin at Taiwan Adventist Hospital for assessing the immunohistochemical analyses.

Conflicts of Interest: C.C. Yen is a manager of Kartigen Biomedical Inc.

References

1. James, S.L.; Abate, D.; Abate, K.H.; Abay, S.M.; Abbafati, C.; Abbasi, N.; Abbastabar, H.; Abd-Allah, F.; Abdela, J.; Abdelalim, A.; et al. Global, regional, and national incidence, prevalence, and years lived with disability for 354 diseases and injuries for 195 countries and territories, 1990–2017: A systematic analysis for the global burden of disease study 2017. *Lancet* **2018**, *392*, 1789–1858. [CrossRef]
2. Curl, W.W.; Krome, J.; Gordon, E.S.; Rushing, J.; Smith, B.P.; Poehling, G.G. Cartilage injuries: A review of 31,516 knee arthroscopies. *Arthrosc. J. Arthrosc. Relat. Surg.* **1997**, *13*, 456–460. [CrossRef]
3. Hjelle, K.; Solheim, E.; Strand, T.; Muri, R.; Brittberg, M. Articular cartilage defects in 1000 knee arthroscopies. *Arthrosc. J. Arthrosc. Relat. Surg.* **2002**, *18*, 730–734. [CrossRef] [PubMed]
4. Årøen, A.; Løken, S.; Heir, S.; Alvik, E.; Ekeland, A.; Granlund, O.G.; Engebretsen, L. Articular cartilage lesions in 993 consecutive knee arthroscopies. *Am. J. Sports Med.* **2004**, *32*, 211–215. [CrossRef] [PubMed]
5. Hunter, W. Of the structure and disease of articulating cartilages. *Clin. Orthop. Relat. Res.* **1995**, *317*, 3–6.
6. Hunter, W. VI. Of the structure and diseases of articulating cartilages. *Philos. Trans.* **1743**, *42*, 514–521.
7. Johnson, L.L. Arthroscopic abrasion arthroplasty: A review. *Clin. Orthop. Relat. Res.* **2001**, *391*, S306–S317. [CrossRef]
8. Mithoefer, K.; Williams, R.J., III; Warren, R.F.; Potter, H.G.; Spock, C.R.; Jones, E.C.; Wickiewicz, T.L.; Marx, R.G. The microfracture technique for the treatment of articular cartilage lesions in the knee: A prospective cohort study. *J. Bone Jt. Surg.* **2005**, *87*, 1911–1920. [CrossRef]
9. Steadman, J.R.; Rodkey, W.G.; Singleton, S.B.; Briggs, K.K. Microfracture technique forfull-thickness chondral defects: Technique and clinical results. *Oper. Tech. Orthop.* **1997**, *7*, 300–304. [CrossRef]
10. Hangody, L.; Kish, G.; Kárpáti, Z.; Udvarhelyi, I.; Szigeti, I.; Bély, M. Mosaicplasty for the treatment of articular cartilage defects: Application in clinical practice. *Orthopedics* **1998**, *21*, 751–756. [CrossRef] [PubMed]
11. Bugbee, W.D.; Convery, F.R. Osteochondral allograft transplantation. *Clin. Sports Med.* **1999**, *18*, 67–75. [CrossRef]

12. Hunziker, E.B. Articular cartilage repair: Are the intrinsic biological constraints undermining this process insuperable? *Osteoarthr. Cartil.* **1999**, *7*, 15–28. [CrossRef] [PubMed]
13. Oussedik, S.; Tsitskaris, K.; Parker, D. Treatment of articular cartilage lesions of the knee by microfracture or autologous chondrocyte implantation: A systematic review. *Arthrosc. J. Arthrosc. Relat. Surg.* **2015**, *31*, 732–744. [CrossRef] [PubMed]
14. Bartha, L.; Vajda, A.; Duska, Z.; Rahmeh, H.; Hangody, L. Autologous osteochondral mosaicplasty grafting. *J. Orthop. Sports Phys. Ther.* **2006**, *36*, 739–750. [CrossRef]
15. Gomoll, A.; Farr, J.; Gillogly, S.; Kercher, J.; Minas, T. Surgical management of articular cartilage defects of the knee. *J. Bone Jt. Surg.* **2010**, *92*, 2470–2490.
16. Knutsen, G.; Engebretsen, L.; Ludvigsen, T.C.; Drogset, J.O.; Grøntvedt, T.; Solheim, E.; Strand, T.; Roberts, S.; Isaksen, V.; Johansen, O. Autologous chondrocyte implantation compared with microfracture in the knee: A randomized trial. *J. Bone Jt. Surg.* **2004**, *86*, 455–464. [CrossRef] [PubMed]
17. Vacanti, J.P. Beyond transplantation: Third annual samuel jason mixter lecture. *Arch. Surg.* **1988**, *123*, 545–549. [CrossRef]
18. Brittberg, M.; Lindahl, A.; Nilsson, A.; Ohlsson, C.; Isaksson, O.; Peterson, L. Treatment of deep cartilage defects in the knee with autologous chondrocyte transplantation. *N. Engl. J. Med.* **1994**, *331*, 889–895. [CrossRef]
19. Harris, J.D.; Siston, R.A.; Pan, X.; Flanigan, D.C. Autologous chondrocyte implantation: A systematic review. *J. Bone Jt. Surg.* **2010**, *92*, 2220–2233. [CrossRef] [PubMed]
20. Basad, E.; Ishaque, B.; Bachmann, G.; Stürz, H.; Steinmeyer, J. Matrix-induced autologous chondrocyte implantation versus microfracture in the treatment of cartilage defects of the knee: A 2-year randomised study. *Knee Surg. Sports Traumatol. Arthrosc.* **2010**, *18*, 519–527. [CrossRef]
21. Iwasa, J.; Engebretsen, L.; Shima, Y.; Ochi, M. Clinical application of scaffolds for cartilage tissue engineering. *Knee Surg. Sports Traumatol. Arthrosc.* **2009**, *17*, 561–577. [CrossRef]
22. Wakitani, S.; Imoto, K.; Yamamoto, T.; Saito, M.; Murata, N.; Yoneda, M. Human autologous culture expanded bone marrow mesenchymal cell transplantation for repair of cartilage defects in osteoarthritic knees. *Osteoarthr. Cartil.* **2002**, *10*, 199–206. [CrossRef] [PubMed]
23. Koh, Y.-G.; Choi, Y.-J.; Kwon, S.-K.; Kim, Y.-S.; Yeo, J.-E. Clinical results and second-look arthroscopic findings after treatment with adipose-derived stem cells for knee osteoarthritis. *Knee Surg. Sports Traumatol. Arthrosc.* **2015**, *23*, 1308–1316. [CrossRef]
24. Sekiya, I.; Muneta, T.; Horie, M.; Koga, H. Arthroscopic transplantation of synovial stem cells improves clinical outcomes in knees with cartilage defects. *Clin. Orthop. Relat. Res.* **2015**, *473*, 2316–2326. [CrossRef] [PubMed]
25. Saw, K.-Y.; Anz, A.; Jee, C.S.Y.; Merican, S.; Ng, R.C.S.; Roohi, S.A.; Ragavanaidu, K. Articular cartilage regeneration with autologous peripheral blood stem cells versus hyaluronic acid: A randomized controlled trial. *Arthrosc. J. Arthrosc. Relat. Surg.* **2013**, *29*, 684–694. [CrossRef] [PubMed]
26. Roberts, S.J.; van Gastel, N.; Carmeliet, G.; Luyten, F.P. Uncovering the periosteum for skeletal regeneration: The stem cell that lies beneath. *Bone* **2015**, *70*, 10–18. [CrossRef]
27. Nejadnik, H.; Hui, J.H.; Feng Choong, E.P.; Tai, B.-C.; Lee, E.H. Autologous bone marrow–derived mesenchymal stem cells versus autologous chondrocyte implantation: An observational cohort study. *Am. J. Sports Med.* **2010**, *38*, 1110–1116. [CrossRef]
28. Wakitani, S.; Okabe, T.; Horibe, S.; Mitsuoka, T.; Saito, M.; Koyama, T.; Nawata, M.; Tensho, K.; Kato, H.; Uematsu, K. Safety of autologous bone marrow-derived mesenchymal stem cell transplantation for cartilage repair in 41 patients with 45 joints followed for up to 11 years and 5 months. *J. Tissue Eng. Regen. Med.* **2011**, *5*, 146–150. [CrossRef]
29. Makris, E.A.; Gomoll, A.H.; Malizos, K.N.; Hu, J.C.; Athanasiou, K.A. Repair and tissue engineering techniques for articular cartilage. *Nat. Rev. Rheumatol.* **2015**, *11*, 21. [CrossRef]
30. Teo, A.Q.A.; Wong, K.L.; Shen, L.; Lim, J.Y.; Toh, W.S.; Lee, E.H.; Hui, J.H.P. Equivalent 10-year outcomes after implantation of autologous bone marrow–derived mesenchymal stem cells versus autologous chondrocyte implantation for chondral defects of the knee. *Am. J. Sports Med.* **2019**, *47*, 2881–2887. [CrossRef] [PubMed]
31. Pittenger, M.F.; Mackay, A.M.; Beck, S.C.; Jaiswal, R.K.; Douglas, R.; Mosca, J.D.; Moorman, M.A.; Simonetti, D.W.; Craig, S.; Marshak, D.R. Multilineage potential of adult human mesenchymal stem cells. *Science* **1999**, *284*, 143–147. [CrossRef] [PubMed]
32. Engler, A.J.; Sen, S.; Sweeney, H.L.; Discher, D.E. Matrix elasticity directs stem cell lineage specification. *Cell* **2006**, *126*, 677–689. [CrossRef]
33. Liu, H.-C. Tissue engineering of cartilage: The road a group of researchers have traveled. *J. Orthop. Sci.* **2008**, *13*, 396. [CrossRef]
34. Liu, H.-C.; Liu, T.-S.T.; Liu, Y.-L.; Wang, J.-H.; Chang, C.-H.; Shih, T.T.-F.; Lin, F.-H. Atelocollagen-embedded chondrocyte precursors as a treatment for grade-4 cartilage defects of the femoral condyle: A case series with up to 9-year follow-up. *Biomolecules* **2021**, *11*, 942. [CrossRef] [PubMed]
35. Zheng, M.-H.; Willers, C.; Kirilak, L.; Yates, P.; Xu, J.; Wood, D.; Shimmin, A. Matrix-induced autologous chondrocyte implantation (maci®): Biological and histological assessment. *Tissue Eng.* **2007**, *13*, 737–746. [CrossRef]
36. Wang, C.-C.; Yang, K.-C.; Lin, K.-H.; Liu, Y.-L.; Liu, H.-C.; Lin, F.-H. Cartilage regeneration in scid mice using a highly organized three-dimensional alginate scaffold. *Biomaterials* **2012**, *33*, 120–127. [CrossRef]
37. Uematsu, K.; Hattori, K.; Ishimoto, Y.; Yamauchi, J.; Habata, T.; Takakura, Y.; Ohgushi, H.; Fukuchi, T.; Sato, M. Cartilage regeneration using mesenchymal stem cells and a three-dimensional poly-lactic-glycolic acid (plga) scaffold. *Biomaterials* **2005**, *26*, 4273–4279. [CrossRef]

38. Chang, C.-H.; Liu, H.-C.; Lin, C.-C.; Chou, C.-H.; Lin, F.-H. Gelatin–chondroitin–hyaluronan tri-copolymer scaffold for cartilage tissue engineering. *Biomaterials* **2003**, *24*, 4853–4858. [CrossRef]
39. Ochiya, T.; Takahama, Y.; Nagahara, S.; Sumita, Y.; Hisada, A.; Itoh, H.; Nagai, Y.; Terada, M. New delivery system for plasmid DNA in vivo using atelocollagen as a carrier material: The minipellet. *Nat. Med.* **1999**, *5*, 707–710. [CrossRef] [PubMed]
40. Uchio, Y.; Ochi, M.; Matsusaki, M.; Kurioka, H.; Katsube, K. Human chondrocyte proliferation and matrix synthesis cultured in atelocollagen® gel. *J. Biomed. Mater. Res.* **2000**, *50*, 138–143. [CrossRef]
41. Matsushita, R.; Nakasa, T.; Ishikawa, M.; Tsuyuguchi, Y.; Matsubara, N.; Miyaki, S.; Adachi, N. Repair of an osteochondral defect with minced cartilage embedded in atelocollagen gel: A rabbit model. *Am. J. Sports Med.* **2019**, *47*, 2216–2224. [CrossRef]
42. Sakai, D.; Mochida, J.; Yamamoto, Y.; Nomura, T.; Okuma, M.; Nishimura, K.; Nakai, T.; Ando, K.; Hotta, T. Transplantation of mesenchymal stem cells embedded in atelocollagen® gel to the intervertebral disc: A potential therapeutic model for disc degeneration. *Biomaterials* **2003**, *24*, 3531–3541. [CrossRef]
43. Li, F.; Carlsson, D.; Lohmann, C.; Suuronen, E.; Vascotto, S.; Kobuch, K.; Sheardown, H.; Munger, R.; Nakamura, M.; Griffith, M. Cellular and nerve regeneration within a biosynthetic extracellular matrix for corneal transplantation. *Proc. Natl. Acad. Sci. USA* **2003**, *100*, 15346–15351. [CrossRef] [PubMed]
44. Kawaguchi, H.; Hirachi, A.; Hasegawa, N.; Iwata, T.; Hamaguchi, H.; Shiba, H.; Takata, T.; Kato, Y.; Kurihara, H. Enhancement of periodontal tissue regeneration by transplantation of bone marrow mesenchymal stem cells. *J. Periodontol.* **2004**, *75*, 1281–1287. [CrossRef]
45. Widjaja, W.; Maitz, P. The use of dermal regeneration template (pelnac®) in acute full-thickness wound closure: A case series. *Eur. J. Plast. Surg.* **2016**, *39*, 125–132. [CrossRef]
46. Steadman, J.R.; Rodkey, W.G.; Rodrigo, J.J. Microfracture: Surgical technique and rehabilitation to treat chondral defects. *Clin. Orthop. Relat. Res.* **2001**, *391*, S362–S369. [CrossRef] [PubMed]
47. U.S. Department of Health and Human Services. *Common Terminology Criteria for Adverse Events (Ctcae), Version 5.0.*; U.S. Department of Health and Human Services: Washington, DC, USA, 2017.
48. Irrgang, J.J.; Anderson, A.F. Development and validation of health-related quality of life measures for the knee. *Clin. Orthop. Relat. Res. (1976–2007)* **2002**, *402*, 95–109. [CrossRef]
49. Brittberg, M.; Peterson, L. Introduction of an articular cartilage classification. *ICRS Newsl.* **1998**, *1*, 5–8.
50. Peterson, L.; Minas, T.; Brittberg, M.; Nilsson, A.; Sjögren-Jansson, E.; Lindahl, A. Two-to 9-year outcome after autologous chondrocyte transplantation of the knee. *Clin. Orthop. Relat. Res. (1976–2007)* **2000**, *374*, 212–234. [CrossRef]
51. Chang, C.H.; Kuo, T.F.; Lin, F.H.; Wang, J.H.; Hsu, Y.M.; Huang, H.T.; Loo, S.T.; Fang, H.W.; Liu, H.C.; Wang, W.C. Tissue engineering-based cartilage repair with mesenchymal stem cells in a porcine model. *J. Orthop. Res.* **2011**, *29*, 1874–1880. [CrossRef]
52. Steadman, J.R.; Rodkey, W.G.; Briggs, K.K. Microfracture to treat full-thickness chondral defects: Surgical technique, rehabilitation, and outcomes. *J. Knee Surg.* **2002**, *15*, 170–176. [PubMed]
53. Knutsen, G.; Drogset, J.O.; Engebretsen, L.; Grøntvedt, T.; Ludvigsen, T.C.; Løken, S.; Solheim, E.; Strand, T.; Johansen, O. A randomized multicenter trial comparing autologous chondrocyte implantation with microfracture: Long-term follow-up at 14 to 15 years. *J. Bone Jt. Surg.* **2016**, *98*, 1332–1339. [CrossRef] [PubMed]
54. Cole, B.J.; Pascual-Garrido, C.; Grumet, R.C. Surgical management of articular cartilage defects in the knee. *J. Bone Jt. Surg.* **2009**, *91*, 1778–1790.
55. Nakagawa, Y.; Mukai, S.; Yabumoto, H.; Tarumi, E.; Nakamura, T. Serial changes of the cartilage in recipient sites and their mirror sites on second-look imaging after mosaicplasty. *Am. J. Sports Med.* **2016**, *44*, 1243–1248. [CrossRef]
56. Alparslan, L.; Winalski, C.S.; Boutin, R.D.; Minas, T. Postoperative Magnetic Resonance Imaging of Articular Cartilage Repair. *Semin. Musculoskelet. Radiol.* **2001**, *5*, 345–364. [CrossRef] [PubMed]
57. Choi, Y.S.; Potter, H.G.; Chun, T.J. Mr imaging of cartilage repair in the knee and ankle. *Radiographics* **2008**, *28*, 1043–1059. [CrossRef]
58. Lee, W.S.; Kim, H.J.; Kim, K.I.; Kim, G.B.; Jin, W. Intra-articular injection of autologous adipose tissue-derived mesenchymal stem cells for the treatment of knee osteoarthritis: A phase iib, randomized, placebo-controlled clinical trial. *Stem Cells Transl. Med.* **2019**, *8*, 504–511. [CrossRef]
59. Miot, S.; Brehm, W.; Dickinson, S.; Sims, T.; Wixmerten, A.; Longinotti, C.; Hollander, A.; Mainil-Varlet, P.; Martin, I. Influence of in vitro maturation of engineered cartilage on the outcome of osteochondral repair in a goat model. *Eur. Cell Mater.* **2012**, *23*, 222–236. [CrossRef]
60. Murphy, S.V.; Atala, A. 3d bioprinting of tissues and organs. *Nat. Biotechnol.* **2014**, *32*, 773–785. [CrossRef] [PubMed]
61. Moroni, L.; Burdick, J.A.; Highley, C.; Lee, S.J.; Morimoto, Y.; Takeuchi, S.; Yoo, J.J. Biofabrication strategies for 3d in vitro models and regenerative medicine. *Nat. Rev. Mater.* **2018**, *3*, 21–37. [CrossRef] [PubMed]
62. Foresti, R.; Rossi, S.; Selleri, S. Bio composite materials: Nano functionalization of 4d bio engineered scaffold. In Proceedings of the 2019 IEEE International Conference on BioPhotonics (BioPhotonics), Taipei, Taiwan, 15–18 September 2019; IEEE: Piscataway, NJ, USA, 2019; pp. 1–2.
63. Foresti, R.; Rossi, S.; Pinelli, S.; Alinovi, R.; Sciancalepore, C.; Delmonte, N.; Selleri, S.; Caffarra, C.; Raposio, E.; Macaluso, G. In-vivo vascular application via ultra-fast bioprinting for future 5d personalised nanomedicine. *Sci. Rep.* **2020**, *10*, 1–13.
64. Cui, X.; Breitenkamp, K.; Finn, M.; Lotz, M.; D'Lima, D.D. Direct human cartilage repair using three-dimensional bioprinting technology. *Tissue Eng. Part A* **2012**, *18*, 1304–1312. [CrossRef] [PubMed]

65. De Mori, A.; Peña Fernández, M.; Blunn, G.; Tozzi, G.; Roldo, M. 3d printing and electrospinning of composite hydrogels for cartilage and bone tissue engineering. *Polymers* **2018**, *10*, 285. [CrossRef] [PubMed]
66. Gonçalves, A.M.; Moreira, A.; Weber, A.; Williams, G.R.; Costa, P.F. Osteochondral tissue engineering: The potential of electrospinning and additive manufacturing. *Pharmaceutics* **2021**, *13*, 983. [CrossRef] [PubMed]
67. O'Conor, C.J.; Case, N.; Guilak, F. Mechanical regulation of chondrogenesis. *Stem Cell Res. Ther.* **2013**, *4*, 1–13. [CrossRef] [PubMed]

polymers

MDPI

Review

Effect of Polymeric Matrix Stiffness on Osteogenic Differentiation of Mesenchymal Stem/Progenitor Cells: Concise Review

Aiah A. El-Rashidy [1,2], Sara El Moshy [2,3], Israa Ahmed Radwan [2,3], Dina Rady [2,3], Marwa M. S. Abbass [2,3], Christof E. Dörfer [4] and Karim M. Fawzy El-Sayed [2,4,5,*]

1 Biomaterials Department, Faculty of Dentistry, Cairo University, Cairo 11562, Egypt; aiah.abdelwahab@dentistry.cu.edu.eg
2 Stem Cells and Tissue Engineering Research Group, Faculty of Dentistry, Cairo University, Cairo 11562, Egypt; sarah.mahmoud@dentistry.cu.edu.eg (S.E.M.); esraa.ahmed@dentistry.cu.edu.eg (I.A.R.); dina.radi@dentistry.cu.edu.eg (D.R.); marwa.magdy@dentistry.cu.edu.eg (M.M.S.A.)
3 Oral Biology Department, Faculty of Dentistry, Cairo University, Cairo 11562, Egypt
4 Clinic for Conservative Dentistry and Periodontology, School of Dental Medicine, Christian Albrechts University, 24105 Kiel, Germany; doerfer@konspar.uni-kiel.de
5 Oral Medicine and Periodontology Department, Faculty of Dentistry, Cairo University, Cairo 11562, Egypt
* Correspondence: karim.fawzy@gmail.com; Tel.: +20-2-236-349-65; Fax: +20-2-236-463-75

Abstract: Mesenchymal stem/progenitor cells (MSCs) have a multi-differentiation potential into specialized cell types, with remarkable regenerative and therapeutic results. Several factors could trigger the differentiation of MSCs into specific lineages, among them the biophysical and chemical characteristics of the extracellular matrix (ECM), including its stiffness, composition, topography, and mechanical properties. MSCs can sense and assess the stiffness of extracellular substrates through the process of mechanotransduction. Through this process, the extracellular matrix can govern and direct MSCs' lineage commitment through complex intracellular pathways. Hence, various biomimetic natural and synthetic polymeric matrices of tunable stiffness were developed and further investigated to mimic the MSCs' native tissues. Customizing scaffold materials to mimic cells' natural environment is of utmost importance during the process of tissue engineering. This review aims to highlight the regulatory role of matrix stiffness in directing the osteogenic differentiation of MSCs, addressing how MSCs sense and respond to their ECM, in addition to listing different polymeric biomaterials and methods used to alter their stiffness to dictate MSCs' differentiation towards the osteogenic lineage.

Keywords: mesenchymal stem cells; polymers; matrix; stiffness; osteoblasts; differentiation

Citation: El-Rashidy, A.A.; El Moshy, S.; Radwan, I.A.; Rady, D.; Abbass, M.M.S.; Dörfer, C.E.; Fawzy El-Sayed, K.M. Effect of Polymeric Matrix Stiffness on Osteogenic Differentiation of Mesenchymal Stem/Progenitor Cells: Concise Review. *Polymers* **2021**, *13*, 2950. https://doi.org/10.3390/polym13172950

Academic Editor: José Miguel Ferri

Received: 16 July 2021
Accepted: 5 August 2021
Published: 31 August 2021

Publisher's Note: MDPI stays neutral with regard to jurisdictional claims in published maps and institutional affiliations.

1. Introduction

Stem/progenitor cells are characterized by their outstanding differentiation potential into multiple types of specialized cell lineages, relying on their pluri- or multipotency, while maintaining their self-replicating ability [1–10]. Among the different stem/progenitor cell types, mesenchymal stem/progenitor cells (MSCs) have been widely used in tissue engineering, cell transplantation, and immunotherapy [11–14]. MSCs were initially isolated from the bone marrow, but can be currently isolated from almost every tissue in the body [15]. MSCs niches are located in different sites, including umbilical cord blood [16], menses blood [17], dental tissues [18], synovial fluid [19], adipose tissues [14], and dental tissues [4]. MSCs reside adjacent to vessel walls, near perivascular regions, on the endosteal surfaces of trabecular bone, and within the interfibrillar spaces [11].

Proliferation and differentiation of MSCs can be triggered by certain growth factors and chemicals, inducing specific genetic events, affecting the release of transcriptional factors, which regulate the differentiation of MSCs into specific lineages [14,17]. Additionally, biomaterial scaffolds can create a microenvironment that provides MSCs with appropriate conditions for directed differentiation [14]. MSCs further can secrete various immunomodulatory molecules, including cytokines, chemokines, and growth factors, which provide the self-regulated regenerative microenvironment for different injured tissues or organs [13,17].

Regeneration and healing of bone injuries, particularly in large bony defects, is a complicated process [11]. Based on the multipotency of MSCs, they can give rise to either osteoblasts, chondrocytes, myoblasts, or adipocytes in response to key transcriptional regulators that control the primary commitment and most of the follow-up differentiation [20]. MSCs further interact with the components of their local microenvironment (niche) of the extracellular matrix (ECM) [21].

ECM was earlier believed to be an inert matrix that only provides physical support to cells; later, the important role of ECM in various cellular processes was introduced [22]. The MSCs niche provides extrinsic signals including growth factors, ECM, and those released due to contact with other cells. Through these signals, the MSCs' niche could regulate the stem/progenitor cells' fate [23,24]. In this context, interactions of MSCs with their niche are reciprocal; thus, MSCs are capable of remodeling the niche in response to signals received from it [24].

Several transcription factors are involved in the osteogenic differentiation pathway, including runt-related transcription factor 2 (Runx2), osterix (Osx, or SP7), Smad, and β-catenin [25–27]. Runx2 expressing cells are defined as pre-osteoblasts, a heterogeneous population of cells that includes all cells transitioning from progenitors to mature osteoblasts. A three-stage differentiation of the pre-osteoblasts then follows. The first stage involves cells' proliferation and expression of transforming growth factor-beta receptor 1 (TGF-βR1), fibronectin, collagen, and osteopontin. The second stage involves the initiation of cellular differentiation and maturation of the ECM with alkaline phosphatase (ALP) and collagen expression. In the final stage, the ECM is enriched with osteocalcin, which promotes matrix mineralization [20]. Runx2 guides MSCs differentiation into osteoblasts besides inhibition of other differentiation pathways, particularly adipogenic and chondrogenic ones [28,29]. Various signaling pathways, such as bone morphogenic proteins (BMPs), Notch, and Wnt signaling pathways, could regulate Runx2 expression.

BMPs are famous for their capability to induce bone formation. They activate intracellular Smad, which translocates to the nucleus and acts as a transcription factor besides promoting Runx2 expression [30]. BMP9 stimulates the activation of Smad1/5/8 in MSCs cells. Moreover, Smad4 knockdown decreases the nuclear translocation of Smad1/5/8 and inhibits osteogenic differentiation [31]. Hence, Smad is of great importance, and its interaction with Runx2 is essential for osteogenic differentiation. Mutation of the C-terminal domain of Runx2 disrupts Runx2–Smad transcriptional activities, which leads to the suppression of osteogenic differentiation [32].

Osx is an essential transcription factor for osteogenic differentiation and subsequent bone formation. In Osx null mice, no bone formation occurred; additionally, in the Runx2 null mice, no expression of Osx was noted, indicating that Osx acts as a downstream of Runx2 and emphasizing its role in MSCs osteogenic differentiation and bone formation [33]. Moreover, Wnt signaling pathway activation in MSCs induces Osx expression and suppresses peroxisome proliferator-activated receptor γ (PPAR-γ) [34]. Moreover, β-catenin has a competitive inhibitory relationship with PPAR-γ, where activation of one of them leads to the degradation of the other [35]. Therefore, Wnt/β-catenin signaling activation shifts MSCs' commitment towards osteogenesis at the expense of adipogenic differentiation [34].

β-catenin further plays a critical role in MSCs' osteogenic differentiation. Its absence blocks the osteogenic differentiation and allows for the chondrogenic differentiation of MSCs [36]. Wnt signaling is essential for the β-catenin function. Wnt signaling accumulates β-catenin in the cytoplasm and translocates it into the nucleus, activating the transcription of downstream genes. The absence of Wnt signaling leads to the degradation of β-catenin and interferes with MSCs' osteogenic differentiation [37]. The sensitivity of β-catenin to matrix stiffness during the differentiation of adipose-derived stromal cells (ASCs) has been demonstrated [38]. β-catenin increased nuclear translocation with increased matrix stiffness and enhanced the expression of Runx2, thus stimulating osteogenesis.

Stem/progenitor cells' behavior is largely affected by extracellular signals from the microenvironment, including chemical and mechanical cues from the ECM [39]. Unlike chemical cues, the mechanical properties of the microenvironment act as signals that are consistent along with time and space, thus providing long-range stimulation to cells over long periods and over relatively long distances. Recent literature has focused on the paramount role of the ECM mechanical properties in controlling stem/progenitor cells' behavior, including maintaining their potency, self-renewal and differentiation, migration, proliferation, and interaction with other cells [39,40]. Matrix-related mechanical stimuli, including strain, shear stress, matrix rigidity, and topography, could impact stem/progenitor cell phenotypes through controlling gene transcription and signaling pathways [40,41].

The extracellular-signal-regulated kinase (ERK) and p38 are members of the mitogen-activated protein kinase (MAPK) enzymes family that is concerned with mechanotransduction pathways [42]. ERK is a potent regulator of MSCs' differentiation, as mechanical stimulation activates ERK through integrin focal adhesion complexes and the initiation of MAPK–ERK signaling cascade [43]. Besides ERK, p38 is involved in MSCs' differentiation. The p38-MAPK signaling activity has been identified as an essential factor for osteoblastic differentiation [44–46]. Ras-mediated signaling has been further presented as a master key that affects multiple intracellular pathways, including ERK, PI3K/AKT, and Smad [47,48]. Inhibition of Ras (RasN17) significantly downregulates AKT, ERK, and Smad1/5/8 activation, as well as osteogenic markers' expression. Conversely, active Ras (RasV12) has little effect on osteogenic markers' expression [49]. Consequently, inducing transcription factors to control and guide MSCs' differentiation has become an essential strategy for guided tissue regeneration [26]. Interference between signaling pathways through interaction between different transcription factors can drive MSCs towards specific cell linage; for example, osteogenic signaling can inhibit the adipogenic signaling pathway, and vice versa [41].

Matrix stiffness has a profound impact on MSCs' behavior. The adhesion, proliferation, and spreading capacity of umbilical cord MSCs varied when cultured on polyacrylamide gels coated with fibronectin with different stiffness (Young's modulus: 13–16, 35–38, 48–53, and 62–68 kPa) [50]. Maximum spreading of MSCs was observed with increased matrix stiffness. The soft matrix promoted adipogenic differentiation with high expression of PPARγ and C/EBPα, while MSCs cultured on the 48–53 kPa matrix differentiate into muscle cells with increased expression of MOYG. On the other hand, MSCs cultured on stiff matrix differentiate into osteoblast with increased expression of ALP, collagen type I, Runx2, and osteocalcin [50]. Additionally, bone-marrow MSCs cultured on fibronectin-coated polyacrylamide hydrogels with different stiffnesses, ranging from 13 to 68 kPa, demonstrated enhanced adhesion, spreading and proliferation upon increasing matrix stiffness [51]. On 62–68 kPa, MSCs exhibited a polygonal morphology with a more extensive spreading area and high expression of Runx2, ALP, and osteopontin. These data highlight the critical role of matrix stiffness in regulating MSCs behavior which aids in the development of new biomaterials for tissue regeneration.

Insights into how stem/progenitor cells sense signals from the ECM and how they respond to these signals at the molecular level have become an area of increasing research [21,22,52]. Lately, stem/progenitor cells were shown to be capable of sensing and responding to the structural and functional cues of the matrix [22,52], such as the topography of the ECM components, adhesive properties of the ECM, and ECM stiffness [24,53]. The cells adhere to the ECM via several specific cell-surface receptors, known as integrins [21,22]. Integrins transmit signals from ECM to the cells, thus affecting the proliferation and differentiation of stem/progenitor cells through mechanotransduction of signals [21,22]. It is suggested that the cells use actomyosin filaments (stress fibers) contractility for reciprocal interactions with their matrix [23]. When cells are grown in vitro, extensive efforts to mimic the in vivo microenvironments have been made to control and direct stem/progenitor cell commitment into specific cell lineages required for regenerative medicine.

Natural and synthetic polymeric materials could offer versatile matrices, which are biocompatible and biodegradable, with tunable characteristics, precise control of their topography, and ease of processing [54,55]. Biomaterial stiffness, which determines the material's resistance to deformation in response to an applied force, is a vital property in tissue engineering. ECM stiffness is calculated by dividing the load by the elastic deformation of the matrix [56], is denoted by the elastic modulus or Young's modulus (E), and represents the resistance that a cell feels when it deforms the ECM [57].

ECM stiffness guides stem/progenitor cells' differentiation down corresponding tissue lineages [58]. Osteogenic differentiation of MSCs was shown to be favored on more rigid substrate, while adipogenic differentiation is favored on softer substrates [21]. Such control of MSCs fate by matrix stiffness was shown to be complementary to, and even synergistic with, the regulatory effects of specialized cell culture media commonly used to direct mesenchymal stem/progenitor cell differentiation into specific lineages [23]. Various biomaterials coupled with different methods of controlling stiffness are employed to develop specific stiffness ranges for regulating MSCs differentiation in vitro. Controlling of substrates' stiffness could be tuned through adjusting the biomaterial composition, the amount/concentration/ratio of material components, the degree of crosslinking, and the reaction conditions [56]. Taking into consideration that the bulk stiffness of most native tissues is much lower than that of plastic or glassware used for in vitro tissue culture [24], the development of biomimetic polymeric matrices of tunable stiffness, mimicking native tissues, allowed new data to reveal more details on the impact of mechanical cues of the microenvironment, especially ECM stiffness, on cellular properties [58].

In this review, we highlight the regulatory role of matrix stiffness in directing the osteogenic differentiation of MSCs, addressing how MSCs sense and respond to their ECM, in addition to listing different polymeric biomaterials commonly used in vitro and methods used to alter their stiffness to dictate MSCs differentiation towards the osteogenic lineage. Moreover, through the current review, we aim to elucidate the effect of ECM stiffness on the MSCs' osteogenic potential and the underlying mechanism, which is of particular importance during the process of designing new materials for bone-tissue regeneration.

2. MSCs and Mechanotransduction

MSCs can sense and assess the stiffness of extracellular substrates [59,60]. The ability of stem/progenitor cells to sense changes in the surrounding environment is known as mechanosensation. Stem/progenitor cells can also transduce mechanical stimuli in the surrounding environment into biochemical signals to induce cellular responses through the process of mechanotransduction [61]. Mechanotransduction is the mechanism underlying the increased osteogenic differentiation of MSCs on stiffer matrix [62]. The process of mechanotransduction in stem/progenitor cells is mediated through focal adhesion, associated integrins, and cellular cytoskeleton, in addition to mechanosensitive ion channels.

2.1. Focal Adhesion and Integrins

Focal adhesions are complexes of highly specialized proteins and macromolecules that can attach cells to the ECM, allowing them to sense and respond to mechanical stimuli [63]. Focal adhesion is composed of the transmembrane protein integrin, which has intracellular and extracellular domains (Figure 1). Integrin's intracellular domain is linked to the actin cytoskeleton via cytoplasmic adapter proteins, which include the actin-binding proteins α-actinin, vinculin, and talin [64]. The integrin extracellular domain binds to ECM components, such as collagen, laminin, and fibronectin, via its extracellular domain, thereby establishing a mechanical connection between ECM and intracellular cytoskeleton components [64–66]. Fifty different proteins have been associated with focal adhesion [67], including intracellular proteins as focal adhesion kinase (FAK) and p130Cas [68,69].

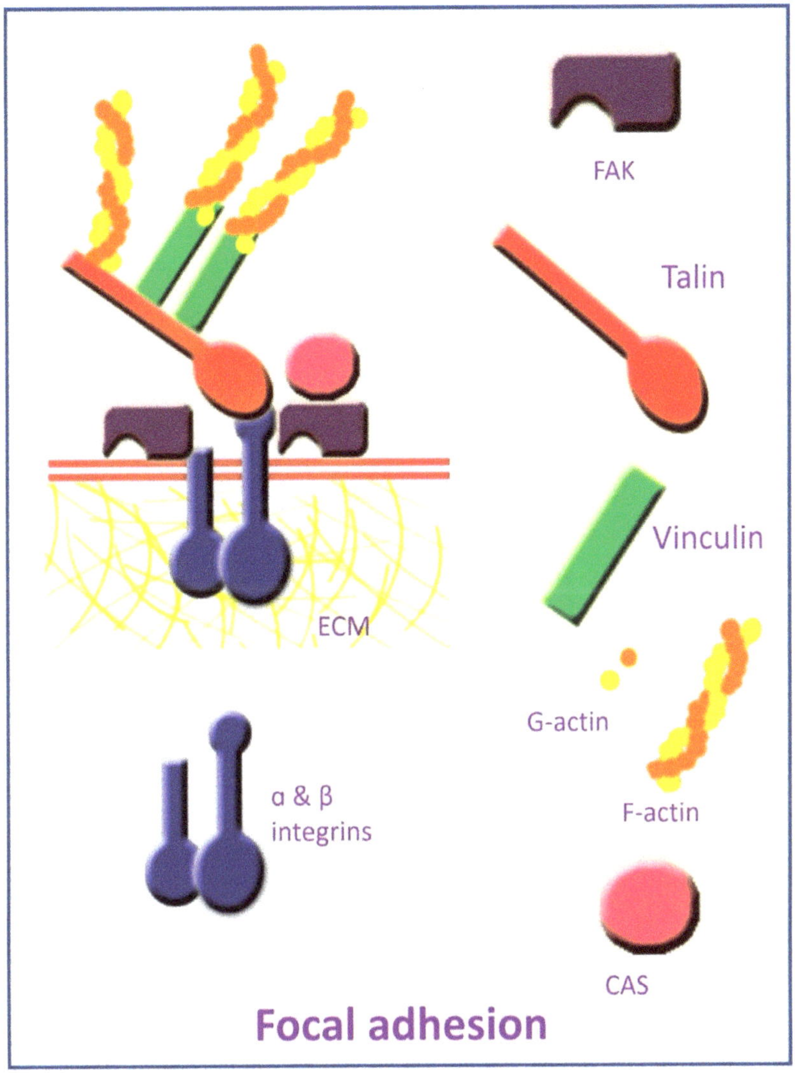

Figure 1. Elements of focal adhesion.

Integrins are alpha and beta subunits heterodimers existing in different combinations [64]. There are 18α and 8β subunits, which account for 24 different integrin heterodimers in mammals specific to an exact set of ECM ligands [70,71]. Through their intracellular and extracellular domains, integrins are capable of joining intracellular cytoskeleton with the external environment, thereby creating mechanical integration between ECM and intracellular cytoskeleton [72]. They can transmit cellular signals to the ECM, and reciprocally can convey signals from the ECM intracellularly [73], triggering an intracellular signaling pathway, resulting in alteration of cellular migration, proliferation, and differentiation [74].

It is noteworthy that MSCs' surface integrins' expression can influence MSCs' lineage commitment [75]. Further, matrix stiffness can influence integrin expression on MSCs, which can dictate and direct stem/progenitor cell fate [76]. Undifferentiated MSCs were found to mostly express α1, α3, αV, β1, and β2 integrins, while α2, α4, α5, α6, β3, β4, and β5 were expressed to a lesser extent [77]. MSCs' osteogenic differentiation was reported to be associated with upregulation of integrin α5 expression on MSCs' surface in response to ECM stimuli [77–80]. Integrin α5 upregulation promotes osteogenesis through activation of FAK via the ERK1/2-MAPKs and PI3K signaling [79]. MSCs' expression of integrin subunits α2 [62,76], α1, αV, and β3 was also upregulated with increased ECM stiffness, favoring osteogenic differentiation [62], while α5 and β1 expression was upregulated in the matrix with lower stiffness [76]. Additionally, activation of MSCs expression of α5β1 and αVβ3 integrin complexes in response to ECM morphology was associated with enhanced osteogenic differentiation [81]. On the other hand, osteogenic differentiation was associated with reduced expression of α1, α3, α4, β3, and β4 integrin subunits [77], while MSCs' adipogenic differentiation was associated with the upregulation of α6 and reduction of α2, α4, α3, β3, and β4 integrin subunits expression [77]. Increased integrin α5 expression can also inhibit both adipogenic and chondrogenic differentiation, while promoting MSCs osteogenic differentiation [82].

The binding of integrin to ECM components triggers the intracytoplasmic assembly of focal adhesion proteins, including talin, FAK, p130Cas, and vinculin, first forming focal complexes, which then grow, giving rise to focal adhesions, linking actin fibers to ECM components [83]. Formation of focal adhesion, with the associated triggering of intracellular signaling pathways, is essential for MSCs migration, proliferation and differentiation [84–86].

MSCs lineage commitment and osteogenic differentiation in response to ECM mechanical cues, including matrix stiffness, involve upregulation of focal adhesion formation. Increasing matrix stiffness can promote number [84] and area of focal adhesions [87]. In turn, upregulation of focal adhesion number and size [60,88,89] has been linked to increased osteogenic differentiation of MSCs. Additionally, tightly packed focal adhesion can stimulate osteogenic differentiation [90].

2.2. Cytoskeleton Elements

Cytoskeletal-related proteins are responsible for the ability of stem/progenitor cells to respond to mechanical cues, including stiffness of the ECM [85]. In addition to their role in providing a cellular structural framework, repolarization of cytoskeleton elements in response to mechanical stimuli, transmits the signals from the ECM to the nucleus, resulting eventually in altered gene expression [64,91,92]. Structural elements of the cellular cytoskeleton include microfilaments, intermediate filaments, and microtubules [64].

The actin cytoskeleton is responsible for the maintenance of cell shape, motility, and contractility. They also act as mechanical sensors for the extracellular environment [93]. It is formed of F-actin, which is a helical polymer of G-actin coupled with actin-binding and actin-bundling proteins, such as α-actinin, vinculin, and talin [94] (Figure 2). Actin cytoskeleton forms a web in association with cellular junctions and forms a core of microvilli, filopodia, and lamellipodia [64]. Actin perinuclear cap is a dome-like structure formed of contractile actin filament and phosphorylated myosin, covering the top of the nucleus and connected to the nucleus through linkers of nucleoskeleton and cytoskeleton (LINC) protein complexes [95,96]. This actin perinuclear cap provides a mechanism through which mechanical signals, transduced through focal adhesion, can reach the nucleus to induce cellular responses [95,97]. Ultimately, the actin perinuclear cap is responsible for conveying signals regarding matrix stiffness to the nucleus [97].

The binding of MSCs to a stiff substrate induces actin polymerization, as evident by an increased ratio of F-actin to G-actin, forming actin stress fibers, which trigger intracellular-signaling pathways [67]. Stress fibers are actomyosin complex composed of F-actin and myosin-2 stabilized by crosslinking proteins [98] (Figure 2). The process of actin polymerization is regulated by the FAK signaling pathway [99]. Actin polymerization and stress fibers formation are essential for establishing cell to ECM interaction [100]. Polymerization of actin dictates lineage commitment of MSCs, as actin depolymerization was noticed during adipogenic differentiation [101]. On the contrary, actin polymerization combined with an increased ratio of F-actin to G-actin upregulated osteogenic differentiation [99,102–105]. On the other hand, disruption of actin polymerization can reduce osteogenic differentiation [103]. Increased osteogenic differentiation on stiffer substrates was also associated with increased expression of F-actin [78], in addition to actin-binding protein (vinculin) [106].

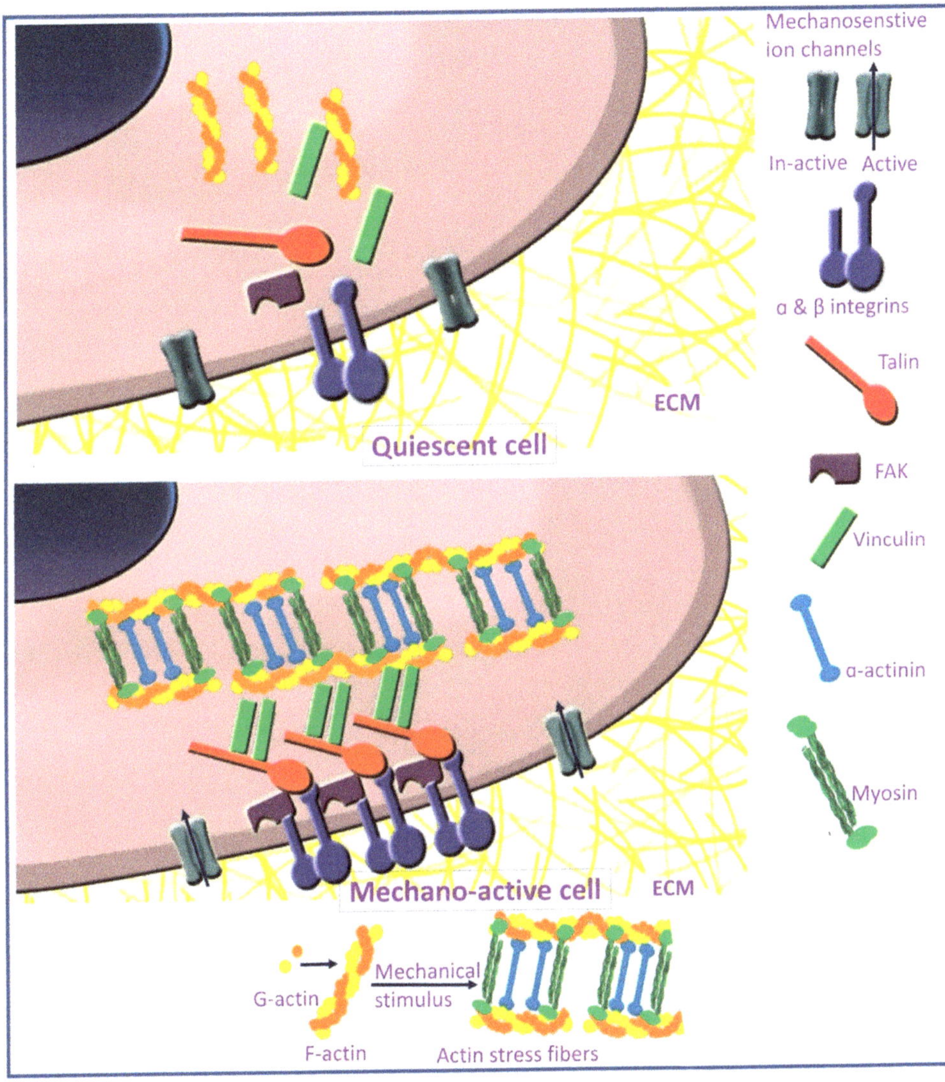

Figure 2. Focal adhesion formation and stress fibers assembly in mechano-active cells.

Actin filaments can also interact with other components of the cellular cytoskeleton as intermediate filaments [64]. Intermediate filaments have a diameter of about 10 nm and have a role in maintaining cell shape and cellular junctions [64]. F-actin promotes intermediate filaments and vinculin assembly and disassembly, which are required for the process of osteogenesis through the transient receptor potential melastatin 7–osterix axis [107].

2.3. Mechanosensitive Ion Channels

Mechanosensitive ion channels are a further mechanism implicated in MSCs' mechanotransduction on stiff matrices. These ion channels are sensitive to substrate stiffness. Upon mechanical stimulation, they allow the intracellular influx of ions and can form complexes with stress fibers, eliciting intracellular signaling pathways [108]. Mechanical stimulation affects cell differentiation through a change in calcium influx through activated channels [109]. The change in the calcium influx results in the activation of the MAPK signaling pathway [110].

2.4. MSCs' Aging and Mechanosensitivity

Different age-dependent changes in MSCs were reported, such as decreased proliferation ability [111] and osteogenic differentiation potential [112–114]. Moreover, age-associated bone loss was linked to the reduced osteogenic potential of MSCs [115]. Aged multipotent progenitor cells lose their sensitivity to alterations in polyacrylamide substrates, while younger multipotent progenitor cells showed a lineage-dependent response to stiffness [116]. The effect of MSCs aging on their mechanosensitivity was investigated by comparing the response of child (11 to 12 years old) and adult MSCs (20–30 years old) to variations in stiffness (10 and 300 kPa) of type I collagen-coated polyacrylamide substrates [117]. Child MSCs revealed more mechanosensitive (increased nuclear-translocation of YAP), improved angiogenesis (enhanced endothelial tubule formation), and osteogenesis (increased alkaline phosphatase activity and mineralization) on stiff substrates as compared to adult MSCs. Based on a customized PCR array, an age-dependent, stiffness-induced upregulation of NOX1, VEGFR1, VEGFR2, WIF1, and JNK3 in child MSCs compared to adults MSCs [117]. Understanding the mechanism behind the age-altered mechanosensitivity of MSCs may open up new avenues to identify potential therapeutic targets to reproduce the enhanced osteogenic and angiogenic potential of adults with bone degeneration and disease.

3. The Role of Matrix Stiffness in Triggering MSCs' Osteogenic Differentiation

Matrix stiffness regulates the MSCs' differentiation into mature specific cells by activating transcription factors that upregulate genes responsible for the initiation and progression of particular cell-linage differentiation. The singling pathways involved in MSCs' osteogenic differentiation are illustrated in (Figure 3). Rigid matrices led to increased MSCs spreading and improved actomyosin contractility, promoting osteogenic differentiation. This enhanced potential was accompanied by increased Runx2, β-catenin, and Smad, implying the significant impact of mechanosensing the matrix stiffness and its role in determining the cell fate [41,50]. The relation between Runx2 expression, owing to mechanosensation with actomyosin contractility, was confirmed by inhibiting myosin, which caused a decrease in Runx2 expression [118]. However, the effect of matrix stiffness on MSCs' differentiation disappeared at the monolayer state [49].

The hippo pathway is one of the signaling pathways involved in MSCs' differentiation and is regulated by intra- and extracellular signals [119]. The downstream effectors of the hippo signaling pathway are yes-associated protein (YAP) and transcriptional co-activator with PDZ-binding motif (TAZ) [120]. YAP and TAZ transduce signals necessary for determining MSCs' fate. The control of the Hippo pathway is through phosphorylation and nuclear translocation of YAP/TAZ [121]. Additionally, matrix stiffness can control the localization and activity of YAP/TAZ [122,123], which is identified through the structural and functional regulation of the cell cytoskeleton to adjust cellular tension [124]. The stresses sensed by MSCs are transmitted to the nucleus and lead to an increase in the nuclear membrane tension, causing expansion of the nuclear pores, which promote sudden nuclear inflow of YAP [125]. In MSCs cultured on a rigid matrix (40 kPa) undergoing osteogenic differentiation, YAP/TAZ has been localized in the nucleus. In comparison, MSCs cultured on a soft matrix (0.7 kPa), YAP/TAZ persisted in the cytoplasm, directing MSCs to undergo adipogenic differentiation [122]. Moreover, YAP knocking-down resulted in inhibition of osteogenesis and enhancement of adipogenesis [126]. During MSCs' osteogenic differentiation, TAZ functions as a co-activator of Runx2 to stimulate osteogenesis and inhibits PPAR-γ, which reduces adipogenic differentiation [127]. These findings highlight the significant role of YAP/TAZ as a potent regulator of stiffness-induced osteogenic differentiation.

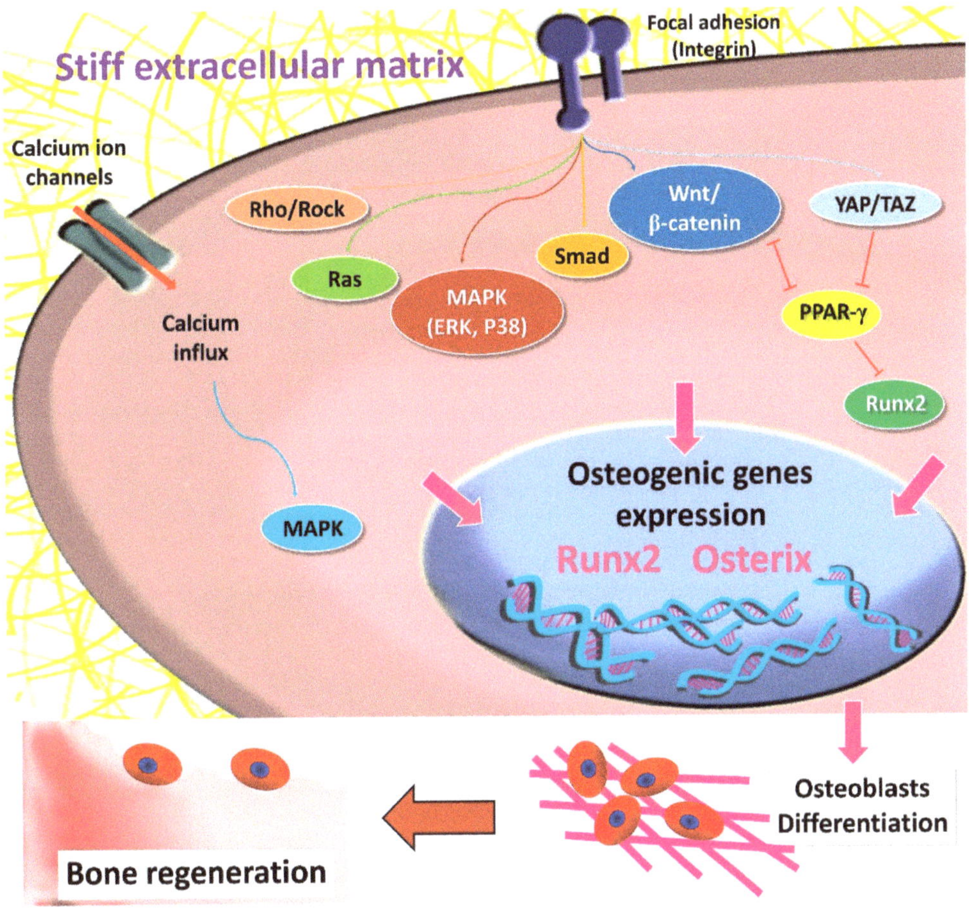

Figure 3. Signaling pathways involved in stiffness induced MSCs' osteogenic differentiation.

MSCs fate is also directed through actomyosin contractility and activated Rho/Rho kinase (ROCK) signaling [88], along with mechanotransduction mediated by focal adhesion and integrin [128]. In response to increased stiffness, activated Rho stimulates actomyosin stress fiber assembly [129], which causes an increase in cell contractility and activation of ERK, promoting osteogenic differentiation [130]. Furthermore, Rho combined with the actin cytoskeleton is essential to maintain nuclear YAP/TAZ in MSCs [122]. Activation of FAK via ROCK signaling led to upregulation of osteogenic marker Runx2, ALP, and matrix mineralization denoting osteogenesis of human adipose stem/progenitor cells [131]. In addition, the inhibition of FAK and ROCK signaling caused an upregulation of adipogenic markers. Furthermore, matrix stiffness modulates MSCs' osteogenic differentiation through the Ras pathway, which is accompanied by an increase in the phosphorylation levels of Smad1/5/8, AKT and ERK [49]. Ras (RasN17) inhibition resulted in a significant decrease of Smad1/5/8, AKT, and ERK activity, as well as osteogenic markers' expression [49].

Cells on stiff matrices develop high cytoskeletal tension, which is evidenced by enhanced actin stress fibers and large spread area. Below a compressive modulus of 25 kPa, regardless of the adhesive ligand presented, there is not enough cytoskeletal tension to promote osteogenic lineage differentiation [88]. Based on these results, it has been postulated that, unless a cell develops cytoskeletal tension exceeding a certain threshold stiffness (substrates with moduli of $\geq$ 25 kPa), osteogenic differentiation will not occur and the cell would need the presence of an osteogenic ligand for Runx2 expression for further differentiation to take place. On the other hand, MyoD1 (a marker for myoblasts) expression demonstrated less ECM dependence compared with Runx2, as it was markedly expressed in cells cultivated on substrates with stiffnesses higher than 9 kPa, regardless of the protein coating [132]. Additionally, on soft poly(acrylamide-co-acrylic acid) substrates (E = 15.4 kPa) that mimic muscle elasticity when grafted with arginine–glycine–aspartate (RGD) peptide sequence, MSCs were directed to a spindle-shaped morphology similar to C2C12 myoblasts, while stiffer matrices (E = 47.5 kPa) that mimic osteoid tissue's crosslinked collagen yield the cells in polygonal morphology, similar to MC3T3-E1 pre-osteoblasts [118].

Inflammation can further counteract the inductive effect of matrix stiffness on osteogenic differentiation. Periodontal ligament stem cells (PDLSCs) cultured with the inflammatory cytokine interleukin (IL)-1β on gelatin/methacrylate hydrogels with different matrix stiffness showed a marked reduction in matrix stiffness-dependent osteogenic differentiation and expression of osteocalcin, as well as Runx2. This was through the activation of p38 signaling pathways, which were activated by IL-1β [133]. Further, macrophages encapsulated in gelatin/methacrylate hydrogels with high stiffness showed a high tendency to polarize toward the pro-inflammatory M1 phenotype, which was associated with a negative impact on the osteogenic differentiation of bone-marrow mesenchymal stem cells (BMMSCs) [134].

4. Matrix-Dependent MSCs' Osteogenic Differentiation

Several natural and synthetic polymeric biomaterials are currently used in tissue engineering and regenerative medicine, serving as biomimetic matrices for in vitro culturing [135]. These biopolymers can be generally divided into two classes: natural and synthetic polymers. Natural polymers include alginate, collagen, gelatin, hyaluronic acid, elastin, actin, keratin, albumin, chitosan, and others. They are characterized by their inherent bioactivity and ability to mimic natural tissues, yet they suffer from possible immunogenicity, structural complexity, and poor mechanical properties. Chitosan can be used for increasing energy storage of α-cobalt molybdate (CoMoO) nano-flakes in the presence of a crosslinking agent such as citric acid [136]. Compared to natural polymers, synthetic polymers have higher mechanical properties, are readily available, with tunable physicochemical properties and degradation rate, but lack natural tissue resemblance [55,137]. Major synthetic polymers used include polyethylene glycol (PEG), polydimethylsiloxane (PDMS), polyesters, polyacrylamide, vinyl polymers, and self-assembling peptides, in addition to poly (lactic acid) (PLA), poly (glycolic acid) (PGA), poly (lactide-co-glycolic acid) (PLGA), and others [55,135].

4.1. Natural Polymers

4.1.1. Alginate

Alginates extracted from seaweeds and algae are composed of β-1,4-linked blocks of β-D-mannuronic acid (M) and its C-5 epimer α-L-guluronic acid [138]. Alginates are widely used polysaccharides for hydrogelation in tissue engineering as they can be gelated easily through the addition of divalent cations [139]. Since alginate could act as a template for binding of manganese ions, the presence of a high concentration of alginate in the electrolytic manganese dioxide altered the morphology from spindle-shaped to cactus-shaped [140].

Spatially modulating the mechanical properties in an alginate bioink, 3D printed constructs were postulated to regulate MSCs' fate. Micro-CT-based imaging with 3D bioprinting and bioreactor system were utilized to fabricate 3D human MSCs-laden porous bone-like scaffolds with varying compressive moduli based on two unmodified polymers (alginate and gelatin). Softer scaffolds with low alginate concentration (0.8% alginate, 0.66 ± 0.08 kPa) revealed accelerated and enhanced osteogenic differentiation with upregulated ALP activity than stiffer scaffolds (1.8% alginate, 5.4 ± 1.2 kPa). In the presence of the osteogenic differentiation medium, cells on soft scaffolds exhibited osteoblastic and early osteocyte-related gene expression and showed a 3D cellular network within the mineralized matrix [141].

Increasing alginate molecular weight, as well as increasing the crosslinking ratio, produces a significantly stiffer bioink. Upon bioprinting cylindrical MSCs laden constructs with spatially variable mechanical stiffness from the core to the periphery, more MSCs underwent osteogenic differentiation within the stiffer regions of the printed constructs as evident by increased ALP staining [142]. In contrast to most studies, an investigation demonstrated that, under basal conditions and in the absence of RGD ligands, alginate hydrogel with bimodal molecular weight distribution (50% LMW and 50% HMW) and 1 wt.% polymer concentration of low-stiffness 3D matrices (tan $\partial \approx 0.4$–0.6) provided a permissive environment for human MSCs osteogenic differentiation and expressed high levels of ALP and osteocalcin as compared to the stiffer 2 wt.% alginate hydrogel with the presence of RGD ligands [143].

4.1.2. Collagen

Collagen is considered to be the most abundant protein in mammals [144]. Being the main ECM protein, collagen, together with soluble factors, may act as a niche for MSCs osteogenesis and bone mineralization [145]. Mechanical properties of collagen fibers vary depending upon their location in different tissues. Thus, the cells can sense local fibrillar microenvironments with different physical cues. Collagen gels were engineered to attain varying fiber stiffness (from 1.1 to 9.3 kPa), while maintaining bulk stiffness below 200 Pa, by changing the polymerization temperature to 4 °C (Col-4), 21 °C (Col-21), and 37 °C (Col-37) without changing the density of the collagen. A polymerization temperature of 4 °C led to shorter, thicker, and stiffer collagen fibers (Col-4), with limited fiber recruitment and force transmission and fewer focal adhesions. Cells grown on Col-4 showed much slower spreading as compared to Col-37 with similar bulk stiffness but with more flexible and longer fibers that can be easily remodeled. Human MSCs cultured on Col-4 revealed a much lower ratio of osteogenic differentiation (21.1%) compared to that on Col-37 with (34.1%) ALP positive reactivity [146].

Matrix stiffness possesses a high impact on cellular bioactivity regardless of the presence or absence of growth factors. This was proved by culturing porcine adipose-derived stem cells (ADSCs) on sequentially integrated benzophenone photo-immobilization and carbodiimide, crosslinking collagen–glycosaminoglycan, with stiffness ranging from 2.85 to 5 MPa, in the presence or absence of covalently immobilized platelet-derived growth factor (PDGF-BB) and BMP-2 [147].

In the presence of osteogenic culture media, human MSCs coated with three bilayers of collagen/alginate nanofilm with relatively high stiffness 24 and 53 MPa revealed an augmented osteogenic differentiation efficiency with a significant increase in ALP by activating transcriptional co-activators with the PDZ binding motif through extracellular signal-related kinase and p38-MAPK signaling [148].

Nanoparticulate mineralized collagen–glycosaminoglycan scaffolds chemically crosslinked with 1-ethyl-3-(3-dimethylaminopropyl) carbodiimide and N-hydroxysuccinimide had a higher range of elastic moduli 3.90 ± 0.36 kPa as compared to non-crosslinked materials. Cultured human MSCs on crosslinked substrates showed higher expression of osteogenic genes and proteins compared to non-crosslinked versions. This was maintained via the mechanotransduction mediators YAP/TAZ and the Wnt signaling pathway [149].

MSCs seeded within a 3D collagen gel with an elastic modulus of ~108 Pa stimulated by vibrations of nanoscale amplitude in a vibrational bioreactor showed increased expression of Runx2, collagen I, ALP, osteopontin, osteocalcin, and BMP-2. This indicated that the viscoelastic properties of the collagen gel allowed the transfer of high-frequency vibrations to the cells seeded in 3D [150].

In the absence of any differentiation supplementation, MSCs grew on stiffer (1.5 kPa) dehydrothermal and 1-ethyl-3-3-dimethyl aminopropyl carbodiimide (EDC) crosslinked collagen–glycosaminoglycan scaffolds showed the greatest level of Runx2 expression, while substrates with lower stiffness (0.5 kPa) resulted in significant elevation of SOX9 expression, indicating that MSCs are directed towards a chondrogenic lineage [151].

Additionally, 3D scaffolds with the highest proportion of collagen in collagen and hydroxyapatite mixture coated on the decellularized cancellous bone with various stiffness (13.00 ± 5.55 kPa, 13.87 ± 1.51 kPa, and 37.7 ± 19.6 kPa) exhibited the highest stiffness that, in turn, promoted higher expressions of osteopontin and osteocalcin [152].

4.1.3. Gelatin

Gelatin is a natural, biocompatible, non-immunogenic, hydrophilic, and biodegradable collagen derivative [153–155]. It is acquired via acid or alkaline hydrolysis of collagen into single molecules [156]. Being derived from a natural source, gelatin is characterized by having RGD cell-binding motifs that can enhance cellular attachment [157]; it can also promote cell proliferation and differentiation [158,159]. One of the major disadvantages of gelatin is its low mechanical properties [153,160], in addition to its thermal instability [161].

Gelatin crosslinking through the addition of chemical groups can help reduce these shortcomings [161]. Stiffness of gelatin hydrogel scaffold can be controlled through changing the ratio of crosslinking agents, such as methacryloyl, giving rise to gelatin–methacrylate (GelMA) [133,162,163], transglutaminase [134], or EDC [164], and through the incorporation of variable additives as starch [162] or polyethylene glycol diacrylate (PEGDA) [163]. Increased gelatin hydrogel scaffold degree of crosslinking and matrix stiffness were positively associated with increased osteogenic differentiation of MSCs [133,134,162–164]. Crosslinked gelatin hydrogel can also be modified to enhance osteogenic potential through incorporation of the bisphosphonate alendronate [163]. GelMA hydrogel with tunable stiffness was constructed by using different GelMA concentrations 10, 12, and 14 wt to yield hydrogel with stiffness 25.75 ± 1.21 kPa, 59.71 ± 8.87 kPa, and 117.82 ± 9.83 kPa, respectively. Osteogenic differentiation of PDLSCs, as well as expression of osteocalcin and Runx2, showed a significant increase with increasing matrix stiffness through activation of ERK1/2 signaling pathway [133].

Gelatin/starch-based hydrogel was fabricated with tunable stiffness from crosslinked gelatin with variable degrees of methacrylation (GelMA; 31%, 72%, and 95%) covalently bound to variable ratios of pentenoates modified starch (10 v% starch and 20 v% starch). Increasing the degree of methacrylation and combining crosslinked gelatin with starch, with subsequent increase in matrix stiffness, effectively promoted osteogenic differentiation of adipose stem cells (ASCs), as was evident by an increased ALP expression. GelMA 95% combined with starch showed the highest degree of osteogenic differentiation, while the highest degree of adipogenic differentiation was observed on the least crosslinked and most flexible gelatin hydrogel (GelMA 31%) [162].

Three-dimensional porous gelatin scaffolds crosslinked using EDC further demonstrated an increase in the elastic modulus from ~0.6 to ~2.5 kPa, without any change in the scaffold internal structure. BMMSCs cultured on EDC-crosslinked gelatin scaffolds with increased stiffness showed an increased osteogenic differentiation as evidenced by increased Runx2 and osteocalcin expression in vitro. Subcutaneous implantation of EDC-crosslinked gelatin scaffold loaded with BMMSCs transfected with adenovirus encoding BMP-2 in mice demonstrated an increased bone formation in vivo, as compared to the control, non-crosslinked scaffold with low stiffness [164].

Transglutaminase-crosslinked gelatin scaffold with variable stiffness was constructed, using gelatin concentrations of 3%, 6%, and 9%. The 9% gelatin gave rise to the highest stiffness (60.54 ± 10.45 kPa), while 3% gelatin resulted in the lowest stiffness (1.58 ± 0.42 kPa). BMMSCs encapsulated in the hydrogel with the highest stiffness demonstrated the highest osteogenic differentiation as revealed by ALP activity, calcified nodule formation, expression of SP7 transcription factor-2, Osx, Runx2, and osteocalcin [134].

Augmentation of GelMA with alendronate and PEGDA showed a positive effect on osteogenic differentiation of BMMSCs. GelMA and alendronate were added at different concentrations and grafted on gelatin molecules, followed by further crosslinking, using 20 wt.% PEGDA to improve hydrogel scaffold stiffness from 4 to 40 kPa. Osteogenic differentiation of grafted BMMSCs was promoted on stiffer hydrogel with higher alendronate concentration, as evident by upregulation of ALP activity, collagen type I, and osteocalcin expression, as well as calcium deposition [163].

4.1.4. Decellularized Matrix and Demineralized Bone

Decellularized cell-derived matrices (dCDMs) could further provide a way to mimic natural tissues. It has been reported that aligned dCDMs could contribute to producing a more homogeneous environment, which resulted in a uniform response of cells to the biophysical environment, displaying a highly homogeneous phenotype and can undergo differentiation if properly stimulated [165]. Substrates displaying linear topographic patterns were obtained by replica molding, using PDMS, while flat PDMS substrates were produced by using a polystyrene dish as a control.

Cell-derived matrices (CDMs) were attained by cultivating MC3T3-E1 pre-osteoblasts in the presence of ascorbic acid for two weeks on linear or flat surfaces. MC3T3-E1 cultivated on nanopatterned substrates produced an aligned fibrillar matrix, whose microarchitectural features remained intact after the decellularization process. Atomic force microscope measurements performed on bare dCDMs revealed very low Young's moduli in the range of (0.01–0.1 kPa) that was increased by genipin crosslinking to reach (i.e., 0.1–1.5 kPa). These matrices were further seeded with murine MSCs and cultivated in the presence of either osteogenic or adipogenic media for two weeks. Both the aligned and random dCDMs promoted murine MSC adhesion and proliferation in their pristine state, while maintaining high levels of stemness markers, with a more homogeneous population of undifferentiated cells, were seen on aligned dCDMs. On the pristine dCDMs, MSCs promptly underwent adipogenic differentiation when stimulated with induction media, while they were minimized in the presence of osteogenic medium, due to very low stiffness. On the contrary, MSCs responded consistently on stiff dCDMs, displaying a significant adipogenic and osteogenic differentiation potential [165].

In another study, 3D demineralized bone matrices with the same 3D microstructure (porosity and pore size), but with various compressive moduli (high: 66.06 ± 27.83 MPa, medium: 26.90 ± 13.16 MPa, and low: 0.67 ± 0.14 MPa), were fabricated by controlling the decalcification duration (1 h, 12 h, and 5 d, respectively). Low-stiffness scaffolds promoted BMMSCs' osteogenic differentiation. Subcutaneous implantation in a rat model further revealed efficient improvement of cells' infiltration and deposition of collagen fibers, in addition to upregulated positive osteocalcin and osteopontin expression, as well as angiogenesis upon utilizing the low-stiffness scaffolds. Further implantation in a femoral condylar defect rabbit model supported the previous findings and revealed that stromal-cell-derived factor-1α/CXC chemokine receptor (SDF-1α/CXCR4) signaling pathway was essential for the stiffness-mediated stem/progenitor recruitment and osteogenic differentiation during bone repair [166].

4.1.5. Hyaluronic Acid

Hyaluronic acid (HA) is a linear non-sulfated polysaccharide made up of repeated disaccharide molecules in alternating patterns (D-glucuronic acid and N-acetyl-D-glucosamine). This pattern is linked through interchanging β-1,4 and β-1,3 glycosidic bonds. HA is a fundamental component of ECM that regulates various cellular biological processes, such as migration, adhesion, proliferation, and differentiation, through binding with a specific receptor on the target cell [167,168]. Owing to its exceptional biocompatible, biodegradable and non-immunogenic properties, HA is clinically used for drug delivery and tissue regeneration [9,169,170]. As being a natural extracellular component, HA mimics the typical ECM and could initiate signaling pathways responsible for osteogenesis [171,172]. Moreover, the physicochemical and biological properties of HA could be altered by chemical modification [173].

Through adjusting the crosslinker (PEGTA) density, a series of hydrogels with different biochemical and biomechanical properties were developed by utilizing a thiol-functionalized HA and a thiol-functionalized recombinant human gelatin. Human BMMSCs were cultured on the hydrogels with different stiffness (storage modulus (G') and corresponding PEGTA concentrations, namely 0.15 kPa (0.25%), 1.5 kPa (1.75%), and 4 kPa (2.5%), in adipogenic and osteogenic conditions. Adipogenic differentiation was confirmed by gene expression of lipoprotein lipase (LPL), as well as PPARγ2, with similar LPL expression levels demonstrated on the hydrogels with varying stiffness, whereas PPARγ2 expression was markedly enhanced upon increasing hydrogel stiffness. Cells exhibited spindle-shape morphology on the 0.15 kPa hydrogel, while displaying elongated and cuboidal appearance, similar to osteoblasts on greater stiffness hydrogels. Human MSCs cultured on the 1.5 kPa hydrogel significantly expressed osteopontin, while those cultured on the 4 kPa hydrogel revealed a significant upregulation in the expression of the late osteogenic gene (bone sialoprotein) [174].

In a novel HA hydrogel platform, ligation of the HAVDI adhesive peptide sequence from the extracellular N-cadherin domain 1 and the RGD adhesive motif from fibronectin led to Rac1-GTP-dependent reductions in the attachment of myosin IIA to the focal adhesions. This lack of myosin IIA incorporation into focal adhesions hindered the maturation of these adhesions with increasing substrate stiffness (E = 5, 10, and 15 kPa) and thereby decreased traction force generation on the underlying substrate. These alterations in the mechanical state of the MSCs further reduced mechanosensitive YAP/TAZ translocation to the nucleus, herewith attenuating the signaling pathways involved in mesenchymal development, including cell proliferation and osteogenic differentiation [175].

An in vitro culture system for osteochondral tissue engineering was developed, using HA gels with various stiffness (G' ranging from 10 to 45 Pa) attained by mixing Glycosil®, a thiol-modified hyaluronan gel with the crosslinking agent PEG at ratios from (1:1 to 7:1). The co-differentiation media (a ratio of 50% chondrogenic:50% osteogenic) proved to be suitable for appropriate chondrogenic and osteogenic differentiation of human MSCs. On the stiffest matrix (HA:PEG construct at a 2:1 ratio), the three chondrogenic markers (aggrecan, collagen II, and sox 9) were expressed by the differentiated human MSCs cultured for 21 days [176].

Moreover, human BMMSCs were initially entrapped in a HA carrying sulfhydryl groups and a hydrophilic polymer bearing both acrylate and tetrazine groups with the shear elastic modulus (G') $=180 \pm 42$ Pa. The stiffness of the matrix was increased ($G' = 520 \pm 80$ Pa) by adding HA conjugated with multiple copies of trans-cyclooctene (TCO) to the human MSCs-laden gel culture media. The 3D matrix tagged with a TCO-cell-adhesive motif promoted the cells to undergo remarkable actin polymerization, changing from a rounded phenotype to a spindle morphology with long processes. After an additional seven days of culture in the modified media, quantitative analysis showed that RGD tagging enhanced cellular expression of matrix metalloproteinase 1, whereas it decreased the expression of tenascin C and collagen I/III. RGD tagging, however, was not sufficient alone to induce chondrogenic, adipogenic, fibroblastic/myofibroblastic, or osteogenic differentiation [177].

Photo-crosslinked methacrylated HA hydrogels incorporating fragmented polycaprolactone (PCL) nanofibers with compression modulus 3122.5 ± 43.7 Pa promoted osteogenic differentiation of adipose-derived stem/progenitor cells incorporated into the composite hydrogel. The biomarkers collage type 1, ALP, and Runx2 were significantly expressed in the hydrogels containing nanofibers. In addition, the results of alizarin red staining confirmed osteogenic differentiation [178].

4.1.6. Fibrin

Fibrin is one of the natural biopolymers that offers many advantages based on its excellent biocompatibility and cell-adhesion properties [179]. However, fibrin has low mechanical properties that can be modified by adjusting the concentration and ionic strength of fibrinogen to obtain a polymeric substrate mimicking native ECM [55]. A high concentration of fibrinogen and thrombin resulted in a stiffer fibrin matrix, as compared to fibrin with lower fibrinogen and thrombin concentration, as altering these two components allowed the tuning of fibrin elasticity. Microfluidic biochips coated with stiff fibrin substrates modified with gold-nanowires-enhanced osteogenic differentiation of human amniotic mesenchymal stem cells (AMSCs) and led to significant elevation in collagen type I levels and matrix mineralization (calcium deposition), while softer fibrin matrices with lower fibrinogen and thrombin concentration enhanced human AMSCs chondrogenic differentiation [180].

4.2. Synthetic Polymers

4.2.1. Polyethylene Glycol

Polyethylene glycol (PEG) is one of the most widely used synthetic polymers in the tissue engineering field. It is characterized by being chemically and biologically inert and by the high hydrophilicity of the polymer backbone. There is a wide range of polymer architectures and lengths that are commercially or synthetically accessible [181].

MSCs cultured on 3D thixotropic PEG-silica nanocomposite gel with high stiffness ($\geq$75 Pa) expressed the highest level of Runx2 and osteocalcin. Additionally, RGD cell-adhesion peptide sequence immobilization in the gel of 75 Pa stiffness promoted ~13% higher expression of the osteogenic transcription factor [182]. Rat BMMSCs cultured on relatively soft (130 kPa) and stiff (3170 kPa) PEG hydrogels with RGD nano-spacings of 49 and 135 nm, incubated in the mixed osteogenic and adipogenic medium, exhibited a higher density of adherent MSCs, and osteogenesis was promoted on stiffer hydrogels. When the hydrogel stiffness was controlled, the large RGD nano-spacing was beneficial for osteogenesis, while the small RGD nano-spacing generated more adipogenesis [183].

Human MSCs were encapsulated in a multilayer PEG-based hydrogel composed of a soft cartilage-like layer of chondroitin sulfate (48 kPa) and low RGD concentrations, a stiff bone-like layer 345 kPa with high RGD concentrations, and an intermediate interfacial layer with 100 kPa. The recorded stiffness of the multilayer hydrogel was 90 kPa. Opposite to static conditions, dynamic mechanical stimulation generated a high expression of collagens with collagen II in the cartilage-like layer, collagen X in the interfacial layer, and collagen I in the bone-like layer with the presence of mineral deposits in the bone layer [184].

PEG/silk fibroin/HA (PEG/SF/HA) scaffold was prepared with varying HA concentrations, which influenced scaffold stiffness (80.98 to 190.51 kPa). PEG/SF/HA containing 50 mg HA cultured with rat BMMSCs enhanced cell adhesion, viability, the expression of all the osteogenesis-related markers in vitro and promoted superior calvarial defect repair in vivo, through modulating gene and protein expression levels [185].

Additionally, human MSCs seeded on regularly and randomly patterned photodegradable PEG hydrogel surfaces with different stiff-to-soft ratios from ~2–3 kPa to ~10–12 kPa displayed higher cell spread, elongated morphologies, and superior YAP activation and osteogenic differentiation on the regularly patterned regions, as compared to those cultured on random patterns [186]. High PEG substrate stiffness (~25 kPa) and $\alpha5\beta1$ integrin signaling stimulated by c(RRETAWA) induced osteogenic differentiation of human MSCs [187].

4.2.2. Polydimethylsiloxane

Polydimethylsiloxane (PDMS) is characterized by its biocompatibility, flexibility, optical clarity, and elastic tunability [188]. Dental follicle stem cells (DFCs) [189], and human exfoliated deciduous teeth (SHED) [190] were cultured on elastic PDMS substrates. Different stiffnesses, ranging from 11 to 93 kPa, were attained by changing the Sylgard®'s crosslinker to base ratios (1:55, 1:45, and 1:35 by weight) [189,190]. Coating PDMS with fibronectin caused a slight increase in ALP-activity of DFCs and continuous expression of cementoblast marker CP23 on standard cell culture dishes [189]. Osteogenic differentiation of SHED and DFCs was not supported by similar grades of ECM stiffness. In a study that involved adding osteogenic differentiation medium to DFCs on PDMS, DFCs revealed a significantly higher ALP activity and accumulation of calcium on the softest substrate (PDMS 1:55) [189], while SHED demonstrated high osteogenic differentiation on PDMS (1:35) stiffer substrate [190].

ASCs were cultured on soft and stiff PDMS substrates with moduli of elasticity ranging from (0.046 $\pm$ 0.02 MPa) and (1.014 $\pm$ 0.15 MPa), respectively. Stiff substrate enhanced the directed differentiation of ASCs into osteogenic lineages as evidenced by positive ALP stain. This enhancement was supplemented with the upregulated expression of Runx2 and Osx transcriptional factors [38].

Osteogenic differentiation of rat MSCs incubated in osteogenic medium grown on PDMS, with stiffness gradients that ranged from 0.19 to 3.10 MPa, utilizing a temperature gradient during curing, was proven to be strongly influenced by substrate stiffness and the ECM macromolecules pre-adsorbed onto the substrates. Calcein Blue (CB)-positive bone-nodule-like colonies were only observed on the stiff end of PDMS coated with fibronectin and gelatin, while oxygen-plasma-treated surfaces were entirely devoid of CB-positive colonies after 1 week of osteoinductive culture [191].

4.2.3. Vinyl Polymers

A variety of functionalized vinyl monomers are commercially available or can be synthetically customized, rendering vinyl polymer-based hydrogels useful as structurally diverse scaffolds [181]. The osteogenic capability of 3D porous scaffolds composed of polytetrafluoroethylene (PTFE) and polyvinyl alcohol (PVA) with and without graphene oxide (GO) nanoparticles was investigated. These two scaffolds were fabricated through chemical crosslinking with small amounts of boric acids and a controlled freeze-drying method. The scaffolds exhibited randomly oriented nanofibers of 2 and 650 nm and compressive moduli of 620 and 130 kPa, respectively. Human ADSCs seeded on stiffer PTFE/PVA/GO scaffolds revealed a significant elevation in ALP activity, calcium deposition, and osteogenic related genes expression as compared to the softer scaffold without graphene oxide [192].

Cylindrical PV alcohol (PVA)/HA hydrogel prepared with a liquid nitrogen–contacting gradual freezing–thawing method to produce hydrogel with a wide range stiffness gradient (between ~20 kPa and ~200 kPa). Human BMMSCs cultured on PVA/HA hydrogel favored certain stiffness ranges to get differentiated into specific cell lineages: ~20 kPa for nerve cell, ~40 kPa for muscle cell, ~80 kPa for chondrocyte, and ~190 kPa for osteoblast [193]. Moreover, a minimal hydrogel matrix stiffness of 4.47 kPa was recognized to activate transcriptional co-activator TAZ and induce MSCs' osteogenic differentiation [194].

4.2.4. Polyesters

Polyesters are popular polymers that contain ester groups in the polymer backbones, enabling them to produce biomedical hydrogels that can undergo biodegradation [181]. Poly(ether-ester-urethane) (PEEU) containing poly (ρ-dioxanone) (PPDO) and PCL segments can be electrospun into fiber meshes. PEEU fiber meshes were tailored by varying the PPDO:PCL weight ratio, thus affecting their stiffness. Human ADSCs cultured on the stiffer fiber meshes (e.g., PEEU70) significantly demonstrated enhanced osteogenic differentiation with higher levels of osteocalcin expression and ALP activity. Moreover, higher levels of HA were detected on the stiffer fiber meshes [195].

Hydrophilic degradable porous 3D nanocomposite scaffolds composed of PCL, Poly (2-hydroxyethylmethacrylate) (PHEMA), and Apacite (apatite–calcite) nanostructures (15 and 25 wt.%) with mechanical values (E ~ 7.109 MPa and σ ~ 0.414 MPa) provided a balanced microenvironment that resulted in osteogenic induction of human BMMSCs. Von Kossa staining, calcium content, and ALP results confirmed the highest bone cells' differentiation on PCLPHEMA/25% Apacite nanocomposites [196].

4.2.5. Polyacrylamide

Polyacrylamide formed from only acrylamide subunits is nonionic. Copolymerizing it with other monomers such as 2-acrylamido-2-methylpropane sulfonate or acrylate forms anionic polyacrylamide, while cationic polyacrylamide could be synthesized upon copolymerization with dimethyl diallyl ammonium. Polyacrylamide substrate is bio-inert; thus, its surface must be conjugating with adhesive ECM proteins to allow for cell attachment [197,198]. Polyacrylamide is widely utilized in literature as a model for investigating the mechanoregulatory role of substrate stiffness combined or uncombined with other parameters in osteogenic differentiation. The stiffness of polyacrylamide hydrogels is commonly modified by altering the concentration of acrylamide monomer or bis-acrylamide crosslinker [62].

Upon seeding human MSCs on 250-Pa polyacrylamide gels coated with a mixture of collagen type 1 and fibronectin, the progression of the cells throughout the cell cycle was prohibited despite the presence of serum. Conversely, the quiescent cells reentered the cell cycle when presented on a stiff polyacrylamide substrate (7.5 kPa). Moreover, the non-proliferative cells revealed an adipogenic differentiation potential upon culturing on 250-Pa gels in adipogenic media or an osteogenic potential into osteoblasts if transferred to a stiff substrate in the presence of osteogenic media [199]. Micropatterned polyacrylamide gels were fabricated with varying stiffness (10 to 40 kPa), using PDMS stamps coated with fibronectin. MSCs cultured on protein-coated gels revealed a stiffness-dependent osteogenic markers' expression (Runx2 and osteopontin) with a maximum expression at 30 kPa [200]. Osteogenic differentiation as revealed by Runx2 expression was upregulated significantly only on collagen I-coated gels with high stiffness (80 kPa), while myogenic differentiation, as ascertained by MyoD1 expression, occurred on all gel–protein coated matrices that had a stiffness of 9 kPa. Peak MyoD1 expression was demonstrated on gels with a modulus of 25 kPa coated with fibronectin. Polyacrylamide hydrogels prepared with variable stiffnesses, ranging from 13 to 68 kPa, through varying the concentrations of bis-acrylamide (0.1%, 0.5%, and 0.7%), showed a difference in the gel morphology. Under scanning electron microscopy, gels with low stiffness (13–16 kPa) appeared flat and non-porous. On the other hand, higher stiffness matrices (48–53 kPa and 62–68 kPa) showed multiple small porosities. Such inherent porosities of polyacrylamide hydrogels could enhance the flow of culture media and better mimic the natural cellular environment, as compared to plastic and glass substrates. Moreover, BMMSCs cultured on 62–68 kPa fibronectin-coated polyacrylamide hydrogels demonstrated a polygonal morphology and revealed an osteogenic phenotype with significantly high levels of ALP, Runx2, and osteopontin [51].

The modulatory effect of extracellular matrix type and density on the mechanotransduction of stem/progenitor cells and the correlated integrin involved in signals translocation were assessed through conjugating each of the four major cell adhesion ECM proteins (fibronectin, collagen I, collagen IV, and laminin) on polyacrylamide hydrogels with tunable stiffness (soft, 3 kPa; and stiff, 38 kPa). The results revealed that increasing ECM ligand density alone can induce YAP nuclear translocation without changing substrate stiffness with a different optimized ligand density. Using antibody-blocking techniques for αvβ3-, α5-, and α2β1-integrins revealed the involvement of αvβ3-, α5-, and α2β1-integrins with fibronectin, while α5-integrin was further associated with collagen type I and IV. On the contrary, laminin was associated with α5- and α2β1-integrins. Moreover, altering ECM type resulted in modulation of human MSC osteogenesis confirmed by quantitative real-time (qRT)-PCR for Runx2 and ALP without changing substrate stiffness [201].

The mechanotransduction role of FAK, α5/β1 integrin and Wnt-signaling pathways mediated by stiff matrices, in regulating osteogenic differentiation of human MSCs cultured on 62–68 kPa fibronectin-coated polyacrylamide hydrogels were further investigated. Throughout osteogenesis, gene and protein expressions of integrin α5/β1 were enhanced, together with the expression of signaling molecules FAK, p-ERK, p-Akt, GSK-3β, p-GSK-3β, and β-catenin. Antibody blocking of integrin α5 significantly downregulated the stiffness-induced expression of osteogenic markers (Runx2, alpha-1 type I collagen, and BGLAP) with associated downregulated expression of ERK, p-ERK, FAK, and β-catenin protein. Reciprocally, GSK-3β, p-GSK-3β, Akt, and p-Akt expressions were upregulated. The presence of the Akt inhibitor Triciribine reduced the expression of p-Akt and p-GSK-3β, whereas Akt, GSK-3β, and β-catenin were unchanged. These results emphasized the role of p-Akt in regulating the expression of p-GSK-3β on 62–68 kPa ECM during osteogenesis [78].

MSCs cultivated on polyacrylamide hydrogels with elasticity (7.0 ± 1.2 and 42.1 ± 3.2 kPa) and coated with type I collagen in osteogenic medium revealed enhanced osteogenic differentiation potential on stiff substrates, with an upregulated expression of Runx2, type I collagen, and osteocalcin genes. On stiff matrices, Western blot analysis revealed an increase in mechanotransducers involved in osteogenic differentiation ROCK, FAK, and ERK1/2, whereas their inhibition resulted in decreased osteogenic markers' expression. Furthermore, α2-integrin was upregulated on stiff matrices during osteogenesis, and its knockdown by siRNA hindered the osteogenic phenotype through FAK, ROCK, and ERK1/2. Therefore, it could be concluded that α2-integrin is involved in osteogenesis mediated by matrix stiffness [62].

Additionally, upon culturing human MSCs on poly acrylamide-co-acrylic acid hydrogels grafted with RGDs, myogenic differentiation occurred at 13–17 kPa, while osteogenic differentiation was revealed at 45–49 kPa stiffness confirmed further with positive protein immunostaining of MyoD, as well as Osx, osteocalcin, and Runx2. Stiffer matrices grafted with BMP-2 mimetic peptide (E = 47.5 kPa) also induced osteoblast lineage commitment, having a similar effect as the ones grafted with RGDs. On the contrary, the osteogenic effect of BMP-2 mimetic peptides on MSCs was inhibited on very soft microenvironments (0.76–3.21 kPa) due to F-actin cytoskeleton reorganization that inhibited BMP-induced smad1/5/8 phosphorylation and subsequent differentiation of the cells into osteoblast-like cells [118].

Umbilical cord (UC) MSCs attained similar behavior upon being cultivated on different stiffness (13–16, 35–38, 48–53, and 62–68 kPa) polyacrylamide gels coated with fibronectin. Quantitative RT-PCR results showed that soft matrices promoted adipogenic differentiation, as evident by upregulated expressions of adipocytic markers (PPARγ and C/EBPα). On the contrary, stiff matrices (48–53 kPa) enhanced the tendency of the cells to differentiate into muscles, as demonstrated by enhanced expression of desmin and MOYG. High stiffness substrates (62–68 kPa) significantly promoted the expression of osteogenic markers, such as Runx2, collagen type I, ALP, and osteocalcin [50].

Moreover, the effect of mechanical loading and biomaterial stiffness on MSCs differentiation was investigated upon cultivating MSCs in osteogenic and adipogenic media on soft (23 ± 0.3 kPa) and stiff (111 ± 2 kPa) polyacrylamide as compared to PDMS (1.5 ± 0.07 MPa) either strained with 8% cyclic strain at 1 Hz or unstrained. Without strain, the expression of ALP was markedly higher on PDMS than on both polyacrylamide types. With 8% cyclic strain, ALP expression was upregulated in all groups, with the highest expression in soft polyacrylamide. Moreover, adipogenesis was the highest on the unstrained soft polyacrylamide, while it was significantly decreased on soft and stiff polyacrylamide when strained [202].

4.2.6. Self-Assembling Peptides

Human MSCs were encapsulated within a 3D culture and grown on top of 2D culture biomimetic self-assembling peptide (SAP) hydrogel containing 1 mg/mL RGDs-functionalized peptide (KFE–RGD) at the shear moduli of 0.25, 1.25, 5, and 10 kPa. Changes in adipogenic and osteogenic gene expression were relatively modest with no visual signs of differentiation as mineral deposition. The cells maintained a fibroblast-like phenotype throughout the culture period. However, on introducing 1:1 mixed adipogenic/osteogenic induction medium, the stiff matrices of 10 kPa induced the most efficient osteogenesis, with alizarin red-stained calcium deposits [203].

4.2.7. Other Polymers

Indirectly 3D-printed "stiffness memory" poly(urea-urethane) (PUU)/POSS elastomeric nano-hybrid scaffolds with thermo-responsive mechanical properties that soften at body temperature by inverse self-assembling have been developed. The initial stiffness and subsequent stiffness relaxation (>10 kPa) of the scaffolds directed the proliferation and differentiation of human BMMSCs towards the osteogenic lineages on stiffer scaffolds over 4 weeks, as measured by immunohistochemistry, histology, ELISA, and qPCR, while soft substrates (<1 kPa) promoted MSCs' chondrogenic differentiation [204]. Table 1 lists the key studies investigating the effect of polymeric matrix stiffness on osteogenic differentiation of mesenchymal stem/progenitor cells.

Table 1. Key studies on the effect of polymeric matrix stiffness on osteogenic differentiation of mesenchymal stem/progenitor cells.

Study	Cell Source	Polymer	Modification	Modulus of Elasticity	Results
			Alginate		
Zhang et al., 2020 [141]	hMSCs	Alginate–gelatin scaffold	3D bioprinted porous scaffolds different alginate concentration (0.8%alg and 1.8%alg) and different initial cell seeding density (1.67, 5, and 15 M cells/mL)	Soft scaffold 0.66 ± 0.08 kPa Stiff scaffold 5.4 ± 1.2 kPa	UpregulatedALP-activity-related, 3D-bone-like-tissue-related, osteoblast-related, and early osteocyte-related gene expression
Freeman and Kelly, 2017 [142]	MSCs	Alginate hydrogel	3D bioprinting matrix with varying alginate molecular weight and cross linker ratio		Osteogenic differentiation with increased ALP staining
Maia et al., 2014 [143]	hMSCs	Alginate hydrogel	3D matrix with bimodal molecular weight distribution at different polymer concentrations (1 and 2 wt.%) and RGD densities (0, 100 or 200 µM)	2 wt.% hydrogels (tan ∂ = 0.2), 1 wt.% hydrogels (tan ∂ = 0.4–0.6).	1 wt.% alginate hydrogel matrices upregulated hMSCs osteogenic differentiation and expressed high levels of ALP and OCN
			Collagen		
Xie et al., 2017 [146]	hMSCs	Collagen gel	Varying polymerization temperature 4, 21, and 37 °C.	Fiber stiffness: 1.1 to 9.3 kPa Bulk stiffness: 16.4 to 151.5 Pa	Collagen gel polymerized at 37 °C resulted in 34.1% ALP positive staining
Banks et al., 2014 [147]	ADSCs	Collagen–glycosaminoglycan (CG)	Chemical Crosslinking with EDAC and NHS Covalent immobilization of PDGF-BB and BMP-2 by benzophenone photolithography	2.85 to 5 MPa	Upregulated expression of collagen 1, ALP, and OCN with increased stiffness
Hwang et al., 2019 [148]	hMSCs	Three bilayers of collagen/alginate nano film		24 and 53 MPa	Increase in alkaline phosphatase activity

Table 1. *Cont.*

Study	Cell Source	Polymer	Modification	Modulus of Elasticity	Results
Zhou et al., 2021 [149]	hMSCs	Nano-particulate mineralized collagen glycosaminoglycan	Chemical crosslinking with EDAC and NHS	3.90 −/+ 0.36 kPa	Increase in expression of ALP, collagen 1, and Runx2
Tsimbouri et al., 2017 [150]	MSCs	Collagen gel	3D collagen gel culture on the vibrational bioreactor	~108 Pa	Increased expression of Runx2, collagen I, ALP, OPN, OCN, and BMP2.
Murphy et al., 2012 [151]	MSCs	Collagen/ glycosaminoglycan	DHT and EDAC crosslinking	0.5, 1, and 1.5 kPa	Osteogenic differentiation with Runx2 expression
Chen et al., 2015 [152]	Rat MSCs	3D scaffold collagen and hydroxyapatite	Coated on decellularized cancellous bone	13.00 ± 5.55 kPa, 13.87 ± 1.51 kPa, and 37.7 ± 19.6 kPa	Highest scaffold stiffness promoted higher expressions of OPN and OC
Chen et al., 2017 [205]	Rat MSCs	Collagen and hydroxyapatite, coated on decellularized cancellous bone	3D oscillatory perfusion bioreactor system	6.74 ± 1.16 kPa-8.82 ± 2.12 kPa-23.61 ± 8.06 kPa	Osteogenic differentiation of MSCs
			Gelatin		
Wan et al., 2019 [133]	PDLSCs	Gelatin	Crosslinked with variable concentrations of methacryloyl	GelMA concentrations of 10, 12, and 14 wt% stiffness 25.75 ± 1.21, 59.71 ± 8.87, and 117.82 ± 9.83 kPa, respectively	Increasing matrix stiffness increased osteogenic differentiation of PDLSCs, with upregulated expression of OCN and Runx2
He et al., 2018 [134]	BMMSCs	Gelatin 3%, 6%, and 9%.	Crosslinked with transglutaminase	9% gelatin gave rise to the highest stiffness (60.54 ± 10.45 kPa), while 3% gelatin resulted in the lowest stiffness (1.58 ± 0.42 kPa)	BMMSCs encapsulated in hydrogel with highest stiffness demonstrated the highest osteogenic differentiation
Van Nieuwenhove et al., 2017 [162]	ADSCs	Gelatin with variable degrees of methacrylation (GelMA 31%, GelMA 72%, and GelMA 95%)	Covalently bound to variable ratios of pentenoates modified starch (10 v% starch and 20 v% starch)		Increase in matrix stiffness promoted osteogenic differentiation of ADSCs
Jiang et al., 2015 [163]	BMMSCs	GelMA encapsulating alendronate	Crosslinked by PEG diacrylate	stiffness increased from 4 to 40 kPa	Increased osteogenic differentiation of BMMSCs on stiffer hydrogel with higher alendronate concentration with upregulated ALP, collagen I, OCN, and calcium deposition

Table 1. *Cont.*

Study	Cell Source	Polymer	Modification	Modulus of Elasticity	Results
Sun et al., 2014 [164]	BMMSCs	Three-dimensional porous gelatin scaffolds	Crosslinked using EDC	Crosslinked scaffold demonstrated an increase in the elastic modulus from w 0.6 to ≈ 2.5 kP without any change in the scaffold internal structure	Increased stiffness increased osteogenic differentiation evidenced by increased Runx2 and OCN in vitro and increased bone formation in vivo
Decellularized matrix and Demineralized Bone					
Ventre et al., 2019 [165]	Murine MSCs	Decellularized MC3T3-E1-cell-derived matrix on replica from PDMS	Genipin crosslinking	Young's modulus increased from (0.01–0.1 kPa) to (0.1–1.5 kPa).	MSCs on stiff dCDMs, revealed significant adipogenic and osteogenic differentiation potentials
Hu et al., 2018 [166]	BMMSC	Demineralized bone matrices	Controlling the decalcification duration (1 h, 12 h, and 5 d, respectively)	High: 66.06 ± 27.83 MPa, Medium: 26.90 ± 13.16 MPa Low: 0.67 ± 0.14 MPa	Low stiffness scaffolds promoted osteogenesis in vitro. Subcutaneous implantation in a rat model and in a femoral condylar defect rabbit model revealed positive OCN and OPN expression
Hyaluronic acid (HA)					
Zhao et al., 2014 [174]	hBMMSCs	Thiol functionalized hyaluronic acid (HA) and thiol functionalized recombinant human gelatin	Crosslinked by poly (ethylene glycol) tetra-acrylate	0.15, 1.5, and 4 kPa	Change in cell morphologies with different stiffness. Cells cultured on the 4 kPa hydrogel revealed an enhanced expression of late osteogenic genes
Cosgrove et al., 2016 [175]	Juvenile bovine MSCs	Methacrylated HA hydrogel	Ligation of the HAVDI adhesive peptide sequence from N-cadherin domain 1 and RGD from fibronectin	5, 10, and 15 kPa	Lack of myosin IIA incorporated into focal adhesions hindered their maturation with increasing substrate stiffness and decreased osteogenesis
Dorcemus et al., 2017 [176]	hMSCs-bone-marrow-derived	Thiol-modified hyaluronan gel	Crosslinked by PEG at ratios ranging from 1:1 to 7:1	Storage moduli from 10 to 45 Pa	Differences between the top (cartilage-forming) and bottom (bone-forming) regions of the scaffold proved its capability for osteochondral engineering

Table 1. *Cont.*

Study	Cell Source	Polymer	Modification	Modulus of Elasticity	Results
Hao et al., 2018 [177]	hMSCs-bone-marrow-derived	HA carrying sulfhydryl groups and a hydrophilic polymer bearing both acrylate and tetrazine groups	Matrix metalloprotease -degradable peptidic crosslinker and adding HA conjugated with multiple copies of trans-cyclooctene (TCO)	(G′) = 180 ± 42 Pa increased to G′ = 520 ± 80 Pa	The 3D matrix tagged with a TCO- motif promoted the cells to undergo change from a rounded to spindle phenotype
			Fibrin		
Hashemzadeh et al., 2019 [180]	hADSCs	Fibrin hydrogels embedding gold nanowires	Altering fibrinogen and thrombin concentration and incorporation of gold nanowires		With high fibrinogen and thrombin concentration, gold nanowires, promoted osteogenic differentiation
			Polyethylene glycol (PEG)		
Pek et al., 2010 [182]	MSCs	Thixotropic polyethylene glycol–silica (PEG–silica) nano composite gel	3D cell culture Cell-adhesion peptide RGD (Arg–Gly–Asp) sequence immobilization	≥75 Pa	Higher expression of the osteogenic transcription factor
Ye et al., 2015 [183]	Rat BMMSCs	PEG	PEG hydrogels with RGD nano-spacings of 49 and 135 nm and incubated in mixed osteogenic and adipogenic medium	Soft hydrogels (130 kPa) and stiff hydrogels (3170 kPa)	Stiff hydrogels promoted osteogenesis. Large RGD nano-spacing promoted osteogenesis
Steinmetz et al., 2015 [184]	hMSCs	Multilayer PEG-based hydrogel	Simple sequential photopolymerization- high RGD concentrations- dynamic mechanical stimulation	345 kPa	Collagen I generation with mineral deposits were evident
Yang et al., 2020 [185]	Rat BMMSCs	PEG/silk fibroin/HA scaffold	Varying HA concentration	80.98 to 190.51 kPa	Expression of all the osteogenesis-related markers in vitro and superior calvarial defect repair in vivo
Yang et al., 2016 [186]	hMSCs	PEG hydrogel	Regularly and randomly patterned photodegradable hydrogel	~10–12 kPa	Osteogenic differentiation of MSCs cultured on random patterns
Gandavarapu et al., 2014 [187]	hMSCs	PEG hydrogels	functionalized with c(RRETAWA) hydrogels through α5 integrins	~25 kPa	Osteogenic differentiation of hMSCs
			Polydimethylsiloxane (PDMS)		
Xie et al., 2018 [38]	ASCs	PDMS		1.014 ± 0.15 MPa	Osteogenic differentiation by ALP stain and upregulation of Runx2 and Osx transcriptional factors
Viale-Bouroncle et al., 2014 [189]	DFCs	PDMS	Coating PDMS with fibronectin and cultured in osteogenic differentiation medium	11 kPa	High ALP activity and accumulation of calcium on the soft substrate

Table 1. *Cont.*

Study	Cell Source	Polymer	Modification	Modulus of Elasticity	Results
Viale-Bouroncle et al., 2012 [190]	SHED	PDMS	Adding osteogenic differentiation medium	93 kPa	High osteogenic differentiation
Wang et al., 2012 [191]	Rat MSCs	PDMS	Osteogenic medium with temperature gradient curing	0.19 to 3.10 MPa	Calcein Blue–positive bone-nodule-like colonies
Vinyl polymers					
Khoramgah et al., 2020 [192]	hADSCs	Poly tetra fluoro ethylene (PTFE) and PVA with and without graphene oxide nanoparticles	3D porous scaffolds-chemical crosslinking with small amounts of boric acids–controlled freeze-drying method	620 and 130 kPa	Elevation in ALP activity, calcium deposition, and osteogenic-related genes expression
Oh et al., 2016 [193]	hBMMSCs	Cylindrical PVA/HA hydrogel	Liquid nitrogen—contacting gradual freezing–thawing method	~20 kPa and ~200 kPa	Stiffness of ~190 kPa led to osteoblast differentiation
Polyesters					
Sun et al., 2019 [195]	hADSCs	Poly(ether-ester-urethane) (PEEU) containing PPDO and PCL segments	Electrospun into fiber meshes with varying PPDO to PCL weight ratios	2.6 ± 0.8 MPa (PEEU40), 3.2 ± 0.9 MPa (PEEU50), 4.0 ± 0.9 MPa (PEEU60) 4.5 ± 0.8 MPa (PEEU70)	Enhanced osteogenic differentiation of hADSCs with higher levels of OCN, ALP, and hydroxyapatite detected on the stiffer fiber meshes
Self-assembling peptides					
Hogrebe and Gooch, 2016 [203]	hMSCs	Biomimetic self-assembling peptide hydrogel containing 1 mg/mL RGD-functionalized peptide (KFE–RGD)	hMSCs were encapsulated within 3D culture and grown on top of 2D culture Adding 1:1 mixed adipogenic/osteogenic induction medium	(G′) 10 kPa	Osteogenesis induction and alizarin red-stained calcium deposits
Other Polymers					
Olivares-Navarrete et al., 2017 [76]	MSCs	Methyl acrylate/methyl methacrylate polymer	Altering monomer concentration.	0.1 MPa to 310 MPa	Chondrogenic and osteogenic differentiation when grown on substrates with less than 10 MPa stiffness
Wu et al., 2018 [204]	hBMMSCs	Poly(urea-urethane) (PUU)/POSS elastomeric nano-hybrid scaffolds	Thermoresponsive scaffolds indirectly 3D printed by inverse self-assembling	>10 kPa	Osteogenic differentiation

5. Conclusions

The world of biomaterials, specifically polymers, lingers to impact the biomedicine field. Various materials are currently under investigation to produce ECM with varying stiffness for tissue engineering. Matrix stiffness regulates the MSCs' differentiation into mature specific cells by activating transcription factors that upregulate the genes responsible for the initiation and progression of particular cell linage differentiation. Rigid matrices lead to increased MSCs spreading and improved actomyosin contractility, promoting osteogenic differentiation. This enhanced potential is accompanied by increased Runx2, β-catenin, and Smad, implying the significant impact of mechanosensing of the matrix stiffness and its role in determining cell fate. The comprehensive signaling mechanisms by which micro-environmental stiffness controls the lineage specification of MSCs are still unknown. Additional research is needed to understand the variety of potential signaling forces involved in MSCs' osteogenic differentiation that can lead to the development of new therapeutic modalities addressing bone disorders. Moreover, we believe that further research on the implicated mechanical and physical factors, such as topographic changes and external mechanical forces affecting cell properties, including cell shape and colony size, can offer a broader understanding of cell-fate determination.

Author Contributions: All authors contributed equally. All authors have read and agreed to the published version of the manuscript.

Funding: This publication was funded by the Christian Albrechts University of Kiel, Germany.

Institutional Review Board Statement: Not applicable.

Informed Consent Statement: Not applicable.

Data Availability Statement: Not applicable.

Conflicts of Interest: The authors declare no conflict of interest.

Abbreviations

3D	Three-dimensional
ADSCs	Adipose derived stromal cells
ALP	Alkaline phosphatase
ASCs	adipose derived stromal cells
BMMSCs	Bone marrow mesenchymal stem cells
BMP-2	Bone morphogenetic protein 2
dCDMs	Decellularized cell derived matrix
DFCs	Dental follicle stem cells
ECM	Extracellular matrix
EDAC	1-ethyl-3-(3-dimethylami-nopropyl) carbodiimide
GC-TRS	Glycol chitin-based thermo-responsive hydrogel scaffold
GelMA	Gelatin with variable degrees of methacrylation
HA	Hyaluronic acid
HA	Hydroxyapatite
hADSCs	Human adipose derived mesenchymal stem cells
hBMMSCs	Human Bone marrow mesenchymal stem cells
hMSCs	Human mesenchymal stem cells
MSCs	Mesenchymal stem cells
NHS	N-hydroxysuccinimide
OCN	Osteocalcin
ON	Osteonectin
OPN	osteopontin
PCL	poly (ε-caprolactone)
PDLSCs	periodontal ligament stem cells
PDMS	Polydimethylsiloxane
PEEU	Poly (ether-ester-urethane)
PEG	Polyethylene glycol
PEGDA	Polyethylene glycol diacrylate
PEGMC	Polyethylene glycol-maleate-citrate
PPDOP	poly (ρ-dioxanone)

PV	polyvinyl
PVA	polyvinyl alcohol
qRT-PCR	Quantitative Reverse transcription real-time polymerase chain reaction
RGD	Arginine-glycine-aspartic acid
Runx2	Runt-related transcription factor-2
SCAP	Stem cells of the apical papilla
SDF-1α	Stromal derived factor-1alpha
SHED	Stem cells isolated from human exfoliated deciduous teeth

References

1. Singh, V.K.; Saini, A.; Kalsan, M.; Kumar, N.; Chandra, R. Describing the Stem Cell Potency: The Various Methods of Functional Assessment and In silico Diagnostics. *Front. Cell Dev. Biol.* **2016**, *4*, 134. [CrossRef]
2. El-Sayed, K.M.; Paris, S.; Graetz, C.; Kassem, N.; Mekhemar, M.; Ungefroren, H.; Fandrich, F.; Dorfer, C. Isolation and characterisation of human gingival margin-derived STRO-1/MACS(+) and MACS(-) cell populations. *Int. J. Oral Sci.* **2015**, *7*, 80–88. [CrossRef]
3. Fawzy El-Sayed, K.M.; Ahmed, G.M.; Abouauf, E.A.; Schwendicke, F. Stem/progenitor cell-mediated pulpal tissue regeneration: A systematic review and meta-analysis. *Int. Endod. J.* **2019**, *52*, 1573–1585. [CrossRef] [PubMed]
4. Fawzy El-Sayed, K.M.; Dorfer, C.; Fandrich, F.; Gieseler, F.; Moustafa, M.H.; Ungefroren, H. Adult mesenchymal stem cells explored in the dental field. *Adv. Biochem. Eng. Biotechnol.* **2013**, *130*, 89–103. [CrossRef] [PubMed]
5. Fawzy El-Sayed, K.M.; Dorfer, C.E. Gingival Mesenchymal Stem/Progenitor Cells: A Unique Tissue Engineering Gem. *Stem. Cells Int.* **2016**, *2016*, 7154327. [CrossRef]
6. Fawzy El-Sayed, K.M.; Elahmady, M.; Adawi, Z.; Aboushadi, N.; Elnaggar, A.; Eid, M.; Hamdy, N.; Sanaa, D.; Dorfer, C.E. The periodontal stem/progenitor cell inflammatory-regenerative cross talk: A new perspective. *J. Periodontal Res.* **2019**, *54*, 81–94. [CrossRef] [PubMed]
7. Fawzy El-Sayed, K.M.; Elsalawy, R.; Ibrahim, N.; Gadalla, M.; Albargasy, H.; Zahra, N.; Mokhtar, S.; El Nahhas, N.; El Kaliouby, Y.; Dorfer, C.E. The Dental Pulp Stem/Progenitor Cells-Mediated Inflammatory-Regenerative Axis. *Tissue Eng. Part B Rev.* **2019**, *25*, 445–460. [CrossRef]
8. Fawzy El-Sayed, K.M.; Jakusz, K.; Jochens, A.; Dorfer, C.; Schwendicke, F. Stem Cell Transplantation for Pulpal Regeneration: A Systematic Review. *Tissue Eng. Part B Rev.* **2015**, *21*, 451–460. [CrossRef]
9. Fawzy El-Sayed, K.M.; Mekhemar, M.K.; Beck-Broichsitter, B.E.; Bahr, T.; Hegab, M.; Receveur, J.; Heneweer, C.; Becker, S.T.; Wiltfang, J.; Dorfer, C.E. Periodontal regeneration employing gingival margin-derived stem/progenitor cells in conjunction with IL-1ra-hydrogel synthetic extracellular matrix. *J. Clin. Periodontol.* **2015**, *42*, 448–457. [CrossRef] [PubMed]
10. Fawzy El-Sayed, K.M.; Paris, S.; Becker, S.T.; Neuschl, M.; De Buhr, W.; Sälzer, S.; Wulff, A.; Elrefai, M.; Darhous, M.S.; El-Masry, M.; et al. Periodontal regeneration employing gingival margin-derived stem/progenitor cells: An animal study. *J. Clin. Periodontol.* **2012**, *39*, 861–870. [CrossRef]
11. Wang, X.; Wang, Y.; Gou, W.; Lu, Q. Role of mesenchymal stem cells in bone regeneration and fracture repair: A review. *Int. Orthop.* **2013**, *37*, 2491–2498. [CrossRef]
12. Zhou, L.L.; Liu, W.; Wu, Y.M.; Sun, W.L.; Dorfer, C.E.; Fawzy El-Sayed, K.M. Oral Mesenchymal Stem/Progenitor Cells: The Immunomodulatory Masters. *Stem. Cells Int.* **2020**, *2020*, 1327405. [CrossRef]
13. Fu, X.; Liu, G.; Halim, A.; Ju, Y.; Luo, Q.; Song, G. Mesenchymal Stem Cell Migration and Tissue Repair. *Cells* **2019**, *8*, 784. [CrossRef] [PubMed]
14. Patel, D.M.; Shah, J.; Srivastava, A.S. Therapeutic potential of mesenchymal stem cells in regenerative medicine. *Stem. Cells Int.* **2013**, *2013*, 496218. [CrossRef]
15. Via, A.G.; Frizziero, A.; Oliva, F. Biological properties of mesenchymal Stem Cells from different sources. *Muscles Ligaments Tendons J.* **2012**, *2*, 154–162.
16. Lee, O.K.; Kuo, T.K.; Chen, W.-M.; Lee, K.-D.; Hsieh, S.-L.; Chen, T.-H. Isolation of multipotent mesenchymal stem cells from umbilical cord blood. *Blood* **2004**, *103*, 1669–1675. [CrossRef]
17. Ding, D.C.; Shyu, W.C.; Lin, S.Z. Mesenchymal stem cells. *Cell Transplant.* **2011**, *20*, 5–14. [CrossRef] [PubMed]
18. Rodríguez-Lozano, F.J.; Bueno, C.; Insausti, C.L.; Meseguer, L.; Ramírez, M.C.; Blanquer, M.; Marín, N.; Martínez, S.; Moraleda, J.M. Mesenchymal stem cells derived from dental tissues. *Int. Endod. J.* **2011**, *44*, 800–806. [CrossRef]
19. Jorgenson, K.D.; Hart, D.A.; Krawetz, R.; Sen, A. Production of Adult Human Synovial Fluid-Derived Mesenchymal Stem Cells in Stirred-Suspension Culture. *Stem. Cells Int.* **2018**, *2018*, 8431053. [CrossRef]
20. Rutkovskiy, A.; Stensløkken, K.-O.; Vaage, I.J. Osteoblast Differentiation at a Glance. *Med. Sci. Monit. Basic Res.* **2016**, *22*, 95–106. [CrossRef] [PubMed]
21. Assis-Ribas, T.; Forni, M.F.; Winnischofer, S.M.B.; Sogayar, M.C.; Trombetta-Lima, M. Extracellular matrix dynamics during mesenchymal stem cells differentiation. *Dev. Biol.* **2018**, *437*, 63–74. [CrossRef] [PubMed]
22. Jhala, D.; Vasita, R. A Review on Extracellular Matrix Mimicking Strategies for an Artificial Stem Cell Niche. *Polym. Rev.* **2015**, *55*, 561–595. [CrossRef]
23. Even-Ram, S.; Artym, V.; Yamada, K.M. Matrix control of stem cell fate. *Cell* **2006**, *126*, 645–647. [CrossRef]

24. Watt, F.M.; Huck, W.T.S. Role of the extracellular matrix in regulating stem cell fate. *Nat. Rev. Mol. Cell Biol.* **2013**, *14*, 467–473. [CrossRef]
25. Gaur, T.; Lengner, C.J.; Hovhannisyan, H.; Bhat, R.A.; Bodine, P.V.; Komm, B.S.; Javed, A.; Van Wijnen, A.J.; Stein, J.L.; Stein, G.S. Canonical WNT signaling promotes osteogenesis by directly stimulating Runx2 gene expression. *J. Biol. Chem.* **2005**, *280*, 33132–33140. [CrossRef]
26. Almalki, S.G.; Agrawal, D.K. Key transcription factors in the differentiation of mesenchymal stem cells. *Differentiation* **2016**, *92*, 41–51. [CrossRef]
27. Long, F. Building strong bones: Molecular regulation of the osteoblast lineage. *Nat. Rev. Mol. Cell Biol.* **2012**, *13*, 27–38. [CrossRef]
28. Komori, T. Regulation of osteoblast differentiation by transcription factors. *J. Cell Biochem.* **2006**, *99*, 1233–1239. [CrossRef] [PubMed]
29. Zhao, Z.; Zhao, M.; Xiao, G.; Franceschi, R.T. Gene transfer of the Runx2 transcription factor enhances osteogenic activity of bone marrow stromal cells in vitro and in vivo. *Mol. Ther.* **2005**, *12*, 247–253. [CrossRef]
30. Fujii, M.; Takeda, K.; Imamura, T.; Aoki, H.; Sampath, T.K.; Enomoto, S.; Kawabata, M.; Kato, M.; Ichijo, H.; Miyazono, K. Roles of bone morphogenetic protein type I receptors and Smad proteins in osteoblast and chondroblast differentiation. *Mol. Biol. Cell* **1999**, *10*, 3801–3813. [CrossRef]
31. Xu, D.-J.; Zhao, Y.-Z.; Wang, J.; He, J.-W.; Weng, Y.-G.; Luo, J.-Y. Smads, p38 and ERK1/2 are involved in BMP9-induced osteogenic differentiation of C3H10T1/2 mesenchymal stem cells. *BMB Rep.* **2012**, *45*, 247–252. [CrossRef]
32. Javed, A.; Bae, J.-S.; Afzal, F.; Gutierrez, S.; Pratap, J.; Zaidi, S.K.; Lou, Y.; Van Wijnen, A.J.; Stein, J.L.; Stein, G.S. Structural coupling of Smad and Runx2 for execution of the BMP2 osteogenic signal. *J. Biol. Chem.* **2008**, *283*, 8412–8422. [CrossRef]
33. Nakashima, K.; Zhou, X.; Kunkel, G.; Zhang, Z.; Deng, J.M.; Behringer, R.R.; De Crombrugghe, B. The novel zinc finger-containing transcription factor osterix is required for osteoblast differentiation and bone formation. *Cell* **2002**, *108*, 17–29. [CrossRef]
34. Kang, S.; Bennett, C.N.; Gerin, I.; Rapp, L.A.; Hankenson, K.D.; MacDougald, O.A. Wnt signaling stimulates osteoblastogenesis of mesenchymal precursors by suppressing CCAAT/enhancer-binding protein α and peroxisome proliferator-activated receptor γ. *J. Biol. Chem.* **2007**, *282*, 14515–14524. [CrossRef]
35. Lecarpentier, Y.; Claes, V.; Duthoit, G.; Hébert, J.-L. Circadian rhythms, Wnt/beta-catenin pathway and PPAR alpha/gamma profiles in diseases with primary or secondary cardiac dysfunction. *Front. Physiol.* **2014**, *5*, 429. [CrossRef]
36. Day, T.F.; Guo, X.; Garrett-Beal, L.; Yang, Y. Wnt/β-catenin signaling in mesenchymal progenitors controls osteoblast and chondrocyte differentiation during vertebrate skeletogenesis. *Dev. Cell* **2005**, *8*, 739–750. [CrossRef] [PubMed]
37. Huelsken, J.; Behrens, J. The Wnt signalling pathway. *J. Cell Sci.* **2002**, *115*, 3977–3978. [CrossRef]
38. Xie, J.; Zhang, D.; Zhou, C.; Yuan, Q.; Ye, L.; Zhou, X. Substrate elasticity regulates adipose-derived stromal cell differentiation towards osteogenesis and adipogenesis through β-catenin transduction. *Acta Biomater.* **2018**, *79*, 83–95. [CrossRef] [PubMed]
39. Kshitiz; Afzal, J.; Chang, H.; Goyal, R.; Levchenko, A. Mechanics of Microenvironment as Instructive Cues Guiding Stem Cell Behavior. *Curr. Stem. Cell Rep.* **2016**, *2*, 62–72. [CrossRef]
40. Kshitiz; Park, J.; Kim, P.; Helen, W.; Engler, A.J.; Levchenko, A.; Kim, D.-H. Control of stem cell fate and function by engineering physical microenvironments. *Integr. Biol.* **2012**, *4*, 1008–1018. [CrossRef] [PubMed]
41. Han, S.B.; Kim, J.K.; Lee, G.; Kim, D.H. Mechanical Properties of Materials for Stem Cell Differentiation. *Adv. Biosyst.* **2020**, *4*, 2000247. [CrossRef]
42. Cargnello, M.; Roux, P.P. Activation and Function of the MAPKs and Their Substrates, the MAPK-Activated Protein Kinases. *Microbiol. Mol. Biol. Rev.* **2011**, *75*, 50–83. [CrossRef]
43. MacQueen, L.; Sun, Y.; Simmons, C.A. Mesenchymal stem cell mechanobiology and emerging experimental platforms. *J. R. Soc. Interface* **2013**, *10*, 20130179. [CrossRef] [PubMed]
44. Greenblatt, M.B.; Shim, J.-H.; Zou, W.; Sitara, D.; Schweitzer, M.; Hu, D.; Lotinun, S.; Sano, Y.; Baron, R.; Park, J.M.; et al. The p38 MAPK pathway is essential for skeletogenesis and bone homeostasis in mice. *J. Clin. Investig.* **2010**, *120*, 2457–2473. [CrossRef]
45. Thouverey, C.; Caverzasio, J. The p38α MAPK positively regulates osteoblast function and postnatal bone acquisition. *Cell Mol. Life Sci.* **2012**, *69*, 3115–3125. [CrossRef]
46. Rodríguez-Carballo, E.; Gámez, B.; Sedó-Cabezón, L.; Sánchez-Feutrie, M.; Zorzano, A.; Manzanares-Céspedes, C.; Rosa, J.L.; Ventura, F. The p38α MAPK function in osteoprecursors is required for bone formation and bone homeostasis in adult mice. *PLoS ONE* **2014**, *9*, e102032. [CrossRef]
47. Chakladar, A.; Dubeykovskiy, A.; Wojtukiewicz, L.J.; Pratap, J.; Lei, S.; Wang, T.C. Synergistic activation of the murine gastrin promoter by oncogenic Ras and beta-catenin involves SMAD recruitment. *Biochem. Biophys. Res. Commun.* **2005**, *336*, 190–196. [CrossRef]
48. Yamashita, M.; Otsuka, F.; Mukai, T.; Otani, H.; Inagaki, K.; Miyoshi, T.; Goto, J.; Yamamura, M.; Makino, H. Simvastatin antagonizes tumor necrosis factor-alpha inhibition of bone morphogenetic proteins-2-induced osteoblast differentiation by regulating Smad signaling and Ras/Rho-mitogen-activated protein kinase pathway. *J. Endocrinol.* **2008**, *196*, 601–613. [CrossRef]
49. Xue, R.; Li, J.Y.; Yeh, Y.; Yang, L.; Chien, S. Effects of matrix elasticity and cell density on human mesenchymal stem cells differentiation. *J. Orthop. Res.* **2013**, *31*, 1360–1365. [CrossRef]
50. Xu, J.; Sun, M.; Tan, Y.; Wang, H.; Wang, H.; Li, P.; Xu, Z.; Xia, Y.; Li, L.; Li, Y. Effect of matrix stiffness on the proliferation and differentiation of umbilical cord mesenchymal stem cells. *Differentiation* **2017**, *96*, 30–39. [CrossRef]

51. Sun, M.; Chi, G.; Li, P.; Lv, S.; Xu, J.; Xu, Z.; Xia, Y.; Tan, Y.; Xu, J.; Li, L.; et al. Effects of Matrix Stiffness on the Morphology, Adhesion, Proliferation and Osteogenic Differentiation of Mesenchymal Stem Cells. *Int. J. Med. Sci.* **2018**, *15*, 257–268. [CrossRef]
52. Ahmed, M.; Ffrench-Constant, C. Extracellular Matrix Regulation of Stem Cell Behavior. *Curr. Stem Cell Rep.* **2016**, *2*, 197–206. [CrossRef]
53. Guilak, F.; Cohen, D.M.; Estes, B.T.; Gimble, J.M.; Liedtke, W.; Chen, C.S. Control of stem cell fate by physical interactions with the extracellular matrix. *Cell Stem Cell* **2009**, *5*, 17–26. [CrossRef]
54. Zhang, N.; Kohn, D.H. Using polymeric materials to control stem cell behavior for tissue regeneration. *Birth Defects Res. Part. C Embryo Today Rev.* **2012**, *96*, 63–81. [CrossRef] [PubMed]
55. Abbass, M.M.S.; El-Rashidy, A.A.; Sadek, K.M.; Moshy, S.E.; Radwan, I.A.; Rady, D.; Dörfer, C.E.; Fawzy El-Sayed, K.M. Hydrogels and Dentin–Pulp Complex Regeneration: From the Benchtop to Clinical Translation. *Polymers* **2020**, *12*, 2935. [CrossRef] [PubMed]
56. Lv, H.; Wang, H.; Zhang, Z.; Yang, W.; Liu, W.; Li, Y.; Li, L. Biomaterial stiffness determines stem cell fate. *Life Sci.* **2017**, *178*, 42–48. [CrossRef] [PubMed]
57. Engler, A.J.; Sen, S.; Sweeney, H.L.; Discher, D.E. Matrix elasticity directs stem cell lineage specification. *Cell* **2006**, *126*, 677–689. [CrossRef]
58. Lv, H.; Li, L.; Sun, M.; Zhang, Y.; Chen, L.; Rong, Y.; Li, Y. Mechanism of regulation of stem cell differentiation by matrix stiffness. *Stem Cell Res. Ther.* **2015**, *6*, 103. [CrossRef]
59. Guo, R.; Lu, S.; Merkel, A.R.; Sterling, J.A.; Guelcher, S.A. Substrate Modulus Regulates Osteogenic Differentiation of Rat Mesenchymal Stem Cells through Integrin β1 and BMP Receptor Type IA. *J. Mater. Chem. B* **2016**, *4*, 3584–3593. [CrossRef]
60. Trappmann, B.; Gautrot, J.E.; Connelly, J.T.; Strange, D.G.; Li, Y.; Oyen, M.L.; Cohen Stuart, M.A.; Boehm, H.; Li, B.; Vogel, V.; et al. Extracellular-matrix tethering regulates stem-cell fate. *Nat. Mater.* **2012**, *11*, 642–649. [CrossRef]
61. Naqvi, S.M.; McNamara, L.M. Stem Cell Mechanobiology and the Role of Biomaterials in Governing Mechanotransduction and Matrix Production for Tissue Regeneration. *Front. Bioeng. Biotechnol.* **2020**, *8*, 597661. [CrossRef]
62. Shih, Y.R.; Tseng, K.F.; Lai, H.Y.; Lin, C.H.; Lee, O.K. Matrix stiffness regulation of integrin-mediated mechanotransduction during osteogenic differentiation of human mesenchymal stem cells. *J. Bone Miner. Res.* **2011**, *26*, 730–738. [CrossRef]
63. Miller, A.E.; Hu, P.; Barker, T.H. Feeling Things Out: Bidirectional Signaling of the Cell–ECM Interface, Implications in the Mechanobiology of Cell Spreading, Migration, Proliferation, and Differentiation. *Adv. Healthc. Mater.* **2020**, *9*, 1901445. [CrossRef] [PubMed]
64. Nanci, A. *Ten Cate's Oral Histology: Development, Structure and Function*, 8th ed.; El Sevier Mosby: St Louis, Hong Kong, 2013; pp. 81–83.
65. Ciobanasu, C.; Faivre, B.; Le Clainche, C. Actomyosin-dependent formation of the mechanosensitive talin-vinculin complex reinforces actin anchoring. *Nat. Commun.* **2014**, *5*, 3095. [CrossRef] [PubMed]
66. Martino, F.; Perestrelo, A.R.; Vinarský, V.; Pagliari, S.; Forte, G. Cellular Mechanotransduction: From Tension to Function. *Front. Physiol.* **2018**, *9*, 824. [CrossRef]
67. Ohashi, K.; Fujiwara, S.; Mizuno, K. Roles of the cytoskeleton, cell adhesion and rho signalling in mechanosensing and mechanotransduction. *J. Biochem.* **2017**, *161*, 245–254. [CrossRef]
68. Chen, H.; Cheng, C.Y. *Focal Adhesion Kinase*; Springer International Publishing: New York, NY, USA, 2018; pp. 1800–1812.
69. Bradbury, P.M.; Turner, K.; Mitchell, C.; Griffin, K.R.; Middlemiss, S.; Lau, L.; Dagg, R.; Taran, E.; Cooper-White, J.; Fabry, B.; et al. The focal adhesion targeting domain of p130Cas confers a mechanosensing function. *J. Cell Sci.* **2017**, *130*, 1263–1273. [CrossRef]
70. Takada, Y.; Ye, X.; Simon, S. The integrins. *Genome Biol.* **2007**, *8*, 215. [CrossRef] [PubMed]
71. Hynes, R.O. Integrins: Bidirectional, allosteric signaling machines. *Cell* **2002**, *110*, 673–687. [CrossRef]
72. Sun, Z.; Guo, S.S.; Fässler, R. Integrin-mediated mechanotransduction. *J. Cell Biol.* **2016**, *215*, 445–456. [CrossRef] [PubMed]
73. Kim, M.; Carman, C.V.; Springer, T.A. Bidirectional transmembrane signaling by cytoplasmic domain separation in integrins. *Science* **2003**, *301*, 1720–1725. [CrossRef]
74. Ellert-Miklaszewska, A.; Poleszak, K.; Pasierbinska, M.; Kaminska, B. Integrin Signaling in Glioma Pathogenesis: From Biology to Therapy. *Int. J. Mol. Sci.* **2020**, *21*, 888. [CrossRef] [PubMed]
75. Petrie, T.A.; Raynor, J.E.; Dumbauld, D.W.; Lee, T.T.; Jagtap, S.; Templeman, K.L.; Collard, D.M.; García, A.J. Multivalent integrin-specific ligands enhance tissue healing and biomaterial integration. *Sci. Transl. Med.* **2010**, *2*, 45ra60. [CrossRef]
76. Olivares-Navarrete, R.; Lee, E.M.; Smith, K.; Hyzy, S.L.; Doroudi, M.; Williams, J.K.; Gall, K.; Boyan, B.D.; Schwartz, Z. Substrate Stiffness Controls Osteoblastic and Chondrocytic Differentiation of Mesenchymal Stem Cells without Exogenous Stimuli. *PLoS ONE* **2017**, *12*, e0170312. [CrossRef]
77. Frith, J.E.; Mills, R.J.; Hudson, J.E.; Cooper-White, J.J. Tailored integrin-extracellular matrix interactions to direct human mesenchymal stem cell differentiation. *Stem. Cells Dev.* **2012**, *21*, 2442–2456. [CrossRef]
78. Sun, M.; Chi, G.; Xu, J.; Tan, Y.; Xu, J.; Lv, S.; Xu, Z.; Xia, Y.; Li, L.; Li, Y. Extracellular matrix stiffness controls osteogenic differentiation of mesenchymal stem cells mediated by integrin α5. *Stem. Cell Res. Ther.* **2018**, *9*, 52. [CrossRef] [PubMed]
79. Hamidouche, Z.; Fromigué, O.; Ringe, J.; Häupl, T.; Vaudin, P.; Pagès, J.C.; Srouji, S.; Livne, E.; Marie, P.J. Priming integrin alpha5 promotes human mesenchymal stromal cell osteoblast differentiation and osteogenesis. *Proc. Natl. Acad. Sci. USA* **2009**, *106*, 18587–18591. [CrossRef] [PubMed]

80. Zhang, D.; Ni, N.; Wang, Y.; Tang, Z.; Gao, H.; Ju, Y.; Sun, N.; He, X.; Gu, P.; Fan, X. CircRNA-vgll3 promotes osteogenic differentiation of adipose-derived mesenchymal stem cells via modulating miRNA-dependent integrin α5 expression. *Cell Death Differ.* **2021**, *28*, 283–302. [CrossRef] [PubMed]
81. Lee, J.; Abdeen, A.A.; Tang, X.; Saif, T.A.; Kilian, K.A. Geometric guidance of integrin mediated traction stress during stem cell differentiation. *Biomaterials* **2015**, *69*, 174–183. [CrossRef]
82. Liu, J.; DeYoung, S.M.; Zhang, M.; Zhang, M.; Cheng, A.; Saltiel, A.R. Changes in integrin expression during adipocyte differentiation. *Cell Metab.* **2005**, *2*, 165–177. [CrossRef] [PubMed]
83. Lock, J.G.; Wehrle-Haller, B.; Strömblad, S. Cell-matrix adhesion complexes: Master control machinery of cell migration. *Semin. Cancer Biol.* **2008**, *18*, 65–76. [CrossRef]
84. Gavazzo, P.; Viti, F.; Donnelly, H.; Oliva, M.A.G.; Salmeron-Sanchez, M.; Dalby, M.J.; Vassalli, M. Biophysical phenotyping of mesenchymal stem cells along the osteogenic differentiation pathway. *Cell Biol. Toxicol.* **2021**, *2021*, 1–19. [CrossRef]
85. Saidova, A.A.; Vorobjev, I.A. Lineage Commitment, Signaling Pathways, and the Cytoskeleton Systems in Mesenchymal Stem Cells. *Tissue Eng. Part B Rev.* **2020**, *26*, 13–25. [CrossRef] [PubMed]
86. Byron, A.; Frame, M.C. Adhesion protein networks reveal functions proximal and distal to cell-matrix contacts. *Curr. Opin. Cell Biol.* **2016**, *39*, 93–100. [CrossRef] [PubMed]
87. Han, S.J.; Bielawski, K.S.; Ting, L.H.; Rodriguez, M.L.; Sniadecki, N.J. Decoupling substrate stiffness, spread area, and micropost density: A close spatial relationship between traction forces and focal adhesions. *Biophys. J.* **2012**, *103*, 640–648. [CrossRef]
88. McBeath, R.; Pirone, D.M.; Nelson, C.M.; Bhadriraju, K.; Chen, C.S. Cell shape, cytoskeletal tension, and RhoA regulate stem cell lineage commitment. *Dev. Cell* **2004**, *6*, 483–495. [CrossRef]
89. Lavenus, S.; Berreur, M.; Trichet, V.; Pilet, P.; Louarn, G.; Layrolle, P. Adhesion and osteogenic differentiation of human mesenchymal stem cells on titanium nanopores. *Eur. Cell Mater.* **2011**, *22*, 84–96. [CrossRef]
90. Frith, J.E.; Mills, R.J.; Cooper-White, J.J. Lateral spacing of adhesion peptides influences human mesenchymal stem cell behaviour. *J. Cell Sci.* **2012**, *125*, 317–327. [CrossRef]
91. Rosenberg, N. The role of the cytoskeleton in mechanotransduction in human osteoblast-like cells. *Hum. Exp. Toxicol.* **2003**, *22*, 271–274. [CrossRef]
92. Wang, N.; Naruse, K.; Stamenović, D.; Fredberg, J.J.; Mijailovich, S.M.; Tolić-Nørrelykke, I.M.; Polte, T.; Mannix, R.; Ingber, D.E. Mechanical behavior in living cells consistent with the tensegrity model. *Proc. Natl. Acad. Sci. USA* **2001**, *98*, 7765–7770. [CrossRef] [PubMed]
93. Harris, A.R.; Jreij, P.; Fletcher, D.A. Mechanotransduction by the actin cytoskeleton: Converting mechanical stimuli into biochemical signals. *Annu. Rev. Biophys.* **2018**, *47*, 617–631. [CrossRef]
94. Pollard, T.D.; Borisy, G.G. Cellular motility driven by assembly and disassembly of actin filaments. *Cell* **2003**, *112*, 453–465. [CrossRef]
95. Kim, D.H.; Wirtz, D. Cytoskeletal tension induces the polarized architecture of the nucleus. *Biomaterials* **2015**, *48*, 161–172. [CrossRef] [PubMed]
96. Khatau, S.B.; Hale, C.M.; Stewart-Hutchinson, P.J.; Patel, M.S.; Stewart, C.L.; Searson, P.C.; Hodzic, D.; Wirtz, D. A perinuclear actin cap regulates nuclear shape. *Proc. Natl. Acad. Sci. USA* **2009**, *106*, 19017–19022. [CrossRef]
97. Kim, D.H.; Khatau, S.B.; Feng, Y.; Walcott, S.; Sun, S.X.; Longmore, G.D.; Wirtz, D. Actin cap associated focal adhesions and their distinct role in cellular mechanosensing. *Sci. Rep.* **2012**, *2*, 555. [CrossRef]
98. Tojkander, S.; Gateva, G.; Lappalainen, P. Actin stress fibers—Assembly, dynamics and biological roles. *J. Cell Sci.* **2012**, *125*, 1855–1864. [CrossRef] [PubMed]
99. Huang, C.H.; Chen, M.H.; Young, T.H.; Jeng, J.H.; Chen, Y.J. Interactive effects of mechanical stretching and extracellular matrix proteins on initiating osteogenic differentiation of human mesenchymal stem cells. *J. Cell Biochem.* **2009**, *108*, 1263–1273. [CrossRef]
100. Svitkina, T. The Actin Cytoskeleton and Actin-Based Motility. *Cold Spring Harb. Perspect. Biol.* **2018**, *10*, a018267. [CrossRef] [PubMed]
101. Chen, L.; Hu, H.; Qiu, W.; Shi, K.; Kassem, M. Actin depolymerization enhances adipogenic differentiation in human stromal stem cells. *Stem Cell Res.* **2018**, *29*, 76–83. [CrossRef]
102. Rodríguez, J.P.; González, M.; Ríos, S.; Cambiazo, V. Cytoskeletal organization of human mesenchymal stem cells (MSC) changes during their osteogenic differentiation. *J. Cell Biochem.* **2004**, *93*, 721–731. [CrossRef]
103. Sonowal, H.; Kumar, A.; Bhattacharyya, J.; Gogoi, P.K.; Jaganathan, B.G. Inhibition of actin polymerization decreases osteogeneic differentiation of mesenchymal stem cells through p38 MAPK pathway. *J. Biomed. Sci.* **2013**, *20*, 71. [CrossRef]
104. Sen, B.; Xie, Z.; Case, N.; Thompson, W.R.; Uzer, G.; Styner, M.; Rubin, J. mTORC2 regulates mechanically induced cytoskeletal reorganization and lineage selection in marrow-derived mesenchymal stem cells. *J. Bone Miner. Res.* **2014**, *29*, 78–89. [CrossRef]
105. Chen, L.; Shi, K.; Frary, C.E.; Ditzel, N.; Hu, H.; Qiu, W.; Kassem, M. Inhibiting actin depolymerization enhances osteoblast differentiation and bone formation in human stromal stem cells. *Stem Cell Res.* **2015**, *15*, 281–289. [CrossRef]
106. Zhang, T.; Lin, S.; Shao, X.; Zhang, Q.; Xue, C.; Zhang, S.; Lin, Y.; Zhu, B.; Cai, X. Effect of matrix stiffness on osteoblast functionalization. *Cell Prolif.* **2017**, *50*. [CrossRef] [PubMed]
107. Khan, A.U.; Qu, R.; Fan, T.; Ouyang, J.; Dai, J. A glance on the role of actin in osteogenic and adipogenic differentiation of mesenchymal stem cells. *Stem Cell Res. Ther.* **2020**, *11*, 283. [CrossRef]

108. Kobayashi, T.; Sokabe, M. Sensing substrate rigidity by mechanosensitive ion channels with stress fibers and focal adhesions. *Curr. Opin. Cell Biol.* **2010**, *22*, 669–676. [CrossRef] [PubMed]
109. Wu, Z.; Wong, K.; Glogauer, M.; Ellen, R.P.; McCulloch, C.A.G. Regulation of Stretch-Activated Intracellular Calcium Transients by Actin Filaments. *Biochem. Biophys. Res. Commun.* **1999**, *261*, 419–425. [CrossRef] [PubMed]
110. Iqbal, J.; Zaidi, M. Molecular regulation of mechanotransduction. *Biochem. Biophys. Res. Commun.* **2005**, *328*, 751–755. [CrossRef]
111. Kretlow, J.D.; Jin, Y.-Q.; Liu, W.; Zhang, W.J.; Hong, T.-H.; Zhou, G.; Baggett, L.S.; Mikos, A.G.; Cao, Y. Donor age and cell passage affects differentiation potential of murine bone marrow-derived stem cells. *BMC Cell Biol.* **2008**, *9*, 1–13. [CrossRef]
112. Stenderup, K.; Justesen, J.; Clausen, C.; Kassem, M. Aging is associated with decreased maximal life span and accelerated senescence of bone marrow stromal cells. *Bone Miner.* **2003**, *33*, 919–926. [CrossRef]
113. Asumda, F.Z.; Chase, P.B. Age-related changes in rat bone-marrow mesenchymal stem cell plasticity. *BMC Cell Biol.* **2011**, *12*, 1–11. [CrossRef]
114. Shen, J.; Tsai, Y.-T.; DiMarco, N.M.; Long, M.A.; Sun, X.; Tang, L. Transplantation of mesenchymal stem cells from young donors delays aging in mice. *Sci. Rep. Cetacean Res.* **2011**, *1*, 1–7. [CrossRef] [PubMed]
115. Lin, T.h.; Gibon, E.; Loi, F.; Pajarinen, J.; Córdova, L.A.; Nabeshima, A.; Lu, L.; Yao, Z.; Goodman, S.B. Decreased osteogenesis in mesenchymal stem cells derived from the aged mouse is associated with enhanced NF-κB activity. *J. Orth Res.* **2017**, *35*, 281–288. [CrossRef]
116. Pelissier, F.A.; Garbe, J.C.; Ananthanarayanan, B.; Miyano, M.; Lin, C.; Jokela, T.; Kumar, S.; Stampfer, M.R.; Lorens, J.B.; LaBarge, M.A. Age-Related Dysfunction in Mechanotransduction Impairs Differentiation of Human Mammary Epithelial Progenitors. *Cell Rep.* **2014**, *7*, 1926–1939. [CrossRef]
117. Barreto, S.; Gonzalez-Vazquez, A.; Cameron, A.R.; Cavanagh, B.; Murray, D.J.; O'Brien, F.J. Identification of the mechanisms by which age alters the mechanosensitivity of mesenchymal stromal cells on substrates of differing stiffness: Implications for osteogenesis and angiogenesis. *Acta Biomater.* **2017**, *53*, 59–69. [CrossRef] [PubMed]
118. Zouani, O.F.; Kalisky, J.; Ibarboure, E.; Durrieu, M.-C. Effect of BMP-2 from matrices of different stiffnesses for the modulation of stem cell fate. *Biomaterials* **2013**, *34*, 2157–2166. [CrossRef] [PubMed]
119. Ma, S.; Meng, Z.; Chen, R.; Guan, K.-L. The Hippo pathway: Biology and pathophysiology. *Annu Rev. Biochem.* **2019**, *88*, 577–604. [CrossRef] [PubMed]
120. Yamaguchi, H.; Taouk, G.M. A Potential role of YAP/TAZ in the interplay between metastasis and metabolic alterations. *Front. Oncol.* **2020**, *10*, 928. [CrossRef] [PubMed]
121. Dobrokhotov, O.; Samsonov, M.; Sokabe, M.; Hirata, H. Mechanoregulation and pathology of YAP/TAZ via Hippo and non-Hippo mechanisms. *J. Transl. Med.* **2018**, *7*, 1–14. [CrossRef]
122. Dupont, S.; Morsut, L.; Aragona, M.; Enzo, E.; Giulitti, S.; Cordenonsi, M.; Zanconato, F.; Le Digabel, J.; Forcato, M.; Bicciato, S. Role of YAP/TAZ in mechanotransduction. *Nature* **2011**, *474*, 179–183. [CrossRef]
123. Zhao, B.; Li, L.; Wang, L.; Wang, C.-Y.; Yu, J.; Guan, K.-L. Cell detachment activates the Hippo pathway via cytoskeleton reorganization to induce anoikis. *Genes Dev.* **2012**, *26*, 54–68. [CrossRef] [PubMed]
124. Aragona, M.; Panciera, T.; Manfrin, A.; Giulitti, S.; Michielin, F.; Elvassore, N.; Dupont, S.; Piccolo, S. A mechanical checkpoint controls multicellular growth through YAP/TAZ regulation by actin-processing factors. *Cell* **2013**, *154*, 1047–1059. [CrossRef] [PubMed]
125. Elosegui-Artola, A.; Andreu, I.; Beedle, A.E.; Lezamiz, A.; Uroz, M.; Kosmalska, A.J.; Oria, R.; Kechagia, J.Z.; Rico-Lastres, P.; Le Roux, A.-L. Force triggers YAP nuclear entry by regulating transport across nuclear pores. *Cell* **2017**, *171*, 1397–1410.e1314. [CrossRef] [PubMed]
126. Lorthongpanich, C.; Thumanu, K.; Tangkiettrakul, K.; Jiamvoraphong, N.; Laowtammathron, C.; Damkham, N.; U-pratya, Y.; Issaragrisil, S. YAP as a key regulator of adipo-osteogenic differentiation in human MSCs. *Stem Cell Res. Ther.* **2019**, *10*, 402. [CrossRef] [PubMed]
127. Hong, J.H.; Hwang, E.S.; McManus, M.T.; Amsterdam, A.; Tian, Y.; Kalmukova, R.; Mueller, E.; Benjamin, T.; Spiegelman, B.M.; Sharp, P.A.; et al. TAZ, a transcriptional modulator of mesenchymal stem cell differentiation. *Science* **2005**, *309*, 1074–1078. [CrossRef]
128. Du, J.; Chen, X.; Liang, X.; Zhang, G.; Xu, J.; He, L.; Zhan, Q.; Feng, X.Q.; Chien, S.; Yang, C. Integrin activation and internalization on soft ECM as a mechanism of induction of stem cell differentiation by ECM elasticity. *Proc. Natl. Acad. Sci. USA* **2011**, *108*, 9466–9471. [CrossRef]
129. Chrzanowska-Wodnicka, M.; Burridge, K. Rho-stimulated contractility drives the formation of stress fibers and focal adhesions. *J. Cell Biol.* **1996**, *133*, 1403–1415. [CrossRef]
130. Arnsdorf, E.J.; Tummala, P.; Kwon, R.Y.; Jacobs, C.R. Mechanically induced osteogenic differentiation–the role of RhoA, ROCKII and cytoskeletal dynamics. *J. Cell Sci.* **2009**, *122*, 546–553. [CrossRef]
131. Hyväri, L.; Ojansivu, M.; Juntunen, M.; Kartasalo, K.; Miettinen, S.; Vanhatupa, S. Focal Adhesion Kinase and ROCK Signaling Are Switch-Like Regulators of Human Adipose Stem Cell Differentiation towards Osteogenic and Adipogenic Lineages. *Stem Cells Int.* **2018**, *2018*, 2190657. [CrossRef]
132. Rowlands, A.S.; George, P.A.; Cooper-White, J.J. Directing osteogenic and myogenic differentiation of MSCs: Interplay of stiffness and adhesive ligand presentation. *Am. J. Physiol. Cell Physiol.* **2008**, *295*, C1037–C1044. [CrossRef]

133. Wan, W.; Cheng, B.; Zhang, C.; Ma, Y.; Li, A.; Xu, F.; Lin, M. Synergistic Effect of Matrix Stiffness and Inflammatory Factors on Osteogenic Differentiation of MSC. *Biophys. J.* **2019**, *117*, 129–142. [CrossRef]
134. He, X.T.; Wu, R.X.; Xu, X.Y.; Wang, J.; Yin, Y.; Chen, F.M. Macrophage involvement affects matrix stiffness-related influences on cell osteogenesis under three-dimensional culture conditions. *Acta Biomater.* **2018**, *71*, 132–147. [CrossRef]
135. Tang, X.; Thankappan, S.K.; Lee, P.; Fard, S.E.; Harmon, M.D.; Tran, K.; Yu, X. Chapter 21—Polymeric Biomaterials in Tissue Engineering and Regenerative Medicine. In *Natural and Synthetic Biomedical Polymers*; Kumbar, S.G., Laurencin, C.T., Deng, M., Eds.; Elsevier: Oxford, UK, 2014; pp. 351–371.
136. Ramkumar, R.; Sundaram, M.M. A biopolymer gel-decorated cobalt molybdate nanowafer: Effective graft polymer cross-linked with an organic acid for better energy storage. *New J. Chem.* **2016**, *40*, 2863–2877. [CrossRef]
137. Yang, J.M.; Olanrele, O.S.; Zhang, X.; Hsu, C.C. Fabrication of Hydrogel Materials for Biomedical Applications. In *Novel Biomaterials for Regenerative Medicine*; Chun, H.J., Park, K., Kim, C.-H., Khang, G., Eds.; Springer Singapore: Singapore, 2018; pp. 197–224.
138. Lee, K.Y.; Mooney, D.J. Alginate: Properties and biomedical applications. *Prog. Polym. Sci.* **2012**, *37*, 106–126. [CrossRef]
139. Lee, H.-p.; Gu, L.; Mooney, D.J.; Levenston, M.E.; Chaudhuri, O. Mechanical confinement regulates cartilage matrix formation by chondrocytes. *Nat. Mater.* **2017**, *16*, 1243–1251. [CrossRef] [PubMed]
140. Wickramaarachchi, K.; Sundaram, M.M.; Henry, D.J.; Gao, X. Alginate Biopolymer Effect on the Electrodeposition of Manganese Dioxide on Electrodes for Supercapacitors. *ACS Appl. Energy Mater.* **2021**, *4*, 7040–7051. [CrossRef]
141. Zhang, J.; Wehrle, E.; Adamek, P.; Paul, G.R.; Qin, X.-H.; Rubert, M.; Müller, R. Optimization of mechanical stiffness and cell density of 3D bioprinted cell-laden scaffolds improves extracellular matrix mineralization and cellular organization for bone tissue engineering. *Acta Biomater.* **2020**, *114*, 307–322. [CrossRef] [PubMed]
142. Freeman, F.E.; Kelly, D.J. Tuning alginate bioink stiffness and composition for controlled growth factor delivery and to spatially direct MSC fate within bioprinted tissues. *Sci. Rep.* **2017**, *7*, 1–12. [CrossRef]
143. Maia, F.R.; Fonseca, K.B.; Rodrigues, G.; Granja, P.L.; Barrias, C.C. Matrix-driven formation of mesenchymal stem cell–Extracellular matrix microtissues on soft alginate hydrogels. *Acta Biomater.* **2014**, *10*, 3197–3208. [CrossRef]
144. Yener, B.; Acar, E.; Aguis, P.; Bennett, K.; Vandenberg, S.L.; Plopper, G.E. Multiway modeling and analysis in stem cell systems biology. *BMC Syst. Biol.* **2008**, *2*, 1–17. [CrossRef]
145. Stein, G.S.; Lian, J.B.; Owen, T.A. Relationship of cell growth to the regulation of tissue-specific gene expression during osteoblast differentiation. *FASEB J.* **1990**, *4*, 3111–3123. [CrossRef]
146. Xie, J.; Bao, M.; Bruekers, S.p.M.; Huck, W.T. Collagen gels with different fibrillar microarchitectures elicit different cellular responses. *ACS Appl. Mater. Interfaces* **2017**, *9*, 19630–19637. [CrossRef] [PubMed]
147. Banks, J.M.; Mozdzen, L.C.; Harley, B.A.; Bailey, R.C. The combined effects of matrix stiffness and growth factor immobilization on the bioactivity and differentiation capabilities of adipose-derived stem cells. *Biomaterials* **2014**, *35*, 8951–8959. [CrossRef] [PubMed]
148. Hwang, J.-H.; Han, U.; Yang, M.; Choi, Y.; Choi, J.; Lee, J.-M.; Jung, H.-S.; Hong, J.; Hong, J.-H. Artificial cellular nano-environment composed of collagen-based nanofilm promotes osteogenic differentiation of mesenchymal stem cells. *Acta Biomater.* **2019**, *86*, 247–256. [CrossRef]
149. Zhou, Q.; Lyu, S.; Bertrand, A.A.; Hu, A.C.; Chan, C.H.; Ren, X.; Dewey, M.J.; Tiffany, A.S.; Harley, B.A.; Lee, J.C. Stiffness of Nanoparticulate Mineralized Collagen Scaffolds Triggers Osteogenesis via Mechanotransduction and Canonical Wnt Signaling. *Macromol. Biosci.* **2021**, *21*, 2000370. [CrossRef]
150. Tsimbouri, P.M.; Childs, P.G.; Pemberton, G.D.; Yang, J.; Jayawarna, V.; Orapiriyakul, W.; Burgess, K.; Gonzalez-Garcia, C.; Blackburn, G.; Thomas, D. Stimulation of 3D osteogenesis by mesenchymal stem cells using a nanovibrational bioreactor. *Nat. Biomed. Eng.* **2017**, *1*, 758–770. [CrossRef]
151. Murphy, C.M.; Matsiko, A.; Haugh, M.G.; Gleeson, J.P.; O'Brien, F.J. Mesenchymal stem cell fate is regulated by the composition and mechanical properties of collagen–glycosaminoglycan scaffolds. *J. Mech. Behav. Biomed. Mater.* **2012**, *11*, 53–62. [CrossRef]
152. Chen, G.; Dong, C.; Yang, L.; Lv, Y. Interfaces 3D scaffolds with different stiffness but the same microstructure for bone tissue engineering. *ACS Appl. Mater.* **2015**, *7*, 15790–15802. [CrossRef] [PubMed]
153. Radhakrishnan, J.; Krishnan, U.M.; Sethuraman, S. Hydrogel based injectable scaffolds for cardiac tissue regeneration. *Biotechnol. Adv.* **2014**, *32*, 449–461. [CrossRef] [PubMed]
154. Echave, M.C.; Saenz del Burgo, L.; Pedraz, J.L.; Orive, G. Gelatin as Biomaterial for Tissue Engineering. *Curr. Pharm. Des.* **2017**, *23*, 3567–3584. [CrossRef] [PubMed]
155. Su, K.; Wang, C. Recent advances in the use of gelatin in biomedical research. *Biotechnol. Lett.* **2015**, *37*, 2139–2145. [CrossRef]
156. Gasperini, L.; Mano, J.F.; Reis, R.L. Natural polymers for the microencapsulation of cells. *J. R. Soc. Interface* **2014**, *11*, 20140817. [CrossRef] [PubMed]
157. Weng, Y.J.; Ren, J.R.; Huang, N.; Wang, J.; Chen, J.Y.; Leng, Y.X.; Liu, H.Q. Surface engineering of Ti–O films by photochemical immobilization of gelatin. *Mater. Sci Eng. C* **2008**, *28*, 1495–1500. [CrossRef]
158. Zhu, Y.; Gao, C.; He, T.; Shen, J. Endothelium regeneration on luminal surface of polyurethane vascular scaffold modified with diamine and covalently grafted with gelatin. *Biomaterials* **2004**, *25*, 423–430. [CrossRef]

159. Salamon, A.; van Vlierberghe, S.; van Nieuwenhove, I.; Baudisch, F.; Graulus, G.-J.; Benecke, V.; Alberti, K.; Neumann, H.-G.; Rychly, J.; Martins, J.C.; et al. Gelatin-Based Hydrogels Promote Chondrogenic Differentiation of Human Adipose Tissue-Derived Mesenchymal Stem Cells In Vitro. *Materials* **2014**, *7*, 1342–1359. [CrossRef]
160. Zhou, D.; Ito, Y. Inorganic material surfaces made bioactive by immobilizing growth factors for hard tissue engineering. *RSC Adv.* **2013**, *3*, 11095–11106. [CrossRef]
161. Sun, M.; Sun, X.; Wang, Z.; Guo, S.; Yu, G.; Yang, H. Synthesis and Properties of Gelatin Methacryloyl (GelMA) Hydrogels and Their Recent Applications in Load-Bearing Tissue. *Polymers* **2018**, *10*, 1290. [CrossRef]
162. Van Nieuwenhove, I.; Salamon, A.; Adam, S.; Dubruel, P.; Van Vlierberghe, S.; Peters, K. Gelatin- and starch-based hydrogels. Part B: In vitro mesenchymal stem cell behavior on the hydrogels. *Carbohydr. Polym.* **2017**, *161*, 295–305. [CrossRef]
163. Jiang, P.; Mao, Z.; Gao, C. Combinational effect of matrix elasticity and alendronate density on differentiation of rat mesenchymal stem cells. *Acta Biomater.* **2015**, *19*, 76–84. [CrossRef] [PubMed]
164. Sun, H.; Zhu, F.; Hu, Q.; Krebsbach, P.H. Controlling stem cell-mediated bone regeneration through tailored mechanical properties of collagen scaffolds. *Biomaterials* **2014**, *35*, 1176–1184. [CrossRef] [PubMed]
165. Ventre, M.; Coppola, V.; Natale, C.F.; Netti, P.A. Aligned fibrous decellularized cell derived matrices for mesenchymal stem cell amplification. *J. Biomed. Mater. Res. A* **2019**, *107*, 2536–2546. [CrossRef]
166. Hu, Q.; Liu, M.; Chen, G.; Xu, Z.; Lv, Y. Demineralized Bone Scaffolds with Tunable Matrix Stiffness for Efficient Bone Integration. *ACS Appl. Mater. Interfaces* **2018**, *10*, 27669–27680. [CrossRef]
167. Lee, J.Y.; Spicer, A.P. Hyaluronan: A multifunctional, megaDalton, stealth molecule. *Curr. Opin. Cell Biol.* **2000**, *12*, 581–586. [CrossRef]
168. Garg, H.G.; Hales, C.A. *Chemistry and Biology of Hyaluronan*; Elsevier: Amsterdam, The Netherlands, 2004.
169. Fawzy El-Sayed, K.M.; Dahaba, M.A.; Aboul-Ela, S.; Darhous, M.S. Local application of hyaluronan gel in conjunction with periodontal surgery: A randomized controlled trial. *Clin. Oral Investig.* **2012**, *16*, 1229–1236. [CrossRef]
170. Xing, F.; Li, L.; Zhou, C.; Long, C.; Wu, L.; Lei, H.; Kong, Q.; Fan, Y.; Xiang, Z.; Zhang, X. Regulation and directing stem cell fate by tissue engineering functional microenvironments: Scaffold physical and chemical cues. *Stem Cells Int.* **2019**, *2019*. [CrossRef]
171. Zhai, P.; Peng, X.; Li, B.; Liu, Y.; Sun, H.; Li, X. The application of hyaluronic acid in bone regeneration. *Int. J. Biol. Macromol.* **2020**, *151*, 1224–1239. [CrossRef]
172. Itano, N. Simple primary structure, complex turnover regulation and multiple roles of hyaluronan. *J. Biochem.* **2008**, *144*, 131–137. [CrossRef] [PubMed]
173. Anseth, K.S.; Burdick, J.A. New directions in photopolymerizable biomaterials. *MRS Bull.* **2002**, *27*, 130–136. [CrossRef]
174. Zhao, W.; Li, X.; Liu, X.; Zhang, N.; Wen, X. Effects of substrate stiffness on adipogenic and osteogenic differentiation of human mesenchymal stem cells. *Mater. Sci. Eng. C* **2014**, *40*, 316–323. [CrossRef] [PubMed]
175. Cosgrove, B.D.; Mui, K.L.; Driscoll, T.P.; Caliari, S.R.; Mehta, K.D.; Assoian, R.K.; Burdick, J.A.; Mauck, R.L. N-cadherin adhesive interactions modulate matrix mechanosensing and fate commitment of mesenchymal stem cells. *Nat. Mater.* **2016**, *15*, 1297–1306. [CrossRef] [PubMed]
176. Dorcemus, D.L.; George, E.O.; Dealy, C.N.; Nukavarapu, S.P. Harnessing External Cues: Development and Evaluation of an In Vitro Culture System for Osteochondral Tissue Engineering. *Tissue Eng. Part A* **2017**, *23*, 719–737. [CrossRef] [PubMed]
177. Hao, Y.; Song, J.; Ravikrishnan, A.; Dicker, K.T.; Fowler, E.W.; Zerdoum, A.B.; Li, Y.; Zhang, H.; Rajasekaran, A.K.; Fox, J.M.; et al. Rapid Bioorthogonal Chemistry Enables in Situ Modulation of the Stem Cell Behavior in 3D without External Triggers. *ACS Appl Mater. Interfaces* **2018**, *10*, 26016–26027. [CrossRef] [PubMed]
178. Patel, M.; Koh, W.-G.J.P. Composite Hydrogel of Methacrylated Hyaluronic Acid and Fragmented Polycaprolactone Nanofiber for Osteogenic Differentiation of Adipose-Derived. *Stem Cells* **2020**, *12*, 902. [CrossRef]
179. Ducret, M.; Montembault, A.; Josse, J.; Pasdeloup, M.; Celle, A.; Benchrih, R.; Mallein-Gerin, F.; Alliot-Licht, B.; David, L.; Farges, J.C. Design and characterization of a chitosan-enriched fibrin hydrogel for human dental pulp regeneration. *Dent. Mater.* **2019**, *35*, 523–533. [CrossRef] [PubMed]
180. Hashemzadeh, H.; Allahverdi, A.; Ghorbani, M.; Soleymani, H.; Kocsis, Á.; Fischer, M.B.; Ertl, P.; Naderi-Manesh, H. Gold Nanowires/Fibrin Nanostructure as Microfluidics Platforms for Enhancing Stem Cell Differentiation: Bio-AFM Study. *Micromachines* **2020**, *11*, 50. [CrossRef]
181. Spicer, C.D. Hydrogel scaffolds for tissue engineering: The importance of polymer choice. *Polym. Chem.* **2020**, *11*, 184–219. [CrossRef]
182. Pek, Y.S.; Wan, A.C.; Ying, J.Y. The effect of matrix stiffness on mesenchymal stem cell differentiation in a 3D thixotropic gel. *Biomaterials* **2010**, *31*, 385–391. [CrossRef] [PubMed]
183. Ye, K.; Wang, X.; Cao, L.; Li, S.; Li, Z.; Yu, L.; Ding, J. Matrix stiffness and nanoscale spatial organization of cell-adhesive ligands direct stem cell fate. *Nano Lett.* **2015**, *15*, 4720–4729. [CrossRef]
184. Steinmetz, N.J.; Aisenbrey, E.A.; Westbrook, K.K.; Qi, H.J.; Bryant, S.J. Mechanical loading regulates human MSC differentiation in a multi-layer hydrogel for osteochondral tissue engineering. *Acta Biomater.* **2015**, *21*, 142–153. [CrossRef] [PubMed]
185. Yang, Y.; Feng, Y.; Qu, R.; Li, Q.; Rong, D.; Fan, T.; Yang, Y.; Sun, B.; Bi, Z.; Khan, A.U. Synthesis of aligned porous polyethylene glycol/silk fibroin/hydroxyapatite scaffolds for osteoinduction in bone tissue engineering. *Stem. Cell Res. Ther.* **2020**, *11*, 1–17. [CrossRef]

186. Yang, C.; DelRio, F.W.; Ma, H.; Killaars, A.R.; Basta, L.P.; Kyburz, K.A.; Anseth, K.S. Spatially patterned matrix elasticity directs stem cell fate. *Proc. Natl. Acad. Sci. USA* **2016**, *113*, E4439–E4445. [CrossRef]
187. Gandavarapu, N.R.; Alge, D.L.; Anseth, K.S. Osteogenic differentiation of human mesenchymal stem cells on α5 integrin binding peptide hydrogels is dependent on substrate elasticity. *Biomater. Sci.* **2014**, *2*, 352–361. [CrossRef]
188. Chou, S.-Y.; Cheng, C.-M.; LeDuc, P.R. Composite polymer systems with control of local substrate elasticity and their effect on cytoskeletal and morphological characteristics of adherent cells. *Biomaterials* **2009**, *30*, 3136–3142. [CrossRef]
189. Viale-Bouroncle, S.; Völlner, F.; Möhl, C.; Küpper, K.; Brockhoff, G.; Reichert, T.E.; Schmalz, G.; Morsczeck, C. Soft matrix supports osteogenic differentiation of human dental follicle cells. *Biochem. Biophys. Res. Commun.* **2011**, *410*, 587–592. [CrossRef]
190. Viale-Bouroncle, S.; Gosau, M.; Küpper, K.; Möhl, C.; Brockhoff, G.; Reichert, T.E.; Schmalz, G.; Ettl, T.; Morsczeck, C. Rigid matrix supports osteogenic differentiation of stem cells from human exfoliated deciduous teeth (SHED). *Differentiation* **2012**, *84*, 366–370. [CrossRef]
191. Wang, P.-Y.; Tsai, W.-B.; Voelcker, N.H. Screening of rat mesenchymal stem cell behaviour on polydimethylsiloxane stiffness gradients. *Acta Biomater.* **2012**, *8*, 519–530. [CrossRef]
192. Khoramgah, M.S.; Ranjbari, J.; Abbaszadeh, H.-A.; Mirakabad, F.S.T.; Hatami, S.; Hosseinzadeh, S.; Ghanbarian, H. Freeze-dried multiscale porous nanofibrous three dimensional scaffolds for bone regenerations. *BioImpacts BI* **2020**, *10*, 73. [CrossRef]
193. Oh, S.H.; An, D.B.; Kim, T.H.; Lee, J.H. Wide-range stiffness gradient PVA/HA hydrogel to investigate stem cell differentiation behavior. *Acta Biomater.* **2016**, *35*, 23–31. [CrossRef]
194. Hwang, J.-H.; Byun, M.R.; Kim, A.R.; Kim, K.M.; Cho, H.J.; Lee, Y.H.; Kim, J.; Jeong, M.G.; Hwang, E.S.; Hong, J.H. Extracellular matrix stiffness regulates osteogenic differentiation through MAPK activation. *PLoS ONE* **2015**, *10*, e0135519. [CrossRef]
195. Sun, X.; Tung, W.; Wang, W.; Xu, X.; Zou, J.; Gould, O.E.; Kratz, K.; Ma, N.; Lendlein, A. The effect of stiffness variation of electrospun fiber meshes of multiblock copolymers on the osteogenic differentiation of human mesenchymal stem cells. *Clin. Hemorheol. Microcirc.* **2019**, *73*, 219–228. [CrossRef]
196. Shams, M.; Karimi, M.; Heydari, M.; Salimi, A. Nanocomposite scaffolds composed of Apacite (apatite-calcite) nanostructures, poly (ε-caprolactone) and poly (2-hydroxyethylmethacrylate): The effect of nanostructures on physico-mechanical properties and osteogenic differentiation of human bone marrow mesenchymal stem cells in vitro. *Mater. Sci. Eng. C Mater. Biol. Appl.* **2020**, *117*, 111271. [CrossRef]
197. Tang, X.; Ali, M.Y.; Saif, M.T.A. A Novel Technique for Micro-patterning Proteins and Cells on Polyacrylamide Gels. *Soft Matter* **2012**, *8*, 7197–7206. [CrossRef] [PubMed]
198. Daliri, K.; Pfannkuche, K.; Garipcan, B. Effects of physicochemical properties of polyacrylamide (PAA) and (polydimethylsiloxane) PDMS on cardiac cell behavior. *Soft Matter* **2021**, *17*, 1156–1172. [CrossRef]
199. Winer, J.P.; Janmey, P.A.; McCormick, M.E.; Funaki, M. Bone marrow-derived human mesenchymal stem cells become quiescent on soft substrates but remain responsive to chemical or mechanical stimuli. *Tissue Eng. Part A* **2009**, *15*, 147–154. [CrossRef]
200. Lee, J.; Abdeen, A.A.; Huang, T.H.; Kilian, K.A. Controlling cell geometry on substrates of variable stiffness can tune the degree of osteogenesis in human mesenchymal stem cells. *J. Mech. Behav. Biomed. Mater.* **2014**, *38*, 209–218. [CrossRef] [PubMed]
201. Stanton, A.E.; Tong, X.; Yang, F. Extracellular matrix type modulates mechanotransduction of stem cells. *Acta Biomater.* **2019**, *96*, 310–320. [CrossRef]
202. Gungordu, H.I.; Bao, M.; van Helvert, S.; Jansen, J.A.; Leeuwenburgh, S.C.G.; Walboomers, X.F. Effect of mechanical loading and substrate elasticity on the osteogenic and adipogenic differentiation of mesenchymal stem cells. *J. Tissue Eng. Regen. Med.* **2019**, *13*, 2279–2290. [CrossRef]
203. Hogrebe, N.J.; Gooch, K.J. Direct influence of culture dimensionality on human mesenchymal stem cell differentiation at various matrix stiffnesses using a fibrous self-assembling peptide hydrogel. *J. Biomed. Mater. Res. A* **2016**, *104*, 2356–2368. [CrossRef]
204. Wu, L.; Magaz, A.; Wang, T.; Liu, C.; Darbyshire, A.; Loizidou, M.; Emberton, M.; Birchall, M.; Song, W. Stiffness memory of indirectly 3D-printed elastomer nanohybrid regulates chondrogenesis and osteogenesis of human mesenchymal stem cells. *Biomaterials* **2018**, *186*, 64–79. [CrossRef]
205. Chen, G.; Xu, R.; Zhang, C.; Lv, Y. Responses of MSCs to 3D scaffold matrix mechanical properties under oscillatory perfusion culture. *ACS Appl. Mater.* **2017**, *9*, 1207–1218. [CrossRef]

polymers

MDPI

Article

Laser-Induced Graphene-Based Enzymatic Biosensor for Glucose Detection

Kalpana Settu *, Pin-Tzu Chiu and Yu-Ming Huang

Department of Electrical Engineering, National Taipei University, New Taipei City 23741, Taiwan; s410787003@gm.ntpu.edu.tw (P.-T.C.); idris.herondale@gmail.com (Y.-M.H.)
* Correspondence: kalpana@mail.ntpu.edu.tw; Tel.: +886-2867-4111-1 (ext. 68806)

Abstract: Laser-induced graphene (LIG) has recently been receiving increasing attention due to its simple fabrication and low cost. This study reports a flexible laser-induced graphene-based electrochemical biosensor fabricated on a polymer substrate by the laser direct engraving process. For this purpose, a 450 nm UV laser was employed to produce a laser-induced graphene electrode (LIGE) on a polyimide substrate. After the laser engraving of LIGE, the chitosan–glucose oxidase (GOx) composite was immobilized on the LIGE surface to develop the biosensor for glucose detection. It was observed that the developed LIGE biosensor exhibited good amperometric responses toward glucose detection over a wide linear range up to 8 mM. The GOx/chitosan-modified LIGE biosensor showed high sensitivity of 43.15 μA mM^{-1} cm^{-2} with a detection limit of 0.431 mM. The interference studies performed with some possible interfering compounds such as ascorbic acid, uric acid, and urea exhibited no interference as there was no difference observed in the amperometric glucose detection. It was suggested that the LIGE-based biosensor proposed herein was easy to prepare and could be used for low-cost, rapid, and sensitive/selective glucose detection.

Keywords: biosensor; laser-induced graphene; polyimide; glucose; enzyme

Citation: Settu, K.; Chiu, P.-T.; Huang, Y.-M. Laser-Induced Graphene-Based Enzymatic Biosensor for Glucose Detection. *Polymers* **2021**, *13*, 2795. https://doi.org/10.3390/polym13162795

Academic Editors: José Miguel Ferri and Claudio Gerbaldi

Received: 5 July 2021
Accepted: 17 August 2021
Published: 20 August 2021

1. Introduction

Numerous novel and cutting-edge technologies and materials are necessary to satisfy the new trends and requisites of analytical systems as the needs for environmental, biomedical, food and beverage analysis are progressing very quickly. The development of biosensors has evolved as one of the most promising research directions to overcome these challenges. Therefore, biosensor-based techniques have recently started being applied for the determination of different clinically, environmentally and biologically active materials [1–3]. In this regard, the design of biosensors in nanoscience/nanotechnology, environmental, medicine and food monitoring has been significantly increased during the past decade for their extensive applications. These advanced technologies have assisted the construction of highly sensitive, selective, customizable, and portable sensors for the determination of various clinically significant materials such as glucose, etc. [4]. The progress of such glucose biosensors has an inordinate significance in diagnosing and controlling diabetes mellitus, which is considered a worldwide public health problem. Diabetes mellitus would increase the risk of heart disease, kidney failure, blindness, postoperative and wound infections [5,6].

Diabetes mellitus has increased worldwide over the past five decades. Diabetes is a medical condition in which patients experience glucose concentration diverging from the normal range of 80–120 mg/dL (4.4–6.6 mM) [7]. In 2019, the International Diabetes Federation (IDF) assessments indicated that approximately 463 million adults have diabetes, and it might rise to 700 million by 2045 [8]. Diabetic patients are required to perform glucose testing several times a day to maintain normal glucose levels. Hence, the rapid quantification of glucose concentration in bodily fluids is vital for diagnosing and treating diabetic patients. For this purpose, the design of an easy, rapid and low-cost technology for the determination of glucose is essential in clinical diagnosis [9,10].

Glucose biosensors have significantly contributed to the detection of glucose levels in diabetic patients [11,12]. Studies have indicated that among various biosensors, glucose oxidase (GOx) enzyme-based electrochemical biosensors were considered to offer good selectivity and sensitivity for glucose detection [13,14]. Amperometry is the widely used electrochemical technique for glucose detection. Amperometric sensors could provide several advantages, such as ease of use, short analysis time, high sensitivity, and higher signal-to-noise ratio compared to other sensors [14–16]. The common idea applied for the development of amperometric biosensors is the efficiency of charge transfer, which can be better enhanced. Additionally, the biocompatibility issues of the sensors could be resolved by modifying electrodes with polymers such as chitosan or hydrogels [17]. In addition, various features of the electrodes could easily be altered by selecting the optimal chemical and electrochemical parameters during the effective electrode modifications [18,19].

The amperometric glucose biosensor generally uses an enzyme glucose oxidase (GOx), which catalyzes glucose oxidation at the electrode and provides high selectivity in glucose detection. Most enzymatic amperometric biosensors are based on disposable screen-printed enzyme electrode strips [20–22]. However, the wastage of materials might occur during the screen-printing process, limiting the applications of screen-printed electrodes.

Graphene, a carbon-based nanomaterial, has gained substantial attention in many areas. In terms of electrochemical properties, graphene could provide high conductivity with a remarkable heterogeneous electron transfer rate [23,24]. In 2014, it was found that polymers such as polyimide (PI) could be directly converted into porous graphene using a CO_2 laser machine with a 10.6 μm wavelength [25]. In addition to infrared CO_2 (10.6 μm) laser, visible laser [26–31] and ultraviolet laser [32] have also been successfully used to synthesize laser-induced graphene (LIG). The laser-irradiation of the PI film caused the photo-thermal generation of the graphene due to the local heating of the film. Upon heating the film, the carbon atoms bonded with oxygen (C–O, C=O) and nitrogen (C–N) atoms via sp^3 and sp^2 hybridization breakdown and rearranged to form several layers of sp^2 hybridized carbon atoms of graphene [25,33]. The laser induction of graphene has been performed in ambient conditions without any material wastage. In addition, the shape/pattern of LIG could also be easily customized by computer design, which holds great promise toward developing glucose biosensors.

Recently, Pereira et al. demonstrated the electrochemical response of GOx adsorbed on a CO_2 laser-scribed LIG [34]. The GOx enzyme adsorbed on LIG remained catalytically active even after running the cyclic voltammetry up to +1.0 V for glucose detection. The LIG electrodes facilitated the direct electron transfer between the GOx and the electrode surface without mediators.

In this study, we fabricated a laser-induced graphene electrode (LIGE) by simple direct laser engraving with the UV laser on polyimide tape. The LIGE surface was immobilized with GOx/Chitosan composite for selective detection on glucose. Amperometric measurement was used to quantify the glucose concentration with the developed LIGE enzymatic biosensor. The novelty of the present work lies in the detection of glucose with enhanced sensitivity using a simple, low-cost LIGE-based biosensor.

2. Materials and Methods

2.1. Chemicals and Instruments

Glucose, uric acid, ascorbic acid, chitosan, and glucose oxidase (GOx, from Aspergillus niger, Type X-S, lyophilized powder, 118,000 units/g solid) were purchased from Sigma-Aldrich Corp. (St. Louis, MO, USA). A single-sided Kapton® polyimide tape with a film thickness of ~30.4 μm and a width of 50 mm was obtained from STAREK Scientific Co., Ltd. (Taipei, Taiwan). Photo/printing paper (HYA300, A4—120 gm^{-2}, 0.15 mm) was purchased from a local book store. All the electrochemical measurements were conducted using a portable potentiostat (PalmSens 4, PalmSens, Houten, The Netherlands). Raman spectroscopic study was conducted using a micro-Raman spectrometer (JASCO NRS-4100; Laser 532 nm) with a spectral resolution of 2 cm^{-1}. Data processing/plotting was performed using Origin 9.1 software (OriginLab Inc., Northampton, MA, USA).

2.2. Fabrication of LIGE Sensor

A 3-electrode system was designed using AutoCAD software with a 3 mm diameter of working electrode and laser-inscribed to graphene-based electrodes. Kapton® polyimide tape was pasted onto a paper substrate and cleaned with isopropanol and deionized water. Then, the designed pattern made in graphic software was inscribed on the surface of the Kapton tape using a laser engraving machine (HANLIN 7WLS, 7 W, 450 nm) to form highly conductive graphene electrodes, as shown in Figure 1. The resistance of the graphene-based electrode was optimized by adjusting the laser power intensity (22% of the machine's maximum power), engraving depth (5%), the distance between the laser head and the polyimide substrate (~13 cm). The duration for fabricating a complete LIGE sensor was 2.8 min.

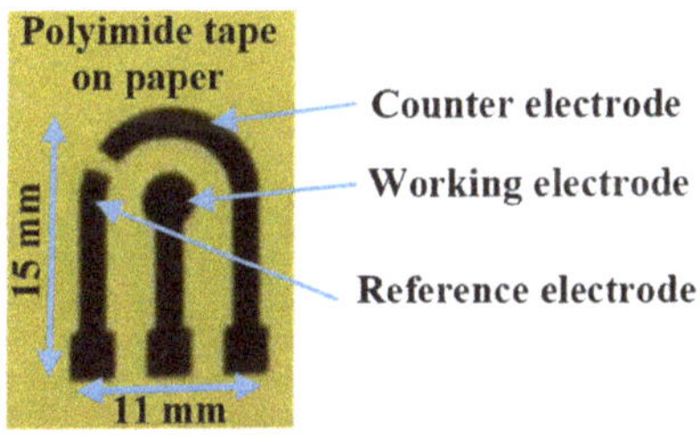

Figure 1. LIG 3-electrode system on polyimide tape fabricated by laser inscribing.

2.3. Immobilization of GOx/Chitosan Composite on the LIGE

The glucose biosensor was prepared by immobilizing the glucose oxidase and chitosan hydrogel homogeneous biocomposite on the LIGE surface. The resulting biocomposite could retain the enzyme bioactivity at considerably extreme conditions [35]. Five milligrams of GOx and three milligrams of chitosan were dissolved in 0.5 mL of deionized water and stirred for 5 min [36]. Subsequently, 5 µL of the mixture was cast onto the surface of the LIGE working electrode. Then, the LIGE sensor was kept in a refrigerator at 4 °C for 24 h.

2.4. Electrochemical Measurements

All the electrochemical measurements were carried out using PalmSens 4 potentiostat (PalmSens, Houten, The Netherlands) at room temperature. The electrochemical redox characteristics of the LIGE were measured by Cyclic voltammetry (CV) with different concentrations of potassium ferri (III)cyanide ($K_3[Fe(CN)_6]$) in 50 mM of phosphate-buffered solution (PBS). CV measurements were performed at a scan rate of 50 mV/s with a potential range from −0.8 to +0.8 V. Chronoamperometry (CA) experiments for glucose detection with LIGE were performed in 50 mM PBS at the fixed applied voltage of 0.8 V for 60 s. The detection principle of glucose is based on the electron transfer mechanism. GOx reacts with glucose in the presence of O_2 and produces gluconolactone and H_2O_2. A change in electrical current occurs at the electrode surface during these reactions due to the electron transfer. Additionally, the resulting current response is proportional to the number of glucose molecules present in the sample.

2.5. Optimization of Applied Potential and pH

CA measurements were used to determine the optimal applied potential and pH for glucose detection. The CA potential was optimized by varying the potential from 0.3 V to 1.3 V (5 mM Glucose, pH 7), and the resulting CA current was sampled at 60 s. CA measurements were performed with a LIGE biosensor at 5 mM glucose solution with the applied potential of 0.8 V by varying the pH of the phosphate-buffered solution from a pH of 5 to 9, and the optimal pH was found.

2.6. Interference Study

The response of the LIGE biosensor for glucose detection was evaluated in the presence of potential interferences such as 0.1 mM ascorbic acid, 0.1 mM uric acid, and 3 mM urea (pH 7; 5 mM glucose; 0.8 V).

3. Results and Discussion

3.1. Characterization of LIGE

3.1.1. Raman Spectra

In this study, a graphene three-electrode system for electrochemical sensing applications was developed by direct laser inscribing on polymer substrate (Polyimide). The prepared LIGE was characterized with Raman spectra, as shown in Figure 2. The Raman spectrum consists of G band at ca. 1592 cm^{-1} related to the E_{2g} phonon of the sp^2 carbon atoms, and D band at ca. 1340 cm^{-1} corresponds to the disordered grain boundaries [37,38]. Two other bands were observed at 2697 and 2900 cm^{-1}. The band at ca. 2700 cm^{-1} is known as the 2D band, an indicator of the number of graphene layers. A sharp peak will appear at ca. 2700 cm^{-1} for monolayer graphene. Here, the broadened band was observed, which would be attributed to the prepared graphene containing many layers with some defects. The band that appeared at 2900 cm^{-1} is called an S3 band, which is a second-order peak derived from the D–G peak combination. The band intensity ratio of S3–2D is proportional to the reduction in defects [38]. This Raman spectra result indicated that the obtained black material on polyimide substrate was carbon-based graphene.

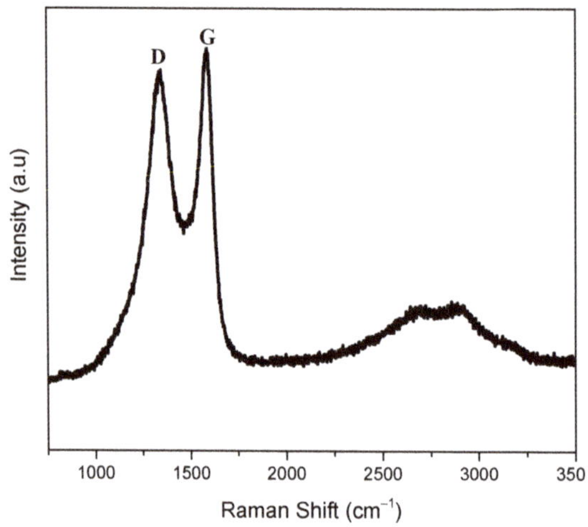

Figure 2. Raman spectra of LIGE.

3.1.2. Electrochemical Characterization

Before developing the glucose biosensor with LIGE, validating the LIGE sensor towards electrochemical sensing was necessary. The ferri/ferrocyanide ($Fe(CN)_6^{3-/4-}$) redox couple is one of the most widely used electron mediators for electrochemical reactions [39]. The performance of an electrochemical sensor towards an electron mediator was considered most relevant to general biochemical sensing applications. Thus, the electrochemical efficacy of the LIGE sensor was evaluated using cyclic voltammetry responses in different concentrations of ferricyanide redox mediator ($K_3[Fe(CN)_6]$), as shown in Figure 3a. As seen from Figure 3a, the oxidation peaks' current increased from 35.495 to 65.043 µA when the ferricyanide concentration ranged from 0.5 to 2.5 mM. The oxidation peak current showed an excellent linear relationship with different ferricyanide concentrations, as shown in Figure 3b. The linear regression equation was $y = 14.54x + 28.69$ $(R^2 = 0.998)$, where y and x are the height of oxidation peak current (μA) and ($K_3[Fe(CN)_6]$) concentration (mM), respectively. The fabricated LIGE provided a favorable response for varying ferricyanide concentrations, indicating excellent electrocatalytic properties. Moreover, the reproducibility of all CV responses was within 5% RSD (relative standard deviation) (n = 4). These results demonstrated the remarkable electrocatalytic response of the fabricated LIGE sensor.

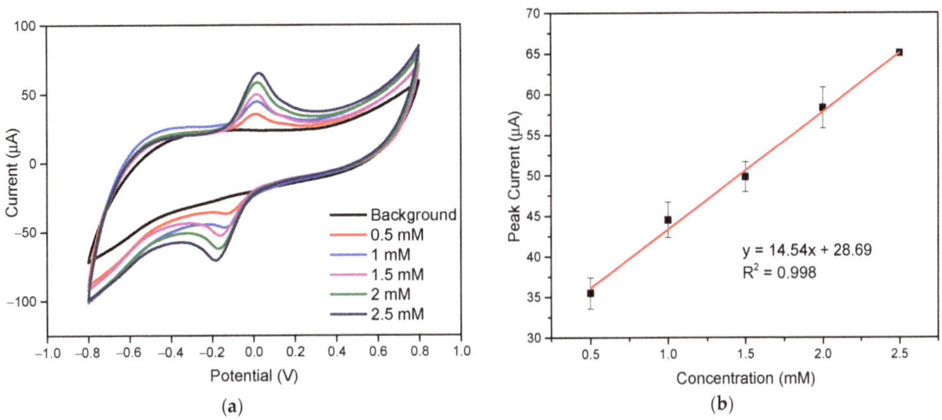

Figure 3. (**a**) CV responses of ferricyanide solutions with varying concentrations; and (**b**) oxidation current peaks vs. concentration. Scan rate was 50 mV s^{-1}.

3.2. Characterization of GOx/Chitosan Immobilized LIGE

Cyclic voltammetry measurement was performed to confirm the LIGE immobilization with GOx/Chitosan. Figure 4 shows the cyclic voltammograms of potassium ferricyanide at bare LIGE and GOx/chitosan composite-modified LIGE. It can be seen that after the immobilization of GOx/chitosan composite onto the LIGE surface, the peak current decreased to 24.325 from 58.336 μA of the bare LIGE. The electron transfer kinetics of $[Fe(CN)6]^{4-}/[Fe(CN)6]^{3-}$ is significantly hindered after the LIGE surface was modified with GOx/chitosan. This result confirmed that the GOx/chitosan was successfully immobilized on the LIGE surface.

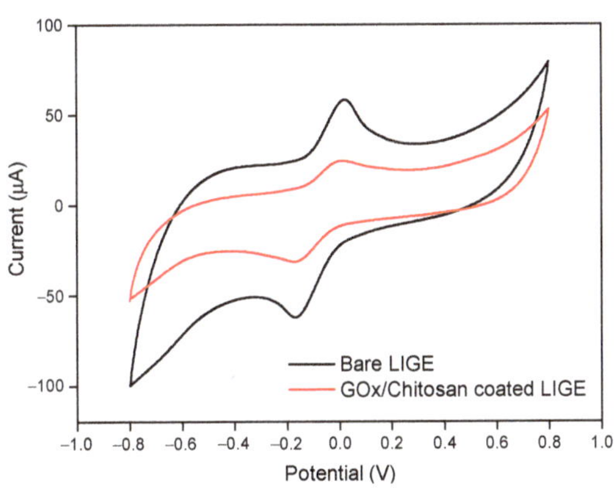

Figure 4. Cyclic voltammograms on bare LIGE and GOx/chitosan-modified LIGE in the presence of 2 mM potassium ferricyanide.

3.3. Amperometric Detection of Glucose by the Proposed LIGE

The chronoamperometry technique was employed to detect glucose using GOx/Chitosan coated LIGE sensor at a constant oxidation potential of +0.8 V. Figure 5a depicts the chronoamperometric responses of the LIGE biosensor with glucose concentrations ranging from 0 to 10 mM. The current response increased with increasing glucose concentrations. The steady-state current response at 60 s was chosen for the detection of glucose concentration. The amperometric current response of the LIGE biosensor exhibited a linear relationship with the glucose concentrations ranging from 0 to 8 mM, and the current began to level off at a glucose concentration higher than 8 mM as shown in Figure 5b. The linear regression equation was $y = 3.05x + 8.54$, with a coefficient of determination $R^2 = 0.97$ and a sensitivity of 43.15 μA mM^{-1} cm^{-2}. The limit of detection was calculated according to the $3s_a/b$ criterion, where b was the slope of the calibration curve, and s_a was the estimated standard deviation of the y-intercepts of the regression line [3]. The detection limit calculated was 0.431 mM. As seen from Figure 5b, the linear part of the calibration curve includes the normal glucose levels (4.4 to 6.6 mM) in the human blood. Thus, this study could offer a simple approach for the clinical glucose measurement with a disposable LIGE-based biosensor. The performance of the proposed biosensor was compared with other reported glucose biosensors, as shown in Table 1. The developed LIGE-based enzymatic glucose biosensor exhibited good analytical characteristics towards glucose detection such as good linearity and high sensitivity. Moreover, the fabrication and detection procedures of the proposed LIGE-based biosensor were also simple, rapid, and cost-effective.

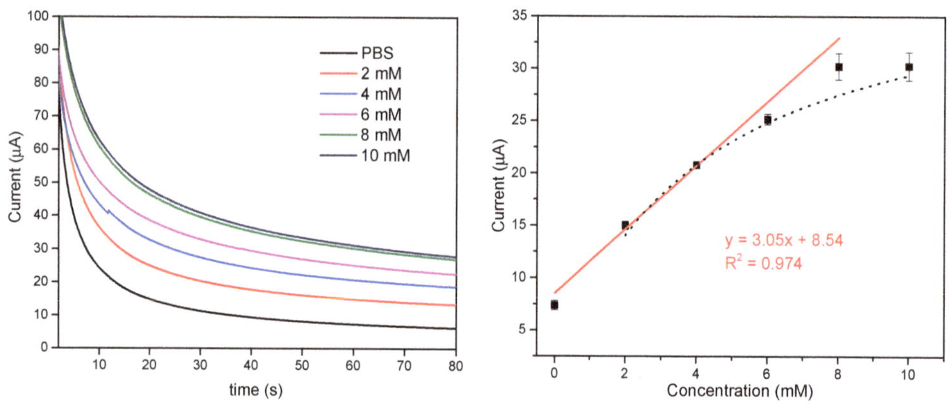

(a) (b)

Figure 5. (**a**) Chronoamperometry response with different concentrations of glucose; (**b**) the relationship between the glucose concentration and the chronoamperometric current response at 60 s.

Table 1. Comparison of the analytical performance of glucose biosensors.

Glucose Biosensor [a]	Sensitivity (μA mM^{-1} cm^{-2})	Linear Range (mM)	LOD (μM)	Reference
GOx/Chitosan-modified LIGE	43.15	0–8	431	This work
GC/MWCNT/Fe_3O_4/PDA–GOx	5.04	2–20	2.25	[40]
LSG/PBSE/PtNPs/GOx	12.64	0.005–3.2	2.57	[41]
MoS_2/Chitosan/GOx-Gelatin/PGE	0.8 (μA mM^{-1})	0.01–0.8	3.18	[42]
CPE/GOx-SiO_2/Lig	0.78	0.5–9	145	[19]
Au–Cys–GA–Gox	2.65	1.5–7	940	[43]
PPy/GOD/SPCE	0.21	0–5	-	[44]

[a] GC—glassy carbon electrode; MWCNT—multi-walled carbon nanotubes; Fe_3O_4/PDA—magnetite/polydopamine; LSG—laser-scribed graphene; PBSE—pyrenebutanoic acid–succinimide ester; PtNPs—platinum nanoparticles; MoS_2—molybdenum disulfide; PGE—pencil graphite electrode; CPE—carbon paste electrode; SiO_2/Lig—silica/lignin; Cys—cysteine; GA—glutaraldehyde; PPy—polypyrrole; SPCE—screen-printed carbon electrodes; GOD/GOx—Glucose oxidase.

3.4. Michaelis–Menten Kinetics

The maximum response current (I_{max}) and the apparent Michaelis–Menten constant (K_m^{app}) were used to analyze the relationship between chronoamperometric signals and enzymatic reaction. As shown in Figure 5b, when glucose concentration exceeds 6 mM, a response plateau was observed with the characteristic of Michaelis–Menten kinetic mechanism. From the calibration plot (Figure 5b dotted line), the current response showed hyperbolic dependence on glucose concentration and was in good agreement with Michaelis–Menten kinetics [45]. The kinetic parameters, the maximum current generated during the enzymatic reaction (I_{max}) and the apparent Michaelis constant (K_m^{app}) are the corresponding a and b parameters of hyperbolic function $y = ax/(b + x)$ [46]. The apparent Michaelis–Menten constant (K_m^{app}) is an indication of enzymatic mimics–substrate kinetics. From the hyperbolic calibration plot (Figure 5b dotted line), the I_{max} and Michaelis constant K_m^{app} were 40.34 μA and 3.75 mM, respectively. The value of K_m^{app} is consistent with the reported value (K_m^{app} = 3.84 mM) for other GOx immobilized on the chitosan complex over triangular silver nanoprisms/platinum biosensor [47].

3.5. Optimization of Applied Potential and Buffer pH

Chronoamperometry measurements were used to determine the optimal applied potential and pH for glucose detection with the developed enzymatic LIGE biosensor. Figure 6a shows the chronoamperometric response of the LIGE biosensor at 60 s with different applied potential values ranging from 0.3 to 1.3 V. The results showed that the current increased with increasing applied potential from 0.3 to 0.8 V and currents tended to level off when the potential increased beyond 0.8 V. Thus, 0.8 V was selected as the optimized potential for amperometric glucose detection. Figure 6b illustrates the chronoamperometry current response of the biosensor as a function of the pH of PBS containing 2 mM glucose. The current responses at pH 5, pH 6, and pH 7 were almost similar. Considering the pH of a physiological buffer, pH 7 was chosen for the glucose detection experiments.

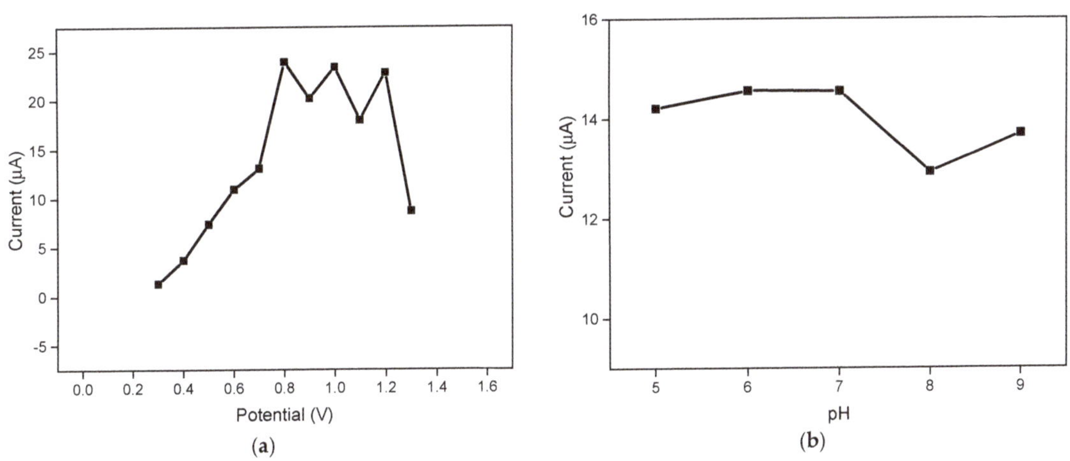

Figure 6. (**a**) Chronoamperometry response at 60 s in different applied potential with 5 mM glucose; and (**b**) Chronoamperometry response at 60 s in different buffer pH with 2 mM glucose.

3.6. Interference Study

The developed LIGE-based enzymatic biosensor was evaluated with possible interferences by comparing the chronoamperometric responses before and after adding some interferents such as ascorbic acid (0.1 mM), uric acid (0.1 mM), and urea (3 mM) in 5 mM glucose. As shown in Figure 7, the chronoamperometric current responses for glucose without and with interferents showed practically no interference. The LIGE was modified with GOx, which is the standard enzyme for biosensors and it has relatively higher selectivity for glucose [48]. Hence, the LIGE biosensor was suggested to possess good selectivity due to the specificity of the GOx enzyme.

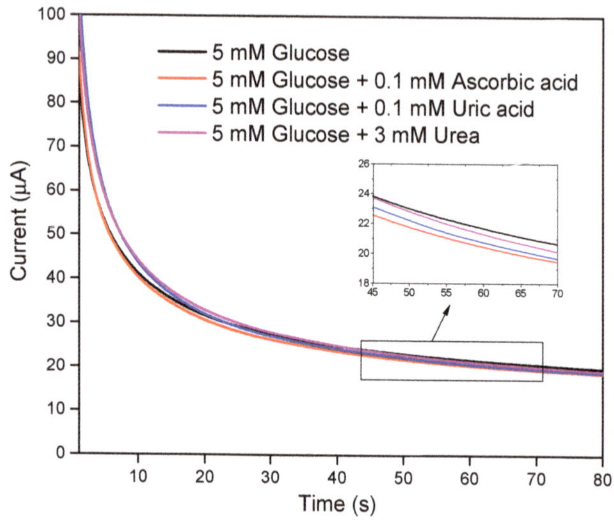

Figure 7. Chronoamperograms of 5 mM glucose with/without interferences. The inset shows the zoomed part of the result from 45 to 70 s.

3.7. Stability and Reproducibility of Biosensor

The stability of the developed GOx/chitosan-modified LIGE biosensor was evaluated by measuring the amperometric current response in the presence of 5 mM glucose over 25 days stored at 4 °C in a refrigerator. The biosensor exhibited ~90% stability for 10 days, and the response remained approximately 72–85% after 10 days. The reproducibility of the developed biosensor was assessed from the current response of different biosensors prepared independently. In this work, all the measurements were taken from at least three independent sensors ($n \geq 3$), and the reproducible signals were obtained with the RSD less than 6%.

4. Conclusions

We developed a simple laser-induced graphene-based enzymatic biosensor for glucose detection. The proposed detection strategy could offer an easy and low-cost route to mass-produce sensitive biosensing electrodes. The chronoamperometric measurements successfully detected the glucose over a linear range from 0 to 8 mM with a detection limit of 0.431 mM. The biosensor response was not affected by interfering compounds (ascorbic acid, uric acid and urea) and demonstrated the high specificity and selectivity of this LIGE biosensor in glucose detection. The proposed LIGE biosensor holds excellent promise in point-of-care diagnosis. Our future study aims to validate the biosensor response in human blood samples for real-life applications.

Author Contributions: K.S.: conceptualization, supervision, validation, writing—review and editing; P.-T.C.: methodology, validation, data curation, writing—original draft; Y.-M.H.: validation, data curation. All authors have read and agreed to the published version of the manuscript.

Funding: This research was supported under Grants MOST 110-2222-E-305-001 and MOST 107-2221-E-305-012-MY3 by the Ministry of Science and Technology, Taiwan, and Grant 109-NTPU_ORDA-F-008 by National Taipei University, Taiwan.

Institutional Review Board Statement: Not applicable.

Informed Consent Statement: Not applicable.

Conflicts of Interest: The authors declare no conflict of interest.

References

1. Ramanavicius, S.; Ramanavicius, A. Charge Transfer and Biocompatibility Aspects in Conducting Polymer-Based Enzymatic Biosensors and Biofuel Cells. *Nanomaterials* **2021**, *11*, 371. [CrossRef] [PubMed]
2. Ramanavicius, S.; Ramanavicius, A. Conducting Polymers in the Design of Biosensors and Biofuel Cells. *Polymers* **2021**, *13*, 49. [CrossRef]
3. Ridhuan, N.S.; Razak, K.A.; Lockman, Z. Fabrication and Characterization of Glucose Biosensors by Using Hydrothermally Grown ZnO Nanorods. *Sci. Rep.* **2018**, *8*, 13722. [CrossRef] [PubMed]
4. German, N.; Ramanaviciene, A.; Ramanavicius, A. Dispersed Conducting Polymer Nanocomposites with Glucose Oxidase and Gold Nanoparticles for the Design of Enzymatic Glucose Biosensors. *Polymers* **2021**, *13*, 2173. [CrossRef]
5. Batool, R.; Rhouati, A.; Nawaz, M.H.; Hayat, A.; Marty, J.L. A review of the construction of nano-hybrids for electro-chemical biosensing of glucose. *Biosensors* **2019**, *9*, 46. [CrossRef]
6. Labib, M.; Sargent, E.H.; Kelley, S.O. Electrochemical Methods for the Analysis of Clinically Relevant Biomolecules. *Chem. Rev.* **2016**, *116*, 9001–9090. [CrossRef] [PubMed]
7. Blunt, B.A.; Barrett-Connor, E.; Wingard, D.L. Evaluation of Fasting Plasma Glucose as Screening Test for NIDDM in Older Adults: Rancho Bernardo Study. *Diabetes Care* **1991**, *14*, 989–993. [CrossRef]
8. *IDF DIABETES ATLAS*, 9th ed.; International Diabetes Federation, 2019; pp. 1–176. Available online: https://diabetesatlas.org/en/resources/ (accessed on 1 July 2021).
9. German, N.; Ramanaviciene, A.; Ramanavicius, A. Formation and Electrochemical Evaluation of Polyaniline and Polypyrrole Nanocomposites Based on Glucose Oxidase and Gold Nanostructures. *Polymers* **2020**, *12*, 3026. [CrossRef]
10. German, N.; Ramanaviciene, A.; Ramanavicius, A. Formation of polyaniline and polypyrrole nanocomposites with em-bedded glucose oxidase and gold nanoparticles. *Polymers* **2019**, *11*, 377. [CrossRef]
11. Gopalan, A.; Muthuchamy, N.; Lee, K. A novel bismuth oxychloride-graphene hybrid nanosheets based non-enzymatic photo-electrochemical glucose sensing platform for high performances. *Biosens. Bioelectron.* **2017**, *89*, 352–360. [CrossRef]
12. Sridara, T.; Upan, J.; Saianand, G.; Tuantranont, A.; Karuwan, C.; Jakmunee, J. Non-Enzymatic Amperometric Glucose Sensor Based on Carbon Nanodots and Copper Oxide Nanocomposites Electrode. *Sensors* **2020**, *20*, 808. [CrossRef]
13. Liu, S.; Ju, H. Reagentless glucose biosensor based on direct electron transfer of glucose oxidase immobilized on colloidal gold modified carbon paste electrode. *Biosens. Bioelectron.* **2003**, *19*, 177–183. [CrossRef]
14. Malitesta, C.; Palmisano, F.; Torsi, L.; Zambonin, P.G. Glucose fast-response amperometric sensor based on glucose oxidase immobilized in an electropolymerized poly(o-phenylenediamine) film. *Anal. Chem.* **1990**, *62*, 2735–2740. [CrossRef]
15. Gorton, L.; Bremle, G.; Csöregi, E.; Jönsson-Pettersson, G.; Persson, B. Amperometric glucose sensors based on immobilized glucose-oxidizing enzymes and chemically modified electrodes. *Anal. Chim. Acta* **1991**, *249*, 43–54. [CrossRef]
16. Koudelka, M.; Gernet, S.; De Rooij, N. Planar amperometric enzyme-based glucose microelectrode. *Sens. Actuators* **1989**, *18*, 157–165. [CrossRef]
17. Tajik, S.; Beitollahi, H.; Nejad, F.G.; Dourandish, Z.; Khalilzadeh, M.A.; Jang, H.W.; Venditti, R.A.; Varma, R.S.; Shokouhimehr, M. Recent Developments in Polymer Nanocomposite-Based Electrochemical Sensors for Detecting Environmental Pollutants. *Ind. Eng. Chem. Res.* **2021**, *60*, 1112–1136. [CrossRef]
18. Andriukonis, E.; Celiesiute-Germaniene, R.; Ramanavicius, S.; Viter, R.; Ramanavicius, A. From microorganism-based amperometric biosensors towards microbial fuel cells. *Sensors* **2021**, *21*, 2442. [CrossRef]
19. Jędrzak, A.; Rębiś, T.; Klapiszewski, Ł.; Zdarta, J.; Milczarek, G.; Jesionowski, T. Carbon paste electrode based on functional GOx/silica-lignin system to prepare an amperometric glucose biosensor. *Sens. Actuators B Chem.* **2018**, *256*, 176–185. [CrossRef]
20. Crouch, E.; Cowell, D.C.; Hoskins, S.; Pittson, R.W.; Hart, J.P. Amperometric, screen-printed, glucose biosensor for analysis of human plasma samples using a biocomposite water-based carbon ink incorporating glucose oxidase. *Anal. Biochem.* **2005**, *347*, 17–23. [CrossRef]
21. Gao, Q.; Guo, Y.; Zhang, W.; Qi, H.; Zhang, C. An amperometric glucose biosensor based on layer-by-layer GOx-SWCNT conjugate/redox polymer multilayer on a screen-printed carbon electrode. *Sens. Actuators B Chem.* **2011**, *153*, 219–225. [CrossRef]
22. Guan, W.-J.; Li, Y.; Chen, Y.-Q.; Zhang, X.-B.; Hu, G.-Q. Glucose biosensor based on multi-wall carbon nanotubes and screen printed carbon electrodes. *Biosens. Bioelectron.* **2005**, *21*, 508–512. [CrossRef]
23. Altuntas, D.B.; Tepeli, Y.; Anik, U. Graphene-metallic nanocomposites as modifiers in electrochemical glucose biosensor transducers. *2D Mater.* **2016**, *3*, 034001. [CrossRef]
24. Pumera, M. Graphene in biosensing. *Mater. Today* **2011**, *14*, 308–315. [CrossRef]
25. Lin, J.; Peng, Z.; Liu, Y.; Ruiz-Zepeda, F.; Ye, R.; Samuel, E.L.G.; Yacaman, M.J.; Yakobson, B.I.; Tour, J.M. Laser-induced porous graphene films from commercial polymers. *Nat. Commun.* **2014**, *5*, 5714. [CrossRef]
26. Zhang, Z.; Song, M.; Hao, J.; Wu, K.; Li, C.; Hu, C. Visible light laser-induced graphene from phenolic resin: A new approach for directly writing graphene-based electrochemical devices on various substrates. *Carbon* **2018**, *127*, 287–296. [CrossRef]

27. Romero, F.J.; Salinas-Castillo, A.; Rivadeneyra, A.; Albrecht, A.; Godoy, A.; Morales, D.P.; Rodriguez, N. In-Depth Study of Laser Diode Ablation of Kapton Polyimide for Flexible Conductive Substrates. *Nanomaterials* **2018**, *8*, 517. [CrossRef]
28. Stanford, M.G.; Zhang, C.; Fowlkes, J.D.; Hoffman, A.; Ivanov, I.N.; Rack, P.D.; Tour, J.M. High-Resolution Laser-Induced Graphene. Flexible Electronics beyond the Visible Limit. *ACS Appl. Mater. Interfaces* **2020**, *12*, 10902–10907. [CrossRef]
29. Tao, L.-Q.; Tian, H.; Liu, Y.; Ju, Z.-Y.; Pang, Y.; Chen, Y.-Q.; Wang, D.-Y.; Tian, X.-G.; Yan, J.-C.; Deng, N.-Q.; et al. An intelligent artificial throat with sound-sensing ability based on laser induced graphene. *Nat. Commun.* **2017**, *8*, 14579. [CrossRef]
30. Bobinger, M.R.; Romero, F.J.; Salinas-Castillo, A.; Becherer, M.; Lugli, P.; Morales, D.P.; Rodríguez, N.; Rivadeneyra, A. Flexible and robust laser-induced graphene heaters photothermally scribed on bare polyimide substrates. *Carbon* **2019**, *144*, 116–126. [CrossRef]
31. Cai, J.; Lv, C.; Watanabe, A. Cost-effective fabrication of high-performance flexible all-solid-state carbon mi-cro-supercapacitors by blue-violet laser direct writing and further surface treatment. *J. Mater. Chem. A* **2016**, *4*, 1671–1679. [CrossRef]
32. Carvalho, A.F.; Fernandes, A.J.S.; Leitao, C.; Deuermeier, J.; Marques, A.; Martins, R.; Fortunato, E.; Costa, F.M. Laser-Induced Graphene Strain Sensors Produced by Ultraviolet Irradiation of Polyimide. *Adv. Funct. Mater.* **2018**, *28*, 1805271. [CrossRef]
33. Tehrani, F.; Bavarian, B. Facile and scalable disposable sensor based on laser engraved graphene for electrochemical detection of glucose. *Sci. Rep.* **2016**, *6*, 1–10. [CrossRef] [PubMed]
34. Pereira, S.; Santos, N.; Carvalho, A.; Fernandes, A.; Costa, F. Electrochemical Response of Glucose Oxidase Adsorbed on Laser-Induced Graphene. *Nanomaterials* **2021**, *11*, 1893. [CrossRef]
35. Luo, X.-L.; Xu, J.-J.; Du, Y.; Chen, H.-Y. A glucose biosensor based on chitosan–glucose oxidase–gold nanoparticles bio-composite formed by one-step electrodeposition. *Anal. Biochem.* **2004**, *334*, 284–289. [CrossRef] [PubMed]
36. Yoon, H.; Nah, J.; Kim, H.; Ko, S.; Sharifuzzaman, M.; Barman, S.C.; Xuan, X.; Kim, J.; Park, J.Y. A chemically modified laser-induced porous graphene based flexible and ultrasensitive electrochemical biosensor for sweat glucose detection. *Sens. Actuators B Chem.* **2020**, *311*, 127866. [CrossRef]
37. Si, Y.; Samulski, E.T. Synthesis of water soluble graphene. *Nano Lett.* **2008**, *8*, 1679–1682. [CrossRef]
38. Johra, F.T.; Lee, J.-W.; Jung, W.-G. Facile and safe graphene preparation on solution based platform. *J. Ind. Eng. Chem.* **2014**, *20*, 2883–2887. [CrossRef]
39. Siprachuabwong, C.; Karuwan, C.; Wisitsorrat, A.; Phokharatkul, D.; Lomas, T.; Sritongkham, P.; Tuantranont, A. Inkjet-printed graphene-PEDOT:PSS modified screen printed carbon electrode for biochemical sensing. *J. Mater. Chem.* **2012**, *22*, 5478–5485. [CrossRef]
40. Kuznowicz, M.; Jędrzak, A.; Rębiś, T.; Jesionowski, T. Biomimetic magnetite/polydopamine/β-cyclodextrins nanocomposite for long-term glucose measurements. *Biochem. Eng. J.* **2021**, *174*, 108127. [CrossRef]
41. Hossain, F.; Slaughter, G. Flexible electrochemical uric acid and glucose biosensor. *Bioelectrochemistry* **2021**, *141*, 107870. [CrossRef]
42. Altuntaş, D.B.; Kuralay, F. MoS2/Chitosan/GOx-Gelatin modified graphite surface: Preparation, characterization and its use for glucose determination. *Mater. Sci. Eng. B* **2021**, *270*, 115215. [CrossRef]
43. Lović, J.; Stevanović, S.; Nikolić, N.D.; Petrović, S.; Vuković, D.; Prlainović, N.; Mijin, D.; Ivić, M.A. Glucose Sensing Using Glucose Oxidase-Glutaraldehyde-Cysteine Modified Gold Electrode. *Int. J. Electrochem. Sci.* **2017**, *12*, 5806–5817. [CrossRef]
44. Weng, B.; Morrin, A.; Shepherd, R.; Crowley, K.; Killard, A.J.; Innis, P.C.; Wallace, G.G. Wholly printed polypyrrole na-noparticle-based biosensors on flexible substrate. *J. Mater. Chem. B* **2014**, *2*, 793–799. [CrossRef] [PubMed]
45. Gao, Z.; Lin, Y.; He, Y.; Tang, D. Enzyme-free amperometric glucose sensor using a glassy carbon electrode modified with poly(vinyl butyral) incorporating a hybrid nanostructure composed of molybdenum disulfide and copper sulfide. *Microchim. Acta* **2017**, *184*, 807–814. [CrossRef]
46. Anusha, J.; Raj, C.J.; Cho, B.-B.; Fleming, A.T.; Yu, K.-H.; Kim, B.C. Amperometric glucose biosensor based on glucose oxidase immobilized over chitosan nanoparticles from gladius of *Uroteuthis duvauceli*. *Sens. Actuators B Chem.* **2015**, *215*, 536–543. [CrossRef]
47. Shi, W.; Ma, Z. Amperometric glucose biosensor based on a triangular silver nanoprisms/chitosan composite film as immobilization matrix. *Biosens. Bioelectron.* **2010**, *26*, 1098–1103. [CrossRef]
48. Bankar, S.; Bule, M.V.; Singhal, R.; Ananthanarayan, L. Glucose oxidase—An overview. *Biotechnol. Adv.* **2009**, *27*, 489–501. [CrossRef]

polymers

Communication

Investigation of the Thermal and Hydrolytic Degradation of Polylactide during Autoclave Foaming

Julia Dreier [1], Christian Brütting [2], Holger Ruckdäschel [2], Volker Altstädt [2] and Christian Bonten [1,*]

[1] Institut für Kunststofftechnik, University of Stuttgart, Pfaffenwaldring 32, 70569 Stuttgart, Germany; julia.dreier@ikt.uni-stuttgart.de
[2] Department of Polymer Engineering, University of Bayreuth, Universitätsstraße 30, 95477 Bayreuth, Germany; christian.bruetting@uni-bayreuth.de (C.B.); ruckdaeschel@uni-bayreuth.de (H.R.); volker.altstaedt@uni-bayreuth.de (V.A.)
* Correspondence: christian.bonten@ikt.uni-stuttgart.de

Abstract: Polylactide (PLA) is one of the most important bioplastics worldwide and thus represents a good potential substitute for bead foams made of the fossil-based Polystyrene (PS). However, foaming of PLA comes with a few challenges. One disadvantage of commercially available PLA is its low melt strength and elongation properties, which play an important role in foaming. As a polyester, PLA is also very sensitive to thermal and hydrolytic degradation. Possibilities to overcome these disadvantages can be found in literature, but improving the properties for foaming of PLA as well as the degradation behavior during foaming have not been investigated yet. In this study, reactive extrusion on a twin-screw extruder is used to modify PLA in order to increase the melt strength and to protect it against thermal degradation and hydrolysis. PLA foams are produced in an already known process from the literature and the influence of the modifiers on the properties is estimated. The results show that it is possible to enhance the foaming properties of PLA and to protect it against hydrolysis at the same time.

Keywords: polylactide; biofoam; hydrolysis; degradation

Citation: Dreier, J.; Brütting, C.; Ruckdäschel, H.; Altstädt, V.; Bonten, C. Investigation of the Thermal and Hydrolytic Degradation of Polylactide during Autoclave Foaming. *Polymers* **2021**, *13*, 2624. https://doi.org/10.3390/polym13162624

Academic Editor: José Miguel Ferri

Received: 29 June 2021
Accepted: 30 July 2021
Published: 6 August 2021

1. Introduction

Degradation describes any type of mechanism in which there is a reduction in molecular weight and consequently a shortening of the polymer chains. These include hydrolysis, enzymatic oxidation, photooxidation and auto-oxidation. Mechanical and thermal as well as UV radiation-induced degradation can also occur. All these processes take place without the presence of microorganisms, which is why they are called abiotic degradation processes. This is their difference from biodegradation, which involves microorganisms. The abiotic degradation processes can lead to fragmentation and the formation of small microparticles, which in turn can be metabolized by microorganisms [1,2].

PLA, as with many other bioplastics, is very sensitive to thermal and hydrolytic degradation, which is characteristic of all polyesters. In this process, cleavage takes place at the hydrolyzable groups, such as esters, by water molecules. The structural formula of PLA is shown below in Figure 1, with the functional groups color-coded.

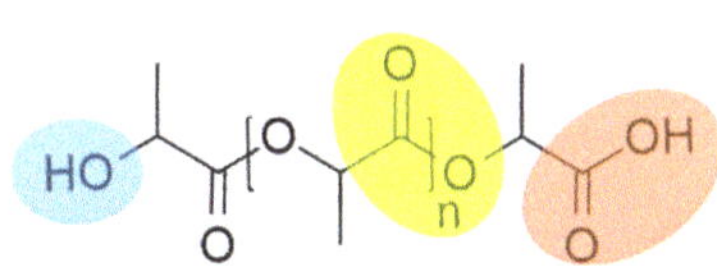

Figure 1. Chemical structure of PLA with its functional groups.

In hydrolysis, a distinction must be made between two different types, acidic and basic ester hydrolysis. Both reactions can occur in PLA and it must be noted whether the reaction takes place within the chain or at the end groups and under which conditions. Hydrolysis is influenced by several parameters, including on the one hand the prevailing environmental conditions such as water activity, temperature, pH and time [3,4]. On the other hand, the degree of crystallization, molar mass, size and geometry of the samples, stereo complex formation, number of acid end groups and hydrophobicity also play a crucial role [5–8]. A special feature in PLA is also the occurrence of the so-called autocatalyzed hydrolysis. The mechanism is similar to that of acid ester hydrolysis. The proton of the carboxyl group catalyzes the hydrolysis reaction. The proton activates the carbonyl group and makes it more susceptible to attack by water molecules. Under neutral conditions, a slower degradation takes place and in alkaline media, a faster degradation takes place than in acidic environments [1,4,9]. Hydrolysis often occurs during processing (for example extrusion or injection molding) under the influence of high temperatures. One way to counteract this is to pre-dry the PLA pellets very well before processing or to add stabilizers during processing [6,7,10]. However, hydrolysis can occur not only at high temperatures, but also at relatively moderate temperatures around 60 °C and under the influence of increased humidity, such as under industrial compost conditions [11].

Hydrolytic degradation does not necessarily end in complete decomposition of the material. However, it must be considered in order to identify and minimize downgrading of the polymer properties. The literature contains numerous papers and patents that counteract undesirable degradation with chemical modifiers [3,6–8,10,12–17]. The substance classes epoxides, carbodiimides and phosphorous acid esters have shown the most promise to date [10,13]. Most of these modifiers react with the end groups of PLA to inhibit the hydrolysis. The determination of the acid value is a method to verify possible reactions between modifiers and PLA and degradation through processing. Although there are some ways to prevent the hydrolysis of PLA, to the best of our knowledge, this was not investigated for foaming. Especially in bead foaming, PLA has to undergo different processing steps, such as compounding, foaming and welding.

Often, biopolymers such as PLA are said to be potential alternatives in packaging applications. Here, expanded PS (EPS) and expanded Polypropylene (EPP) are the market leaders due to their possible complex geometries compared with their low densities. In order to compete with these materials, PLA bead foams have to be made and fused together. Standau et al. [18] showed different ways of producing bead foams. As an example, expanded PLA (EPLA) can be made by a stirring autoclave process described by Nofar [19]. This process is characterized by a water-polymer mixture which is processed at temperatures far above 100°C, which leads to a tremendous degradation and therefore a loss in mechanical stability. To be competitive with polyolefine bead foams, the used polylactides need to be modified in order to prevent degradation during processing.

As a conclusion, PLA suffers from hydrolytic degradation during processing. Therefore, several modifiers have been used to increase the molecular structure and prevent the PLA from degradation during processing. Modified samples have been processed in a stirring autoclave process according to the literature to evaluate the stabilizing effect of the modifiers.

2. Experimental

2.1. Materials

In this study, PLA Ingeo 2003 D from NatureWorks Ltd. (Minnetonka, USA) was modified with two different chemical modifiers. Dicumyl peroxide (DCUP) from Sigma Aldrich was used as a melt strength enhancer and poly carbodiimide from Lanxess, Cologne, Germany as a hydrolysis stabilizer. Both modifiers were incorporated by reactive extrusion with a ZSK 26 twin screw extruder (Coperion GmbH, Stuttgart, Germany). The concentrations were selected based on pre-trials with 0.1 and 0.2 wt% for the peroxide and 0.5 wt% for the poly carbodiimide.

2.2. Methods

2.2.1. Rheological Investigation

Using a Discovery HR-2 plate-to-plate rheometer from TA Instruments, USA (plate diameter 25 mm; plate spacing 1 mm), the linear viscoelastic deformation behavior of PLA and the modified grades was determined. Prior to measurement, the materials were dried overnight at 40 °C in a vacuum oven. The viscosity measurements were then carried out under a nitrogen atmosphere. To determine the thermal stability and to investigate if the chemical reactions were completed, time-sweep measurements were performed at a constant angular velocity of 1 rad/s, a shear stress amplitude of 5 % and a measurement temperature of 180 °C. The measurement time was 1800 s. A frequency sweep was performed to determine the viscoelastic flow behavior. Here, at a shear stress amplitude of 5% and a temperature of 180 °C, the angular velocity is varied from 500 to 0.01 rad/s.

2.2.2. Size Exlcusion Chromatography (SEC)

An Agilent 1260 (Agilent, Waldbronn, Germany), equipped with a PSS-SDV precolumn (10 µm) 50 mm × 8 mm and three linear PSS-SDV columns (10 µm) 300 mm × 8 mm (with pore sizes of 103 Å, 105 Å and 106 Å), from PSS (Mainz, Germany) was used to determine the molar mass. Measurements were made in chloroform at a temperature of 30 °C and a flow rate of 1.0 mL/min. Using a PSS SECurtiy 1260 differential detector (PSS, Mainz, Germany), the molar mass could be calculated. In order to investigate the influence of the modification and the chemical constitution, GPC measurements were performed using a PSS DVD 1260 viscosity detector (PSS, Mainz, Germany).

2.2.3. Acid Value Determination

The acid number was determined according to EN ISO 2114:2000 to verify whether the acid end group had been modified by reactive extrusion. For the subsequent determination of the acid number, 3 or 6 g PLA were dissolved in 100 mL chloroform (Carl Roth, Karlsruhe/Germany) and stirred for at least three hours. The glass electrode (type 6.0229.100 from Methrom (Filderstadt, Germany)) was then placed in the analysis vessel. While stirring, the methanolic KOH solution was added stepwise (0.5 mL steps or 0.2 mL steps around the equivalence point) and the voltage value of the potentiometer ("AL15" from Tintometer GmbH—Lovibond Water Testing, Dortmund, Germany) was noted. The measurement was stopped as soon as the voltage value of the potentiometer changed only slightly after the addition of the methanolic KOH solution. The potentiometer "AL15" was used (Tintometer GmbH—Lovibond Water Testing, Dortmund, Germany).

2.2.4. Foaming

A self-made stirring autoclave was used for processing the polymer granules. Here, the materials were processed as it is known for EPLA bead foams in the literature [19]. Deionized water and the pellets were put in the autoclave. Afterwards, it was sealed and a pressure of 50 bar of CO_2 was applied. The setup was heated up to a temperature of 120 °C, saturated for 30 min and released out of the autoclave. The gained materials after this processing were used to evaluate the degradation behavior of the samples.

3. Results

3.1. SEC Results

The used PLA was modified on a twin-screw extruder with DCUP (0.1 wt%/0.2 wt%) and PDCI (0.5 wt%). After the reactive modification, the effectiveness of the modifiers was verified. One way to investigate this is by the method of SEC, which can be used to determine the molar mass and a molar mass distribution. The principle is based on the separation of molecules according to their molecular size. Figure 2 shows the molar mass and polydispersity index (PDI) of all produced compounds.

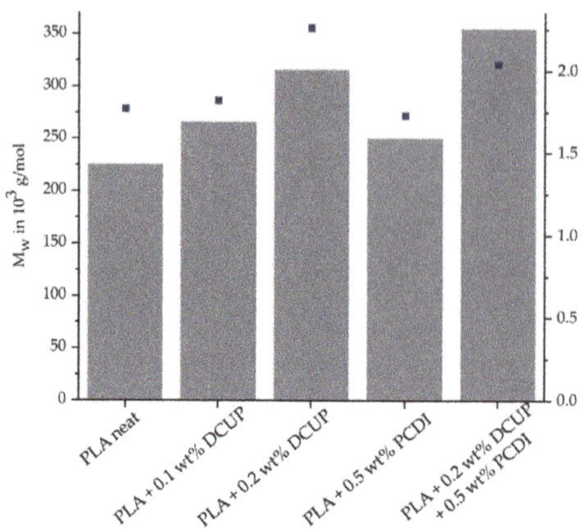

Figure 2. SEC-Results of all PLA compounds.

The organic peroxide led, as well as the PDCI, to an increased molar mass. As the amount of DCUP increases, this effect is enhanced. The addition of PCDI leads also to an increase in molar mass but is less pronounced than DCUP. The reason why the molar mass does not increase that much is that PCDI only reacts with the end groups of PLA. The combination of DCUP and PCDI showed the highest increase in molar mass. The results show on the one hand that a successful modification took place during the reactive extrusion on the twin-screw extruder and thus a chemical reaction between the two components has occurred. Furthermore, the increase in molar mass indicates a change in the linear structure of the PLA, which will be considered in more detail in the following rheological investigations.

3.2. *Rheological Investigation*

Figure 3 shows the measured frequency sweeps of the unmodified PLA compared to the modified ones.

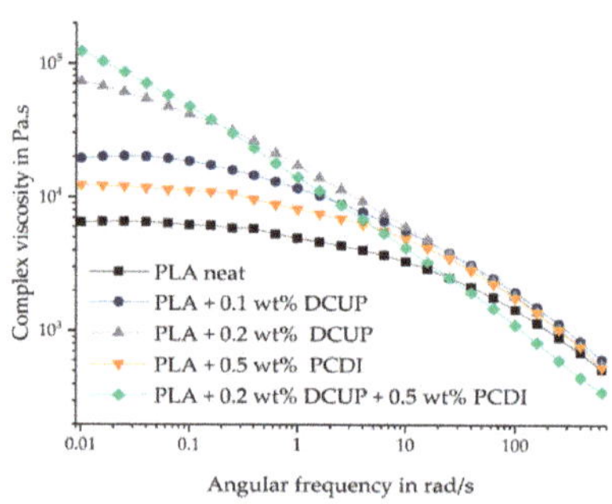

Figure 3. Rheological measurements of the PLA compounds.

All materials show a typical shear thinning behavior. Pure, unmodified PLA shows a chain progression typical for linear polymers. Modification with DCUP leads to an increase in zero viscosity, but the curves still show typical shear thinning behavior and form a Newtonian plateau in the low-frequency range, except PLA with 0.2. wt% DCUP and 0.5 wt% PCDI. With higher amount of DCUP, a higher zero shear viscosity is observed. The increase in viscosity can be explained by the change in the molecular chain structure of the modified PLA. DCUP decomposes during reactive extrusion to form radicals that can attack the PLA chain leading to a chain extension or/and branching. It has already been described in the literature that DCUP can lead to branching or cross-linking of polymer chains [12]. The longer chains and branching in the modified PLA result in entanglements that act like a physical network, leading to longer relaxation times. This leads to the fact that the Newtonian range is only indicated at low frequencies. The addition of 0.2 wt% DCUP is more effective than that of 0.1 wt% because a higher number of possible reactive groups are available which leads to an amount of extensions. PCDI was added as a hydrolysis stabilizer. As seen with the SEC results before, the addition of 0.5 wt% PCDI leads to an increase in molar mass und therefore also to an increase in the complex viscosity compared to neat PLA. Consequently, the combination of DCUP with PCDI results in the highest complex viscosity of all materials and to a change in the curve flow due to a complementation of the chain extension and the prevention to degradation.

Timesweep measurements were used to observe the thermal stability. The thermal stability is crucial for the processing, especially for the foaming process of polymers. In addition, information is obtained about the conversion of the chemical reaction during the reactive extrusion. Degradation reactions as well as post-reactions are generally undesirable and should be avoided or at least taken into account. The normalized complex viscosity as a function of time is shown in Figure 4.

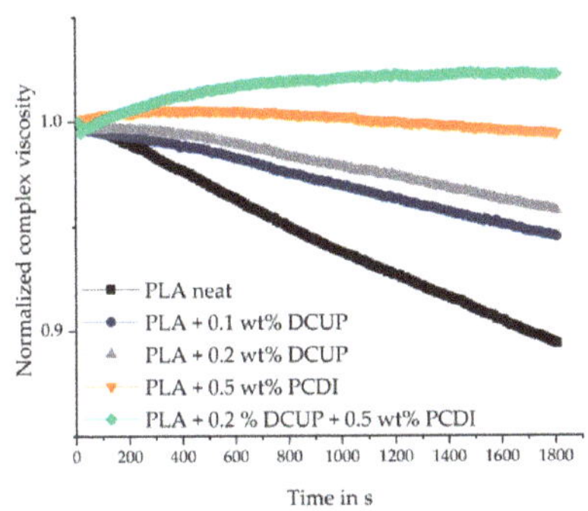

Figure 4. Normalized time-dependent viscosity of the compounds.

The time sweep was done for 30 min at isothermal temperature of 180 °C for all investigated compounds. It can be seen that the neat PLA has the highest degree of degradation, followed by the different peroxide-containing compounds. The PCDI stabilized material and the modified stabilized modifications show the highest relative viscosity indicating no or even a low degradation during the timesweep. After the first 400 s, for all materials except those with PCDI, the normalized complex viscosity decreases due to degradation processes. The use of PCDI led to the desired stabilization process. However, only a slight decrease in complex viscosity for the PLA with 0.5 wt% PCDI can be observed after 30 min. Additionally, the reaction regarding the chain extension proceeded completely and no post-reactions could be observed. The combination of DCUP and PCDI has the highest complex viscosity and is stable over the time of 1800 s.

3.3. Acid Value Determination

The determination of the acid value is a method to verify if chemical reactions at the end groups of PLA occurred. Table 1 shows the results of the acid value of the different PLA compounds before and after the foaming.

Table 1. Determined acid values in mmol/kg.

Compound	Acid Value before Foaming	Acid Value after Foaming
PLA neat	18.78 ± 0.04	30.60 ± 0.13
PLA + 0.1 wt% DCUP	20.80 ± 0.25	31.87 ± 1.20
PLA + 0.2 wt% DCUP	26.63 ± 0.19	34.83 ± 0.31
PLA + 0.5 wt% PCDI	0.72 ± 0.03	2.03 ± 0.98
PLA + 0.2 wt% DCUP + 0.5 wt% PCDI	1.15 ± 0.20	5.67 ± 0.31

The higher the acid value, the higher the amount of carboxyl end groups of PLA and therefore more possibilities for a hydrolytic reaction are given. PCDI can react with these end groups and lead to a small acid value. This means that PCDI is capturing the carboxyl end groups, which hinders the hydrolytic reaction. The results before indicated that a reaction between PLA and PCDI happened. The acid values before foaming were determined after the compounding step. The addition of 0.5 wt% PCDI leads to the smallest acid value. It confirms the other results and the expected reaction of PLA and PCDI. The combination of both modifiers has a higher acid value of 1.15 but it is still in a range where hardly any end groups exist in the whole polymer chains. As a comparison the neat PLA showed an acid value of about 18.8 mmol/Kg. The conditions during the foaming led for all materials to an increase of the acid value indicating a chain scission during the processing. This is not surprising, looking at the processing conditions (PLA pellets in water at 120 °C for 30 min). As it is known from literature, hydrolysis of PLA takes place around 60 °C and under humidity. Therefore, it can be concluded that the processing conditions for the foaming contain the ideal surroundings for hydrolysis.

4. Conclusions

PLA was successfully modified via reactive extrusion. By using 0.2 wt% of an organic peroxide, the molar mass and complex viscosity could be increased. To prevent PLA from undergoing hydrolysis, a stabilizer was used. To receive good foaming properties as well as a thermally and hydrolytically stable PLA, the two modifiers were combined. The combination of 0.2 wt% DCUP and 0.5 wt% PCDI leads to the highest molar mass and the best thermal stabilization. By using the acid value determination, it could be shown that the stabilizer reacts with the end groups of PLA and that several processing steps such as extrusion and foaming can lead to a degradation of PLA. As a result, it is indispensable to prevent PLA from hydrolysis to ensure that there is no downgrade of the properties of PLA. In future works, different modifiers and stabilizers could be investigated and also the following processing steps of bead foams, such as welding.

Author Contributions: Conceptualization, J.D.; methodology, J.D. and C.B. (Christian Brütting); software, J.D., validation, J.D. and C.B. (Christian Brütting); formal analysis, J.D.; investigation, J.D. and C.B. (Christian Brütting); resources, J.D.; data curation, J.D.; writing—original draft preparation, J.D. and C.B. (Christian Brütting); writing—review and editing, J.D. and C.B. (Christian Brütting); visualization, J.D.; supervision, V.A., H.R. and C.B. (Christian Bonten); funding acquisition, V.A., H.R. and C.B. (Christian Bonten). All authors have read and agreed to the published version of the manuscript.

Funding: This research was funded by German Research Foundation (DFG) (392851979).

Institutional Review Board Statement: Not applicable.

Informed Consent Statement: Not applicable.

Data Availability Statement: Not applicable.

Acknowledgments: The authors kindly like to thank the scientific staff members who contributed with inspiring discussions and all technicians and students for their support during the experiments.

Conflicts of Interest: The authors declare no conflict of interest.

References

1. Lucas, N.; Bienaime, C.; Belloy, C.; Queneudec, M.; Silvestre, F.; Nava-Saucedo, J.-E. Polymer biodegradation: Mechanisms and estimation techniques. *Chemosphere* **2008**, *73*, 429–442. [CrossRef] [PubMed]
2. Farah, S.; Anderson, D.G.; Langer, R. Physical and mechanical properties of PLA, and their functions in widespread applications—A comprehensive review. *Adv. Drug Deliv. Rev.* **2016**, *107*, 367–392. [CrossRef] [PubMed]
3. Elsawy, M.A.; Kim, K.-H.; Park, J.-W.; Deep, A. Hydrolytic degradation of polylactic acid (PLA) and its composites. *Renew. Sustain. Energy Rev.* **2017**, *79*, 1346–1352. [CrossRef]
4. Haider, T.P.; Völker, C.; Kramm, J.; Landfester, K.; Wurm, F.R. Kunststoffe der Zukunft? Der Einfluss von bioabbaubaren Polymeren auf Umwelt und Gesellschaft. *Angew. Chem.* **2019**, *131*, 50–63. [CrossRef]
5. Lim, L.-T.; Auras, R.; Rubino, M. Processing technologies for poly(lactic acid). *Prog. Polym. Sci.* **2008**, *33*, 820–852. [CrossRef]
6. Al-Itry, R.; Lamnawar, K.; Maazouz, A. Improvement of thermal stability, rheological and mechanical properties of PLA, PBAT and their blends by reactive extrusion with functionalized epoxy. *Polym. Degrad. Stab.* **2012**, *97*, 1898–1914. [CrossRef]
7. Yang, L.; Chen, X.; Jing, X. Stabilization of poly(lactic acid) by polycarbodiimide. *Polym. Degrad. Stab.* **2008**, *93*, 1923–1929. [CrossRef]
8. Janorkar, A.V.; Metters, A.T.; Hirt, D.E. Modification of Poly(lactic acid) Films: Enhanced Wettability from Surface-Confined Photografting and Increased Degradation Rate Due to an Artifact of the Photografting Process. *Macromolecules* **2004**, *37*, 9151–9159. [CrossRef]
9. Gorrasi, G.; Pantani, R. Effect of PLA grades and morphologies on hydrolytic degradation at composting temperature: Assessment of structural modification and kinetic parameters. *Polym. Degrad. Stab.* **2013**, *98*, 1006–1014. [CrossRef]
10. Ehrenstein, G.W.; Pongratz, S. *Beständigkeit von Kunststoffen*; Edition Kunststoffe; Hanser: München, Germany, 2007; ISBN 978-3-446-21851-2.
11. Mohanty, A.K.; Drzal, L.T.; Misra, M. *Natural Fibers, Biopolymers, and Biocomposites*; Taylor & Francis: Boca Raton, FL, USA, 2005; ISBN 9781280400148.
12. Badia, J.D.; Ribes-Greus, A. Mechanical recycling of polylactide, upgrading trends and combination of valorization techniques. *Eur. Polym. J.* **2016**, *84*, 22–39. [CrossRef]
13. Cicero, J.A.; Dorgan, J.R.; Dec, S.F.; Knauss, D.M. Phosphite stabilization effects on two-step melt-spun fibers of polylactide. *Polym. Degrad. Stab.* **2002**, *78*, 95–105. [CrossRef]
14. Göttermann, S.; Standau, T.; Weinmann, S.; Altstädt, V.; Bonten, C. Effect of chemical modification on the thermal and rheological properties of polylactide. *Polym. Eng. Sci.* **2017**, *57*, 1242–1251. [CrossRef]
15. Karst, D.; Yang, Y. Molecular modeling study of the resistance of PLA to hydrolysis based on the blending of PLLA and PDLA. *Polymer* **2006**, *47*, 4845–4850. [CrossRef]
16. Oliveira, M.; Santos, E.; Araújo, A.; Fechine, G.J.; Machado, A.V.; Botelho, G. The role of shear and stabilizer on PLA degradation. *Polym. Test.* **2016**, *51*, 109–116. [CrossRef]
17. Rasal, R.M.; Janorkar, A.V.; Hirt, D.E. Poly(lactic acid) modifications. *Prog. Polym. Sci.* **2010**, *35*, 338–356. [CrossRef]
18. Standau, T.; Zhao, C.; Castellón, S.M.; Bonten, C.; Altstädt, V. Chemical Modification and Foam Processing of Polylactide (PLA). *Polymers* **2019**, *11*, 306. [CrossRef] [PubMed]
19. Nofar, M.; Ameli, A.; Park, C.B. A novel technology to manufacture biodegradable polylactide bead foam products. *Mater. Des.* **2015**, *83*, 413–421. [CrossRef]

polymers

MDPI

Review

Applications of Biopolymers for Drugs and Probiotics Delivery

Roxana Gheorghita [1,2], Liliana Anchidin-Norocel [1], Roxana Filip [3], Mihai Dimian [2,4] and Mihai Covasa [1,5,*]

1 Department of Health and Human Development, Stefan cel Mare University of Suceava, 720229 Suceava, Romania; roxana.puscaselu@usm.ro (R.G.); liliana.norocel@usm.ro (L.A.-N.)
2 Integrated Center for Research, Development and Innovation in Advanced Materials, Nanotechnologies, and Distributed Systems for Fabrication and Control, Stefan cel Mare University of Suceava, 720229 Suceava, Romania; dimian@usm.ro
3 Hipocrat Clinical Laboratory, 720003 Suceava, Romania; roxana.filip@usm.ro
4 Department of Computers, Electronics and Automation, Stefan cel Mare University of Suceava, 720229 Suceava, Romania
5 Department of Basic Medical Sciences, College of Osteopathic Medicine, Western University of Health Sciences, Pomona, CA 91766, USA
* Correspondence: mcovasa@westernu.edu

Abstract: Research regarding the use of biopolymers has been of great interest to scientists, the medical community, and the industry especially in recent years. Initially used for food applications, the special properties extended their use to the pharmaceutical and medical industries. The practical applications of natural drug encapsulation materials have emerged as a result of the benefits of the use of biopolymers as edible coatings and films in the food industry. This review highlights the use of polysaccharides in the pharmaceutical industries and as encapsulation materials for controlled drug delivery systems including probiotics, focusing on their development, various applications, and benefits. The paper provides evidence in support of research studying the use of biopolymers in the development of new drug delivery systems, explores the challenges and limitations in integrating polymer-derived materials with product delivery optimization, and examines the host biological/metabolic parameters that can be used in the development of new applications.

Keywords: drug; controlled release; polysaccharide; probiotics

Citation: Gheorghita, R.; Anchidin-Norocel, L.; Filip, R.; Dimian, M.; Covasa, M. Applications of Biopolymers for Drugs and Probiotics Delivery. *Polymers* **2021**, *13*, 2729. https://doi.org/10.3390/polym13162729

Academic Editors: José Miguel Ferri, Vicent Fombuena Borràs and Miguel Fernando Aldás Carrasco

Received: 30 June 2021
Accepted: 11 August 2021
Published: 15 August 2021

1. Introduction

Biopolymers are generated by living organisms [1–5] and are defined as biologically degradable polymers [6]. They represent possible materials for the replacement of synthetic plastics due to an increased interest in developing environmental sustainability [7]. Biopolymers have a structural backbone with carbon, oxygen, and nitrogen atoms which makes them easily biodegradable. Biodegradation breaks them down into carbon dioxide, water, humic matter (organic macromolecular material), biomass, and other natural substances; thus, these materials are naturally recycled through biological processes [3].

A classification system based on their origin, synthesis and processing of different biodegradable polymers [8–12] is presented in Figure 1 in the form of a social network analysis. It divides the biopolymers in four major categories: extracted from biomass products (agrobiopolymers), from microorganisms, and from biotechnological and petrochemical products. Biopolymers from biomass products have diverse compounds such as polysaccharides (starches, celluloses, alginates, pectins, gums, and chitosan) [13]; proteins of animal origin (whey, collagen, and gelatin); proteins of vegetal origin (zein, soya, and wheat gluten) [14,15]; and lipids (bees wax, carnauba wax, and free fatty acids) [16,17]. Most biopolymers can be extracted from natural sources such as plants, animals, and microorganisms including algae and agro-wastes [18]. Agro-sources of biopolymers include bananas, maize, potatoes, tapioca, yams, rice, corn, wheat, cotton, sorghum, and barley [19,20], while animal sources are derived from cattle, pigs, and other products. Agro-waste-based sources include apple pomace [21], tomato pomace, pineapple [22], orange and lemon peels, wheat straw, rice husks [23,24], paper wastes, crops, wood, and green wastes, while the marine sources are sponges [25], corals, lobsters, fishes, and shrimps [26]. Biomaterials manufactured from these products are described as stretchy, soft, and gel-like, with many characteristics of both solids and fluids. It is known that biopolymers can be smart and flexible materials even in living organisms [6] because they have a structure that is constantly manipulated either in response to environmental changes or by enzymes throughout different stages of the organism's lifecycle [27].

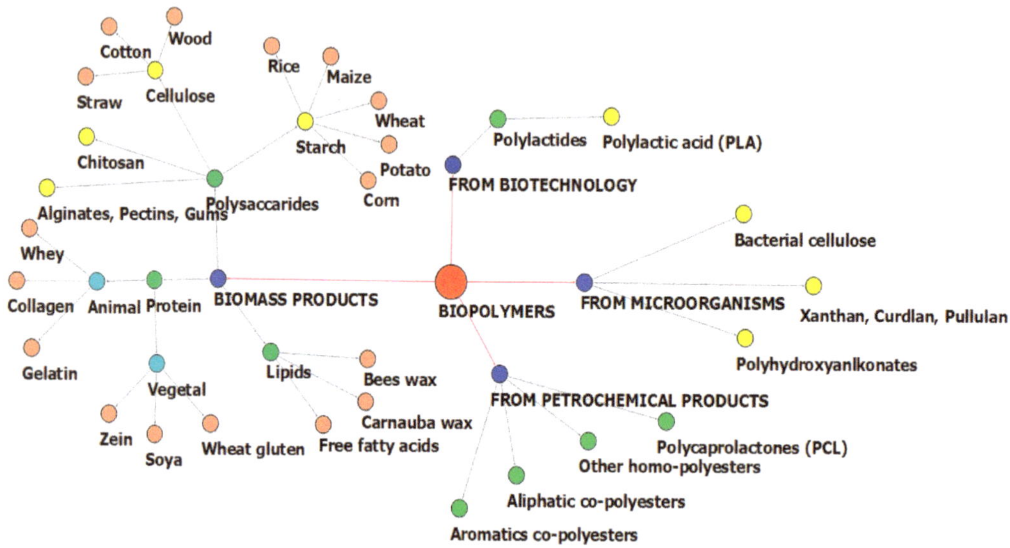

Figure 1. A social network graphical illustration of biopolymers classification.

The biopolymer composites can be prepared by several methods such as extrusion, electrospinning, grafting, different types of molding [28], solvent casting, melt blending, intercalation, filament winding, phase separation [29], laser printing, and film stacking [5].

Currently, the manufactured design and optimization of biopolymers through mathematical models are very advantageous because they improve their physical, chemical, electrical, and mechanical properties in order to increase resistance in humid, warm, or cold storage conditions and for applications that require specific features [30].

Although significant work has been done on the use of a wide range of biopolymers, most of them have been based on polysaccharides due to their improved properties compared to other categories such as proteins or lipids. Thus, this review focuses on the ability of biopolymers to be used successfully in the pharmaceutical industry as encapsulating agents, particularly for delivery of drugs and probiotics. Specifically, the paper describes the use of alginate, chitosan, agar, starch, and cellulose by focusing on the properties and characteristics that make them suitable candidates for product delivery, offering advantages over the chemically derived polymers. The following will also be presented: the development of encapsulated substances based on biopolymers; challenges and limitations such as the encapsulation process, shelf life, controlled release of embedded drugs, protection, and viability of live strains; and the rate of release at different pH mediums of the gastrointestinal fluids.

2. Biopolymers vs. Conventional Synthetic Materials

Several studies have been conducted concerning the utilization of biopolymers with the aim of developing sustainable packaging materials. Although significant improvements have been made, there is still considerable debate over economic considerations, environmental concerns, and product packaging performance [31].

Living organisms produce a variety of polymers as a significant part of their morphological, cellular, and dry matter. These biopolymers play vital roles in the life cycle of organisms including the preservation and expression of genetic information, catalysis of reaction [32], energy or other nutrients, sensing of abiotic and biotic factors, protecting against the attack of other cells, storage of carbon, and negotiation of the adhesion to the surface of other organisms [4].

Biopolymers present important features such as biodegradability [33,34], biocompatibility [35], sustainability [36], bioresorbability [37], flexibility [38], antibacterial activity [6], renewability [39], and stability [2]. They are also less toxic [40], non-immunogenic [41], non-carcinogenic, non-thrombogenic, carbon neutral, and have the advantage of easy extraction [42]. These properties are directly influenced by parameters such as the type of material used as the structural matrix (charge distribution, molecular mass, and conformation), film developing conditions (concentration, pH, solvent, temperature, etc.) and category and concentration of additives (antimicrobials, crosslinking agents, plasticizers, antioxidants, etc.) [43].

Until recently, conventional synthetic materials have become part of most materials in our lives, including those present in beverages and food, clothes, daily used instruments, and baby-toys, and even in biomedical applications such as surgical equipment, drug delivery systems, and cosmetic personal care materials. Some studies have associated these materials with potential adverse health problems, particularly in pregnant women and newborn infants. To this end, hormonally active agents are a group of polymeric chemicals that have been associated with critical health issues such as cancerous tumors, congenital disabilities, and other disorders [40]. Furthermore, people have become more aware of the effects of chemically derived compounds and are more cautious in their use of conventional synthetic materials due to their effects on health and the environment. Today's consumers are more informed and sophisticated in their preferences and choices, increasingly looking for natural and vegan alternate products with high biocompatibility and low environmental implications. Furthermore, increasing efforts and research on the management of plastic waste on Earth are aimed towards finding eco-friendly alternatives to plastics [5]. Such eco-friendly alternatives can be represented by biopolymers, which are disposed in the environment and easily degradable through the enzymatic actions of microorganisms [44].

Compared with conventional synthetic materials that have a simpler and more random structure, these biopolymers are complex molecular assemblies that adopt defined and precise 3D configuration and structures [45]. Based on the composition and chemical structure of biopolymers, they are almost identical to the macromolecules of the native extracellular environment [46]. Many characteristics differentiate between the two types of materials, which are summarized in Table 1. Biopolymers have multiple advantages over conventional plastics due to their low/no toxicity, biodegradability, sustainability, biocompatibility, and extreme hydrophilicity. Furthermore, their morphology and chemical modifications can have a significant impact on their rate of biodegradation [47], an important feature in the development of new applications for food, biomedical, and pharmaceutical industries. Conversely, synthetic materials have a low cost and high thermal and mechanical properties that make them more usable than biopolymers.

Some applications of biopolymers have used mixtures with synthetic materials (such as polyethylene and polyvinyl alcohol), plasticizers (sorbitol and glycerin), nitrogenous bases, and others, thus obtaining a partially biodegradable material [7].

Table 1. Characteristics of biopolymers vs. synthetic polymers.

Characteristic of Materials	Biopolymers	Synthetic Polymers	References
Main source	Agro-resources	Petroleum and gas	[48]
Biodegradability/environmentally friendly	YES	NO/slow	[49,50]
Structure	Well defined	Stochastic	[48]
Chemical backbone structure	Carbon, oxygen, and nitrogen	Mostly carbon	[48]
Dispersity	Unity	>1	[51]
Physicochemical resistance	Low	High	[52]
Toxicity	Low	High	[41]
Thermal stability	Low	High	[52]
Mechanical properties	Low	High	[53]
Sustainability	High	Low	[52]
Availability	High	Decreasing	[52]
Cost	High (depends on the type)	Low	[54]

Although biopolymers have many advantages, there are a number of limitations in their processing, starting from the extraction and all the way up to the final product. First, being a completely natural product, biopolymers' final properties depend largely on the raw material. This can vary greatly due to the origin, climatic conditions, location, harvesting, and processing. Therefore, the world production of biopolymers cannot always maintain the same sustainability. To date, no universal acceptable procedures have been developed for the collection and manufacturing of biopolymer powders from vegetable materials. This is important both for the safety as well as the quality and performance of the final product. Second, because the production of biopolymers is still in its infancy, the production costs are quite high [55]. However, the elimination of recycling and waste taxes through world legislation mitigates some of the high costs. Third, the production of biopolymers necessitates special equipment other than those currently used. The development of such equipment and protocols requires time, additional costs, and trained staff. However, given that biopolymer processing technology is relatively easy and accessible, some existing equipment has been adapted for this purpose, thus reducing the costs [56]. Finally, the lower performance of biopolymers compared to conventional materials may limit their use, although continuous research advances in the technology and material combinations show great improvements in their characteristics that are comparable to conventional materials currently used [57].

3. Applications of Biopolymers

Recent research demonstrates the potential applications of biopolymers as materials for manufacturing medical devices [58]. The most suitable characteristics for suggesting these biomaterials are molecular weight, lubricity, material chemistry, water absorption degradation, shape and structure, solubility [59], hydrophilicity/hydrophobicity [60], erosion mechanism [61], and surface energy [62]. Besides these, other applications of biopolymers, such as those presented in Figure 2, are found in industries such as pharmaceutical preparations with encapsulation; food (edible film packaging and emulsifier) [63]; agriculture, which includes sustainable activities, methods for water recovery, and materials used as soil conditioner; the cosmetics industry (especially hydrogels) [64]; and water treatment substances, biosensors, and even data storage elements [1]. In these industries, polysaccharide-based materials have been developed under different forms such as films, membranes, fibers, hydrogels, food casing [65], sponges, and air gels [66].

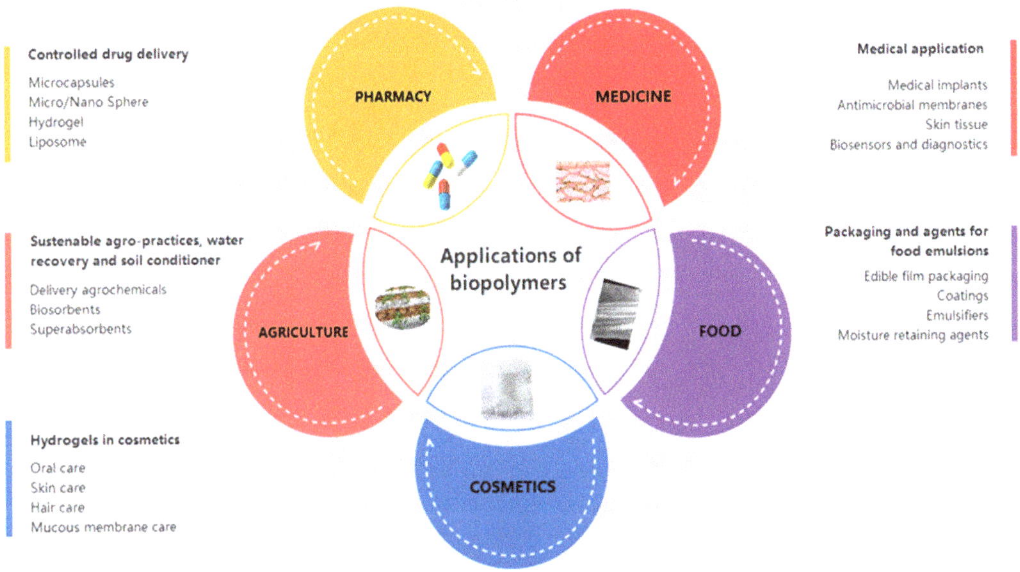

Figure 2. Applications of biopolymers.

Packaging in the bio-medical industry is a method that allows for the closure of a pharmaceutical product from its fabrication to its end use. In pharmaceutical packaging, biopolymers are used to protect pills, nutraceuticals, drugs, surgical devices, powders, and liquids [63]. Pharmaceutical packaging has an impact on the isolation and ensures the safety, identity, and convenience of using the products. Packaging should be compatible with the patient's condition, be free of adverse effects on his/her health, and be environmentally safe [64].

As mentioned above, biopolymers can be also used in the preparation of edible packaging films for food products [65]. These films made from biomaterials can be ingested with the food because they are prepared from polysaccharides and proteins. Edible films have received special attention in the last years because of their alternative potential to replace synthetic materials, which could minimize packaging waste and reduce environmental pollution [66]. As a food packaging material, it can also improve the antimicrobial effect of packaging [67], shelf-life heat resistance, flexibility, mechanical strength, and barrier properties [68]. Edible films/coatings are currently used in a variety of other applications including collagen casings for sausages, chocolate coatings for fruits, and coatings for chocolates and other items [69]. Furthermore, biopolymers are used as emulsifiers and as both thickening and moisture-retaining agents in the food industry with the goal of improving the stability and physicochemical properties of food emulsions [70,71]. Finally, biopolymers have been extensively used in the delivery of bioactive compounds such as probiotics that are susceptible to degradation during preparation, storage, or under the adverse environmental conditions of the human gut. Similarly, they have been used for applications in the pharmaceutical industries as a delivery agent to improve drug stability and bioavailability. In the following sections, we will discuss the applications of biopolymers in drugs and probiotics delivery.

3.1. Biopolymers in the Pharmaceutical Industry

Due to their special properties, biopolymers have slowly begun to replace conventional materials. Whereas in the beginning they were mainly used in the food industry, their application in other related industries took place relatively quickly. In the pharmaceutical industry, they were initially used for the same purpose as in the food industry, which is as thickening and emulsifying agents, host molecules, bulking agents, or fibers. In addition, their use in cosmetics has increased substantially. According to existing data, it is estimated that the world biopolymer market will reach approximately USD 10 billion by the end of 2021, an increase by approximately 17% between 2017 and 2021. The largest market segment is owned by Western Europe with approximately 41.5% of the global market [72]. In biomedicine, polymers have been used successfully both experimentally and in in vivo applications, wound dressing, tissue engineering, drug delivery, or in medical devices such as electronics, sensors, and batteries. Furthermore, due to their physical, thermal, mechanical, and optical properties, biopolymers are ideal materials widely used for food and pharmaceutical applications [5] (Figure 3).

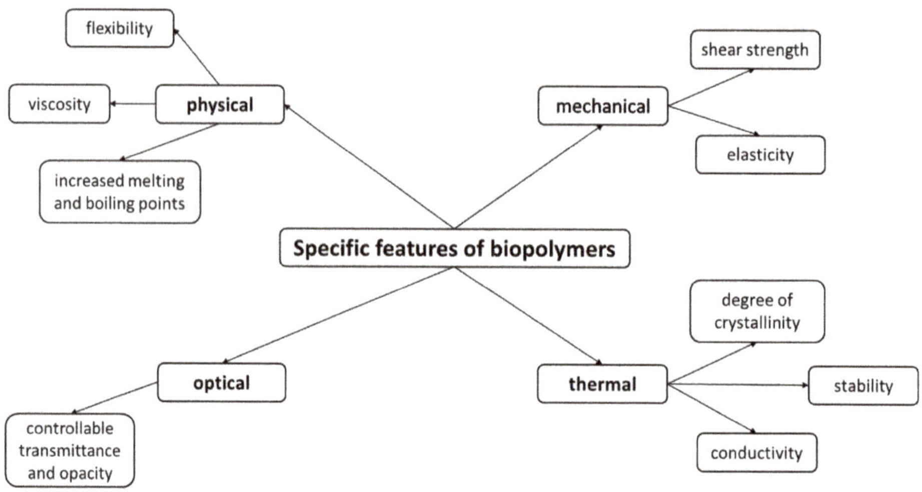

Figure 3. Specific features of biopolymers.

The composition and matrix of biopolymers can be manipulated in order to obtain the appropriate functional properties such as microstructure, permeability, and chargeability that are dependent on the internal structure of the polymer. Electrical characteristics influence the bonding of particles in the biopolymer matrix and their capacity to aggregate. The biopolymer fractions that prevent aggregation are the ones with a high electrical charge [58]. Based on these properties, biopolymers are used successfully to obtain nanoparticles, nanoemulsions, nanogels, or hydrogels with applications in the biomedical industry as carrier systems. Among them, polysaccharides are the most widely used category of biopolymers, either individually or in mixtures with other biopolymers to replace the synthetic materials or exist in addition to them.

3.1.1. Biopolymers for Controlled Drug Release

Encapsulation involves the protection of living cells from destruction by entrapment in biopolymer membranes and it is applied in micro and macrocapsules [73]. It is the procedure by which one or more materials, representing the active part or core material, is embedded or coated with another material or system, which is actually a mantle, shell, carrier, or encapsulant [74] (Figure 4).

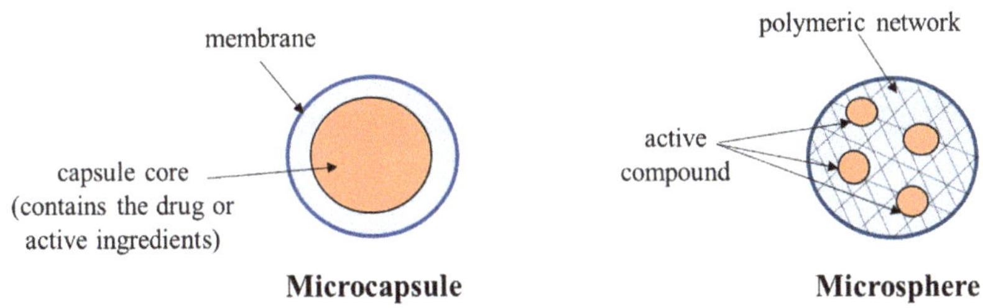

Figure 4. Graphical representation of drug encapsulation (adapted from Madene et al. [74] with permission from the publisher).

A specific feature of macrocapsules is the relatively large difference between the surface area and the volume. Thus, it is necessary to use a large number of nutrients to obtain an appropriate diffusion gradient for the entry of nutrients. This aspect overlaps with the necessary nutrition for the cells. In macrocapsules, living cells are entrapped in large diffusion chambers formed as flat sheets, hollow fibers, and disks with semi-permeable properties [75]. Macrocapsules can be used as intra or extra-vascular devices [76]. In intravascular devices, cells are connected to the bloodstream as a shunt, oriented outside the artificial capillaries. They are found in the vicinity of blood circulation, assisting with the rapid transfer of therapeutic and nutritional substances such as oxygen. The biggest disadvantage is the potential for developing thrombosis. Therefore, research is moving towards their use as extravascular devices with cells entrapped within semi-permeable diffusion chambers and placed transdermally or in the peritoneal cavity without the need for direct circulatory access. This involves a minor surgery and permits a quick and easy substitution in case of graft failure or when the transplant has to be replaced for other reasons [73]. Microcapsules allow for a fast transfer of beneficial substances and accurately mimics the release of substances such as glucose or insulin. Due to their benefits, most studies focused on developing microcapsules with low or non-inflammatory responses. This feature is used successfully in the treatment of endocrine diseases [77].

Many biocompatible polymers have been used as encapsulation materials. For this, a biopolymer must meet certain criteria: (i) stable and not interacting with the drug it contains; (ii) not interfering with the function and cellular viability; non-toxic, inexpensive, and biodegradable; (iii) both the biopolymer and its degradation products must be non-antagonistic to the host; (iv) molecular weight, solubility characteristics, glass transition temperature, microstructure, and chemical functionality should allow for proper drug diffusion and release; (v) biosafe and biocompatible; and (vi) when biocompatibility needs to be improved, the biopolymer should be combined with other compounds for a synergistic effect.

Depending on the mechanism that controls the release of the active agent from the delivery system, the controlled-release modalities may be different. Thus, biopolymer erosion, diffusion, and swelling, followed by diffusion or degradation, may occur [78]. The erosion mechanisms involve: (i) hydrolysis of hydrogels, an important feature for the controlled release of macromolecules; (ii) solubilization of water-insoluble biopolymers by reactions with groups pendant from the polymer covalently bonded atoms; and (iii) cleavage of hydrolytically labile bonds within the biopolymer covalently bonded atoms. The diffusion process occurs when an encapsulated drug or other active agent crosses the outer membrane of the capsule through the biopolymer used for the controlled-release device. In the case of diffusion-controlled systems, the drug delivery system must be stable in the biological environment and must maintain its size and shape through the swelling or degradation [79]. For example, when biopolymers are combined with other bioactive agents, the drug must be able to diffuse through their molecular structure or through pores when it reaches the biological environment. At this stage, it is very important that there are no changes to the biopolymer itself. Swelling-controlled release devices are those systems that, although dry in the initial phase, will swell when they reach the body and come into contact with fluids or water. The swelling ability of the biopolymers can be triggered by changing the environmental conditions of the delivery system. This is one of the most important and useful features of the biopolymers because, by changing the pH or temperature, the release of drugs or incorporated active substances can be controlled [80]. Finally, the biodegradation of a biopolymer in the body is a natural process through which the active ingredient is completely eliminated.

Synthetic polymers have long been of interest for use as encapsulating agents of various therapeutic substances. Although they show improved pharmacokinetics compared to small molecule drugs, their accumulation in the body has raised toxicity issues [81]. With the reorientation of the medical industry towards the use of biopolymers, the major issue is the selection of the right compounds based on the need and desired effects. Not all biopolymers are suitable as encapsulating agents for drugs. It is important that they release the active substance to the target area at the right time in a safe manner and without side effects, especially considering that the predominant routes are oral or intravenous administration [82]. The most used biopolymers and which are the focus of this review are those based on polysaccharides, such as sodium alginate, chitosan, agar, starch, and cellulose. They react synergistically with other biopolymers and polymers, have low toxicity and non-immunogenic behavior, and are compatible with tissues and cells. These polysaccharides are stable in vitro and in vivo, and are used in the development of microcapsules, microspheres, or nanocapsules. When tested in vivo, they showed high biocompatibility and biodegradability, facilitating treatment, minimizing side effects, and improving the health condition. Their high solubility is a plus for their use as disintegrants in water-soluble tablets. For example, when used in tablets, the coating of chitosan and starch improved their visual appearance, protected the drug from degradation, and masked the unpleasant taste of the incorporated substance [83]. When used as capsule material, gelatin was replaced with alginate, a vegan version, or with cellulose, for hard capsules. The main biopolymers that are widely used and presented in this review are alginate, chitosan, agar, starch, and cellulose.

Alginate and Its Use for Drug Delivery

Alginate is probably the most widely used biopolymer in the pharmaceutical and medical industries. This is due to its special encapsulation properties and role in wound healing. It was first isolated in the 1980s and since then it became a multifunctional compound in many applications. Thus, it is obtained at low cost and is a renewable and readily available, biodegradable, non-toxic, biocompatible, and, importantly, mucoadhesive and non-immunogenic compound [84]. It is recognized in the pharmaceutical industry as an excipient and is used to treat reflux esophagitis [85].

Structurally, alginate is a hidrosoluble polysaccharide formed from alternative blocks of 1–4 linked α-L-guluronic acid and β-D-mannuronic acid residues. It contains varying lengths of G-blocks, M-blocks, and/or MG/GM-blocks. High G content alginates have the ability to form stiffer, brittle, and more porous gels, but with increased strength, while high M content alginates tend to obtain more elastic and weaker gels [86].

Alginate is obtained from brown algae and is found as alginic acid sodium, calcium, and magnesium salts. The algae species used for the extraction of trading alginates are *Macrocystis pyrifera*, *Laminaria hyperborea*, *Saccharina japonica*, and *Ascophyllum nodosum*. It can be synthesized by various species of bacteria such as *Azotobacter vinelandii* and various *Pseudomonas* species, although they are not commercially available [72]. Alginate extraction is achieved by a relatively simple process. First, the raw material from the algae is ground and washed with acid followed by the extraction with hot alkali. The alginic acid is obtained after the extract has been filtered, precipitated with calcium, and acidified. The required salt form of alginate is obtained by treating insoluble alginic acid with metallic carbonates, oxides, or hydroxides [87]. Alginate biocompatibility has been extensively studied, with data showing that oral administration of alginate does not trigger many immune responses, in addition to the finding that it is non-toxic and biodegradable [88].

By contrast, intravenous administration of most commercial alginates can lead to adverse body reactions and fibrosis [89].

As an encapsulating agent, alginate was first used in the treatment of diabetes in the encapsulation of pancreatic islet cells [90]. Since then, it has been used for both macro and microencapsulation, and for other endocrine and recombinant cells for the release of therapeutic gene products such as growth hormones or human clotting factor IX (Table 2). It is also used in bioartificial organs such as they kidneys or for the protection of hepatocytes or parathyroids. Alginate gels cannot provide immunoprotection because they are too porous. Therefore, in most applications, alginate gels must be coated with cationic polymers of synthetic origin. For alginate-based coatings, the most used cationic polymers are poly-L-lysine and poly-L-ornithine, but lately polyethylene glycol (PEG), glutaraldehyde, chitosan, and agarose have also been applied. Occasionally, other substances are used to reduce permeability, to ensure mechanical stability, and to increase the durability of the capsules; however, PEG remains the most used coating material. Another way to stabilize alginate gels is through the application of covalent crosslinking molecules, although this method of encapsulation interferes with the functional viability of the cells and can lead to cell toxicity [91].

Table 2. Alginate use for drug delivery.

Biopolymers	Entrapped Substances	Applications	Benefits	References
Alginate	Rifampicin	Drug delivery carriers	Nanoparticles are pH sensitive with the highest release of the active substance occurring at a pH of 7.4. Toxicity and safety tests were excellent with no systemic toxicity after oral administration of nanoparticles.	[92]
Alginate	Ibuprofen	Drug delivery system	Controlled drug release was maintained for 4 h (67.53% of the drug formulation).	[93]
Alginate and methylcellulose	Indomethacin	Drug delivery carrier	Controlled drug release. There was no interaction between the loaded drug and the polymers.	[94]
Alginate	Metformin hydrochloride	Drug delivery system	Good release time; microspheres may be used in the treatment of diabetes.	[95]
Alginate	Diclofenac sodium	Controlled-release microparticles	The drug: polymer (ratio 1:3) was obtained by emulsification and the drug release followed zero order kinetics, optimum for controlled drug release delivery.	[96]
Alginate and sodium carboxymethylcellulose	Ceftriaxone sodium	Multiarticulate beads	The use of the biopolymer matrix decreased drug release in gastric conditions but sustained it at intestinal pH. The beads swelled at pH 1.2 but particle diffusion and erosion occurred at pH 6.8.	[97]
Alginate	Furosemide	Controlled drug delivery beads	Drug release was controlled due to the thicker membrane and reduced beads swelling. Release of Furosemide depends on the conditions of the coating treatment.	[98]
Alginate	Isoniazid	Oral drug delivery	Microspheres were present in the intestinal lumen 4 h after administration and were detectable in the intestine after 24 h of oral administration. Approximately 26% of the drug was released in the gastrointestinal fluid (pH 1.2) in 6 h and 71.25% in the simulated intestinal fluid (pH 7.4) in 30 h.	[99]

Table 2. *Cont.*

Biopolymers	Entrapped Substances	Applications	Benefits	References
Alginate	Nicotinic acid	Aerogels	The release of the drug was prolonged when the core was surrounded by several alginate-based membranes.	[100]
Alginate, carboxymethylcellulose, and chitosan	Amoxicillin	Coated beads	Due to coating, 50% of the drug was released within 4 h. In gastric pH conditions, the drug release was prolonged from 61 min to up to 8 h.	

Thus, based on its characteristics, alginate seems to be the most suitable biopolymer used for drug encapsulation (Table 2). This is due to its specific properties, especially as a matrix for controlled drug delivery devices. In addition, alginate is cheap and readily available, is accepted for consumption in quantum statis doses, is nontoxic, and ensures the protection of the mucous membranes of the upper gastrointestinal tract [101].

Chitosan and Its Use for Drug Delivery

Chitosan is a polysaccharide found in shellfish, fungi, annelids, mollusks, and insects. It is the second most outspread natural polysaccharide on Earth, after cellulose. Commercially, it is produced from chitin, being a poly β (1 → 4) -2-amino-2-deoxy-β-D-glucan deacetylated chitin. It has a strong affinity for polyanions, contains reactive NH_3^+ and OH^- groups, and is soluble in acidic aqueous solutions. It is nontoxic, odorless, bio-compatible, and biodegradable. Due to its antibacterial properties, chitosan is used for microencapsulation, in particular for cells that require a cationic environment. Numerous applications in drug delivery include drug targeting systems for oral, nasal, ocular, and transdermal routes [102]. For this purpose, chitosan has been used in the development of gels, films, oral tablets, beads, and microspheres [103].

The capacity of chitosan as an encapsulating agent is greatly influenced by its molecular weight, degree of deacetylation and crystallinity, and extent of ionization/the free amino group. Thus, when the amino group at the 2-position of glucosamine units of chitosan is the main site for the immobilization of thiol groups, it results in thiolated chitosan. The thiolated chitosan derivatives are chitosan-cysteine, chitosan-thiolactic acid, chitosan-thioglycolicacid, chitosan-homocystenine, chitosan-N-acetylcysteine, and chitosan-glutathione. The thiolated chitosan has been used for anticancer drugs because it offers efficient mucoadhesivity, membrane permeation, and an enhancing capability and improved inhibition for P-glycoprotein [104]. Phosphorylated chitosan and its derivatives have different features such as high hidrosolubility and a metal chelating tendency, used in tissue regeneration, drug delivery intermediates, fuel cells, and in the food industry [105].

Structurally, chitosan is composed of free amine groups in media with a pH over 7.5 and protonated amines are formed in media with a lower pH. These pH-sensitive characteristics make chitosan-based compounds suitable in controlled-release technologies. Under well-established conditions, chitosan microcapsules containing the drug as an active ingredient permits its slow release at the target site [106]. For example, when encapsulated in chitosan, lipophilic drugs were effectively released into the intestinal tract [107]. When used as a vehicle to encapsulate vaccines, it allowed for their controlled release and delivery to targeted sites [108].

There are a number of advantages in using chitosan in the pharmaceutical industry for drug delivery, such as: (i) the controlled release of encapsulated substances; (ii) the elimination of toxic agents in the development process (due to dissolution in aqueous solution); (iii) crosslinking readily available free amino groups; and (iv) improved membrane absorption by mixing cationic chitosan with an anionic material [109]. When used as a coating agent for nanoparticles for the treatment of brain disease, chitosan protects against enzyme degradation, controls release, and improves bioavailability. In addition, it enhances drug permeability across the blood–brain barrier by affecting tight junctions [110].

Equally important are its hemostatic, bacteriostatic, anticholesterol, anticarcinogenic, and fungal characteristics.

In addition, chitosan has good bioadhesive properties and slows down the drug release in the nasal cavity, thus increasing bioavailability and the transfer of drugs from the nasal cavity to the brain [111]. When chitosan was used in membrane development, it increased permeability to acidic drugs. It is insoluble at a pH greater than 6.5 and prevents the burst effect of the release in the first segments of the gastrointestinal tract [112]. It has also been used successfully for antiviral and antibiotic encapsulation, as seen from Table 3.

Table 3. Chitosan use for drug delivery.

Biopolymers	Entrapped Substances	Applications	Benefits	References
Chitosan	Acyclovir	Drug delivery system	The grafting efficiency was 94% and the slow release of the drug was prolonged to 12 h.	[113]
Chitosan	Chlorhexidine diacetate	Buccal tablets	The tablets dissolved more quickly in vitro than chlorhexidine powder and both maintained and even improved the antimicrobial activity of the drug, particularly against *Candida albicans*, due to antimicrobial activity of the polymer itself.	[114]
Chitosan aspartate, glutamate, and hydrochloride	Vancomycin	Peptidic model drug	The sustained release from the microspheres minimized its solubilization in the upper gastrointestinal tract.	[115]
Chitosan	Tetracycline	Controlled drug system	The concentration of drug released was above the minimum limit required for the inhibition of *Staphylococcus aureus*.	[116]
Chitosan and oleic acid	Camptothecin	Controlled drug system	The encapsulation efficiency was about 78%. When its effectiveness in the simulated gastrointestinal fluids was tested, the drug was released slowly into the gastric environment. In intestinal fluids, the drug release was controlled. The drug embedded in chitosan was 75% protected from hydrolysis.	[117]
Chitosan	Satranidazole	Subgingival films for periodontitis	The drug was released for 96 h; the population of gram-positive bacteria was reduced.	[118]
Chitosan and alginate	Amygdalin	Drug delivery system	The controlled amygdalin release was performed for 10 h; the maximum amygdalin released was 70.46% at pH 3.1, 81.86% at pH 5.0, and 86.03% at pH 7.4.	[119]
Chitosan and graphene	Isosfamide	Sustained drug microspheres	The drug diffusion was the most controlled for when isosfamide was entrapped in microspheres.	[120]
Chitosan and xanthan gum	Ciprofloxacin	Controlled-release hydrogel	The entrapment efficiency of the prepared hydrogel increased with the drug increasing concentration and the maximum was reached at 93.8%.	[121]
Chitosan	Interferon-α	Nanoparticles for oral delivery	In mice, the nanoparticles were found in plasma at 1 h after administration, unlike the commercial interferon which could not be detected.	[122]

Chitosan has been extensively used as a matrix for extended drug release, especially due to the simple obtaining procedure, low cost, and biocompatibility. The biocompatibility of chitosan also derives from the fact that it is already part of the human food chain due to its presence in numerous fungi [123]. Chitosan increases the solubility of insoluble drugs when used in mixtures with inorganic nanoparticles, forming a stable complex with safe delivery to the specific site. It was effective when encapsulated hemoglobin, astaxanthin, quercetin, vaccines, or vitamins. Besides its applicability in drug delivery, chitosan is also used in wound dressing, tissue engineering, bioimaging, biosensors, and packaging, among other uses [124].

Agar and Its Use for Drug Delivery

Agar is a long-chain biopolymer obtained from species of algae from the Rhodophyceae class, most commonly found in *Gelidium sp.* and *Gracilaria sp.* It represents the supporting structure of algae and is composed of a mixture of agarose and agaropectin, the gelling and the non-gelling fraction, respectively [125]. Agaropectin is usually removed during processing in order to obtain an agar powder with higher gel strength. Agarose is composed of repetitive units of D-galactose and 3-6, anhydro-L-galactose, linked by alternating α- (1 → 3) and β- (1 → 4) glycosidic bonds.

The ratio of agarose to agaropectin depends largely on seaweed growth, the environmental condition of seaweed growth, extraction methods, and rheological and gelling properties. These changes affect the final mechanical properties of the gels [126]. Agar quality can be significantly improved by modification, which is the most widely used chemical method. It involves hydroxypropylation, acetylation, etherification, and oxidation, the last one being the most commonly used [127]. Due to its gelling capacity, gel reversibility, and high hysteresis, agar is intensely used in various applications, mainly in the food industry, due to its ability to form gel and have an odorless taste. The most important agar evaluation index is gel strength, an important feature for pricing and developing new applications. The easiest way to improve agar characteristics is to remove the sulfate groups with hydrogen peroxide. Thus, after modification, the viscosity, ash content, and sulfate content decreased. Conversely, the gel strength, whiteness, and transparency increased after modification, in contrast to gelling, melting, and dissolving temperatures that decreased after modification [128]. Unlike other biopolymers, agar has been widely used as an encapsulating agent for probiotics since 1988 [129]. The method followed a simple way of encapsulation, which involves the use of drug microparticles and their dispersing at high temperatures in a hydrophilic liquid vehicle. After cooling, due to the transition to ambient temperature, the beads solidify. The same encapsulation method is currently used both for agar and other biopolymers used for this purpose [130]. The method depends on dropping a hot hydrophilic polymeric solution on the top of a cooled organic liquid, such as ethyl acetate which is a non-toxic compound, during which the polymer and the incorporated drug are insoluble. Usually, when only agar is used as an encapsulating agent, the release of the drug occurs in two phases. The first and faster phase leads to the release of 10–20% of the drug, based on the agar content of the beads. The second is a slower and more prolonged phase and becomes even slower as dissolution proceeds. In the first phase, the drug presents in a molecular state on the surface and is released in the outer layer of the bead so that in the second phase, its release is due to dissolution from the solid core. When used as an encapsulating agent, the larger the mass of agar in the beads, the denser the matrix formed and the lower the transfer of drug molecules through the beads. Similarly, the beads that contain a lower percentage of agar in the composition have a higher water content, which explains the rapid rate of drug release [131]. Therefore, agar can be used for the development of sustained-release dosage systems because it is a natural, inert, non-toxic, renewable, biocompatible, and inexpensive material.

Starch and Its Use for Drug Delivery

Starch is one of the most abundant renewable biopolymers on Earth and is non-allergenic, GRAS (generally recognized as safe), and cheap [132]. It is found in peas, corn, rice, wheat, potato, and beans [133]. Starch granules vary in size, shape, particle size distribution, and in the amylose–amylopectin ratio depending on the botanical origin and maturity [134]. The high encapsulation efficiency was reached when the amylose:amylopectin ratio was 25:75 [135]. Starch granules are composed of amylose and amylopectin, free fatty acids and lysophopholipids, proteins, phosphate esters, and water [136]. Amylose is the linear fraction and is composed of glucopyranose units linked by α-(1,4)-glycosidic linkages, while amylopectin is a highly branched polymer with short α-(1,4)-glycosidic chains linked by α-(1,6)-glycosidic branching points [137]. Although amylopectin has a high viscosity and is a good thickening agent, it produces very weak gels with poor mechanical properties [138]. Starch is a biopolymer available in the form of powders, hydrogels, films, and sponges [139]. Due to its low cost, physicochemical features, biodegradability, and biocompatibility, native and modified starch has been widely used in the food, chemical, pharmaceutical, and environmental industries [140].

In the pharmaceutical and medical industries, starch has been used as a pharmaceutical excipient, a tablet super disintegrant (immediate release tablet formulations), and a controlled/sustained-release polymer or as plasma volume expander, useful for patients suffering from trauma, heavy blood loss, or in cancer treatment [141]. Research has focused on the ability of native starch to be dissolved by pancreatic enzymes after oral ingestion, followed by absorption from the small intestine into the systemic circulation. There is also a resistant part of starch that is not digested in the small intestine and is fermented by colonic bacteria. When used as an encapsulant for drugs, it is combined with other biopolymers precisely to limit or attenuate enzymatic degradation in the stomach, thus facilitating the absorption of an adequate amount of the therapeutic agent [142]. In pharmacotherapy, the main objective of such a system is to provide controlled drug release and prevent fluctuations of active substances in the blood in order to maintain drug plasma concentration within the optimal range, in accordance with therapeutic recommendations.

Starch with high crystallinity levels has been explored as an encapsulation matrix. In order to be used successfully in drug delivery and other industries, starch can be modified so that the physicomechanical properties are adjusted to maximize its use. Starch can be modified by chemical, physical, enzymatic, and genetic processes. Of these, chemical process is used most frequently due to its non-disintegrating nature and potential increase in the functionality of the modified starch. The applications of starch as an encapsulating agent of active substances are presented in Table 4.

Table 4. Applications of starch for drug delivery.

Biopolymers	Entrapped Substances	Applications	Benefits	References
Corn starch	Chlorhexidine gluconate	Long-term drug delivery system	In vitro drug release was observed for 21 days and it inhibited *Staphylococcus aureus* growth.	[134]
Corn starch	Clonidine	Transdermal patches	Transdermal patches with a single dose of 30 μg hydrogel had an effect 15 min after application in treated mice.	[143]
Corn starch	Salicylic acid	Hydrogel membrane	The diffusion of the drug through the membrane was 4.11×10^{-6} cm^2/s. The hydrogel was tested as an artificial skin for transferring nutrients or medicines, or for healing substances to the target area.	[144]
Corn starch/ethylene vinyl alcohol blend	Non-steroid anti-inflammatory agent	Drug delivery carriers	In vitro tests showed an immediate burst effect, followed by a slower, controlled release of the drug that lasted up to 10 days.	[145]
Potato starch	Ibuprofen, benzocaine, and sulphapyridine	Starch-based stable carriers	Encapsulation altered starch digestion; resistant starch was available in the colon for fermentation.	[146]
Glutinous rice starch, sodium alginate, and calcium chloride	Metformin hydrochloride	Hydrogel beads for controlled drug delivery	The initial drug entrapment efficiency was very low for the metformin hydrochloride because of its high solubility. Encapsulation improved it when combined with pre-gelatinized starch gel.	[147]
Ensete ventricosum starch	Epichlorohydrin	Drug-release sustaining pharmaceutical excipient	The in vitro drug release profile showed a minimum burst release, followed by a sustained release for 12 h.	[148]
Starch-clay composites	Tramadol	Tablet formulations	The controlled drug release of tramadol from starch-clay biocomposites was achieved in approximately 350 min.	[149]
Starch-chitosan	Hydroxyurea	Cancer therapy	The drug release was sensitive to pH and increased in the acid environment. The drug/starch/chitosan had a toxicity effect and, at certain concentrations, killed cancer cells.	[150]
PVA-corn starch hydrogel	Erythromycin	Wound dressing	The release of erythromycin from the PVA/corn starch network was higher than the drug containing PVA hydrogel (after 1800 min, released 76.7 mg of the total drug).	[151]

Table 4. *Cont.*

Biopolymers	Entrapped Substances	Applications	Benefits	References
Corn starch-sponge matrix	Uranine, indomethacin, and nifedipine	Sustained-release capsule	After intraduodenal administration, 2.5% of the capsule exhibited a sustained release of the drug in the plasma.	[152]
Starch-poly-ε caprolactone	Dexamethasone	Drug delivery and tissue engineering applications	The drug from the outermost layer of the microparticles was quickly released. In vitro tests showed a sustained-release pattern for 30 days.	[153]
High-amylose starch-microcrystalline cellulose	Ranitidine hydrochloride	Gastric-floating drug delivery systems	In vitro tests indicated that the system with 3:7 (wt./wt.) starch/cellulose ratio maintained the buoyancy for more than a day; the drug release was 45.87% in the first hour, followed by a sustained release for up to 10 h.	[154]
Maize starch	Probiotics, e.g., *Lactobacillus plantarum*	Microencapsulated probiotic	In low acid environments, *L. plantarum* encapsulated in the starch matrix was more stable. After simulated digestion and heating treatments, the cells maintained their high viability, unlike formulations with native starch	[155]

Most starch-based drug delivery systems have been developed with starches extracted from potato, maize, corn, cassava, and wheat [118]. As shown in Table 4, starch is a viable source of biopolymer, used as an encapsulating agent for controlled drug delivery systems. In its unmodified form, starch is not as effective as a drug delivery system due to poor mechanical properties, such as low shear stress resistance or high retrogradation and syneresis, thermal decomposition, reduced processability, and solubility in common organic solvents [156]. However, after modification, starch can be used successfully for this purpose. For example, modifying starch in order to obtain resistant starch has led to its use for improving the gut microbiota population with a role in modulating signaling pathways associated with anti-inflammation, anti-diabetes, and anti-obesity [157]. Resistant starch, due to its high amylose content and low amylopectin, has been recognized as a healthy food for humans and animals. It can be considered prebiotic and may reach the colon due to its resistance to digestion by pancreatic enzymes in the small intestine [158]. Therefore, encapsulation has been suggested as the best approach to improve prebiotic–probiotic symbiosis.

Cellulose and Its Use for Drug Delivery

Cellulose, a natural polymer, is the most renewable and abundant polysaccharide. Cellulose has been used as an immunoprotective macrocapsule because it does not form a hydrogel and it is mostly applied in inert diffusion chambers. As an encapsulating agent, it is beneficial for cytotoxic epithelial cells in the treatment of pancreatic cancer, insulin-producing cell lines (HIT-T15), embryonic kidney cells, and hybridoma cells. It is recognized as a new nanovehicle for oral colorectal cancer treatment with high drug release at a neutral pH compared to acid pH, being proposed as a safe oral delivery system for controlled colon cancer treatment [159].

Cellulose is the structural part of the cell wall of green plants, algae, or oomycetes. It is part of the polysaccharide group and is composed of a linear chain of β (1 → 4)-linked D-glucose units. Considering it has an amphiphilic character, it can be used as a surfactant and/or stabilizer at the water–oil interface in pickering emulsions [160]. Cellulose is insoluble in water and most organic solvents. The cellulose derivative, carboxy-methyl-cellulose (CMC), contains carboxymethyl groups bound to the OH-groups of glucopyranose monomers on the cellulose backbone. CMC is mostly applied as a matrix molecule and, in order to ensure mechanical stability and immunoprotection, requires surface coating. CMC has been used as an encapsulating agent for probiotics, but due to the hydrophilicity of the cellulose derivatives, physical degradation occurs when passing through the digestive system. Combined with alginate, it provides a better medium system for probiotics with enhanced tolerance at low pH and a more durable delivery of probiotic cells. Long-term storage stability depends on low water activity and low temperature. The most used dehydration methods to reduce water activity are freeze-drying, spray-drying, vacuum-drying, convective air-drying, and fluidized bed-drying. Among all, freeze-drying is the best method for preserving cells' viability because it reduces the damage to biological structures by eliminating water through sublimation [155].

Cellulose crystals have been used in combination with chitosan to encapsulate vitamin C. Stability of vitamin C is highly dependent on light, pH, and the dissolved oxygen in the environment, but is maintained due to the encapsulation with cellulose and chitosan crystals, and this may be a way to preserve highly unstable compounds during long-term storage in functional systems [161]. Similarly, nanofibrillated cellulose, combined with soybean oil-in-water emulsion and whey protein isolate, was used to encapsulate vitamin D3. Vitamin D3 encapsulation efficiency has improved with increasing emulsifier concentrations. Increasing the concentration of nanofibrillated cellulose has improved the stability and efficiency of encapsulation against environmental stresses (pH changes, salt addition, and thermal processing). The procedure may be the basis for more suitable encapsulation technologies for liposoluble vitamins in emulsion-based food products [162].

In addition to encapsulating vitamins, cellulose and cellulose derivatives have been used as agents to encapsulate drugs and probiotics with active substances. For example, ethyl cellulose nanoparticles have been shown to be effective in encapsulating clarithromycin (3:1 weight ratio of ethyl cellulose:clarithromycin). Once encapsulated, clarithromycin was more effective against *Helicobacter pylori* gastric infections. Tests performed in vivo on laboratory mice have clearly indicated better elimination of bacteria from the stomach by encapsulated clarithromycin compared to the nonencapsulated drug [163]. Ethyl cellulose and microcrystalline cellulose were also used for the encapsulation of antihypertensive drugs. This is important considering that, in the standard method of manufacturing microspheres involving emulsification and solvent evaporation, the solvents used are usually dichloromethane or chloroform, which are hazardous for the environment. Therefore, less toxic substances such as ethyl acetate are used to prepare the microspheres. Furthermore, the drug release from the microspheres is faster than the tableted ones, suggesting that tableting of the microparticulate systems may be optimal [163].

3.2. Biopolymers in Probiotic Encapsulation and Delivery

Due to their special properties, biopolymers have been used to encapsulate probiotics. Probiotics are living organisms with benefits on the hosts' health if ingested in adequate amounts [164]. According to the International Scientific Association for Probiotics and Prebiotics (ISAPP), a sufficient amount of probiotics with a beneficial effect on the hosts' health involves ingesting 1×10^9 CFU per serving [165]. Unlike probiotics, prebiotics are nutrients, usually high-fiber foods, providing the substrate that is selectively utilized by the hosts' microorganisms, conferring a health benefit [165]. Most probiotics in the human body form the commensal intestinal microbiota with a role in increasing resistance to infections and boosting host immune system, glucose and lipid metabolism, degradation of complex carbohydrates, and synthesis of vitamins and bile acid [166]. Although the effects of probiotics on various diseases is still debatable, several studies showed beneficial effects in the treatment and prevention of infectious diseases. For example, strains of *Lactobacillus plantarum*, *Lactobacillus casei*, or *Lactobacillus paracasei* had antifungal, antibacterial, and antioxidant effects. Other strains have been shown to have anti-inflammatory effects, to lessen the risk of osteoporosis, maintain cholesterol levels, and prevent the proliferation of cancer cells [167]. Therefore, due to their beneficial effects and health claims, there has been a worldwide explosion of probiotic-based health products in the form of dietary supplements [168]. As such, the global probiotic market is soon reaching USD 50 billion and, with that, the range of probiotics-containing products and associated health claims continue to expand rapidly. Currently, in Europe, the probiotic market is subject to regulatory requirements and compliance with rules and regulations in order to meet certain standards for product registration and use [169].

The use of encapsulation technologies of probiotics has been intensely studied in order to increase probiotics' viability throughout manipulation, storage, commercialization, and incorporation in food and pharmaceutical products so that these cells are viable during their transit and residence in the gastrointestinal tract. Therefore, improving probiotic survival and resistance to adverse conditions through encapsulation is paramount to their effectiveness in health and disease conditions. To prove their effectiveness, encapsulated probiotic strains have been incorporated into a wide range of food products such as yoghurt [170], cheeses [171], frozen dairy desserts [172], beverages [173], and meat products [174], increasing their therapeutic effects [168]. Once encapsulated, probiotics embedded in food matrices maintained their viability even two months under refrigeration [175]. They can also be mixed in a single microcapsule or in dual core capsules with separation microcompartments [176] and a combination of at least two strains can improve their effect [174]. In order to scale up production, several industrial partnerships between food producers and probiotic companies have been formed. For example, Christian Hansen and Dos Pinos developed probiotic ice cream, Balchem Encapsulates and Rosell Institute developed probiotic raisins and bars, and Dannon uses probiotics encapsulated in their products [177]. Encapsulated probiotics have been effective in irritable bowel syndrome [178], colitis, abdominal pain [179], and other gut or metabolic conditions characterized by microbiota dysbiosis [180].

Several encapsulation methods have been developed and used.

(1) Microencapsulation represents a physicochemical or mechanical process used to trap a substance (active agent) into a coating material (defined as wall material). In this way, spherical particles have a thin and strong but semipermeable membrane with a diameter from nanometers to a few millimeters [174]. The purpose of the procedure is to protect compounds or viable cells against environmental agents that can destroy the core [181].
(2) Spray-drying technique is suitable for industrial applications on a large scale, involving atomization of a liquid mixture and the solvent is evaporated at contact with hot air or gas.
(3) Lyophilization involves freezing the cells with the material used for encapsulation (usually at freezing temperatures), followed by vacuum elimination of water at a pressure between 0.05 to 0.1 mBar and temperature between −50 °C to −30 °C. To preserve and stabilize the activity of lyophilized probiotics, cryoprotectants are added, such as lactose, trehalose, sorbitol, sucrose, milk protein, or skim milk.
(4) Extrusion is the most common technique to use biopolymers as encapsulation materials. The method involves obtaining a hydrocolloid solution, followed by the addition of microorganisms, formation of droplets using a syringe needle (pilot scale) or an extruder (industrial scale), and their release into a hardening solution (typically calcium chloride) [182].
(5) Emulsion is when a small volume of a hydrocolloid suspension containing microorganisms (discontinuous phase) is added to a larger volume of vegetable oil (continuous phase). Using an emulsifier, the mixture is homogenized. After emulsion formation, it can be insolubilized to manufacture gel capsules. The big disadvantage of this method is that the particles obtained vary greatly in shape and size, although bead sizes can be reduced by mechanical homogenization [183].
(6) Spray–freeze-drying is a combined procedure that involves steps used in lyophilization (freeze-drying) and spray-drying. The advantage is that it provides capsules with a controlled size and higher specific surface area, unlike those obtained by spray-drying. The disadvantages of the method refers to high costs (approximately 50 times higher than the classic spray-drying version), long processing times, and the high-energy requirement.

(7) Layer-by-layer is technology based on alternating coating layers of cationic (e.g., chitosan) with anionic (e.g., alginate) biopolymers on cells via electrostatic interaction [184]. It has the advantage of enhanced bacterial viability throughout the gastrointestinal tract, along with the survival of probiotic cells against acidic and bile salt insults, mucoadhesion and growth on intestinal tissues, and in vivo survival [179].

In addition to ensuring cell viability along the gastrointestinal tract, the stability of probiotics during storage is also very important. In this regard, encapsulation has proved to be an effective method. For this, the material used to encapsulate microorganisms is the first and most important factor in maintaining their viability [185]. It improves the survival of probiotics during manufacturing processes, especially heat processing [186] and storage [187]. An important aspect of this process is cytotoxicity. According to ISO 10993-5 [188], a material used for encapsulation is potentially cytotoxic when cell viability decreases below 70% after exposure [189]. In this regard, polysaccharides such as alginate, starch, chitosan, and cellulose, as well as other biopolymers or chemicals, have low or no toxicity and do not affect cell viability. On the contrary, they maintained cellular stability for a long time, particularly when kept in refrigerated or frozen conditions.

Although the encapsulation method has many advantages, there are still several aspects that must be consider. These are: (i) biosafety concerns preventing clinical translation of the cell microencapsulation; (ii) concerns regarding the manipulation and extraction procedures that must be refined in order to be as minimally invasive as possible; (iii) concerns regarding the optimization of cost effectiveness; and (iv) concerns regarding the consideration of internationally accepted regulations for the use of probiotics. Therefore, applications of biopolymers for the coating of encapsulated strains for the purpose of protection in the intestinal gastrointestinal tract or as carriers for direct encapsulation of microorganisms should involve procedures that facilitate high bacterial viability.

3.2.1. Alginate's Use for Probiotic Delivery

Among biomaterials used for encapsulation, alginate is the most widely used due to its strong gelling properties and ability to coat within a short time. Additionally, as a dietary fiber, alginate strengthens the functionality of probiotics used in several diseases such as diabetes or obesity [190]. As seen from Table 5, alginate has proven to be a good microencapsulation agent by extending cell viability in refrigeration and freezing conditions, as well as in adverse gastric and intestinal environments.

Table 5. Utilization of alginate as a probiotic encapsulating material.

Biopolymers	Encapsulated Strain	Encapsulation Method	Benefits	References
Alginate and gelatin	*Lactobacillus rhamnosus*	Extrusion	The cells of *L. rhamnosus* survived in beads with 10^5 CFU/g after four months (initially 10^9 CFU/g).	[191]
Alginate	*Lactococcus lactis* spp. *cremoris*	Extrusion	No release of bacteria in the stomach simulated condition (first 120 min) or the survival in the intestinal fluid until 240 min.	[192]
Alginate	*Bifidobacterium pseudocatenulatum*	Extrusion	None of the uncoated probiotic cells survived after immersion in the simulated small intestine fluid. By contrast, 5.6 log10 CFU/g of viable probiotic cells remained in the tested microgels.	[193]
Alginate	*Staphylococcus succinus* and *Enterococcus fecium*	Extrusion	The encapsulated cells showed 98.75–88.75% of viability in simulated gastric fluids. Survival was constant throughout the storage time and decreased from 8.1 log CFU/mL to 7.9 log CFU/mL after 30 days of storage at 4 °C.	[194]
Alginate and milk	*Lactobacillus bulgaricus*	Extrusion	The viability of the encapsulated probiotic was the same after 120 min of incubation in an acid medium (simulated gastric fluid with pH 2.5). The viability of encapsulated *L. bulgaricus* was kept at 8 log CFU/g after 120 min of incubation at pH 2.0. Stability of the encapsulated probiotic can be preserved for one month after storage at 4 °C.	[195]

Table 5. *Cont.*

Biopolymers	Encapsulated Strain	Encapsulation Method	Benefits	References
Alginate and starch	*Lactobacillus fermentum*	Lyophilization	The survival rate of the probiotic was significantly higher for microparticles blended with starch than those with no starch.	[196]
Alginate, chitosan, and locust beam	*Lactobacillus rhamnosus*	Freeze-drying	In contrast to the alginate-based capsules, the alginate locust beam capsules improved stress tolerance (6× for freeze-drying, 100× for thermotolerance, and 10× for acid).	[197]
Alginate and chitosan	*Saccharomyces cerevisiae* Y235	Emulsification	The viable microencapsulated cells were kept at 7.00 log CFU/g after six months at −20 °C and remained 6.29 log CFU/g after incubation in SGF for 2 h and in SIF for 12 h, reaching the standard value (10^6–10^7 CFU/g).	[198]
Alginate and chitosan	*Bifidobacterium pseudocatenulatum*		The highest stability of *B. pseudocatenulatum* was at the highest concentrations of alginate (4.41 g/100 mL) and chitosan (0.56 g/100 mL). Resistance of alginate–chitosan capsule in SGF was better than in SIF.	[199]
Alginate and chitosan	*Bifidobacterium breve*	Layer-by-layer	Three-layer coated matrix was the best method to increase viability from <3 log CFU/mL, seen in encapsulated cells, up to a maximum of 8.84 ± 0.17 log CFU/mL upon exposure to in vitro gastric conditions. Multilayer-coated alginate released their loads to the intestine with a gradual delivery over 240 min.	[200]
Alginate, starch, and chitosan	*Lactobacillus acidophilus*	Extrusion	Biopolymers ensured better stability of probiotics after exposure to SGF and SIF with 6.35 log CFU/g, while lower counts were noticed for freeze-dried microcapsules. During storage, cell viability of the probiotics stored in the freeze-dried form was up to six logs for 30 days and 135 days in the moist form when kept at room temperature.	[201]
Alginate, chitosan, and xanthan gum	*Lactobacillus plantarum*	Extrusion	Sequential incubation of biopolymers in SGF and SIF facilitated high survival of *L. plantarum* (95%) at pH < 2. Encapsulation improved storage stability of *L. plantarum* at 4 °C.	[202]

Abbreviations: CFU/g, colony-forming unit per gram; CFU/mL, colony-forming unit per milliliter; SGF, simulated gastric fluids; and SIF, simulated intestinal fluids.

Alginate is hemocompatible, does not accrue in organs, is water soluble, biodegradable, and can form gels under mild conditions. It develops gel at ambiental temperatures and prevents the destruction of the activity of thermolabile drugs. By cross-linking with other agents, it forms insoluble gel that delays the drug release. However, alginates have low mechanical properties, therefore they must be reinforced by combining with other biopolymers or with various conjugates in order to obtain both ionically and covalently cross-linkable capsules.

3.2.2. Chitosan Use for Probiotic Delivery

Like other polymers, chitosan has been used to encapsulate probiotics. The best-performing formulas were identified in combinations of chitosan with other biopolymers such as alginate, agar, or gelatin. The most used combination is with alginate, in which chitosan is used as a final layer of microcapsules (Table 6). This is because at pH 7, chitosan that is positively charged develops strong bonds with gelatin and agar, which are negatively charged [203]. What sets chitosan apart is its antibacterial properties as it is a cationic polysaccharide. The disadvantage of chitosan, however, is the need for solubilization in an acidic environment. Usually, acetic acid is used to solubilize the powder and obtain the coating-forming solution without turbidity, which occurs when a compound has not been fully solubilized [204]. Chitosan cannot be used individually as an encapsulating agent with a role in maintaining cell viability. This is because it increases cell membrane permeability, leading, in the end, to cell loss [205]. Due to this, chitosan is mixed with other natural substances when used as an encapsulating agent.

Table 6. Chitosan use for probiotic encapsulation.

Biopolymers	Encapsulated Strain	Encapsulation Method	Benefits	References
Chitosan, agar, and gelatin	*Lactobacillus plantarum*	Emulsification	Particles with a diameter of approximately 6 mm did not solubilize in SGF 20 min after exposure. Cell viability in the biopolymer-free formula decreased completely after 2 h, unlike coated particles whose viability was 9.2 CFU/g after 2 h.	[206]
Chitosan and alginate	vaccine with *Lactobacillus plantarum*	Extrusion	The oral vaccine containing *L. plantarum*, used against spring viremia of carp virus, was effective even after 56 days due to the encapsulation.	[207]
Chitosan and xanthan gum	*Pediococcus acidilactici*	Extrusion	The encapsulated cells maintained their cell viability for 8 h in the gastrointestinal fluid with maximum release occurring after 24 h. The encapsulated cells maintained their viability for three days when tested in deionized water.	[208]
Chitosan and alginate	*Bifidobacterium breve*	Extrusion	In an acidic medium (pH 2), cell viability was maintained for 1 h. As pH increased (4 and 5), cell viability increased to 120 min. After 2 h, the swelling ratio decreased, a sign that the microcapsules began to disintegrate. Chitosan maintained cellular stability at pH 4 and 5, and alginate at pH 2.	[209]
Chitosan and alginate	*Lactobacillus reuteri DSM 17938*	Vibration technology	Unencapsulated cells were more labile to gastrointestinal stress conditions (reduction by 2.09 log cycles after 3 h). The encapsulated ones resisted better with a reduction of 0.82 log cycles.	[210]
Chitosan and alginate	*Saccharomyces boulardii*	Extrusion	Encapsulation of strains with chitosan and alginate facilitated maintenance of cell viability up to 6 h after administration in mice.	[211]
Chitosan and alginate	*L. acidophilus and L. casei*	Extrusion	Galactooligosaccharides potentiated the effect of microencapsulation. Cell viability was reduced by 3.1 logs for *L. acidophilus* and 2.9 logs for *L. casei* when tested at a very low pH (1.55) of SIF.	[212]
Chitosan and alginate	*Bacteria strain 4.1.Z (B. amyloliquefaciens, B. subtilis,* and *B. methylotrophicus)*	Vibration and extrusion	After lyophilization, the microcapsules maintained their viability (10^6–10^7 CFU/g) for about two months under refrigeration. Chitosan maintained the integrity of capsules for 24 h.	[213]
Chitosan and alginate	*Lactobacillus reuteri KUB-AC5*	Emulsification	The viability of non-encapsulated cells decreased in 40 min from 8 logs CFU/mL to <4 log CFU/mL, being completely eliminated after 1 h. The encapsulated cells were much more stable with a reduction of 1 log CFU/mL after 180 min at pH 1.8.	[214]
Chitosan and hydrochloride-alginate	*Bacillus licheniformis*	Orifice-polymerization method	The chitosan coating protected the microcapsules; cell release (6.19 CFU/mL) in 1 h in SGF (pH 2) and 4 h in the simulated intestinal fluid (pH 6).	[215]

Abbreviations: CFU/g, colony-forming unit per gram; CFU/mL, colony-forming unit per milliliter; SGF, simulated gastric fluids; and SIF, simulated intestinal fluids.

Although the effectiveness of chitosan encapsulation has been demonstrated, it does not appear to be the best biopolymer for probiotic encapsulation. Besides the fact that it cannot be used individually and produces turbidity, the obtained microparticles are usually larger in size, more porous, wrinkled [213], sticky, and the aggregation is worsened [211]. However, utilization of chitosan should not be limited as chitosan improves potential bioadhesion and facilitates the controlled release of bacteria [213].

3.2.3. Agar Use for Probiotic Delivery

Agar is one of the polysaccharides intensively used to obtain tablets or other formulas of drugs released in the gastrointestinal tract, but is less used as an encapsulating agent for probiotics. Although there have been attempts to microencapsulate with agar, when used in combination with other biopolymers (Table 6), research on its use is limited. This is due to the higher ability to obtain films and lesser ability to facilitate the development of coatings. These applications are mainly due to the ability of agar to form viscous solutions by solubilizing the powder in water at very high temperatures (over 90 °C) in order to obtain a termoreversible gel. Agar cannot produce gel at lower temperatures and high temperatures affect the viability of microorganisms. To date, no working method that could involve the solubilization of agar powder in liquids with lower temperatures has been developed. When used as an encapsulating material for essential oils, the temperature of the film-forming solution is lowered to 40 °C, after which essential oils are incorporated [216,217]. Agar-based films have low mechanical properties such as low tensile strength and poor elasticity. Therefore, in the development of films, it is preferred to mix it with other polysaccharides, proteins, or lipids.

3.2.4. Starch Use for Probiotic Delivery

Starch can be used as an encapsulating agent but it has weaker characteristics than alginate, chitosan, or cellulose. In the pharmaceutical field, starch is used mainly for encapsulating drugs or active substances when molded into tablets or oral formulations. This is because starch is strongly hydrophilic and is easily dissolved in liquids at ambient temperatures. However, it has the ability to form very small microspheres constituted in resistant aggregates, allowing for a better protection of the core [218]. For example, when used as an encapsulating agent for *L. plantarum*, rice starch maintained cell viability both at 4 °C (refrigeration conditions) and at temperatures above 50 °C. However, as an encapsulating agent, starch proved to be more effective when mixed with other compounds. For example, the combination of starch and alginate resulted in microcapsules with increased probiotics resistance to simulated gastric conditions (Table 7).

Table 7. Starch as an encapsulation material for probiotics.

Biopolymers	Encapsulated Strain	Encapsulation Method	Benefits	References
Rice starch	*Lactobacillus casei, Lactobacillus brevis,* and *Lactobacillus plantarum*	Extrusion	The viability of encapsulated cells (8.27/8.46/7.65 log CFU/g) was kept constant for two months at refrigeration. In contrast, non-encapsulated cells lost their viability by approximately 3 log CFU/g during storage.	[219]
Starch and pectin	*Lactobacillus plantarum*	Extrusion	Cell viability was reduced from 10 log CFU/g to 1 log CFU/g for free cells maintained for 2 h in gastric conditions (pH 1.5–3). Cells encapsulated in pectin had higher viability (4.6 log CFU/g) but the best protection was observed with the addition of starch to which the viability increased to 6.94 log CFU/g.	[220]
Starch from corn and rice	*Lactobacillus plantarum*	Freeze-drying	Encapsulated cells showed thermal stability and maintained their integrity for 35 min at 55 °C. Unencapsulated cells subjected to the same treatment lost their viability by 63% after only 10 min of exposure to 55 °C.	[221]
Starch, alginate, chitosan, and inulin	*Lactobacillus casei* and *Bifidobacterium bifidum*	Emulsification	Encapsulated *L. casei* and *B. bifidum* lost their viability when subjected to simulated gastric conditions for 120 min. Cell viability decreased from 25.10×10^{10} CFU/mL to 6.30×10^{6} CFU/mL for *L. casei*. Encapsulated *B. bifidum* lost 4.65 log/mL of the bacterial culture, while the unencapsulated form had undetectable cell viability after 90 min.	[222]
Starch and alginate	*Lactobacillus fermentum*	Emulsification	Encapsulated in the matrix, cells maintained viability when stored at 4 °C for 45 days. In environmental conditions, however, cells showed a decrease of 1.7 log after 24 h, with complete loss after 2 weeks.	[196]

Table 7. *Cont.*

Biopolymers	Encapsulated Strain	Encapsulation Method	Benefits	References
Starch	*Lactobacillus paracasei*	Electrospinning	Tested at different storage temperatures (4, 25, and 37 °C), *L. paracasei* cells maintained their initial viability of 13.6 × 10 CFU/mL when stored for three weeks at 4 °C and 25 °C but not at 37 °C. Unencapsulated cells lost about 90% of their viability regardless of the storage temperature.	[223]
Maize starch, maltodextrin, and gum arabic	*Lactobacillus acidophilus*	Spray-drying	After 30days of storage at room temperature, only strains encapsulated with maltodextrin, namely gum arabic, maintained their cell viability of 10^6 CFU/g. After 60 days, no encapsulating material prevented the loss of cell viability. Of the tested coatings, starch least protected the bacterial strains.	[224]
Taro and rice starch	*Lactobacillus paracasei*	Spray-drying	When stored, the taro–starch encapsulated strains were more stable; cells maintained their viability for a month, both at temperatures of 4 °C and 25 °C.	[218]
Cassava starch and alginate	*Lactobacillus brevis*	Emulsification	Encapsulation efficiency was higher than 89%. In gastrointestinal conditions, cell viability was better for microcapsules than free cells (96.07% compared to 76.51%). After 5 h of maintenance in the same conditions, viability of *L. brevis* encapsulated cells was 8.69 log CFU/mL, unlike the non-encapsulated ones with 6.87 log CFU/mL.	[225]
Starch and alginate	*Lactobacillus casei*	Extrusion	The addition of 2% starch to the alginate-based film-forming solution increased cell viability from 4 × 10^8 to 3.1 × 10^{11}. Increasing starch did not change the results. Tested under simulated gastrointestinal conditions, cell viability was maintained for up to 6 h.	[226]

3.2.5. Cellulose Use for Probiotic Delivery

Carboxymethyl cellulose (CMC) is one of the most widely used forms of cellulose due to the fact that it is the most affordable in terms of spread and cost. Although it is a compound that prevents lipid oxidation and reduces oxygen permeability due to the small number of hydroxyl groups in the structure, CMC is a highly water-soluble compound. This limits its use as an encapsulation material for probiotics as it is degraded in the digestive system. Therefore, when used in combination with other biopolymers such as polysaccharides (carrageenan, alginate, chitosan, and starch), proteins (gelatin), or other natural compounds (inulin), it increased the viability of encapsulated anaerobic probiotics by 36% [227]. The microencapsulation characteristics of CMC have been improved due to the addition of other substances such as gelatin or carrageenan. In general, the encapsulated probiotics maintained their viability for 120 min in simulated gastric conditions, regardless of the type of probiotic encapsulated. When lyophilization was used as the encapsulation method, cellulose and alginate maintained the viability of *L. plantarum* for approximately 160 days in refrigeration conditions (Table 8).

Table 8. Cellulose as a probiotic encapsulating material.

Biopolymers	Encapsulated Strain	Encapsulation Method	Benefits	References
CMC and gelatin	*Lactobacillus rhamnosus*	Emulsification	After 120 min of exposure to SGF and SIF, cell viability was maintained at approximately 77.5% (4 log CFU/mL) in capsules and 60% (5 log CFU/mL) in free cells.	[228]

Table 8. *Cont.*

Biopolymers	Encapsulated Strain	Encapsulation Method	Benefits	References
CMC and κ-carrageenan	*Lactobacillus plantarum*	Extrusion	Cellular stability was greatly improved for encapsulated samples: in an acidic medium (pH 2), it decreased from 10 log CFU/g to 0 after 90 min (non-encapsulated cells) and to about 8 log CFU/g after 120 min (encapsulated cells). During storage for 30 days at 4 °C, cell stability changed from 10 log CFU/g to 2 log CFU/g (free cells) and from 10 log CFU/g to 7 log CFU/g (encapsulated cells).	[229]
Cellulose and pectin	lactic acid bacteria	High-pressure microfluidization	Viability of non-encapsulated cells decreased from 9.56 to 5.29 log CFU/mL in an acid medium, while encapsulation protected cells (decrease of 1.88 log CFU/mL after 2 h in the same conditions).	[230]
CMC and inulin	*Lactobacillus plantarum*	Casting	Cell viability decreased during storage whether or not probiotics were encapsulated.	[227]
CMC and rice bran	*Lactobacillus reuteri*	Emulsification	After heat exposure (85 °C, 25 s), cell viability decreased by more than 57%, although *L. reuteri* is a thermotolerant bacterium. However, the survival rate of encapsulated cells was approximately 6 log CFU/g.	[231]
CMC and chitosan	*Lactobacillus rhamnosus*	Extrusion	Microencapsulated strains were stable at pH 2–4; at the highest pH value tested (12.5), all microcapsules disintegrated.	[232]
Cellulose, alginate, starch, and lecithin	*Lactobacillus rhamnosus*	Extrusion	Under gastric conditions, viability of encapsulated cells was 37% higher than that of free ones. Encapsulation had a positive effect on storage, in which viability decreased by 1.23 log (25 degrees) and 1.08 log (4 degrees), unlike free cells in which stability decreased by 3.17 and 1.93.	[233]
Cellulose and alginate	*Lactobacillus plantarum*	Extrusion and lyophilization	Lyophilized encapsulated cells showed the best stability in the simulated gastrointestinal conditions: gradual release of 2.6×10^6 CFU/mL for 210 min. When refrigerated, encapsulated cells maintained viability for up to 160 days.	[155]
Cellulose and alginate	*Lactobacillus plantarum*	Extrusion-dripping	After 120 min in SGF, viability of non-encapsulated cells decreased by 66.6%, while encapsulated strains had a 58.4% better viability than that of free cells. The addition of cellulose protected the capsules from the action of pH.	[234]

Abbreviations: CMC, carboxymethyl cellulose; SGF, simulated gastric fluids; SIF, simulated intestinal fluids; CFU/g, colony-forming unit per gram; and CFU/mL, colony-forming unit per milliliter.

4. Challenges and Limitations

The rapid adoption of the use of biopolymers is still hindered by several factors. First, research thus far has been conducted primarily in vitro; therefore, more in vivo and clinical trials are needed to demonstrate the health benefits of biopolymers and the biocompatibility in various biomedical applications, in particular when used as encapsulation materials for drug delivery. When used for the treatment of various diseases, more studies are needed to assess the appropriate compound: either alone or in combination to achieve the desired payload in a highly regulated and site-specific manner at therapeutically relevant concentrations [227].

A second challenge is to obtain materials with properties similar or better to synthetic products by improving end-use mechanical properties, kinetics and release, thermal resistance, and barrier properties. For example, some products exhibit low mechanical properties, rapid degradation, and high hydrophilic capacity especially in humid or adverse environments, rendering their application unviable.

Although numerous studies have examined the use of encapsulated probiotics, there is a need for in depth interdisciplinary research that includes microbiologists, medical doctors, and biomaterial, food, agro and chemical engineers. This will lead to better and more efficient prototypes of probiotic encapsulating formulations, to the identification of the most specific/effective probiotic strains, and to the most suitable polymeric carriers applied for product manufacturing. In addition, this will lead to the optimization of the entire process based on the natural characteristics and sensitivity of the selected strain, and will identify ways on how to develop the best formula based on in vitro, in vivo, and pre-clinical techniques considering the release, manufacturing, packaging, transportation, and storage of capsules. Third, the challenges related to costs, economic aspects, and the gap between policy and implementation of the new technologies on a global level need to be addressed in this rapidly emerging field.

5. Conclusions

Although research in this area has revolutionized the biomedical and pharmaceutical industries, significant work still lies ahead if we are to effect changes not only at the individual level but also to bring sustained and affordable environmental changes. Biopolymers have proven effective as encapsulation materials for controlled drug release systems. Significant progress has been made on the biocompatibility, biodegradability, and mechanical and thermal properties of the materials involved. However, challenges still remain in developing target-specific carriers that are biocompatible with various delivery routes for providing sustainable release at the target site. Strategies have been proposed to improve stability of polymers, product kinetics, and release time as well as clinical efficiency. For biomedical applications, it is important to develop uniform guidelines for polymer applications in order to improve versatility and safety and avoid contamination. From the evidence presented thus far, it is obvious that among biopolymers, polysaccharides-based applications are the most used in the field due to their protective, physicochemical, and low immunogenicity characteristics. Although they present some limitations, the ability to react synergistically with other biopolymers or other natural or synthetic substances make their applications widely used. More studies evaluating the technical parameter optimization, efficiency of encapsulation with different formulations, and product-loading capacity concerning viability and metabolic activity should be undertaken. For example, studies examining the functional interactions between the polymer networks and the coating materials, in order to improve capsule stability, product metabolic activity, release time, and viability, should be high on the list. Notwithstanding current limitations from the host perspective, the use of polymers, particularly polysaccharides-based, will continue to expand with an eye towards improving polysaccharide-drug interactions, the optimization of pharmacokinetics and pharmacodynamics, and the compatibility of the polysaccharide with the target tissue. Nevertheless, while more research is needed, polymer-based applications are of great benefit for delivering small molecules that are highly effective, biopotent, and safe.

Author Contributions: All authors have contributed equally to this work. All authors have read and agreed to the published version of the manuscript.

Funding: The work was supported by the project titled "The analysis of interrelationship between gut microbiota and the host with applications in the prevention and control of type 2 diabetes", co-financed by the European Regional Development Fund through Competitiveness Operational Program under the contract number 120/16.09.2016.

Institutional Review Board Statement: Not applicable.

Informed Consent Statement: Not applicable.

Data Availability Statement: The data presented in this study are available upon request from the corresponding author.

Conflicts of Interest: The authors declare no conflict of interest.

References

1. Rebelo, R.; Fernandes, M.; Fangueiro, R. Biopolymers in medical implants: A brief review. *Procedia Eng.* **2017**, *200*, 236–243. [CrossRef]
2. Bala, I.A.; Abdullahi, M.R.; Bashir, S.S. A review on formulation of enzymatic solution for biopolymer hydrolysis. *J. Chem.* **2017**, *6*, 9–13.
3. Yadav, P.; Yadav, H.; Shah, V.G.; Shah, G.; Dhaka, G. Biomedical biopolymers, their origin and evolution in biomedical sciences: A systematic review. *J. Clin. Diagnostic Res.* **2015**, *9*, 21–25. [CrossRef]
4. Aggarwal, J.; Sharma, S.; Kamyab, H.; Kumar, A. The Realm of Biopolymers and Their Usage: An Overview. Available online: http://www.jett.dormaj.com (accessed on 24 June 2021).
5. Udayakumar, G.P.; Muthusamy, S.; Selvaganesh, B.; Sivarajasekar, N.; Rambabu, K.; Banat, F.; Sivamani, S.; Sivakumar, N.; Hosseini-Bandegharaei, A.; Show, P.L. Biopolymers and composites: Properties, characterization and their applications in food, medical and pharmaceutical industries. *J. Environ. Chem. Eng.* **2021**, *9*, 105322. [CrossRef]
6. Pattanashetti, N.A.; Heggannavar, G.B.; Kariduraganavar, M.Y. Smart biopolymers and their biomedical applications. *Procedia Manuf.* **2017**, *12*, 263–279. [CrossRef]
7. Rendón-Villalobos, R.; Ortíz-Sánchez, A.; Tovar-Sánchez, E.; Flores-Huicochea, E. The role of biopolymers in obtaining environmentally friendly materials. In *Composites from Renewable and Sustainable Materials*; InTech: Houston, TX, USA, 2016.
8. Nath, K.; Bhattacharyya, S.K.; Das, N.C. Biodegradable polymeric materials for EMI shielding. In *Materials for Potential EMI Shielding Applications*; Elsevier: Amsterdam, The Netherlands, 2020; pp. 165–178.
9. Kumar, S.; Thakur, K. Bioplastics—Classification, production and their potential food applications. *J. Hill Agric.* **2017**, *8*, 118. [CrossRef]
10. Prajapati, S.K.; Jain, A.; Jain, A.; Jain, S. Biodegradable polymers and constructs: A novel approach in drug delivery. *Eur. Polym. J.* **2019**, *120*, 109191. [CrossRef]
11. Arora, S. Biopolymers as Packaging Material in Food and Allied Industry Value Addition of Makhana and Its by-Products View Project. 2018. Available online: https://www.researchgate.net/publication/342765641 (accessed on 1 February 2021).
12. Díez-Pascual, A.M. Synthesis and applications of biopolymer composites. *Int. J. Mol. Sci.* **2019**, *20*, 2321. [CrossRef] [PubMed]
13. Qasim, U.; Osman, A.I.; Al-Muhtaseb, A.H.; Farrell, C.; Al-Abri, M.; Ali, M.; Vo, D.V.N.; Jamil, F.; Rooney, D.W. Renewable cellulosic nanocomposites for food packaging to avoid fossil fuel plastic pollution: A review. *Environ. Chem. Lett.* **2021**, *19*, 613–641. [CrossRef]
14. Hamouda, T. Sustainable packaging from coir fibers. In *Biopolymers and Biocomposites from Agro-Waste for Packaging Applications*; Elsevier: Amsterdam, The Netherlands, 2021; pp. 113–126.
15. Shivam, P. Recent developments on biodegradable polymers and their future trends. *Int. Res. J. Sci. Eng.* **2016**, *4*, 17–26.
16. Chen, H.; Wang, J.; Cheng, Y.; Wang, C.; Liu, H.; Bian, H.; Pan, Y.; Sun, J.; Han, W. Application of protein-based films and coatings for food packaging: A review. *Polymers* **2019**, *11*, 2039. [CrossRef]
17. Swain, S.K.; Pattanayak, A.J.; Sahoo, A.P. *Functional Biopolymer Composites*; Springer: Berlin, Germany, 2018; pp. 159–182.
18. Varma, K.; Gopi, S. Biopolymers and their role in medicinal and pharmaceutical applications. In *Biopolymers and Their Industrial Applications*; Elsevier: Amsterdam, The Netherlands, 2021; pp. 175–191.
19. Carvalho, A.J.F. Starch: Major sources, properties and applications as thermoplastic materials. In *Monomers, Polymers and Composites from Renewable Resources*; Elsevier: Amsterdam, The Netherlands, 2008; pp. 321–342. ISBN 9780080453163.
20. Temesgen, S.; Rennert, M.; Tesfaye, T.; Nase, M. Review on spinning of biopolymer fibers from starch. *Polymers* **2021**, *13*, 1121. [CrossRef] [PubMed]
21. Gustafsson, J.; Landberg, M.; Bátori, V.; Åkesson, D.; Taherzadeh, M.J.; Zamani, A. Development of bio-based films and 3D objects from apple pomace. *Polymers* **2019**, *11*, 289. [CrossRef]
22. Liu, W.; Misra, M.; Askeland, P.; Drzal, L.T.; Mohanty, A.K. "Green" composites from soy based plastic and pineapple leaf fiber: Fabrication and properties evaluation. *Polymer* **2005**, *46*, 2710–2721. [CrossRef]
23. Maraveas, C. Production of sustainable and biodegradable polymers from agricultural waste. *Polymers* **2020**, *12*, 1127. [CrossRef] [PubMed]
24. Heredia-Guerrero, J.A.; Heredia, A.; Domínguez, E.; Cingolani, R.; Bayer, I.S.; Athanassiou, A.; Benítez, J.J. Cutin from agro-waste as a raw material for the production of bioplastics. *J. Exp. Bot.* **2017**, *68*, 5401–5410. [CrossRef] [PubMed]
25. Khrunyk, Y.; Lach, S.; Petrenko, I.; Ehrlich, H. Progress in modern marine biomaterials research. *Mar. Drugs* **2020**, *18*, 589. [CrossRef]
26. Kaur, S.; Dhillon, G.S. The versatile biopolymer chitosan: Potential sources, evaluation of extraction methods and applications. *Crit. Rev. Microbiol.* **2014**, *40*, 155–175. [CrossRef]
27. Torres, F.G.; Troncoso, O.P.; Pisani, A.; Gatto, F.; Bardi, G. Natural polysaccharide nanomaterials: An overview of their immunological properties. *Int. J. Mol. Sci.* **2019**, *20*, 5092. [CrossRef]
28. Velu, R.; Calais, T.; Jayakumar, A.; Raspall, F. A comprehensive review on bio-nanomaterials for medical implants and feasibility studies on fabrication of such implants by additive manufacturing technique. *Materials* **2020**, *13*, 92. [CrossRef]
29. Chaitanya, S.; Singh, I. Processing of PLA/sisal fiber biocomposites using direct- and extrusion-injection molding. *Mater. Manuf. Processes* **2017**, *32*, 468–474. [CrossRef]

30. Khanna, S.; Srivastava, A.K. A simple structured mathematical model for biopolymer (PHB) production. *Biotechnol. Prog.* **2005**, *21*, 830–838. [CrossRef] [PubMed]
31. Ncube, L.K.; Ude, A.U.; Ogunmuyiwa, E.N.; Zulkifli, R.; Beas, I.N. Environmental impact of food packaging materials: A review of contemporary development from conventional plastics to polylactic acid based materials. *Materials* **2020**, *13*, 4994. [CrossRef]
32. Wankhade, V. Animal-derived biopolymers in food and biomedical technology. In *Biopolymer-Based Formulations: Biomedical and Food Applications*; Elsevier: Amsterdam, The Netherlands, 2020; pp. 139–152. ISBN 9780128168981.
33. Parker, G. Measuring the environmental performance of food packaging: Life cycle assessment. In *Environmentally Compatible Food Packaging*; Elsevier: Amsterdam, The Netherlands, 2008; pp. 211–237. ISBN 9781845691943.
34. Shankar, S.; Rhim, J.-W. Bionanocomposite films for food packaging applications. In *Reference Module in Food Science*; Elsevier: Amsterdam, The Netherlands, 2018.
35. Hassan, M.E.; Bai, J.; Dou, D.Q. Biopolymers; Definition, classification and applications. *Egypt. J. Chem.* **2019**, *62*, 1725–1737. [CrossRef]
36. Soldo, A.; Miletić, M.; Auad, M.L. Biopolymers as a sustainable solution for the enhancement of soil mechanical properties. *Sci. Rep.* **2020**, *10*, 1–13. [CrossRef] [PubMed]
37. Song, J.; Winkeljann, B.; Lieleg, O. Biopolymer-based coatings: Promising strategies to improve the biocompatibility and functionality of materials used in biomedical engineering. *Adv. Mater. Interfaces* **2020**, *7*, 2000850. [CrossRef]
38. Gabor, D.; Tita, O. Biopolymers used in food packaging: A review. *Acta Univ. Cibiniensis Ser. E Food Technol.* **2012**, *16*, 3–19.
39. Sadasivuni, K.K.; Saha, P.; Adhikari, J.; Deshmukh, K.; Ahamed, M.B.; Cabibihan, J.J. Recent advances in mechanical properties of biopolymer composites: A review. *Polym. Compos.* **2020**, *41*, 32–59. [CrossRef]
40. Jummaat, F.; Bashir Yahya, E.; Khalil HPS, A.; Adnan, A.S.; Mohammed Alqadhi, A.; Abdullah, C.K.; Sofea, A.A.; Olaiya, N.G.; Abdat, M.; Hps, K. The role of biopolymer-based materials in obstetrics and gynecology applications: A review. *Polymers* **2021**, *13*, 633. [CrossRef]
41. Reddy, M.S.B.; Ponnamma, D.; Choudhary, R.; Sadasivuni, K.K. A comparative review of natural and synthetic biopolymer composite scaffolds. *Polymers* **2021**, *13*, 1105. [CrossRef]
42. Olivia, M.; Jingga, H.; Toni, N.; Wibisono, G. Biopolymers to improve physical properties and leaching characteristics of mortar and concrete: A review. In *IOP Conference Series: Materials Science and Engineering*; Institute of Physics Publishing: Bristol, UK, 2018; Volume 345, p. 012028.
43. Gurgel Adeodato Vieira, M.; Altenhofen da Silva, M.; Oliveira dos Santos, L.; Masumi Beppu, M. Natural-based plasticizers and biopolymer films: A review. *Eur. Polym. J.* **2011**, *47*, 254–263. [CrossRef]
44. Polman, E.M.N.; Gruter, G.J.M.; Parsons, J.R.; Tietema, A. Comparison of the aerobic biodegradation of biopolymers and the corresponding bioplastics: A review. *Sci. Total Environ.* **2021**, *753*, 131953. [CrossRef]
45. Mohan, S.; Oluwafemi, O.S.; Kalarikkal, N.; Thomas, S.; Songca, S.P. Biopolymers—Application in nanoscience and nanotechnology. In *Recent Advances in Biopolymers*; InTech: Houston, TX, USA, 2016.
46. Imre, B.; Pukánszky, B. Compatibilization in bio-based and biodegradable polymer blends. *Eur. Polym. J.* **2013**, *49*, 1215–1233. [CrossRef]
47. Jayanth, D.; Kumar, P.S.; Nayak, G.C.; Kumar, J.S.; Pal, S.K.; Rajasekar, R. A review on biodegradable polymeric materials striving towards the attainment of green environment. *J. Polym. Environ.* **2018**, *26*, 838–865. [CrossRef]
48. Cziple, F.A.; Velez Marques, A.J. Biopolymers Versus Synthetic Polymers. Available online: https://creativepegworks.wordpress.com/2015/12/01/biopolymers-vs-synthetic-polymers/ (accessed on 26 June 2021).
49. Simionescu, B.C.; Ivanov, D. Natural and synthetic polymers for designing composite materials. In *Handbook of Bioceramics and Biocomposites*; Springer: Berlin, Germany, 2016; pp. 233–286. ISBN 9783319124605.
50. Gowthaman, N.S.K.; Lim, H.N.; Sreeraj, T.R.; Amalraj, A.; Gopi, S. Advantages of biopolymers over synthetic polymers. In *Biopolymers and Their Industrial Applications*; Elsevier: Amsterdam, The Netherlands, 2021; pp. 351–372.
51. Luh, T.-Y.; Yang, H.-C.; Lin, N.-T.; Lin, S.-Y.; Lee, S.-L.; Chen, C.-H. OMCOS for functional polymers-double-stranded DNA-like polymers. *Pure Appl. Chem.* **2008**, *80*, 819–829. [CrossRef]
52. Colmenares, J.C.; Kuna, E. Photoactive hybrid catalysts based on natural and synthetic polymers: A comparative overview. *Molecules* **2017**, *22*, 790. [CrossRef] [PubMed]
53. Ryder, K.; Ali, M.A.; Carne, A.; Billakanti, J. The potential use of dairy by-products for the production of nonfood biomaterials. *Crit. Rev. Environ. Sci. Technol.* **2017**, *47*, 621–642. [CrossRef]
54. Maraveas, C. Environmental sustainability of greenhouse covering materials. *Sustainability* **2019**, *11*, 6129. [CrossRef]
55. Mtibe, A.; Motloung, M.P.; Bandyopadhyay, J.; Ray, S.S. Synthetic biopolymers and their composites: Advantages and limitations—An overview. *Macromol. Rapid Commun.* **2021**, *42*, 2100130. [CrossRef] [PubMed]
56. Ortelli, S.; Costa, A.L.; Torri, C.; Samorì, C.; Galletti, P.; Vineis, C.; Varesano, A.; Bonura, L.; Bianchi, G. Innovative and sustainable production of biopolymers. In *Factories of the Future: The Italian Flagship Initiative*; Springer: Berlin, Germany, 2019; pp. 131–148. [CrossRef]
57. Adeyeye, O.A.; Sadiku, E.R.; Reddy, A.B.; Ndamase, A.S.; Makgatho, G.; Sellamuthu, P.S.; Perumal, A.B.; Nambiar, R.B.; Fasiku, V.O.; Inrahim, I.D.; et al. The use of biopolymers in food packaging. In *Green Polymers and Their Nanocomposites*; Springer: Singapore, 2019; pp. 137–158. [CrossRef]

58. Wróblewska-Krepsztul, J.; Rydzkowski, T.; Michalska-Pożoga, I.; Thakur, V.K. Biopolymers for biomedical and pharmaceutical applications: Recent advances and overview of alginate electrospinning. *Nanomaterials* **2019**, *9*, 404. [CrossRef]
59. Ferraris, S.; Spriano, S.; Scalia, A.C.; Cochis, A.; Rimondini, L.; Cruz-Maya, I.; Guarino, V.; Varesano, A.; Vineis, C. Topographical and biomechanical guidance of electrospun fibers for biomedical applications. *Polymers* **2020**, *12*, 2896. [CrossRef]
60. Dhandayuthapani, B.; Yoshida, Y.; Maekawa, T.; Kumar, D.S. Polymeric scaffolds in tissue engineering application: A review. *Int. J. Polym. Sci.* **2011**, *2011*, 290602. [CrossRef]
61. Ogueri, K.S.; Jafari, T.; Escobar, I.J.L.; Laurencin, C.T. Polymeric biomaterials for scaffold-based bone regenerative engineering. *Regen. Eng. Transl. Med.* **2019**, *5*, 128–154. [CrossRef]
62. Bhatia, S. Natural polymers vs. synthetic polymer. In *Natural Polymer Drug Delivery Systems*; Springer: Berlin, Germany, 2016; pp. 95–118.
63. Innovative Packaging of Medicines-IJAHR. Available online: https://www.ipinnovative.com/journals/IJAHR/article-full-text/10648 (accessed on 27 June 2021).
64. Zadbuke, N.; Shahi, S.; Gulecha, B.; Padalkar, A.; Thube, M. Recent trends and future of pharmaceutical packaging technology. *J. Pharm. Bioallied Sci.* **2013**, *5*, 98–110. [CrossRef]
65. Gheorghita Puscaselu, R.; Amariei, S.; Norocel, L.; Gutt, G. New edible packaging material with function in shelf life extension: Applications for the meat and cheese industries. *Foods* **2020**, *9*, 562. [CrossRef]
66. Janes, M.E.; Dai, Y. Edible films for meat, poultry and seafood. In *Advances in Meat, Poultry and Seafood Packaging*; Elsevier: Amsterdam, The Netherlands, 2012; pp. 504–521. ISBN 9781845697518.
67. Avila-Sosa, R.; Palou, E.; López-Malo, A. Essential oils added to edible films. In *Essential Oils in Food Preservation, Flavor and Safety*; Elsevier: Amsterdam, The Netherlands, 2016; pp. 149–154. ISBN 9780124166448.
68. Zhao, Y.; Mcdaniel, M. Sensory quality of foods associated with edible film and coating systems and shelf-life extension. In *Innovations in Food Packaging*; Elsevier: Amsterdam, The Netherlands, 2005; pp. 434–453. ISBN 9780123116321.
69. Min, S.; Krochta, J.M. Antimicrobial films and coatings for fresh fruit and Vegetables. In *Improving the Safety of Fresh Fruit and Vegetables*; Elsevier: Amsterdam, The Netherlands, 2005; pp. 454–492. ISBN 9781855739567.
70. Mcclements, D.J. Biopolymers in food emulsions. In *Modern Biopolymer Science*; Elsevier: Amsterdam, The Netherlands, 2009; pp. 129–166. ISBN 9780123741950.
71. Ahmed, Z.; Ahmad, A. Biopolymer produced by the lactic acid bacteria: Production and practical application. In *Microbial Production of Food Ingredients and Additives*; Elsevier: Amsterdam, The Netherlands, 2017; pp. 217–257.
72. Martau, G.A.; Mihai, M.; Vodnar, D.C. The use of chitosan, alginate, and pectin in the biomedical and food sector-biocompatibility, bioadhesiveness, and biodegradability. *Polymers* **2019**, *11*, 1837. [CrossRef]
73. De Vos, P.; Lazarjani, H.A.; Poncelet, D.; Faas, M.M. Polymers in cell encapsulation from an enveloped cell perspective. *Adv. Drug Deliv. Rev.* **2014**, *67–68*, 15–34. [CrossRef]
74. Madene, A.; Jacquot, M.; Scher, J.; Desobry, S. Flavour encapsulation and controlled release—A review. *Int. J. Food Sci. Technol.* **2006**, *41*, 1–21. [CrossRef]
75. Therapeutic Applications of Cell Microencapsulation—Google Cărți. Available online: https://books.google.ro/books?id=7u9d2fkYHvMC&pg=PA1&hl=ro&source=gbs_toc_r&cad=3#v=onepage&q&f=false (accessed on 9 August 2021).
76. Biopolymer Nanoparticles: A Review of Prospects for Application as Carrier for Therapeutics and Diagnostics—Pharmaceutical Research and Allied Sciences. Available online: https://ijpras.com/article/biopolymer-nanoparticles-a-review-of-prospects-for-application-as-carrier-for-therapeutics-and-diagnostics (accessed on 9 August 2021).
77. Lee, M.K.; Bae, Y.H. Cell transplantation for endocrine disorders. *Adv. Drug Deliv. Rev.* **2000**, *42*, 103–120. [CrossRef]
78. Sharma, K.; Singh, V.; Arora, A. Natural biodegradable polymers as matrices intransdermal drug delivery. *Int. J. Drug Dev. Res.* **2011**, *3*, 85–103.
79. DJ, M. Designing biopolymer microgels to encapsulate, protect and deliver bioactive components: Physicochemical aspects. *Adv. Colloid Interface Sci.* **2017**, *240*, 31–59. [CrossRef]
80. Altomare, L.; Bonetti, L.; Campiglio, C.E.; de Nardo, L.; Draghi, L.; Tana, F.; Farè, S. Biopolymer-based strategies in the design of smart medical devices and artificial organs. *Int. J. Artif. Organ* **2018**, *41*, 337–359. [CrossRef]
81. Schmaljohann, D. Thermo- and pH-responsive polymers in drug delivery. *Adv. Drug Deliv. Rev.* **2006**, *58*, 1655–1670. [CrossRef]
82. George, A.; Shah, P.A.; Shrivastav, P.S. Natural biodegradable polymers based nano-formulations for drug delivery: A review. *Int. J. Pharm.* **2019**, *561*, 244–264. [CrossRef] [PubMed]
83. Polymers in Pharmaceutical Drug Delivery System: A Review. Available online: https://www.researchgate.net/publication/285986312_Polymers_in_pharmaceutical_drug_delivery_system_A_review (accessed on 26 July 2021).
84. Paques, J.P.; Van Der Linden, E.; Van Rijn, C.J.M.; Sagis, L.M.C. Preparation methods of alginate nanoparticles. *Adv. Colloid Interface Sci.* **2014**, *209*, 163–171. [CrossRef]
85. Severino, P.; da Silva, C.F.; Andrade, L.N.; de Lima Oliveira, D.; Campos, J.; Souto, E.B. Alginate nanoparticles for drug delivery and targeting. *Curr. Pharm. Des.* **2019**, *25*, 1312–1334. [CrossRef]
86. De, S.; Robinson, D. Polymer relationships during preparation of chitosan-alginate and poly-l-lysine-alginate nanospheres. *J. Control. Release* **2003**, *89*, 101–112. [CrossRef]
87. Smith, A.M.; Miri, T. Alginates in Foods. In *Practical Food Rheology: An Interpretive Approach*; John Wiley&Sons: Hoboken, NJ, USA, 2010; pp. 113–132. ISBN 9781405199780.

88. Espevik, T.; Otterlei, M.; Skjåk-Braek, G.; Ryan, L.; Wright, S.D.; Sundan, A. The involvement of CD14 in stimulation of cytokine production by uronic acid polymers. *Eur. J. Immunol.* **1993**, *23*, 255–261. [CrossRef]
89. De Vos, P.; De Haan, B.; Pater, J.; Van Schilfgaarde, R. Association between capsule diameter, adequacy of encapsulation, and survival of microencapsulated rat islet allografts. *Transplantation* **1996**, *62*, 893–899. [CrossRef] [PubMed]
90. Lim, F.; Sun, A.M. Microencapsulated islets as bioartificial endocrine pancreas. *Science* **1980**, *210*, 908–910. [CrossRef]
91. Gattás-Asfura, K.M.; Stabler, C.L. Chemoselective cross-linking and functionalization of alginate via Staudinger ligation. *Biomacromolecules* **2009**, *10*, 3122–3129. [CrossRef] [PubMed]
92. Thomas, D.; KurienThomas, K.; Latha, M.S. Preparation and evaluation of alginate nanoparticles prepared by green method for drug delivery applications. *Int. J. Biol. Macromol.* **2020**, *154*, 888–895. [CrossRef] [PubMed]
93. Design and Characterization of Ibuprofen Loaded Alginate Microspheres Prepared by Ionic Gelation Method. Available online: https://www.researchgate.net/publication/327743699_Design_and_Characterization_of_Ibuprofen_Loaded_Alginate_Microspheres_Prepared_by_Ionic_Gelation_Method (accessed on 24 July 2021).
94. Bose, P.S.; Reddy, P.S.; Nagaraju, R. Preparation and evaluation of indomethacin loaded alginate microspheres. *Ceska Slov. Farm.* **2016**, *65*, 104–110.
95. Hariyadi, D.M.; Hendradi, E.; Erawati, T.; Jannah, E.N.; Febrina, W. Influence of drug-polymer ratio on physical characteristics and release of metformin hydrochloride from metformin alginate microspheres. *Trop. J. Pharm. Res.* **2018**, *17*, 1229–1233. [CrossRef]
96. Ahmed, M.M.; El-Rasoul, S.A.; Auda, S.H.; Ibrahim, M.A. Emulsification/internal gelation as a method for preparation of diclofenac sodium–sodium alginate microparticles. *Saudi Pharm. J.* **2013**, *21*, 61–69. [CrossRef] [PubMed]
97. Patel, N.; Lalwani, D.; Gollmer, S.; Injeti, E.; Sari, Y.; Nesamony, J. Development and evaluation of a calcium alginate based oral ceftriaxone sodium formulation. *Prog. Biomater.* **2016**, *5*, 117–133. [CrossRef]
98. Setty, C.M.; Sahoo, S.S.; Sa, D.B. Alginate-coated alginate-polyethyleneimine beads for prolonged release of furosemide in simulated intestinal fluid. *Drug Dev. Ind. Pharm.* **2008**, *31*, 435–446. [CrossRef] [PubMed]
99. Rastogi, R.; Sultana, Y.; Aqil, M.; Ali, A.; Kumar, S. Alginate microspheres of isoniazid for oral sustained drug delivery. *Int. J. Pharm.* **2007**, *334*, 71–77. [CrossRef]
100. Veronovski, A.; Knez, Ž.; Novak, Z. Preparation of multi-membrane alginate aerogels used for drug delivery. *J. Supercrit. Fluids* **2013**, *79*, 209–215. [CrossRef]
101. Lee, S.; Kim, Y.C.; Park, J.H. Zein-alginate based oral drug delivery systems: Protection and release of therapeutic proteins. *Int. J. Pharm.* **2016**, *515*, 300–306. [CrossRef] [PubMed]
102. Varshosaz, J. The promise of chitosan microspheres in drug delivery systems. *Expert Opin. Drug Deliv.* **2007**, *4*, 263–273. [CrossRef]
103. Bakshi, P.S.; Selvakumar, D.; Kadirvelu, K.; Kumar, N.S. Chitosan as an environment friendly biomaterial—A review on recent modifications and applications. *Int. J. Biol. Macromol.* **2020**, *150*, 1072–1083. [CrossRef]
104. Negm, N.A.; Hefni, H.H.H.; Abd-Elaal, A.A.A.; Badr, E.A.; Abou Kana, M.T.H. Advancement on modification of chitosan biopolymer and its potential applications. *Int. J. Biol. Macromol.* **2020**, *152*, 681–702. [CrossRef]
105. Jayakumar, R.; Reis, R.L.; Mano, J.F. Chemistry and applications of phosphorylated chitin and chitosan. *E Polym.* **2006**, *6*, 1–16. [CrossRef]
106. Raza, Z.A.; Khalil, S.; Ayub, A.; Banat, I.M. Recent developments in chitosan encapsulation of various active ingredients for multifunctional applications. *Carbohydr. Res.* **2020**, *492*, 108004. [CrossRef] [PubMed]
107. Hussain, S.A.; Abdelkader, H.; Abdullah, N.; Kmaruddin, S. Review on micro-encapsulation with Chitosan for pharmaceuticals applications. *MOJ Curr. Res. Rev.* **2018**, *1*, 77–84. [CrossRef]
108. Koppolu, B.P.; Smith, S.G.; Ravindranathan, S.; Jayanthi, S.; Suresh Kumar, T.K.; Zaharoff, D.A. Controlling chitosan-based encapsulation for protein and vaccine delivery. *Biomaterials* **2014**, *35*, 4382–4389. [CrossRef]
109. Prabaharan, M. Review paper: Chitosan derivatives as promising materials for controlled drug delivery. *J. Biomater. Appl.* **2008**, *23*, 5–36. [CrossRef] [PubMed]
110. Yu, S.; Xu, X.; Feng, J.; Liu, M.; Hu, K. Chitosan and chitosan coating nanoparticles for the treatment of brain disease. *Int. J. Pharm.* **2019**, *560*, 282–293. [CrossRef] [PubMed]
111. Wang, X.; Chi, N.; Tang, X. Preparation of estradiol chitosan nanoparticles for improving nasal absorption and brain targeting. *Eur. J. Pharm. Biopharm.* **2008**, *70*, 735–740. [CrossRef]
112. Panos, I.; Acosta, N.; Heras, A. New drug delivery systems based on chitosan. *Curr. Drug Discov. Technol.* **2008**, *5*, 333–341. [CrossRef]
113. Rokhade, A.P.; Patil, S.A.; Aminabhavi, T.M. Synthesis and characterization of semi-interpenetrating polymer network microspheres of acrylamide grafted dextran and chitosan for controlled release of acyclovir. *Carbohydr. Polym.* **2007**, *67*, 605–613. [CrossRef]
114. Giunchedi, P.; Juliano, C.; Gavini, E.; Cossu, M.; Sorrenti, M. Formulation and in vivo evaluation of chlorhexidine buccal tablets prepared using drug-loaded chitosan microspheres. *Eur. J. Pharm. Biopharm.* **2002**, *53*, 233–239. [CrossRef]
115. Cerchiara, T.; Luppi, B.; Bigucci, F.; Petrachi, M.; Orienti, I.; Zecchi, V. Controlled release of vancomycin from freeze-dried chitosan salts coated with different fatty acids by spray-drying. *J. Microencapsul.* **2003**, *20*, 473–478. [CrossRef]
116. Govender, S.; Pillay, V.; Chetty, D.J.; Essack, S.Y.; Dangor, C.M.; Govender, T. Optimisation and characterisation of bioadhesive controlled release tetracycline microspheres. *Int. J. Pharm.* **2005**, *306*, 24–40. [CrossRef]

117. Almeida, A.; Araújo, M.; Novoa-Carballal, R.; Andrade, F.; Gonçalves, H.; Reis, R.L.; Lúcio, M.; Schwartz, S.; Sarmento, B. Novel amphiphilic chitosan micelles as carriers for hydrophobic anticancer drugs. *Mater. Sci. Eng. C* **2020**, *112*, 110920. [CrossRef]
118. Nair, S.; Anoop, K. Design and in vitro evaluation of controlled release Satranidazole subgingival films for periodontitis therapy. *Recent Res. Sci. Technol.* **2014**, *2*, 6–11.
119. Sohail, R.; Abbas, S.R. Evaluation of amygdalin-loaded alginate-chitosan nanoparticles as biocompatible drug delivery carriers for anticancerous efficacy. *Int. J. Biol. Macromol.* **2020**, *153*, 36–45. [CrossRef] [PubMed]
120. Shariatinia, Z.; Mazloom-Jalali, A. Chitosan nanocomposite drug delivery systems designed for the ifosfamide anticancer drug using molecular dynamics simulations. *J. Mol. Liq.* **2019**, *273*, 346–367. [CrossRef]
121. Hanna, D.H.; Saad, G.R. Encapsulation of ciprofloxacin within modified xanthan gum-chitosan based hydrogel for drug delivery. *Bioorg. Chem.* **2019**, *84*, 115–124. [CrossRef] [PubMed]
122. Cánepa, C.; Imperiale, J.C.; Berini, C.A.; Lewicki, M.; Sosnik, A.; Biglione, M.M. Development of a drug delivery system based on chitosan nanoparticles for oral administration of interferon-α. *Biomacromolecules* **2017**, *18*, 3302–3309. [CrossRef]
123. Sathiyaseelan, A.; Saravanakumar, K.; Anand Mariadoss, A.V.; Wang, M.H. Biocompatible fungal chitosan encapsulated phytogenic silver nanoparticles enhanced antidiabetic, antioxidant and antibacterial activity. *Int. J. Biol. Macromol.* **2020**, *153*, 63–71. [CrossRef] [PubMed]
124. Ali, A.; Ahmed, S. A review on chitosan and its nanocomposites in drug delivery. *Int. J. Biol. Macromol.* **2018**, *109*, 273–286. [CrossRef]
125. Mostafavi, F.S.; Zaeim, D. Agar-based edible films for food packaging applications—A review. *Int. J. Biol. Macromol.* **2020**, *159*, 1165–1176. [CrossRef]
126. Bertasa, M.; Dodero, A.; Alloisio, M.; Vicini, S.; Riedo, C.; Sansonetti, A.; Scalarone, D.; Castellano, M. Agar gel strength: A correlation study between chemical composition and rheological properties. *Eur. Polym. J.* **2020**, *123*, 109442. [CrossRef]
127. Zhang, C.; An, D.; Xiao, Q.; Weng, H.; Zhang, Y.; Yang, Q.; Xiao, A. Preparation, characterization, and modification mechanism of agar treated with hydrogen peroxide at different temperatures. *Food Hydrocoll.* **2020**, *101*, 105527. [CrossRef]
128. Xiao, Q.; An, D.; Zhang, C.; Weng, H.; Zhang, Y.; Chen, F.; Xiao, A. Agar quality promotion prepared by desulfation with hydrogen peroxide. *Int. J. Biol. Macromol.* **2020**, *145*, 492–499. [CrossRef] [PubMed]
129. El-Helw, A.E.R.; El-Said, Y. Preparation and characterization of agar beads containing phenobarbitone sodium. *J. Microencapsul.* **1988**, *5*, 159–163. [CrossRef]
130. Evandro, M.; Denis, R.; Zenia, A.; Emre, K.; Denis, P. Oil encapsulation in core-shell alginate capsules by inverse gelation. I: Dripping methodology. *J. Microencapsul.* **2017**, *34*, 82–90. [CrossRef]
131. Rivadeneira, J.; Audisio, M.; Gorustovich, A. Films based on soy protein-agar blends for wound dressing: Effect of different biopolymer proportions on the drug release rate and the physical and antibacterial properties of the films. *J. Biomater. Appl.* **2018**, *32*, 1231–1238. [CrossRef]
132. Zhu, F. Encapsulation and delivery of food ingredients using starch based systems. *Food Chem.* **2017**, *229*, 542–552. [CrossRef] [PubMed]
133. Queiroz, V.M.; Kling, I.C.S.; Eltom, A.E.; Archanjo, B.S.; Prado, M.; Simão, R.A. Corn starch films as a long-term drug delivery system for chlorhexidine gluconate. *Mater. Sci. Eng. C* **2020**, *112*, 110852. [CrossRef]
134. Qi, X.; Tester, R.F. Starch granules as active guest molecules or microorganism delivery systems. *Food Chem.* **2019**, *271*, 182–186. [CrossRef] [PubMed]
135. Hoyos-Leyva, J.D.; Bello-Pérez, L.A.; Alvarez-Ramirez, J.; Garcia, H.S. Microencapsulation using starch as wall material: A review. *Food Rev. Int.* **2018**, *34*, 148–161. [CrossRef]
136. Santana, Á.L.; Angela, M.; Meireles, A. New starches are the trend for industry applications: A review. *Food Public Health* **2014**, *2014*, 229–241. [CrossRef]
137. Chen, J.; Wang, Y.; Liu, J.; Xu, X. Preparation, characterization, physicochemical property and potential application of porous starch: A review. *Int. J. Biol. Macromol.* **2020**, *148*, 1169–1181. [CrossRef]
138. Šárka, E.; Dvořáček, V. New processing and applications of waxy starch (a review). *J. Food Eng.* **2017**, *206*, 77–87. [CrossRef]
139. Hemamalini, T.; Giri Dev, V.R. Comprehensive review on electrospinning of starch polymer for biomedical applications. *Int. J. Biol. Macromol.* **2018**, *106*, 712–718. [CrossRef]
140. Wang, B.; Sui, J.; Yu, B.; Yuan, C.; Guo, L.; Abd El-Aty, A.M.; Cui, B. Physicochemical properties and antibacterial activity of corn starch-based films incorporated with *Zanthoxylum bungeanum* essential oil. *Carbohydr. Polym.* **2021**, *254*, 117314. [CrossRef]
141. Singh, A.; Nath, L.; Singh, A. Pharmaceutical, food and non-food applications of modified starches: A critical review. *Electron. J. Environ. Agric. Food Chem.* **2010**, *9*, 1214–1221.
142. Rodrigues, A.; Emeje, M. Recent applications of starch derivatives in nanodrug delivery. *Carbohydr. Polym.* **2012**, *87*, 987–994. [CrossRef]
143. Saboktakin, M. In-vitro and in-vivo characterizations of novel modified starch transdermal drug delivery system. *MOJ Proteom. Bioinform.* **2017**, *5*, 132–135. [CrossRef]
144. Pal, K. Starch based hydrogel with potential biomedical application as artificial skin. *Afr. J. Biomed. Res.* **2006**, *9*, 23–29. [CrossRef]
145. Malafaya, P.B.; Elvira, C.; Gallardo, A.; San Román, J.; Reis, R.L. Porous starch-based drug delivery systems processed by a microwave route. *J. Biomater. Sci. Polym. Ed.* **2001**, *12*, 1227–1241. [CrossRef]

146. Janaswamy, S. Encapsulation altered starch digestion: Toward developing starch-based delivery systems. *Carbohydr. Polym.* **2014**, *101*, 600–605. [CrossRef] [PubMed]
147. Sachan, N.; Bhattacharya, A. Modeling and characterization of drug release from glutinous rice starch based hydrogel beads for controlled drug delivery. *Int. J. Health Res.* **2010**, *2*, 55398. [CrossRef]
148. Tesfay, D.; Abrha, S.; Yilma, Z.; Woldu, G.; Molla, F. Preparation, optimization, and evaluation of epichlorohydrin cross-linked enset (ensete ventricosum (welw.) cheeseman) starch as drug release sustaining excipient in microsphere formulation. *BioMed Res. Int.* **2020**, *2020*, 2147971. [CrossRef]
149. Alabi, C.O.; Singh, I.; Odeku, O.A. Evaluation of starch-clay composites as a pharmaceutical excipient in tramadol tablet formulations. *Polim. Med.* **2020**, *50*, 33–40. [CrossRef]
150. Starch-Chitosan Modified Blend as Long-Term Controlled Drug Release For Cancer Therapy. Available online: http://staff.uokufa.edu.iq/publication_details.php?khalidahk.abbas&recordID=6799 (accessed on 26 June 2021).
151. Mehrizi, A.A. Synthesis and Characterization of Erythromycin Loaded PVA-Corn Starch Hydrogel Wound Dressing. Available online: https://www.academia.edu/15627564/Synthesis_and_characterization_of_Erythromycin_loaded_PVA_corn_starch_hydrogel_wound_dressing (accessed on 26 June 2021).
152. Shibata, N.; Nishumura, A.; Naruhashi, K.; Nakao, Y.; Miura, R. Preparation and pharmaceutical evaluation of new sustained-release capsule including starch-sponge matrix (SSM). *Biomed. Pharmacother.* **2010**, *64*, 352–358. [CrossRef]
153. Balmayor, E.R.; Tuzlakoglu, K.; Azevedo, H.S.; Reis, R.L. Preparation and characterization of starch-poly-ε-caprolactone microparticles incorporating bioactive agents for drug delivery and tissue engineering applications. *Acta Biomater.* **2009**, *5*, 1035–1045. [CrossRef]
154. Xu, J.; Tan, X.; Chen, L.; Li, X.; Xie, F. Starch/microcrystalline cellulose hybrid gels as gastric-floating drug delivery systems. *Carbohydr. Polym.* **2019**, *215*, 151–159. [CrossRef] [PubMed]
155. Li, W.; Liu, L.; Tian, H.; Luo, X.; Liu, S. Encapsulation of *Lactobacillus plantarum* in cellulose based microgel with controlled release behavior and increased long-term storage stability. *Carbohydr. Polym.* **2019**, *223*, 115065. [CrossRef]
156. Ab'lah, N.; Wong, T.W. Starch as oral colon-specific nano- and microparticulate drug carriers. In *Polymer Science and Innovative Applications*; Elsevier: Amsterdam, The Netherlands, 2020; pp. 287–330.
157. Yang, X.; Darko, O.K.; Huang, Y.; He, C.; Yang, H.; He, S.; Li, J.; Li, J.; Hocker, B.; Yin, Y. Resistant starch regulates gut microbiota: Structure, biochemistry and cell signalling. *Cell Physiol. Biochem.* **2017**, *42*, 306–318. [CrossRef] [PubMed]
158. Fuentes-Zaragoza, E.; Sánchez-Zapata, E.; Sendra, E.; Sayas, E.; Navarro, C.; Fernández-Lõpez, J.; Pérez-Alvarez, J.A. Resistant starch as prebiotic: A review. *Starch Staerke* **2011**, *63*, 406–415. [CrossRef]
159. Pooresmaeil, M.; Behzadi Nia, S.; Namazi, H. Green encapsulation of LDH(Zn/Al)-5-Fu with carboxymethyl cellulose biopolymer; new nanovehicle for oral colorectal cancer treatment. *Int. J. Biol. Macromol.* **2019**, *139*, 994–1001. [CrossRef]
160. Li, Y.; Yu, S.; Chen, P.; Rojas, R.; Hajian, A.; Berglund, L. Cellulose nanofibers enable paraffin encapsulation and the formation of stable thermal regulation nanocomposites. *Nano Energy* **2017**, *34*, 541–548. [CrossRef]
161. Baek, J.; Ramasamy, M.; Willis, N.C.; Kim, D.S.; Anderson, W.A.; Tam, K.C. Encapsulation and controlled release of vitamin C in modified cellulose nanocrystal/chitosan nanocapsules. *Curr. Res. Food Sci.* **2021**, *4*, 215–223. [CrossRef]
162. Mitbumrung, W.; Suphantharika, M.; Mcclements, D.J.; Winuprasith, T. Encapsulation of vitamin D3 in pickering emulsion stabilized by nanofibrillated mangosteen cellulose: Effect of environmental stresses. *J. Food Sci.* **2019**, *84*, 3213–3221. [CrossRef]
163. Pan-In, P.; Banlunara, W.; Chaichanawongsaroj, N.; Wanichwecharungruang, S. Ethyl cellulose nanoparticles: Clarithomycin encapsulation and eradication of H. pylori. *Carbohydr. Polym.* **2014**, *109*, 22–27. [CrossRef]
164. Serna-Cock, L.; Vallejo-Castillo, V. Probiotic encapsulation. *Afr. J. Microbiol. Res.* **2013**, *7*, 4743–4753. [CrossRef]
165. Salminen, S.; Collado, M.C.; Endo, A.; Hill, C.; Lebeer, S.; Quigley, E.M.M.; Sanders, M.E.; Shamir, R.; Swann, J.R.; Szajewska, H.; et al. The international scientific association of probiotics and prebiotics (ISAPP) consensus statement on the definition and scope of postbiotics. *Nat. Rev. Gastroenterol. Hepatol.* **2021**, *2021*, 1–19. [CrossRef]
166. Silva, D.R.; Sardi, J.C.O.; Pitangui, N.S.; Roque, S.M.; da Silva, A.C.B.; Rosalen, P.L. Probiotics as an alternative antimicrobial therapy: Current reality and future directions. *J. Funct. Foods* **2020**, *73*, 104080. [CrossRef]
167. Misra, S.; Pandey, P.; Mishra, H.N. Novel approaches for co-encapsulation of probiotic bacteria with bioactive compounds, their health benefits and functional food product development: A review. *Trends Food Sci. Technol.* **2021**, *109*, 340–351. [CrossRef]
168. Kailasapathy, K. Microencapsulation of Probiotic Bacteria 39 Microencapsulation of Probiotic Bacteria: Technology and Potential Applications. Available online: https://www.caister.com/bacteria-plant (accessed on 27 July 2021).
169. Burgain, J.; Gaiani, C.; Linder, M.; Scher, J. Encapsulation of probiotic living cells: From laboratory scale to industrial applications. *J. Food Eng.* **2011**, *104*, 467–483. [CrossRef]
170. Kailasapathy, K. Encapsulation technologies for functional foods and nutraceutical product development. *CAB Rev.* **2009**, *4*, 33. [CrossRef]
171. Viability and Survival of Free and Encapsulated Probiotic Bacteria in Cheddar Cheese | Western Sydney University ResearchDirect. Available online: https://researchdirect.westernsydney.edu.au/islandora/object/uws%3A5276 (accessed on 30 July 2021).
172. Homayouni, A.; Azizi, A.; Ehsani, M.R.; Yarmand, M.S.; Razavi, S.H. Effect of microencapsulation and resistant starch on the probiotic survival and sensory properties of synbiotic ice cream. *Food Chem.* **2008**, *111*, 50–55. [CrossRef]
173. Mcmaster, L.D.; Kokott, S.A. Micro-encapsulation of Bifidobacterium lactis for incorporation into soft foods. *World J. Microbiol. Biotechnol.* **2005**, *21*, 723–728. [CrossRef]

174. Muthukumarasamy, P.; Holley, R.A. Survival of Escherichia coli O157:H7 in dry fermented sausages containing micro-encapsulated probiotic lactic acid bacteria. *Food Microbiol.* **2007**, *24*, 82–88. [CrossRef] [PubMed]
175. Frakolaki, G.; Giannou, V.; Kekos, D.; Tzia, C. A review of the microencapsulation techniques for the incorporation of probiotic bacteria in functional foods. *Crit. Rev. Food Sci. Nutr.* **2021**, *61*, 1515–1536. [CrossRef]
176. Zhao, C.; Zhu, Y.; Kong, B.; Huang, Y. Dual-core prebiotic microcapsule encapsulating probiotics for metabolic syndrome. *ACS Appl. Mater. Interfaces* **2020**, *12*, 42586–42594. [CrossRef] [PubMed]
177. Zhang, Y.; Lin, J.; Zhong, Q. The increased viability of probiotic *Lactobacillus salivarius* NRRL B-30514 encapsulated in emulsions with multiple lipid-protein-pectin layers. *Food Res. Int.* **2015**, *71*, 9–15. [CrossRef]
178. Whorwell, P.J.; Morel, J.; Duane, C. Efficacy of an encapsulated probiotic Bifidobacterium infantis 35624 in women with irritable bowel syndrome. *Am. J. Gastroenterol.* **2006**, *101*, 1581–1590. [CrossRef]
179. Anselmo, A.C.; Mchugh, K.J.; Webster, J.; Langer, R.; Jaklenec, A. Layer-by-layer encapsulation of probiotics for delivery to the microbiome. *Adv. Mater.* **2016**, *28*, 9486–9490. [CrossRef] [PubMed]
180. Zheng, D.-W.; Li, R.-Q.; An, J.-X.; Xie, T.-Q.; Han, Z.-Y.; Xu, R.; Fang, Y.; Zhang, X.-Z. Prebiotics-encapsulated probiotic spores regulate gut microbiota and suppress colon cancer. *Adv. Mater.* **2020**, *32*, 2004529. [CrossRef] [PubMed]
181. Solanki, H.K.; Pawar, D.D.; Shah, D.A.; Prajapati, V.D.; Jani, G.K.; Mulla, A.M.; Thakar, P.M. Development of microencapsulation delivery system for long-term preservation of probiotics as biotherapeutics agent. *BioMed Res. Int.* **2013**, *2013*, 620719. [CrossRef]
182. Martín, M.J.; Lara-Villoslada, F.; Ruiz, M.A.; Morales, M.E. Microencapsulation of bacteria: A review of different technologies and their impact on the probiotic effects. *Innov. Food Sci. Emerg. Technol.* **2015**, *27*, 15–25. [CrossRef]
183. Capela, P.; Hay, T.K.C.; Shah, N.P. Effect of homogenisation on bead size and survival of encapsulated probiotic bacteria. *Food Res. Int.* **2007**, *40*, 1261–1269. [CrossRef]
184. Razavi, S.; Janfaza, S.; Tasnim, N.; Gibson, D.L.; Hoorfar, M. Nanomaterial-based encapsulation for controlled gastrointestinal delivery of viable probiotic bacteria. *Nanoscale Adv.* **2021**, *3*, 2699–2709. [CrossRef]
185. Darjani, P.; Hosseini Nezhad, M.; Kadkhodaee, R.; Milani, E. Influence of prebiotic and coating materials on morphology and survival of a probiotic strain of *Lactobacillus casei* exposed to simulated gastrointestinal conditions. *LWT Food Sci. Technol.* **2016**, *73*, 162–167. [CrossRef]
186. Anal, A.K.; Singh, H. Recent advances in microencapsulation of probiotics for industrial applications and targeted delivery. *Trends Food Sci. Technol.* **2007**, *18*, 240–251. [CrossRef]
187. Tripathi, M.K.; Giri, S.K. Probiotic functional foods: Survival of probiotics during processing and storage. *J. Funct. Foods* **2014**, *9*, 225–241. [CrossRef]
188. ISO 10993-5:2009(en), Biological Evaluation of Medical Devices—Part 5: Tests for In Vitro Cytotoxicity. Available online: https://www.iso.org/obp/ui/#iso:std:iso:10993:-5:ed-3:v1:en (accessed on 27 June 2021).
189. Singh, P.; Medronho, B.; dos Santos, T.; Nunes-Correia, I.; Granja, P.; Miguel, M.G.; Lindman, B. On the viability, cytotoxicity and stability of probiotic bacteria entrapped in cellulose-based particles. *Food Hydrocoll.* **2018**, *82*, 457–465. [CrossRef]
190. Puscaselu, R.G.; Lobiuc, A.; Dimian, M.; Covasa, M. Alginate: From food industry to biomedical applications and management of metabolic disorders. *Polymers* **2020**, *12*, 2417. [CrossRef] [PubMed]
191. Lopes, S.; Bueno, L.; De Aguiar Júnior, F.; Finkler, C.L.L. Preparation and characterization of alginate and gelatin microcapsules containing *Lactobacillus rhamnosus*. *An. Acad. Bras. Cienc.* **2017**, *89*, 1601–1613. [CrossRef] [PubMed]
192. Ramos, P.E.; Silva, P.; Alario, M.M.; Pastrana, L.M.; Teixeira, J.A.; Cerqueira, M.A.; Vicente, A.A. Effect of alginate molecular weight and *M/G* ratio in beads properties foreseeing the protection of probiotics. *Food Hydrocoll.* **2018**, *77*, 8–16. [CrossRef]
193. Zhang, Z.; Gu, M.; You, X.; Sela, D.A.; Xiao, H.; Mcclements, D.J. Encapsulation of bifidobacterium in alginate microgels improves viability and targeted gut release. *Food Hydrocoll.* **2021**, *116*, 106634. [CrossRef]
194. Sathyabama, S.; Kumar, M.R.; Devi, P.B.; Vijayabharathi, R.; Priyadharisini, V.B. Co-encapsulation of probiotics with prebiotics on alginate matrix and its effect on viability in simulated gastric environment. *LWT Food Sci. Technol.* **2014**, *57*, 419–425. [CrossRef]
195. Shi, L.E.; Li, Z.H.; Li, D.T.; Xu, M.; Chen, H.Y.; Zhang, Z.L.; Tang, Z.X. Encapsulation of probiotic lactobacillus bulgaricus in alginate'milk microspheres and evaluation of the survival in simulated gastrointestinal conditions. *J. Food Eng.* **2013**, *117*, 99–104. [CrossRef]
196. Martin, M.J.; Lara-Villoslada, F.; Ruiz, M.A.; Morales, M.E. Effect of unmodified starch on viability of alginate-encapsulated *Lactobacillus fermentum* CECT5716. *LWT Food Sci. Technol.* **2013**, *53*, 480–486. [CrossRef]
197. Cheow, W.S.; Kiew, T.Y.; Hadinoto, K. Controlled release of *Lactobacillus rhamnosus* biofilm probiotics from alginate-locust bean gum microcapsules. *Carbohydr. Polym.* **2014**, *103*, 587–595. [CrossRef]
198. Song, H.; Yu, W.; Liu, X.; Ma, X. Improved probiotic viability in stress environments with post-culture of alginate-chitosan microencapsulated low density cells. *Carbohydr. Polym.* **2014**, *108*, 10–16. [CrossRef]
199. Kamalian, N.; Mirhosseini, H.; Mustafa, S.; Manap, M.Y.A. Effect of alginate and chitosan on viability and release behavior of Bifidobacterium pseudocatenulatum G4 in simulated gastrointestinal fluid. *Carbohydr. Polym.* **2014**, *111*, 700–706. [CrossRef] [PubMed]
200. Cook, M.T.; Tzortzis, G.; Khutoryanskiy, V.V.; Charalampopoulos, D. Layer-by-layer coating of alginate matrices with chitosan-alginate for the improved survival and targeted delivery of probiotic bacteria after oral administration. *J. Mater. Chem. B* **2013**, *1*, 52–60. [CrossRef] [PubMed]

201. de Araújo Etchepare, M.; Raddatz, G.C.; de Moraes Flores, É.M.; Zepka, L.Q.; Jacob-Lopes, E.; Barin, J.S.; Ferreira Grosso, C.R.; de Menezes, C.R. Effect of resistant starch and chitosan on survival of *Lactobacillus acidophilus* microencapsulated with sodium alginate. *LWT Food Sci. Technol.* **2016**, *65*, 511–517. [CrossRef]
202. Fareez, I.M.; Lim, S.M.; Mishra, R.K.; Ramasamy, K. Chitosan coated alginate-xanthan gum bead enhanced pH and thermotolerance of *Lactobacillus plantarum* LAB12. *Int. J. Biol. Macromol.* **2015**, *72*, 1419–1428. [CrossRef]
203. Prata, A.S.; Grosso, C.R.F. Production of microparticles with gelatin and chitosan. *Carbohydr. Polym.* **2015**, *116*, 292–299. [CrossRef]
204. Morrison, I.D.; Ross, S. *Colloidal Dispersions: Suspensions, Emulsions, and Foams*; John Wiley&Sons: Hoboken, NJ, USA, 2002; p. 616.
205. Kwiecień, I.; Kwiecień, M. Application of polysaccharide-based hydrogels as probiotic delivery systems. *Gels* **2018**, *4*, 47. [CrossRef] [PubMed]
206. Albadran, H.A.; Monteagudo-Mera, A.; Khutoryanskiy, V.V.; Charalampopoulos, D. Development of chitosan-coated agar-gelatin particles for probiotic delivery and targeted release in the gastrointestinal tract. *Appl. Microbiol. Biotechnol.* **2020**, *104*, 5749–5757. [CrossRef]
207. Jia, S.; Zhou, K.; Pan, R.; Wei, J.; Liu, Z.; Xu, Y. Oral immunization of carps with chitosan–alginate microcapsule containing probiotic expressing spring viremia of carp virus (SVCV) G protein provides effective protection against SVCV infection. *Fish Shellfish Immunol.* **2020**, *105*, 327–329. [CrossRef]
208. Argin, S.; Kofinas, P.; Lo, Y.M. The cell release kinetics and the swelling behavior of physically crosslinked xanthan–chitosan hydrogels in simulated gastrointestinal conditions. *Food Hydrocoll.* **2014**, *40*, 138–144. [CrossRef]
209. Cook, M.T.; Tzortis, G.; Charalampopoulos, D.; Khutoryanskiy, V.V. Production and evaluation of dry alginate-chitosan microcapsules as an enteric delivery vehicle for probiotic bacteria. *Biomacromolecules* **2011**, *12*, 2834–2840. [CrossRef] [PubMed]
210. De Prisco, A.; Maresca, D.; Ongeng, D.; Mauriello, G. Microencapsulation by vibrating technology of the probiotic strain *Lactobacillus reuteri* DSM 17938 to enhance its survival in foods and in gastrointestinal environment. *LWT Food Sci. Technol.* **2015**, *61*, 452–462. [CrossRef]
211. Graff, S.; Hussain, S.; Chaumeil, J.-C.; Charrueau, C. Increased intestinal delivery of viable saccharomyces boulardii by encapsulation in microspheres. *Pharm. Res.* **2008**, *25*, 1290–1296. [CrossRef] [PubMed]
212. Krasaekoopt, W.; Watcharapoka, S. Effect of addition of inulin and galactooligosaccharide on the survival of microencapsulated probiotics in alginate beads coated with chitosan in simulated digestive system, yogurt and fruit juice. *LWT Food Sci. Technol.* **2014**, *57*, 761–766. [CrossRef]
213. Mirtič, J.; Rijavec, T.; Zupančič, Š.; Zvonar Pobirk, A.; Lapanje, A.; Kristl, J. Development of probiotic-loaded microcapsules for local delivery: Physical properties, cell release and growth. *Eur. J. Pharm. Sci.* **2018**, *121*, 178–187. [CrossRef]
214. Rodklongtan, A.; La-ongkham, O.; Nitisinprasert, S.; Chitprasert, P. Enhancement of *Lactobacillus reuteri* KUB-AC5 survival in broiler gastrointestinal tract by microencapsulation with alginate–chitosan semi-interpenetrating polymer networks. *J. Appl. Microbiol.* **2014**, *117*, 227–238. [CrossRef]
215. Wu, Q.X.; Xu, X.; Xie, Q.; Tong, W.Y.; Chen, Y. Evaluation of chitosan hydrochloride-alginate as enteric micro-probiotic-carrier with dual protective barriers. *Int. J. Biol. Macromol.* **2016**, *93*, 665–671. [CrossRef]
216. Çakmak, H.; Özselek, Y.; Turan, O.Y.; Fıratlıgil, E.; Karbancioğlu-Güler, F. Whey protein isolate edible films incorporated with essential oils: Antimicrobial activity and barrier properties. *Polym. Degrad. Stab.* **2020**, *179*, 109285. [CrossRef]
217. Zhang, B.; Liu, Y.; Wang, H.; Liu, W.; Cheong, K.; Teng, B. Characterization of seaweed polysaccharide-based bilayer films containing essential oils with antibacterial activity. *LWT Food Sci. Technol.* **2021**, *150*, 111961. [CrossRef]
218. Alfaro-Galarza, O.; López-Villegas, E.O.; Rivero-Perez, N.; Tapia- Maruri, D.; Jiménez-Aparicio, A.R.; Palma-Rodríguez, H.M.; Vargas-Torres, A. Protective effects of the use of taro and rice starch as wall material on the viability of encapsulated *Lactobacillus paracasei* subsp. Paracasei. *LWT Food Sci. Technol.* **2020**, *117*, 108686. [CrossRef]
219. Ashwar, B.A.; Gani, A.; Gani, A.; Shah, A.; Masoodi, F.A. Production of RS4 from rice starch and its utilization as an encapsulating agent for targeted delivery of probiotics. *Food Chem.* **2018**, *239*, 287–294. [CrossRef]
220. Dafe, A.; Etemadi, H.; Dilmaghani, A.; Mahdavinia, G.R. Investigation of pectin/starch hydrogel as a carrier for oral delivery of probiotic bacteria. *Int. J. Biol. Macromol.* **2017**, *97*, 536–543. [CrossRef]
221. Benavent-Gil, Y.; Rodrigo, D.; Rosell, C.M. Thermal stabilization of probiotics by adsorption onto porous starches. *Carbohydr. Polym.* **2018**, *197*, 558–564. [CrossRef]
222. Zanjani, M.A.K.; Ehsani, M.R.; Ghiassi Tarzi, B.; Sharifan, A. Promoting *Lactobacillus casei* and Bifidobacterium adolescentis survival by microencapsulation with different starches and chitosan and poly L-lysine coatings in ice cream. *J. Food Process. Preserv.* **2018**, *42*, e13318. [CrossRef]
223. Lancuški, A.; Aiman, A.A.; Avrahami, R.; Vilensky, R.; Vasilyev, G.; Zussman, E. Design of starch-formate compound fibers as encapsulation platform for biotherapeutics. *Carbohydr. Polym.* **2017**, *158*, 68–76. [CrossRef]
224. Reyes, V.; Chotiko, A.; Chouljenko, A.; Sathivel, S. Viability of *Lactobacillus acidophilus* NRRL B-4495 encapsulated with high maize starch, maltodextrin, and gum arabic. *LWT Food Sci. Technol.* **2018**, *96*, 642–647. [CrossRef]
225. Thangrongthong, S.; Puttarat, N.; Ladda, B.; Itthisoponkul, T.; Pinket, W.; Kasemwong, K.; Taweechotipatr, M. Microencapsulation of probiotic *Lactobacillus brevis* ST-69 producing GABA using alginate supplemented with nanocrystalline starch. *Food Sci. Biotechnol.* **2020**, *29*, 1475–1482. [CrossRef] [PubMed]

226. Sultana, K.; Godward, G.; Reynolds, N.; Arumugaswamy, R.; Peiris, P.; Kailasapathy, K. Encapsulation of probiotic bacteria with alginate–starch and evaluation of survival in simulated gastrointestinal conditions and in yoghurt. *Int. J. Food Microbiol.* **2000**, *62*, 47–55. [CrossRef]
227. Zabihollahi, N.; Alizadeh, A.; Almasi, H.; Hanifian, S.; Hamishekar, H. Development and characterization of carboxymethyl cellulose based probiotic nanocomposite film containing cellulose nanofiber and inulin for chicken fillet shelf life extension. *Int. J. Biol. Macromol.* **2020**, *160*, 409–417. [CrossRef] [PubMed]
228. Singh, P.; Medronho, B.; Miguel, M.G.; Esquena, J. On the encapsulation and viability of probiotic bacteria in edible carboxymethyl cellulose-gelatin water-in-water emulsions. *Food Hydrocoll.* **2018**, *75*, 41–50. [CrossRef]
229. Dafe, A.; Etemadi, H.; Zarredar, H.; Mahdavinia, G.R. Development of novel carboxymethyl cellulose/k-carrageenan blends as an enteric delivery vehicle for probiotic bacteria. *Int. J. Biol. Macromol.* **2017**, *97*, 299–307. [CrossRef] [PubMed]
230. Chen, B.; Lin, X.; Lin, X.; Li, W.; Zheng, B.; He, Z. Pectin-microfibrillated cellulose microgel: Effects on survival of lactic acid bacteria in a simulated gastrointestinal tract. *Int. J. Biol. Macromol.* **2020**, *158*, 826–836. [CrossRef]
231. Chitprasert, P.; Sudsai, P.; Rodklongtan, A. Aluminum carboxymethyl cellulose–rice bran microcapsules: Enhancing survival of *Lactobacillus reuteri* KUB-AC5. *Carbohydr. Polym.* **2012**, *90*, 78–86. [CrossRef] [PubMed]
232. Singh, P.; Medronho, B.; Alves, L.; da Silva, G.J.; Miguel, M.G.; Lindman, B. Development of carboxymethyl cellulose-chitosan hybrid micro- and macroparticles for encapsulation of probiotic bacteria. *Carbohydr. Polym.* **2017**, *175*, 87–95. [CrossRef]
233. Huq, T.; Fraschini, C.; Khan, A.; Riedl, B.; Bouchard, J.; Lacroix, M. Alginate based nanocomposite for microencapsulation of probiotic: Effect of cellulose nanocrystal (CNC) and lecithin. *Carbohydr. Polym.* **2017**, *168*, 61–69. [CrossRef] [PubMed]
234. Fareez, I.M.; Lim, S.M.; Zulkefli, N.A.A.; Mishra, R.K.; Ramasamy, K. Cellulose Derivatives Enhanced Stability of Alginate-Based Beads Loaded with *Lactobacillus plantarum* LAB12 against Low pH, High Temperature and Prolonged Storage. *Probiotics Antimicrob. Proteins* **2017**, *10*, 543–557. [CrossRef]

polymers

MDPI

Article

Effect of Crystallinity on the Properties of Polycaprolactone Nanoparticles Containing the Dual FLAP/mPEGS-1 Inhibitor BRP-187

Antje Vollrath [1,2], Christian Kretzer [3], Baerbel Beringer-Siemers [1], Blerina Shkodra [1,2], Justyna A. Czaplewska [1,2], Damiano Bandelli [1,2], Steffi Stumpf [1,2], Stephanie Hoeppener [1,2], Christine Weber [1,2], Oliver Werz [2,3] and Ulrich S. Schubert [1,2,*]

[1] Laboratory of Organic Chemistry and Macromolecular Chemistry (IOMC), Friedrich Schiller University, Humboldtstraße 10, 07743 Jena, Germany; antje.vollrath@uni-jena.de (A.V.); baerbel.beringer-siemers@uni-jena.de (B.B.-S.); Blerina.Shkodra-pula@uni-jena.de (B.S.); justyna.czaplewska@uni-jena.de (J.A.C.); damianobandelli@gmail.com (D.B.); steffi-stumpf@uni-jena.de (S.S.); s.hoeppener@uni-jena.de (S.H.); christine.weber@uni-jena.de (C.W.)

[2] Jena Center for Soft Matter (JCSM), Friedrich Schiller University, Philosophenweg 7, 07743 Jena, Germany; oliver.werz@uni-jena.de

[3] Department of Pharmaceutical/Medicinal Chemistry, Institute of Pharmacy, Friedrich Schiller University, Philosophenweg 14, 07743 Jena, Germany; christian.kretzer@uni-jena.de

* Correspondence: ulrich.schubert@uni-jena.de

Citation: Vollrath, A.; Kretzer, C.; Beringer-Siemers, B.; Shkodra, B.; Czaplewska, J.A.; Bandelli, D.; Stumpf, S.; Hoeppener, S.; Weber, C.; Werz, O.; et al. Effect of Crystallinity on the Properties of Polycaprolactone Nanoparticles Containing the Dual FLAP/mPEGS-1 Inhibitor BRP-187. *Polymers* **2021**, *13*, 2557. https://doi.org/10.3390/polym13152557

Academic Editors: José Miguel Ferri, Vicent Fombuena Borràs and Miguel Fernando Aldás Carrasco

Received: 29 June 2021
Accepted: 29 July 2021
Published: 31 July 2021

Abstract: Seven polycaprolactones (PCL) with constant hydrophobicity but a varying degree of crystallinity prepared from the constitutional isomers ε-caprolactone (εCL) and δ-caprolactone (δCL) were utilized to formulate nanoparticles (NPs). The aim was to investigate the effect of the crystallinity of the bulk polymers on the enzymatic degradation of the particles. Furthermore, their efficiency to encapsulate the hydrophobic anti-inflammatory drug BRP-187 and the final in vitro performance of the resulting NPs were evaluated. Initially, high-throughput nanoprecipitation was employed for the εCL and δCL homopolymers to screen and establish important formulation parameters (organic solvent, polymer and surfactant concentration). Next, BRP-187-loaded PCL nanoparticles were prepared by batch nanoprecipitation and characterized using dynamic light scattering, scanning electron microscopy and UV-Vis spectroscopy to determine and to compare particle size, polydispersity, zeta potential, drug loading as well as the apparent enzymatic degradation as a function of the copolymer composition. Ultimately, NPs were examined for their potency in vitro in human polymorphonuclear leukocytes to inhibit the BRP-187 target 5-lipoxygenase-activating protein (FLAP). It was evident by Tukey's multi-comparison test that the degree of crystallinity of copolymers directly influenced their apparent enzymatic degradation and consequently their efficiency to inhibit the drug target.

Keywords: polycaprolactone (PCL); polyesters; hydrophobic-hydrophilic balance (HHB); nanoparticle formulation; nanoparticle crystallinity; FLAP antagonist; BRP-187

1. Introduction

With the first clinical approval of a polymer-based nano-drug in 1995 [1], interest in developing polymers as nanocarriers of (bio)pharmaceutical drugs has been steadily growing [2,3]. Due to their favorable characteristics, e.g., adjustable physical and mechanical properties, it is not surprising that polymer-based nanomaterials are now established in many areas of bionanotechnology. Polymers are widely used in delivery systems for therapeutics, in matrices for tissue engineering and, among others, in polymer-based composites for biomedical purposes [4–6].

The main criteria in designing polymers for therapeutic use have been based mainly on the biocompatibility and the biodegradability of the polymer backbone as well as the suitability of the polymer to be processed into a stable pharmaceutical formulation [7].

However, there are other parameters of equal importance to be considered to optimize a polymer for its application as a delivery vehicle. In fact, parameters such as molar mass, functional end-groups, hydrophobic-hydrophilic balance (HHB), melting temperature (T_m) and crystallinity strongly influence the drug loading and the drug release kinetics from the polymer matrix [8].

Independent investigations of the influence of the polymer crystallinity on the resulting particle characteristics (e.g., particle formation and degradation) while keeping the key properties of the system constant are rare or only provide partial conclusions typically due to influences of a third variable [9]. In particular, alterations of HHB are frequently accompanied by changes in crystallinity [10]. It is hence currently not fully understood if degradation or general performance of hydrophobic pharmapolymers in aqueous media are, in fact, strongly influenced by polymer crystallinity or if the hydrophobicity is the dominating factor.

Today, the most commonly used polymers for biomedical applications are polyesters, such as polylactide (PLA), poly(lactide-*co*-glycolide) (PLGA) and polycaprolactone (PCL) [11]. They are easy to access and offer a range of interesting advantages; i.e., (i) a complete hydrolytic and/or enzymatic biodegradation, (ii) a facile and controlled synthesis to obtain defined molar masses, (iii) various modification possibilities of the polymer structure, and (iv) commercial availability [12]. The advantages of PCL compared to other aliphatic polyesters include interesting thermal properties, higher durability and manufacturability, and a good compatibility with other polymers [13,14]. Thus, PCL represents a promising candidate to design materials with tailor-made properties [13,15,16]. Bandelli et al. recently demonstrated that copolymerization of the constitutional isomers ε-caprolactone (εCL) and δ-caprolactone (δCL) with a varying ratio of εCL and δCL can generate a library of five copolyesters featuring a constant HHB and similar molar masses in the range of 7 to 10 kDa, but the copolymers showed a varying crystallinity [17]. They are hence suitable materials to study the sole influence of crystallinity on the particle properties and performance. In this study, we utilized this library of poly(εCL-*ran*-δCL) to formulate drug-loaded nanoparticles (NPs). The aim was to investigate, firstly, whether such polymers provide suitable properties (particle size and polydispersity) to form an NP-based drug delivery system, and secondly, to study the effect of the crystallinity of the bulk polymers on the enzymatic degradation and the in vitro performance of the resulting NPs.

The anti-inflammatory drug BRP-187 (4-(4-chlorophenyl)-5-[4-(quinoline-2-ylmethoxy)-phenyl] isoxazol-3-carboxylic acid) is a dual inhibitor of the 5-lipoxygenase-activating protein (FLAP) and microsomal prostaglandin E2 synthase-1 (mPGES-1), which are crucial proteins within arachidonic acid (AA) metabolism. Inhibition of mPGES-1 and FLAP prevents the biosynthesis of pro-inflammatory prostaglandin (PG)E_2 and leukotrienes (LTs), respectively [18]. Several in vitro and in vivo studies with inhibitors of FLAP and/or mPGES-1 have demonstrated their efficient anti-inflammatory activity while exhibiting fewer adverse effects compared to the conventional non-steroidal anti-inflammatory drugs (NSAIDs) [19,20]. These observations suggest that dual inhibition of FLAP and/or mPGES-1, rather than blocking cyclooxygenase-1 or -2 pathways, might be a better strategy for intervention with inflammation. However, BRP-187 is a fatty acid-like molecule with poor water solubility and a strong tendency to bind plasma proteins [18]. Molecules exhibiting such properties typically cause challenges in reaching a sufficient bioavailability in vivo and require technological solutions to improve their pharmacokinetic drawbacks. We have previously demonstrated that encapsulating BRP-187 into PLGA NPs and acetalated dextran NPs enhanced its enzyme inhibition efficacy in vitro [21].

In the present study, we initially performed a high-throughput (HT) nanoprecipitation approach for the homopolymers PεCL and PδCL to screen a range of polymer and surfactant concentrations for the preparation of empty (unloaded) PCL particles. Once optimal formulation conditions were established, and drug-loaded NPs were prepared by batch nanoprecipitation of the PεCL and PδCL homopolymers as well as of the poly(εCL-*ran*-δCL) copolymers with BRP-187. PCL NPs with and without BRP-187 were characterized for their critical quality attributes, namely their particle size, polydispersity index (PDI), surface charge and drug loading. Other particle properties i.e., nanodispersion stability and NP degradation behavior were also investigated. Ultimately, BRP-187-NPs were studied in vitro in human polymorphonuclear leukocytes (PMNL) for their efficiency to inhibit the drug target FLAP, in comparison to the free BRP-187.

2. Methods

2.1. Materials

The PCL homopolymers and P(εCL-*ran*-δCL) copolymers were synthesized as previously reported [17]. Key characterization data are listed in Table S1 in the Supplementary Information (SI). For further details on their synthesis, the reader is referred to literature reports [17]. Polyvinylalcohol (PVA) (Mowiol 4-88), tetrahydrofuran (THF), dimethylsulfoxide (DMSO) and lipase from the yeast *Candida rugosa* were purchased from Sigma-Aldrich (Germany). BRP-187 was synthesized according to a published protocol [18]. Further materials are described in the specific experimental sections.

2.2. Automated High-Throughput Nanoprecipitation

Automated high-throughput nanoprecipitation was performed in a 96-well plate (Greiner Bio-One GmbH, Frickenhausen, Germany) utilizing a FasTrans liquid handling robot (Analytik Jena GmbH, Jena, Germany). Starting with a polymer stock solution of 10 mg mL^{-1} in THF, a dilution series with varying concentrations (0.25, 0.5, 1, 2, 3, 4, 5, 6, 7, 8, 9 and 10 mg mL^{-1}) was prepared. The polymer solutions (40 μL) were then automatically pipetted into 200 μL of either purified water (GenPure ultrapure water system, Thermo Scientific, Waltham, MA, USA) or PVA surfactant-containing aqueous solutions with a concentration of 0.25%, 0.5% or 1.0% (*w/v*). The resulting NP dispersions were mixed by pipetting up and down three times and then left for two hours for solvent evaporation. Each formulation was prepared twice. After solvent evaporation, the samples were diluted with pure water (1:2 ratio for a polymer concentration up to 4 mg ml^{-1} and 1:10 ratio for all NPs prepared with a polymer concentration above 4 mg mL^{-1}) and investigated via dynamic light scattering (as described in Section 2.4) [22].

2.3. Batch Nanoprecipitation

Polymer solutions with 5 mg mL^{-1} or 2.5 mg mL^{-1} were prepared in THF via batch nanoprecipitation. For the drug-loaded particles, 10 mg mL^{-1} of BRP-187 dissolved in DMSO were mixed with the polymer solution prior to formulation, which corresponded to 3% (*w/w*) of the drug to polymer mass. The drug stock solution was sonicated in an ultrasound water bath for 15 min at room temperature to ensure good dissolution. Particle formulation was carried out by injecting the polymer/drug solution into an aqueous phase containing 0.3% (*w/v*) PVA using a syringe pump (Aladdin AL1000-220, World Precision Instruments, Berlin, Germany) with a flow rate of 2 mL min^{-1} while stirring at 800 rpm. The solvent/non-solvent ratio was set to 1:8. The resulting particle suspensions were stirred for 24 h at room temperature for solvent evaporation and then centrifuged at 12.851 x g for 60 min at 20 °C using a Rotina 380 R centrifuge (Hettich Lab Technology, Tuttlingen, Germany). The supernatant was removed, and the NPs were redispersed in 2.5 mL pure water, vortexed and sonicated in an ultrasonic water bath for 30 min. The NPs were stored overnight at 4 °C and lyophilized in aliquots of 200 μL. After lyophilization, the mass of the NPs was determined using a precise analytical balance (MYA 11.4Y, Radwag Waagen, Hilden, Germany). The yield was calculated as follows: (mass of NPs recovered − mass of found PVA)/(mass of polymer + mass of drug) in the formulation × 100. To check reproducibility, five individual batches of the drug-loaded PCL particles were prepared and analyzed individually. The data provided represent the average values and the standard deviation of these five batches.

2.4. Dynamic Light Scattering (DLS) and Electrophoretic Light Scattering (ELS)

DLS measurements were performed utilizing a Nano ZS (Malvern Panalytical, Malvern, United Kingdom) with a laser wavelength of λ = 633 nm with non-invasive back-scatter (NIBS) technology [22]. The particle size is reported as the hydrodynamic diameter (d_H). The particle size distribution (PDI) was measured using pure water as a dispersant with a refractive index RI of 1330 and a viscosity of 0.8872 cP at 25 °C. Samples obtained from the automated HT-nanoprecipitation were measured at 25 °C in a micro cuvette (Brand GmbH, Wertheim, Germany) without any filtering step with the following settings: measurements of each sample were repeated three times for 10 sec at 25 °C. The samples obtained from batch nanoprecipitation were measured at a dilution of 1:10 up to 1:100 utilizing the following settings: five repeated measurements, each with five runs of 30 s. The zeta-potential of the lyophilized NPs was investigated by ELS using the same instrument at 25 °C with three repeated measurements.

The apparent degradation behavior of the NPs was analyzed by DLS by monitoring changes in the mean count rate at fixed measurement settings: measuring position at 4.65, attenuator factor 7 at 37 °C [21]. Before investigating, NPs were mixed with the enzyme solution (a lipase from Candida rugosa) in a 1:4 mass ratio of polymer to enzyme and incubated at 37 °C for pre-determined timepoints.

2.5. UV-Vis Spectroscopy Measurements

UV-Vis spectroscopy measurements were performed with the Infinite M200 Pro plate reader (Tecan Group, Männedorf, Switzerland). For determination of the encapsulation efficiency (EE) and the loading capacity (LC) of the BRP-187 in the PCL particles, lyophilized NPs were dissolved in DMSO, and the solutions were investigated in a flat-transparent 96-well quartz plate (Hellma, Jena, Germany) at λ = 316 nm with 3 $\times$ 3 multiple reads per well and a 2000 µm well border. A calibration curve of BRP-187 was obtained for each batch in the concentration range of 1.2 to 312.5 µg mL^{-1} with R^2 = 0.9997. The LC was calculated as follows: LC = (mass of drug recovered)/(mass of particle recovered) x 100. The EE was calculated as follows: EE = LC found/(mass of drug used) $\times$ 100. The determination of PVA in the NPs (%, *w/w*) was performed according to the published protocol [23].

2.6. Scanning Electron Microscopy (SEM)

A Sigma VP Field Emission Scanning Electron Microscope (Carl-Zeiss, Jena, Germany) equipped with an InLens detector with an accelerating voltage of 6 kV was used for electron microscopy imaging. Before the measurement, the samples were coated with a thin layer of platinum (4 nm) via sputter coating (CCU-010 HV, Safematic, Zizers, Switzerland).

2.7. Cell Isolation

The leukocytes isolation was performed according to a published protocol [21]. Leukocyte concentrates were prepared from peripheral blood obtained from healthy human adult donors that received no anti-inflammatory treatment for the last ten days (Institute of Transfusion Medicine, University Hospital Jena). The approval for the protocol was given by the ethical committee of the University Hospital Jena, and all methods were performed in accordance with the relevant guidelines and regulations. To isolate PMNL, the leukocyte concentrates were mixed with dextran (*Leuconostoc* spp. MW ~40,000, Sigma Aldrich, Taufkirchen, Germany) for sedimentation of erythrocytes and the supernatant was centrifuged on lymphocyte separation medium (Histopaque®-1077, Sigma Aldrich, Taufkirchen, Germany). Contaminating erythrocytes in the pelleted neutrophils were removed by hypotonic lysis (water). PMNL were then washed twice in ice-cold phosphate-buffered saline (PBS) and finally resuspended in PBS plus 0.1% of glucose and 1 mM $CaCl_2$.

2.8. Determination of FLAP-Dependent 5-LO Product Formation in PMNL

The evaluation of the effects on FLAP was performed according to our established protocol [21]. We assessed FLAP-dependent 5-LO product formation in human PMNL, cells (5×10^6 mL^{-1}) were pre-incubated with BRP-187 or NPs for indicated timepoints at 37 °C. The cells were stimulated with 2.5 µM Ca^{2+}-ionophore A23187 (Cayman, Ann Arbor, USA) for 10 min, and the incubation was stopped with 1 mL ice-cold methanol containing 200 ng mL^{-1} PGB_1 as an internal standard. Samples were subjected to solid phase extraction, and the formed lipid mediators (leukotriene B_4 (LTB_4), trans-isomers of LTB_4, 5-hydroxyeicosatetraenoic acid (5-HETE)) were separated and analyzed by reverse-phase high-performance liquid chromatography (RP-HPLC) as previously described [24]. Statistical analysis was performed with log-transformed values to obtain Gaussian-distributed data sets. Experiments were analyzed via one-way ANOVA and Tukey's multicomparison test with GraphPad Prism 9.1.2 (GraphPad, La Jolla, CA, USA).

2.9. Cell Viability

Freshly isolated PMNL were incubated with a control sample with 0.1% DMSO, BRP-187 (10 µM) or NPs containing the respective amount of BRP-187 (10 µM) at 37 °C in PBS containing 0.1% of glucose. After 5 h the cell suspension was subjected to a Vi-CELL XR cell counter (Beckman Coulter, Lahntal, Germany), for determination of cell viability by trypan blue staining.

3. Results and Discussion

In our previous study, five poly(εCL-*ran*-δCL) copolymers, herein named ε87-δ13, ε81-δ19, ε75-δ25, ε61-δ39 and ε45-δ55, and the two respective homopolymers PεCL and PδCL, herein referred to as ε100-δ0 and ε0-δ100, were synthesized exhibiting a constant HHB [17]. It was demonstrated that the HHB of the bulk polymers correlated with the HHB of the corresponding NPs when particles were prepared in THF using a polymer concentration of 1 mg mL^{-1} [17]. In the present study, the particle formation of the ε100-δ0 and ε0-δ100 was investigated over a wider range of polymer concentrations in THF ranging from 0.25 to 10 mg mL^{-1} using an automated pipetting robot that was adapted for the HT-nanoprecipitation [25]. Particles were formulated without surfactant as well as with PVA of different concentrations (0.25 to 1% (*w/v*)). Previous studies revealed that PVA of less than 0.5% (*w/v*) generated stable drug-loaded PLGA NPs, and it could be demonstrated that even concentrations of up to 5% (*w/v*) were generally non-toxic in vitro [26]. ε100-δ0 and ε0-δ100 homopolymers both formed NPs up to the highest tested polymer concentration of 10 mg mL^{-1} when PVA was used as a surfactant (SI, Figure S1). Even the lowest tested PVA concentration of 0.25% (*w/v*) was sufficient to obtain stable particle dispersions and ε100-δ0 NPs with a size of 150 to 300 nm and ε0-δ100 NPs with a particle size of 120 to 280 nm with PDI < 0.3. However, ε0-δ100 NPs prepared without surfactant failed to produce stable NP dispersions above concentrations of 0.5 mg mL^{-1} as indicated by a strong aggregation of the particles. This is not surprising since ε0-δ100 is above its glass transition temperature at room temperature, which could disturb the particle formation in the absence of a stabilizer. It is well-known that several factors influence the final NP properties, including the polymer concentration, the solvent used to dissolve the polymer and the type and the concentration of the surfactant [26–28]. THF was demonstrated to be a suitable solvent in the HT-screening, resulting in stable particle formation within a broad polymer concentration range when PVA was used as a surfactant. Hence, it was selected as solvent for the subsequently performed BRP-187 encapsulation experiments. All other formulation parameters for the preparation of PCL[BRP-187] NP were adapted from our previous study that described the encapsulation of BRP-187 into PLGA NPs [21]. The first batch nanoprecipitation with the drug and a polymer concentration of 5 mg mL^{-1} in THF yielded large particles with a diameter (d_H) of 400 to 600 nm with high LC values (SI, Table S2). However, the particles revealed significant aggregation after centrifugation and lyophilization, as indicated by the higher PDI values of 0.3 to 0.6. Hence, the initial polymer concentration was reduced to 2.5 mg mL^{-1} to optimize the dispersion stability and to decrease the particle size [28]. Particles within a size range of 200 to 260 nm and PDI values below 0.3 were obtained for all PCLs using a polymer concentration of 2.5 mg mL^{-1}

(Table 1, SI Table S4). It was further observed that empty NPs were approximately 30 to 50 nm smaller compared to the BRP-187-loaded NPs (SI, Table S3). The particle size of the empty NPs increased by approximately 40 to 80 nm when NPs were lyophilized and subsequently reconstituted in water (SI, Table S3). Similar tendencies were also observed for the PCL[BRP-187] NPs, although here the difference in size was on average only about 30 to 50 nm (Table 1), presumably caused by the strong affinity of the hydrophobic drug with the polymer matrix [29]. The particles were also investigated via SEM (Figure 1), which revealed individual or clustered particle populations within the particle size range as indicated by DLS measurements.

Table 1. Overview of PCL[BRP-187] NP properties prepared in THF using a polymer concentration of 2.5 mg mL^{-1}.

εCL/δCL (mol %)	T_m (°C)	X_c [a] (%)	d_H [b] (nm)	PDI [b]	ZP [c] (mV)	d_H [c] (nm)	PDI [c]	PVA % (*w*/*w*)	Yield [d] (%)	LC [e] (%)
ε100-δ0	69	73	229 ± 13	0.08 ± 0.02	−50 ± 1	268 ± 21	0.27 ± 0.09	4.5	87	1.5 ± 0.1
ε87-δ13	54	44	211 ± 5	0.08 ± 0.02	−38 ± 2	251 ± 13	0.30 ± 0.14	4.5	76	1.4 ± 0.5
ε81-δ19	52	38	218 ± 13	0.08 ± 0.02	−41 ± 1	267 ± 24	0.37 ± 0.27	4.7	81	1.4 ± 0.2
ε75-δ25	42	28	225 ± 13	0.16 ± 0.11	−34 ± 1	260 ± 23	0.42 ± 0.20	5.0	67	1.9 ± 0.6
ε61-δ39	24	4	209 ± 13	0.06 ± 0.12	−40 ± 1	223 ± 16	0.16 ± 0.10	8.2	61	1.7 ± 0.1
ε45-δ55	/ *	0	200 ± 13	0.10 ± 0.12	−32 ± 1	237 ± 61	0.18 ± 0.08	6.7	52	1.4 ± 0.2
ε0-δ100	/ *	8	259 ± 32	0.28 ± 0.14	−45 ± 2	262 ± 20	0.26 ± 0.26	5.5	54	3.2 ± 1.2

d_H represents the intensity-weighted distribution (n ≥ 4 batches) and zeta-potential (ZP) (n = 3 ELS measurements) * Amorphous or near amorphous polymers with glass transition temperature T_g below 37 °C [17]. [a] Bulk degree of crystallinity as determined by wide-angle X-ray scattering (WAXS) at room temperature. [b] NPs measured after purification. [c] NPs measured after lyophilization and subsequent resuspension in water. [d] Yield = (mass of NPs recovered − mass of found PVA)/(mass of polymer + mass of drug) in the formulation × 100. [e] Determined by UV-VIS spectroscopy at λ = 316 nm (n = 4) and calculated using LC = (mass of drug recovered)/(mass of particle recovered) × 100.

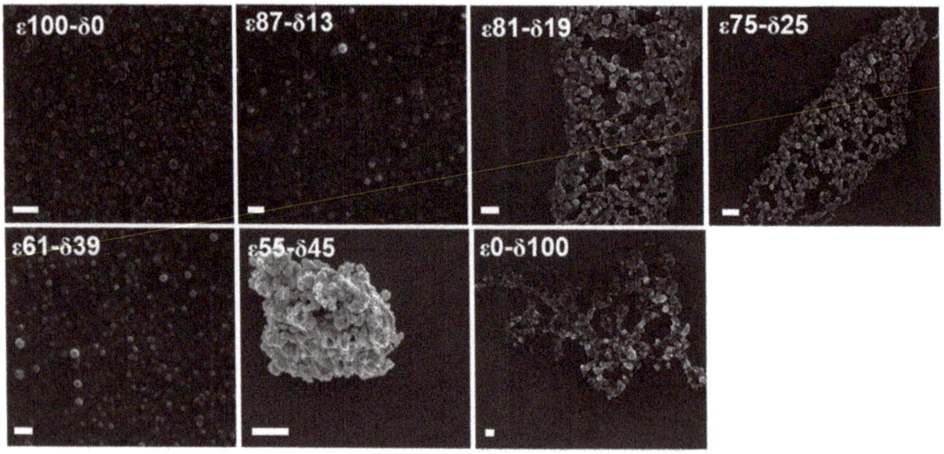

Figure 1. SEM micrographs of PCL[BRP-187] particles consisting of the homo- or copolymers with a varying composition. Scale bar = 1 μm.

The average LC of the PCL[BRP-187] NPs was between 1.4 and 1.9% for ε100-δ0 and the poly(εCL-*ran*-δCL) copolymers (Table 1) and similar to the LC values of PLGA NPs encapsulating the same drug [21]. The only exception was the ε0-δ100 homopolymer with an LC of 3.2%, probably due to its almost liquified state at room temperature. This resulted in a viscous dispersion with emulsion-like properties in which the drug was apparently entrapped during the purification process.

In general, the yield of both empty and drug-loaded PCL NPs decreased with increasing molar fraction of δCL (Figure 2A). In other words, NP yield increased with the degree of crystallinity of the polyester materials. Amorphous materials are frequently utilized as excipients in pharmaceutical formulations since they are known to increase the dissolution rate of insoluble drugs and to enhance their bioavailability [30]. However, their major disadvantage is seen in the fact that they exhibit high energy states at a molecular level and thus are prone to physical instabilities. In particular, such tendencies were observed with the NPs of the amorphous PδCL homopolymer, which displayed a higher polydispersity and the lowest yield. In technical terms, the low yield of the copolymers with a higher fraction of δCL could have resulted from their near-molten state at room temperature causing them to sediment at a lower rate due to their lower density. Thus, after 60 min of centrifugation, a lower amount of the NPs was recovered.

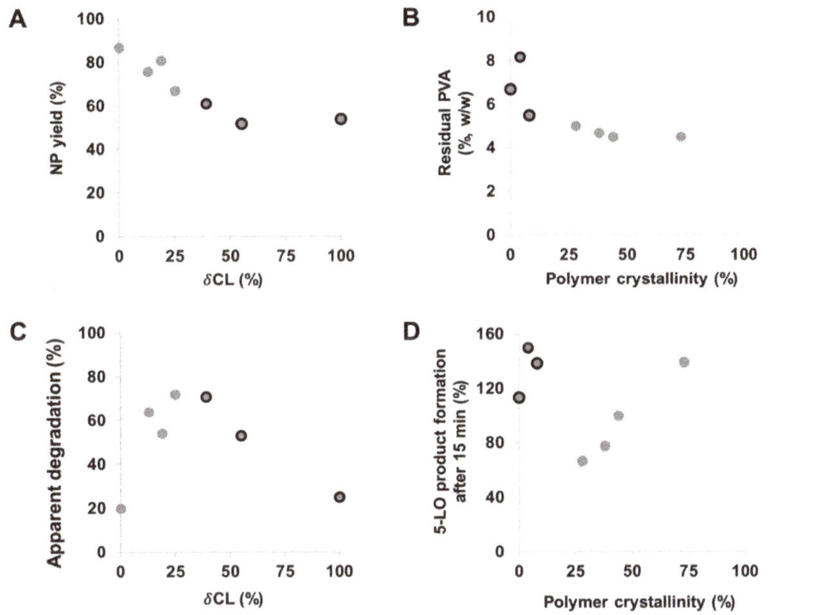

Figure 2. Influence of the δCL fraction on the yield of drug-loaded PCL NPs (**A**), influence of the polymer crystallinity on the residual PVA content of drug-loaded PCL NPs (**B**), apparent degradation represented by the normalized relative count rate (%) after 20 h plotted against the δCL fraction of the copolymers (**C**) and influence of polymer crystallinity on the efficiency of drug-loaded PCL NPs to inhibit 5-LO product formation (**D**). Black-circled data points represent PCL polymers with a degree of crystallinity below 10% and a $T_g < 37$ °C.

Furthermore, it was observed that the residual amount of PVA in the drug-loaded NPs was higher compared to the empty NPs for all PCL copolymers (Table 1 and SI, Table S3). As mentioned before, such differences are typically a result of strong drug–polymer interactions [31], and in this case, the interactions of the BRP-187 with the chains of PVA polymer. Moreover, the residual PVA content was noticeably higher for less crystalline copolymers with a higher δCL fraction and highest for the particles consisting of the PδCL homopolymer (Figure 2B). Apparently, the surfactant molecules tended to stick to the surface or were even incorporated into the particles formed from amorphous polyesters that are above their glass transition temperature during formulation. As soon as the materials were semicrystalline and below T_m, the degree of crystallinity did not influence the amount of residual PVA anymore. Besides providing dispersion stability, surfactants also influence the degradation rate of NPs since they adsorb at the surface of the particles forming a layer that protects from enzymatic hydrolysis to some degree [32]. Additional characterization experiments of the PCL[BRP-187] NPs were performed to investigate the degradation kinetics as well as the biological evaluation of the NP efficiency to inhibit the drug targets in vitro.

3.1. Degradation Studies

Among the aliphatic polyesters that are most commonly investigated for drug delivery applications, PCL has a superior thermal stability, with a decomposition temperature of 100 °C higher above that of the typical PLA- and PGA-based polymers [15]. Due to its high durability, PεCL has found a wide range of applications mainly for implantable medical devices [33,34], in which degradation occurs over two to four years [13]. However, to tailor their application for drug delivery purposes, faster degradation kinetics of the PεCL are desirable and can be achieved by copolymerization of εCL with its isomer δCL [9]. Introducing δCL repeating units to the PεCL polymer decreases its degree of crystallinity [17], and as such, it increases its rate of degradation as confirmed by investigations of films [35]. Figure 3 shows the enzymatic degradation of the PCL[BRP-187] particles incubated for 24 h at 37 °C as monitored by DLS. The apparent NPs degradation was inferred by monitoring changes in the sample concentration over time, as indicated by the count rate on the DLS under constant measurement settings [21]. In agreement with literature reports regarding film degradation, Figure 3 reveals that the degradation of the most crystalline ε100-δ0 was the slowest in the nanoparticulate state. ε100-δ0 is predominantly a semicrystalline material with a melting point considerably higher than the experimental temperature of 37 °C. It was noticed that except for the ε0-δ100 homopolymer, which degraded only about 25% after 24 h, the NP degradation rate generally increased with the amount of the δCL (Figure 3A). This was expected since the long-range order and the compact structure of crystalline materials requires higher levels of energy for degradation compared to the less organized molecular arrangement of amorphous materials [29,36].

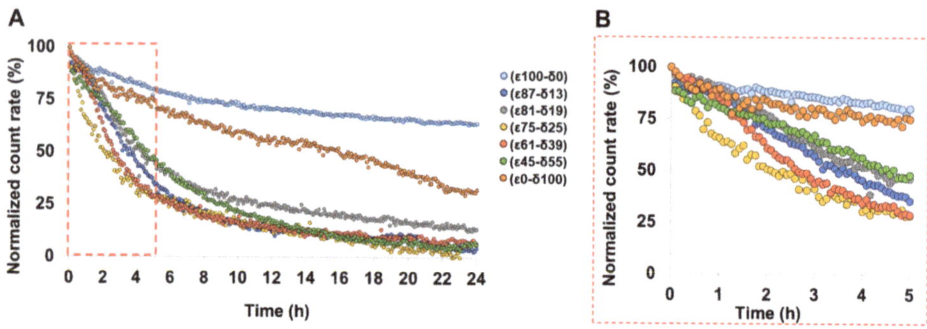

Figure 3. Normalized count rate of BRP-187-loaded PCL NPs incubated with *Candida rugosa* as measured by DLS for 24 h (**A**). (**B**) depicts a zoomed-in area into the data until 5 h.

This observation is further confirmed by other studies that have also demonstrated that the amorphous regions within bulk and/or films of PεCL polymer degraded faster compared to the crystalline regions [35,37,38]. Another study with similar observations argued that polyesters with higher crystallinity exhibit a slower degradation because in a densely packed crystal, it is more difficult for the enzymes to reach the cleavable bonds [39]. In general, our results revealed that all copolyester NPs featured an apparent degradability above 50% within 5 h (Figure 3B). A faster initial degradation was particularly observed for the ε75-δ25 and ε61-δ39 copolymers since they exhibit melting points (42 °C and 24 °C, respectively [17]) that are closer to the experimental temperature of 37 °C, which was chosen to simulate the conditions of the human body (Figures 2C and 3B) [8].

3.2. In vitro Performance of NPs

Although clear influences of the polymer crystallinity and physical state on NP formulation and enzymatic degradation were found, other effects might come into play in the more complex environment of a cell. The PCL[BRP-187] particles were hence tested in human PMNL for bioactivity. PMNL are the most abundant leukocytes in the blood and are a major source for FLAP-dependent 5-LO product biosynthesis, thus they are suitable cells for evaluation of various anti-LT agents. Note that FLAP as helper-protein of 5-LO has no enzyme activity that can be experimentally assessed, but instead assists 5-LO in LT formation by facilitating the access towards the substrate for the 5-LO enzyme. At first, the PCL[BRP-187] NPs were compared to the free drug for their influence on the cell viability of PMNL (cytotoxicity). No cytotoxic effects of the particles were found within a 5 h incubation as shown in the SI (Figure S2). These results were in agreement with previous studies that demonstrated PCL NPs to be biocompatible [40,41]. Considering their good biocompatibility, all PCL[BRP-187] particles as well as the free drug were studied for their efficiency to inhibit the drug target FLAP in PMNL and, thus, to prevent 5-LO product formation [42]. Therefore, a drug concentration of 0.3 μM was chosen for free and encapsulated BRP-187, which were investigated at different preincubation times (15 min, 1 h, 2 h and 5 h, respectively). As shown in Figure 4A, 5-LO product formation was clearly suppressed after 15 min of incubation with the PCL[BRP-187] particles to variant degrees, but essentially the particles performed as efficiently as the free drug. Apart from this, there was no significant difference between the different PCL[BRP-187] polymers at longer incubation time points (i.e., 1–5 h; Figure 4B and SI, Figure S3). More specifically, the NPs prepared with ε75-δ25 prevented the 5-LO product formation most after 15 min of incubation (Figure 4A). This observation also correlated with the fastest apparent degradation of the ε75-δ25 copolymer (Figure 3B), which might be promoted by its melting point of 42 °C, which is around the temperature of cell incubation (i.e., 37 °C). Karavelidis et al. reported that other polyesters with melting points around 37 °C exhibited a faster drug release [8]. It can be inferred that the rapid degradation of ε75-δ25 led to an accelerated release of the BRP-187, thereby considerably preventing the 5-LO product formation at early time points (Figure 2C,D). NPs formed from PCL with higher εCL fraction and hence higher T_m, as well as a higher degree of crystallinity, were less effective. As shown in Figure 2D, the 5-LO product formation was almost linearly dependent on the polymer crystallinity if only the semicrystalline materials are considered. The better performance of the NPs with lower crystallinity could be explained by two effects based on two different release mechanisms. Firstly, less crystalline materials with a larger fraction of amorphous domains enable a faster diffusion of the drug through the polymer matrix without barriers formed by crystalline domains [11,43]. Secondly, if the drug release is promoted through polymer degradation, these amorphous domains would most likely be more accessible for enzymes catalyzing the polyester hydrolysis [11].

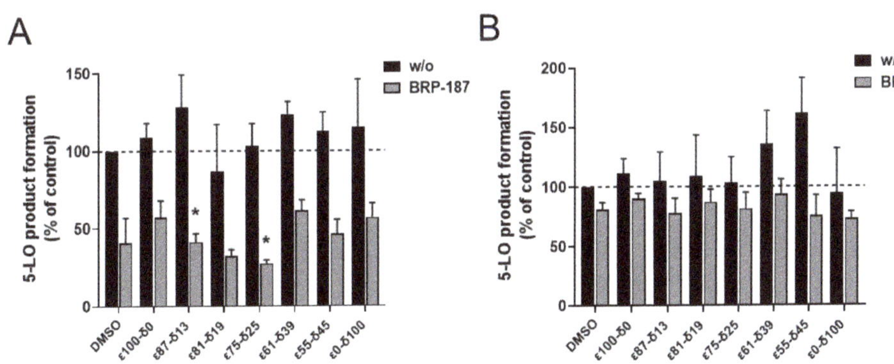

Figure 4. Measurement of 5-LO product formation as an indicator for the inhibition of the drug target FLAP by BRP-187 [36]. PMNL preincubated with DMSO, BRP-187 (0.3 μM), empty PCL particles (labeled as w/o) or PCL particles with BRP-187 (labeled with BRP-187; 0.3 μM respective BRP-187) for 15 min (**A**) or 5 h (**B**) at 37 °C. Values are given as 5-LO products as a percentage of control (DMSO) (n = 3). Statistical analysis was performed via one-way ANOVA and Tukeys multi comparison test with logarithmic trans-formed data (* $p < 0.05$).

Based on the apparent enzymatic degradation of the PCL NPs, a burst release of the drug is conceivable considering the immediate decrease in the count rate of at least 5 to 10% of all PCL NPs (Figure 3B). The slightly higher efficiency of the NPs to inhibit 5-LO product formation in PMNL within 15 min supports this idea (Figure 4A). For the polymers with a degree of crystallinity below 10% and $T_g < 37$ °C, namely ε61-δ39, ε55-δ45 and ε00-δ100, inhibition of 5-LO product formation is less apparent after 15 min of incubation (Figure 2D, black-circled data points). This is presumably because these polymers are molten and more viscous at 37 °C. As a consequence, they could delay the release of the drug and therefore hamper the drug action in the cells.

Furthermore, the coating effect of PVA might reduce the influence of the crystallinity of NPs or their intracellular drug release. It is reported that increasing amounts of residual surfactant decrease the cellular uptake of the NPs [44,45]. This could explain why the PCL copolymers with a higher fraction of δCL containing more residual PVA (Figure 2B) were less efficient to inhibit 5-LO product formation after 15 min of incubation (Figure 2C, Figure 4A) when compared to the PCL copolymers with a higher εCL fraction containing less residual PVA. However, no correlation was observed between suppression of 5-LO product formation and the PVA content in the formulation, showing that the trend cannot be generalized (SI, Figure S4).

4. Conclusions

A library of poly(εCL-*ran*-δCL) copolymers with a constant HHB but different degrees of crystallinity were used to encapsulate BRP-187 into polymer NPs. PCL[BRP-187] particles with a diameter of 200 to 300 nm were successfully produced, whereby a comparable drug-loading was observed with LC between 1.4% and 1.9%, with the exception of the PδCL homopolymer, which revealed a higher LC. It was evident that the degree of crystallinity directly influenced the enzymatic degradation rate of the PCL copolymer, whereby the degradation increased with an increasing fraction of δCL repeating units. In addition, increasing the amount of δCL in the polymer increased the amount of residual surfactant in the NP formulation but decreased the final NP yield. The release of bioactive BRP-187 from the PCL NPs was demonstrated in vitro in PMNL by inhibiting FLAP-dependent 5-LO product formation, whereby the inhibition efficiency was dependent on the degree of crystallinity of the copolymers used for the particle formulation. The NPs of ε75-δ25 revealed the fastest degradation and inhibited the 5-LO product formation more than the other copolymers after 15 min of incubation in PMNL; longer preincubation times (1 to 5 h) reduced the potency. In conclusion, although all PCL copolymers were suited to produce NPs, the ε75-δ25 copolymer can be considered as a more promising candidate to be further investigated for both its physicochemical properties and its performance in more complex biological models. When designing superior materials for NP-mediated drug delivery, it hence seems promising to rely on polymers that are in a solid state of matter at 37 °C but feature a low degree of crystallinity. However, it is not yet clear if these observations can be applied to other systems. Thus, our future research will concentrate on the encapsulation of other anti-inflammatory drugs in the polymer library with constant HHB to determine if the effect of polymer crystallinity of the present PCL systems can be transferred to other actives. In addition, we are currently establishing similar libraries mimicking the HHB of PLA to understand if our findings can be generalized in the field of polyester-based drug carrier materials.

Supplementary Materials: The following are available online at https://www.mdpi.com/article/10.3390/polym13152557/s1, Figure S1: Hydrodynamic diameter (intensity-weighted distribution, circles) and PDI (bars) of the homopolymer ε100-δ0 and ε0-δ100 NPs over a range of PVA concentration used in the formulation, Figure S2: Cell viability measured with a Beckman ViCell XR cell counter by trypan blue staining. A total of 1 × 107 PMNL were diluted in PBS plus 0.1% of glucose and incubated with DMSO, BRP-187 (10 μM), empty PCL particles (labeled as w/o) or PCL particles with BRP-187 (labeled with BRP-187; respective amount to 10 μM BRP-187) for 5 h at 37 °C. Values are given as 5-LO products as a percentage of control (DMSO) (n = 3), Figure S3: Measurement of 5-LO product formation as indicator for the inhibition of the drug target 5-lipoxygenase-activating protein (FLAP) by BRP-187.[37] A total of 5 × 106 polymorphonuclear leukocytes (PMNL) diluted in PBS containing 0.1% glucose and 1mM $CaCl_2$ were preincubated with DMSO, BRP-187 (0.3 μM), empty PCL particles (labeled as w/o) or PCL particles with BRP-187 (labeled as BRP-187; 0.3 μM respective BRP-187) for 1 h (A) and 2 h (B) at 37 °C and further stimulated with 2.5 μM A23187 for 10 min. The reaction was stopped with 1 mL ice-cold methanol containing 200 ng mL-1 PGB1 as internal standard. Lipid mediators were extracted via solid-phase extraction (SPE) and analyzed with HPLC. Values are given as 5-LO products (LTB4, its trans-isomers 4 and 5-HETE) as a percentage of control (DMSO) (n = 3), Figure S4: Influence of the residual PVA on the efficiency of drug-loaded PCL NPs on 5-LO inhibition. Black-circled data points represent PCL polymers with bulk degree of crystallinity below 10% and glass transition temperature Tg < 37 °C. Table S1: Molar mass and composition of the (co)polyesters. Details are described in a previous publication, Table S2: Properties of PCL[BRP-187] NPs formulated from THF utilizing polymer concentration of 5 mg mL^{-1} (n = 1 batch), Table S3: Particle properties of empty PCL NPs prepared in THF with c = 2.5 mg mL 1 (n = 2 batches) obtained by DLS and ELS measurements after purification and after lyophilization and subsequent resuspension (n = 2 for purified NPs, n = 1 for lyophilized NPs), Table S4: DLS intensity-weighted size distribution of PCL[BRP-187] NPs of one formulation round after purification, as well as after lyophilization and resuspension in water.

Author Contributions: Performance of experiments, A.V., C.K., B.B.-S., S.S.; writing —review and editing, A.V., C.K., B.S., C.W.; polymer synthesis, D.B.; synthesis of BRP-187, J.A.C.; supervision, S.H., C.W., O.W., U.S.S.; project administration, A.V. All authors have read and agreed to the published version of the manuscript.

Funding: This work was supported by the Deutsche Forschungsgemeinschaft (DFG), Collaborative Research Center SFB 1278 "PolyTarget", (project number 316213987, projects A04, A06 and Z01). The SEM facilities of the Jena Center for Soft Matter (JCSM) were established with a grant from the DFG (INST 275/307-1 FUGG).

Institutional Review Board Statement: Not applicable.

Informed Consent Statement: Not applicable.

Data Availability Statement: The data presented in this study are available on request from the corresponding author.

Conflicts of Interest: The authors declare no conflict of interest.

References

1. Barenholz, Y. Doxil[(R)]—the first fda-approved nano-drug: Lessons learned. *J. Control. Release* **2012**, *160*, 117–134. [CrossRef]
2. Mitragotri, S.; Burke, P.A.; Langer, R. Overcoming the challenges in administering biopharmaceuticals: Formulation and delivery strategies. *Nat. Rev. Drug Discov.* **2014**, *13*, 655–672. [CrossRef]
3. Murthy, S.K. Nanoparticles in modern medicine: State of the art and future challenges. *Int. J. Nanomed.* **2007**, *2*, 129–141.
4. Guo, Z.C.; Poot, A.A.; Grijpma, D.W. Advanced polymer-based composites and structures for biomedical applications. *Eur. Polym. J.* **2021**, *149*, 110388. [CrossRef]
5. Keshvardoostchokami, M.; Majidi, S.S.; Huo, P.P.; Ramachandran, R.; Chen, M.L.; Liu, B. Electrospun nanofibers of natural and synthetic polymers as artificial extracellular matrix for tissue engineering. *Nanomaterials* **2021**, *11*, 21. [CrossRef]
6. Shkodra, B.; Vollrath, A.; Schubert, U.S.; Schubert, S. Polymer-based nanoparticles for biomedical applications. In *Colloids for Nanobiotechnology—Synthesis, Characterization and Potential Applications*, 1st ed.; Parak, W., Feliu, N., Eds.; Elsevier: Amsterdam, The Netherlands, 2020; Volume 16, pp. 233–252.
7. Englert, C.; Brendel, J.C.; Majdanski, T.C.; Yildirim, T.; Schubert, S.; Gottschaldt, M.; Windhab, N.; Schubert, U.S. Pharmapolymers in the 21st century: Synthetic polymers in drug delivery applications. *Prog. Polym. Sci.* **2018**, *87*, 107–164. [CrossRef]
8. Karavelidis, V.; Karavas, E.; Giliopoulos, D.; Papadimitriou, S.; Bikiaris, D. Evaluating the effects of crystallinity in new biocompatible polyester nanocarriers on drug release behavior. *Int. J. Nanomed.* **2011**, *6*, 3021–3032.
9. Song, Q.L.; Xia, Y.N.; Hu, S.Y.; Zhao, J.P.; Zhang, G.Z. Tuning the crystallinity and degradability of PCL by organocatalytic copolymerization with delta-hexalactone. *Polymer* **2016**, *102*, 248–255. [CrossRef]
10. Schneiderman, D.K.; Hillmyer, M.A. Aliphatic polyester block polymer design. *Macromolecules* **2016**, *49*, 2419–2428. [CrossRef]
11. Kamaly, N.; Yameen, B.; Wu, J.; Farokhzad, O.C. Degradable controlled-release polymers and polymeric nanoparticles: Mechanisms of controlling drug release. *Chem. Rev.* **2016**, *116*, 2602–2663. [CrossRef]
12. Danhier, F.; Ansorena, E.; Silva, J.M.; Coco, R.; Le Breton, A.; Preat, V. PLGA-based nanoparticles: An overview of biomedical applications. *J. Control. Release* **2012**, *161*, 505–522. [CrossRef]

13. Woodruff, M.A.; Hutmacher, D.W. The return of a forgotten polymer-polycaprolactone in the 21st century. *Prog. Polym. Sci.* **2010**, *35*, 1217–1256. [CrossRef]
14. Hakkarainen, M.; Hoglund, A.; Odelius, K.; Albertsson, A.C. Tuning the release rate of acidic degradation products through macromolecular design of caprolactone-based copolymers. *J. Am. Chem. Soc.* **2007**, *129*, 6308–6312. [CrossRef] [PubMed]
15. Engelberg, I.; Kohn, J. Physicomechanical properties of degradable polymers used in medical applications—A comparative-study. *Biomaterials* **1991**, *12*, 292–304. [CrossRef]
16. Lee, I.H.; Palombo, M.S.; Zhang, X.P.; Szekely, Z.; Sinko, P.J. Design and evaluation of a cxcr4 targeting peptide 4dv3 as an hiv entry inhibitor and a ligand for targeted drug delivery. *Eur. J. Pharm. Biopharm.* **2019**, *138*, 11–22.
17. Bandelli, D.; Muljajew, I.; Scheuer, K.; Max, J.B.; Weber, C.; Schacher, F.H.; Jandt, K.D.; Schubert, U.S. Copolymerization of caprolactone isomers to obtain nanoparticles with constant hydrophobicity and tunable crystallinity. *Macromolecules* **2020**, *53*, 5208–5217. [CrossRef]
18. Garscha, U.; Voelker, S.; Pace, S.; Gerstmeier, J.; Emini, B.; Liening, S.; Rossi, A.; Weinigel, C.; Rummler, S.; Schubert, U.S.; et al. BRP-187: A potent inhibitor of leukotriene biosynthesis that acts through impeding the dynamic 5-lipoxygenase/5-lipoxygenase-activating protein (FLAP) complex assembly. *Biochem. Pharmacol.* **2016**, *119*, 17–26. [CrossRef]
19. Koeberle, A.; Werz, O. Natural products as inhibitors of prostaglandin e-2 and pro-inflammatory 5-lipoxygenase-derived lipid mediator biosynthesis. *Biotechnol. Adv.* **2018**, *36*, 1709–1723. [CrossRef]
20. Koeberle, A.; Zettl, H.; Greiner, C.; Wurglics, M.; Schubert-Zsilavecz, M.; Werz, O. Pirinixic acid derivatives as novel dual inhibitors of microsomal prostaglandin e-2 synthase-1 and 5-lipoxygenase. *J. Med. Chem.* **2008**, *51*, 8068–8076. [CrossRef]
21. Shkodra-Pula, B.; Kretzer, C.; Jordan, P.M.; Klemm, P.; Koeberle, A.; Pretzel, D.; Banoglu, E.; Lorkowski, S.; Wallert, M.; Hoppener, S.; et al. Encapsulation of the dual FLAP/mPEGs-1 inhibitor brp-187 into acetalated dextran and plga nanoparticles improves its cellular bioactivity. *J. Nanobiotechnol.* **2020**, *18*, 73. [CrossRef] [PubMed]
22. Malvern Panalytical; Zetasizer Nano User Manual man0485. Available online: https://www.Malvernpanalytical.Com/de/learn/knowledge-center/user-manuals/man0485en (accessed on 8 July 2021).
23. Spek, S.; Haeuser, M.; Schaefer, M.M.; Langer, K. Characterisation of pegylated PLGA nanoparticles comparing the nanoparticle bulk to the particle surface using UV/Vis spectroscopy, SEC, H-1 nmr spectroscopy, and x-ray photoelectron spectroscopy. *Appl. Surf. Sci.* **2015**, *347*, 378–385. [CrossRef]
24. Werz, O.; Burkert, E.; Samuelsson, B.; Radmark, O.; Steinhilber, D. Activation of 5-lipoxygenase by cell stress is calcium independent in human polymorphonuclear leukocytes. *Blood* **2002**, *99*, 1044–1052. [CrossRef] [PubMed]
25. Perevyazko, I.Y.; Vollrath, A.; Pietsch, C.; Schubert, S.; Pavlov, G.M.; Schubert, U.S. Nanoprecipitation of poly(methyl methacrylate)-based nanoparticles: Effect of the molar mass and polymer behavior. *J. Polym. Sci. Pol. Chem.* **2012**, *50*, 2906–2913. [CrossRef]
26. Shkodra-Pula, B.; Grune, C.; Traege, A.; Vollrath, A.; Schuber, S.; Fischer, D.; Schubert, U.S. Effect of surfactant on the size and stability of plga nanoparticles encapsulating a protein kinase c inhibitor. *Int. J. Pharm.* **2019**, *566*, 756–764. [CrossRef] [PubMed]
27. Beck-Broichsitter, M.; Nicolas, J.; Couvreur, P. Solvent selection causes remarkable shifts of the "ouzo region" for poly(lactide-co-glycolide) nanoparticles prepared by nanoprecipitation. *Nanoscale* **2015**, *7*, 9215–9221. [CrossRef]
28. Homs, M.; Caldero, G.; Monge, M.; Morales, D.; Solans, C. Influence of polymer concentration on the properties of nano-emulsions and nanoparticles obtained by a low-energy method. *Colloids Surf. A Physicochem. Eng. Asp.* **2018**, *536*, 204–212. [CrossRef]
29. Sanchez, A.; Mejia, S.P.; Orozco, J. Recent advances in polymeric nanoparticle-encapsulated drugs against intracellular infections. *Molecules* **2020**, *25*, 3760. [CrossRef]
30. Newman, A.; Hastedt, J.E.; Yazdanian, M. New directions in pharmaceutical amorphous materials and amorphous solid dispersions. *AAPS Open* **2017**, *3*, 7. [CrossRef]
31. Du, S.; Li, W.S.; Wu, Y.R.; Fu, Y.; Yang, C.Q.; Wang, J. Comparison of the physical and thermodynamic stability of amorphous azelnidipine and its coamorphous phase with piperazine. *RSC Adv.* **2018**, *8*, 32756–32764. [CrossRef]
32. Lee, S.C.; Oh, J.T.; Jang, M.H.; Chung, S.I. Quantitative analysis of polyvinyl alcohol on the surface of poly(d,l-lactide-co-glycolide) microparticles prepared by solvent evaporation method: Effect of particle size and PVA concentration. *J. Control. Release* **1999**, *59*, 123–132. [CrossRef]
33. Hollander, J.; Genina, N.; Jukarainen, H.; Khajeheian, M.; Rosling, A.; Makila, E.; Sandler, N. Three-dimensional printed PCL-based implantable prototypes of medical devices for controlled drug delivery. *J. Pharm. Sci.* **2016**, *105*, 2665–2676. [CrossRef] [PubMed]
34. Manoukian, O.S.; Arul, M.R.; Sardashti, N.; Stedman, T.; James, R.; Rudraiah, S.; Kumbar, S.G. Biodegradable polymeric injectable implants for long-term delivery of contraceptive drugs. *J. Appl. Polym. Sci.* **2018**, *135*, 46068. [CrossRef]
35. Eldsater, C.; Erlandsson, B.; Renstad, R.; Albertsson, A.C.; Karlsson, S. The biodegradation of amorphous and crystalline regions in film-blown poly(epsilon-caprolactone). *Polymer* **2000**, *41*, 1297–1304. [CrossRef]
36. Pitt, G.G.; Gratzl, M.M.; Kimmel, G.L.; Surles, J.; Sohindler, A. Aliphatic polyesters ii. The degradation of poly (dl-lactide), poly (ε-caprolactone), and their copolymers in vivo. *Biomaterials* **1981**, *2*, 215–220. [CrossRef]
37. Cook, W.J.; Cameron, J.A.; Bell, J.P.; Huang, S.J. Scanning electron-microscopic visualization of biodegradation of polycaprolactones by fungi. *J. Polym. Sci. Pol. Lett.* **1981**, *19*, 159–165. [CrossRef]
38. Tilstra, L.; Johnsonbaugh, D. The biodegradation of blends of polycaprolactone and polyethylene exposed to a defined consortium of fungi. *J. Environ. Polym. Degrad.* **1993**, *1*, 10. [CrossRef]

39. Bandelli, D.; Helbing, C.; Weber, C.; Seifer, M.; Muljajew, I.; Jandt, K.D.; Schubert, U.S. Maintaining the hydrophilic hydrophobic balance of polyesters with adjustable crystallinity for tailor-made nanoparticles. *Macromolecules* **2018**, *51*, 5567–5576. [CrossRef]
40. Lukasiewicz, S.; Mikolajczyk, A.; Blasiak, E.; Fic, E.; Dziedzicka-Wasylewska, M. Polycaprolactone nanoparticles as promising candidates for nanocarriers in novel nanomedicines. *Pharmaceutics* **2021**, *13*, 191. [CrossRef]
41. Ortiz, R.; Prados, J.; Melguizo, C.; Arias, J.L.; Ruiz, M.A.; Alvarez, P.J.; Caba, O.; Luque, R.; Segura, A.; Aranega, A. 5-fluorouracil-loaded poly(epsilon-caprolactone) nanoparticles combined with phage e gene therapy as a new strategy against colon cancer. *Int. J. Nanomed.* **2012**, *7*, 95–107.
42. Radmark, O.; Werz, O.; Steinhilber, D.; Samuelsson, B. 5-lipoxygenase, a key enzyme for leukotriene biosynthesis in health and disease. *Biochim. Biophys. Acta* **2015**, *1851*, 9. [CrossRef] [PubMed]
43. Tallury, P.; Alimohammadi, N.; Kalachandra, S. Poly(ethylene-co-vinyl acetate) copolymer matrix for delivery of chlorhexidine and acyclovir drugs for use in the oral environment: Effect of drug combination, copolymer composition and coating on the drug release rate. *Dent. Mater.* **2007**, *23*, 404–409. [CrossRef] [PubMed]
44. Prabha, S.; Labhasetwar, V. Effect of residual polyvinyl alcohol on nanoparticle-mediated gene transfection in breast cancer cells. *Mol. Ther.* **2003**, *7*, S67.
45. Sahoo, S.K.; Panyam, J.; Prabha, S.; Labhasetwar, V. Residual polyvinyl alcohol associated with poly (d,l-lactide-co-glycolide) nanoparticles affects their physical properties and cellular uptake. *J. Control. Release* **2002**, *82*, 105–114. [CrossRef]

polymers

MDPI

Review

Recent Advances in Antioxidant Polymers: From Sustainable and Natural Monomers to Synthesis and Applications

Chrysanthos Maraveas [1,*], Ilker S. Bayer [2,*] and Thomas Bartzanas [1]

1 Department of Natural Resources and Agricultural Engineering, Agricultural University of Athens, 11855 Athens, Greece; t.bartzanas@aua.gr
2 Smart Materials, Istituto Italiano di Tecnologia, 16163 Genova, Italy
* Correspondence: c.maraveas@maraveas.gr (C.M.); Ilker.Bayer@iit.it (I.S.B.)

Abstract: Advances in technology have led to the production of sustainable antioxidants and natural monomers for food packaging and targeted drug delivery applications. Of particular importance is the synthesis of lignin polymers, and graft polymers, dopamine, and polydopamine, inulin, quercetin, limonene, and vitamins, due to their free radical scavenging ability, chemical potency, ideal functional groups for polymerization, abundance in the natural environment, ease of production, and activation of biological mechanisms such as the inhibition of the cellular activation of various signaling pathways, including NF-κB and MAPK. The radical oxygen species are responsible for oxidative damage and increased susceptibility to cancer, cardiovascular, degenerative musculoskeletal, and neurodegenerative conditions and diabetes; such biological mechanisms are inhibited by both synthetic and naturally occurring antioxidants. The orientation of macromolecules in the presence of the plasticizing agent increases the suitability of quercetin in food packaging, while the commercial viability of terpenes in the replacement of existing non-renewable polymers is reinforced by the recyclability of the precursors (thyme, cannabis, and lemon, orange, mandarin) and marginal ecological effect and antioxidant properties. Emerging antioxidant nanoparticle polymers have a broad range of applications in tumor-targeted drug delivery, food fortification, biodegradation of synthetic polymers, and antimicrobial treatment and corrosion inhibition. The aim of the review is to present state-of-the-art polymers with intrinsic antioxidant properties, including synthesis scavenging activity, potential applications, and future directions. This review is distinct from other works given that it integrates different advances in antioxidant polymer synthesis and applications such as inulin, quercetin polymers, their conjugates, antioxidant-graft-polysaccharides, and polymerization vitamins and essential oils. One of the most comprehensive reviews of antioxidant polymers was published by Cirillo and Iemma in 2012. Since then, significant progress has been made in improving the synthesis, techniques, properties, and applications. The review builds upon existing research by presenting new findings that were excluded from previous reviews.

Keywords: antioxidant; antioxidant polymers; lignin polymers; graft polymers; dopamine; polydopamine; inulin; quercetin; limonene; vitamins

Citation: Maraveas, C.; Bayer, I.S.; Bartzanas, T. Recent Advances in Antioxidant Polymers: From Sustainable and Natural Monomers to Synthesis and Applications. *Polymers* **2021**, *13*, 2465. https://doi.org/10.3390/polym13152465

Academic Editor: José Miguel Ferri

Received: 5 July 2021
Accepted: 23 July 2021
Published: 27 July 2021

1. Introduction

1.1. Definition of Antioxidants. Their Importance in Food Preservation

Antioxidants are a class of naturally occurring or synthetic compounds. The naturally occurring antioxidants include Vitamin C and Vitamin E (tocotrienols and tocopherols in general) [1–3]. Other classes are phenolic compounds and carotenoids [4]. Synthetic antioxidant molecules include α-lipoic acid, N-acetyl cysteine, melatonin, gallic acid, captopril, taurine, catechin, and quercetin [5]; these compounds are indispensable in the scavenging of free radical species. For example, gallic acid and poly Trolox ester polymers scavenge free radicals in the cytosolic cellular compartment [6]. In contrast, β-carotene, and vitamin E (tocopherol), are most effective against lipid peroxidation.

Wattamwar et al. [6] attribute the formation of reactive oxygen species to cellular processes, including inflammation—a process that results in the activation of endothelial cell macrophages, which in turn trigger the development of Nicotinamide adenine dinucleotide phosphate (NADPH) oxidase complex [6]. The NADPH complex is responsible for the conversion of molecular oxygen into O_2, a superoxide reactive anion, which is converted into H_2O_2 (hydrogen peroxide) following contact with superoxide dismutase [6]. The hydrogen peroxide molecule is a potent reactant, which binds with copper and iron metal ions, releasing hydroxyl radicals. Alternatively, $ONOO^-$ (peroxynitrite) is formed if the superoxide reactive anion is formed in a NO-rich environment. The reactive oxygen and nitrogen species formed through a sequence of reactions triggered by NADPH are responsible for cellular oxidative and lipid damage—phenomena that result in the release of oxidative stress markers (4-hydroxy-2-trans-nonenal, 3-nitrotyrosine (3NT), protein carbonyl). The primary function of melatonin, gallic acid, captopril, taurine, catechin, and quercetin, vitamin C, and E analogs is to bind to the markers to prevent structural damage to cellular proteins.

The biological importance of different antioxidants in food preservation and augmentation of body defense mechanisms depends on chemical properties. Pryor et al. [3] evaluated the performance of both natural and synthetic antioxidants in sodium dodecyl sulfate micelle solutions; it was noted that hydrogen bonding on the para and ether O_2 atoms predicted the reactivity of the antioxidants towards peroxyl radicals. Beyond H-bonding affinity towards reactive oxygen species was predicted by the presence of bulky tert-butyl groups [3]. Even though synthetic and natural antioxidants exhibit similar potency towards ROS, the selection of the chemicals in industrial applications is predicted by regulatory standards. European Parliament and Council Directive No.1333/2008 restricts the use of synthetic antioxidants in food preservation [4], except for 2, 4-dichlorophenoxyacetic acid (2,4-DA), 2-naphthol (2NL), 4-phenyl phenol (OPP), tert-butyl hydroquinone (TBHQ), butylated hydroxytoluene (BHT), and butylated hydroxyanisole (BHA) [4]. The listed synthetic antioxidants have found broad application in the food industry, particularly in preserving fruits and vegetables, due to their low cost, stability, performance, and widespread availability [3–5,7].

Vitamin C has been proven to exhibit superior activity in the quenching of reactive oxygen species and free radicals, resulting in the formation of ascorbyl radicals. The latter is a less potent radical compared to ROS based on the potential for oxidative damage [2]. Vitamin E exhibits a similar mechanism of action as Vitamin C. However, in the former case, the main mechanism of action involves protecting biological liquid compartments and the cleavage of the lipid peroxidation chain reactions. Alternatively, tocotrienols and tocopherols, in general, have been proven to contribute to the inactivation of ROS before regeneration by ascorbate [2]. Experimental evidence has also demonstrated the practical benefits of chain-breaking tocopherols in mitigating the auto-oxidation of polyunsaturated fatty acids. This process is responsible for cancer cell growth and atherosclerosis, as well as other life-threatening conditions [3]. Other unique biological functions associated with naturally occurring antioxidants include the modulation of the activity of specific enzymes, including mitogen-activated protein kinase, protein tyrosine kinase (PTK), protein tyrosine phosphatase (PTP), protein phosphatase 2A (PP2A), and protein kinase [2].

The commercial suitability of naturally occurring antioxidants is informed by various considerations beyond regulatory standards (European Parliament and Council Directive No.1333/2008) on food preservation, encompassing antimicrobial and antifungal activities [4], lipid oxidation reactions, and chemical reactions related to anti-oxidative effects [3–5]. The synergistic impact of different considerations helps explain why phenolic antioxidants are often preferred in food preservation (see Table 1).

The extraction techniques predict the selection of natural antioxidants for food preservation. Complex extraction techniques such as microwave-assisted extraction (MAE), ultrasound-assisted extraction, pressurized liquid extraction (PLE), high hydrostatic pressure (HHP), and supercritical fluid extraction (SFE) and methanol, ethanol, acetone, and water predict the cost of the extraction process and industrial application [1] (see Table 2).

Table 1. Active compounds, natural sources of antioxidants, and their use in different food matrices [4].

Natural Source	Main Active Compound	Food Matrix
Fennel and chamomile aqueous extracts	Phenolic compounds	Biscuits Cottage cheese Yogurt
Olive leaf and cakes extracts byproducts	Phenolic compounds	Antioxidant film
Litchi fruit pericarp extract	Phenolic compounds	Cooked nuggets
Green tea extract	Polyphenols	Sunflower oil
Cloves and cinnamon	Phenolic compounds	Meat samples
Tomato pomace extract	Carotenoids	Lamb steaks packaged
Ginkgo biloba leaves extract	Polyphenols	Pork meat
Cloudberry, beetroot, and willow herb	Flavonoid	Cooked pork patties
Canola olive oils, rice bran, and walnut	Polyphenols, vitamins E and B	Pork frankfurters
Coffee	Chlorophylls and carotenoids	Not defined
Wine	Phenolic compounds	Meat, fish, cereals

Table 2. Extraction techniques for different classes of naturally occurring antioxidants [4].

Extraction Process	Source	Antioxidant Extracted
Organic Solvents:		
Ethanol, dichloromethane, hexane	Coffee leaves	Chlorophylls and carotenoids
Ethanol, acetone, and water	*Baccharides* species	Phenolic content
	Sweet potato	Polyphenols and anthocyanins
Methanol, ethanol, and acetone	Spent grain	Phenolic content
	Spice herb	Phenolic content
	Peel of eggplant	Total phenolics, flavonoids, tannins, and anthocyanins
Supercritical fluid extraction (SFE)	Mango peel	Carotenoids
	Apple pomace	Phenolic compounds
	Myrtle leaves and berries	Phenolic acids, flavonoids, and anthocyanins
	Green algae	Carotenoids and phenolic compounds
	Cape gooseberry	Phenolic compounds and β-carotene
High Hydrostatic Pressure (HHP)	Red macroalgae	Proteins, polyphenols, and polysaccharides
	Tomato pulp	Flavonoids and lycopene
	Watercress	Phenolic acids and flavonoids from watercress
	Papaya seeds	Phenolic content
Pressurized liquid extraction (PLE)	Peppermint	Phenolic compounds and essential oils
	Carrot byproducts	Carotenoids
Ultrasound-assisted extraction	Green propolis	Phenolic compounds
Microwave-assisted extraction (MAE)	Pomegranate peels	Phenolic compounds
	Phalera macrocarpa fruit peel	Phenolic compounds

Natural and synthetic antioxidants are integral to reducing oxidative stress and imbalances in the cell redox reactions associated with reactive oxygen species. The formation of ROS is linked to the suppression of the innate antioxidant defense systems or the overproduction of oxygen species. Both mechanisms are common, considering that oxygen is ubiquitous in the environment, especially in biological processes involving aerobic organisms [2]. The visual illustration suggests that free radical scavenging ability is predicted by chemical structure and active components.

Supercritical Carbon Dioxide-Based Techniques

New evidence shows that the development of active packaging films is augmented by supercritical carbon dioxide (SC$-CO_2$) impregnation [8]. The observations made by Franco et al. [8] were collaborated by Lukic et al. [9] and Trucilo et al. [10], who documented the development of active food packaging products made of PLA/PCL combined with thymol and/or carvacrol and liposomes. The antioxidant and packaging properties of the active packaging materials were augmented by supercritical carbon dioxide (SC-CO_2) impregnation [9]. The choice of the supercritical CO_2 is grounded in its ability to yield good material properties based on its near-zero surface tension, low viscosity, and high diffusion coefficient [10]. The suitable material properties partly explain why SC-CO_2 has been extensively applied in membrane formation, desorption, micronization, and extraction. However, in contrast to Franco et al. [8], Lukic et al. [9], Trucilo et al. [10], Ozkan et al. [11] and Cejudo et al. [12] have argued that the utility of SC-CO_2 in active packaging films was dependent on the role of the material (co-solvent, anti-solvent, or swelling agent). On the one hand, SC-CO_2 can function as a supercritical anti-solvent (this is critical in micronization or co-precipitation processes). On the other hand, it can function as a solvent or co-solvent. Following the appraisal of the two techniques, it was clear that the SC-CO_2 supercritical anti-solvent process (SAS) is highly appropriate in packaging given it yields products with customized properties such as spherical nanoparticles, and nanostructured filaments with controlled mean size and particle size distribution.

1.2. Natural Phenolic Polymers (Lignin and Proanthocyanidins (PAs))

Lignin is a naturally occurring polymer/heterogeneous biomacromolecule that supports the connective tissues in plants [13–15]. The rigidity, tensile strength, resistance to chemical degradation, and pressure-resistance of lignin are linked to the phenyl propane created using coniferyl alcohol, p-coumaryl alcohol, and sinapyl alcohol units [16]—precursor-specific structural modifications of lignin translate to unique, different structural modifications. The microscopic-level changes improve the utility of the material in the production of plastics, paints, films, PU-based forms, nutritional supplements, resins and coatings, beverage additives, food additives, adhesive binders, carbon fiber composites, and enhancement of the structural properties of existing materials [17]. The application of lignin-based compounds in dopamine polymerization, synthesis of polydopamine and copolymers, polymerization of inulin, enzyme-catalyzed polyphenol, polymerization, antioxidant quercetin polymers, polyquercetin and quercetin copolymers, and antioxidant properties, and antioxidant terpene polymers is explored under Sections 2 and 3.

Proanthocyanidins are a class of natural phenolic compounds derived from natural phenols (flavan-3-ols) and are fundamental polyphenolic components of the human diet, including red wine, cocoa, dark chocolate, orange juices, grapefruit, parsley, black and green tea, onions, olives, broccoli, and apple skin [18,19]. The PAs are critical to the investigation of sustainable and natural monomers, synthesis, and applications of antioxidant polymers considering that they are the second most naturally occurring and abundant plant polyphenols after lignin [18]. In addition, PA compounds possess ideal chemical properties for industrial and biological applications. For example, cinnamon-derived polyphenols have been proven to have superior anti-inflammatory characteristics–a factor that is instrumental in the treatment of inflammation-related conditions, including inflammatory bowel disease, rheumatoid arthritis, asthma, cancer, diabetes, cardiovascular diseases, degenerative musculoskeletal diseases, and neurodegenerative conditions. The need for plant-based alternatives is supported by the limitations of traditional non-steroidal anti-inflammatory drugs (NSAIDs) [20], the manifestation of appropriate antioxidant properties [21,22], and a variety of naturally occurring precursors, including agro-wastes.

1.3. Recovery of Antioxidants from Agro-Wastes

The recovery of antioxidants from agro-wastes is a facile and scalable method in the manufacturing of a broad array of lignin-containing compounds for industrial and biological applications. Martinez-Avila et al. [23] noted that the fermentation of fruit peel waste generated high-quality phenolic antioxidants for industrial application. However, the recovery of antioxidants from agro-wastes is impacted by production-related constraints—traditional processes are associated with significant hazards to human health, low yields, and negative ecological effects; this explains why solid-state fermentation and solid-state shear pulverization have become the techniques of choice in the recovery of polyphenols from agro-wastes [23,24]. From a manufacturing perspective, the technical constraints are offset by the cost advantage. The cost of agro-waste is about USD (United stated dollars) 0.05–0.10, which is significantly low compared to synthetic antioxidants, which cost > USD 6 per kg (Table 3) [24]. Future advances in technology might resolve the low yields, hazards to human health, and negative ecological effects associated with toxic products.

Table 3. Cost of recovering antioxidants from selected agro-wastes and the phenol content [24].

Agro-Waste	Annual production (thousand tons)	Cost ($/kg)	Calculated Total Cost	Antioxidant	Total Phenol content (mg GA/g)
Red Grape	100	0.050	5	αtocopherol, Gallic acid	~0.02
Turmeric				Curcumin	~3.1
Coffee grounds	0.005	0.050	0.00025	Melanoidin, Chalcogenic acid	17
Orange peel	20	0.050	1	Ascorbic acid and Flavonoids	~1.3
Irganox I10	Not known	6.36		Four substituted 2,6 di-tertiary butyl phenols	

In theory, it is feasible to extract antioxidants from nearly all commonly available agricultural wastes. Nonetheless, there is a strong preference for antioxidant-rich precursors such as orange peel, coffee grounds, turmeric shavings, and waste and grape pomace wastes, which exhibit superior performance in the reinforcement of plastic properties.

The antioxidants are incorporated into polyolefins to enhance their material properties, primarily recyclability, weathering resistance, and high-temperature processability [23,24], which vary in line with the chemical properties of the precursors/agro-waste materials. The presence of hydrogen-donating hindered phenols in the agro-waste materials is integral to the capture of alkyl peroxide radicals, a process that impedes the propagation of free radical reactions [24]. The precursor-specific chemical behavior underscores the need to select appropriate natural antioxidant molecules for antimicrobial treatment, biodegradation of polymers, tumor-targeted drug delivery, and food fortification.

1.4. Natural Antioxidant Macromolecules

Antioxidant polymers are a broad class of polymers, including polylactones [25], polymer nanoparticles integrated with living polymerization techniques [26], poly(b-malic acid) derivatives [27], lignin graft polymers, and polyphenols, which offer unparalleled capabilities in tumor-targeted drug delivery, food fortification [28], biodegradation of polymers, and antimicrobial treatment [29]. Emerging research evidence indicates that hydrophobic antioxidant polymers are effective corrosion inhibitors in steel structures.

The development of green, biocompatible, and biodegradable antioxidants is critical in polymer processing industries, packaging, health, cosmetics, textiles, and agricultural sectors. The transition to green materials is supported by growing environmental awareness. The need to explore antioxidant molecules is supported by widespread availability in nature, such as lignin and polyphenol adduct polysaccharides. Additionally, certain natural essential oils such as limonene also feature antioxidant properties and can be polymerized into renewable polylimonene while preserving their antioxidant properties. The antioxidant polymers (including polyfurfuryl alcohol and its derivatives) can be directly sourced from agricultural wastes. Polyphenols drawn from a variety of sources, including fruits and vegetables, dopamine, pyrogallol, tannic acid, and green tea catechins, undergo controlled oxidation and bio-inspired polymerization, which is vital in biomedical applications. For example, quercetin (a potent antioxidant) is polymerized to encapsulate cancer treatment medicines. Recently, edible flavors such as vanillin (phenolic aldehyde) have also been polymerized into an antioxidant copolymer known as polyoxalate co-vanillyl alcohol (PVAX) for biomedical applications. Vitamin C and other poly vitamins with potent antioxidant properties can facilitate the synthesis of ADMET, acyclic diene metathesis polymerization. The procedure involves a step-growth polymerization where α, ω-diene monomer is transformed in the presence of a ruthenium catalyst.

In contrast to synthetic antioxidants, natural antioxidants are naturally occurring compounds generated by plants, fungi, bacteria, and mammals, and are environmentally benign and largely biocompatible. Additionally, natural polymers are preferred due to the diminished risk of harmful byproduct formation in the course of their use relative to synthetic antioxidants [30]. Common antioxidant polymers include essential vitamins such as retinoids, ascorbic acid, tocopherols, and polyphenolic compounds epigallocatechin gallate (EGCG), quercetin, curcumin, and resveratrol [31]. The natural antioxidants exhibit unique material properties, including the ability to scavenge reactive oxygen species, targeted delivery, and long-term functionality/reactivity [32]. Such properties are vital agricultural, textiles, cosmetics, health, and packaging industries. Future applications require the development of materials with customized properties for specialized applications.

1.5. A Review of Antioxidant Molecule-Incorporating Polymers

The encapsulation of antioxidant molecules in conventional polymers is integral to the regulation of the potential toxicity, lability, solubility, and diffusion, and other antioxidant properties; this is particularly true for natural antioxidant polymers, which achieve high antioxidizing effectiveness at relatively low concentrations [33]. In other cases, the incorporation of antioxidant molecules into polymers has been proven useful in autosynergism and heterosynergism [33]. Lith and Ameer [32] and Sawant et al. [1] reported the successful encapsulation of vitamin C antioxidant polymer in PEO (or poly(ethylene glycol), PEG) micelles. The incorporation of vitamin C into PEG/PEO micelles was beneficial in the treatment of cancer. Preliminary in vivo experiments using mouse models demonstrated selective necrosis of breast cancer cells; the outcomes demonstrate the unique capabilities of antioxidant compounds embedded into polymers [1,32]. However, the potency of the antioxidant polymers was dependent on concentration—higher concentrations were established to be highly effective in overcoming cancer antioxidant defense mechanisms. Wang et al. [26] reported successful incorporation of paclitaxel into polyethylene glycol-b-poly(D, L-lactide) (PEG-PDLLA) micelle for tumor-targeted drug delivery. Preliminary clinical data show that the technique offers promising prospects in the treatment of tumors.

Beyond targeted drug delivery, the encapsulation of antioxidant molecules in conventional polymers was proven beneficial in agricultural and food packaging applications. The incorporation of primary phenolic antioxidants into polyolefins and polyethylene has been proven useful in the protection of the thermo-oxidative stability of polymers used in packing [30]. The stabilization of synthetic polymers exposed to natural elements, including UV radiation, shear forces, and higher temperatures, is integral to their durability post-production. However, there are ecological implications associated with the incorporation of natural phenolic antioxidants into polymers [30]. Some studies have suggested that the incorporation of antioxidant polymers into polymers poses significant environmental and health hazards. For example, the study found traces of plastics (polyethylene) in drinking water [34]. The problem was attributed to the byproducts of phenolic antioxidant additives used in pipeline production [34]. The compounds migrated to waterways via diffusion. The presence of polymers in drinking water elevates the risk of health complications associated with reactive oxygen species [32]. Even though the latter findings point to the negative effects of antioxidant molecules in conventional polymers, there are tangible benefits associated with the embedding of antioxidant species. The positive benefits in the agricultural, textiles, cosmetics, health and packaging industries help to offset the adverse effects.

2. Lignin Polymers and Antioxidant Properties

The review of lignin polymers as potential antioxidants for biodegradation of polymers and antimicrobial treatment [29], tumor-targeted drug delivery, food fortification [28], and diabetes treatment [15] is supported by the chemical composition of lignin structures, abundance in the natural environment, and ease of production; conservative estimates suggest that at least 70 million tons of lignin are generated from pulp [35]—of which 2% is converted to value-added products including lignin for biological, pharmaceutical and chemical industry applications [35]. The widespread availability of lignin in plant species predicts cost. The availability of lignin is predicted by the rate of lignin biosynthesis, which is responsible for biotic and abiotic stress management, organ/tissue development, and growth [36]; these processes are mediated by a broad class of enzymes and genes. Higher expression of CCR, CCoAOMT B, C3H, 4CL, and F5H was associated with pronounced tissue growth [37]. Since the genes predict the presence of G, S, and H units [16], the genetic composition of the different plant species strongly predicts the utility of the plant lignin in tumor-targeted drug delivery, food fortification [28], biodegradation of polymers, and antimicrobial treatment [29]. The observations made by Liu et al. [16] are valid, considering that the G, S, and H units which comprise sinapyl alcohol, coniferyl alcohol, and p-coumaryl alcohol, respectively, predict the rate of lignin monomer copolymerization. The function of different genes on plant species is heightened in Table 4.

Table 4. Function of selected genes [16].

Genes	Plant Species	Functions
4CL	*Oryza sativa* (*OsAAE3*)	Decreased lignin accumulation and increased sensitivity to rice blast
CCR	*Paspalum dilatatum* (*PdCCR1*)	Increased lignin content and altered lignin deposition
CAD	*Populus trichocarpa* (*PtrCAD1*)	Modified lignin content and structure
MOMT	*Arabidopsis thaliana* (*MOMT4*)	Depressed lignin biosynthesis and increased saccharification yields
COMT	*Sorghum bicolor* (*SbCOMT*)	Methylate the tricin precursors and participated in S lignin biosynthesis

The focus on selected plant species as precursors for lignin-based polyurethanes, dopamine polymerization, polydopamine and copolymers, polymerization of inulin, enzyme-catalyzed polyphenol polymerization, antioxidant quercetin polymers, polyquercetin and quercetin copolymers, and antioxidant properties, antioxidant terpene polymers, polylimonine synthesis and properties, antioxidant random copolymers, and gallic acid grafted polymers is beyond the scope of this research inquiry. This means that the desired chemical properties of lignin-based compounds are considered in general without a specific emphasis on a particular plant species.

Lignin molecules have vast antioxidant properties due to the presence of certain functional groups such as p-hydroxy acetophenone extracted from oil palm fronds [38]. The antioxidant properties of lignin-derived polymers are predicted by the precursors. Different precursors have different antioxidant properties, which are predicted by the presence of carbohydrates, G + S phenols, Pi-conjugated carbons, ArC (Py-products with primary, secondary and tertiary carbons on the side chains), carboxylic groups, OH groups, and phenols [14]. The number of functional groups that predict chemical behavior considering the chemical structure of lignin is poorly understood—the aliphatic OH and phenol groups increase the probability for functionalization of the compound [10] and biological and pharmacological function. The antioxidant properties and higher binding affinity of the chemical functional groups in lignin enable the material to bind to bile acids in the intestines—a process that facilitates serum control.

The antioxidant properties are also integral to tumor suppression—animal models suggest that lignin reduces the adverse effects associated with different carcinogens, including 3,2-dimethyl-4-ami-biphenyl [15]. The phenolic compounds isolated from the different lignin structures have been proven to inhibit microbial growth through the inhibition of oxygen-mediated reactions, ATP depletion, and interference with the intracellular pH [35].

In other cases, organic functional groups within the lignin structure (carvacrol, thymol, and cinnamaldehyde) trigger bacterial lysis and damage the cell membrane [35]. The lignin compounds extracted from corn Stover have exhibited appropriate antioxidant properties in eliminating free radical initiators in the red blood cells such as AAPH (2,2-azobis (2-amidinopropane)) [35]. The trends and applications of lignin as a natural antioxidant supersede the antimicrobial effects; this means that natural lignin best functions as a free radical scavenger/borrower [35]. The utility of the antioxidant properties transcends pharmaceutical applications—the chemical antioxidant properties offer protective benefits to the skin and eyes; this means that topical formulations can be prepared from lignin extracts for cosmetic applications. Other potential commercial applications can be explored due to the heterogeneity of lignin structures drawn from diverse sources. Moreover, research has demonstrated that it is possible to customize the behavior of lignin in different applications to stimulate the desired biological function [35].

Even though diverse plants are a suitable source of lignin, only selected sources are used for commercial applications due to wide variations in the lignin composition by weight 10–40 wt%, the need to balance resource use and promote green economy [39]. In other cases, the herbaceous weight composition of biomass is lower 15–25% (w/w) [15]; this explains why most lignin is derived from byproducts during pulp production [15,39]. The production-related requirements have practical implications on the choice of lignin precursors for biological, polymer processing industries, agricultural, cosmetics, textiles, and agricultural applications. This observation is supported by the link between the precursor and chemical properties and production methods [35]. Certain sources are associated with a higher cost of production, low yield extraction, higher energy use, poor solubility, and presence of chemical impurities; this means that the choice of fractionation methods of technical lignin predicts commercial utility [35].

Even though precursor-specific chemical functional groups might be ideal in the development of lignin-based antioxidants for pharmaceutical, medical and packaging applications, other factors have to be taken into account, including the availability of the plant species and lignin content. The *Quercus* plant species have an extremely low lignin content (3.8%); this is in contrast to 54% in *Acacia auricuriformis* [36]. The high concentration of lignin in the latter justifies its use as a preferred source of antioxidants for a broad array of applications. However, there is significant information asymmetry about the performance of different plants with variable concentrations of lignin—most studies focus on selected plant species [14], with proven benefits. The bias towards selected plant species could have practical implications in commercial applications.

2.1. *Lignin Graft Polymers and Copolymers*

The development of lignin graft polymers and copolymers is integral to the sustainability of modern civilizations and human life. However, there are critical constraints in the synthesis of lignin-based high-performance materials [40]. Most lignin-based materials have been used in the development of low-value products such as pesticides, animal feeds, surfactants, binders, dyestuff dispersants, concrete additives, and materials for dust control, which account for about 2% of the total fraction of lignin produced globally [40]; this means that there is a clear mismatch between the rates of lignin production and the development of value-added products. A feasible approach for the synthesis of value-added lignin products is the derivatives that entail the development of lignin graft polymers and copolymers.

Modern plastics are primarily sourced from petroleum byproducts, but there are valid and emerging concerns about the environmental implications, greenhouse emissions, and destruction of marine and terrestrial environments [41–43]. The growing environmental concerns have validated the need to develop sustainable methods for reducing the carbon footprint associated with plastics; this has, in turn, triggered the development of bio-based plastics for agricultural and packaging applications [44–46]. A fundamental constraint includes the shortage of suitable precursors for bio-based plastics. Chitosan and corn are less abundant, pose a direct threat to commercial agriculture, and commercial viability is challenging [45,47], making it unfeasible to upscale operations. The highlighted challenges demonstrate the need for alternative and complementary sources of sustainable and biodegradable plastics; this informs the development of lignin graft polymers and copolymers for sustainable plastic materials. The focus on lignin in place of other forms of biomass is grounded on the natural abundance in plant and tree species [36,37]. The relative composition varies between 10 and 40 wt% [15,39], depending on the tree species, which is significant considering that about 70 million tons of pulp are generated [35]. The pulping byproducts are rich in lignin; this means that the precursor can be extracted using a scalable and low-cost method. Affordability is integral to the gradual phasing out of synthetic plastics with biodegradable materials in emerging and developed economies.

Lignin graft polymers have been synthesized using a variety of techniques, including blending natural lignin extracted from plants with commercial-off-the-shelf polymers [13] to augment the mechanical and thermal properties of the polymer blends. Copolymer blends have other desirable properties, including superior gene delivery properties, surfactant properties, UV absorber properties, super-plasticizer capabilities, and enhanced elasticity [16]. The realization of these properties is dependent on controlled polymerization and the selection of the backbone structure/polymers and the branch polymers [13,16,48]. The two are connected via covalent bonding.

The properties of the blends are predicted by a broad range of factors, including the polymerization techniques, click reactions, reversible-addition-fragmentation chain-transfer (RAFT) polymerization, and radical-mediated lignin-graft copolymerization [13,16,48]. Other feasible synthetic techniques include single-electron transfer living radical polymerization (SET-LRP), atom transfer radical polymerization (ATRP), macromolecular design via the interchange of xanthates (MADIX), and single-electron transfer-degenerative transfer living radical polymerization (SET-DTLRP) [49]. Even though there are multiple feasible copolymerization techniques, ATRP and RAFT are widely used.

The attainment of the desired properties during polymerization is contingent on graft density, the length of the grafts, and the functional groups on the graft polymers. The most commonly used grafting techniques (Figure 1) include grafting-through, grafting into, and grafting from [16]. The graft-from technique is characterized by the grafting of the polymers from active sites at the backbone polymer (lignin).

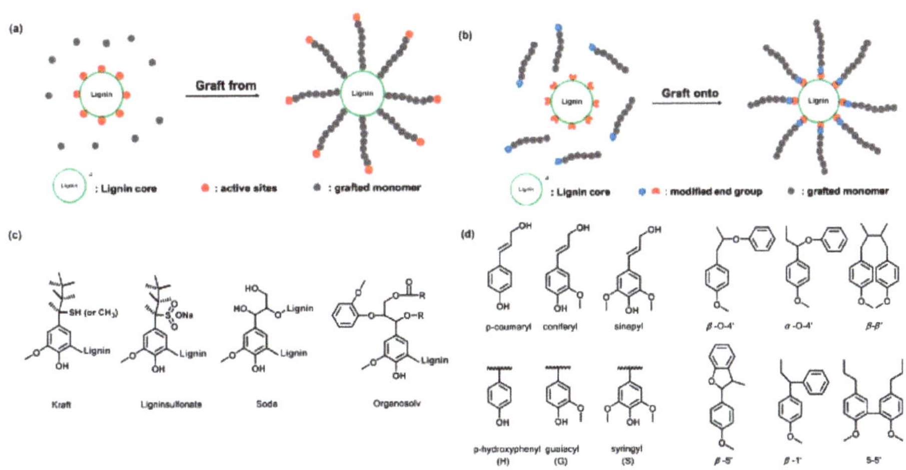

Figure 1. (**a**) An illustration of the graft-from synthesis of lignin copolymers; (**b**) a demonstration of graft-onto methods; (**c**) industrial lignin products; (**d**) sinapyl alcohol, coniferyl alcohol, and p-coumaryl alcohol units [16].

The grafted polymers emerge from the backbone as a result of radical polymerization, ring-open polymerization, RAFT, and atom transfer radical polymerization (ATRP) [16,48] (see Figure 2). Each polymerization technique has its benefits and constraints. For example, ATRP relies on an alkyl halide initiator, which is then radicalized, resulting in the polymerization of monomers [13]; this process is augmented by a transition metal catalyst. In contrast, RAFT polymerization is augmented by radicalization from the addition of a radical-generating species [13,16], while ROP is characterized by the opening of a cyclic molecule in the polymerization process; this results in the formation of a reactive species that undergo a chemical reaction with the next cyclic molecule [13]. The choice of either copolymerization process is predicted by a broad range of factors, including cost, the quality of the desired product. For example, Liu et al. [16] argue that RAFT is an ideal technique for newly made monomers, and ATRP is appropriate for commercially available acrylate monomers. In the former case, the lignin backbone is modified through the incorporation of an ester linkage to create a RAFT agent moiety; this procedure is subsequently followed by the polymerization of the vinyl monomers.

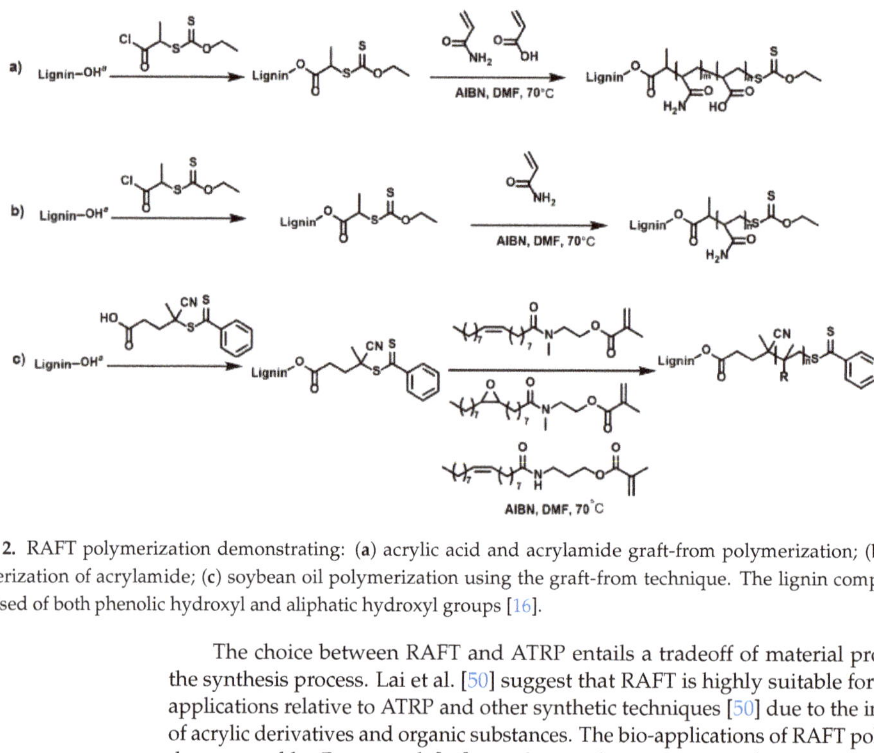

Figure 2. RAFT polymerization demonstrating: (**a**) acrylic acid and acrylamide graft-from polymerization; (**b**) RAFT polymerization of acrylamide; (**c**) soybean oil polymerization using the graft-from technique. The lignin compound is comprised of both phenolic hydroxyl and aliphatic hydroxyl groups [16].

The choice between RAFT and ATRP entails a tradeoff of material properties and the synthesis process. Lai et al. [50] suggest that RAFT is highly suitable for commercial applications relative to ATRP and other synthetic techniques [50] due to the incorporation of acrylic derivatives and organic substances. The bio-applications of RAFT polymerization documented by Boyer et al. [49] corroborate the assessment of Lai et al. [50] of the key benefits associated with RAFT. For instance, RAFT features most of the desirable aspects of traditional free radical polymerization, namely facile reaction conditions, tolerance of most functionalities, site-specific functionality, control of molecular weight distribution and molecular weight, and compatibility with a wide array of monomers. In contrast to traditional living radical polymerization (LRP), RAFT features site-specific functionality [50]. The commercial significance is offset by synthetic requirements.

The RAFT polymerization technique depends on the ability of RAFT agents to transport an ideal leaving group bonded to an S-atom to the reaction site; it is often challenging to attain such reaction specificity. In view of these constraints, the RAFT polymerization has low yields. There are also concerns about the synthesis of toxic compounds, namely

Grignard reagents [50]. The inadequacy of RAFT and ATRP techniques demonstrates that new methods are required to meet the need for the synthesis of next-generation lignin graft polymers and copolymers.

2.2. Lignin-Based Polyurethanes

The synthesis of lignin-based thermoplastic polyurethanes has been demonstrated through copolymerization [40]. Lignin precursor is incorporated into the backbone of the polyurethane structure, but most of the products are rarely recyclable or processable due to extensive crosslinking and networking within the polymer structure [7,40,51]; this means that the polydisperse nature of lignin molecules is a critical impediment to the synthesis of thermoplastic polyurethanes. Saito et al. [40] suggested that the challenge could be offset through the incorporation of tie molecules to link the hard segments or crystals. The synthesis of recyclable thermoplastic lignin-based polyurethanes could also be augmented by bridging the soft segment (characterized by a low glass transition temperature T_g) with the hard segments containing lignin to ensure that the final product features the mechanical rigidity of lignin and the soft rubbery properties of the soft segments. The weight ratio of lignin in polyurethane has a direct impact on the material properties—a higher lignin content translated to better stress–strain behavior (see Figure 3) [40]. The higher stress and strain tolerance and other desirable mechanical properties make lignin-based polyurethane materials appropriate for high-strength applications.

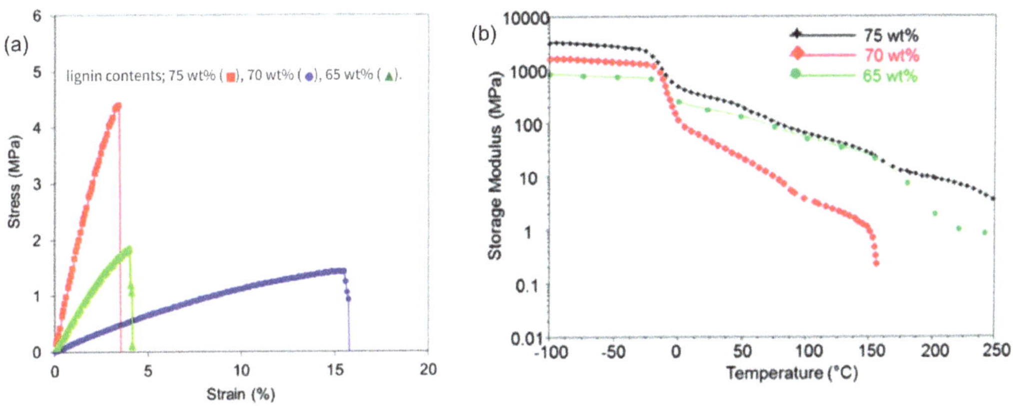

Figure 3. (**a**,**b**) Stress–strain behavior and storage modulus of lignin-based polyurethane thermoplastics [40].

Saito et al. [40] reported the successful synthesis of lignin-based polyurethanes made of telechelic polybutadiene soft segments. Additionally, Da Silva et al. [51] used a kraft lignin precursor to synthesize vanillin and lignin-based polyurethanes, while Cateto et al. [7] synthesized lignin-based polyurethanes using 4,40-methylene-diphenylene diisocyanate (MDI), polycaprolactone (PCL). The successful synthesis of lignin-based polyurethanes from a wide variety of precursors demonstrates that existing synthetic routes can be modified to suit different commercial applications, with or without a catalyst. However, specific adjustments must be made in the synthesis process, including the chemical modification via etherification and esterification reactions [7]. Additionally, oxypropylated lignin materials have been proven useful relative to non-oxypropylated materials in the formation of rigid polyurethane foams. However, there are critical constraints associated with chemical modification. For example, the costs are higher after esterification and etherification; this means that the industrial production of high-value products from lignin precursors is often offset by cost constraints [7], which can be partly offset through the development of new synthetic routes [7,41,51]. The cost constraints reported by Cateto et al. [7] were also acknowledged by Da Silva et al. [51], who argued that cost was not the only constraint.

Other issues relating to the food-versus-fuel dilemma and technical know-how must be resolved before the "novel systems become more competitive against the current industrial alternatives based on petrochemical resources" (p. 1273). The observations made by Cateto et al. [7], Da Silva et al. [51], and Saito et al. [40] illustrate that for lignocellulose feedstock (LCF), bio-refineries are preferred from a sustainability perspective. However, the future development of bio-based products would be catalyzed by conversion products, which offer competitive benefits relative to petrochemical products. This observation is supported by the chemical properties of lignin materials depicted in Table 5.

Table 5. Second-order kinetics of lignin-based materials [7].

Sample	*p* (30 min)	CV (%)	Linear Fit	The Domain of Validity (pNCO)	[NCO]0	Kapp 10^3 (L/mol/s)
(PCL400/1/0)	0.93	0.41	1 + 0.525t	0.93	4.72	1.87
A(PCL400/1/10)	0.91	0.19	1 + 0.491t	0.87	3.52	2.32
A(PCL400/1/15)	0.87	0.67	1 + 0.461t	0.82	3.54	2.16
A(PCL400/1/20)	0.84	0.43	1 + 0.469t	0.71	3.57	2.20
A(PCL400/1/25)	0.77	0.94	1 + 0.377t	0.66	3.59	1.76
A(PCL750/1/15)	0.84	1.30	1 + 0.350t	0.73	2.47	2.36
A(PCL750/1/20)	0.79	1.33	1 + 0.323t	0.66	2.56	2.08
A(PCL750/1/25)	0.74	1.18	1 + 0.327t	0.56	2.66	2.07
(PCL1000/1/0)	0.79	1.12	1 + 0.123t	0.79	1.73	1.16
A(PCL1000/1/10)	0.82	0.40	1 + 0.212t	0.75	1.98	1.77
A(PCL1000/1/15)	0.78	0.42	1 + 0.247t	0.63	2.11	1.98
A(PCL1000/1/20)	0.75	1.04	1 + 0.260t	0.57	2.23	1.94
A(PCL1000/1/25)	0.70	1.55	1 + 0.257t	0.48	2.35	1.84

The physicochemical properties and kinetic properties affirm the suitability of lignin antioxidants in a broad range of engineering and non-engineering applications. However, the formation of the products is dependent on a broad range of factors, including the yield from dopamine polymerization and the polydopamine and copolymers. The antioxidant properties of lignin-based polyurethanes are linked to the presence of an aromatic ring with hydroxyl and methoxy functional groups [50,51]. The functional groups are integral to the oxidation propagation reaction, which is inhibited by hydrogen donation [51]. According to the literature, the performance of lignin polymers was comparable to 2,2-diphenyl-1-picrylhydrazyl (DPPH), and butylated hydroxyl-toluene, among other commercial antioxidant polymers [50].

3. Dopamine Polymerization

The self-assembly of polymer molecules has gained significant research attention in research and development because of the versatility of the process, its practicability, and its ability to form products with a broad range of morphologies, including hierarchical assemblies, cylinders, vesicles, and micelles [52]. Polydopamine is a versatile organic molecule that can be deposited onto any substrate at the nanoscale; this means that the material might act as a suitable coat or primer [52–54]. The versatile chemistry of polydopamine is linked to the molecule's ability to form strong covalent bonds [54], physical interactions associated with π-interactions, hydrogen bonding, presence of different surface functional groups including amines, and thiols [54]. However, surface functionalization in isolation does not explain the broad commercial application of dopamine and polydopamine; the material exhibits appropriate antibacterial activities through gelation [54].

The extensive conjugation influences the application of polydopamine coatings in polymers, ceramics, and metals [53]. The extended application of polydopamine in the coating is augmented by extended bio-conjugation, which is integral in the bio-adhesion of protein molecules, adsorption-resistant surfaces, drug delivery systems, contrast agents, and bioadhesives. The adhesion properties of polydopamine are also vital in cardiovascular, diagnostic, and neurotechnology applications [53]. In other cases, the polydopamine coatings have been proven useful in the surface functionalization at the nanoscale to enhance the chemical and electronic properties [55].

The oxidative polymerization of dopamine primarily occurs in the presence of aminoethyl and catechol groups, which are catalysts for oxidative polymerization [56]—a process that is integral to the formation of PDA nano-coatings and polydopamine coatings [57] through the formation of the aromatic rings of dopamine. Recently, there has been a growing demand for nano-coatings developed using dopamine polymerization techniques; this could be attributed to the capability for secondary modification, generalizability, and simplicity of the synthetic process relative to other techniques [57].

Other unique applications of dopamine polymerization include the electrochemical analysis of dopamine, uric acid, and ascorbic acid using hollow nitrogen-doped carbon microspheres (HNCMS)-based glassy electrodes [58]. The application of dopamine polymerization in biosensors and bioelectronics documented by Xiao et al. [58] was corroborated by Kalimuthu and John [59], who reported successful electrochemical determination of xanthine, ascorbic acid, dopamine, and uric acid using 2-amino-1, 3, 4-thiadiazole (p-ATD)-modified glassy carbon electrodes.

The application of dopamine polymerization in biosensors and bioelectronics and the development of the desired products in manufacturing applications depends on a broad array of factors such as size control of polydopamine nodules and chemical composition of the precursors (carboxylic acid-containing compounds) [60]. The carboxylic-acid-containing compounds introduce an acidic environment, which yields products with unique chemical characteristics compared to dopamine polymerization under basic conditions.

Even though the findings reported by Chen et al. [61] seem to favor acidic dopamine polymerization, basic polymerization techniques have yielded stable products through the customization of the synthesis process. Du et al. [57] observed that the challenges associated with dopamine polymerization under basic environments could be offset by light-triggered regulation of light initiation and termination of dopamine polymerization; this was achieved through the incorporation of small quantities of antioxidant Vitamin C (sodium ascorbate). Vitamin C contributed to the inhibition of the polymerization process under basic conditions—a process that has remained a challenge in traditional synthesis. Du et al. [57] attributed the superior performance to Vitamin C's ability to delay/inhibit dopamine polymerization and reduce the reactive dopamine quinone. The inhibition process can be halted through UV irradiation. Once vitamin C is exposed to UV radiation, instantaneous dopamine polymerization is achieved. The customization of the process using a natural antioxidant and UV radiation helps to explain why it was feasible to attain optimal dopamine polymerization under basic conditions through a facile, scalable, and environmentally benign process.

Fichman and Schneider [54] reported a facile synthetic route for dopamine polymerization under basic conditions in the absence of Vitamin C. Optimal polymerization was achieved in this case through gelation of 1 wt% MAX1 peptide; this experiment was conducted in the presence of 10 mM dopamine at neutral pH [54], and molecular oxygen. The process resulted in the spontaneous polymerization of dopamine. The temperature was adjusted to room temperature to trigger a hydrophobic effect, which was instrumental in promoting MAX1 assembly. Considering that the final product exhibited a 66-fold improvement in mechanical rigidity relative to other materials synthesized using alternative synthetic routes [54], sodium ascorbate is not a prerequisite.

The reports of the positive synthesis of polydopamine and dopamine under basic and acidic conditions by Chen et al. [61], Du et al. [57], Fichman and Schneider [54], Kwon and Bettinger [53], and Qui et al. [52] do not address other emergent challenges such as the relationship between film thickness and solution pH, dopamine concentration and self-polymerization time optimization. The critical requirements for the process underscore the need to select a suitable pH range for the controlled synthesis of materials with the desired film thickness, mechanical rigidity, and industrial/biological performance.

3.1. Polydopamine and Copolymers

The chemistry of polydopamine (PDA) copolymers predicts the synthesis of derivative materials and promising applications in biomedicine, environmental, and energy applications. The application of PDA is predicted by the presence of the different functional groups (azido, alkyl-thiol, amino, carboxyl derivative groups, and carboxyl and alkyl groups) and the synthesis process [62]. There are three feasible mechanisms for polydopamine synthesis from derivatives, namely oxidative copolymerization of the mixtures (comprising of copolymers and monomers), and oxidative polymerization of DA analogs with additional functional groups (see Figure 4).

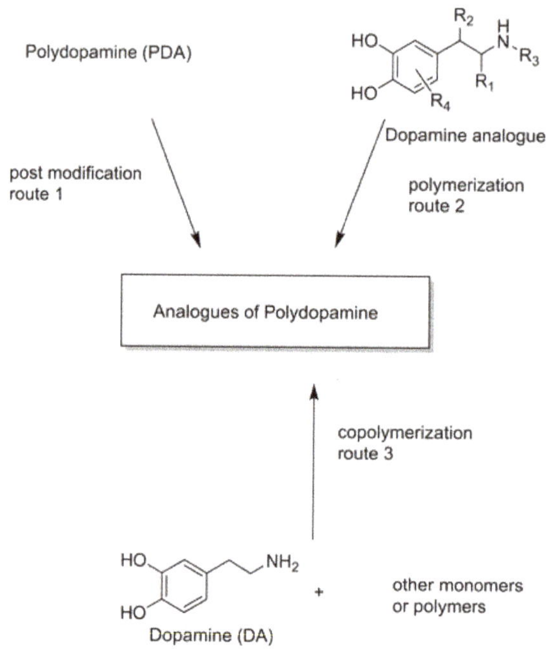

Figure 4. Polydopamine polymerization from different analogs [62].

Melanoidins

Melanoidin compounds are byproducts of the Maillard reaction and the reduction of amino acids, sugars and proteins during food preservation and processing [61]. In most cases, the compounds are drawn from coffee processing—the roasting of coffee bean components results in the generation of brown-colored nitrogenous compounds with high molecular weight [62], referred as melanoidins.

4. Polymerization of Inulin

Inulin is categorized as a fructan carbohydrate, which functions as a primary ingredient in industrial food processing. Inulin hydrogels have been employed as carriers for colonic drug targeting; such processes are fundamental to prebiotic fermentation, chocolate, and cheese production [63–65], as noted in the preceding sections. From a nutritional perspective, the functional properties of inulin and health beneficial effects include prebiotic effects, high dietary fiber content, and lower calorie value [65]. The physiological benefits linked to the consumption of inulin-rich foods are presented in Table 2. In industrial food production, inulin has been employed in fat replacement as a bulking agent [65], stabilizer, and fat replacer in cheese production. The observations made by Karimi et al. [65] are in agreement with Aidoo et al. [63], who reported the incorporation of inulin in sugar-free chocolate production (inulin functions as an alternative sweetener), and Guimarães et al. [64] research on inulin's role in prebiotic fermentation (especially in beverage stabilization). In other cases, inulin has been proven useful in the retro-gradation and gelatinization of wheat starch [66].

The potential applications of inulin are not confined to food production, given that inulin has been proven to be an ideal in vitro drug carrier of lignin derivatives. Vervoot et al. [67] noted that inulin was an ideal carrier for colonic drug targeting [67]; this is achieved through the surface functionalization of inulin with N, N, N′, N′-tetramethyl-ethylenediamine, and ammonium persulphate [67]. Colonic drug delivery is critical to the treatment of colon cancer and metabolic/bowel complications. Similar to Vervoot et al. [67], Kumar et al. [68] noted that inulin was a suitable bioactive polymer for pathogen-mimicking vaccine delivery systems. The biomedical applications of inulin in targeted drug delivery offer promising prospects in the treatment and management of life-threatening conditions, including cancer, Ebola, HIV/AIDS, malaria, and tuberculosis. The biomedical application of inulin is mediated by free-radical polymerization—a process that contributes to the incorporation of vinyl functional groups on the surface of inulin [67].

Beyond industrial applications, the polymerization of inulin has broader implications on the health of mammals. Wada et al. observed that the fermentation of indigestible carbohydrates in the diet by bowel microflora predicts the formation of short-chain fatty acids [69]. Acetate, propionate, and butyrate are the primary short-chain fatty acids (SCFA) derived from the digestion and fermentation of undigested carbohydrates and inulin in the human colon [70].

The formation of both short-chain and long-chain fatty acids has long-term effects on bowel physiology. The observations reflect long-standing beliefs and evidence concerning the relationship between improved health outcomes and high-fiber diets, causal epidemiological evidence, and metagenomics studies linking metabolic diseases to variations in the gut microbe and molecular signaling [71].

Even though there is growing use of inulin in biomedical and food-related industries, the extent of application depends on a broad range of factors, including the polymerization of inulin; this underscores the need for customized synthetic processes, including the reaction between N,N-dimethylformamide with glycidyl methacrylate in the presence of a catalyst (4-dimethylaminopyridine) [67]. The polymerization of inulin is integral to industrial application, primarily in the food and biomedical sectors, but there are critical technical constraints that impede successful polymerization.

5. Enzyme-Catalyzed Polyphenol Polymerization

Similar to the antioxidants for food production and biopharmaceutical applications, enzyme-catalyzed polyphenol polymerization reactions have a broad range of practical applications, including antioxidant, antibacterial, anticancer, cardioprotective, and neuroprotective properties [72]. However, the progression of most reactions is impeded under natural conditions; this justifies the need for enzyme-mediated reactions to attain the desired effects. In a recent study, Oliver et al. [72] noted that the impressive therapeutic activity of polyphenols in therapeutic agents and other clinical applications could not be realized without proper adjustment of the production processes and chemical properties. Under natural conditions, polyphenols exhibit poor membrane permeability, rapid metabolism, and poor bioavailability and UV degradation. The challenges have been partly resolved through amidification, esterification, free-radical grafting [73], step-growth, free radical, and enzyme-catalyzed reactions (direct polymerization of polyphenol monomers), and enzyme grafting-mediated conjugation with macromolecules.

Even though multiple techniques have been proposed to improve the industrial application of polyphenols, enzyme grafting and catalyzation of the reaction were given preference given polyphenol polymerization under natural conditions is constrained by the reaction mechanisms and chemical composition of the reactants, which impede synthesis using conventional methods [74,75]. The forward polymerization process is augmented by the inclusion of enzymes. The utility of laccase (an enzyme) in catechol, resorcinol, and hydroquinone selective polymerization was confirmed by Sun et al. [75] (see Figure 5). The process was phased; the initial phases entailed the formation of quinone intermediates through laccase catalysis, followed by oxidation and formation of covalent bonds. In the subsequent steps, carbon-carbon (C-C) and ether bonds were formed, linking catechol units, resorcinol, and hydroquinone units.

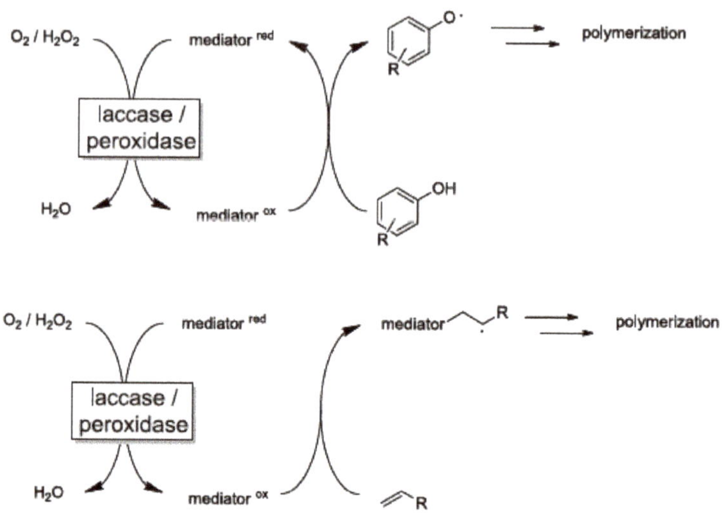

Figure 5. Illustration of laccase and peroxidase mediator systems (PMSs) in polymerization reactions [76].

A key benefit associated with the reaction process is the formation of reaction products under mild conditions and the elimination of toxic byproducts such as formaldehyde [74]. The merits of the laccase-catalyzed polymerization process are offset by the complexity of the reaction mechanism [76]. Hollman and Arends [76] argue that even though catalysts are critical to the progression of reactions, which are unfeasible using conventional methods, there are practical constraints such as unfavorable steric interactions between the polyphenols and the active sites within the enzyme. Alternatively, the suitability of the enzymes in reaction processes could be impaired by unfavorable redox reactions. Since enzymes in isolation do not address the challenges and limitations associated with the synthesis of new products, alternative processes must be developed to improve the yields in enzyme-catalyzed polyphenol polymerization reactions.

Other experiments have reported the practical benefits of using alternative enzymes, primarily hydrolases, to catalyze bond cleavage reactions and the reverse action of hydrolysis [74]. Similar to laccase hydrolyses, they catalyze polyphenol polymerization via bond-forming reaction. However, hydrolyses are suitable in oxidative polymerization reactions that are specific to peroxidase and laccase catalysts in the presence of phenol derivatives. Other factors that inform the choice of enzymes for polyphenol polymerization include inhibition of laccase-initiated polymerization and peroxidase-initiated polymerization and the use of molecular oxygen in place of hydrogen peroxide [76]. The focus on molecular oxygen was validated by the following considerations.

Even though hydrolyses, laccase, and peroxidase polymers have proven to be useful in polyphenol polymerization, the concerns associated with polyphenol polymerization should be addressed. Various mechanisms were proposed by Hollman and Arends [76] to address the catalyst-related problems; these include the ingenious use of a mediator as a radical transfer catalyst. However, the radical is not incorporated into the end product. The feasibility of this reaction has been demonstrated in the use of phenothiazines to address constraints associated with the sterically hindered/congested phenols and consequently catalyze the polymerization process. In other cases, the production constraints can be offset by the incorporation of ABTS or transition metal ions, particularly $Mn^{2+/3+}$, to facilitate the pyrrole polymerization process. If transition metal ion species and ABTS prove less useful, polyoxometallates can be employed. The complexities of polyphenol catalysis underscore the challenges that impact the synthesis and application of sustainable and natural monomers and antioxidants in food packaging, biopharmaceuticals, and agricultural applications.

The synthesis-related constraints raise fundamental questions about the commercial exploitation of naturally occurring antioxidants such as tocotrienols and tocopherols [1–3], phenolic compounds, carotenoids [4]. Even though polymerization of natural polyphenols is impacted by the generation of formaldehyde byproducts, unfavorable steric interactions between the polyphenols and the active sites within the enzyme, inhibition of laccase-initiated polymerization, the process offers practical benefits relative to the use of artificial antioxidant molecules such as α-lipoic acid, N-acetyl cysteine, melatonin, gallic acid, captopril, taurine, catechin, and quercetin [5], or the replacement of catalysts with alternative processes such as amidification, esterification, free-radical grafting [73], step-growth, and free radical-catalyzed reactions to improve the percentage yield (see Table 6).

Table 6. Yields, MW, and PD from acrylamide and styrene polymerization as initiators [76].

Initiator: Styrene Polymerization		
Yield %	MW [$\times 10^{-3}$ g·mol^{-1}]	PD
17	27	2.1
59	68	2.0
14	80	2.0
14	97	2.2
14	57	1.6
Initiator: Acrylamide Polymerization		
Yield %	MW [$\times 10^{-3}$ g·mol^{-1}]	PD
93	124	2.5
84	56	2.9
76	5	4.4
72	27	3.3
78	85	2.7
38	10.5	3.9

The outcomes documented by Hollman and Arends [76] validate the polymerization of polyphenols, quercetin, and other organic molecules with suitable antioxidant properties. A critical challenge concerns the shortcomings and tradeoffs of existing polymerization processes.

6. Antioxidant Quercetin Polymers

The polymerization of quercetin is part of a broader effort in recent years to enhance the performance of natural and synthetic antioxidant polymers via covalent insertion of antioxidant species (such as vitamins, curcumin, and quercetin) into polymeric chains [77]. The polymerization products have demonstrated appropriate properties in food packaging (as preserving agents) [78], biomedical and pharmaceutical applications, especially in the production of antioxidant-laden nanoparticles for cancer treatment [77,79,80], or contact lenses and hemodialysis membranes. In both applications, the potency of the antioxidants is contingent on successful polymerization. A distinct advantage of bioconjugates is the exploitation of unique benefits of different bio-conjugates, the slow rates of macromolecular system degradation, and better chemical and cellular stability.

Quercetin is a plant-derived aglycone flavonoid with antioxidant properties that are suitable in the food packaging industry and the production of targeted therapies for cancer, heart, liver, and lung complications [78,80,81]. The primary sources include tea, red wine, and common fruits and vegetables such as onions, berries, apples, red grapes, broccoli, and cherries [80]. The antioxidant properties influence medical and nutritional applications as nutraceutical/nutritional supplement to boost the immune system and elevate protection against cardiovascular and lung conditions, osteoporosis, and tumor growth [80,81]; this is achieved through the neutralization of reactive oxygen species and the mediation of the transduction pathways, enzymatic activity, and function of glutathione. The specific mechanisms for the prevention of liver damage include the inhibition of the cellular activation of various signaling pathways, including NF-κB and MAPK [80]. Moreover, the antioxidant inhibits the release and expression of apoptosis-related proteins triggered by LPS/d-GalN, suppresses oxidation marker-mediated production of LPS/d-GalN. Following the inhibition of the mechanisms listed above, quercetin catalyzes the antioxidant signal transduction pathways (particularly Nrf2/GCL/GSH) and other processes that increase the concentration of glutathione in the cells [79].

The chemical structure and properties of quercetin contribute to its bioavailability and solubility in cellular fluids and moderation of biochemical pathways. However, the commercial exploitation of the antioxidant properties of quercetin depends on the success of the polymerization process, which can be tailored to generate bioactive compounds such as template quercetin (QCT) nanoparticles for free-radical scavenging. The cellular function of the QCT nanoparticles was demonstrated through material characterization using transmission electron microscopy (TEM), dynamic light scattering, H-NMR spectroscopy, Fourier-transform infrared spectroscopy, and UV-Visible spectroscopy [80]. The demonstrated biopharmaceutical application of oxidation-triggered self-polymerization quercetin in the production of targeted therapies by Xu et al. [80] is in line with the findings of Sunoqrot et al. [79]. The latter study documented the successful bio-inspired polymerization of quercetin—a process that resulted in the synthesis of cancer therapies. The nanotechnology-based quercetin–curcumin therapeutics exhibited superior cytotoxic behavior against cancer tumors. However, the superior biological function was dependent on precision in the synthetic process.

In the current case, the quercetin–curcumin-loaded nano-medicine was synthesized in the presence of curcumin and thiol-terminated poly (ethylene glycol) (PEG)-mediated surface functionalization; this was integral in facilitating steric stabilization in a single reaction step. The reactants were exposed to dimethyl sulfoxide [79]—a universal solvent to obtain a homogenous solution. The next polymerization steps entailed the gradual addition of water. The single-step synthesis process preferred by Sunoqrot et al. [79] was confirmed to be efficient by Pouci et al. [77], who noted that the one-step synthesis route was integral to the synthesis of polymer antioxidant conjugates (that are less susceptible to degradation) without emitting toxic byproducts.

Even though Sunoqrot et al. [79] documented successful synthesis and clinical application of quercetin–curcumin-loaded nanoparticles for targeted cancer drug synthesis, the approach does not align with Zhang et al.'s [81] research on the antioxidant properties of quercetin. In contrast to Sunoqrot et al. [79], Zhang et al. [81] advocated the use of quercetin based on its greater reduction potential relative to curcumin. The reduction potential helps to predict the total antioxidant capacity (TAC), which is fourfold higher for quercetin compared to curcumin. The unique chemical and biological properties help explain the cellular action against LPS-induced reactive oxygen species. Considering that experimental data predict clinical applications, further research is necessary to assess the effectiveness of quercetin only and quercetin–curcumin-loaded nanoparticles for targeted cancer tumor treatment. The need to compare antioxidants' potential benefits in combined or individual treatments is consistent with the research of Zhang et al. [81].

The biomedical applications of quercetin documented by Xu et al. [80] are but a microcosm of the potential industrial applications considering that quercetin is indispensable in the food production industry, where it is incorporated as an additive in bio-polyether (PEO), bio-polyester (PLA), and commercial starch-based polymer (Mater-Bio) [78]; this depends on the extent of photo-stabilization and artificial photo-stabilization. The oxidation-triggered self-polymerization is depicted in Figure 6. The illustration demonstrates the role of incubation, oxidizing agent, and universal organic solvents in the synthesis of nanoparticles.

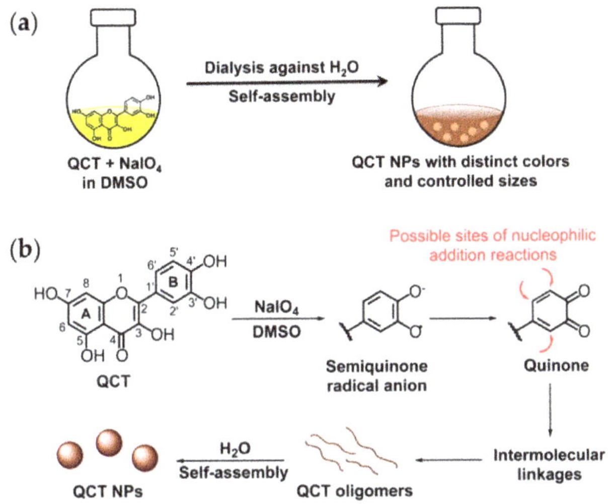

Figure 6. Synthesis process for template quercetin (QCT) nanoparticles [80] (**a**) Schematic of synthesis and (**b**) Potential chemical pathways of synthesis.

The impact of quercetin additives on the tensile strength, Young's Modulus, and elongation at break of bio-polyether, bio-polyester, and commercial starch-based polymer (Mater-Bio) is depicted in Table 7. The data demonstrate the influence of quercetin on stabilized and un-stabilized systems. The mechanical testing outcomes showed a significant improvement in the elastic modulus, tensile strength, and elongation at break after stabilization with quercetin additives (Q): Cyasorb® and synthetic Light Stabilizers (LS). In particular, a 5 wt% increase in elongation at break and the tensile strength resulted in a 10–20% improvement in the elastic modulus, tensile strength, and elongation at break. However, the microscale changes in the mechanical properties were attributed to different mechanisms.

Table 7. Changes in the mechanical properties (elastic modulus, tensile strength, and elongation at break) before and after stabilization with quercetin additives (Q): Cyasorb and synthetic Light Stabilizers (LS) [78].

Samples	E, MPa	TS, MPa	EB,%
PLA	1556 ±52	47.1 ± 2.5	9.3 ± 1.5
PLA/Q	1684 ±56	44.8 ± 2.2	14.8 ±2.0
PLA/LS	1843 ±62	51.2 ± 2.5	19.5 ±2.5
MB	105 ±5	24.2 ± 1.2	435 ±22
MB/Q	119 ±6	27.0 ± 1.3	470 ±24
MB/LS	120 ±6	24.6 ± 1.2	440 ±22
PEO	60.5 ± 2.3	3.5 ± 0.3	105 ± 5.0
PEO/Q	55.2 ± 2.5	3.1 ± 0.2	110 ± 6.0
PEO/LS	54.2 ± 2.3	3.2 ± 0.3	115 ± 7.5

On the one hand, the changes in the elongation at break of the stabilized bio-polyether, bio-polyester, and commercial starch-based polymer have been linked to the plasticizing effects triggered by the low MW stabilizing molecules such as quercetin [78]. On the other hand, the improvements in the tensile strength and elongation at break have been linked to higher rigidity of the stabilized films in relation to the machine direction [78]. The contrasting changes at the micro scale show that quercetin has unique effects on biofilms for food packaging—a phenomenon that introduces new challenges and benefits in the polymerization of quercetin for industrial applications. On a positive note, the higher orientation of the macromolecules in the presence of the plasticizing agent increases the suitability of quercetin in food packaging. It is deduced that other challenges could be addressed through innovations in polymerization.

The improvements in elastic modulus, tensile strength, and elongation at the break following stabilization with Q and LS by Morici et al. [78] are significant considering that the most common forms of quercetin polymerization are inspired by nature. The polymerization of polyquercetin offers further practical opportunities for developing customized therapeutics that influence glutathione (GSH) and enzymatic activity, moderate the signal transduction pathways, and diminish the availability of reactive oxygen species (ROS).

Polyquercetin and Quercetin Copolymers and Antioxidant Properties

Similar to quercetin, polyquercetin has superior antioxidant properties, depending on the synthetic route and ingredients. A facile and scalable micro-emulsion polymerization/crosslinking technique was developed by Sahiner [82], involved reacting quercetin molecules with glycerol diglycidyl ether (GDE) in the presence of l-α lecithin and cyclohexane, which act as the surfactant and organic phase, respectively. The process was proven to be useful in the synthesis of polyquercetin and quercetin copolymers with customizable antioxidant properties. In contrast to the methods proposed by Sahiner [82], Pivec et al. [83] explored an alternative technique comprising 0.5 g of flavonoid rutin (RU) hydrate in ultra-pure water. The mixture was subjected to enzymatic polymerization through the incorporation of Trametes Versicolor-based laccase (500 mg). The mixing was performed at room temperature, away from UV light for 24 h, after which ammonium sulfate was incorporated to enhance the precipitation process; this was followed by centrifugation, removal of the supernatant solution, and dialysis. A comparison of the two techniques shows that the microemulsion polymerization/crosslinking technique developed by Sahiner [82] was more facile and scalable compared to the lengthy procedures undertaken by Pivec et al. [83]. The focus on the synthetic route is grounded on the chemical structure–antioxidant activity relationship discussed in the preceding sections [77–79,82–85].

In contrast to quercetin molecules, which have found broad applications in targeted drug delivery and food packaging, polyquercetin has mainly been used as an electrode component for glass carbon electrodes modified by multi-walled carbon nanotubes; this was the case in studies conducted by Ziyatdinova et al. [84,85]. In both studies, the oxidation potentials of polyquercetin influenced the chronoamperometric determination of the antioxidant capacity and the quantification of gallic acid, catechin, and epigallocatechin gallate (EGCG) [84,85]. The variable industrial applications of quercetin and polyquercetin demonstrate the influence of polymerization and the choice of the crosslinking method.

7. Antioxidant Terpene Polymers

Similar to polyphenols, tocotrienols, tocopherols, quercetin, α-lipoic acid, N-acetyl cysteine, melatonin, polyquercetin, gallic acid, captopril, taurine, vitamins, and catechin terpenes are antioxidant polymers synthesized via natural and artificial polymerization processes [86–90]. However, the synthesis and commercial application of terpenes as antioxidants in food packaging and biomedical industries has attracted less research interest compared to polyphenols, vitamin C and D, and quercetin polymers—a phenomenon that is linked to the unique function and features of different terpenes compounds (D-limonene, β-ionone, geraniol, eugenol, myrcene) [91]. For example, each of the listed terpenes functions as an ideal insecticide (in citral mixtures) due to the antibacterial and antifungal properties (see Table 8); there are no notable side effects on human health given the diminished neurotoxicity risks.

Table 8. Classifications of terpenes and their potential application as pesticides [91].

Terpene Type	Function	Features
Limonene	Limonene enhances the properties of other terpenes.	Redistilled limonene has less odor, more stable than d-limonene
Beta-ionone	Antibacterial and antifungal properties	Beta-ionone has prophylactic value.
Geraniol	Like beta-ionone, antibacterial and antifungal properties.	Geraniol gives a pleasant fragrance.
Eugenol	Clove oil; anesthetic (anti-itch), antibacterial antifungal properties.	Contain a distinct fragrance which is like geraniol
Myrcene	Possesses antifungal, antibacterial properties	Famous for its fragrance properties

Despite the potential for application in various manufacturing applications, there has been minimal industry interest in terpene and terpene derivatives. Winnacker et al. [88] attributed this phenomenon to inadequate understanding of how terpenes could be combined with other natural and synthetic polymers to augment material characteristics (such as tensile strength, photo-degradation, elongation at break, antimicrobial properties, and biocompatibility) [88]. Moreover, even though the feedstock is cheap and widely available, there are gaps in knowledge concerning effective chemical processes to reduce material costs, and determining potential new applications for terpene derivatives and monomers remain underexplored [88].

The most widely applied procedures include radical initiator-free addition and solvation of thiols to terpenes such as R +/− and S−/− limonene and -β-pinene; chain and step-growth polymerization techniques involving acrylic monoterpenes myrcene (7-methyl-3-methylene-1,6-octadiene) and ocimenes and Bayer–Villiger oxidation of menthone into methide [89]. The application potential of terpenes includes the synthesis of terpene/fatty acid-based polyesters for industrial applications [89]. The reaction parameters, including conversion efficiency for polyester synthesis via copolymerization of propylene oxide (PO), are depicted in Table 9 [88].

Table 9. Mediated copolymerization of PO, time conversion, MW and Tg [88].

Entry Monomer	Cat.	[PO]/[2][b]	[ah-1 c]/[cat.-2][b]	Time [h]	Conversion[c] [%]	Mn[d] [kDa]	Mw/Mn[d]	cis[e] [%]	Tg[f] [8C]
1	cat.-2 a	500	100	0.75	86	11.4	1.25	>99	105
2	cat.-2 a	500	100	2.0	>99	18.8	1.66	52	91
3[g]	cat.-2 a	100	150	3.0	>99[h]	12.2	1.22	>99	102
4	cat.-2 b	500	100	2.0	76	9.0	1.18	>99	101
5	cat.-2 b	500	100	3.0	>99	14.6	1.72	10	78
6	cat.-2 c	500	100	1.0	63	7.4	1.11	>99	97
7	cat.-2 c	500	100	2.5	>99	10.2	1.16	97	103
8	cat.-2 c	500	100	5.0	>99	9.8	1.16	95	102
9	cat.-2 c	1000	200	4.0	>99	22.5	1.28	88	104
10	cat.-2 c	2000	400	9.0	>99	36.7	1.30	94	107
11	cat.-2 c	4000	800	20.5	>99	55.4	1.29	98	109

In other cases, terpenoid-based polymeric resins have been employed in stereolithographic 3D printing [87] and as starting materials in the production of sustainable polymers and advanced materials [88]. The renewed industry attention on terpenes and their derivatives is validated by the progress made in controlled polymerization, interesting structures, and the broad scope of industrial application, renewability, and abundant feedstock.

Polylimonine Synthesis and Properties

The synthesis of polylimonine has practical implications on the production of terpenes, given that the former classes of chemical compounds are optically active terpenes, which exhibit antioxidant behavior [92]. Commercial applications of limonene and polylimonene include food preservation, enhancing the flavor and aroma of foods [93]. A fundamental constraint in the development of polymeric products globally is the recyclability of the precursor and the ecological effects; this remains a priority concern in polylimonene synthesis, considering that most precursors are sourced from fossil fuels [94]. In light of this challenge, the synthesis of non-toxic and renewable plastics is a priority for manufacturers. Such plastics would be ideal substitutes for existing non-renewable polymers. Traditionally, ideal precursors for renewable polymers are starch, essential oils, and wood. However, there have been growing research interests in the production of renewable and non-toxic plastics from terpenes.

Terpenes are naturally found in citrus plants and essential oils. Other sources of terpenes include tea, thyme, cannabis, Spanish sage, and citrus fruits (lemon, orange, mandarin) [91]. Beyond the ubiquity of the precursors, terpenes (D-limonene in particular) are ideal based on their chemical structure and the presence of isoprene units; these units are key enablers for addition-based polymerization processes. The choice of D-limonene as a precursor in the current case is supported by the worldwide production capacity [94] and compatibility with microbial mediated synthesis processes, including PMD, Codon optimized MVK, and PMK (see Table 10).

Table 10. Microbial-based synthetic routes for polylimonene through microbial mediated synthesis processes [93].

Host	Engineering Design	Limonene Synthase Origin, (+/−)-Limonene, Accession Number
E. coli BLR (DE3) codon +	• Abies grandis tGPPS	*Mentha spicata*, (−), L13459
E. coli DH1 ΔacrAB	• HMGS and tHMGR from Staphylococcus aureus • Codon optimized MVK, PMK, and PMD from *S. cerevisiae* • AACT and IDI from *E. coli* • tGPPS from Abies grandis • One plasmid containing the mevalonate pathway genes and one plasmid with tGPPS and tLS • A. Borkumensis efflux pump	*M. spicata*, (−), L13459, codon-optimized
E. coli DH1	• HMGS and tHMGR from Staphylococcus aureus • Codon optimized MVK, PMK, and PMD from *S. cerevisiae* • AACT and IDI from *E. coli* • tGPPS from Abies grandis • One plasmid	*M. Spicata*, (−), accession number not clear, codon-optimized
E. coli BL21(DE3)	• HMGS and tHMGR, MVK, PMK, and PMD from *S. cerevisiae* • AACT and IDI from *E. coli* • tGPPS from Abies grandis/GPPS from Streptomyces sp. strain KO-3988	*M. Spicata*, (−), L13459, codon-optimized
Synechocystis sp. PCC 6803	• DXS, IDI, and CrtE from Synechocystis	*Schizonepeta tenuifolia*, enantio-selectivity not clear, AF282875
Synechococcus sp. PCC 7002	• Wild-type and ΔglgC background were compared	*M. Spicata*, (−), Q40322, codon-optimized
S. cerevisiae AE9	• Yeast FPPS (ERG20 K197G) mutated to partly produce GPP	*Perilla frutescens*, (−), KM015220 and Citrus limon, (+), AF514287
S. cerevisiae EPY210C	• tHMGR from *S. cerevisiae* • upc2–1 transcription factor	*M. spicata*, (−), L13459 and C. Limon, (+), AF514287, codon-optimized

The biotechnology-based production of limonene from microorganisms does not address production-related constraints, including the molar mass loss/lack of sufficient molar mass, selection of an acceptable monomer for conversion. Most of the precursors for terpenes have practical drawbacks and limitations—a factor influencing the choice of orange peel-based precursors—a precursor proven useful in the synthesis of polylimonene through recycling polymerization [94]. Other sources of terpenes include tea, thyme, cannabis, Spanish sage, and citrus fruits (lemon, orange, mandarin) [91]. Beyond the ubiquity of the precursors, terpenes (D-limonene in particular) are ideal.

Under standard conditions, Coelho and Vieira synthesized polylimonene by placing D-limonene in a solution containing Ziegler-Natta catalysts and Lewis acids in the first stage. The molar mass of the mixed species was below 1000 g/mol [94]. The second stage entailed the initiation of free-radical polymerization using benzoyl peroxide. The concentration of the free radical initiator is adjusted accordingly to attain the desired conversion efficiency. In the current case, the desired conversion efficiency was 12% [94]. The reduced copolymerization rates demonstrate there is an inverse link between the D-limonene ratio, molar masses, and concentration of styrene, methyl methacrylate, and n-butyl acrylate).

8. Antioxidant Random Copolymers

The synthesis of antioxidant copolymers such as dual-function amphiphilic random copolymers is critical to the suppression of medical-related infections through Fe^{2+} chelating ability and supra-macromolecular re-arrangement and morphological changes in hydrous environments and biocompatibility [95]. The polymerization of the tertiary amine-containing cationic monomer with hydroxyl-tyrosol yielded copolymers with ideal amphiphilic balances, molar ratios, and functional groups (especially the aromatic rings) for biomedical device coatings. In other cases, polypropylene (PP) antioxidant random copolymers have been proven useful in hot water applications in the presence of nucleating agents. Even though the utility of nucleating agents was not demonstrated by Taresco et al. [95], Grabmann et al. [96] argued that the compounds contributed to the modification of the crystalline structure—a factor that translated to higher thermal stability, mechanical strength (toughness and creep); this was observed in β-nucleated PP-R grades. In contrast, non-nucleated grades with α-crystal structure did not exhibit improvements in thermal and mechanical properties.

The utility of antioxidant random copolymers transcends biomedical devices to encompass food packaging applications, but the suitability of the materials for such applications is offset by the rapid depletion of antioxidants and chemical changes induced by microwave heating. Shahidi [97] observed that the viability of polymers for food preservation depends on the ability of natural antioxidants, additives, and preservatives to prevent rancidity—a process that impacts the odor and sensory appeal of foods.

Alin and Hakkarainen [98] explored the suitability of propylene-ethylene copolymer (PP-C), propylene-ethylene random copolymer (PP-R), and polypropylene homopolymer (PP) in plastic packaging containers labeled as microwave safe. The microwave testing of the polymers established the following. First, excessive heating (and subsequent analysis using microwave-assisted extraction (MAE) and HPLC) triggered the migration of the antioxidant molecules, particularly Irgafos 168 and 100 from the propylene-ethylene copolymer (PP-C), propylene-ethylene random copolymer (PP-R), and polypropylene homopolymer (PP) [98]. The migration resulted in unequal distribution of the antioxidants diminishing the utility of the polymer packaging materials in food packaging. A notable observation was the influence of polypropylene materials on the migration rate of the antioxidants and the inverse relationship with crystallinity. Better migration resistance was noted in the polypropylene homo-polymer (PP) potentially due to higher crystallinity [98]; this pattern remained unchanged despite contact with fatty acid stimulants. The influence of acetic acid and ethanol/iso-octane concentration on the rate of migration is depicted in Table 11.

Table 11. Antioxidant migration under microwaving and heating conditions.

	Microwaving		Heating	
Sample	**I168**	**I1010**	**I168**	**I1010**
		90/10 iso-octane/ethanol, 80 °C		
PPC	509	502	686 ± 31 a	726 ± 64 a
PPR	4735	1283	4955 ± 204 a	1342 ± 217 a
PP	615	190	748 ± 85 a	277 ± 45 a
		Iso-octane, 80 °C		
PPC	–	–	777	765
PPR	–	–	4909	1477
PP	–	–	780	429
		3% acetic acid, 100 °C		
PPC	b	9	b	b
		3% acetic acid, 120 °C		
PPC	b	9	b	b
		10% ethanol, 120 °C		
PPC	b	6	b	b

In light of the preliminary observations made by Alin and Hakkarainen [98], polypropylene homo-polymer (PP) is an ideal alternative compared to propylene-ethylene copolymer (PP-C), propylene-ethylene random copolymer (PP-R), considering that the migration of antioxidants promotes toxicity and emergence of undesirable flavors [99]. A fundamental constraint moving forward was the development of packaging materials that are suitable and resistant to antioxidant migration, regardless of the fatty acid concentration in food. The experimental data shows that the composition of the food items, exposure to heat, and microwave energy predicts the chemical composition of antioxidant polymers for food packaging.

8.1. Gallic Acid Grafted Polymers

Similar to lignin, polyphenol, and quercetin polymers, gallic acid grafted polymers are ideal for targeted drug delivery, packaging, and commercial, industrial applications. Cho et al. [100] documented successful synthesis and characterization of gallic acid-grafted-chitosans, which demonstrated superior antioxidant scavenging ability against various free radicals such as hydrogen peroxide (93% effectiveness) and 2,2-diphenyl-1-picrylhydrazyl (92% effectiveness) at low concentrations (50 μg/mL) [101]. The outcomes affirm the superiority of gallic acid grafted copolymers in free radical scavenging. In other studies, gallic acid grafted chitosan-casein phosphopeptide nanoparticles have demonstrated superior anticancer and antioxidant properties under simulated gastrointestinal conditions [102]. The observations made by alternatively, the incorporation of chitosan on resveratrol via free radical-induced grafting proven effective in the synthesis of resveratrol modified species [73]. Cho et al. [100] contrast with Curcio et al. [79], who documented successful free radical-induced grafting of gallic acid-chitosan and catechin-chitosan conjugates for antioxidant applications.

Considering that the in-vitro models were confined to the two free radicals (2,2-diphenyl-1-picrylhydrazyl, and H_2O_2, it remains unclear whether gallic acid-grafted-chitosans would retain the high antioxidant scavenging ability at higher concentrations. The outcomes documented by Cho et al. are in agreement with Spizzirri et al., who documented successful synthesis of antioxidant–protein conjugates (gallic acid (GA) and catechin (CT) attached on gelatin) via grafting reaction [103]. Similar to gallic acid grafted polymers, the Gallic Acid-Catechin-Gelatin copolymers exhibited superior scavenging ability against linoleic acid peroxide hydroxyl radicals and 2,2′-diphenyl-1-picrylhydrazyl.

8.2. Resveratrol Conjugated and Grafted Polymers

Resveratrol exhibits comparable antioxidant activity as quercetin, polyphenols, Vitamin C, and E, phenolic compounds, carotenoids, α-lipoic acid, melatonin, gallic acid, and

N-acetyl cysteine, captopril. However, the material is superior to other antioxidants based on its broad biological activity, including anticancer, anti-inflammatory and chemoprevention properties, and rapid metabolism within the body due to its shortened half-life [104]. The mechanical, biological, and chemical properties can be enhanced through modifications in the synthetic routes. For example, Siddalingappa et al. [104] reported the successful synthesis of Resveratrol-PEG ethers via alkali-mediated etherification. The plasma stability of the moieties was augmented by aggregation behavior and polymer hydrophobicity, and the antioxidant demonstrated ideal antioxidant behavior in mice models. Further customized changes were viable through carbodiimide coupling reaction-mediated surface and chemical modifications with α-methoxy poly (ethylene glycol)-co-polylactide succinyl ester resveratrol, MeO-PEG succinyl ester resveratrol, and α-methoxy-ω-carboxylic acid poly(ethylene glycol) succinylamide resveratrol [104].

9. Antioxidant-Polysaccharides Graft Polymers

Polysaccharide graft polymers play an integral role in biomedical (targeted drug delivery systems) and commercial engineering applications given their unique properties and wide range of sources (microbial and natural). In contrast to other precursors, polysaccharides are cheap and widely available [105]. Moreover, the materials have comparable properties as antioxidant terpene polymers in terms of biodegradation ability, water solubility, and non-toxicity. Grafted polysaccharides have attracted significant research attention based on a broad scope of application, chemical stability, efficiency, biocompatibility, tailored properties, and biodegradation ability. Recent advances in drug delivery research have demonstrated that the rate of in vivo drug delivery could be predicted by the extent of crosslinking and grafting [105]. The enhanced functionality is linked to chemical bonding between the grafted polymer chains and polymeric substrate. The free radicals are introduced onto the backbone structure through irradiation and chemical initiation.

Despite the promising material prospects, the polysaccharide graft polymers have not conclusively addressed challenges associated with traditional drug delivery systems, such as sustaining the desired drug dosage in tissues for an extended period (length of treatment). However, there are promising prospects that the uncontrolled drug release kinetics would be resolved through better absorption rates [105]. The progress made in targeted drug delivery by Pal and Das [105] is in line with Lemarchand et al. [106] research on polyester-polysaccharide nanoparticles for drug delivery systems. In particular, the nanoparticles graft polymers from poly (D, L-lactide) or poly(ε–caprolactone) (PCL) demonstrated ideal pharmacokinetic behavior linked to the synthesis route—emulsion solvent evaporation. The material exhibited superior stability with or without emulsions, and strong amphiphilic properties, which eliminated the need for surfactants.

The emulsion solvent evaporation technique's ability to influence the targeted drug delivery is comparable to RAFT (reversible addition-fragmentation chain transfer polymerization), reversible deactivation radical polymerization, and atom transfer radical polymerization; these two procedures were employed by Garcia-Valdez in the modification of polysaccharides to attain ideal control over the molecular weight distribution, and carefully control the macromolecular shapes, and sizes [107]. From an engineering perspective, the experimental outcomes reported following the adoption of different synthetic routes introduce unique challenges because the outcomes were based on laboratory-scale models; there are concerns that the procedures could lead to different outcomes in large experiments. The highlighted concerns could be resolved through advances in synthesis technologies, including lab-on-a-chip.

10. Poly (Trolox Ester) Polymers

Poly Trolox ester polymers are useful antioxidants in human cells based on their ability to relieve oxidative stress in injured cells and resolve iron particle toxicity at the nanoscale [107,108]. Even though there is a broad class of compounds that exhibits antioxidant behavior (including α-lipoic acid, N-acetyl cysteine, melatonin, gallic acid, captopril, taurine, catechin, and quercetin), the biological function of most is impaired by oxidative stress-linked biological incompatibility at the nanoscale [108]. Wattamwar et al. claimed that poly Trolox ester polymers were able to overcome these barriers by inhibiting the response of free radicals while retaining higher compatibility with medical devices. In contrast to quercetin, lignin, and polydopamine, poly Trolox ester polymers are a class of synthetic analogs of vitamin E (with superior hydrophilic, tunable particle sizes, antioxidant properties) [108]. The materials can be synthesized via a facile single emulsion technique; the nanoparticle concentration is varied over time to achieve the desired nanoparticle sizes.

Resolving iron particle toxicity is a priority in biomedical applications, given that the risk is elevated by the frequency of medical diagnosis and treatment (iron supplements for anemia and MRI contrast agents). Iron particle toxicity was mitigated through targeted delivery of functionalized poly Trolox ester polymers to the platelet endothelial cell adhesion molecules (PECAM-1) [101]. In theory, the modification of the pro-oxidant and antioxidant activity of poly Trolox ester polymers could be applied in virtually all high-oxidative-stress environments within the cells. However, in reality, this might not be feasible considering that poly (Trolox ester) nanoparticles' ability to induce/inhibit cellular oxidative stress depends on advanced programming of the molecules, based on knowledge of the molecular compounds being oxidized and the cumulative oxidative status of the cells [6]. If the listed variables are not known, the cellular function of poly (Trolox ester) nanoparticles might be impaired. Such complexities limit the clinical application of Trolox esters.

11. Polymerization of Vitamins

Reversible addition-fragmentation chain transfer (RAFT) is suitable for the Vitamin C and B2 mediated polymerization of synthetic monomers such as methacrylates, acrylates, and acrylamides under LED irradiation. The efficiency of the polymerization process is contingent on the optimization of the irrational wavelength and oxygen. Zhang et al. linked the high throughput in vitamin-mediated polymerization to the low reaction volume platform, molecular weight distribution, low reaction volumes and the concentrations of Vitamin B_2 and C [109]. The costs of the synthetic process can be offset by the inclusion of off-the-shelf vitamin supplements in place of reagent/analytical grade Vitamin C and B_2. The free radical-initiated chain polymerization of biomolecules has critical implications on disease prevalence, given that the progression of Alzheimer's disease and other neurodegenerative diseases can be linked to free radical damage of cells. On a positive note, this process can be offset by anti-oxygenic nutrients, such as vitamin-rich (C, E and co-enzyme Q) fruits and vegetables, which offer protective benefits against free radical damage [110]. The polymerization of compounds with complex architectures has been augmented by advances in technology, particularly photo-induced electron/energy transfer–reversible addition-fragmentation chain-transfer (PET-RAFT) polymerization.

The PET-RAFT process is remarkably superior relative to RAFT in isolation due to higher oxygen tolerance [111]. However, other issues remain unresolved, including the experimental testing of oxygen tolerant species. Most of the documented experiments were conducted in water—a factor that limits the utility of the process in the presence of organic solvents. The initiation of vitamin polymerization in the presence of organic solvents is critical in biological processes, given that vitamins and biological molecules comprise organic functional groups. The constraint was partially resolved by Gormley et al. [109], who documented successful polymerization of oxygen resistant compounds in organic solvents, dimethyl sulfoxide (DMSO). Nonetheless, it is of note that the initiation of the process requires a zinc tetraphenyl porphyrin (ZnTPP) catalyst. From an industrial and commercial perspective, the additional costs linked to catalysis pale in comparison relative to the additional benefits associated with successful PET-RAFT in organic solvents. The findings documented by Zhang et al. [110], Tappel et al. [111], and Gormley et al. [109] demonstrate the excellent potential of oxygen tolerant PET-RAFT and RAFT vitamin-based photo-initiation mechanism in biological applications.

12. Conclusions

The published literature has demonstrated the antioxidant properties of various naturally occurring compounds, such as quercetin, polyphenols, vitamins, and synthetic molecules including α-lipoic acid, N-acetyl cysteine, melatonin, gallic acid, lignin, captopril, taurine, and catechin; the free radical scavenging abilities predicted the potency of the antioxidants (hydrogen bonding within the ether O_2 atoms in peroxide radical) and the location of bulky tert-butyl functional groups. The extraction of antioxidants from agro-wastes generates high-quality phenolic antioxidants. On the downside, the benefits are offset by low yields and negative ecological effects, as well as synthesis-related trade-offs between functionality, chemical stability, and biocompatibility. Other constraints are related to the chemical properties of the products—polyphenols exhibit poor membrane permeability, rapid metabolism, and poor bioavailability and UV degradation, despite free radical grafting [73], step growth, free radical, and enzyme-catalyzed reactions. A vast majority of research in antioxidant materials so far has been focused on the entrapment or encapsulation of low-molecular-weight and enzymatic antioxidant compounds in polymersomes, micelles, or nanoparticles that often exhibit poor long-term stability. The structural incorporation of antioxidant moieties in polymers or direct polymerization while maintaining their intrinsic antioxidant properties has been attracting increasing interest from the research community. Polymerized antioxidants can be produced through grafting or backbone incorporation, and offer the potential advantages of protecting the antioxidant from inactivation, delivering continuous antioxidant protection as long as the polymer exists, as well as a relatively high mass of antioxidant payload.

Potential industrial applications of antioxidants in the food production industry are diverse, especially in the forms of blends with bio-polyether (PEO), bio-polyester (PLA), and commercial starch-based polymer (Mater-Bi) for photo-stabilization. The blended products exhibit better tensile strength, photostability, and elongation at break and modulus of elasticity. Active polymer packaging is becoming increasingly important as an emerging technology [112] that can significantly improve the quality and stability of food, eliminating direct addition of chemicals and the need to make significant changes in production processes. If antioxidant polymers are brought into packaging systems, major improvements in maintaining the stability of oxidation-sensitive food products can be made. Antioxidant polymers have been proven to be scavengers that function by reducing the presence of reactive oxygen species, which act as initiators of oxidation processes. In this review, we have seen that antioxidant polymers can be specifically designed and optimized for each specific product or application and in brief, they can be implemented by the food industry as new emerging sustainable polymeric materials.

Author Contributions: C.M. writing—original draft preparation, I.S.B. review and editing, T.B. review and editing. All authors have read and agreed to the published version of the manuscript.

Funding: This research received no external funding.

Institutional Review Board Statement: Not applicable.

Informed Consent Statement: Not applicable.

Data Availability Statement: The data presented in this study are available on request from the corresponding author.

Conflicts of Interest: The authors declare no conflict of interest.

References

1. Sawant, R.; Vaze, O.; D'Souza, G.; Rockwell, K.; Torchilin, V. Palmitoyl ascorbate-loaded polymeric micelles: Cancer cell targeting and cytotoxicity. *Pharm Res.* **2011**, *28*, 301–308. [CrossRef]
2. Augustyniak, A.; Bartosz, G.; Čipak, A.; Duburs, G.; Horáková, L.; Łuczaj, W.; Žarković, N. Natural and synthetic antioxidants: An updated overview. *Free Radic. Res.* **2010**, *44*, 1216–1262. [CrossRef] [PubMed]
3. Pryor, W.A.; Strickland, T.; Church, D.F. Comparison of the efficiencies of several natural and synthetic antioxidants in aqueous SDS [sodium dodecyl sulfate] micelle solutions. *J. Am. Chem. Soc.* **1988**, *110*, 2224–2229. [CrossRef]

4. Lourenço, S.C.; Moldão-Martins, M.; Alves, V.D. Antioxidants of Natural Plant Origins: From Sources to Food Industry Applications. *Molecules* **2019**, *24*, 4132. [CrossRef]
5. Flora, S.; Shrivastava, R.; Mittal, M. Chemistry and Pharmacological Properties of Some Natural and Synthetic Antioxidants for Heavy Metal Toxicity. *Curr. Med. Chem.* **2013**, *20*, 4540–4574. [CrossRef] [PubMed]
6. Wattamwar, P.P.; Hardas, S.S.; Butterfield, D.A.; Anderson, K.W.; Dziubla, T.D. Tuning of the pro-oxidant and antioxidant activity of trolox through the controlled release from biodegradable poly(trolox ester) polymers. *J. Biomed. Mater Res. Part A* **2011**, *99*, 184–191. [CrossRef]
7. Cateto, C.A.; Barreiro, M.F.; Rodrigues, A.E.; Belgacem, M.N. Kinetic study of the formation of lignin-based polyurethanes in bulk. *React. Funct. Polym.* **2011**, *71*, 863–869. [CrossRef]
8. Franco, P.; Incarnato, L.; De Marco, I. Supercritical CO_2 impregnation of α-tocopherol into PET/PP films for active packaging applications. *J. CO_2 Util.* **2019**, *34*, 266–273. [CrossRef]
9. Lukic, I.; Vulic, J.; Ivanovic, J. Antioxidant activity of PLA/PCL films loaded with thymol and/or carvacrol using $scCO_2$ for active food packaging. *Food Packag. Shelf Life* **2020**, *26*, 100578. [CrossRef]
10. Trucillo, P.; Campardelli, R.; Reverchon, E. Production of liposomes loaded with antioxidants using a supercritical CO_2 assisted process. *Powder Technol.* **2018**, *323*, 155–162. [CrossRef]
11. Ozkan, G.; Franco, P.; Capanoglu, E.; De Marco, I. PVP/flavonoid coprecipitation by supercritical antisolvent process. *Chem. Eng. Process. Process. Intensif.* **2019**, *146*, 107689. [CrossRef]
12. Bastante, C.C.; Cardoso, L.C.; Serrano, C.M.; de la Ossa, E.M. Supercritical impregnation of food packaging films to provide antioxidant properties. *J. Supercrit. Fluids* **2017**, *128*, 200–207. [CrossRef]
13. Grossman, A.; Vermerris, W. Lignin-based polymers and nanomaterials. *Curr. Opin. Biotechnol.* **2019**, *56*, 112–120. [CrossRef] [PubMed]
14. Ponomarenko, J.; Lauberts, M.; Dizhbite, T.; Lauberte, L.; Jurkjane, V.; Telysheva, G. Antioxidant activity of various lignins and lignin-related phenylpropanoid units with high and low molecular weight. *Holzforschung* **2015**, *69*, 795–805. [CrossRef]
15. Vinardell, M.P.; Mitjans, M. Lignins and Their Derivatives with Beneficial Effects on Human Health. *Int. J. Mol. Sci.* **2017**, *18*, 1219. [CrossRef]
16. Liu, Q.; Luo, L.; Zheng, L. Lignins: Biosynthesis and Biological Functions in Plants. *Int. J. Mol. Sci.* **2018**, *19*, 335. [CrossRef]
17. Abhilash, M.; Thomas, D. 15-Biopolymers for Biocomposites and Chemical Sensor Applications. In *Biopolymer Composites in Electronics*; Sadasivuni, K.K., Ponnamma, D., Kim, J., Cabibihan, J.-J., Eds.; AlMaadeed MABT-BC in Elsevier: Amsterdam, The Netherlands, 2017; pp. 405–435. [CrossRef]
18. Mena, P.; Calani, L.; Bruni, R.; Del Rio, D. Chapter 6-Bioactivation of High-Molecular-Weight Polyphenols by the Gut Microbiome. In *Diet-Microbe Interactions in the Gut*; Tuohy, K., Ed.; Academic Press: San Diego, CA, USA, 2015; pp. 73–101. [CrossRef]
19. Pizzorno, J.; Murray, M.T. *Textbook of Natural Medicine*, 4th ed.; Churchill Livingstone: London, UK, 2012; Available online: https://www.elsevier.com/books/textbook-of-natural-medicine/pizzorno/978-1-4377-2333-5 (accessed on 25 July 2021).
20. Gunawardena, D.; Govindaraghavan, S.; Münch, G. Anti-Inflammatory Properties of Cinnamon Polyphenols and their Monomeric Precursors. *Polyphen. Hum. Health Dis.* **2013**, *1*, 409–425. [CrossRef]
21. Gonçalves, C.; Dinis, T.; Batista, M.T. Antioxidant properties of proanthocyanidins of Uncaria tomentosa bark decoction: A mechanism for anti-inflammatory activity. *Phytochemistry* **2005**, *66*, 89–98. [CrossRef]
22. Chai, W.-M.; Ou-Yang, C.; Huang, Q.; Lin, M.-Z.; Wang, Y.-X.; Xu, K.-L.; Pang, D.D. Antityrosinase and antioxidant properties of mung bean seed proanthocyanidins: Novel insights into the inhibitory mechanism. *Food Chem.* **2018**, *260*, 27–36. [CrossRef]
23. Martinez-Avila, G.C.G.; Aguilera, A.F.; Saucedo, S.; Rojas, R.; Rodriguez, R.; Aguilar, C.N. Fruit Wastes Fermentation for Phenolic Antioxidants Production and Their Application in Manufacture of Edible Coatings and Films. *Crit. Rev. Food Sci. Nutr.* **2014**, *54*, 303–311. [CrossRef]
24. Iyer, K.A.; Zhang, L.; Torkelson, J.M. Direct Use of Natural Antioxidant-rich Agro-wastes as Thermal Stabilizer for Polymer: Processing and Recycling. *ACS Sustain. Chem. Eng.* **2016**, *4*, 881–889. [CrossRef]
25. Kricheldorf, H.R.; Kreiser-Saunders, I. Polymerizations of Lactide. *Polymer* **1994**, *35*, 4175–4180. [CrossRef]
26. Wang, C.E.; Stayton, P.S.; Pun, S.H.; Convertine, A.J. Polymer nanostructures synthesized by controlled living polymerization for tumor-targeted drug delivery. *J. Control. Release* **2015**, *219*, 345–354. [CrossRef]
27. Coulembier, O.; Degée, P.; Hedrick, J.L.; Dubois, P. From controlled ring-opening polymerization to biodegradable aliphatic polyester: Especially poly(β-malic acid) derivatives. *Prog. Polym. Sci.* **2006**, *31*, 723–747. [CrossRef]
28. Garbetta, A.D.; Antuono, I.; Melilli, M.G.; Sillitti, C.; Linsalata, V.; Scandurra, S. Inulin enriched durum wheat spaghetti: Effect of polymerization degree on technological and nutritional characteristics. *J. Funct. Foods* **2020**, *71*, 104004. [CrossRef]
29. Nanni, G.; Heredia-Guerrero, J.A.; Paul, U.C.; Dante, S.; Caputo, G.; Canale, C.; Athanassiou, A.; Fragouli, D.; Bayer, I.S. Poly (furfuryl alcohol)-polycaprolactone blends. *Polymers* **2019**, *11*, 1069. [CrossRef] [PubMed]
30. Kirschweng, B.; Tátraaljai, D.; Földes, E.; Pukánszky, B. Natural antioxidants as stabilizers for polymers. *Polym. Degrad. Stab.* **2017**, *145*, 25–40. [CrossRef]
31. Cirillo, G.; Iemma, F. *Antioxidant Polymers: Synthesis, Properties, and Applications*; John Wiley & Sons: Hoboken, NJ, USA, 2012.
32. Van Lith, R.; Ameer, G.A. Antioxidant Polymers as Biomaterial. In *Oxidative Stress Biomater*; Elsevier: Amsterdam, The Netherlands, 2016; pp. 251–296.

33. Quilez-Molina, A.I.; Heredia-Guerrero, J.A.; Armirotti, A.; Paul, U.C.; Athanassiou, A.; Bayer, I.S. Comparison of physicochemical, mechanical and antioxidant properties of polyvinyl alcohol films containing green tealeaves waste extracts and discarded balsamic vinegar. *Food Packag. Shelf Life* **2020**, *23*, 100445. [CrossRef]
34. Brocca, D.; Arvin, E.; Mosbæk, H. Identification of organic compounds migrating from polyethylene pipelines into drinking water. *Water Res.* **2002**, *36*, 3675–3680. [CrossRef]
35. Mahmood, Z.; Yameen, M.; Jahangeer, M.; Riaz, M.; Ghaffar, A.; Javid, I. *Lignin as Natural Antioxidant Capacity. Lignin-Trends Appl*; Intechopen: Budapest, Hungary, 2017.
36. Rahman, M.M.; Tsukamoto, J.; Rahman, M.; Yoneyama, A.; Mostafa, K.M. Lignin and its effects on litter decomposition in forest ecosystems. *Chem. Ecol.* **2013**, 540–553. [CrossRef]
37. Llerena, J.P.P.; Figueiredo, R.; Brito, M.S.; Kiyota, E.; Mayer, J.L.S.; Araujo, P. Deposition of lignin in four species of Saccharum. *Sci. Rep.* **2019**, *9*, 5877. [CrossRef]
38. Latif, N.H.A.; Rahim, A.A.; Brosse, N.; Hussin, M.H. The structural characterization and antioxidant properties of oil palm fronds lignin incorporated with p-hydroxyacetophenone. *Int. J. Biol. Macromol.* **2019**, *130*, 947–957. [CrossRef]
39. Kabir, A.S. Effects of Lignin as a Stabilizer or Antioxidant in Polyolefins. Electronic Thesis and Dissertation Repository. 2017, p. 4796. Available online: https://ir.lib.uwo.ca/etd/4796 (accessed on 25 July 2021).
40. Saito, T.; Perkins, J.H.; Jackson, D.C.; Trammel, N.E.; Hunt, M.A.; Naskar, A.K. Development of lignin-based polyurethane thermoplastics. *RSC Adv.* **2013**, *3*, 21832–21840. [CrossRef]
41. Cui, S.; Borgemenke, J.; Liu, Z.; Li, Y. Recent advances of "soft" bio-polycarbonate plastics from carbon dioxide and renewable bio-feedstocks via straightforward and innovative routes. *J. CO_2 Util.* **2019**, *34*, 40–52. [CrossRef]
42. Turner, D.A.; Williams, I.D.; Kemp, S. Resources, Conservation and Recycling Greenhouse gas emission factors for recycling of source-segregated waste materials. *Resour. Conserv. Recycl.* **2015**, *105*, 186–197. [CrossRef]
43. Zheng, J.; Suh, S. Strategies to reduce the global carbon footprint of plastics. *Nat. Clim. Chang.* **2019**, *9*. [CrossRef]
44. Nisticò, R. Aquatic-derived biomaterials for a sustainable future: A European opportunity. *Resources* **2017**, *6*, 65. [CrossRef]
45. Fernandez, J.G.; Ingber, D.E. Manufacturing of large-scale functional objects using biodegradable chitosan bioplastic. *Macromol. Mater. Eng.* **2014**, *299*, 932–938. [CrossRef]
46. Lackner, M. Bioplastics-Biobased plastics as renewable and/or biodegradable alternatives to petroplastics. In *Encyclopedia of Chemical Technology*; Othmer, K., Ed.; Wiley Publishing, Inc.: Hoboken, NJ, USA, 2015.
47. Sapei, L.; Padmawijaya, K.S.; Sijayanti, O.; Wardhana, P.J. The effect of banana starch concentration on the properties of chitosan-starch bioplastics. *Mater Sci.* **2015**, *7*, 101–105.
48. Liu, H.; Chung, H. Lignin-Based Polymers via Graft Copolymerization. *J. Polym. Sci. A Polym. Chem.* **2017**, *55*, 3515–3528. [CrossRef]
49. Boyer, C.; Bulmus, V.; Davis, T.P.; Ladmiral, V.; Liu, J.; Perrier, S. Bioapplications of RAFT polymerization. *Chem. Rev.* **2009**, *109*, 5402–5436. [CrossRef] [PubMed]
50. Lai, J.T.; Filla, D.; Shea, R. Functional Polymers from Novel Carboxyl-Terminated Trithiocarbonates as Highly Efficient RAFT Agents. *Macromolecules* **2002**, *35*, 6754–6756. [CrossRef]
51. da Silva, E.B.; Zabkova, M.; Araújo, J.; Cateto, C.; Barreiro, M.; Belgacem, M.; Rodrigues, A. An integrated process to produce vanillin and lignin-based polyurethanes from Kraft lignin. *Chem. Eng. Res. Des.* **2009**, *87*, 1276–1292. [CrossRef]
52. Qiu, Y.; Zhu, Z.; Miao, Y.; Zhang, P.; Jia, X.; Liu, Z.; Zhao, X. Polymerization of dopamine accompanying its coupling to induce self-assembly of block copolymer and application in drug delivery. *Polym. Chem.* **2020**, *11*, 2811–2821. [CrossRef]
53. Kwon, I.S.; Bettinger, C.J. Polydopamine nanostructures as biomaterials for medical applications. *J. Mater. Chem. B* **2018**, *6*, 6895–6903. [CrossRef]
54. Fichman, G.; Schneider, J. Dopamine Self-Polymerization as a Simple and Powerful Tool to Modulate the Viscoelastic Mechanical Properties of Peptide-Based Gels. *Molecules* **2021**, *26*, 1363. [CrossRef] [PubMed]
55. Barclay, T.G.; Hegab, H.M.; Clarke, S.R.; Ginic-Markovic, M. Versatile Surface Modification Using Polydopamine and Related Polycatecholamines: Chemistry, Structure, and Applications. *Adv. Mater. Interfaces* **2017**, *4*. [CrossRef]
56. Lakshminarayanan, R.; Madhavi, S.; Sim, C. Oxidative Polymerization of Dopamine: A High-Definition Multifunctional Coatings for Electrospun Nanofibers-An Overview. In *Dopamine-Health and Disease*; IntechOpen: London, UK, 2018. [CrossRef]
57. Du, X.; Li, L.; Behboodi-Sadabad, F.; Welle, A.; Li, J.; Heissler, S.; Zhang, H.; Plumeré, N.; Levkin, P.A. Bio-inspired strategy for controlled dopamine polymerization in basic solutions. *Polym. Chem.* **2017**, *8*, 2145–2151. [CrossRef]
58. Xiao, C.; Chu, X.; Yang, Y.; Li, X.; Zhang, X.; Chen, J. Hollow nitrogen-doped carbon microspheres pyrolyzed from self-polymerized dopamine and its application in simultaneous electrochemical determination of uric acid, ascorbic acid and dopamine. *Biosens. Bioelectron.* **2011**, *26*, 2934–2939. [CrossRef]
59. Kalimuthu, P.; John, A. Simultaneous determination of ascorbic acid, dopamine, uric acid and xanthine using a nanostructured polymer film modified electrode. *Talanta* **2010**, *80*, 1686–1691. [CrossRef]
60. Kohri, M.; Nannichi, Y.; Kohma, H.; Abe, D.; Kojima, T.; Taniguchi, T.; Kishikawa, K. Size control of polydopamine nodules formed on polystyrene particles during dopamine polymerization with carboxylic acid-containing compounds for the fabrication of raspberry-like particles. *Colloids Surf. A Physicochem. Eng. Asp.* **2014**, *449*, 114–120. [CrossRef]
61. Chen, T.-P.; Liu, T.; Su, T.-L.; Liang, J. Self-Polymerization of Dopamine in Acidic Environments without Oxygen. *Langmuir* **2017**, *33*, 5863–5871. [CrossRef] [PubMed]
62. Mrówczyński, R.; Markiewicz, R.; Liebscher, J. Chemistry of polydopamine analogues. *Polym. Int.* **2016**, *65*, 1288–1299. [CrossRef]

63. Aidoo, R.P.; Depypere, F.; Afoakwa, E.O.; Dewettinck, K. Industrial manufacture of sugar-free chocolates–Applicability of alternative sweeteners and carbohydrate polymers as raw materials in product development. *Trends Food Sci. Technol.* **2013**, *32*, 84–96. [CrossRef]
64. Guimarães, J.T.; Silva, E.K.; Lelis, A.L.R.C.; Cunha, R.; Freitas, M.Q.; Meireles, M.A.A.; da Cruz, A.G. Manufacturing a prebiotic whey beverage exploring the influence of degree of inulin polymerization. *Food Hydrocoll.* **2018**, *77*, 787–795. [CrossRef]
65. Karimi, R.; Azizi, M.H.; Ghasemlou, M.; Vaziri, M. Application of inulin in cheese as prebiotic, fat replacer and texturizer: A review. *Carbohydr. Polym.* **2015**, *119*, 85–100. [CrossRef]
66. Olkku, J.; Rha, C. Gelatinisation of starch and wheat flour starch—A review. *Food Chem.* **1978**, *3*, 293–317. [CrossRef]
67. Vervoort, L.; Mooter, G.V.D.; Augustijns, P.; Busson, R.; Toppet, S.; Kinget, R. Inulin hydrogels as carriers for colonic drug targeting: I. Synthesis and characterization of methacrylated inulin and hydrogel formation. *Pharm. Res.* **1997**, *14*, 1730–1737. [CrossRef]
68. Kumar, S.; Kesharwani, S.S.; Kuppast, B.; Bakkari, M.A.; Tummala, H. Pathogen-mimicking vaccine delivery system designed with a bioactive polymer (inulin acetate) for robust humoral and cellular immune responses. *J. Control. Release* **2017**, *261*, 263–274. [CrossRef]
69. Ito, H.; Wada, T.; Ohguchi, M.; Sugiyama, K.; Kiriyama, S.; Morita, T. The Degree of Polymerization of Inulin-Like Fructans Affects Cecal Mucin and Immunoglobulin A in Rats. *J. Food Sci.* **2008**, *73*, H36–H41. [CrossRef]
70. Boets, E.; Deroover, L.; Houben, E.; Vermeulen, K.; Gomand, S.V.; Delcour, J.A.; Verbeke, K. Quantification of In Vivo Colonic Short Chain Fatty Acid Production from Inulin. *Nutrients* **2015**, *7*, 8916–8929. [CrossRef]
71. Morrison, D.J.; Preston, T. Formation of short chain fatty acids by the gut microbiota and their impact on human metabolism. *Gut Microbes* **2016**, *7*, 189–200. [CrossRef]
72. Oliver, S.; Vittorio, O.; Cirillo, G.; Boyer, C. Enhancing the therapeutic effects of polyphenols with macromolecules. *Polym. Chem.* **2016**, *7*, 1529–1544. [CrossRef]
73. Curcio, M.; Puoci, F.; Iemma, F.; Parisi, O.I.; Cirillo, G.; Spizzirri, U.G.; Picci, N. Covalent Insertion of Antioxidant Molecules on Chitosan by a Free Radical Grafting Procedure. *J. Agric. Food Chem.* **2009**, *57*, 5933–5938. [CrossRef] [PubMed]
74. Uyama, H.; Kobayashi, S. Enzyme-catalyzed polymerization to functional polymers. *J. Mol. Catal. B Enzym.* **2002**, *19–20*, 117–127. [CrossRef]
75. Sun, X.; Bai, R.; Zhang, Y.; Wang, Q.; Fan, X.; Yuan, J.; Cui, L.; Wang, P. Laccase-Catalyzed Oxidative Polymerization of Phenolic Compounds. *Appl. Biochem. Biotechnol.* **2013**, *171*, 1673–1680. [CrossRef]
76. Hollmann, F.; Arends, I.W.C.E. Enzyme Initiated Radical Polymerizations. *Polymers* **2012**, *4*, 759–793. [CrossRef]
77. Puoci, F.; Morelli, C.; Cirillo, G.; Curcio, M.; Parisi, O.I.; Maris, P.; Sisci, D.; Picci, N. Anticancer activity of a quercetin-based polymer towards HeLa cancer cells. *Anticancer Res.* **2012**, *32*, 2843–2847. [PubMed]
78. Morici, E.; Arrigo, R.; Dintcheva, N.T. Quercetin as natural stabilizing agent for bio-polymer. *AIP Conf. Proc.* **2014**, *1599*, 314–317. [CrossRef]
79. Sunoqrot, S.; Al-Debsi, T.; Al-Shalabi, E.; Ibrahim, L.H.; Faruqu, F.N.; Walters, A.; Palgrave, R.G.; Al-Jamal, K.T. Bioinspired Polymerization of Quercetin to Produce a Curcumin-Loaded Nanomedicine with Potent Cytotoxicity and Cancer-Targeting Potential in Vivo. *ACS Biomater. Sci. Eng.* **2019**, *5*, 6036–6045. [CrossRef]
80. Xu, D.; Hu, M.-J.; Wang, Y.-Q.; Cui, Y.-L. Antioxidant Activities of Quercetin and Its Complexes for Medicinal Application. *Molecules* **2019**, *24*, 1123. [CrossRef]
81. Zhang, M.; Swarts, S.G.; Yin, L.; Liu, C.; Tian, Y.; Cao, Y.; Swarts, M.; Yang, S.; Zhang, S.B.; Zhang, K.; et al. Antioxidant Properties of Quercetin. *Adv. Exp. Med. Biol.* **2011**, *701*, 283–289. [CrossRef]
82. Sahiner, N. One step poly(quercetin) particle preparation as biocolloid and its characterization. *Colloids Surf. A Physicochem. Eng. Asp.* **2014**, *452*, 173–180. [CrossRef]
83. Pivec, T.; Kargl, R.; Maver, U.; Bračič, M.; Elschner, T.; Žagar, E.; Gradišnik, L.; Kleinschek, K.S. Chemical Structure–Antioxidant Activity Relationship of Water–Based Enzymatic Polymerized Rutin and Its Wound Healing Potential. *Polymers* **2019**, *11*, 1566. [CrossRef] [PubMed]
84. Ziyatdinova, G.; Kozlova, E.; Budnikov, H. Polyquercetin/MWNT-modified Electrode for the Determination of Natural Phenolic Antioxidants. *Electroanalysis* **2017**, *29*, 2610–2619. [CrossRef]
85. Ziyatdinova, G.K.; Kozlova, E.; Budnikov, H.C. Chronoamperometric evaluation of the antioxidant capacity of tea on a polyquercetin-modified electrode. *J. Anal. Chem.* **2017**, *72*, 382–389. [CrossRef]
86. Firdaus, M.; Espinosa, L.M.; De Meier, M.A.R. Terpene-Based Renewable Monomers and Polymers via Thiol À Ene Additions. *Macromolecules* **2011**, *44*, 7253–7262. [CrossRef]
87. Weems, A.C.; Chiaie, K.R.D.; Worch, J.; Stubbs, C.J.; Dove, A.P. Terpene- and terpenoid-based polymeric resins for stereolithography 3D printing. *Polym. Chem.* **2019**, *10*, 5959–5966. [CrossRef]
88. Winnacker, M.; Rieger, B. Recent Progress in Sustainable Polymers Obtained from Cyclic Terpenes: Synthesis, Properties, and Application Potential. *ChemSusChem* **2015**, *8*, 2455–2471. [CrossRef]
89. Singh, A.; Kamal, M. Synthesis and characterization of polylimonene: Polymer of an optically active terpene. *J. Appl. Polym. Sci.* **2012**, *125*, 1456–1459. [CrossRef]
90. Zhao, J.; Schlaad, H. Synthesis of Terpene-Based Polymers. In *Biosynthetic-Synthetic Polymer Conjugates*; Schlaad, H., Ed.; Springer: Berlin/Heidelberg, Germany, 2011.

91. Cox-Georgian, D.; Ramadoss, N.; Dona, C.; Basu, C. Therapeutic and Medicinal Uses of Terpenes. *Med. Plants* **2019**, *2019*, 333–359. [CrossRef]
92. Yang, Z.; Peng, H.; Wang, W.; Liu, T. Crystallization behavior of poly(ε-caprolactone)/layered double hydroxide nanocomposites. *J. Appl. Polym. Sci.* **2010**, *116*, 2658–2667. [CrossRef]
93. Jongedijk, E.; Cankar, K.; Buchhaupt, M.; Schrader, J.; Bouwmeester, H.; Beekwilder, J. Biotechnological production of limonene in microorganisms. *Appl. Microbiol. Biotechnol.* **2016**, *100*, 2927–2938. [CrossRef]
94. Coelho, F.M.; Vieira, R.P. Synthesis of Renewable Poly(limonene): A Kinetic Modeling Study to Improve the Polymerization. *Braz. Arch. Biol. Technol.* **2020**, *63*, 1–10. [CrossRef]
95. Taresco, V.; Crisante, F.; Francolini, I.; Martinelli, A.; D'Ilario, L.; Vitiani, L.R.; Buccarelli, M.; Pietrelli, L.; Piozzi, A. Antimicrobial and antioxidant amphiphilic random copolymers to address medical device-centered infections. *Acta Biomater.* **2015**, *22*, 131–140. [CrossRef] [PubMed]
96. Grabmann, M.K.; Wallner, G.M.; Maringer, L.; Buchberger, W.; Nitsche, D. Hot air aging behavior of polypropylene random copolymers. *J. Appl. Polym. Sci.* **2019**, *136*. [CrossRef]
97. Shahidi, F. *Handbook of Antioxidants for Food Preservation*; Elsevier B.V: Amsterdam, The Netherlands, 2015.
98. Alin, J.; Hakkarainen, M. Type of Polypropylene Material Significantly Influences the Migration of Antioxidants from Polymer Packaging to Food Simulants During Microwave Heating. *J. Appl. Polym. Sci.* **2010**, *118*, 1084–1093. [CrossRef]
99. Alin, J.; Hakkarainen, M. The significant effect of polypropylene material on the migration of antioxidants from food con-tainer to food simulants. In Proceedings of the 43rd IUPAC World Polymer Congress, Glasgow, UK, 11–16 July 2010.
100. Cho, Y.-S.; Kim, S.-K.; Ahn, C.-B.; Je, J.-Y. Preparation, characterization, and antioxidant properties of gallic acid-grafted-chitosans. *Carbohydr. Polym.* **2011**, *83*, 1617–1622. [CrossRef]
101. Hu, B.; Wang, Y.; Xie, M.; Hu, G.; Ma, F.; Zeng, X. Polymer nanoparticles composed with gallic acid grafted chitosan and bioactive peptides combined antioxidant, anticancer activities and improved delivery property for labile polyphenols. *J. Funct. Foods* **2015**, *15*, 593–603. [CrossRef]
102. Spizzirri, U.G.; Iemma, F.; Puoci, F.; Cirillo, G.; Curcio, M.; Parisi, O.I. Synthesis of antioxidant polymers by grafting of gallic acid and catechin on gelatin. *Biomacromolecules* **2009**, *10*, 1923–1930. [CrossRef]
103. Siddalingappa, B.; Benson, H.A.E.; Brown, D.H.; Batty, K.T.; Chen, Y. Stabilization of Resveratrol in Blood Circulation by Conjugation to mPEG and mPEG-PLA Polymers: Investigation of Conjugate Linker and Polymer Composition on Stability, Metabolism, Antioxidant Activity and Pharmacokinetic Profile. *PLoS ONE* **2015**, *10*, e0118824. [CrossRef] [PubMed]
104. Pal, S.; Das, R. Polysaccharide-Based Graft Copolymers for Biomedical Applications. *Polysacch. Based Graft Copolym.* **2013**, 325–345. [CrossRef]
105. LeMarchand, C.; Couvreur, P.; Besnard, M.; Costantini, D.; Gref, R. Novel Polyester-Polysaccharide Nanoparticles. *Pharm. Res.* **2003**, *20*, 1284–1292. [CrossRef]
106. Garcia-Valdez, O.; Champagne, P.; Cunningham, M. Graft modification of natural polysaccharides via reversible deactivation radical polymerization. *Prog. Polym. Sci.* **2018**, *76*, 151–173. [CrossRef]
107. Cochran, D.B.; Wattamwar, P.P.; Wydra, R.; Hilt, J.Z.; Anderson, K.W.; Eitel, R.E.; Dziubla, T.D. Suppressing iron oxide nanoparticle toxicity by vascular targeted antioxidant polymer nanoparticles. *Biomaterials* **2013**, *34*, 9615–9622. [CrossRef]
108. Wattamwar, P.P.; Mo, Y.; Wan, R.; Palli, R.; Zhang, Q.; Dziubla, T.D. Antioxidant Activity of Degradable Polymer Poly(trolox ester) to Suppress Oxidative Stress Injury in the Cells. *Adv. Funct. Mater.* **2010**, *20*, 147–154. [CrossRef]
109. Gormley, A.J.; Yeow, J.; Ng, G.; Conway, Ó.; Boyer, C.; Chapman, R. An Oxygen-Tolerant PET-RAFT Polymerization for Screening Structure–Activity Relationships. *Angew. Chem. Int. Ed.* **2018**, *57*, 1557–1562. [CrossRef]
110. Zhang, T.; Yeow, J.; Boyer, C. A cocktail of vitamins for aqueous RAFT polymerization in an open-to-air microtiter plate. *Polym. Chem.* **2019**, *10*, 4643–4654. [CrossRef]
111. Tappel, A.; Tappel, A. Oxidant free radical initiated chain polymerization of protein and other biomolecules and its relationship to diseases. *Med. Hypotheses* **2004**, *63*, 98–99. [CrossRef] [PubMed]
112. Tran, T.N.; Heredia-Guerrero, J.A.; Mai, B.T.; Ceseracciu, L.; Marini, L.; Athanassiou, A.; Bayer, I.S. Bioelastomers based on cocoa shell waste with antioxidant ability. *Adv. Sustain. Syst.* **2017**, *1*, 1700002. [CrossRef]

polymers

MDPI

Article

An Innovative Computational Strategy to Optimize Different Furnish Compositions of Tissue Materials Using Micro/Nanofibrillated Cellulose and Biopolymer as Additives

Flávia P. Morais [1,*], Ana M. M. S. Carta [2], Maria E. Amaral [1] and Joana M. R. Curto [1,3,*]

1 Fiber Materials and Environmental Technologies Research Unit (FibEnTech-UBI), University of Beira Interior, Rua Marquês d'Ávila e Bolama, 6201-001 Covilhã, Portugal; mecca@ubi.pt
2 Forest and Paper Research Institute (RAIZ), R. José Estevão, 3800-783 Eixo, Portugal; ana.carta@thenavigatorcompany.com
3 Chemical Process Engineering and Forest Products Research Centre (CIEPQPF), University of Coimbra, R. Sílvio Lima, Polo II, 3004-531 Coimbra, Portugal
* Correspondence: flavia.morais@ubi.pt (F.P.M.); jmrc@ubi.pt (J.M.R.C.)

Abstract: The furnish management of tissue materials is fundamental to obtain maximum quality products with a minimum cost. The key fiber properties and fiber modification process steps have a significant influence on the structural and functional properties of tissue paper. In this work, two types of additives, a commercial biopolymer additive (CBA) that replaces the traditional cationic starch and micro/nanofibrillated cellulose (CMF), were investigated. Different formulations were prepared containing eucalyptus fibers and softwood fibers treated mechanically and enzymatically and both pulps with these two additives incorporated independently and simultaneously with drainage in the tissue process range. The use of these additives to reduce the percentage of softwood fibers on tissue furnish formulations was investigated. The results indicated that a maximum of tensile strength was obtained with a combination of both additives at the expense of softness and water absorbency. With a reduction of softwood fibers, the incorporation of additives increased the tensile strength and water absorbency with a slight decrease in HF softness compared with a typical industrial furnish. Additionally, a tissue computational simulator was also used to predict the influence of these additives on the final end-use properties. Both additives proved to be a suitable alternative to reduce softwood fibers in the production of tissue products, enhancing softness, strength and absorption properties.

Keywords: absorbency; commercial biopolymer additive (CBA); micro/nanofibrillated cellulose (CMF); softness; strength; tissue paper materials

Citation: Morais, F.P.; Carta, A.M.M.S.; Amaral, M.E.; Curto, J.M.R. An Innovative Computational Strategy to Optimize Different Furnish Compositions of Tissue Materials Using Micro/Nanofibrillated Cellulose and Biopolymer as Additives. *Polymers* **2021**, *13*, 2397. https://doi.org/10.3390/polym13152397

Academic Editors: José Miguel Ferri, Vicent Fombuena Borràs and Miguel Fernando Aldás Carrasco

Received: 26 May 2021
Accepted: 17 July 2021
Published: 21 July 2021

1. Introduction

In recent years, there has been great growth potential in the tissue paper market. Factors such as the abundance of fibrous raw materials, the high demand for hygiene and health products and the evolution of socio-economic living standards have promoted this global growth. In addition, advances in manufacturing technologies for these materials have made the market increasingly competitive [1]. Tissue paper materials are increasingly used in consumer goods products such as napkins, facial tissues, paper towels, toilet papers, diapers and tissue masks. Tissue papers are produced to meet the desired properties for a specific application in the market such as a low basis weight, softness, strength and absorption [2]. To optimize the process, a great deal of effort been made to improve the properties of the raw materials used in the production of these papers [1–4]. The highest quality tissue products are produced with a blend of hardwood and softwood fiber pulps. The production of tissue papers with 100% eucalyptus fibers generally presents a higher softness and absorption; however, the strength is degraded during the creping and converting processes in the tissue manufacture [3,4]. Softwood fibers are introduced because these pulps provide strength to the papers and ensure the paper machine's runnability [5–8]. The fibers can also be subject to modification processes such as enzymatic and mechanical modifications. Refining is an important process in the development of functional tissue properties; however, it is associated with a large energy consumption [9,10]. The application of enzymes enhances not only the development of the tissue properties but also the reduction of energy costs associated with the refining process [11,12].

In addition to the modification of fiber pulps, chemical additives are used to modify properties or improve the production process for different materials. Different chemical additives used in the tissue paper industry are designed and applied according to different needs such as dry strength agents, wet strength agents, softeners and lotions [13,14]. Usually, these additives are cationic as the fibers carry negative charges and they are also more effective in much lower concentrations than anionic polymers. Nevertheless, these cationic additives normally adhere in excess to the paper, negatively affecting the creping process and the product quality [14]. The addition of strength agents can ensure that tissue products remain integrated when applied in wet conditions while the addition of debonding agents can improve the softness and bulk of these products [15–21]. However, reaching a balance between these two properties is a challenge because the conditions that maximize the softness are those that minimize strength properties. This balance can be achieved by selecting fibers, refining treatments and adding chemical polymers. Therefore, the tissue manufacturer must balance these factors in terms of cost and effect. The tissue paper industry provides the design, development and production of many chemical and polymeric additives [22].

The application of innovative and versatile additives such as biopolymers and micro/nanofibrillated cellulose (CMF), replacing synthetic additives currently used, is a sustainable strategy to be considered in order to produce high-quality tissue premium products with innovative features. Biopolymers as additives can be new natural and green approaches to new possibilities and advances in modifying or improving the final end-use tissue properties [23]. The same is true for CMF as an additive that has been extensively investigated for its application in tissue products due to its high specific surface area, excellent mechanical properties, flexibility, biocompatibility, biodegradability and lack of toxicity, among other properties [24–27]. The furnish formation is also affected by the CMF incorporation as an additive. CMF decreases porosity and air permeability when added to the tissue structures. CMF content and structural changes present a proportional behavior. These decreases are caused by the CMF bonding with the cellulose fibers in the 3D network structures, closing the porous structure. Furthermore, this reduction in porosity is also correlated with the increase in the structure's apparent density [28].

The presence of CMF can present an impact on the structure porosity because of the increasing number of hydrogen bonds present in this nanocellulose. The increase of these bonds is related to its increase in the surface/volume ratio due to the nanosized scale [28]. By filling the large pores created by the 3D fiber network, the CMF incorporation contributes to smaller pore dimensions and a narrower distribution of the structures, providing abundant pores with a uniform size and, consequently, a decrease in the properties of water absorption capacity and capillarity penetration [24,25]. The flexibility of CMF can be a filler for large pores in structures [25]. Although the mechanical properties of tissue products are improving, CMF also brings a difficulty to the dewatering of papermaking furnishes. The fibrillation degree influences the dewatering capacity of pulp suspensions. The drainage time increases depending on the CMF fibrillation degree. Particles resulting from fibrillation are more easily incorporated into the fiber network, partially closing the pores between the fibers and consequently limiting the wet network's ability to drain water. This drainage difficulty is also caused by the increased water retention capacity of CMF. However, this dewatering of pulp suspensions containing CMF can be improved with the use of retention aids and also with cationic starch-treated CMF with colloidal silica [28]. In addition to a negative impact on dewatering, the accumulation of CMF particles in the closed circuit of the tissue machine can increase the viscosity of the recirculating water, thus affecting processability. Therefore, there is a need to study the feasibility of using CMF to reduce the fiber content. Our study [24] confirmed that the use of 1 to 8% of CMF incorporated into tissue paper formulations, along with a reduction of hardwood and softwood fiber content, achieves a trade-off between the final end-use tissue properties and fiber blend costs, with drainability between 26 and 32 °SR (Schopper–Riegler degree). In this range of drainability, the furnish formulations are able to produce high-quality tissue papers without difficulties in tissue paper machine runnability and efficiency [2,24].

The application of these additives leads to the optimization of strength properties without compromising the softness and absorbency properties of these types of papers. These strategies can allow a reduction of softwood fibers in the production of tissue paper without affecting the strength performance and decrease the overall tissue manufacturing cost [24,27–30]. The aim of this work was to perform a comparative study between different tissue furnish compositions and the incorporation of two additives, a commercial tissue biopolymer additive (CBA) as a cationic starch substitute and CMF, and their effect on the properties of the final end-use tissue paper. For this purpose, the fiber and suspension drainability characterizations of different investigated samples were performed. Laboratory isotropic structures with a light basis weight of 20 g/m^2 were prepared and the softness, tensile strength and absorbency properties were evaluated.

2. Materials and Methods

2.1. Materials

An industrial, never-dried, bleached eucalyptus kraft pulp and an industrial, air-dried, bleached pinus kraft pulp were used in this work after disintegration according to ISO 5263-1. The additives used in this study were CBA and CMF. This commercial tissue additive, CBA, is a cationic polymeric solution specially designed as a replacement for cationic starch. CBA presents a viscosity of 500–900 cP, a pH of 2.1–4.0 and a density of 1.030 ± 0.040 g/cm^3 at 20 °C (data supplied by a commercial enterprise). CMF was supplied by a research institute and obtained by mechanical (refining and homogenization) and enzymatic (cellulase) treatments using an industrial bleached eucalyptus kraft pulp. This sample was characterized in our previous work, which can be used for more detailed information [24]. CMF (Figure 1) presents an intrinsic viscosity of 1577 mL/g (ISO 5351), a carboxyl group content of 8.4 ± 0.2 mmol/100 g (TAPPI T 237 om-93), a suspension drainability degree superior to 80 °SR (Schopper–Riegler) and 56% of particles below 1000 nm, representing 10% of particles below 200 nm [24].

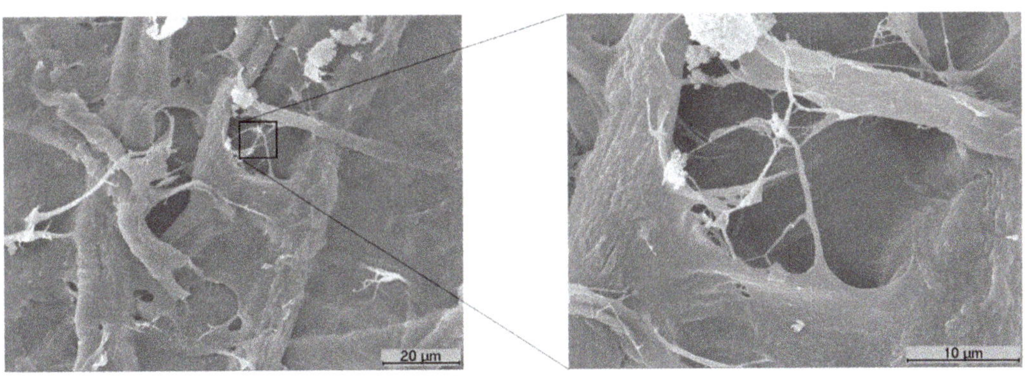

Figure 1. SEM image of the CMF additive, highlighting a high magnification image (3000×) where it is possible to verify the micro and nanofibrils of this sample.

2.2. Enzymatic and Mechanical Treatments

The softwood pulp was beaten in a PFI mill at 4000 revolutions under a refining intensity of 3.33 N/mm (ISO 5264-2) followed by an enzymatic treatment. The enzymatic treatment was performed according to the method described by Morais et al. [12]. The treatment was carried out at a consistency of 4% at pH 7 at 40 °C with continuous mechanical agitation to ensure an efficient mass transfer for 60 min and 30 mg of enzyme/g of pulp.

2.3. Tissue Formulations

Table 1 presents the various tissue formulations (investigated samples) carried out in this study. The different formulations were performed to study the effect of additives in the tissue structure properties and understand their implication in the reduction of softwood pulp usually used in this type of paper. The first formulation presents a similar combination of eucalyptus fibers and softwood fibers typically used in tissue mills. All of the samples presented similar drainability properties around 25 °SR, which is recommended for tissue paper products and means that drainage and processability are in the range for the tissue process and, consequently, processability is not negatively impacted [2].

Table 1. Tissue formulations used in the study.

	Eucalyptus Pulp	Softwood Pulp with Mechanical and Enzymatic Treatment	CBA (Commercial Biopolymer Additive)	CMF (Micro/Nanofibrillated Cellulose)
Formulation 1	75%	25%	-	-
Formulation 2	75%	25%	2%	-
Formulation 3	75%	25%	-	2%
Formulation 4	75%	25%	2%	2%
Formulation 5	90%	10%	-	-
Formulation 6	90%	10%	2%	-
Formulation 7	90%	10%	-	2%
Formulation 8	90%	10%	2%	2%

2.4. Preparation of Tissue Structures

The tissue structures were produced in a batch laboratory sheet former according to the method described by Morais et al. [6,7]. An adaptation to ISO 5269-1 was used, namely, the suppression of the pressing operation and the production of structures with a basis weight of 20 g/m^2. The tissue structures were prepared with the different formulations described in Table 1. Finally, all samples were conditioned at 23 $\pm$ 1 °C, with a relative humidity of 50 $\pm$ 2%, according to ISO 187.

2.5. Characterization of the Morphological, Drainability and Chemical Properties

The morphological properties of pulp fibers and formulations were evaluated automatically by an image analysis using a fiber analyzer, the MorFi equipment (TECHPAP, Grenoble, France).

In order to characterize the surfaces of the different formulations, the obtained tissue structures were coated with gold using a Sputter Quorum Q 15 OR ES (Laughton, East Sussex, UK) to be analyzed by scanning electron microscopy (SEM) using a Hitachi S-2700 model (Tokyo, Japan) with an accelerating voltage of 20 kV at different magnifications. ImageJ, an image analysis software, was also used to characterize the pore properties of the samples. The SEM images were processed and analyzed using a methodology with defined criteria for the stabilization of the average of the measured values [31].

The chemical functional groups were analyzed using Fourier transform infrared spectroscopy with attenuated total reflectance (FTIR-ATR) at a wavelength of 600–4000 cm^{-1} with a resolution of 4 cm^{-1}.

The drainability of pulp suspensions and formulations was evaluated by the °SR method (ISO 5267/1). Triplicate assays were performed for both methods.

2.6. Characterization of the Tissue Properties

2.6.1. Structural Properties

The structures produced were characterized in terms of basis weight (ISO 12625-6), thickness and bulk (ISO 12525-3). The basis weight was obtained by the quotient between the average mass of each structure and the respective area (0.02138 m^2). The thickness was obtained using a Frank-PTI micrometer. The bulk was obtained by the quotient between the thickness and the basis weight of the structures. Additionally, the porosity (apparent theoretical porosity) of the structure was also calculated according to the following equation:

$$Porosity\ (\%) = 100 \times \left(1 - \frac{\rho_{structures}}{\rho_{cellulose}}\right) \quad (1)$$

where $\rho_{cellulose}$ was the density of the cellulose (1.5 g/cm^3) [12] and $\rho_{structure}$ was the apparent density of the structures (g/cm^3).

2.6.2. Softness Properties

Softness was assessed using a Tissue Softness Analyzer, the TSA (Emtec). A two-step measurement was used. The first was a noise measurement where the real softness (TS7) was measured. The second step was a deformation measurement where stiffness, plasticity, elasticity and hysteresis were measured. The handfeel (HF) parameter value was calculated using both steps.

2.6.3. Strength Properties

Strength properties were determined according to ISO 12625-4 using the tensile index and calculated according to the following equation:

$$Tensile\ Index\ (\mathrm{Nm/g}) = \frac{\frac{break}{W}}{G} \quad (2)$$

where *break* was the breaking force (N), *W* was the sample width (50 mm) and *G* was the sample grammage or basis weight (g/m^2).

2.6.4. Absorbency Properties

The absorbency properties were determined by the methods of water absorbency capacity per unit of mass and Klemm capillary rise.

The immersion absorption method was used to determine the water absorption capacity per unit of mass of the samples. The assay was carried out according to an adaptation of ISO 12625-8. The sample was immersed for 30 s. At the end of that time, the sample was removed from the water and left to run on a support that guaranteed an amplitude of 30° for approximately 1 min. Finally, the amount of water that the sample was able to absorb during this time was determined. The assay was performed in triplicate for each sample.

The Klemm capillary rise method was measured through an adaptation of ISO 8787. The assay consisted of the water rising in the sample (50 mm wide) at a certain time. The capillary rise was recorded at 10, 20, 30, 60, 180, 300 and 600 s. Four readings were performed for each sample. All of the assays were performed in a controlled temperature and humidity laboratory.

2.7. Computational Studies

SimTissue, a tissue simulator, was developed specifically for furnish management and optimization by FibEnTech-UBI members, in the scope of Project InPaCTus. This tissue simulator was used to compare different end-use final tissue properties and to access them for the different formulations to those performed experimentally [24].

A 3D fiber-based computational simulator was used to model the tissue structures as planar random networks. This simulator is open-source software, and the code is available on GitHub (https://github.com/eduardotrincaoconceicao/voxelfiber, accessed on April 2021). A more detailed description of this 3D simulator can be found in Conceição et al. and Curto et al. [32,33]. The fibers are represented by a voxel chain and modeled according to fiber dimensions and properties such as the fiber length and width ratio, flexibility, fiber wall thickness and lumen dimensions [34]. The structural properties were obtained following this computational study and the resulting 3D computational structures. The computational studies were performed using MATLAB® (R 2020a, 9.8.0.1323502, MathWorks, Natick, MA, USA).

3. Results and Discussion

3.1. Characterization of Pulp Fibers and Tissue Formulations

In a first approach, the characterization of the raw materials used in the study was carried out. The eucalyptus pulp presented a length weighted by length of 0.729 ± 0.003 mm, a width of 19.1 ± 0.0 μm, a coarseness of 6.31 ± 0.04 mg/100 m and fine elements of 38.5 ± 0.4% in length. The softwood pulp presented a length weighted by length of 1.889 ± 0.052 mm, a width of 30.1 ± 0.2 μm, a coarseness of 20.52 ± 3.23 mg/100 m and fine elements of 36.6 ± 4.2% in length. In this study, only softwood pulp was subjected to fiber modification treatments, namely, refining and enzymatic treatments, as this eucalyptus pulp is already an industrial pulp suitable for the production of tissue papers. These treatments were carried out in order to improve fiber flexibility and inter-fiber bond properties with a greater strength three-dimensional matrix [35]. Analyzing the treatments separately, the refining effect was more pronounced than the enzymatic treatment. The results indicated that refining decreased the length weighted by length by 8% and coarseness by 30% and increased the fiber width by 7% and the fine content by 17% while the enzymatic treatment decreased the fiber length weighted by length by 2% and the coarseness by 18% and increased the fiber width by 23% and the fine content by 4%. The combination of these two treatments decreased the fiber length weighted by length by 10% and the coarseness by 26% and increased the fiber width by 7% and the fine content by 11% compared with the untreated pulp. Figure 2 presents an analysis of coarseness as a function of the length/width ratio (slenderness ratio) for the softwood samples. The results indicated that two groups can be considered with the fiber modification treatments applied. In the first one, the softwood fibers without treatment (63 of length/width ratio and 20.52 mg/100 m of coarseness) and with enzymatic treatment (61 of length/width ratio and 16.92 mg/100 m of coarseness) presented a higher slenderness ratio and higher coarseness. In the second group, the refined softwood fibers (54 of length/width ratio and 15.81 mg/100 m of coarseness) and with both treatments (52 of length/width ratio and 15.09 mg/100 m of coarseness) presented a lower slenderness ratio and lower coarseness. This suggests that the first group was more susceptible to presenting better tissue properties of softness and absorbency than the second group, which was more susceptible to presenting higher strength properties [7]. Additionally, the first group presented °SR between 12–14 and the second group presented 55–66 °SR. The mechanical refining and the combination of both treatments promoted a more efficient fiber fibrillation and consequently decreased the suspension drainability.

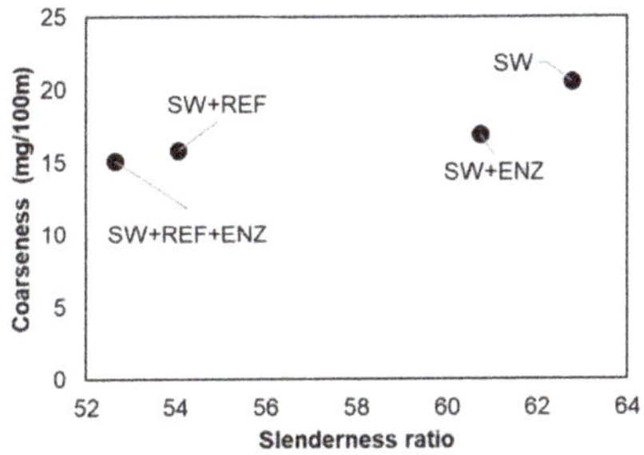

Figure 2. Coarseness as a function of the slenderness ratio of softwood fiber pulp (SW) with enzymatic treatment (SW + ENZ), with refining (SW + REF) and with the combination of both fiber modification treatments (SW + REF + ENZ).

The softwood pulp with the refining and enzymatic treatment was also assessed through SEM images (Figure 3a). This treatment made it possible to modify the properties of the softwood pulp, promoting inter-fiber bonds and the combination of key structural properties such as the pore dimension and distribution. The pore properties depend on the fiber properties and the mechanical and enzymatic processes to which they are subjected [36]. The results indicated that these treatments had an impact on the fiber dimension and flexibility, resulting in a structure with a surface porosity of 83 ± 1% and a pore diameter distribution of 35% between 2 and 10 μm and 65% between 10 and 27 μm (Figure 3b).

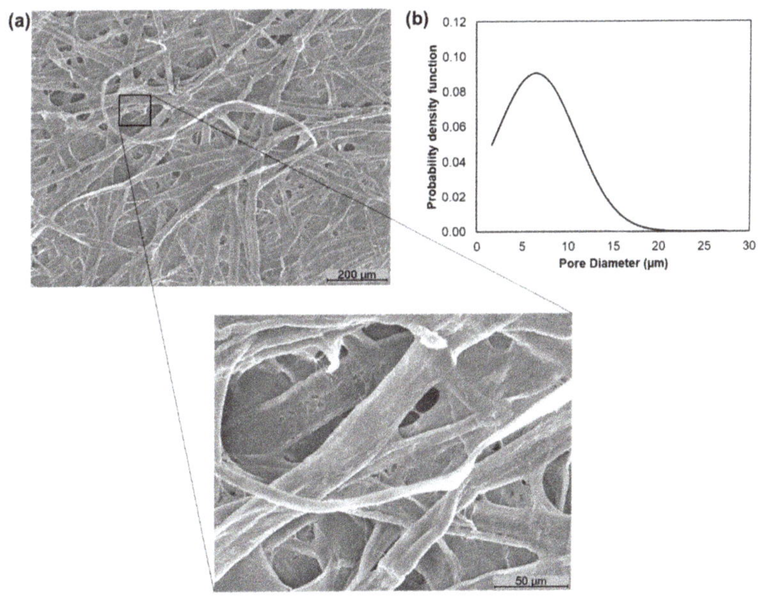

Figure 3. SEM image of the softwood pulp with refining and enzymatic treatment with a high magnification image (500×) (**a**) and its pore diameter distribution (**b**).

In a second approach, the characterization of the tissue formulations was carried out. As previously reported, this study compares two versatile additives: CBA and CMF. The MorFi analyzer is only able to evaluate the fiber suspension properties so the presence of the CBA does not affect its fibrous composition. The formulations with 75% eucalyptus pulp and 25% softwood pulp-treated and with incorporated additives presented a length weighted by length between 0.717 and 0.729 mm, a width between 19.4 and 19.9 μm, coarseness between 7.78 and 8.27 mg/100 m and fine elements between 32.7 and 35.7% in length. With a reduction of the softwood fibers, the formulations with 90% eucalyptus pulp and 10% softwood pulp-treated and with incorporated additives presented a length weighted by length between 0.719 and 0.722 mm, a width between 19.1 and 20.1 μm, coarseness between 6.77 and 7.61 mg/100 m and fine elements between 30.4 and 32.8% in length. Additionally, these formulations presented an °SR degree between 20 and 25 °SR. Although the softwood pulp presented low drainability, its combination with eucalyptus pulp and with different additives allowed the tissue formulations to present a range of °SRs suitable of producing premium tissue materials [2] without compromising the processability. Figure 4 presents the SEM image of the formulation 4 (75% eucalyptus pulp + 25% softwood pulp-treated + 2% CBA + 2% CMF) as well as its pore diameter distribution. The 3D structure matrix resulting from this formulation was presented as a multi-structured material with bonding between CMF fibrils and pulp fibers. The CBA addition also promoted a more closed structure with a surface porosity of 88 ± 1% and a pore diameter distribution of 40% between 1 and 10 μm and 60% between 10 and 22 μm (Figure 4b).

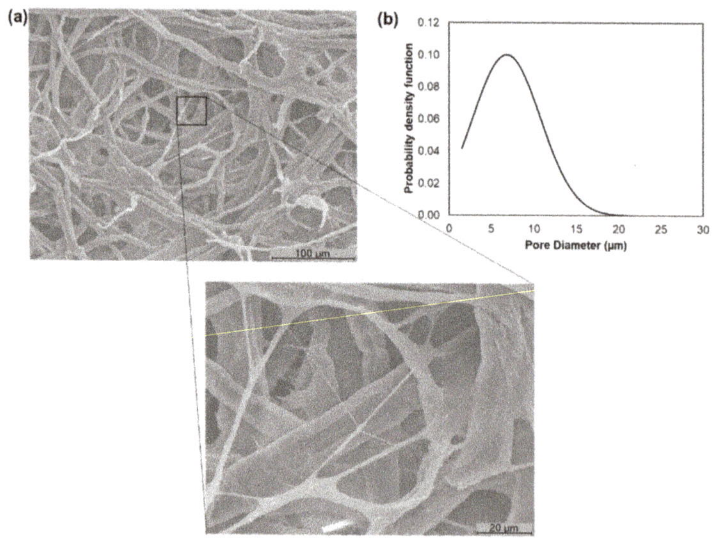

Figure 4. SEM image of tissue formulation 4 (75% eucalyptus pulp + 25% softwood pulp-treated + 2% CBA + 2% CMF) with a high magnification image (1000×) where is possible to verify the inter-fiber bonding between CMF and the structure in more detail (**a**) and its pore diameter distribution (**b**).

To complement this study, the FTIR-ATR technique allowed us to chemically characterize, identify and quantify the formulations with CBA and CMF, evaluating the existence of physical retentions in the structure without new chemical bonds between these additives and the cellulose eucalyptus and softwood fiber 3D matrix. Figure 5 shows the FTIR spectrum of formulation 1 (a typical furnish mixture at a tissue mill), formulation 2 (with CBA incorporation) and formulation 3 (with CMF incorporation). The typical absorption bands of hydroxyl groups between 3000 and 3500 cm^{-1} indicative of the –OH stretching of the intra- and inter-molecular interactions of hydrogen bonds were verified for the three formulations. A –OH band with a higher area was observed for the structures with CBA. The biggest differences in the FTIR spectrum were found in the bands between 2800 and 3000 cm^{-1}. This range corresponds with the =C–H stretching in the methyl groups of cellulose and hemicellulose. The band with the highest intensity in this range was observed for the formulation with CMF followed by the formulation with CBA and formulation 1. Additionally, the bands between 2200 and 2400 cm^{-1} correspond with the –OH asymmetrical stretching vibration of the carboxylic acid due to the enzymatic treatment applied to the materials [12]. The bands between 1500 to 1700 cm^{-1} were due to the C=O stretch of hemicelluloses. Other characteristics of cellulose were also observed between 800 and 1500 cm^{-1} such as the angular deformation of C-H and primary alcohol C-O bonds, the absorption band of C–O–C bonds, the β-glycosidic bonds between glucose units, C=H stretching, C=O stretch vibration in the syringyl ring and carboxylate ion group vibration.

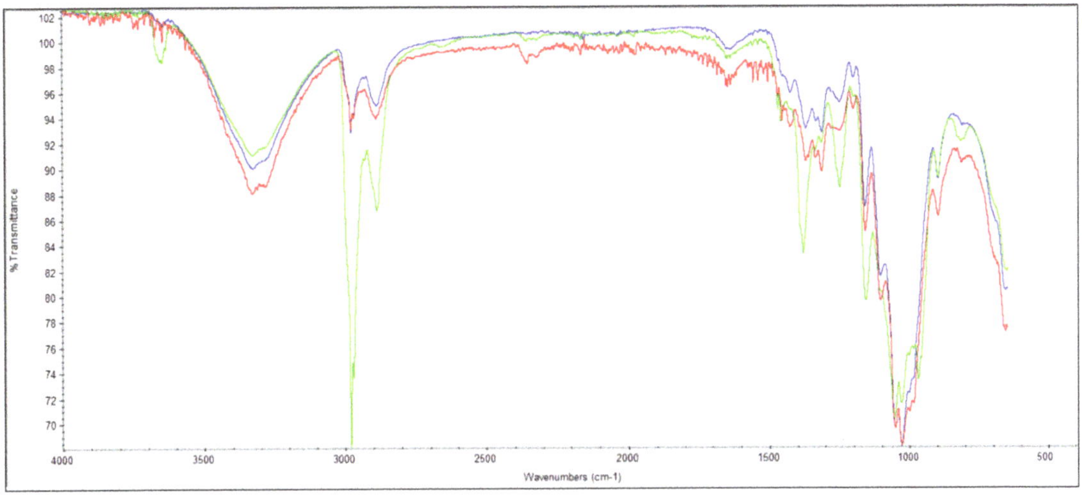

Figure 5. FTIR-ATR spectrum of formulation 1 (blue), formulation 2 (red) and formulation 3 (green).

3.2. *Characterization of the Properties of the Final End-Use Tissue*

3.2.1. Structural Properties

Bulk and porosity are essential parameters for the quality of tissue products that influence the softness and liquid absorption properties [6,7]. Figure 6a shows, for the different tissue formulations, the behavior of the structural properties. The addition of CBA in formulation 2 promoted a decrease in bulk and porosity by 1% and 0.1%, respectively, compared with formulation 1. The addition of CMF (formulation 3) showed a tissue structure with lower bulk (<4%) and porosity (<0.5%) compared with formulation 1. The combination of these versatile additives (formulation 4) also promoted an increase of hydrogen and inter-fiber bonds, inducing a decrease of 11% in bulk and 1% in porosity of the tissue structures in comparison with formulation 1. This densification effect can be observed in the cross-section of the structures, as shown in Figure 7. With the reduction of the reinforcement fibers by 15% in the formulations, bulk and porosity were increased. This reduction promoted an increase of 20% in bulk and 2% in porosity when comparing formulations 1 and 5. The same trend in formulations 1–4 was observed in formulations 5–8; i.e., the combination of both additives presented a higher influence on the structural properties followed by CMF and CBA. Formulations 6, 7 and 8 promoted a reduction in bulk by 4, 10 and 22%, respectively, and a reduction in porosity by 0.4, 1 and 3%, respectively, compared with the control (formulation 5). The addition of both CBA and CMF in formulations with a softwood fiber reduction (90% eucalyptus fiber and 10% softwood fiber) allowed the obtaining of structures with increased structural properties compared with typical industrial structures (75% of eucalyptus fiber and 25% softwood fiber). The results indicated that formulations 6 and 7 increased the structure bulk by 16% and 9%, respectively, compared with formulation 1 (Figure 8). However, the combination of these two additives promoted a 6% reduction in bulk compared with this formulation typically used industrially.

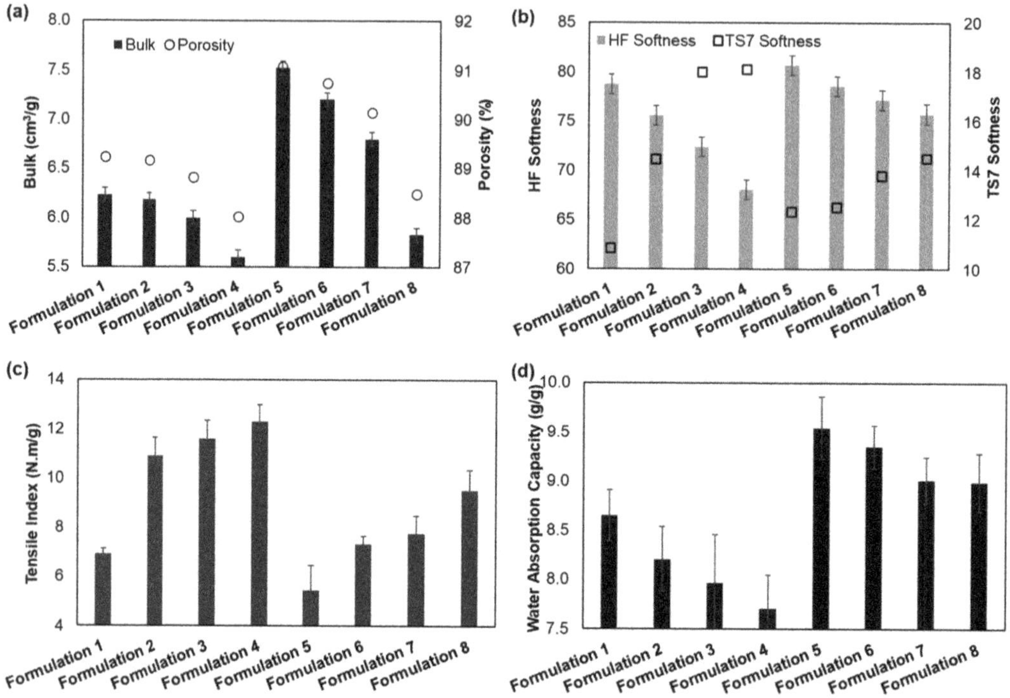

Figure 6. Properties of bulk and porosity (**a**), softness (**b**), tensile index (**c**) and water absorption capacity per unit of gram (**d**) of all tissue formulations.

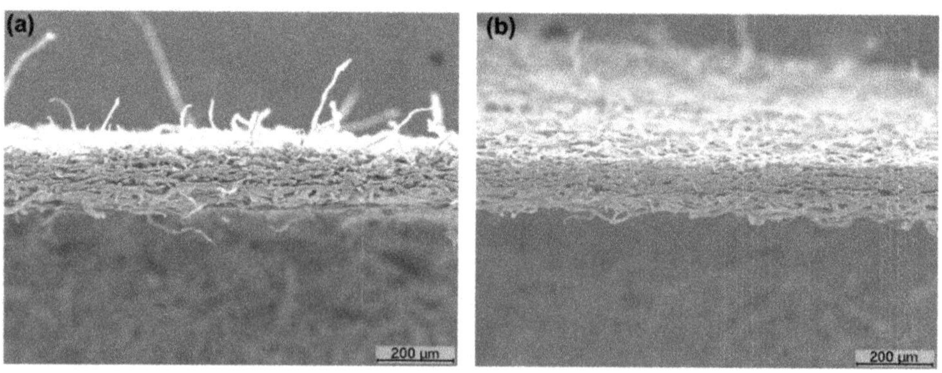

Figure 7. SEM images of the cross-section of the tissue structures prepared with formulation 1 (**a**) and formulation 4 (**b**).

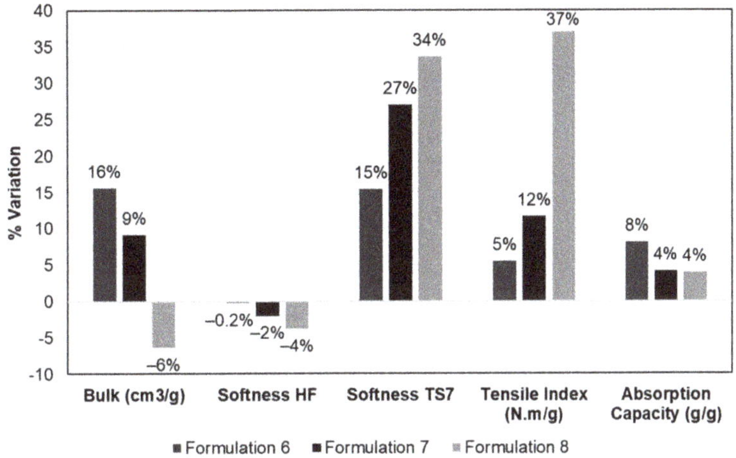

Figure 8. Percentage of variation in the properties of bulk, softness HF, softness TS7, tensile index and water absorption capacity of formulations 6, 7 and 8 compared with formulation 1 obtained with SimTissue.

3.2.2. Softness Properties

Softness is also a characteristic that is highly appreciated by consumers, resulting in the combination of structure flexibility, bulk and surface softness. Softness properties of the different formulations are shown in Figure 6b. The HF (handfeel) values decreased with the additives loaded in the paper 3D matrix. The increase in fiber flexibility, fine elements and, consequently, the number of inter-fiber bonds promoted a more closed and compact 3D matrix with less bulk, resulting in a structure with less softness. The addition of CBA in formulation 2 promoted a decrease in softness HF by 4% compared with formulation 1. The addition of CMF (formulation 3) showed a tissue structure with lower softness compared with formulation 1 with a decrease of 8%. The combination of these versatile additives (formulation 4) also promoted a decrease of 14% in the softness HF of the tissue structure compared with formulation 1. The reduction of the reinforcement fibers in the formulations increased softness properties by 2% compared with formulations 1 and 5. As with bulk properties, the same trend was also observed for softness properties. Formulations 6, 7 and 8 promoted a reduction in softness by 3%, 4% and 6%, respectively, compared with formulation 5. Regarding Figure 8, the softness HF values decreased by 0.2%, 2% and 4% with the incorporation of CBA, CMF and the combination of both, respectively. The production of structures with a low content of SW fibers and versatile additives promoted a small reduction in softness, improving strength properties, as reported below, contributing to the formation of a fibrous structure with less stiffness [37].

Softness TS7 is inversely related to softness HF values; therefore, it is influenced by the presence of fibers in the structure surface. TS7 values increased with the addition of additives in the structure (Figure 6b). This result indicated that the fibers in the structure presented higher bonds, causing structure densification and rigidity. The fibers in the z-direction were more adherent to the paper structure, reducing the number of free fibers on the tissue surface compared with the structures without additives (Figure 7). Compared with formulation 1, softness TS7 of the tissue structure increased by 33%, 65% and 67% with the incorporation of CBA, CMF and both, respectively (formulations 2–4). With the reduction of the reinforcement fibers in the formulations, this property also increased by 13% compared with formulations 1 and 5. Additionally, formulations 6, 7 and 8 promoted an increase in softness TS7 by 2%, 12% and 18%, respectively, compared with formulation 5. Compared with formulation 1 (75:25), softness TS7 increased by 15%, 27% and 34% with the incorporation of CBA, CMF and both, respectively, in formulations with a softwood fiber reduction (90:10) (Figure 8).

3.2.3. Strength Properties

The tensile index is an essential parameter not only for the product quality but also to ensure the paper machine runnability, which depends on the fiber intrinsic strength and the inter-fiber bonds. The results of the tensile index for each tissue formulation are shown in Figure 6c. The incorporation of additives promoted an increase in the tensile index of the tissue structures. Differences in strength improvements varied with the performance of CBA and CMF and became more noticeable with the combination of these additives. Formulation 3 with the addition of CMF showed improvements in the tensile index (+67% compared with formulation 1) related to formulation 2 with the addition of CBA (+57% compared with formulation 1). Formulation 4, with the incorporation of both additives, presented higher structure strength improvements by 77% compared with formulation 1. This result was corroborated by the results of the literature as the addition of CMF and chemical additives in the tissue structures improve strength properties of these products [13,25,27]. These additives promoted an increase in the surface area of the tissue structures due to the micro/nanofibrils and consequently formed stronger and more stable structures through inter-fiber interaction and hydrogen bonding [27]. This fact can also be verified by the densification of the tissue structures with the incorporation of both additives (Figure 7b).

Reinforcement fibers play a fundamental role in strength properties; therefore, the reduction of these fibers in formulation 5 promoted a 21% reduction of the tensile index compared with formulation 1. Additionally, formulations 6, 7 and 8 promoted an increase in the tensile index by 34%, 42% and 74%, respectively, compared with formulation 5. It is important to highlight that the strength gains obtained with the addition of CMF and CBA and the softwood reduction in the tissue structures were higher when compared with formulation 1. Despite the structures presenting a low basis weight and no pressing, these additives favored strength properties of these tissue structures. Additionally, as shown in Figure 8, the tensile index increased by 5%, 12% and 37% with the incorporation of CBA, CMF and both, respectively.

The commercial biopolymer additive used in this work was a cationic polymeric solution that is especially intended to be a substitute for cationic starch. Generally, cationic starch is used in the printing and writing paper industry with functions of a dry strength additive, emulsification of sizing agents, retention and drainage. This starch improves the formation, internal bonding, surface strength, tensile strength, writing and printing surface and energy consumption [38]. However, in the tissue paper industry, this starch has several inconveniences as the increase in strength properties promotes an abrupt reduction in softness properties. A solution found in the tissue industry was to design a practically constant biopolymer that would increase strength properties while preserving softness properties with a reduction of softwood fiber content. Furthermore, this additive allows the reduction of the contaminating organic load and costs as a result of reducing the use of cationic starch [38]. Therefore, the results of this work indicated that, compared with formulation 1, a typical industrial mixture can produce tissue structures containing 2% of the CBA and 10% of softwood fibers with 5% tensile index increases while maintaining the softness (Figure 8). This cationic polymeric solution also improved the inter-fiber bonding and CMF particle retention. This retention was proven by the 38% increase in strength properties with both additives (formulation 8) compared with formulation 1. The incorporation of CMF alone (formulation 7) only increased strength properties by 12% compared with formulation 1 (Figure 8).

In addition, an inverse relationship between the properties of softness and strength can also be observed in these formulations (Figure 9); i.e., strength improvements present a negative impact on the tissue structure softness. The formulations promoted an increase of 1.52 units of softness HF for each unit with a decreased tensile index ($y = -1.52x + 89.51$;

R^2 = 0.89). In our previous studies, this relationship was also observed in formulations with only eucalyptus fibers and without additives in which the softness HF increased by 2.4 units per unit of the decreased tensile index [6,7]. This can contribute to the reduction of costs associated with the use of softwood fibers in industrial furnishes as there is a balance between the properties of softness and strength in the tissue structures with the use of these versatile additives.

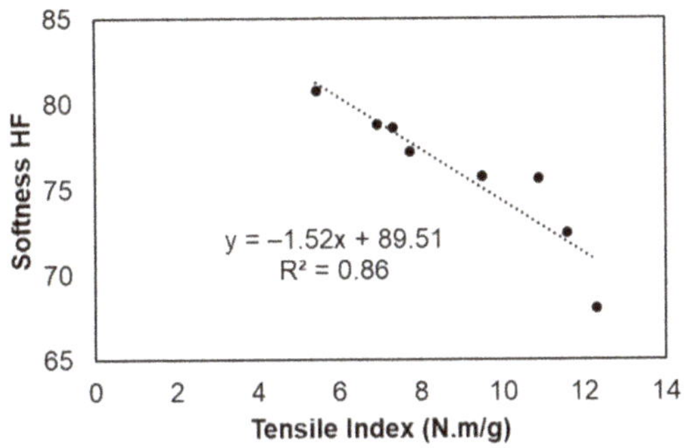

Figure 9. Correlation between the tensile index and softness HF of tissue formulations.

3.2.4. Absorbency Properties

Absorbency properties are essential in tissue products, mainly in paper towels and napkin s, depending on their fibrous structure. Tissue products with a more porous 3D matrix will present better water absorption as empty spaces for water interaction are more available [12]. Figure 6d presents the water absorption capacity of the tissue formulations in the study. Overall, the incorporation of CBA and CMF into the tissue formulations decreased the water absorption properties compared with the control. This result was also in line with the decrease in bulk and porosity properties (Figure 6a). This effect was observed from the SEM images in Figures 4a and 7b in which the formulations with these additives presented a less porous surface with a more compact cross-section. The addition of CBA and CMF (formulations 2 and 3) presented a decrease of 5% and 8% of this property, respectively, compared with formulation 1. The incorporation of both additives presented higher absorption decreases by 11% compared with formulation 1. The reduction of the reinforcement fibers in the formulations also increased the absorption properties by 10% compared with the formulations 1 and 5. Additionally, formulations 6, 7 and 8 promoted a reduction in absorption by 2%, 6% and 6%, respectively, compared with formulation 5. Compared with formulation 1, absorption increased by 8%, 4% and 4% with the incorporation of CBA, CMF and both, respectively, in formulations with a softwood fiber reduction (90:10) (Figure 8). In this work, we also found a good linear correlation (R^2 = 0.78) between the water absorption properties and the softness of the tissue structures made from these formulations (Figure 10a). The formulations promoted an increase of 5.39 units of softness HF for each unit of increased water absorption capacity ($y = -5.39x + 29.06$). The same inverse trend observed for the softness and strength properties was also verified for the absorbency and strength properties (Figure 10b). The tissue formulations promoted an increase of 3.37 units of water absorption capacity for each unit of decreased tensile index ($y = -3.37x + 38.19$; R^2 = 0.81).

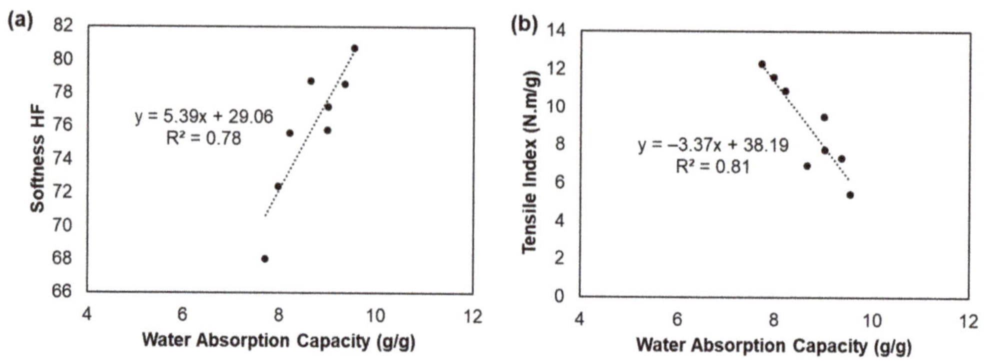

Figure 10. Correlation between the water absorption capacity and softness HF (**a**) and tensile index (**b**) of the tissue formulations.

The capillary rise of the structures depends on the adhesion and cohesion forces of the water molecules that ascend through the fiber walls or between the pores. Therefore, there must be a balance between the pore dimensions to promote the capillary rise efficiency. Inter-molecular bonds between fibers and water molecules in structures with higher pores can impair the capillary rise. On the other hand, in the case of channels that are too narrow, the water progression will be strangled by the limited space for capillary rise [39]. Figure 11 presents the Klemm capillary rise of the tissue formulations in this study. The same trend in the water absorption properties was observed in the capillary rise of tissue formulations. The formulations with a content of fewer softwood fibers presented better capillary ascents and the additives promoted the capillary rise decrease. The structures with a higher bulk and a more porous 3D matrix promoted a higher capillary rise. Comparatively of both additives, CBA promoted a better water affinity compared with CMF, corresponding with water capillary rise improvements.

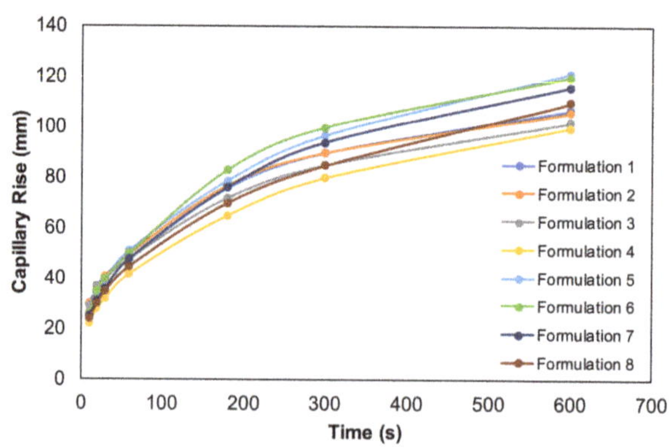

Figure 11. Klemm capillary rise method as a function of time for all tissue formulations.

These results indicated that a compromise could be found between the properties of softness, strength and absorbency with the incorporation of versatile additives in tissue formulations, reducing the incorporation of softwood fibers in an industrial furnish. This effect could promote a reduction in costs associated with the use of these fibers and also the mechanical and enzymatic refining applied to softwood fibers as the incorporation of these additives promoted structure strength improvements while maintaining the softness.

3.3. Computational Prediction of the CBA and CMF Performance in Formulations with 100% Eucalyptus Fibers

The high manufacturing costs associated with tissue paper materials are due to the market softwood pulp prices and the energy consumption associated with the fiber modification processes such as refining or beating. Finding strategies to reduce these processes is one of the focuses of the research on the management and optimization of the furnish and each type of tissue product. For this reason, we developed a tissue simulator, SimTissue, with the ability to predict the final end-use properties of different hypothetical scenarios of industrial interest. Our approach in this study was to estimate improvements in tissue properties with the incorporation of CBA and CMF, separately, in formulations with 100% eucalyptus fibers compared with formulations containing eucalyptus fibers and softwood fibers (Figure 12). Note that these simulations were carried out with a reinforcing fiber with high strength properties, according to a set of pulps studied previously [6,7]. It is important to present the simulations of these fiber mixtures as they represent the limits achievable in the tissue properties.

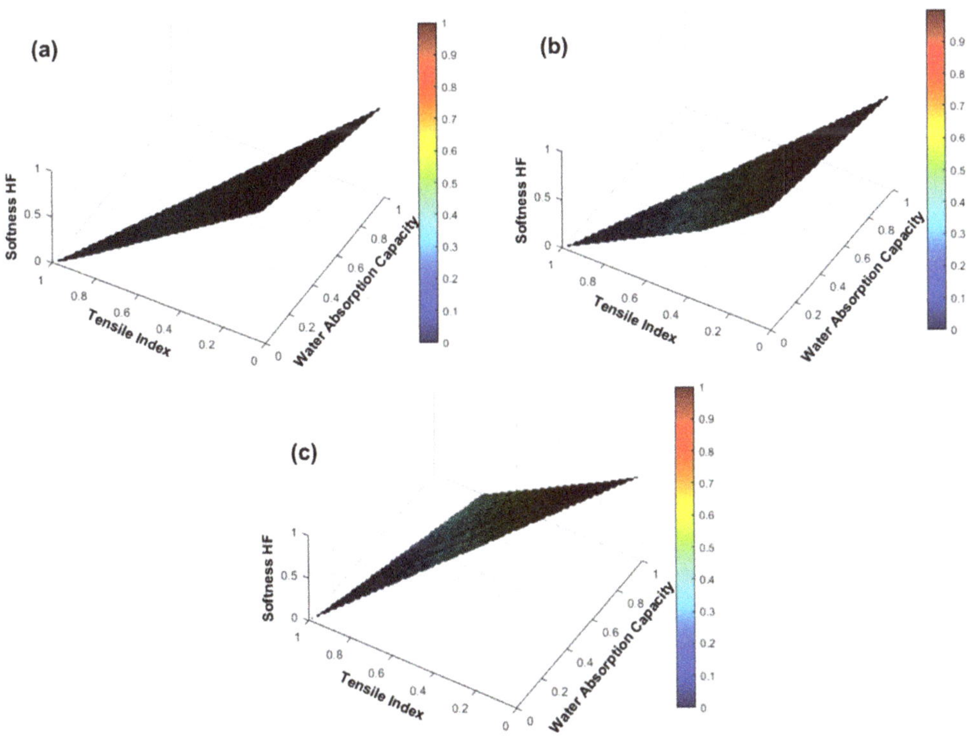

Figure 12. Evolution of the end-use tissue properties of softness HF, tensile index and water absorption capacity with the incorporation of CBA (**a**), CMF (**b**) and softwood fibers (**c**) in formulations with 100% eucalyptus fibers. All variables were normalized to present the same scale range.

As expected, the incorporation of CBA, CMF and softwood fibers promoted an improvement in strength properties with adverse effects on softness and absorbency. According to the tissue simulator, a formulation with 90% eucalyptus fibers and 10% softwood fibers promoted the production of a tissue product with a 68 softness HF, 15 Nm/g tensile index and 8 g/g water absorption capacity [24]. In order to replace this softwood fiber, the incorporation of 3% CBA allowed us to obtain improved tissue properties in the same range (72 softness HF, 15 Nm/g tensile index and 8 g/g water absorption capacity). The same trend was also verified for the incorporation of 14% CMF, presenting 70 softness HF, 15 Nm/g tensile index and 7 g/g water absorption capacity. However, in a previous work, we found that formulations with more than 10% CMF incorporation are not economically viable. The trade-off between the cost and the effectiveness of the formulations was achieved with the incorporation of CMF between 1 to 8%, presenting optimized properties at an industrial range [24].

This approach saves the consumption of softwood fibers in tissue production and is aimed at the use of biocompatible and biodegradable additives in the industrial production process. The simulator's predictive capacity allowed the optimization and development of innovative formulations, saving laboratory and industrial resources.

4. Conclusions

This work presented experimental and computational approaches to evaluate and predict the effects of the incorporation of CBA and CMF with eucalyptus and softwood fibers on the softness, strength and absorbency properties of tissue paper materials. Overall, the addition of both additives enhanced tensile strength properties at the expense of reduced softness and absorbency properties, followed by CMF and CBA. This cationic polymeric solution also improved the inter-fiber bonding and CMF particle retention. With the softwood fiber reduction in the formulations, the incorporation of additives promoted a strength and absorbency increase with a low softness decrease. It was possible to obtain a trade-off between the tissue end-use properties with CBA and CMF additions separately. With this work, it was possible to produce tissue furnish formulations using these additives with a reduction of the softwood fiber percentage in a drainage range of the tissue industrial process. This suggested that these additives presented a good potential to be introduced in the production of high-quality tissue materials such as tissue masks or premium napkins and facial papers. These biocompatible and biodegradable additives can promote the maximization of eucalyptus fibers in industrial tissue production, optimizing the process and associated manufacturing costs.

Author Contributions: Conceptualization, F.P.M. and J.M.R.C.; methodology, F.P.M. and J.M.R.C.; software, F.P.M. and J.M.R.C.; validation, F.P.M. and J.M.R.C.; formal analysis, F.P.M. and J.M.R.C.; investigation, F.P.M. and J.M.R.C.; resources, M.E.A., A.M.M.S.C. and J.M.R.C.; data curation, F.P.M. and J.M.R.C.; writing—original draft preparation, F.P.M. and J.M.R.C.; writing—review and editing, F.P.M., M.E.A., A.M.M.S.C. and J.M.R.C.; visualization, F.P.M. and J.M.R.C.; supervision, M.E.A. and J.M.R.C.; project administration, J.M.R.C.; funding acquisition, M.E.A., A.M.M.S.C. and J.M.R.C. All authors have read and agreed to the published version of the manuscript.

Funding: This research was supported by Project InPaCTus—Innovative Products and Technologies from eucalyptus, Project No 21 874 funded by Portugal 2020 through European Regional Development Fund (ERDF) in the frame of COMPETE 2020 no 246/AXIS II/2017.

Institutional Review Board Statement: Not applicable.

Informed Consent Statement: Not applicable.

Data Availability Statement: The raw/processed data required to reproduce these findings cannot be shared at this time due to legal or ethical reasons.

Acknowledgments: The authors are also very grateful for the support given by the Fiber Materials and Environmental Technologies Research Unit (FibEnTech-UBI) on the extent of the project reference UIDB/00195/2020.

Conflicts of Interest: The authors declare no conflict of interest.

References

1. Raunio, J.-P.; Löyttyniemi, T.; Ritala, R. Online quality evaluation of tissue paper structure on new generation tissue machines. *Nord. Pulp Pap. Res. J.* **2018**, *33*, 133–141. [CrossRef]
2. de Assis, T.; Reisinger, L.W.; Pal, L.; Pawlak, J.; Jameel, H.; Gonzalez, R.W. Understanding the Effect of Machine Technology and Cellulosic Fibers on Tissue Properties—A Review. *BioResources* **2018**, *13*, 4593–4629. [CrossRef]

3. Boudreau, J.; Germgård, U. Influence of various pulp properties on the adhesion between tissue paper and Yankee cylinder surface. *BioResources* **2014**, *9*, 2107–2114. [CrossRef]
4. de Assis, T.; Pawlak, J.; Pal, L.; Jameel, H.; Reisinger, L.W.; Kavalew, D.; Campbell, C.; Pawlowska, L.; Gonzalez, R.W. Comparison between uncreped and creped handsheets on tissue paper properties using a creping simulator unit. *Cellulose* **2020**, *27*, 5981–5999. [CrossRef]
5. Fišerová, M.; Gigac, J.; Stankovská, M.; Opálená, E. Influence of Bleached Softwood and Hardwood Kraft Pulps on Tissue Paper Properties. *Cellul. Chem. Technol.* **2019**, *53*, 469–477. [CrossRef]
6. Morais, F.P.; Bértolo, R.A.C.; Curto, J.M.R.; Amaral, M.E.C.C.; Carta, A.M.M.S.; Evtyugin, D.V. Comparative characterization of eucalyptus fibers and softwood fibers for tissue papers applications. *Mater. Lett. X* **2019**, *4*, 100028. [CrossRef]
7. Morais, F.P.; Bértolo, R.A.C.; Curto, J.M.R.; Amaral, M.E.C.C.; Carta, A.M.M.S.; Evtyugin, D.V. Characterization data of pulp fibres performance in tissue papers applications. *Data Brief* **2020**, *29*, 105253. [CrossRef]
8. Stankovská, M.; Gigac, J.; Fišerová, M.; Opálená, E. Blending impact of hardwood pulps with softwood pulp on tissue paper properties. *Wood Res.* **2020**, *65*, 447–458. [CrossRef]
9. Gigac, J.; Fišerová, M. Influence of pulp refining on tissue paper properties. *Tappi J.* **2008**, *7*, 27–32.
10. Chang, C.-H.; Yu, S.-T.; Perng, Y.-S. Effects of furnish and refining on properties of household paper. *Cellul. Chem. Technol.* **2018**, *52*, 433–440.
11. Demuner, B.J.; Junior, N.P.; Antunes, A.M.S. Technology prospecting on enzymes for the pulp and paper industry. *J. Technol. Manag. Innov.* **2011**, *6*, 148–158. [CrossRef]
12. Morais, F.P.; Carta, A.M.M.S.; Amaral, M.E.; Curto, J.M.R. Cellulose Fiber Enzymatic Modification to Improve the Softness, Strength, and Absorption Properties of Tissue Papers. *BioResources* **2021**, *16*, 846–861. [CrossRef]
13. Park, J.Y.; Melani, L.; Lee, H.; Kim, H.J. Effect of chemical additives on softness components of hygiene paper. *Nord. Pulp Pap. Res. J.* **2019**, *34*, 173–181. [CrossRef]
14. Valencia, C.; Valencia, Y.; Tovar, C.D.G. Synthesis and Application of a Cationic Polyamine as Yankee Dryer Coating Agent for the Tissue Paper-Making Process. *Polymers* **2020**, *12*, 173. [CrossRef]
15. Chan, E.; Woods, B.M.; Salaam, L.T.E. Soft Tissue Paper Having a Polyhydroxy Compound and Lotion Applied onto a Surface Thereof. U.S. Patent 7,972,475 B2, 5 July 2011.
16. Siqueira, E.J.; Salon, M.-C.B.; Belgacem, M.N.; Mauret, E. Carboxymethylcellulose (CMC) as a model compound of cellulose fibers and polyamideamine epichlorohydrin (PAE)-CMC interactions as a model of PAE-fibers interactions of PAE-based wet strength papers. *J. Appl. Polym. Sci.* **2015**, *132*, 42144. [CrossRef]
17. Adhikari, B.B.; Appadu, P.; Kislitsin, V.; Chae, M.; Choi, P.; Bressler, D.C. Enhancing the Adhesive Strength of a Plywood Adhesive Developed from Hydrolyzed Specified Risk Materials. *Polymers* **2016**, *8*, 285. [CrossRef]
18. Miller, J.H.; Sumnicht, D.W.; Oriaran, T.P.; Schuh, B.J.; Ramirez, A.J.; Lee, J.A. High Softness, High Durability Bath Tissues with Temporary Wet Strength. U.S. Patent 9,309,627 B2, 12 April 2016.
19. Miller, J.H.; Sumnicht, D.W.; Oriaran, T.P.; Schuh, B.J.; Ramirez, A.J.; Lee, J.A. Multi-Ply Bath Tissue with Temporary Wet Strength Resin and/or a Particular Lignin Content. U.S. Patent 9,879,382, 30 January 2018.
20. Chen, Z.; Zhang, L.; He, Z. Rethinking the determination of wet strength of paper. *BioResources* **2018**, *13*, 2184–2186. [CrossRef]
21. Zhang, Y.; Li, N.; Chen, Z.; Ding, C.; Zheng, Q.; Xu, J.; Meng, Q. Synthesis of High-Water-Resistance Lignin-Phenol Resin Adhesive with Furfural as a Crosslinking Agent. *Polymers* **2020**, *12*, 2805. [CrossRef]
22. Ginebreda, A.; Guilén, D.; Barceló, D.; Darbra, R.M. Additives in the Paper Industry. In *Global Risk-Based Management of Chemical Additives I: Production, Usage and Environmental Occurrence*, 1st ed.; Bilitewski, B., Darbra, R.M., Barceló, D., Eds.; Hdb Env Chem; Springer: Berlin/Heidelberg, Germany, 2012; pp. 11–34.
23. Shen, J.; Fatehi, P.; Ni, Y. Biopolymers for surface engineering of paper-based products. *Cellulose* **2014**, *21*, 3145–3160. [CrossRef]
24. Morais, F.P.; Carta, A.M.M.S.; Amaral, M.E.; Curto, J.M.R. Micro/nano-fibrillated cellulose (MFC/NFC) fibers as an additive to maximize eucalyptus fibers on tissue paper production. *Cellulose* **2021**, *28*, 6587–6605. [CrossRef]
25. Guan, M.; An, X.; Liu, H. Cellulose nanofiber (CNF) as a versatile filler for the preparation of bamboo pulp based tissue paper handsheets. *Cellulose* **2019**, *26*, 2613–2624. [CrossRef]
26. Patiño-Masó, J.; Serra-Parareda, F.; Tarrés, Q.; Mutjé, P.; Espinach, F.X.; Delgado-Aguilar, M. TEMPO-Oxidized Cellulose Nanofibers: A Potential Bio-Based Superabsorbent for Diaper Production. *Nanomaterials* **2019**, *9*, 1271. [CrossRef] [PubMed]
27. Zambrano, F.; Wang, Y.; Zwilling, J.D.; Venditti, R.; Jameel, H.; Rojas, O.; Gonzalez, R. Micro- and nanofibrillated cellulose from virgin and recycled fibers: A comparative study of its effects on the properties of hygiene tissue paper. *Carbohyd. Polym.* **2021**, *254*, 117430. [CrossRef] [PubMed]
28. Zambrano, F.; Starkey, H.; Wang, Y.; de Assis, C.A.; Venditti, R.; Pal, L.; Jameel, H.; Hubbe, M.A.; Rojas, O.J.; Gonzalez, R. Using Micro- and Nanofibrillated Cellulose as a Means to Reduce Weight of Paper Products: A Review. *BioResources* **2020**, *15*, 4553–4590. [CrossRef]
29. González, I.; Boufi, S.; Pèlach, M.A.; Alcalà, M.; Vilaseca, F.; Mutjé, P. Nanofibrillated cellulose as paper additive in Eucalyptus pulps. *BioResources* **2012**, *7*, 5167–5180. [CrossRef]
30. de Assis, T.C.A.; Iglesias, M.C.; Bilodeau, M.; Johnson, D.; Phillips, R.; Peresin, M.S.; Bilek, E.M.; Rojas, O.J.; Venditti, R.; Gonzalez, R. Cellulose micro- and nanofibrils (CMNF) manufacturing—financial and risk assessment. *Biofuels Bioprod. Biorefining* **2018**, *12*, 251–264. [CrossRef]

31. Hotaling, N.A.; Bharti, K.; Kriel, H.; Simon, C.G., Jr. DiameterJ: A Validated Open Source Nanofiber Diameter Measurement Tool. *Biomaterials* **2015**, *61*, 327–338. [CrossRef]
32. Conceição, E.L.T.; Curto, J.M.R.; Simões, R.M.S.; Portugal, A.T.G. Coding a simulation model of the 3D structure of paper. In *Computational Modeling of Objects Represented in Images. CompIMAGE 2010. Lecture Notes in Computer Science*, 1st ed.; Barneva, R.P., Brimkov, V.E., Hauptman, H.A., Natal Jorge, R.M., Tavares, J.M.R.S., Eds.; Springer: Berlin/Heidelberg, Germany, 2010; Volume 6026, pp. 299–310.
33. Curto, J.M.R.; Conceição, E.L.T.; Portugal, A.T.G.; Simões, R.M.S. Three dimensional modeling of fibrous materials and experimental validation. *Mater. Werkst.* **2011**, *42*, 370–374. [CrossRef]
34. Morais, F.P.; Carta, A.M.M.S.; Amaral, M.E.; Curto, J.M.R. 3D Fiber Models to Simulate and Optimize Tissue Materials. *BioResources* **2020**, *15*, 8833–8848. [CrossRef]
35. de Assis, T.; Huang, S.; Driemeier, C.E.; Donohoe, B.S.; Kim, C.; Kim, S.H.; Gonzalez, R.; Jameel, H.; Park, S. Toward an understanding of the increase in enzymatic hydrolysis by mechanical refining. *Biotechnol. Biofuels* **2018**, *11*, 289. [CrossRef]
36. Martins, V.D.F.; Cerqueira, M.A.; Fuciños, P.; Garrido-Maestu, A.; Curto, J.M.R.; Pastrana, L.M. Active bi-layer cellulose-based films: Development and characterization. *Cellulose* **2018**, *25*, 6361–6375. [CrossRef]
37. Trepanier, R. Pulp fiber quality and the relationship with paper tissue properties. In Proceedings of the issue Conference & Expo 2017: The Power of TAPPI & RISI, Miami Beach, FL, USA, 3–6 October 2017; pp. 1–4.
38. Maurer, H.W. Starch in the Paper Industry. In *Starch—Chemistry and Technology. Food Science and Technology*, 3rd ed.; BeMiller, J., Whistler, R., Eds.; Academic Press: Cambridge, MA, USA, 2009; pp. 657–713.
39. Beuther, J.C.; Veith, M.W.; Zwick, K.J. Characterization of Absorbent Flow Rate in Towel and Tissue. *J. Eng. Fibers Fabr.* **2010**, *5*, 1–7. [CrossRef]

polymers

MDPI

Article

Comparative Study of the Properties of Plasticized Polylactic Acid with Maleinized Hemp Seed Oil and a Novel Maleinized Brazil Nut Seed Oil

Aina Perez-Nakai [1], Alejandro Lerma-Canto [1], Ivan Dominguez-Candela [2], Daniel Garcia-Garcia [1], Jose Miguel Ferri [1] and Vicent Fombuena [1,*]

[1] Technological Institute of Materials (ITM), Universitat Politècnica de València (UPV), Plaza Ferrándiz y Carbonell 1, 03801 Alcoy, Spain; aipena@epsa.upv.es (A.P.-N.); allercan@epsa.upv.es (A.L.-C.); dagarga4@epsa.upv.es (D.G.-G.); joferaz@upvnet.upv.es (J.M.F.)

[2] Instituto de Seguridad Industrial, Radiofísica y Medioambiental (ISIRYM), Universitat Politècnica de València (UPV), Plaza Ferrándiz y Carbonell s/n, 03801 Alcoy, Spain; ivdocan@doctor.upv.es

* Correspondence: vifombor@upv.es

Abstract: In this study, for the first time, Brazil nut seed oil was chemically modified with maleic anhydride to obtain maleinized Brazil nut seed oil (MBNO). The same process was developed to obtain maleinized hemp seed oil (MHO). The use of MBNO and MHO was studied as bio-based plasticizers by incorporating them with different contents ranging from 0 to 10 phr in a polylactic acid (PLA) matrix. By means of mechanical, thermal and thermomechanical characterization techniques, the properties of the different formulations were studied to evaluate the plasticizing effect of the MBNO and MHO. With the addition of both plasticizers, a significant increase in ductile properties was observed, reaching an increase in elongation at break of 643% with 7.5 phr MBNO and 771% with 10 phr MHO compared to neat PLA. In addition, it has been observed that the mechanical resistant properties do not decrease, since the oils enhance the crystallization of PLA by increasing the free volume between its chains and counteracting the effect. Finally, a disintegration test was carried out under thermophilic conditions at 58 °C for 27 days, demonstrating that the incorporation of MHO and MBNO does not significantly affect the biodegradability of neat PLA.

Keywords: maleinized hemp seed oil; maleinized Brazil nut seed oil; bio-plasticizers; polylactic acid

Citation: Perez-Nakai, A.; Lerma-Canto, A.; Domingez-Candela, I.; Garcia-Garcia, D.; Ferri, J.M.; Fombuena, V. Comparative Study of the Properties of Plasticized Polylactic Acid with Maleinized Hemp Seed Oil and a Novel Maleinized Brazil Nut Seed Oil. *Polymers* **2021**, *13*, 2376. https://doi.org/10.3390/polym13142376

Academic Editors: Andreia F. Sousa, José Miguel Ferri, Vicent Fombuena Borràs and Miguel Fernando Aldás Carrasco

Received: 30 June 2021
Accepted: 16 July 2021
Published: 20 July 2021

1. Introduction

Since the end of the 20th century, environmental concerns and the development of sustainable strategies have been on the rise [1]. This concern carries over to the present day, where plastic represents a large part of human life, with up to 368 million tonnes of plastic produced globally in 2019 [2]. The problem arises mainly because of the negative impact on the ecosystems generated by the amount of plastic waste, since only 173 million tonnes are collected for recycling or landfill [3], but also by the exploitation of non-renewable natural resources, such as petroleum. The problem is magnified in the packaging industry, as it is the largest consumer of plastics, accounting for 39.6% of total plastic consumption in Europe [2]. In view of this situation, research is mainly focused on obtaining polymeric materials with formulations based on renewable resources or even with the property of being biodegradable. A wide variety of biopolymers or bio-based materials have now been obtained that contribute to this sustainable development.

Bio-based polymers are polymers that are made from biological substances, i.e., non-fossil materials [4]. These materials may or may not be biodegradable depending on their ability to be broken down by microorganisms [4]. Currently, some of the most promising biopolymers are extracted from biomass (starch, cellulose, protein, chitin, etc.), such as thermoplastic starches (TPSs), given the abundance of polysaccharides in the biomass. Another promising alternative are also those obtained by microbial production, such as polyhydroxyalkanoates (PHAs) and polyhydroxybutyrate (PHB) or polylactic acid (PLA) [5–7].

In 2020, bioplastics accounted for only 1.2 million tonnes, less than 1% of all plastics produced [3]. However, the market for these polymers has been growing for years and the trend is expected to increase, with production estimated to increase by 11.28% from 2019 to the end of 2025 [8]. Specifically, PLA is one of the biopolymers with the greatest potential for industrial use as a substitute for petroleum-based polymers [9,10]. This is due to its great potential due to the combination of its mechanical properties, easy processability and its price compared to other biopolymers [11,12]. PLA has a suitable thermal stability and resistance to being industrially processed by injection molding, welding, thermoforming or extrusion [13]. Moreover, as it is biodegradable, one of its most interesting applications is mainly in the packaging sector [14]. In terms of mechanical properties, this material is comparable to non-degradable polymers in the "commodity" range [14,15]. However, there is a major limitation of PLA, its brittleness, which is a major drawback in the packaging industry [16]. To alleviate this disadvantage, there are several proposals. One solution to improve the flexibility of this polymer is its blending with other polymers such as polyethylene glycol (PEG), thermoplastic starch (TPS) or polybutylene succinate-co-adipate (PBSA), among others. However, the lack of miscibility between the components makes it difficult to achieve this improvement in toughness [5,11,17,18]. For this reason, some authors proposed modified vegetable oils (MVOs) as compatibilizing agents or even as plasticizing additives [12,19]. The reason for this is that they are a renewable and sustainable substitute for synthetic modifiers [20,21], which are also respectful of human health due to their non-toxicity, as they do not generate the migration of substances such as Bisphenol A (BPA), as is the case with conventional epoxy resins used in consumer products [22].

Several techniques can be used for chemical modification, such as epoxidation [23,24], maleinization [5,10,25], acrylation or hydroxylation. The MVOs available on the market today are mainly epoxidized soybean oil (ESBO) [25] and epoxidized linseed oil (ELO) [26]. Authors such as Garcia-Garcia et al. or Chieng et al. [26,27] have reported the improvement of PLA stiffness properties with these oils. Additionally, other authors investigated other modified oils for PLA formulations such as epoxidized cottonseed oil (ECSO) [24] or epoxidized palm oil (EPO) [27]. Another commercial option is maleinized linseed oil, which has provided excellent properties to PLA, as reported Ferri et al. [28].

In this work, the chemical modification of process of oils carried out is the maleinization. Maleinization is a chemical process, usually carried out in a single step, which consists of incorporating maleic anhydride molecules into the triglycerides that make up vegetable oils when conjugated carbon–carbon double bonds are present. For this purpose, there are several methods that can be used, such as the so-called "ene" reaction, Diels-Alder addition and free radical copolymerization, the first being the most favorable and the one used in this work [5,29]. This procedure requires a temperature of about 200 °C to result in the addition of the anhydric groups at the allylic positions of the fatty acid, as shown in Figure 1.

Both Brazil nut (*Bertholletia excelsa*) and hemp seed (*Cannabis sativa* L.) are interesting as MVOs, since both have an interesting lipid profile for functionalization. The Brazil nut is a brown fruit generally cultivated in the Amazon [30]. Its oil is an interesting object of study, since it has between 60.8% and 72.5% lipids [30,31]. Among them, it has 75.6% unsaturated fatty acids (UFA) [30], and both monounsaturated (MUFA) and polyunsaturated (PUFA), which means a significant amount of double bonds that allow its functionalization by chemical processes such as maleinization. On the other hand, hemp seed oil is also attractive for this purpose, as it is high in linoleic acid (55.3%) and linolenic acid (20.3%) [32]. Both allow their oil to be extracted by cold pressing with a good yield and also have good oxidative stability [31,33]. Thereupon, this work evaluates the potential of MBNO and MHO as a bio-based plasticizer to improve the ductile properties of PLA related to its brittleness and the comparison of these with commercial maleinized linseed oil (MLO).

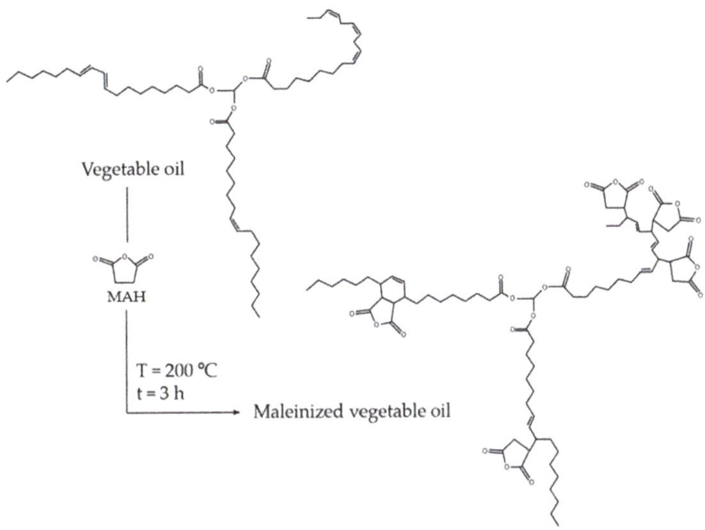

Figure 1. Schematic representation of the maleinization process of the triglyceride presents in vegetable oil.

2. Materials and Methods

2.1. Materials

The Brazil nuts used in this work were of Brazilian origin and were purchased from FrutoSeco (Alicante, Spain). On the other hand, the hemp seeds were obtained from a local market in Callosa de Segura (Alicante, Spain). The oil from both seeds was extracted by cold pressing with an extruder model DL-ZYJ05 purchased from Nanchang Dulong Industrial Company (Zucheng, China). The density of Brazil nut oil was 916 kg·m^{-3}, its iodine value (IV) between 99.5 and 102.6 g 100 g^{-1} and its acid value (AV) 0.2 mg·KOH g^{-1}. The density of hemp oil was 935 kg·m^{-3}, its IV is 163.8 g·100g^{-1} and its AV is 8 mg·KOH g^{-1}. For both cases, density, IV and AV values were obtained according to ISO 1675, ISO 3961 and ISO 660, respectively. For the maleinization of both oils, maleic anhydride supplied by Sigma Aldrich (Madrid, Spain), with a purity of 98% was used. The PLA used for mixing with the oils was a commercial-grade Ingeo Biopolymer 2003D from NatureWorks LLC (Minnetonka, MN, USA) in pellet form.

2.2. Maleinization of Brazil Nut (Bertholletia excelsa) and Hemp (Cannabis sativa L.*) Oils*

The same method was used for the maleinization of Brazil nut and hemp seed oil. For this purpose, a round flask with three necks with a capacity of 500 mL was used, where a stirrer was placed in the central neck and used at a speed of 300 rpm. A digital thermometer was placed in the second neck to measure the temperature during the process, and the extraction of samples and the introduction of maleic anhydride was carried out through the third neck. The maleinization ratio used was 2.4:1, following recommendations from previous works [5]. A total of 9 g of maleic anhydride per 100 g of each virgin oil was introduced. During the maleinization process, three temperature stages were used, at 180 °C, 200 °C and 220 °C, and samples were taken every 30 min. Initially, 300 g of oil was introduced into the round flask and when the first temperature of 180 °C was reached, 1/3 of the quantity of the total amount of maleic anhydride was added. This temperature was maintained for 1 h and increased to repeat the same at 200 °C and 220 °C. Finally, the mixture was allowed to cool to room temperature. The degree of maleinization of the oil was determined according to ISO 660:2009 using the following expression:

$$Acid\ Value = \frac{56.1 \times c \times V}{m} \quad (1)$$

where c is the exact concentration of the KOH standard solution used (mol·L^{-1}), V the volume of the KOH standard solution used (mL) and m the mass analyzed (g).

2.3. Manufacturing of PLA with Maleinized Vegetable Oils

Several PLA compositions were made with different phr of MBNO and MHO, which are summarized in Table 1, where the codes assigned to each of them are indicated. Prior to mixing, the PLA was dried for 24 h at 40 °C. The required amounts of PLA and oil were weighed and added, to be mixed manually. The mixtures were processed with a co-rotating twin-screw extruder (D = 30 mm; L/D = 20:1) from DUPRA (Alicante, Spain), at a rotation speed of 40 rpm with a temperature profile of 165 °C, 168.5 °C, 172.5 °C and 175 °C from the feed zone to die, respectively. The compositions obtained were cooled to room temperature, pelletized and dried at 40 °C for 24 h. Finally, each composition was shaped by injected molding from Mateu and Solé (Barcelona, Spain) at temperature profiles of 170 °C, 180 °C, 190 °C and 200 °C from the feed section to injection nozzle.

Table 1. Composition of plasticized PLA with different contents (phr) of maleinized vegetable oils and labeling of the formulations.

Reference	PLA	Maleinized Brazil Nut Oil—MBNO (phr)	Maleinized Hemp Seed Oil—MHO (phr)
PLA	100	0	0
PLA + 2.5 MBNO	100	2.5	0
PLA + 5 MBNO	100	5	0
PLA + 7.5 MBNO	100	7.5	0
PLA + 10 MBNO	100	10	0
PLA + 2.5 MHO	100	0	2.5
PLA + 5 MHO	100	0	5
PLA + 7.5 MHO	100	0	7.5
PLA + 10 MHO	100	0	10

2.4. Mechanical Characterization of PLA Formulations Plasticized with MBNO and MHO

Several standardized tensile, flexural, impact and hardness tests were carried out to determine the mechanical properties of the different PLA compositions manufactured with the different types and proportions of MVOs. The tensile and flexural tests were performed according to ISO 527 and ISO 178, respectively, with an Ibertest ELIB 30 universal testing machine from SAE Ibertest (Madrid, Spain). For the tensile test, a crosshead speed of 10 mm·min^{-1} was used with a 5 kN load cell. In the case of flexural test, the deformation rates were studied at 5 mm·min^{-1}. Both tests were performed on at least 5 different samples of each material and the average values of the elongation at break (%), tensile and flexural strength, and Young's modulus and flexural modulus were calculated. To determine the impact resistance by energy absorption, a 6 J Charpy pendulum from Metrotec S.A. (San Sebastian, Spain) was used. According to ISO 179, rectangular samples of 80 × 10 × 4 mm^3 without notches were employed. Finally, a Shore D hardness tester model 673-D from J. Bot S.A. (Barcelona, Spain) was used to determine the hardness according to ISO 868. A minimum of 5 samples were used and the results shown are the obtained average.

2.5. Thermal Analysis of PLA Formulations Plasticized with MBNO and MHO

The thermal properties of PLA and plasticized PLA compositions were determined by differential scanning calorimetry (DSC) and thermogravimetric analysis (TGA). The glass transition temperature (T_g), cold crystallization temperature (T_{cc}) and melting temperature

(T_m) were obtained by DSC with a Mettler Toledo DSC 821 (Schwerzenbach, Switzerland). The test conditions were under nitrogen atmosphere (flow rate 66 mL·min^{-1}) with a thermal program to remove thermal history consisting of a first heating from 30 °C to 200 °C at 10 °C·min^{-1}, followed by a cooling to 30 °C at 2 °C·min^{-1} and a last heating to 350 °C at 2 °C·min^{-1}. Thermal transitions were determined from the second heating. The percentage crystallinity of each material was calculated using the following equation:

$$X_c\ (\%) = \frac{\Delta H_m - \Delta H_c}{w \Delta H_m^o} \times 100 \tag{2}$$

where ΔH_m is the enthalpy of fusion, ΔH_c the cold crystallization enthalpy, w the mass fraction of the material and ΔH_m^o the enthalpy of fusion for a theoretical pure crystalline PLA structure, which was assumed to be 93 J g^{-1} [34].

TGA tests were carried out to determine the initial degradation temperature (T_0) and the maximum degradation temperature (T_{max}). The equipment used was a TGA/SDTA 851 from Mettler Toledo (Schwerzenbach, Switzerland) and the samples (7–10 mg) were tested under nitrogen atmosphere (flow rate 66 mL·min^{-1}) and heating from 30 °C to 700 °C at 20 °C·min^{-1}. T_0 was determined at 5% mass loss, while T_{max} was calculated from the first derivate of the TGA curves (DTG).

2.6. Thermomechanical Characterization of PLA Formulations Plasticized with MBNO and MHO

To analyze the changes in storage modulus (G') and damping factor, a dynamic thermo-mechanical analysis (DMTA) was performed with an oscillatory rheometer AR G2 from TA Instruments (New Castle, DE, EEUU), equipped with a clamp attachment for solid samples. The 40 × 10 × 4 mm^3 rectangular samples were tested in torsion mode and subjected to a thermal program from 30 °C to 130 °C with a heating rate of 2 °C·min^{-1}. The frequency was set at 1 Hz and the maximum deformation (γ) was 0.1%.

2.7. Degree of Disintegration under Composting Conditions of PLA Formulations Plasticized with MBNO and MHO

The degree of disintegration of each compound was carried out according to the recommendations of the ISO 20200 standard. For this purpose, 7 different square samples for each compound of dimensions 25 × 25 × 1 mm^3 were buried in controlled soil. Previously, samples were dried at 40 °C for 24 h. To evaluate the disintegration process, samples were unburied on days 3, 7, 14, 17, 21, 24 and 27, washed with distilled water, dried at 45 °C for 24 h and weighed in an analytical balance. The degree of disintegration was calculated with the following equation given in ISO 20200:

$$D = \frac{m_i - m_r}{m_i} \times 100 \tag{3}$$

where m_i is the initial mass of the sample before the test and m_r is the weight of the sample extracted from compost soil on different days after drying. Photographs were also taken of each sample recovered on the days indicated to also qualitatively assess the disintegration process.

3. Results

3.1. Synthesis of Maleinized Hemp Seed Oil and Brazil Nut Oil

The evolution of the AV values along the three temperature stages of 180 °C, 200 °C and 220 °C used throughout the maleinization process of BNO and HO has been plotted in Figure 2. Initially, the AV of BNO is 0.20 mg KOH g^{-1} and 8 mg KOH g^{-1} for the HO. In both cases, it can be observed how these values increase significantly and analogously after the first hour at 180 °C to approximately 47 mg KOH g^{-1} and 50 mg KOH g^{-1} for the BNO and HO, respectively, indicating that maleinization is occurring. After two hours, at the second temperature of 200 °C, another significant increment can be observed up to values close to 90 mg KOH g^{-1} in the case of BNO and up to 80 mg KOH g^{-1} in the case of HO. Finally, at 220 °C, after three hours, there is a drastic increase in the AV up to 105 mg KOH g^{-1} in HO and even up to 130 mg KOH g^{-1} in BNO. In the two oils, the AV follows a very similar trend in which a clear increase is observed as the temperature increases. This is due to the fact that at temperatures close to 200 °C, the "ene" reaction is favored, in which maleic anhydride can easily bind to an allylic position of the unsaturated fatty acid [35]. As the temperature of 220 °C is maintained over time, the acidity index tends to stabilize, in the case of HO at 105 mg KOH g^{-1} and at 130 mg KOH g^{-1} for BNO. Both values are in complete agreement with the AV of commercial MLO reported by Quiles-Carrillo et al. [36], which indicates that it ranges between 105 and 130 mg KOH g^{-1}, suggesting that both the MBNO and MHO developed have similar characteristics to one of the few maleinized oils available on the market today.

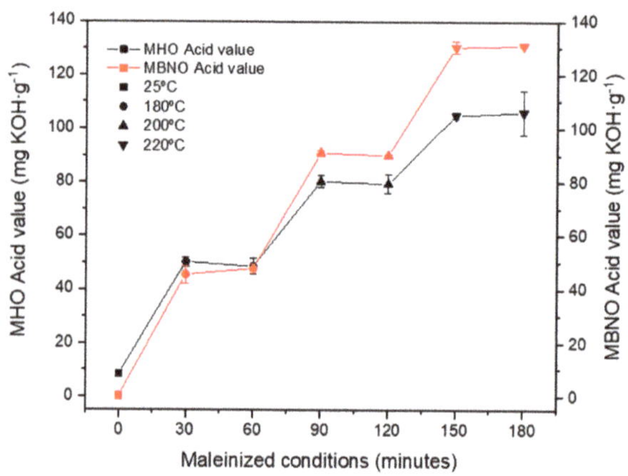

Figure 2. Effect of temperature and time on the efficiency of the maleinization of HO and BNO with maleic anhydride.

Another consequence observed as a result of the maleinization process is an important change in the color of the oils. Figure 3 shows a progressive color change of BNO, in which, at 180 °C, the color is light yellow, similar to the color of the original oil, and changes to a reddish color as maleinization occurs. The same occurs with HO. Ernzen et al. [37] reported a similar change during the maleinization process of soybean oil, in which the oil changed from a yellow to an orange–reddish color.

(a) (b) (c)

Figure 3. Influence of the reaction temperature and time on the color during maleinization of BNO, (**a**) 180 °C–0 min, (**b**) 200 °C–120 min and (**c**) 220 °C–180 min.

3.2. Effect of MBNO and MHO on Mechanical Properties of Plasticized PLA Formulations

The results obtained from the mechanical properties of the different PLA formulations with both MBNO and MHO show a significant reduction in PLA stiffness, indicating that both oils are effective as renewable PLA plasticizers. Figure 4 shows the tensile mechanical properties of PLA with MBNO and MHO. Unplasticized PLA has a Young's modulus close to 3000 MPa (2977 ± 21 MPa) and a tensile strength of 35.8 ± 7.3 MPa, relatively high values in the thermoplastic commodity range. However, its toughness is low because it shows an elongation at break of 7.4 ± 7%. By incorporating MBNO into the PLA matrix, an increase in elongation is observed from 2.5 phr, reaching the maximum value of 52% ± 3.0% at 7.5 phr, which represents an increase of 643% with respect to neat PLA. From this concentration of MBNO, a decrease in elongation occurs, probably due to plasticizer saturation, a phenomenon that some authors call an anti-plasticization effect [38]. On the other hand, the effect of MHO on the elongation at break is observed from 7.5 phr MHO where it increases drastically up to 42% ± 6% and even increases up to 61% ± 3.0% with the addition of 10 phr MHO, i.e., 771% more than neat PLA. These results obtained are close to the results reported by Ferri et al. [28], in which an elongation of 78.4% was obtained by incorporating commercial MLO into PLA. In Figure 5, it is possible to observe the changes in the appearance of the different tensile specimens after the test. Figure 5a shows specimens made with different MBNO contents. It can be seen how from 7.5 phr of MBNO content, the specimen has a lower final elongation. Figure 5b shows the specimens with different MHO contents after the tensile test. The main difference with respect to MBNO is observed in the PLA + 10% MHO sample, whose elongation is higher than that obtained with 7.5 phr MHO. This increase in elongation at break is due to the enhancement of molecular mobility, which is explained by several plasticization theories. The first of these, the lubricity theory, holds that the plasticizer functions as a molecular lubricant of the polymer. On the other hand, gel theory, applied to amorphous thermoplastics such as the grade of PLA used in this work, suggests that the plasticizer molecules are placed between the polymer chains and weaken the interactions between them. Finally, the free volume theory argues that the plasticizer increases the free volume and thus decreases the interactions between the polymer chains [39].

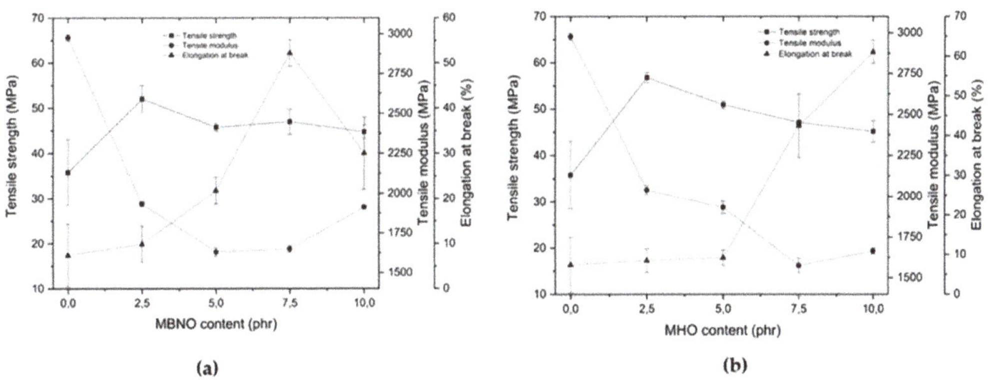

Figure 4. Plot of evolution of tensile properties of PLA formulations plasticized with various contents of MBNO (**a**) and MHO (**b**).

In addition, with the incorporation of both oils, an increase in tensile strength is observed with 2.5 phr, with respect to neat PLA, and then a decreasing trend as the percentage of oil incorporation increases. Several authors report a similar effect, but with lower strength values than PLA as an oil content increases. Garcia-Garcia et al. [40] reported a decrease in strength from 46.5 MPa to 42.2 MPa with the addition of 10% epoxidized karanja oil—EKO. In the case of MBNO and MHO, by incorporating between 2.5 and 10 phr, higher strengths were obtained than those of neat PLA, counterbalanced, in turn, by an improvement in its elongation. This phenomenon can be explained by the fact that MHO and MBNO, by improving the mobility between PLA chains, also facilitate crystallization [41], which at the same time generates an increase in its stiffness. Finally, Young's modulus follows a decreasing trend in both cases, obtaining the lowest value with the incorporation of 5 phr MBNO of up to 45.2% and with 7.5 phr MHO of up to 47%. From these contents in phr, the modulus tends to increase as the elongation tends to decrease. This decreasing evolution is in line with the results reported by authors such as Carbonell-Verdu et al. [24], who with the incorporation of ECSO to PLA obtained a reduction in the tensile modulus of 47.5% with respect to neat PLA. The high decrease in tensile modulus obtained with MBNO and MHO is related to the decrease in molecular interactions that facilitate movement.

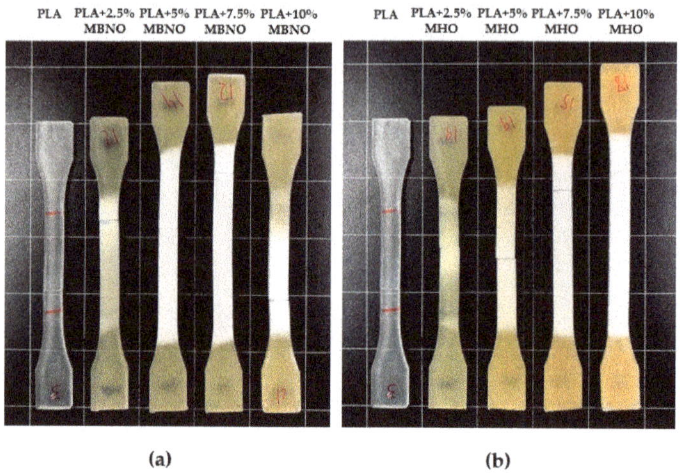

Figure 5. Plot of evolution of tensile properties of PLA formulations plasticized with various contents of MBNO (**a**) and MHO (**b**).

In parallel, Figure 6 shows a decrease in strength and flexural modulus as MBNO and MHO content increases. In the case of MBNO, a maximum reduction in flexural strength of 37.7% and flexural modulus of 20% is obtained with 10 phr and 39.6% and 18.4%, respectively, with 10 phr MHO content. However, as with the tensile properties, it is at 5 phr MBNO and 7.5 phr MHO that the decrease in strength and modulus tends to stabilize, suggesting plasticizer saturation. Ferri et al. [28] reported a decrease in the flexural strength of PLA with commercial MLO of up to 24%, and also a saturation around 15 phr MLO. The antiplasticizing effect of polymers depends, above all, on the molecular weight and the concentration of the diluent, so that in each formulation it is produced with a specific percentage of plasticizer [42].

On the other hand, Figure 7 summarizes the energy absorbed by Charpy's impact test and the Shore D hardness of neat PLA and formulations. The energy absorbed by Charpy's impact test is closely related to the toughness of the material; therefore, it is also representative for evaluating the effectiveness of MBNO and MHO as a plasticizer. Due to its brittleness, neat PLA has a relatively low absorbed energy (around 35.5 $kJ \cdot m^{-2}$). By adding plasticizers, an increase in absorbed energy is observed, in the case of MBNO up to 20% higher and in the case of MHO up to 46% higher. The energy absorbed by a material depends both on the deformation capacity linked to the ductility properties and on the breaking strength, which is related to the mechanical properties [28]; therefore, the results are in full agreement with the obtained values of elongation, modulus and breaking strength. As for hardness, the value decreases progressively as more plasticizer is incorporated, although it is not as strongly visualized as with other properties. With MBNO the Shore D hardness decreased from 80 to 71.7 at 7.5 phr MBNO, and with 10 phr MHO to 72.9. Other studies reported a similar decrease from 75.6 to 59.6 with the addition of 22.5% commercial ELO to PLA/hazelnut shell flour (HSF) blends [43].

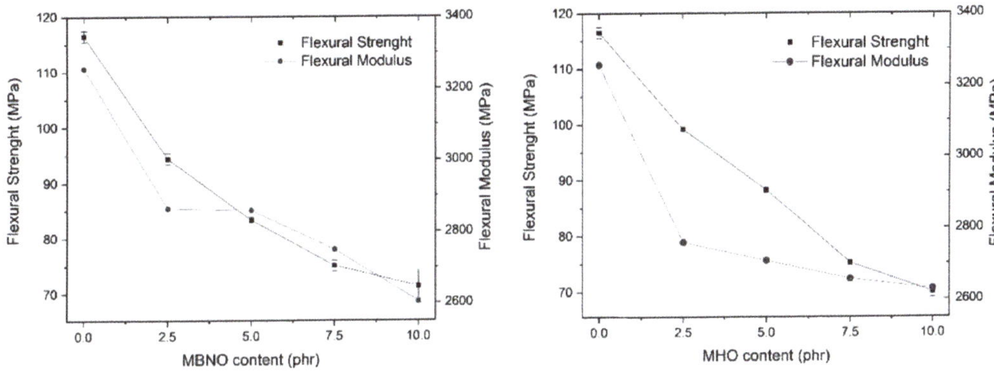

(a) (b)

Figure 6. Plot of evolution of flexural properties of PLA formulations plasticized with various contents of MBNO (**a**) and MHO (**b**).

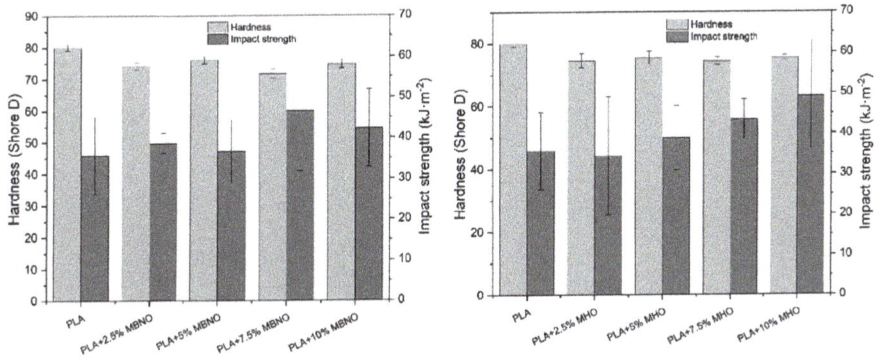

(a) (b)

Figure 7. Graphic representation of impact strength and hardness (Shore D) of PLA formulations plasticized with various contents of MBNO (**a**) and MHO (**b**).

3.3. Effect of MBNO and MHO on Thermal Properties of Plasticized PLA Formulations

Thermal stability of PLA and plasticized formulations with different content of MBNO and MHO was performed by thermogravimetry analysis (TGA). Figure 8 show the weight loss versus temperature curves for each sample (TG) and their corresponding first derivative curves (DGT). Table 2 shows some characteristic thermal parameters of the thermograms, such as the degradation onset temperature ($T_{5\%}$), which indicates the temperature at which a 5% weight loss occurs, and the maximum degradation temperature (T_{max}), which corresponds to the peak of the first derivative curve. Neat PLA possesses good thermal stability with a $T_{5\%}$ of 347.9 °C and a T_{max} of 386 °C. As can be seen, adding MBNO or MHO to PLA causes a slight decrease in $T_{5\%}$ as they are incorporated in higher quantities. In this case, the addition of 10 phr MNBO results in a $T_{5\%}$ reduction of 4.5 °C, while with 10 phr MHO, the reduction is 10.1 °C. Regarding the maximum degradation temperature, it can be observed that both plasticizers lead to a decrease in temperature, obtaining the lowest value at 10 phr, with a reduction of 16.4 °C for both plasticizers. Similar behavior was observed by Garcia-Campo et al. [44] for PLA/PHB/PCL blends compatibilized with epoxidized soybean oil (ELO). The authors observed how the addition of ELO into the blend resulted in a reduction in $T_{5\%}$ by 19.1 °C and T_{max} by 20 °C.

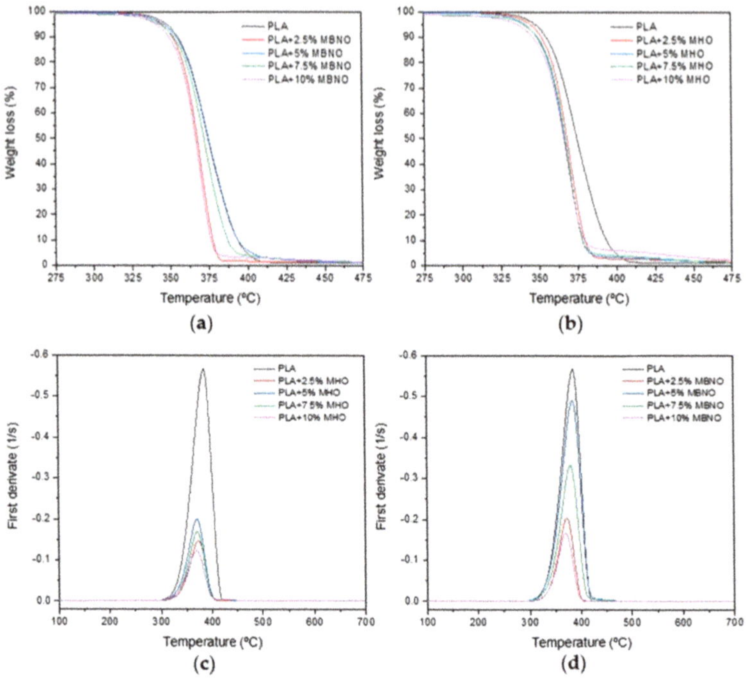

Figure 8. TGA (**a**,**b**) and DTGA (**c**,**d**) of unplasticized PLA and PLA plasticized with different content of MBNO and MHO.

Table 2. Summary of the TGA and DSC thermal parameters of PLA unplasticized and plasticized with different amounts of MBNO and MHO content.

Samples	TGA Parameters		DSC Parameters					
	$T_{5\%}$ (°C) [a]	T_{max} (°C)	T_g (°C)	T_{cc} (°C)	ΔH_c (Jg^{-1})	T_m (°C)	AH_m (Jg^{-1})	X_{PLA} (%)
PLA	347.9	386.2	61.0	123.64	1.2	150.7	3.2	1.7
PLA + 2.5%MBNO	346.6	372.1	58.0	100.8	24.9	153.3	26.8	2.1
PLA + 5%MBNO	346.5	384.6	57.9	120.8	9.2	150.3	15.0	5.8
PLA + 7.5%MBNO	344.5	379.3	58.7	123.7	3.3	151.2	10.0	6.8
PLA + 10%MBNO	343.4	369.6	57.8	123.5	2.6	151.0	3.2	7.3
PLA + 2.5%MHO	346.9	373.1	57.8	103.1	21.0	153.3	22.6	1.6
PLA + 5% MHO	342.9	370.6	56.8	101.8	20.8	153.6	28.4	7.5
PLA + 7.5% MHO	342.2	370.6	57.3	113.9	13.5	150.5	19.5	6.0
PLA + 10% MHO	337.8	369.6	56.9	109.9	9.7	149.2	20.3	10.6

[a] $T_{5\%}$, calculated at 5% mass loss.

Table 2 shows the main thermal properties obtained by DSC. On the other hand, Figure 9 shows the plot representation of the dynamic curve of DSC obtained with the unplasticized PLA and PLA plasticized with different contents of MBNO and MHO. As can be seen, both maleinized oils have a direct effect on some thermal properties of PLA. Unplasticized PLA has a glass transition temperature (T_g) located at 61 °C, a cold crystallization temperature (T_{cc}) at 123.6 °C and a melting temperature located at 150.7 °C. Both plasticizers, MBNO and MHO, decrease the T_g and T_{cc} of PLA, which is indicative of an increase in the mobility of the polymer chains at lower temperatures, evidencing the plasticizing effect of both maleinized oils [45]. The same evolution was reported by Dominguez-Candela et al. [46], who employed epoxidized chia oil (ECO) in PLA. On the other hand, no significant variations were observed in T_m, showing that after the addition of the plasticizers, the T_m remained at values around 150–153 °C for all the samples. As it is possible to observe in Figure 9, the addition of lower contents of maleinized oils provides two small melting temperature peaks, which are associated to the formation of two regions with different crystallinities, a result of the compatibilization process between the PLA matrix and the plasticizer. The addition of higher oil contents broadens these two peaks, producing an overlap between them. Regarding crystallinity, it is observed that PLA without plasticizing has a crystallinity of 1.7%, whereas as MBNO or MHO is added, the crystallinity increases progressively. As can be seen in Table 2, with a 10 phr MBNO, the crystallinity of PLA increases to 7.3%, while with the 10 phr MHO, it increases to 10.6%. A similar increase was reported by Carbonell-Verdu et al. [5], who observed that the addition of 10 phr commercial MLO and MCSO to PLA increased its crystallinity up to 11.6% and 19.1%, respectively. The increase in crystallinity with the addition of plasticizers, in this case MBNO and MHO, is caused by the facilitation of the laminar rearrangement of the amorphous zones of PLA, which is caused by the increased mobility of the polymer chains [47].

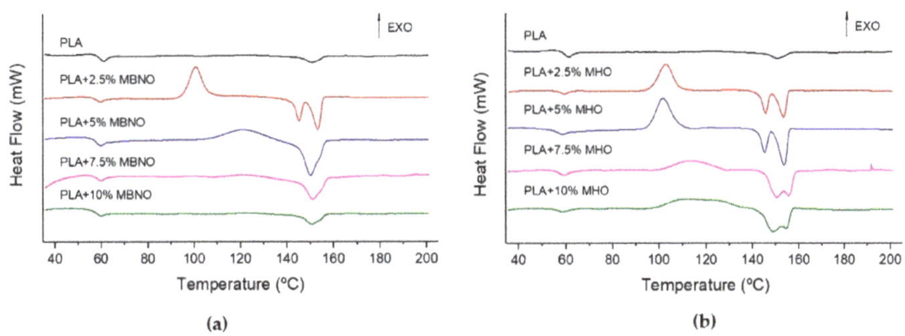

Figure 9. Dynamic DSC curve of PLA unplasticized and plasticized with different amounts of MBNO (**a**) and MHO content (**b**).

3.4. Effect of MBNO and MHO on Thermomechanical Properties of Plasticized PLA Formulations

The evolution of the storage modulus (G′) as a function of temperature is shown in Figure 10a,b. Neat PLA, at room temperature, has a storage modulus above 1000 MPa and this remains constant up to 60 °C. With the incorporation of MBNO and MHO in the PLA matrix, no significant changes are observed in terms of G′ at room temperature, since in all of them the values remain between 900 and 1100 MPa, although it is observed that with the incorporation of the plasticizers, the G′ tends to decrease slightly, showing its plasticizer effect. This is followed by a drop in G′ of up to three orders of magnitude, a change that is related to T_g. In this case, a significant difference can be observed between the plasticized samples and neat PLA. As can be seen after the addition of 10 phr MBNO, the T_g decreases from 60 °C to 48 °C and to almost 50 °C with 7.5 phr MHO. The almost 10 °C shift in T_g with both plasticizers is due, as described above, to the fact that the plasticizer generates an increase in the free volume, which improves the mobility of the chain due to a lower interaction between them, being able to move with lower energy [45]. Neat PLA above 75 °C behaves like an elastomer up to 95–100 °C. Above this temperature, the energetic conditions favor and promote crystallization, improving the elastic behavior and leading to an increase in the storage modulus up to 60 MPa. Here, again, differences are observed when incorporating the plasticizers into PLA. As can be seen, this temperature decreases by around 10 °C in all the MBNO-plasticized formulations and by around 5 °C in the MHO-plasticized samples with respect to neat PLA. This decrease in temperature is due to the higher mobility of the chains, which allows their rearrangement into a packed structure with lower energy [48].

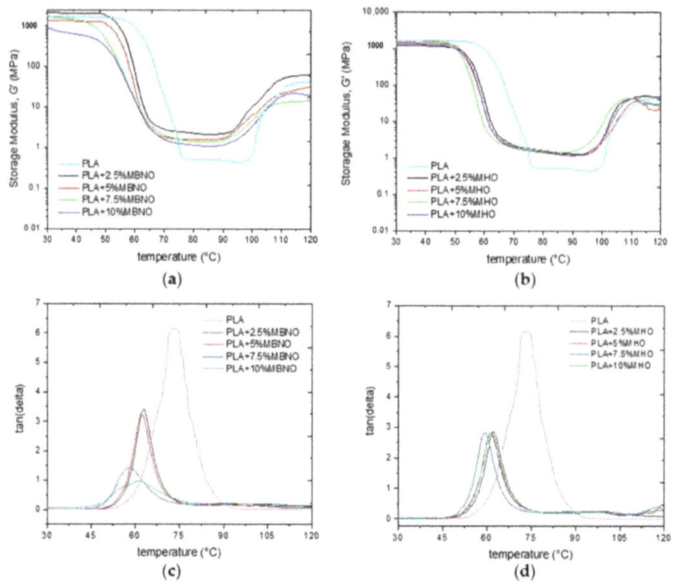

Figure 10. Storage modulus (**a**,**b**) and damping factor (**c**,**d**) of unplasticized PLA and PLA plasticized with different content of MBNO and MHO as a function of temperature.

The damping factor (tan δ), which represents the energy lost due to viscous behavior relative to the energy stored due to elastic behavior, is shown in Figure 10c,d. The peak tan δ, is a way to obtain the T_g value of the materials and, as can be seen, while that of neat PLA is around 73 °C, in the formulations plasticized with MBNO and MHO, this temperature decreases to 58 °C and 60 °C for samples plasticized with 7.5 phr MBNO and MHO, respectively. Santos et al. [49] observed a similar decrease in T_g with the addition to 10 wt.% and 20 wt.% of oligoesters obtained from sunflower oil biodiesel in PLA. The authors observed that the T_g of PLA decreased from 62 °C to 52 °C and 44 °C, respectively.

3.5. Disintegration under Composting Condition of PLA Formulations

The disintegration process of PLA formulations with MBNO and MHO in compost soil is shown visually in Figures 11 and 12, respectively. Initially, the samples were translucent in all formulations; however, after 3 days of incubation, a change in visual appearance to opaque was observed. This may be due to several factors. One cause can be attributed to the 50% relative humidity test condition, since possible hydrolytic degradation due to water absorption affects the refractive index [50]. On the other hand, taking into account the fact that the test was carried out under thermophilic conditions at 58 °C, the opacity may also be due to crystallization, since this temperature is close to the T_g obtained in the thermal study by DSC. As can be seen in Figure 13, where the weight loss with respect to the initial mass at different periods of incubation for two PLA formulations with MBNO (a) and MHO (b) after 7 days was buried, the samples began to lose mass, which led to increased embrittlement, which is observed in the images with the appearance of cracks. However, it was not until day 14 that significant weight loss and inconsistency of the samples was observed. In the case of neat PLA, a faster degradation than in the plasticized PLA was observed, since on day 17 it had already exceeded 90% mass loss, the degree of disintegration determined by the ISO 20200 standard for considering a material to be disintegrable. In the case of plasticized PLA, in the formulations with 2.5 phr and 5 phr of both MBNO and MHO, 90% mass loss was reached at day 27, while with the 7.5 phr and 10 phr formulations, it was reached at day 24. Although the difference in the disintegration time of PLA formulations with plasticizer is small, and in all cases, the time is longer than with neat PLA, and a slight increase in the disintegration rate is observed when more plasticizer is added. This delay in disintegration with respect to neat PLA when introducing plasticizer is due to the fact that the PLA grade used in this work is very amorphous, as can be seen in the thermal analysis; when introducing plasticizer, crystallinity increases, making it difficult for microorganisms to act in the degradation, which act faster in amorphous domains [51,52]. A similar trend was reported by Balart et al. [53], who observed an increase in disintegration time upon the incorporation of ELO into the PLA matrix. Therefore, in view of the results, as a conclusion of the disintegration study, it has been demonstrated that the maleinized oils slightly retard the disintegration of PLA samples. However, PLA compounds plasticized with MBNO and MHO can be considered equally biodegradable by composting.

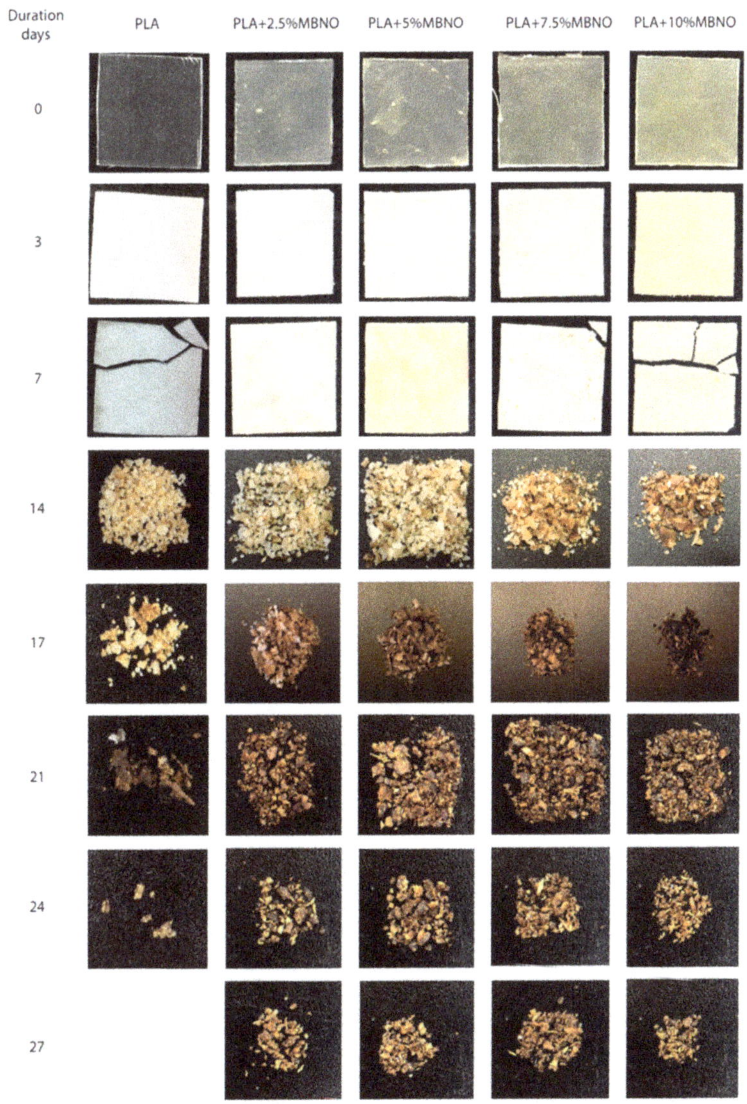

Figure 11. Visual appearance of disintegration under composting conditions of PLA and PLA plasticized with different contents of MBNO.

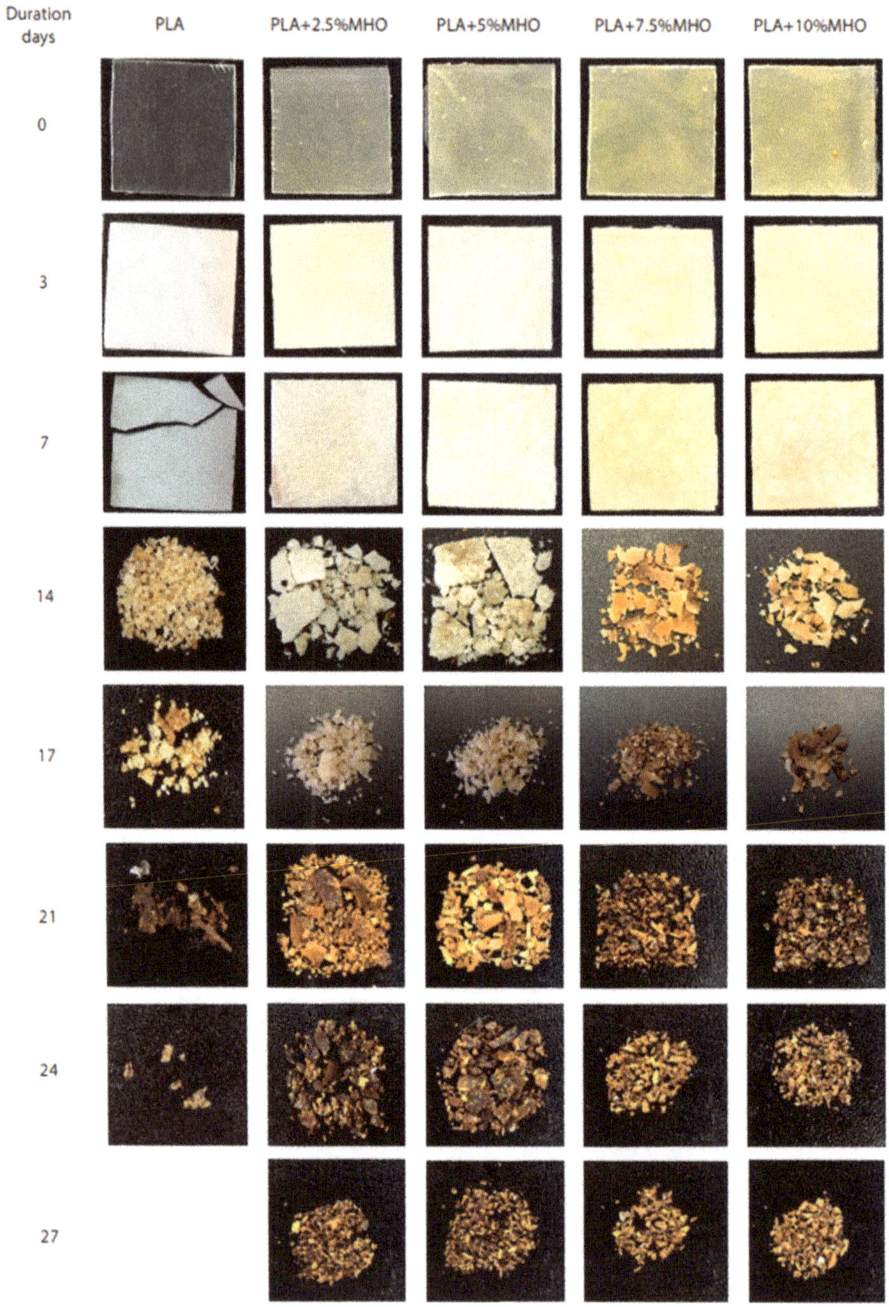

Figure 12. Visual appearance of disintegration under composting conditions of PLA and PLA plasticized with different contents of MHO.

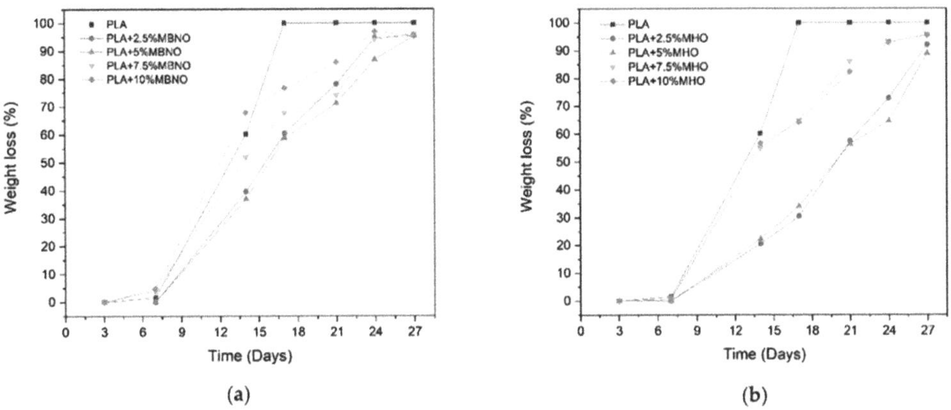

Figure 13. Weight loss recorded during disintegration test of PLA formulations with MBNO (**a**) and MHO (**b**).

4. Conclusions

This research work has developed for the first time a maleinized Brazil nut oil (MBNO). This, as well as a maleinized hemp oil (MHO), were introduced into the PLA matrix to study and analyze their effect as bio-based plasticizers. In addition, the results obtained show MBNO and MHO provided a similar performance compared with a commercially available maleinized linseed oil (MLO), demonstrating its potential. For example, the elongation at break of neat PLA is 7.4%, which is quite low given its brittleness. With the addition of MBNO and MHO, an improvement of 643% and 771%, respectively, was achieved. On the other hand, with values higher than 7.5 phr MBNO, a decrease in the elongation at break was observed due to the anti-plasticizing effect produced by saturation in the mixture. In terms of absorbed impact energy, a 20% higher value was obtained with the addition of 7.5 phr MBNO and 46% higher with 10 phr MHO compared to unplasticized PLA. In addition, mechanical properties such as tensile strength do not decrease, as both maleinized oils provide an improvement in the mobility of PLA chains and an increase in free volume, increasing the degree of crystallinity and thus counteracting the plasticizing effect that would decrease these properties. Thermal parameters, such as T_g, show a decreasing influence with the presence of these bio-plasticizers developed, but not drastically. Finally, the disintegration test under composting conditions showed that the addition of MBNO and MHO, although slightly delaying the process, did not lead to a loss of the biodegradability of PLA. Therefore, both MBNO and MHO are shown as potential bio-plasticizers of organic origin to increase the ductility of PLA without affecting its mechanical properties and not affecting its biodegradability, which makes them two bio-plasticizers of interest for industrial applications.

Author Contributions: Conceptualization, A.P.-N. and V.F.; methodology, A.P.-N., A.L.-C. and I.D.-C.; validation, D.G.-G., J.M.F. and V.F.; formal analysis, A.P.-N., A.L.-C. and I.D.-C.; investigation, A.P.-N. and A.L.-C.; resources, D.G.-G. and V.F.; data curation, D.G.-G.; writing—original draft preparation, A.P.-N., J.M.F. and D.G.-G.; writing—review and editing, V.F.; visualization, D.G.-G.; supervision, J.M.F. and V.F.; project administration, V.F.; funding acquisition, J.M.F., D.G.-G. and V.F. All authors have read and agreed to the published version of the manuscript.

Funding: I.D.-C. wants to thank Universitat Politècnica de València for his FPI grant (PAID-2019-SP20190013) and Generalitat Valenciana (GVA) for his FPI grant (ACIF/2020/233). J.M.F. thanks the postdoc contract (APOSTD/2019/122) Generalitat Valenciana (2019–2021).

Institutional Review Board Statement: Not applicable.

Informed Consent Statement: Not applicable.

Acknowledgments: The authors want to thank the Cátedra FACSA-FOVASA of Water, Waste and Circular Economy for their support, which promotes and supports training, dissemination, innovation, social responsibility and entrepreneurship activities in the field of circular economy and has kindly provided the necessary chemical reagents.

Conflicts of Interest: The authors declare no conflict of interest.

References

1. *Report of the World Commission on Environment and Development: Our common future*; Development and International Co-operation: Environment, Oxford University Press: London, UK, 1987; Volume 10, p. 300.
2. *An Analysis of European Plastics Production, Demand and Waste Data*; Plastics Europe: Brussels, Belgium, 2020; p. 64.
3. Rai, P.; Mehrotra, S.; Priya, S.; Gnansounou, E.; Sharma, S. Recent advances in the sustainable design and applications of biodegradable polymers. *Bioresour. Technol.* **2021**, *325*, 124739. [CrossRef]
4. Pasaoglu, M.E.; Koyuncu, I. Substitution of Petroleum-Based Polymeric Materials Used in the Electrospinning Process with Nanocellulose: A Review and Future Outlook. *Chemosphere* **2020**, *269*, 128710. [CrossRef]
5. Carbonell-Verdu, A.; Garcia-Garcia, D.; Dominici, F.; Torre, L.; Sanchez-Nacher, L.; Balart, R. PLA films with improved flexibility properties by using maleinized cottonseed oil. *Eur. Polym. J.* **2017**, *91*, 248–259. [CrossRef]
6. Mittal, V.; Akhtar, T.; Matsko, N. Engineering. Mechanical, thermal, rheological and morphological properties of binary and ternary blends of PLA, TPS and PCL. *Macromol. Mater. Eng.* **2015**, *300*, 423–435. [CrossRef]
7. Chivrac, F.; Pollet, E.; Averous, L. Progress in nano-biocomposites based on polysaccharides and nanoclays. *Mater. Sci. Eng. R Rep.* **2009**, *67*, 1–17. [CrossRef]
8. *Biodegradable Plastics Market Is Expected to Grow USD 8971.32 Million by the End of 2025*; Valuates Reports; PR Newswire: Bangalore, India, 2020.
9. Yu, Y.; Cheng, Y.; Ren, J.; Cao, E.; Fu, X.; Guo, W. Plasticizing effect of poly(ethylene glycol)s with different molecular weights in poly(lactic acid)/starch blends. *J. Appl. Polym. Sci.* **2015**, *132*, 132. [CrossRef]
10. Arrieta, M.P.; Fortunati, E.; Dominici, F.; Rayón, E.; López, J.; Kenny, J.M. PLA-PHB/cellulose based films: Mechanical, barrier and disintegration properties. *Polym. Degrad. Stab.* **2014**, *107*, 139–149. [CrossRef]
11. Fortunati, E.; Armentano, I.; Iannoni, A.; Kenny, J.M. Development and thermal behaviour of ternary PLA matrix composites. *Polym. Degrad. Stab.* **2010**, *95*, 2200–2206. [CrossRef]
12. Ferri, J.M.; Garcia-Garcia, D.; Sánchez-Nacher, L.; Fenollar, O.; Balart, R. The effect of maleinized linseed oil (MLO) on mechanical performance of poly (lactic acid)-thermoplastic starch (PLA-TPS) blends. *Carbohydr. Polym.* **2016**, *147*, 60–68. [CrossRef] [PubMed]
13. Carrasco, F.; Gámez-Pérez, J.; Santana, O.; Maspoch, M. Processing of poly (lactic acid)/organomontmorillonite nanocomposites: Microstructure, thermal stability and kinetics of the thermal decomposition. *Chem. Eng. J.* **2011**, *178*, 451–460. [CrossRef]
14. Nampoothiri, K.M.; Nair, N.R.; John, R. An overview of the recent developments in polylactide (PLA) research. *Bioresour. Technol.* **2010**, *101*, 8493–8501. [CrossRef]
15. Wee, Y.-J.; Kim, J.-N.; Ryu, H.-W. Biotechnological production of lactic acid and its recent applications. *Biotechnology* **2006**, *44*, 163–172.
16. Chieng, B.W.; Ibrahim, N.A.; Yunus, W.M.Z.W.; Hussein, M.Z. Poly (lactic acid)/poly (ethylene glycol) polymer nanocomposites: Effects of graphene nanoplatelets. *Polymers* **2014**, *6*, 93–104. [CrossRef]
17. Müller, P.; Bere, J.; Fekete, E.; Móczó, J.; Nagy, B.; Kállay, M.; Gyarmati, B.; Pukánszky, B. Interactions, structure and properties in PLA/plasticized starch blends. *Polymers* **2016**, *103*, 9–18. [CrossRef]
18. Malwela, T.; Ray, S.S. Enzymatic degradation behavior of nanoclay reinforced biodegradable PLA/PBSA blend composites. *Int. J. Biol. Macromol.* **2015**, *77*, 131–142. [CrossRef] [PubMed]
19. Ghorui, S.; Bandyopadhyay, N.; Ray, D.; Sengupta, S.; Kar, T. Industrial Crops and Products. Use of maleated castor oil as biomodifier in unsaturated polyester resin/fly ash composites. *Ind. Crop. Prod.* **2011**, *34*, 893–899. [CrossRef]
20. Meier, M.A.; Metzger, J.O.; Schubert, U.S.J. Plant oil renewable resources as green alternatives in polymer science. *Chem. Soc. Rev.* **2007**, *36*, 1788–1802. [CrossRef] [PubMed]
21. Maisonneuve, L.; Lebarbé, T.; Grau, E.; Cramail, H.J. Structure—Properties relationship of fatty acid-based thermoplastics as synthetic polymer mimics. *Polym. Chem.* **2013**, *4*, 5472–5517. [CrossRef]
22. Liao, C.; Liu, F.; Alomirah, H.; Loi, V.D.; Mohd, M.A.; Moon, H.-B.; Nakata, H.; Kannan, K.J.E.s. Bisphenol S in urine from the United States and seven Asian countries: Occurrence and human exposures. *Technology* **2012**, *46*, 6860–6866. [CrossRef]
23. Janković, M.R.; Govedarica, O.M.; Sinadinović-Fišer, S.V.J. The epoxidation of linseed oil with in situ formed peracetic acid: A model with included influence of the oil fatty acid composition. *Ind. Crop. Prod.* **2020**, *143*, 111881. [CrossRef]
24. Carbonell-Verdu, A.; Samper, M.D.; Garcia-Garcia, D.; Sanchez-Nacher, L.; Balart, R.J. Plasticization effect of epoxidized cottonseed oil (ECSO) on poly (lactic acid). *Ind. Crop. Prod.* **2017**, *104*, 278–286. [CrossRef]
25. Stanzione, J., III; La Scala, J. Sustainable polymers and polymer science: Dedicated to the life and work of Richard P. Wool. *J. Appl. Polym. Sci.* **2016**, *133*. [CrossRef]
26. Garcia-Garcia, D.; Ferri, J.M.; Montanes, N.; Lopez-Martinez, J.; Balart, R.J. Plasticization effects of epoxidized vegetable oils on mechanical properties of poly (3-hydroxybutyrate). *Polym. Int.* **2016**, *65*, 1157–1164. [CrossRef]

27. Chieng, B.W.; Ibrahim, N.A.; Then, Y.Y.; Loo, Y.Y.J. Epoxidized vegetable oils plasticized poly (lactic acid) biocomposites: Mechanical, thermal and morphology properties. *Molecules* **2014**, *19*, 16024–16038. [CrossRef]
28. Ferri, J.M.; Garcia-Garcia, D.; Montanes, N.; Fenollar, O.; Balart, R.J. The effect of maleinized linseed oil as biobased plasticizer in poly (lactic acid)-based formulations. *Polym. Int.* **2017**, *66*, 882–891. [CrossRef]
29. Quiles-Carrillo, L.; Blanes-Martínez, M.; Montanes, N.; Fenollar, O.; Torres-Giner, S.; Balart, R.J. Reactive toughening of injection-molded polylactide pieces using maleinized hemp seed oil. *Eur. Polym. J.* **2018**, *98*, 402–410. [CrossRef]
30. Chunhieng, T.; Goli, T.; Piombo, G.; Pioch, D.; Brochier, J.; Montet, D. Recent analysis of the composition of Brazil nut Bertholletia excelsa. *Amaz. Bois Trop.* **2004**, *280*.
31. Cornelio-Santiago, H.P.; Bodini, R.B.; de Oliveira, A.L.J. Potential of Oilseeds Native to Amazon and Brazilian Cerrado Biomes: Benefits, Chemical and Functional Properties, and Extraction Methods. *J. Am. Oil Chem. Soc.* **2021**, *98*, 3–20. [CrossRef]
32. Mikulcová, V.; Kašpárková, V.; Humpolíček, P.; Buňková, L.J. Formulation, characterization and properties of hemp seed oil and its emulsions. *Molecules* **2017**, *22*, 700. [CrossRef]
33. Abuzaytoun, R.; Shahidi, F.J. Oxidative stability of flax and hemp oils. *J. Am. Oil Chem. Soc.* **2006**, *83*, 855–861. [CrossRef]
34. Ferri, J.M.; Fenollar, O.; Jorda-Vilaplana, A.; García-Sanoguera, D.; Balart, R.J. Effect of miscibility on mechanical and thermal properties of poly (lactic acid)/polycaprolactone blends. *Polym. Int.* **2016**, *65*, 453–463. [CrossRef]
35. Eren, T.; Küsefoğlu, S.H.; Wool, R.J. Polymerization of maleic anhydride–modified plant oils with polyols. *J. Appl. Polym. Sci.* **2003**, *90*, 197–202. [CrossRef]
36. Quiles-Carrillo, L.; Montanes, N.; Sammon, C.; Balart, R.; Torres-Giner, S.J. Compatibilization of highly sustainable polylactide/almond shell flour composites by reactive extrusion with maleinized linseed oil. *Ind. Crop. Prod.* **2018**, *111*, 878–888. [CrossRef]
37. Ernzen, J.R.; Bondan, F.; Luvison, C.; Henrique Wanke, C.; De Nardi Martins, J.; Fiorio, R.; Bianchi, O.J. Structure and properties relationship of melt reacted polyamide 6/malenized soybean oil. *J. Appl. Polym. Sci.* **2016**, *133*, 133. [CrossRef]
38. Mikus, P.-Y.; Alix, S.; Soulestin, J.; Lacrampe, M.; Krawczak, P.; Coqueret, X.; Dole, P.J. Deformation mechanisms of plasticized starch materials. *Carbohydr. Polym.* **2014**, *114*, 450–457. [CrossRef]
39. Daniels, P.H.J. Brief overview of theories of PVC plasticization and methods used to evaluate PVC-plasticizer interaction. *J. Vinyl Addit. Technol.* **2009**, *15*, 219–223. [CrossRef]
40. Garcia-Garcia, D.; Carbonell-Verdu, A.; Arrieta, M.; López-Martínez, J.; Samper, M.J. Stability. Improvement of PLA film ductility by plasticization with epoxidized karanja oil. *Polym. Degrad. Stab.* **2020**, *179*, 109259. [CrossRef]
41. Cui, L.; Zhang, R.; Wang, Y.; Zhang, C.; Guo, Y.J. Effect of plasticizer poly (ethylene glycol) on the crystallization properties of stereocomplex-type poly (lactide acid). *Wuhan Univ. J. Nat. Sci.* **2017**, *22*, 420–428. [CrossRef]
42. Moraru, C.; Lee, T.C.; Karwe, M.; Kokini, J.J. Plasticizing and Antiplasticizing Effects of Water and Polyols on a Meat-Starch Extruded Matri. *J. Food Sci.* **2002**, *67*, 3396–3401. [CrossRef]
43. Balart, J.; Fombuena, V.; Fenollar, O.; Boronat, T.; Sánchez-Nacher, L.J. Processing and characterization of high environmental efficiency composites based on PLA and hazelnut shell flour (HSF) with biobased plasticizers derived from epoxidized linseed oil (ELO). *Compos. Part B Eng.* **2016**, *86*, 168–177. [CrossRef]
44. Garcia-Campo, M.J.; Quiles-Carrillo, L.; Masia, J.; Reig-Pérez, M.J.; Montanes, N.; Balart, R.J. Environmentally friendly compatibilizers from soybean oil for ternary blends of poly (lactic acid)-PLA, poly (ε-caprolactone)-PCL and poly (3-hydroxybutyrate)-PHB. *Materials* **2017**, *10*, 1339. [CrossRef]
45. Dobircau, L.; Delpouve, N.; Herbinet, R.; Domenek, S.; Le Pluart, L.; Delbreilh, L.; Ducruet, V.; Dargent, E.J. Molecular mobility and physical ageing of plasticized poly (lactide). *Polym. Eng. Sci.* **2015**, *55*, 858–865. [CrossRef]
46. Dominguez-Candela, I.; Ferri, J.M.; Cardona, S.C.; Lora, J.; Fombuena, V.J. Dual Plasticizer/Thermal Stabilizer Effect of Epoxidized Chia Seed Oil (Salvia hispanica L.) to Improve Ductility and Thermal Properties of Poly (Lactic Acid). *Ploymers* **2021**, *13*, 1283. [CrossRef] [PubMed]
47. Xiong, Z.; Li, C.; Ma, S.; Feng, J.; Yang, Y.; Zhang, R.; Zhu, J.J. The properties of poly (lactic acid)/starch blends with a functionalized plant oil: Tung oil anhydride. *Carbohydr. Polym.* **2013**, *95*, 77–84. [CrossRef] [PubMed]
48. Silverajah, V.; Ibrahim, N.A.; Zainuddin, N.; Yunus, W.M.Z.W.; Hassan, H.A.J. Mechanical, thermal and morphological properties of poly (lactic acid)/epoxidized palm olein blend. *Molecules* **2012**, *17*, 11729–11747. [CrossRef]
49. Santos, E.F.; Oliveira, R.V.; Reiznautt, Q.B.; Samios, D.; Nachtigall, S.M.J. Sunflower-oil biodiesel-oligoesters/polylactide blends: Plasticizing effect and ageing. *Polym. Test.* **2014**, *39*, 23–29. [CrossRef]
50. Fukushima, K.; Tabuani, D.; Abbate, C.; Arena, M.; Ferreri, L.J. Effect of sepiolite on the biodegradation of poly (lactic acid) and polycaprolactone. *Polym. Degrad. Stab.* **2010**, *95*, 2049–2056. [CrossRef]
51. Ray, S.S.; Yamada, K.; Ogami, A.; Okamoto, M.; Ueda, K.J. New polylactide/layered silicate nanocomposite: Nanoscale control over multiple properties. *Macromol. Rapid Commun.* **2002**, *23*, 943–947. [CrossRef]
52. Tokiwa, Y.; Calabia, B.P.J. Biodegradability and biodegradation of poly (lactide). *Appl. Microbiol. Biotechnol.* **2006**, *72*, 244–251. [CrossRef]
53. Balart, J.; Montanes, N.; Fombuena, V.; Boronat, T.; Sánchez-Nacher, L.J. Disintegration in compost conditions and water uptake of green composites from poly (lactic acid) and hazelnut shell flour. *J. Polym. Environ.* **2018**, *26*, 701–715. [CrossRef]

polymers MDPI

Article

Effects of Poloxamer Content and Storage Time of Biodegradable Starch-Chitosan Films on Its Thermal, Structural, Mechanical, and Morphological Properties

Abril Fonseca-García [1,2], Carolina Caicedo [3], Enrique Javier Jiménez-Regalado [1], Graciela Morales [1] and Rocio Yaneli Aguirre-Loredo [1,2,*]

1 Centro de Investigación en Química Aplicada (CIQA), Blvd. Enrique Reyna Hermosillo 140, Saltillo, Coahuila 25294, Mexico; abril.fonseca@ciqa.edu.mx (A.F.-G.); enrique.jimenez@ciqa.edu.mx (E.J.J.-R.); graciela.morales@ciqa.edu.mx (G.M.)
2 CONACYT-CIQA, Blvd. Enrique Reyna Hermosillo 140, Saltillo, Coahuila 25294, Mexico
3 Grupo de Investigación en Química y Biotecnología (QUIBIO), Facultad de Ciencias Básicas, Universidad Santiago de Cali, Pampalinda, Santiago de Cali 760035, Colombia; carolina.caicedo03@usc.edu.co
* Correspondence: yaneli.aguirre@ciqa.edu.mx

Citation: Fonseca-García, A.; Caicedo, C.; Jiménez-Regalado, E.J.; Morales, G.; Aguirre-Loredo, R.Y. Effects of Poloxamer Content and Storage Time of Biodegradable Starch-Chitosan Films on Its Thermal, Structural, Mechanical, and Morphological Properties. *Polymers* **2021**, *13*, 2341. https://doi.org/10.3390/polym13142341

Academic Editors: José Miguel Ferri, Vicent Fombuena Borràs and Miguel Fernando Aldás Carrasco

Received: 25 June 2021
Accepted: 10 July 2021
Published: 17 July 2021

Abstract: Biodegradable packaging prepared from starch is an alternative to fossil-based plastic packaging. However, the properties of starch packaging do not comply with the necessary physicochemical properties to preserve food. Hence, in a previous study, we reported the preparation of a composite polymer material based on starch-chitosan-pluronic F127 that was found to be an adequate alternative packaging material. In this study, we modified the physicochemical properties of this material by storing it for 16 months under ambient conditions. The results indicate that the incorporation of pluronic F127 in the blend polymer can help avoid the retrogradation of starch. Moreover, at higher concentrations of pluronic F127, wettability is reduced. Finally, after storage, the materials exhibited surface modification, which is related to a color change and an increase in solubility, as well as a slight increase in stiffness.

Keywords: biodegradable packaging; biopolymer; starch; storage conditions; thermal properties; physicochemical properties

1. Introduction

Synthetic-origin plastics have shaped various areas of our life, bringing with them numerous new and improved applications. However, their advantages, the environmental pollution resulting from their excessive use, slight or prolonged degradation, and inadequate waste disposal are critical. Studies have been focused on the development of polymeric materials of biological and biodegradable origin using biopolymers such as starch, cellulose, chitosan, and proteins that do not contribute to the accumulation of CO_2 at the end of their life cycles but instead promote sustainable development in the face of any emerging ecological crisis [1–4].

Starch is an excellent choice for developing biodegradable packaging materials, owing to its abundance in nature and higher economical viability. As a nontoxic polymer, it can be safely used for food packaging applications; moreover, it does not impart flavor to the packaged food. Therefore, there is no risk of affecting the organoleptic properties of the packed foods [5].

However, biodegradable packaging based on starch and other biopolymers, such as chitosan or gelatin, are susceptible to hydration and thus do not resist high relative humidity conditions, direct contact with liquid water, or foods with high moisture content [6–8]. Therefore, through previous research, mixtures of starch with other biopolymers have been developed with better barrier properties against moisture [9–12]. A mixture of cornstarch, chitosan, and different proportions of a poloxamer, known as pluronic F127, was developed [11]. The study evaluated the effects of the pluronic concentration and its mechanical and thermal properties, permeability to water vapor, and solubility in water of its starch-chitosan films. The presence of the poloxamer considerably improved the moisture resistance of biodegradable packaging materials when made using starch in a concentration range of 1−5%. The packaging materials are less susceptible to water with a higher the concentration of poloxamer [11].

Storage time and conditions also significantly affect the properties of starch-based packaging materials due to the retrogradation exhibited by starch. Starch is composed of amylose and amylopectin. Amylose is a molecule with a mainly linear structure, whereas amylopectin is a highly branched molecule. The amylose-to-amylopectin ratio in corn starch granules is reported to be approximately 25/75 [13]. The proportion, size, and molecular organization of amylose and amylopectin, as well as the concentration of solids, significantly affects the retrogradation rate of materials made using starch [14,15]. The retrogradation of native starch mainly occurs due to the rearrangement of the amylose chains [16]. After storing for a long time, the amorphous molecules of the gelatinized starch tend to recrystallize to again form the ordered structures of double-helix crystallites. Thus, in gelatinized starch, both components, amylose and amylopectin, crystallize to form double helices generated from the external branches of amylopectin or amylose molecules [14,17].

The results obtained in a previous study were promising, suggesting this material, made using starch-chitosan-poloxamer, to be a viable alternative for producing packaging materials [11]. However, determining the behavior of this material during storage period is important.

This study hypothesized that new poloxamer-containing starch-chitosan formulations could maintain their structural integrity for several months, significantly delaying the starch retrogradation process that typically occurs in short periods. Therefore, in this study, the thermal, structural, mechanical, and morphological behaviors of corn starch-chitosan and poloxamer F127 stored for zero and sixteen months at room temperature were evaluated using differential scanning calorimetry (DSC), X-ray, FTIR-attenuated total reflection (ATR), tensile stress, wettability by the contact angle, color, and scanning electron microscopy (SEM) techniques.

2. Materials and Methods

2.1. Materials

The following biopolymers were used: corn starch from KMC (30% amylose; Brande, Denmark); shrimp shell chitosan (practical grade, deacetylation degree $\geq$ 75%; Sigma-Aldrich, Saint Louis, MO, USA); pluronic F127 (70% ethylene oxide; Sigma-Aldrich, Saint Louis, MO, USA). Glacial acetic acid (Productos Químicos Monterrey SA, Monterrey, Mexico) and glycerol (reagent grade; Meyer, Mexico City, Mexico) were also used.

2.2. Film Preparation

The methodology proposed by Fonseca-García, Jiménez-Regalado [11] was followed for producing biodegradable films. The polymers were used in various concentrations: starch 5% (*w/w*), chitosan 1% (*w/w*), and poloxamer (0%, 1%, 3%, and 5% *w/w* of biopolymers). The polymer solutions were mixed in a ratio of 60:20:20, and glycerol was added as a plasticizer at 25% (*w* of glycerol/*w* of total polymers). Fifty milliliters of each polymer dispersion was poured into a plastic mold and dried at 60 °C (Duo-Vac Oven, Lab-Line Instruments, Inc., Melrose Park, IL, USA) for four hours. The dried films were peeled off and stored in airtight storage bags for 16 months in the dark and at ambient conditions with a relative humidity (RH) of 32.73% and temperature of 25 °C.

2.3. Color

Color was determined using a portable spectrophotometer (Mini Scan EZ 4500L, HunterLab, Reston, VA, USA) based on the CIELab scale with the coordinates L^*, a^*, and b^* [18]. The films were placed on a white copy paper used as standard with color coordinate values of $L^* = 91.76$, $a^* = 2.14$, and $b^* = -10.64$.

2.4. Contact Angle

The contact angle was determined using the video goniometer (Rame-Hart instrument co., Succasunna, NJ, USA) to determine the hydrophobicity of the films [10]. A 5-μL drop of deionized water was placed on the opaquest side of a film sample (the side in contact with the mold) with dimensions of 3 cm × 1.5 cm, using a micropipette, and six measurements were recorded. Subsequently, with a program created using LabVIEW 2014 software, the image was captured, and angles formed by the water droplets on the film were calculated using ImageJ software.

2.5. Scanning Electron Microscopy

To analyze the morphology of the packaging films, SEM (JCM-6000, JEOL, Tokyo, Japan) was performed with a voltage of 10 kV. The films were coated with a gold-palladium layer (DESK II coating system, Denton Vacuum Inc., Moorestown, NJ, USA) [11].

2.6. Fourier Transform Infrared Spectroscopy (FTIR)

The vibrational modes of principal functional groups present in the packaging film were analyzed by FTIR spectrum using a Nicolet iS50 FTIR (Thermo Fisher Scientific, Madison, WI, USA) by ATR technique [10].

2.7. X-ray Diffraction (XRD) Analysis

The XRD of biodegradable films was determined using an X-ray diffractometer (D500, Siemens Aktiengesellschaft, Munich, Germany) with a voltage of 35 kV, current of 25 mA, and Cu Kα radiation of 1.5406 Å over a Bragg angular range (2θ) of 5–35° [11].

2.8. Differential Scanning Calorimetry

DSC was used to measure the gelatinization peak (T_P), melting temperatures (T_{m1} and T_{m2}), and the melting enthalpy (ΔH_{m2}). The analyses were performed at a temperature range of 25–300 °C under a nitrogen atmosphere (50 cm^3 min^{-1}) using a thermogravimetric analyzer (DSC 2 STAR System, Mettler-Toledo AG, Schwerzenbach, Switzerland), according to the standard ASTM D3418-15 [19].

2.9. Water Solubility

To determine the susceptibility of biodegradable films in regards to contact with liquid water at room temperature, the methodology proposed by Fonseca-García, Jiménez-Regalado [11] was followed without modifications.

2.10. Mechanical Properties

The tensile strength and percentage of elongation at break were determined using the methodology proposed by Fonseca-García, Jiménez-Regalado [11]. The experimental data regarding the mechanical properties were analyzed using OriginPro 8.5.0 SR1 software (OriginLab Corporation, Northampton, MA, USA) through an analysis of variance and the Tukey test with a significance level of $p < 0.05$.

3. Results and Discussion

3.1. Appearance and Color

All biodegradable films developed in this study appeared to be homogeneous. Figure 1 shows the starch-chitosan composite films with different proportions of the poloxamer pluronic F127 stored for 0 and 16 months; the images were captured on a black background for contrast. All films presented a translucent white color, as seen in Figure 1, and exhibited no significant visible changes during the 16-month storage process. They did not become brittle or more brittle to handle with time in storage. No visibly considerable color change was noted in the films when compared to the initial time (0) with the samples stored for 16 months. However, when evaluating the film color using a colorimeter, it was observed that the samples did change their coloration after 16 months of storage when compared to the initial time, as indicated by the values of the L^*, a^*, and b^* color parameters in Table 1.

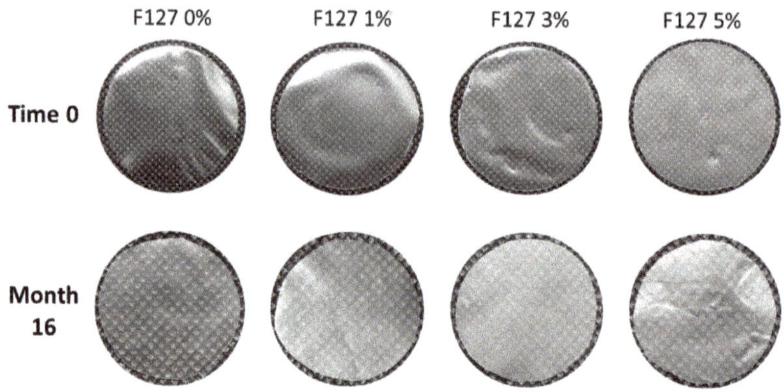

Figure 1. Biodegradable films based on corn starch and chitosan with pluronic F127 at 0%, 1%, 3%, and 5%; images were captured in the initial storage time (time 0) and after 16 months of storage at room temperature; the photos of the films at 16 months were taken at higher magnification.

Table 1. Effect of poloxamer content and storage time on water solubility, color, contact angle, and mechanical properties of corn starch-chitosan biodegradable films.

Content of F127	Water Solubility (%S)	Tensile Strength (MPa)	Elongation at Break (%)	Hunter Color Values [1]			Contact Angle (°)
				L^*	a^*	b^*	ϴ
Time 0 months							
F127 0%	42.58 [§]	4.24 [a,§]	149.05 [c,§]	93.74	2.29	−11.50	99.93
F127 1%	7.81 [§]	6.49 [a,b,§]	53.30 [a,§]	93.52	1.87	−9.31	33.22
F127 3%	4.02 [§]	4.96 [a,§]	95.20 [b,§]	94.22	2.31	−11.56	43.88
F127 5%	3.34 [§]	3.68 [a,§]	84.96 [b,§]	94.53	2.10	−10.54	46.35
Time 16 months							
F127 0%	24.40	5.18 [b]	68.53 [c]	90.88	1.53	−8.05	67.66
F127 1%	17.66	7.01 [c]	60.60 [c]	90.81	1.47	−7.75	31.89
F127 3%	24.17	5.96 [b]	30.67 [b]	91.09	1.41	−7.79	46.09
F127 5%	56.98	3.27 [a]	12.40 [a]	89.53	1.30	−6.02	50.74

Mean values of at least three replicates. [§] Experimental data and significant difference taken from Fonseca-García, Jiménez-Regalado [11].
[1] Hunter color parameters, L^*: color coordinate, lightness (0 = black, 100 = white); a^*: color coordinate, greenness and redness (+ = red, − = green); b^*: color coordinate, blueness to yellowness (+ = yellow, − = blue). L^* standard= 91.76, a^* standard = 2.14, and b^* standard = −10.64. [a,b,c]: values with different letter in a same column denote significant difference.

3.2. Morphology by SEM

Figure 2 shows a comparison of the films with different amounts of poloxamer evaluated at time 0 and 16 months of storage. The evaluation of the initial time (zero) showed that in the biodegradable films formulated from starch and chitosan, the presence of the poloxamer significantly improved the morphology of the surface of these materials. In the film formulated without F127, an irregular surface was observed without cracks. This appearance was gradually reduced until a smoother surface was obtained with an increase in the content of F127.

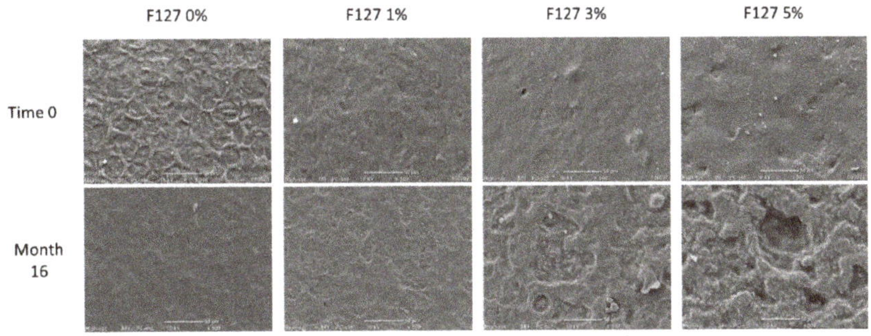

Figure 2. SEM images for the surface (500×) of composite films of corn starch-chitosan with pluronic F127 at ratios of 0%, 1%, 3%, and 5% stored for 0 and 16 months.

A significant change was observed in the morphology of all the composite biodegradable films after 16 months of storage under ambient conditions; however, the most noticeable or drastic changes were observed in the materials containing F127 at concentrations of 3% and 5% (Figure 2). A less homogeneous morphology was observed in these films, which may be due to a change in the organizational structure of the starch chains. Garalde, Thipmanee [20] reported a similar morphology in thermoplastic starch/poly (butylene adipate-co-terephthalate) (PBAT) films, which occurred when the starch was selectively removed from the composite film.

3.3. Contact Angle

The wettability of the biodegradable films was evaluated in terms of the contact angle; Table 1 shows the contact angle values of the starch-chitosan-poloxamer (F127) films. In films F127 0% and F127 1%, after 16 months of storage, the hydrophilicity increased as the contact angle decreased from 99.93° to 67.66° and 33.22° to 31.89°, respectively. In the case of F127 3% and F127 5%, the hydrophilicity decreased as the contact angle increased from 43.88° to 46.09° and 46.25° to 50.74°, respectively. The results suggest that pluronic F127 has a considerable effect on the wettability of biodegradable starch-chitosan films.

3.4. FTIR

Figure 3 shows the spectra of the films stored for different durations. After 16 months of storage, the FTIR-ATR analysis showed modifications in the absorption in the region of 1300 cm^{-1} to 900 cm^{-1} for all materials (F127 0%, F127 1%, F127 3%, and F127 5%); the bands in this region are sensitive to the gelation of starch because of their association with C–O stretching of the ring, linkages (C–O–C), and COH groups. Notably, the band at 1017 cm^{-1} is reported as sensitive to amorphous starch, which is constant in the spectra after 16 months of storage [21–23]. Casu and Reggiani [24] reported that the band at 3300 cm^{-1}, attenuated after 16 months of storage, can be assigned to the O-H stretching of the groups in amorphous amylose. Moreover, the molecule of water showed that the absorption bands at 3300 and 1646 cm^{-1} are associated with OH stretching and deformation vibrations, respectively [24]. In addition, the absorption band at 2883 cm^{-1} was associated with C-H bond stretching, which did not show a modification in the four materials after 16 months [25]. Finally, another important phenomenon identified in the spectra of all films after storage was the disappearing of the absorption band at 1150 cm^{-1}, which was associated with C-O stretching of C–O–C in glycosidic linkage; this result suggests the depolymerization of the starch molecule [26].

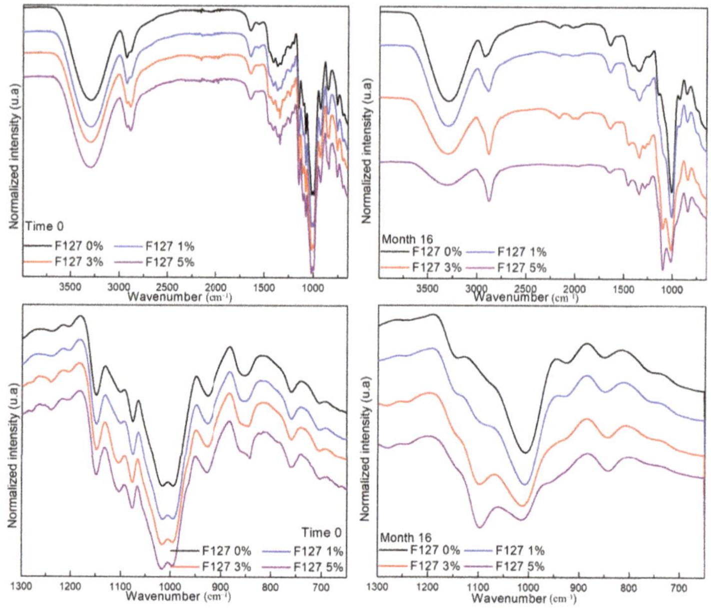

Figure 3. FTIR spectra of biodegradable films of corn starch-chitosan with pluronic F127 at ratios of 0%, 1%, 3%, and 5% stored for 0 and 16 months.

3.5. X-ray Diffraction

To determine if there was retrogradation or modification in the conformation of the structural matrix of the biodegradable films, Figure 4 shows the XRD patterns of biodegradable films F127 0%, F127 1%, F127 3%, and F127 5%. The patterns show that there was a higher atomic ordering in the biodegradable film in F127 0% after 16 months of storage. Peaks at 16°, 19.65°, and 22° were defined after storage; these peaks were related to the crystalline starch, which indicates the retrogradation of starch in this film. However, in films F127 1%, F127 3%, and F127 5%, there was no increase in the atomic ordering after 16 months of storage, as peaks related to starch could not be identified. In F127 3% and F127 5% films, there were two peaks at 19° and 23°, which were less intense after 16 months; these peaks were associated with pluronic 127 [11]. The X-ray patterns indicate that pluronic F127 avoided retrogradation after the storage period of 16 months. Mina Hernandez [27] reported a significant increase in the rearrangement of the polymer chains of mixtures of thermoplastic starch and polycaprolactone during short storage periods (5 and 26 days). Furthermore, the author found that the retrogradation of starch-based polymers occurs more rapidly when the materials are stored in conditions of high RH, which also negatively impacts their mechanical performance.

3.6. Thermal Behavior by DSC

The results of the thermal behavior of the biodegradable films are presented in Table 2. DSC experiments have shown that the melting temperature (T_m) of the neat corn starch film decreases with a significant increase in enthalpy (ΔH_m), indicating a gain in the ordering of the polymeric structures involved. In contrast, chitosan films exhibited an increase of 15 °C after 16 months of storage. The enthalpy values for this biopolymer decreased as a function of the evaluated time. Notably, the mixture of starch and chitosan (F127 0%) exhibited a thermal behavior in which starch transitions predominated. The incorporation of the poloxamer in the starch-chitosan matrix ensured gelatinization, with the absence of the band at ~62 °C (T_P, gelatinization peak). A decrease was observed in the melting temperature (56.1 °C) corresponding to neat F127. The increase in poloxamer content resulted in an increase in the melting temperature and a decrease in enthalpy after 16 months, with respect to its nonfunctionalized homologous (F127 0%). This allowed for demonstrating the plasticizing effect generated by the poloxamer due to the destabilization of the crystalline regions. The fusion of intra- and inter-molecular double helices and the partial recovery of the crystalline structure of amylopectin (Figure S1, Supplementary Material) can be observed [28]. The crystals formed during the starch retrogradation process are less orderly and homogeneous than native starch; this is reflected in their lower melting temperatures. This effect is similar to that observed in other starches, such as sago [29]. The DSC parameters of starch in biodegradable films are summarized in Table 2, and this result is in accordance with that obtained from the XRD analysis.

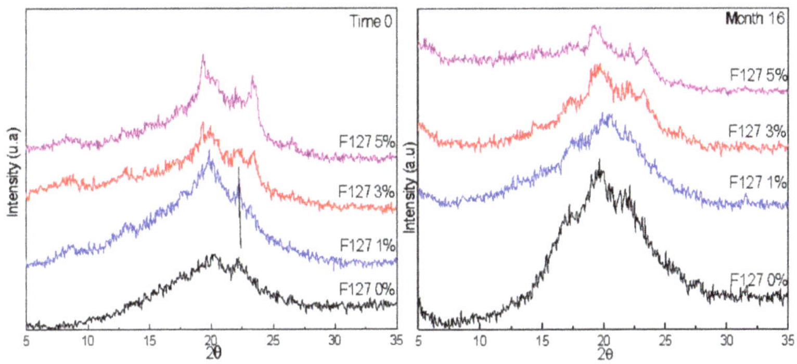

Figure 4. X-ray diffraction patterns of biodegradable films of corn starch-chitosan with pluronic F127 at ratios of 0%, 1%, 3%, and 5% stored for 0 and 16 months.

Table 2. DSC parameters of biodegradable films of corn starch-chitosan with pluronic F127 at ratios of 0%, 1%, 3%, and 5% stored for 0 and 16 months.

Film Sample	T_p (°C)	T_{m1} (°C)		T_{m2} (°C)		ΔH_{m2} (J g^{-1})	
	Month 0	Month 0	Month 16	Month 0	Month 16	Month 0	Month 16
Corn starch	63.13	–	–	141.03	117.75	6.0	18.7
Chitosan	–	–	–	104.68	119.47	28.0	19.6
F127		56.10	56.10				
F127 0%	62.06	–	–	138.96	106.63	6.4	10.3
F127 1%	–	45.52	51.52	137.63	112.94	8.4	9.1
F127 3%	–	51.95	51.68	–	108.05	–	6.3
F127 5%	–	47.84	51.46	114.36	115.22	8.5	4.0

– negligible.

3.7. Water Solubility (WS)

The WS of biodegradable materials is a crucial parameter because it is important that they can decompose in both terrestrial and aquatic environments after being discarded, where they can potentially contaminate the environment on several occasions. A significant increase in the water solubility capacity of the starch-chitosan-based materials was observed (Table 1) at all poloxamer contents (1%, 3%, and 5%) after 16 months of storage with water solubility values of 18%, 24%, and 57%, respectively, compared to the materials evaluated at time zero [11]. Shaker, Elbadawy [30] found that by using different types of poloxamers, such as pluronic F127 and F68, the degree of solubility and the dissolution rate of drugs improved, thus facilitating their usage in pharmaceutical production.

3.8. Mechanical Properties

The mechanical performance of corn starch and chitosan films with various poloxamer concentrations was significantly affected with storage time, as shown in Table 1. A slight increase in the stiffness of the materials was observed, presenting an increase in the tensile stress values. In comparison, the percentage of elongation of the materials decreased when the content of poloxamer was >3%. This mechanical behavior may be associated with two changes in the films. First, SEM micrographs showed that when the poloxamer content was >3% in the films, a higher superficial modification was observed. Second, based on the FTIR spectra of the films, the starch contained in these materials was presumed to depolymerize after storage for 16 months. Furthermore, hydrogen bonds are the predominant ones in these biodegradable films; however, this type of bond is known to be weak, and the tenacity is affected in this case. This mechanical property is diminished by lower values of elongation, without resistance being sacrificed.

4. Conclusions

Storage time can significantly modify the physicochemical, morphological, and structural properties of biodegradable films comprising starch, chitosan, and poloxamer. This information is important to ascertain the state of packaging materials of biodegradable origin when stored at room temperature. Based on physicochemical characterization, there is evidence suggesting that poloxamer avoids the retrogradation of cornstarch. The biodegradable films showed a slight modification in their surface morphology after 16 months, similar to erosion, and this phenomenon can modify the color, WS, mechanical behavior, and wettability of films. Lower melting temperatures and enthalpies in the poloxamer films, as well as storage for 16 months, indicate that the retrogradation of amylopectin only partially recovers the crystalline structure of the native starch. The use in the pharmaceutical, medical, and packaging industry of a biodegradable material such as the one developed in this study may be possible, owing to the favorable properties observed in its mechanical performance and acceptable solubility in liquid water, which can be sufficiently solubilized in a short period of time with the presence of the poloxamer.

Supplementary Materials: The following are available online at https://www.mdpi.com/article/10.3390/polym13142341/s1, Figure S1: DSC thermograms of biodegradable films of corn starch-chitosan with pluronic F127 at ratios of 0%, 1%, 3%, and 5% and stored for 0 and 16 months.

Author Contributions: Conceptualization, A.F.-G., E.J.J.-R. and R.Y.A.-L.; methodology, A.F.-G. and R.Y.A.-L.; formal analysis, A.F.-G., C.C. and R.Y.A.-L.; investigation, A.F.-G., E.J.J.-R. and R.Y.A.-L.; visualization, A.F.-G., E.J.J.-R. and R.Y.A.-L.; supervision, E.J.J.-R. and R.Y.A.-L.; project administration, E.J.J.-R. and R.Y.A.-L.; writing—original draft preparation, A.F.-G., C.C., E.J.J.-R., G.M. and R.Y.A.-L.; writing—review and editing, A.F.-G., C.C., E.J.J.-R., G.M. and R.Y.A.-L. All authors have read and agreed to the published version of the manuscript.

Funding: This research was funded by Centro de Investigación en Química Aplicada (CIQA) under internal projects 6521 and 6610, and by Dirección General de Investigaciones (DGI) of Universidad Santiago de Cali under call no. 01-2021.

Institutional Review Board Statement: Not applicable.

Informed Consent Statement: Not applicable.

Data Availability Statement: The data presented in this study are available on request from the corresponding author.

Acknowledgments: Thank you to M.C. Maria Guadalupe Mendez-Padilla and M.C. Myrna Salinas Hernández, M.C. Blanca Huerta Martinez, and LNMG-CIQA for DSC, DRX, and SEM characterizations, respectively. Fonseca-García and Aguirre-Loredo thank CONACYT for their nominations as a researchers assigned to CIQA.

Conflicts of Interest: The authors declare no conflict of interest.

References

1. Gómez-Aldapa, C.A.; Velazquez, G.; Gutierrez, M.C.; Castro-Rosas, J.; Jiménez-Regalado, E.J.; Aguirre-Loredo, R.Y. Characterization of Functional Properties of Biodegradable Films Based on Starches from Different Botanical Sources. *Starch Stärke* **2020**, *72*, 1900282. [CrossRef]
2. Wang, H.; Ding, F.; Ma, L.; Zhang, Y. Edible films from chitosan-gelatin: Physical properties and food packaging application. *Food Biosci.* **2021**, *40*, 100871. [CrossRef]
3. Yadav, M.; Behera, K.; Chang, Y.-H.; Chiu, F.-C. Cellulose Nanocrystal Reinforced Chitosan Based UV Barrier Composite Films for Sustainable Packaging. *Polymers* **2020**, *12*, 202. [CrossRef]
4. Abera, G.; Woldeyes, B.; Demash, H.D.; Miyake, G. The effect of plasticizers on thermoplastic starch films developed from the indigenous Ethiopian tuber crop Anchote (Coccinia abyssinica) starch. *Int. J. Biol. Macromol.* **2020**, *155*, 581–587. [CrossRef]
5. Samsudin, H.; Hani, N.M. Chapter 8—Use of Starch in Food Packaging. In *Starch-Based Materials in Food Packaging*; Villar, M.A., Barbosa, S.E., García, M.A., Castillo, L.A., López, O.V., Eds.; Academic Press: Cambridge, MA, USA, 2017; pp. 229–256.
6. Aguirre-Loredo, R.Y.; Rodriguez-Hernandez, A.I.; Morales-Sánchez, E.; Gomez-Aldapa, C.; Velazquez, G. Effect of equilibrium moisture content on barrier, mechanical and thermal properties of chitosan films. *Food Chem.* **2016**, *196*, 560–566. [CrossRef]
7. Beigmohammadi, F.; Barzoki, Z.M.; Shabaniyan, M. Rye Flour and Cellulose Reinforced Starch Biocomposite: A Green Approach to Improve Water Vapor Permeability and Mechanical Properties. *Starch Stärke* **2020**, *72*, 1900169. [CrossRef]
8. Meng, W.; Shi, J.; Zhang, X.; Lian, H.; Wang, Q.; Peng, Y. Effects of peanut shell and skin extracts on the antioxidant ability, physical and structure properties of starch-chitosan active packaging films. *Int. J. Biol. Macromol.* **2020**, *152*, 137–146. [CrossRef] [PubMed]
9. Aguirre-Loredo, R.Y.; Rodríguez-Hernández, A.I.; Chavarría-Hernández, N. Physical properties of emulsified films based on chitosan and oleic acid. *CyTA J. Food* **2014**, *12*, 305–312. [CrossRef]
10. Caicedo, C.; Loredo, R.Y.A.; García, A.F.; Ossa, O.H.; Arce, A.V.; Pulgarin, H.L.C.; Torres, Y. Ávila Rheological, Thermal, Superficial, and Morphological Properties of Thermoplastic Achira Starch Modified with Lactic Acid and Oleic Acid. *Molecules* **2019**, *24*, 4433. [CrossRef]
11. Fonseca-García, A.; Jiménez-Regalado, E.J.; Aguirre-Loredo, R.Y. Preparation of a novel biodegradable packaging film based on corn starch-chitosan and poloxamers. *Carbohydr. Polym.* **2021**, *251*, 117009. [CrossRef] [PubMed]
12. Gómez-Aldapa, C.A.; Velazquez, G.; Gutiérrez, M.C.; Rangel-Vargas, E.; Castro-Rosas, J.; Aguirre-Loredo, R.Y. Effect of polyvinyl alcohol on the physicochemical properties of biodegradable starch films. *Mater. Chem. Phys.* **2020**, *239*, 122027. [CrossRef]
13. Panaitescu, D.M.; Frone, A.N.; Ghiurea, M.; Chiulan, I. Influence of storage conditions on starch/PVA films containing cellulose nanofibers. *Ind. Crop. Prod.* **2015**, *70*, 170–177. [CrossRef]
14. Roman, L.; Yee, J.; Hayes, A.; Hamaker, B.R.; Bertoft, E.; Martinez, M.M. On the role of the internal chain length distribution of amylopectins during retrogradation: Double helix lateral aggregation and slow digestibility. *Carbohydr. Polym.* **2020**, *246*, 116633. [CrossRef] [PubMed]
15. Yu, S.; Ma, Y.; Sun, D.-W. Impact of amylose content on starch retrogradation and texture of cooked milled rice during storage. *J. Cereal Sci.* **2009**, *50*, 139–144. [CrossRef]
16. Leyva, J.D.H.; Alonso, L.; Rueda-Enciso, J.; Yee-Madeira, H.; Bello-Perez, L.; Alvarez-Ramirez, J. Morphological, physicochemical and functional characteristics of starch from Marantha ruiziana Koern. *LWT* **2017**, *83*, 150–156. [CrossRef]
17. Eliasson, A.C. *10—Gelatinization and Retrogradation of Starch in Foods and Its Implications for Food Quality, in Chemical Deterioration and Physical Instability of Food and Beverages*; Skibsted, L.H., Risbo, J., Andersen, M.L., Eds.; Woodhead Publishing: Cambridge, UK, 2010; pp. 296–323.
18. Gómez-Aldapa, C.A.; Díaz-Cruz, C.A.; Castro-Rosas, J.; Jiménez-Regalado, E.J.; Velazquez, G.; Gutiérrez, M.C.; Aguirre-Loredo, R.Y. Development of Antimicrobial Biodegradable Films Based on Corn Starch with Aqueous Extract of *Hibiscus sabdariffa* L. *Starch Stärke* **2021**, *73*, 2000096. [CrossRef]
19. ASTM. *D3418. Standard Test Method for Transition Temperatures and Enthalpies of Fusion and Crystallization of Polymers by Differential Scanning Calorimetry*; ASTM International: Philadelphia, PA, USA, 2015.
20. Garalde, R.A.; Thipmanee, R.; Jariyasakoolroj, P.; Sane, A. The effects of blend ratio and storage time on thermoplastic starch/poly(butylene adipate-co-terephthalate) films. *Heliyon* **2019**, *5*, e01251. [CrossRef]
21. Rindlava, Å.; Hulleman, S.H.; Gatenholma, P. Formation of starch films with varying crystallinity. *Carbohydr. Polym.* **1997**, *34*, 25–30. [CrossRef]
22. Rubens, P.; Snauwaert, J.; Heremans, K.; Stute, R. In situ observation of pressure-induced gelation of starches studied with FTIR in the diamond anvil cell. *Carbohydr. Polym.* **1999**, *39*, 231–235. [CrossRef]
23. Van Soest, J.J.G.; De Wit, D.; Tournois, H.; Vliegenthart, J.F.G. Retrogradation of Potato Starch as Studied by Fourier Transform Infrared Spectroscopy. *Starch Stärke* **1994**, *46*, 453–457. [CrossRef]
24. Casu, B.; Reggiani, M. Infrared spectra of amylose and its oligomers. In *Journal of Polymer Science Part C: Polymer Symposia*; Wiley Subscription Services, Inc., A Wiley Company: New York, NY, USA, 1964.
25. Kahvand, F.; Fasihi, M. Plasticizing and anti-plasticizing effects of polyvinyl alcohol in blend with thermoplastic starch. *Int. J. Biol. Macromol.* **2019**, *140*, 775–781. [CrossRef] [PubMed]
26. Pawlak, A.; Mucha, M. Thermogravimetric and FTIR studies of chitosan blends. *Thermochim. Acta* **2003**, *396*, 153–166. [CrossRef]

27. Hernandez, J.H.M. Effect of the Incorporation of Polycaprolactone (PCL) on the Retrogradation of Binary Blends with Cassava Thermoplastic Starch (TPS). *Polymers* **2020**, *13*, 38. [CrossRef]
28. Vamadevan, V.; Bertoft, E. Impact of different structural types of amylopectin on retrogradation. *Food Hydrocoll.* **2018**, *80*, 88–96. [CrossRef]
29. Li, C.; Hamaker, B.R. Effects of different storage temperatures on the intra- and intermolecular retrogradation and digestibility of sago starch. *Int. J. Biol. Macromol.* **2021**, *182*, 65–71. [CrossRef] [PubMed]
30. Shaker, M.A.; Elbadawy, H.M.; Shaker, M.A. Improved solubility, dissolution, and oral bioavailability for atorvastatin-Pluronic® solid dispersions. *Int. J. Pharm.* **2020**, *574*, 118891. [CrossRef]

polymers

MDPI

Article

Prediction of the Miscibility of PBAT/PLA Blends

Shen Su [1,2]

[1] Department of Circular and Bio-Based Plastics, Fraunhofer UMSICHT, Institute for Environment, Safety and Energy Technology, Osterfelder Str. 3, 46047 Oberhausen, Germany; shen.su@umsicht.fraunhofer.de or shen.su@ruhr-uni-bochum.de; Tel.: +49-208-8598-1422

[2] Department of Mechanical Engineering, Ruhr-University Bochum, Universitaetsstr. 150, 44780 Bochum, Germany

Abstract: Designing polymer structures and polymer blends opens opportunities to improve the performance of plastics. Blending poly(butylene adipate-co-terephthalate) (PBAT) and polylactide (PLA) is a cost-effective approach to achieve a new sustainable material with complementary properties. This study aimed to predict the theoretical miscibility of PBAT/PLA blends at the molecular level. First, the basic properties and the structure of PBAT and PLA are introduced, respectively. Second, using the group contribution methods of van Krevelen and Hoy, the Hansen and Hildebrand solubility parameters of PBAT and PLA were calculated, and the effect of the molar ratio of the monomers in PBAT on the miscibility with PLA was predicted. Third, the dependence of the molecular weight on the blend miscibility was simulated using the solubility parameters and Flory–Huggins theory. Next, the glass transition temperature of miscible PBAT/PLA blends, estimated using the Fox equation, is shown graphically. According to the prediction and simulation, the blends with a number-average molecular weight of 30 kg/mol for each component were thermodynamically miscible at 296 K and 463 K with the possibility of spinodal decomposition at 296 K and 30% volume fraction of PBAT. This study contributes to the strategic synthesis of PBAT and the development of miscible PBAT/PLA blends.

Keywords: Flory–Huggins; free energy of mixing; glass transition temperature; group contribution; molecular weight; miscibility prediction; PBAT; PLA; simulation; solubility parameter

Citation: Su, S. Prediction of the Miscibility of PBAT/PLA Blends. *Polymers* **2021**, *13*, 2339. https://doi.org/10.3390/polym13142339

Academic Editors: José Miguel Ferri, Vicent Fombuena Borràs and Miguel Fernando Aldás Carrasco

Received: 20 June 2021
Accepted: 14 July 2021
Published: 16 July 2021

1. Introduction

Physical blending is an economical approach to achieve complementary material properties without developing a new polymer. Poly(butylene adipate-co-terephthalate) (PBAT) and polylactide (PLA) currently account for 13.5 and 18.8 percent of global bioplastic production capacity, respectively [1]. Blends made of commercially available PBAT and PLA tend to macrophase-separate and exhibit two glass transition temperatures (T_g) [2–4]. Therefore, unmodified PBAT/PLA blends have poor miscibility.

Experimentally, PBAT/PLA blends are often prepared by melt blending. At an elevated temperature for a sufficiently long time, transesterification can take place between the two polyesters, resulting in enhanced miscibility [5]. Two criteria often used to evaluate the blend miscibility [6–8] include (1) phase morphology: whether it is homogeneous down to the molecular level; (2) glass transition temperature: whether a binary blend exhibits a single T_g. The morphological investigation depends on the measuring points of the samples in scanning electron microscopy (SEM). A blend may not be truly miscible at the molecular level, even though it shows a homogeneous phase if examined on sufficiently large length scales [6]. Thermal analysis using differential scanning calorimetry (DSC) can only determine the miscibility of polymer blends with well-separated T_g values [9]. Furthermore, different heating rates can lead to a variation in T_g values to some extent. Therefore, both the blend preparation and the analysis can practically affect the actual blend miscibility.

To exclude the experimental influences, a theoretical study of the intrinsic miscibility is necessary. The concept of group contributions has been applied to calculate the solubility parameters and the blend miscibility [10]. According to the literature research, Dil et al. indicate that PBAT and PLA have the Hildebrand solubility parameters (HiSP) of 22.2 $MPa^{1/2}$ and 21.9 $MPa^{1/2}$, respectively [11]. Ding et al. report that the δ (PBAT) and δ(PLA) are 21.9 $MPa^{1/2}$ and 20.7 $MPa^{1/2}$, respectively [12]. As a result, the difference of solubility parameters δ(PBAT-PLA) is 0.3 or 1.2 $MPa^{1/2}$. Materials with similar HiSP have a high affinity with each other [13]. However, the PBAT/PLA blend miscibility is still difficult to determine directly from the difference in solubility parameters. One reason for this is the irregular structure of PBAT molecules that consist of two randomly arranged monomers (BA and BT). Their molar ratio can differ due to the synthesis and degradation. If PBAT were degraded, the resulting lower molecular PBAT-chains might have different ratios of BA and BT. Another reason is that the intermolecular interaction depends on the molecular weights of both polymers, blend ratio, and temperature [14]. To the author's best knowledge, only Park et al. [15] have studied the miscibility of blends (PETG/PLA) with some similarity in molecular structure to PBAT/PLA blends from the theoretical aspect. Therefore, PBAT/PLA blends need more investigation regarding the solubility parameters and miscibility from the molecular level and the thermodynamic aspect. A good understanding of the intrinsic miscibility is not only of academic interest but also contributes to designing miscible PBAT/PLA blends without adding expensive additives.

This study aimed to explore the dependence of the PBAT structure and the molecular weights of PBAT and PLA with different weight ratios on the blend miscibility. First, group contributions theories were used to estimate the effect of the molar ratio of the two monomers in PBAT on the solubility parameter and the difference of solubility parameter between both polymers. Second, the Flory–Huggins model was used to establish phase diagrams and spinodal curves. Furthermore, the composition-dependent glass transition of miscible PBAT/PLA blends was estimated. The first novelty was to find a correlation between the molar ratios of the two monomers in PABT and the solubility parameter between PBAT and PLA. The second novelty was to simulate the phase diagrams and spinodal curves of PBAT/PLA blends, taking into account parameters including temperature, molecular weights, and the component ratios, as well as the calculated solubility parameters.

2. Materials

Poly(butylene adipate-co-terephthalate) (PBAT) (Figure 1) is a fully biodegradable polyester with two types of dimers: BT and BA. One dimer is the rigid section consisting of 1,4-butanediol and terephthalic acid monomers. The other dimer is the flexible section consisting of 1,4-butanediol and adipic acid monomers [16], resulting in high flexibility and high ductility [4]. The largest manufacturer, BASF (Ludwigshafen, Germany), produces PBAT under the brand name Ecoflex® (e.g., Ecoflex F Blend C1200). This polymer possesses a density of 1.26 g/mol, a number average molecular weight (M_n) of 52.1 kg/mol, and a polydispersity index of 2.0.

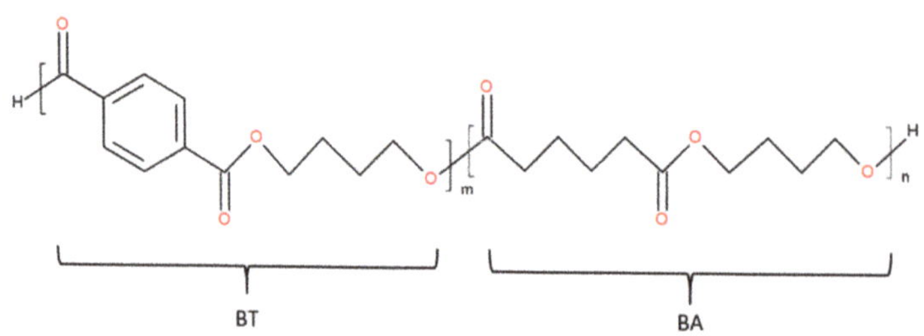

Figure 1. Chemical structure of PBAT.

Poly(lactic acid) or polylactide (PLA) is a biodegradable and renewable aliphatic polyester [17]. Poly(lactic acid) is produced by direct polycondensation of lactic acid. Polylactide can be produced by ring-opening polymerization of cyclic lactide. PLA possesses two optical active and crystallizable isomeric forms: PDLA and PLLA (Figure 2). PLA exhibits many favorable features such as high modulus of elasticity, high strength and high transparency (in the amorphous state), and good processability [18]. The largest manufacturer, NatureWorks (Minnetonka, MN, USA), produces PLA under the trade name Ingeo^{TM} (for example, Ingeo^{TM} Biopolymer 2003D). This polymer has a density of 1.24 g/mol, a number average molecular weight (M_n) of 127.0 kg/mol, a polydispersity index of 1.6, and a D-isomer content of approximately 4.4%.

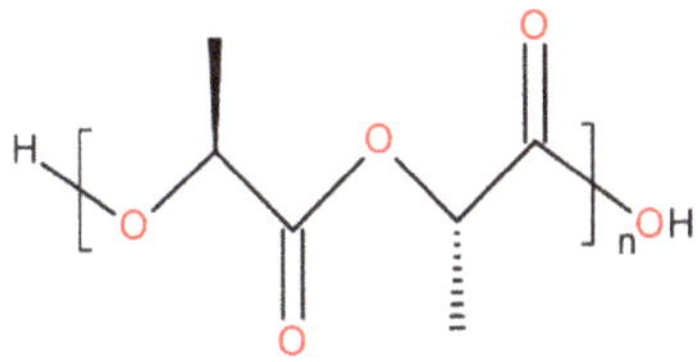

Figure 2. Chemical structure of PLLA.

3. Prediction of Solubility Parameters

The basis of a molecular group contribution method is additivity. This assumes that the physical property of a polymer is calculable by the additive contributions from the individual structural and functional groups in its repeating unit. Using a group contribution method enables us to establish a correlation between the chemical structure of polymers and their interaction. The term "solubility parameter" quantifies the intermolecular interaction.

Coleman et al. has reported a group contribution method to calculate the one-dimensional Hildebrand solubility parameter (HiSP) for polymers with T_g values below room temperature [19]. However, PLA possesses a glass transition temperature higher than room temperature (approx. 60 °C). Therefore, Coleman's method was inappropriate in this miscibility study of PBAT/PLA blends. Group contribution methods developed by van Krevelen [10] and Hoy [20] were used to estimate the three-dimensional HSP and one-dimensional HiSP for polymers. Compared with the real system, the group contribution methods neglect the possible reactions between the blend components, such as transesterification while melt blending, which may change the molecular structures.

In this section, these two different group contribution methods were applied to calculate the solubility parameters of PBAT and PLA. As mentioned in the Introduction, PBAT can consist of different ratios of its monomers (BA and BT). Furthermore, the sequence in the chain may change due to, e.g., polymer degradation. Considering these variations, the estimation of its solubility parameter was performed using three different assumptions:

(1) PBAT (alternating): alternating arrangement with a molar ratio of 1/1.
(2) (PBT): only monomers of BT constitute the polymer.
(3) (PBA): only monomers of BA constitute the polymer.

3.1. Solubility Parameter Calculation According to van Krevelen's Method

The group contributions F_{di}, F_{pi}, and E_{hi} [10] are applied to calculate HSP, including dispersion interactions (δ_d), polar interactions (δ_p), and hydrogen bond interactions (δ_h), as well as the total Hildebrand solubility parameters (δ_t or HiSP). The following equations are used for the calculation:

$$\delta_d = \frac{\sum F_{di}}{V} \quad (1)$$

$$\delta_p = \frac{\sqrt{\sum F_{pi}^2}}{V} \quad (2)$$

$$\delta_h = \sqrt{\frac{\sum E_{hi}}{V}} \tag{3}$$

$$\delta_t = \sqrt{\delta_d^2 + \delta_p^2 + \delta_h^2} \tag{4}$$

where F_{di} represents the group contributions of type i to the dispersion component F_d of the molar attraction constant; F_{pi} represents the group contributions to the polar component F_p; E_{hi} is the hydrogen-bonding energy per structural group i; V is the molar volume. Details of the calculation are described in the Supplementary Material on the sheet "van Krevelen". The calculated HSP and HiSP using van Krevelen's method are listed (Table 1).

Table 1. Calculated solubility parameters using van Krevelen's method.

Polymer Type	δ_d [MPa$^{1/2}$]	δ_p [MPa$^{1/2}$]	δ_h [MPa$^{1/2}$]	HiSP [MPa$^{1/2}$]
alternating PBAT	18.21	5.89	9.17	21.22
PBT	19.63	6.19	9.37	22.62
PBA	14.99	5.00	8.45	17.92
PLA	15.33	8.44	10.98	20.66

Due to the random arrangement of the copolymers BT and BA and possible different molar ratios, the solubility parameters for each segment is different. In the case of a BT-rich PBAT (molar ratio of BT $\geq$ 50%), the HiSP of PBAT was between 21.22 and 22.62, which are the values of alternating PBAT with BT/BA = 1/1 and PBT. In the case of a BA-rich PBAT (molar ratio of BA $\geq$ 50%), the HiSP of PBAT was between 17.92 and 21.22, the values of PBA and alternating PBAT with BA/BT =1/1. The calculated HiSP of PBAT (alternating) and PLA was 21.22 [MPa$^{1/2}$] and 20.66 [MPa$^{1/2}$], respectively, so the difference of total solubility parameter $\Delta\delta$ between them was 0.56 [MPa$^{1/2}$]. Since the HiSP of PLA was in the range of BA-rich PBAT (Figure 3), varying the molar ratio of the monomers in PBAT may lead to a smaller difference in solubility parameters and an increase in the intermolecular interaction between PBAT and PLA.

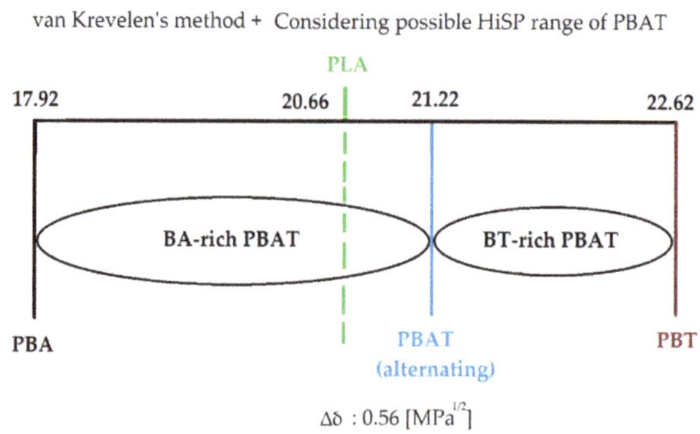

Figure 3. Values and ranges of calculated solubility parameters using van Krevelen's method.

3.2. Solubility Parameter Calculation According to Hoy's Method

The methods of van Krevelen and Hoy, based on the same basic assumption of additivity, have the same order of accuracy. These two are of the same order of accuracy for predicting the solubility parameters [10]. However, van Krevelen and Hoy have applied different ways of calculation. Van Krevelen's method first predicts the partial solubility parameters, which are then used to calculate the total solubility parameter [10,21]. In contrast, Hoy's method first determines the total solubility parameter using the molar attraction functions and auxiliary equations. From the total, the three partial ones are calculated using additive molar functions and expressions for the components [20,22]. The equations used in Hoy's method are below.

$$\alpha(P) = 777\Delta_T(P)/V \tag{5}$$

$$n = 0.5/\Delta_T(P) \tag{6}$$

$$\delta_t = (F_i + \frac{B}{\overline{n}})/V \tag{7}$$

$$\delta_P = \delta_t(\frac{1}{\alpha(P)}\frac{F_P}{F_t + B/\overline{n}})^{1/2} \tag{8}$$

$$\delta_h = \delta_t[(\alpha(P) - 1/\alpha(P))]^{1/2} \tag{9}$$

$$\delta_d = (\delta_t^2 - \delta_P^2 - \delta_h^2)^{1/2} \tag{10}$$

where $\alpha(P)$ is the molecular aggregation number of a polymer; $\Delta_T(P)$ is the Lydersen correction for polymers derived by Hoy; V is the molar volume; B is a base value of 277; F_t is the molar attraction function; F_P is the polar component of molar attraction function; n is the molecular aggregation number. For explanations of variables used before, see Section 3.1 "van Krevelen's method".

The three-dimensional HSP and the overall one-dimensional HiSP were calculated (Table 2). (For details of the calculation, see the Supplementary Material on the sheet "Hoy").

Table 2. Calculated solubility parameters using Hoy's method.

Polymer Type	δ_d [MPa$^{1/2}$]	δ_p [MPa$^{1/2}$]	δ_h [MPa$^{1/2}$]	HiSP [MPa$^{1/2}$]
alternating PBAT	15.94	11.43	9.35	21.73
PBT	15.64	12.82	11.32	23.18
PBA	16.13	10.44	7.98	20.80
PLA	14.02	12.73	9.77	21.31

From the table above, it can be seen that BA-rich PBAT had a HiSP in the range of 20.8–21.73, while BT-rich PBAT possessed a HiSP in the range of 21.73 to 23.18 [MPa$^{1/2}$]. The calculated HiSP for PBAT and PLA was 21.73 [MPa$^{1/2}$] and 21.31 [MPa$^{1/2}$], respectively, indicating a solubility parameter difference of 0.42 [MPa$^{1/2}$]. Furthermore, Hoy's method shows the same tendency as van Krevelen's method: that the HiSP of PLA was in the range of the values of BA-rich PBAT (Figure 4).

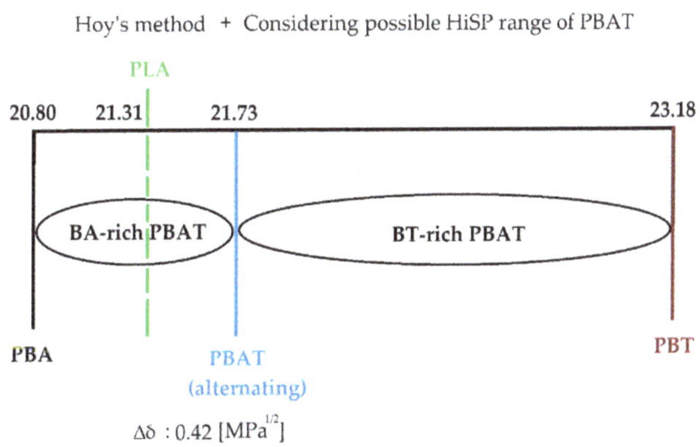

Figure 4. Values and ranges of calculated solubility parameters using Hoy's method.

According to the two group contribution methods, the mean value of the difference in calculated total solubility parameters between alternating PBAT and PLA was 0.49 MPa$^{1/2}$.

The two different methods show consistency in the total solubility parameter and that PLA was in the range of the one of BA-rich PBAT. Furthermore, the same tendency of the solubility parameter can be seen in both methods:

δ(PBA) < δ(PLA) < δ(alternating PBAT) < δ(PBT)

4. Simulation of the Blend Miscibility

Based on the solubility parameters, the miscibility of PBAT/PLA blends can be further studied from the aspect of thermodynamics. The thermodynamic criterion of solubility of two dissimilar components is described by the equation:

$$\Delta G_M = \Delta H_M - T\Delta S_M \tag{11}$$

where ΔG_m is the free energy of mixing; the ΔH_M is the enthalpy of mixing (heat of mixing); T is the absolute temperature; ΔS_M is the entropy of mixing.

A negative value of ΔG_m is generally required to obtain a miscible system. For low molecular weight materials, an increasing temperature generally results in an increase in miscibility as the $T\Delta S_M$ term increases, thus driving G_M to be more negative [8,23]. However, both PBAT and PLA are high molecular weight molecules, implying that the negative contribution from the $T\Delta S_M$ term is small. An equation based on the Flory–Huggins theory correlates the miscibility of a polymer blend with several parameters [24].

$$\frac{\Delta G_M}{RTV} = \frac{\phi_1}{V_1} ln\phi_1 + \frac{\phi_2}{V_2} ln\phi_2 + \frac{\phi_1\phi_2}{RT}(\delta_1 - \delta_2)^2 \tag{12}$$

where ΔG_M is the free enthalpy of mixing; R the gas constant of 8.31 kg·m^2/s^2·mol·K; T the absolute temperature; V the volume of the system; ϕ_1 and ϕ_2 the volume fraction of component, respectively; δ_1 and δ_2 are the solubility parameters; V_1 and V_2 are the molar volumes, respectively. Since the densities of PBAT and PLA are very close, for simplicity, we assumed that both polymers had the same density of 1.25 g/mol. Based on this assumption, for PBAT/PLA blends, the volume fraction is equal to the mass fraction.

The phase diagram (Figure 5) offers a simulation of the PBAT/PLA blend miscibility with assumptions at room temperature (296 K, approximately 23 °C), which is important for the blend preparation by solution blending. PBAT structure was alternating; the HiSP difference was 0.49 MPa$^{1/2}$; the density of both polymers was 1.25 kg/mol; the blend had the molecular weights: M_n52/127, M_n52/60, M_n52/30, and M_n30/30. As an example, M_n52/127 implies M_n(PBAT) is 52 kg/mol and M_n(PLA) is 127 kg/mol, respectively. Details of the calculation are in the Supplementary Material on the sheet "Flory–Huggins".

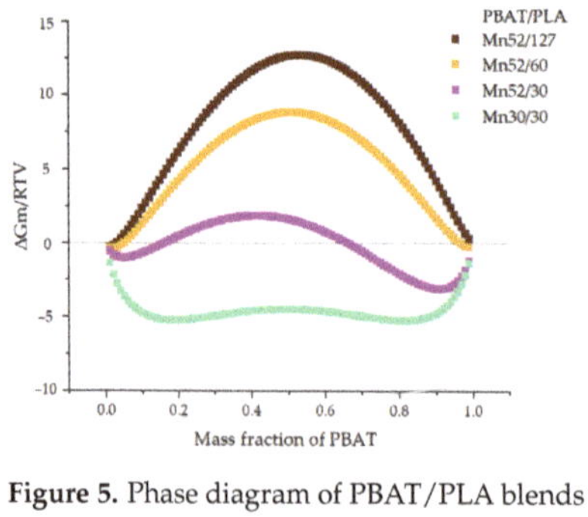

Figure 5. Phase diagram of PBAT/PLA blends with various molecular weights at 296 K.

From the phase diagram, the curve of the PBAT/PLA blends M_n52/127 with the respective molecular weight of 52 and 127 kg/mol is above the zero line, implying blend immiscibility made of the commercially available polymers. A similar trend is shown by the curve M_n52/60, representing the blend made of the original PBAT and PLA with molecular weights of 60 kg/mol. When the molecular weight of PLA was reduced to 30 kg/mol, M_n52/30 showed an interesting curve: in the region of low PBAT content (<15 wt.%), the value of free energy of mixing was slightly negative; this was followed by the curve with a PBAT mass fraction of 15–70 wt.%. Starting with a PBAT content of approximately 70 wt.%, the curve was in the negative region again. The PBAT/PLA blends with the molecular weight of 30 kg/mol (M_n30/30) showed negative values of the free energy of mixing in the whole range. However, it can be seen that in the middle range (40–60 wt.% PBAT), this curve exhibited a slight increase, while the curve shows even lower negative values at the PBAT/PLA ratio of about 20/80 or 80/20. Generally, a lower negative value would lead to a higher probability of miscibility.

Since PBAT/PLA blends are often melt-blended at an elevated temperature, the blend miscibility at 463 K, approx. 190 °C, was simulated (Figure 6).

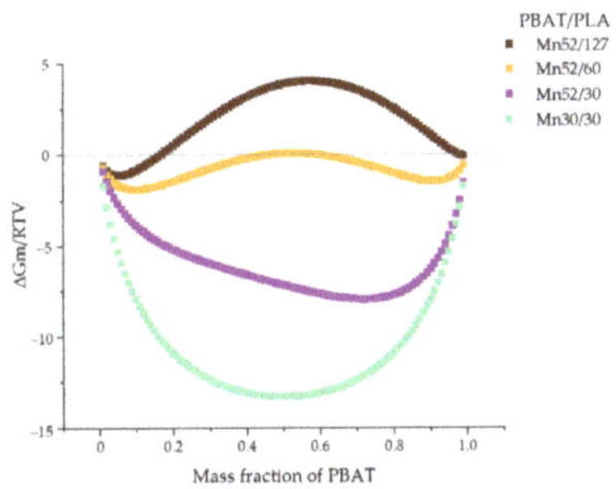

Figure 6. Phase diagram of PBAT/PLA blends with various molecular weights at 463 K.

Even at this higher temperature, M_n52/127 (PBAT/PLA blends consisting of original commercial polymers) showed a curve mostly close to or above the zero line, implying poor miscibility. With the decreased molecular weights of PLA, the curve of M_n52/60 displays a maximum slightly above the zero line (when PBAT content is approx. 50–60 wt.%) and two minima slightly below the zero line (when PBAT content is around 10 wt.% or 90 wt.%), indicating a relatively unstable state of miscibility. A small fluctuation of the temperature or composition may change the theoretical miscibility. When the molecular weights of PLA were 30 kg/mol, M_n52/30 blends showed negative values of the free energy of mixing in the whole range, indicating the blend should be miscible at 463 K. The blends with an M_n of 30 kg/mol for each component demonstrated even lower values, indicating thermodynamically miscibility in the molten state at 463 K.

It is concluded that the mixing behavior depends strongly on the molecular weights of both components, their ratios, and the temperature. As shown in both phase diagrams, the miscibility of PBAT/PLA generally increases with decreasing molecular weights. At 296 K, the blends M_n52/127, M_n52/60, and M_n52/30 each showed a maximum positive value of ΔG_M in the middle range (PBAT content: approx. 50 wt.%) in the curves, indicating poor miscibility of these blends, especially at the blend ratio of 50/50. Generally, in addition to the negative value of ΔG_M, the second derivative of ΔG_M with respect to the volume fraction of the second blend components was a necessary condition of the blend miscibility. At 296 K, the blends M_n30/30 showed negative values of ΔG_M in the whole range of compositions while displaying two minima (at the ratio of about 10/90 and 90/10) and one maximum (at the ratio of 50/50). According to the second derivative of ΔG_M (eq.12 = 0, at 30% volume fraction of PBAT), spinodal decomposition can occur at 296 K due to the negative value of free energy of mixing but beginning of the convex curve of the spinodal (Figure 7).

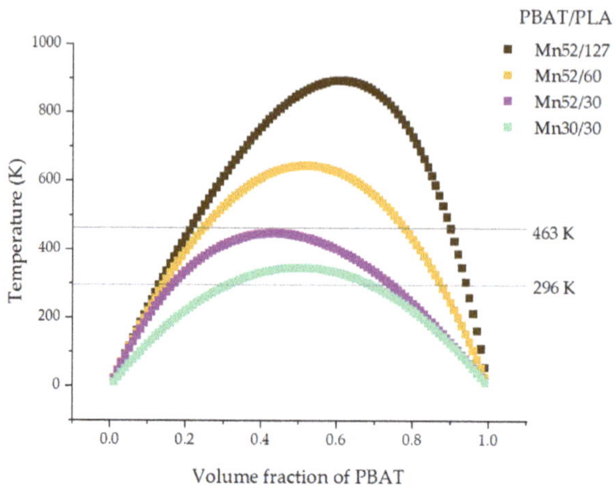

Figure 7. Spinodal curves for PBAT/PLA blends.

At 463 K (e.g., for melt-blending), the curve of the blends $M_n30/30$ shows negative values in the entire range of compositions indicating good miscibility. Even the blends $M_n52/30$ would be miscible in the molten state while melt-blending at 463 K. After that, if the PBAT/PLA blends $M_n52/30$ were stored at a lower temperature between −28 °C and 61 °C (corresponding to the T_g values of both polymers) for a sufficiently long time, the PBAT chains would behave like rubber. For this reason, the blends $M_n52/30$ could change the miscibility from miscible in the molten state to partially miscible or immiscible at a lower temperature after a long enough time. However, the PBAT/PLA with an M_n of 30 kg/mol for each component should be miscible both at 296 K and 463 K, according to the simulation.

5. Estimation of the Glass Transition Temperature

T_g can reflect whether a polymer blend is completely miscible. A fully miscible PBAT/PLA blend exhibits a single T_g. Generally, the T_g value of binary miscible PBAT/PLA blends is predictable based on the Fox equation:

$$\frac{1}{T_g(\text{Blend})} = \frac{m_{\text{PBAT}}}{T_g(\text{PBAT})} + \frac{1 - m_{\text{PBAT}}}{T_g(\text{PLA})} \tag{13}$$

where T_g(Blend) is the predicted glass transition temperature of the miscible PBAT/PLA blend; m_{PBAT} is the mass (weight) fraction of PBAT; T_g(PBAT) and T_g(PLA) are the glass transition temperature of neat PBAT and neat PLA, respectively.

The value of T_g(PBAT) and T_g(PLA) was −28.3 °C and 61.6 °C, respectively [4]. For details of the T_g(Blend) calculation, see the Supplementary Material on the sheet "Fox". The composition-dependent T_g of miscible PBAT/PLA blends is shown graphically (Figure 8).

T_g(Blend) tends to decrease with an increasing mass fraction of PBAT. The glass transition temperature is slightly below 40 °C for PBAT/PLA (20/80). This value is about 10 °C for PBAT/PLA (50/50). Moreover, this value decreases to approximately −15 °C when the ratio of PBAT/PLA is 80/20. To the author's best knowledge, the glass transition temperatures have not been studied for fully miscible PBAT/PLA blends without compatibilizers. Unmodified PBAT/PLA blends with a wide range of ratios (0/100, 10/90, ... 90/10, 100/0) have been reported to have two almost unchanged glass transition temperatures at about −30 °C and 61 °C, which correspond to the T_g of neat PBAT and PLA in the DSC [4].

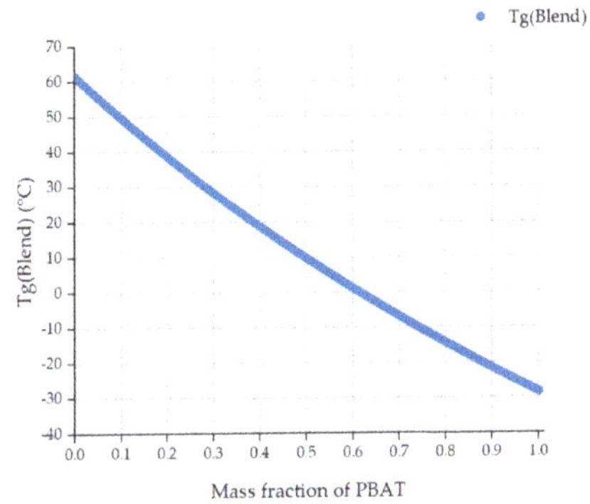

Figure 8. Glass transition temperature of miscible PBAT/PLA blends.

6. Conclusions

In this study, the blend miscibility of PBAT/PLA blends was predicted. The solubility parameters calculated using the group contribution methods of van Krevelen and Hoy had a mean difference of 0.49 $MPa^{1/2}$ between alternating PBAT and PLA. To a certain extent, a higher affinity would be possible between the two polymers when the monomers BA and BT reach a molar ratio exceeding 1 to 1. In this way, the structural optimization of PBAT will fundamentally improve the solubility of PBAT and PLA. Furthermore, a simulation of the miscibility of PBAT/PLA blends was established by using the calculated HiSP and different parameters. According to the simulation, the state of a PBAT/PLA blend can vary from immiscible to miscible, depending strongly on the molecular weights and weight ratio of both polymers at a constant temperature. Generally, the higher the molecular weights, the lower the predicted probability of the blend miscibility. Another tendency is that the higher the temperature, the higher the probability of the blend miscibility. The blends $M_n52/30$ displayed negative values in the whole range of compositions at 463 K. If $M_n52/30$ were melt-blended and then stored at a temperature above the T_g of PBAT (−28 °C, approx. 245 K) for enough long time, the miscibility could change from miscible to partially miscible or immiscible, due to the mobility of PBAT chains. The blends $M_n30/30$ showed negative values of ΔG_M both at 296 K and 463 K, according to the simulation. However, spinodal decomposition of $M_n30/30$ can appear at 296 K (at 30% volume fraction of PBAT) due to the negative value of ΔG_M and the curvature of the spinodal. Moreover, the glass transition temperature of miscible PBAT/PLA blends was calculated using the Fox equation. A single T_g would show at about 40 °C, 10 °C, and −15 °C for PBAT/PLA blends with the composition of (20/80), (50/50), and (80/20), respectively. This study gives the theoretical prediction of the miscibility for PBAT/PLA blends. The next scientific challenge will be the experimental discovery of to what extent the theoretical prediction is consistent with the practical results, especially the molecular weight-dependent miscibility of PBAT and PLA.

Supplementary Materials: The following are available online at https://www.mdpi.com/article/10.3390/polym13142339/s1, Excel table with the sheets of "vanKrevelen", "Hoy", "Flory–Huggins" and "Fox".

Funding: This research received no external funding.

Institutional Review Board Statement: Not applicable.

Informed Consent Statement: Not applicable.

Data Availability Statement: For details of the data used in this paper, see the supplementary material.

Acknowledgments: The author gives sincere thanks to Rodion Kopitzky (Fraunhofer UMSICHT) for discussions.

Conflicts of Interest: The authors declare no conflict of interest.

References

1. European Bioplastics e.V. Bioplastics Market Data. Available online: https://www.european-bioplastics.org/market/ (accessed on 24 May 2021).
2. Liu, T.-Y.; Lin, W.-C.; Yang, M.-C.; Chen, S.-Y. Miscibility, thermal characterization and crystallization of poly(l-lactide) and poly(tetramethylene adipate-co-terephthalate) blend membranes. *Polymer* **2005**, *46*, 12586–12594. [CrossRef]
3. Farsetti, S.; Cioni, B.; Lazzeri, A. Physico-Mechanical Properties of Biodegradable Rubber Toughened Polymers. *Macromol. Symp.* **2011**, *301*, 82–89. [CrossRef]
4. Su, S.; Duhme, M.; Kopitzky, R. Uncompatibilized PBAT/PLA Blends: Manufacturability, Miscibility and Properties. *Materials* **2020**, *13*, 4897. [CrossRef] [PubMed]
5. Kim, Y.J.; Park, O.O. Miscibility and Biodegradability of Poly(Butylene Succinate)/Poly(Butylene Terephthalate) Blends. *J. Polym. Environ.* **1999**, *7*, 53–66. [CrossRef]
6. Arrighi, V.; Cowie, J.M.; Fuhrmann, S.; Youssef, A. Miscibility Criterion in Polymer Blends and its Determination. In *Encyclopedia of polymer blends. Volume 1, Fundamentals*; Isayev, A.I., Ed.; Wiley-VCH: Weinheim, Germany, 2010; pp. 153–198, ISBN 9783527805204.
7. Imre, B.; Renner, K.; Pukanszky, B. Interactions, structure and properties in poly(lactic acid)/thermoplastic polymer blends. *Express Polym. Lett.* **2014**, *8*, 2–14. [CrossRef]
8. Hamad, K.; Kaseem, M.; Ayyoob, M.; Joo, J.; Deri, F. Polylactic acid blends: The future of green, light and tough. *Prog. Polym. Sci.* **2018**, *85*, 83–127. [CrossRef]
9. Fekete, E.; Földes, E.; Pukánszky, B. Effect of molecular interactions on the miscibility and structure of polymer blends. *Eur. Polym. J.* **2005**, *41*, 727–736. [CrossRef]
10. van Krevelen, D.W.; Nijenhuis, K.t. *Properties of Polymers: Their Correlation with Chemical Structure; Their Numerical Estimation and Prediction from Additive Group Contributions/D.W. van Krevelen*, 4th, completely rev. ed.; Revised by K. Te Nijenhuis; Elsevier B.V.: Amsterdam, The Netherlands; Boston, MA, USA, 2009; ISBN 9780080915104.
11. Dil, E.J.; Carreau, P.J.; Favis, B.D. Morphology, miscibility and continuity development in poly(lactic acid)/poly(butylene adipate-co-terephthalate) blends. *Polymer* **2015**, *68*, 202–212. [CrossRef]
12. Ding, Y.; Feng, W.; Huang, D.; Lu, B.; Wang, P.; Wang, G.; Ji, J. Compatibilization of immiscible PLA-based biodegradable polymer blends using amphiphilic di-block copolymers. *Eur. Polym. J.* **2019**, *118*, 45–52. [CrossRef]
13. Hansen, C.M. *Hansen Solubility Parameters: A User's Handbook*, 2nd ed.; CRC Press: Boca Raton, FL, USA, 2007; ISBN 9781420006834.
14. Nedoma, A.J.; Robertson, M.L.; Wanakule, N.S.; Balsara, N.P. Measurements of the Composition and Molecular Weight Dependence of the Flory−Huggins Interaction Parameter. *Macromolecules* **2008**, *41*, 5773–5779. [CrossRef]
15. Park, J.Y.; Hwang, S.Y.; Yoon, W.J.; Yoo, E.S.; Im, S.S. Compatibility and physical properties of poly(lactic acid)/poly(ethylene terephthalate glycol) blends. *Macromol. Res.* **2012**, *20*, 1300–1306. [CrossRef]
16. Kijchavengkul, T.; Auras, R.; Rubino, M.; Selke, S.; Ngouajio, M.; Fernandez, R.T. Biodegradation and hydrolysis rate of aliphatic aromatic polyester. *Polym. Degrad. Stab.* **2010**, *95*, 2641–2647. [CrossRef]
17. Farah, S.; Anderson, D.G.; Langer, R. Physical and mechanical properties of PLA, and their functions in widespread applications—A comprehensive review. *Adv. Drug Deliv. Rev.* **2016**, *107*, 367–392. [CrossRef] [PubMed]
18. Su, S.; Kopitzky, R.; Tolga, S.; Kabasci, S. Polylactide (PLA) and Its Blends with Poly(butylene succinate) (PBS): A Brief Review. *Polymers* **2019**, *11*, 1193. [CrossRef] [PubMed]
19. Coleman, M.M.; Serman, C.J.; Bhagwagar, D.E.; Painter, P.C. A practical guide to polymer miscibility. *Polymer* **1990**, *31*, 1187–1203. [CrossRef]
20. Hoy, K.L. Solubility Parameter as a Design Parameter for Water Borne Polymers and Coatings. *J. Coat. Fabr.* **1989**, *19*, 53–67. [CrossRef]
21. Stefanis, E.; Panayiotou, C. Prediction of Hansen Solubility Parameters with a New Group-Contribution Method. *Int. J. Thermophys.* **2008**, *29*, 568–585. [CrossRef]
22. Kitak, T.; Dumičić, A.; Planinšek, O.; Šibanc, R.; Srčič, S. Determination of Solubility Parameters of Ibuprofen and Ibuprofen Lysinate. *Molecules* **2015**, *20*, 21549–21568. [CrossRef] [PubMed]
23. Robeson, L.M. *Polymer Blends: A Comprehensive Review*; Carl Hanser Verlag: Munich, Germany; Cincinnati, OH, USA, 2007; ISBN 1569904081.
24. Jost, V.; Kopitzky, R. Blending of Polyhydroxybutyrate-co-valerate with Polylactic Acid for Packaging Applications—Reflections on Miscibility and Effects on the Mechanical and Barrier Properties. *Chem. Biochem. Eng. Q.* **2015**, *29*, 221–246. [CrossRef]

polymers

MDPI

Article

Rheological Characterization and Modeling of Thermally Unstable Poly(3-hydroxybutyrate-co-3-hydroxyvalerate) (PHBV)

Silvia Lajewski *, Annika Mauch, Kalman Geiger and Christian Bonten

Institut für Kunststofftechnik, University of Stuttgart, Pfaffenwaldring 32, 70569 Stuttgart, Germany; a_mauch@t-online.de (A.M.); kalman.geiger@ikt.uni-stuttgart.de (K.G.); christian.bonten@ikt.uni-stuttgart.de (C.B.)
* Correspondence: silvia.lajewski@ikt.uni-stuttgart.de; Tel.: +49-711-685-62831

Abstract: Presently, almost every industry uses conventional plastics. Its production from petroleum and extensive plastic pollution cause environmental problems. More sustainable alternatives to plastics include bioplastics such as poly(3-hydroxybutyrate-co-3-hydroxyvalerate) (PHBV), which is produced by bacteria and is biodegradable even in seawater. High temperature sensitivity as well as massive thermal degradation cause difficulties during the processing of PHBV. The aim of this work is to create a detailed rheological characterization and master curves to gain deeper knowledge about the material and its processing parameters. The rheological characterization was performed with frequency sweeps in the range of 0.1 rad/s to 628 rad/s and time sweeps over 300 s. Creating master curves at the reference temperature of 180 °C with the software IRIS delivers Carreau and Arrhenius parameters. These parameters allow for a calculation of the master curves for all other temperatures by means of the temperature shift factor. Moreover, the rheological measurements reveal a minimum rheological measurement temperature of 178 °C and a surprisingly high activation energy of 241.8 kJ/mol.

Keywords: bioplastics; PHBV; rheological characterization; plastics processing

Citation: Lajewski, S.; Mauch, A.; Geiger, K.; Bonten, C. Rheological Characterization and Modeling of Thermally Unstable Poly(3-hydroxybutyrate-co-3-hydroxyvalerate) (PHBV). *Polymers* **2021**, *13*, 2294. https://doi.org/10.3390/polym13142294

Academic Editors: José Miguel Ferri, Vicent Fombuena Borràs and Miguel Fernando Aldás Carrasco

Received: 28 June 2021
Accepted: 8 July 2021
Published: 13 July 2021

1. Introduction

Presently, plastics are used for every imaginable application. Because of the industry's focus on the benefits of the processing and application of plastics, important aspects of plastics have been neglected. In addition to the problematic use of finite resources such as petroleum, masses of plastic waste and the creation of plastic patches in oceans produce a major environmental threat [1]. In order to solve these problems without losing the positive properties of conventional plastics, the industry is in search of better alternatives. A very up-to-date and detailed overview of current research trends in biobased and biodegradable polymers can be found in [2]. A promising biopolymer that can make a small but important contribution along the way in the fight against the ocean garbage patches is poly(3-hydroxybutyrate-co-3-hydroxyvalerate) (PHBV).

PHBV belongs to the group of polyhydroxyalkanoates (PHA) and is a copolymer of polyhydroxybutyrate (PHB) [3]. The linear molecular structure of PHBV consists of a linear carbonyl chain with randomly distributed methyl and ethyl side-chains [4]. PHBV is a biodegradable, biocompatible, hydrophobic, and non-toxic polymer [5,6]. It is produced, for instance, by the Gram-negative bacterium Ralstonia eutropha as an energy and carbon store. The production takes place under a lack of oxygen, nitrogen, and phosphor as well as a surplus of carbon [7]. To produce the characteristic side-chains of PHBV, propionic acid and valeric acid are added [8]. The brittle and stiff thermoplastic polymer shows viscoelastic properties [9]. In addition to the application of PHBV for films and packaging, there is also research on applications in medical technology. However, due to high production costs as well as its brittle and temperature-sensitive behavior, PHBV is not currently established in industry and PHA only accounted for 1.7% of the global production capacities of bioplastics in 2020 [10].

Depending on the composition, the melting point of PHBV varies between 50 °C and 180 °C [11]. Temperatures above the melting point cause unstable behavior of PHBV. Thus, the influence of high temperatures leads to a strong degradation of the polymer chains. The thermal degradation processes with α-deprotonation, β-elimination, and random chain scission proceed via several intermediates. At the end of thermal degradation, propylene, acetaldehyde, ketene, acetone, and carbon dioxide occur [12,13].

Viscosity curves are suitable tools for the analysis of the rheological properties of PHBV. These measurements are carried out in the molten state and can therefore accurately reflect the processing behavior of the plastics considered in the application. Even though PHBV has so far been used mainly in blends, the rheological characterization of the pure material is also important in order to obtain a statement of the expected material behavior. Only in this way can the pros and cons for the use of this biopolymer be assessed.

The viscosity η is plotted double-logarithmically over the angular frequency ω or the shear rate $\dot{\gamma}$ (cf. Figure 1). A strong linear decrease at high frequencies is called the pseudo-plastic flow area or shear thinning behavior. At low frequencies, there is a constant viscosity called zero shear viscosity η_0. Between the areas of zero shear viscosity and shear thinning, a transition occurs. The measured shear viscosity function of polymers with linear molecular structure (PS, PE-HD, PC, or even PHBV) is precisely approached with the Carreau model. The Carreau model, shown in Equation (1), calculates the viscosity η as a function of the shear rate $\dot{\gamma}$ with three parameters: the zero shear viscosity A_0, the reciprocal transient shear rate A_1, and the gradient of the pseudo-plastic flow area A_2.

$$\eta(\dot{\gamma}) = \frac{A_0}{(1 + A_1\dot{\gamma})^{A_2}} \tag{1}$$

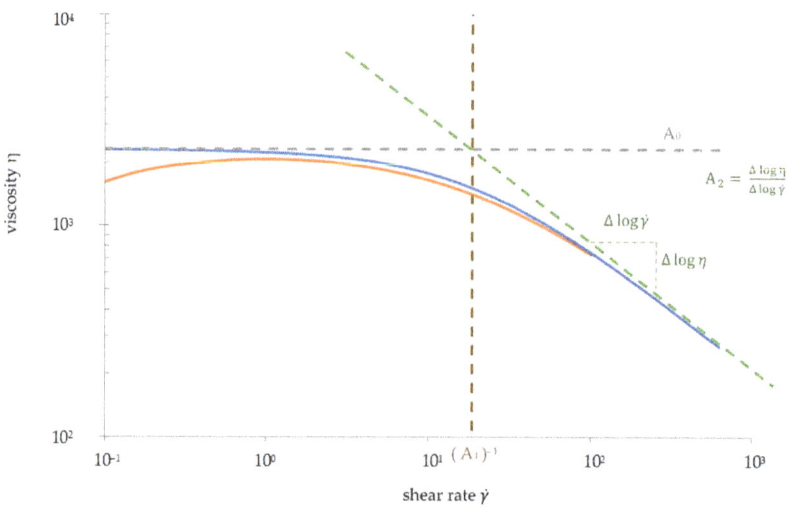

Figure 1. Viscosity curve of thermoplastics. The blue curve is an approximated viscosity curve derived from the Carreau model, and the orange curve is from the rheological measurement of PHBV.

Because of the temperature sensitivity of thermoplastics, the viscosity curve shifts with the change of temperature. At higher temperatures, the viscosity curve shifts by

−45° [14,15]. According to Equation (2), the temperature shift can be calculated in the Carreau model (3) with the temperature shift factor a_T:

$$\eta(\dot{\gamma}) = \frac{a_T \cdot A_0}{(1 + a_T \cdot A_1 \dot{\gamma})^{A_2}} \quad (2)$$

The nondimensional temperature shift factor a_T results from the Arrhenius model as the best approach for the temperature-dependent viscosity function of semi-crystalline polymers like PHBV and can be calculated following Equation (3) [15–17]:

$$a_T = \exp\left(\frac{E_0}{R}\left(\frac{1}{T} - \frac{1}{T_0}\right)\right) \quad (3)$$

To plot viscosity curves, rheological measurements are necessary. Important values of a rheological measurement are the storage modulus G′ and the loss modulus G″. G′ contains the reversible elastic energy, whereas G″ represents the irreversible viscous energy component.

The complex viscosity can be used for an easier analysis of rheological measurements. As shown in Equation (4), the complex viscosity is calculated via the ratio of shear stress τ and the shear rate $\dot{\gamma}$ or the angular frequency ω [18].

$$\eta^* = \frac{\tau(t)}{\dot{\gamma}(t)} = \frac{\tau_0 \sin(\omega t + \delta)}{\gamma_0 \omega \cos(\omega t)} \quad (4)$$

The empirical approach of Cox–Merz [19] in Equation (5) specifies the composition between the viscosity η and the amount of the complex viscosity η*. At low frequencies, the viscosity corresponds to the amount of the complex viscosity if the shear rate takes the same value as the angular frequency [16].

$$\eta(\dot{\gamma}) = |\eta^*(\omega)| \text{ if } \dot{\gamma} = \omega \quad (5)$$

Thus, the graph of the complex viscosity and angular frequency can be linked with the graph of the viscosity and the shear rate.

In a preliminary work, viscosity curves of PHBV were plotted from rheological measurements. A frequency sweep (FS) from 100 rad/s to 0.1 rad/s was performed to obtain the viscosity curves. The viscosity curves did not fulfill the expected and discussed model curves. As shown in Figure 1, the viscosity decreases at frequencies lower than 10 rad/s where a constant zero shear viscosity is expected. This decrease can be explained with thermal degradation.

Due to the extreme thermal degradation already occurring just above the melting temperature, it is difficult to accurately predict viscosity curves at temperatures below or above the measurement temperature. Moreover, the reasonable range of temperature and frequency as well as the zero shear viscosity at a certain temperature are unknown. However, for further rheological measurements, applications, and processing of PHBV, it is necessary to gain this knowledge. Therefore, the aim of this work is to determine important rheological parameters of PHBV. Additionally, we try to better understand the thermal degradation of PHBV by comparing and analyzing measured and calculated viscosity curves.

2. Materials and Methods

2.1. Material

PHBV pellets from TianAn Biopolymer, Ningbo, China, of type ENMAT Y1000P were used for all measurements. According to the manufacturer, this type has a valerate content of only 1–2%. The molar mass was described in the literature as 485,000 g/mol [20]. The pellets were dried overnight in a vacuum drier from Binder GmbH, Tuttlingen, Germany, at 40 °C.

2.2. Methods

Rheological measurements were performed on the computer-controlled plate–plate rheometer Discovery HR-2 from TA Instruments, New Castle, Great Britain. A plate–plate rheometer measures torque with a rotatable upper plate. Using torque, angular frequency, and geometry data, the complex viscosity and angular frequency can be calculated [14,21]. Additional data from the rheological measurement are the storage and loss modulus.

At the beginning of the measurement, the pellets were placed at the lower plate and molded for three minutes. After three minutes, the gap of 1 mm was approached and the excess mass was removed. Three minutes later, the measurement was started as a frequency sweep or a time sweep. In the case of the frequency sweeps, the measurement was made from high to low frequencies. The parameters for the rheological measurement are presented in Table 1.

Table 1. Measurement parameters of the rheological measurements.

	Frequency Sweep	Time Sweep
preparation/molding time in min	6	6
plate diameter in mm	25	25
gap in mm	1	1
temperature in °C	177–190	180–190
angular frequency in rad/s	628.0–1.0 628.0–0.1	0.1 1.0 10.0 100.0
time in s	approx. 102	300
strain in %	5	5

The software IRIS Rheo-Hub 2020 from IRIS Development LLC, Amherst, MA, USA was used to evaluate the measured data. The horizontal temperature shift is calculated with the "t-T-Shift mode" function. For the viscosity shift with the Carreau model, the function "Viscosity Fit -> 2nd Carreau" is used. All shifting parameters and data can be exported.

3. Results and Discussion

3.1. Preliminary Measurement for Error Analysis

In order to analyze the fluctuations at several rheological measurements, a time sweep for 300 s at a temperature of 180 °C and a frequency of 1 rad/s was set five times. Figure 2 shows the semi-logarithmical plot of the complex viscosity over time.

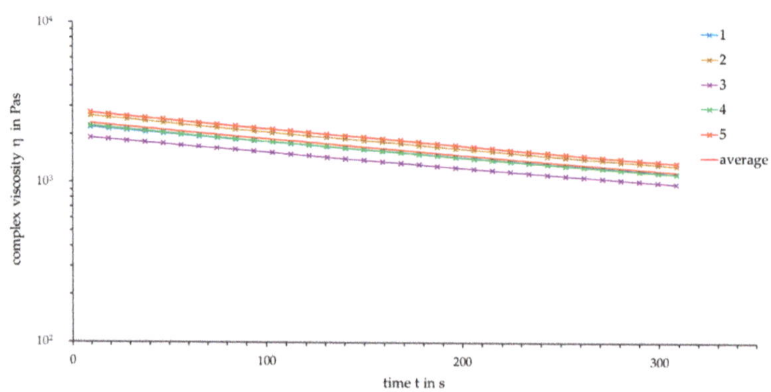

Figure 2. Error analysis, time sweep (ϑ = 180 °C, ω = 1 rad/s).

High deviations between the measurements occurred, even though all measurements were set under the same conditions. At the beginning of the measurements, the standard deviation amounted to 12.55% and decreased to 10.27% after five minutes. The decrease of the standard deviation can be explained by the decreasing viscosity and longer measurement time. Further measurements at 190 °C show a lower standard deviation between 10.39% and 7.94%. Thus, the high standard deviation can be explained not only by the material properties of PHBV but also by the rheological measurement process, which is susceptible to inaccuracies in handling. For the following discussions and analysis, the standard deviation is given as 12.5%.

3.2. Minimum Rheological Measurement Temperature

Thermal degradation has a huge impact on the viscosity curves of PHBV. Therefore, it is helpful to work with low measurement temperatures to minimize degradation effects. In order to investigate the impact of temperature on PHBV and to find the minimum measurement temperature, frequency sweeps were performed at different temperatures. Each frequency sweep at a certain temperature was measured three times, and the average was taken. For a clear illustration, error bars are omitted in Figure 3.

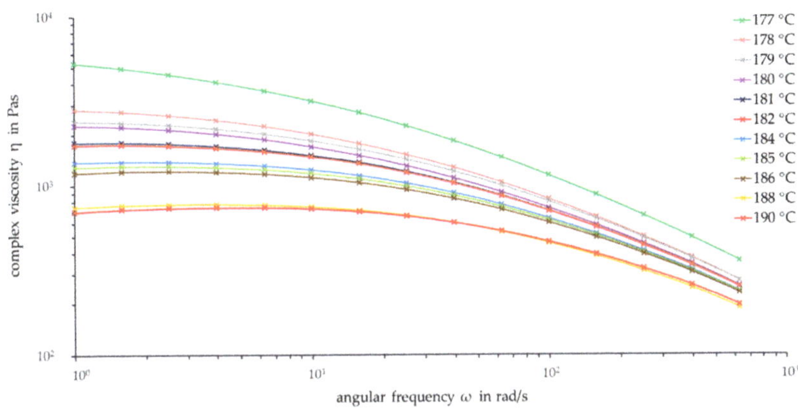

Figure 3. Frequency sweeps at different temperatures (ϑ = 177–190 °C, ω = 1–628 rad/s).

As can be seen, a higher measurement temperature resulted in a lower complex viscosity. Despite the high standard deviation, the viscosity curves are perfectly ordered according to temperature. The viscosity curves for 188 °C and 190 °C are visibly lower than the other curves. With the decrease in the complex viscosity between 10 rad/s and 1 rad/s, the thermal degradation of PHBV for high temperatures is illustrated. In the temperature range between 186 °C and 178 °C, the slopes of the viscosity curves above 100 rad/s are comparable. This can be explained by the assumption that in an ideal case the viscosity curves shift with the temperature by −45°. Consequently, the slope of −1, which is equal to a shift of −45°, shows a temperature invariance in the form of the viscosity curve. In Figure 3, the viscosity curves with different temperatures approach this temperature invariance with their behavior in the pseudo-plastic flow area. At lower frequencies, between 10 rad/s and 1 rad/s, the viscosity curves differentiate further. A zero shear viscosity has not yet been reached yet, especially at lower temperatures.

The viscosity curve at 177 °C is remarkably higher than the curve at 178 °C. Due to the low temperature of 177 °C, the PHBV pellets cannot be completely melted with the given rheological measurement setup. Other measurements with longer melting times up to eight minutes showed the same effect. As is common with thermoplastic polymers, PHBV does not have a precise melting point due to the entangled polymer chains but rather a melting range where the polymer plastifies. Further research with differential scanning calorimetry (DSC) confirms this behavior: PHBV shows a pronounced melting peak with a peak temperature of 177.4 °C. Only complete melting of the sample material allows meaningful rheological measurements. Therefore, the minimum suitable temperature for rheological measurements of the PHBV pellets is 178 °C, which is an important parameter for its later processing.

It is worth mentioning that even a change in the measurement temperature of only 1 °C effects visible changes of the complex viscosity. This behavior indicates a high temperature dependency on the viscosity of PHBV and is a challenge in the processing and application of PHBV.

3.3. Master Curve at the Reference Temperature of 180 °C

With the measurement data from temperatures between 178 °C and 181 °C, it is possible to calculate a master curve for the reference temperature of 180 °C via IRIS. This master curve delivers the Carreau parameters A_0 = 2297.3 Pas, A_1 = 0.054 s, and A_2 = 0.605, as well as the Arrhenius parameter E_0/R = 29,077.3 K, which includes the activation energy E_0 = 241,763.2 J/mol. This value is unusually high. In the literature, data of 25–80 kJ/mol can be found for polymer melts in general [15].

After calculating the temperature shift factors a_T with Equation (3), all master curves for other temperatures can be calculated by applying Equation (2). Figure 4 shows the calculated Carreau model as well as the IRIS viscosity fit and the measurement data for the reference temperature 180 °C.

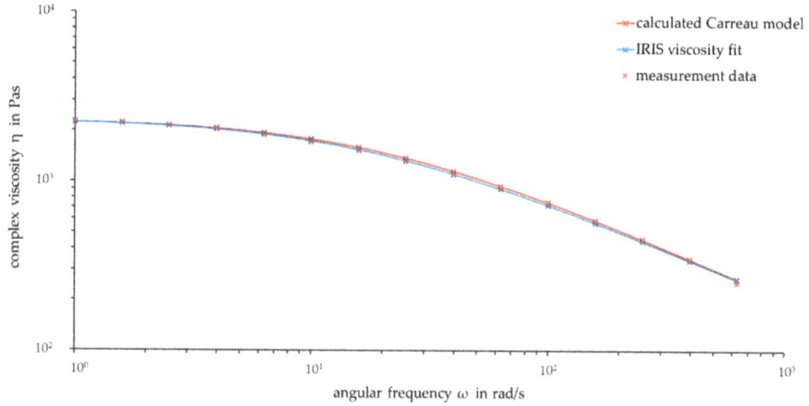

Figure 4. Frequency sweep: Calculated Carreau model, IRIS viscosity fit, and measurement data from reference temperature (ϑ = 180 °C, ω = 1–628 rad/s).

The viscosity curve of the calculated Carreau model is similar to the IRIS viscosity fit. As the temperatures of the two curves correspond to each other, this behavior was expected. Between a frequency of 10 rad/s and 250 rad/s, the viscosity curve of the calculated Carreau model is slightly higher than the curve fitted with IRIS.

It is interesting to compare the calculated values at 180 °C with the values of a time sweep at 180 °C (not shown here). The Carreau model does not consider thermal degradation explicitly, so the calculated Carreau model values are considered ideal. Comparing the start data with the data of the time sweep after the same time as the frequency sweep underlines the temporal and thermal degradation of PHBV. These degradation processes intensify with increasing time and decreasing frequency.

At a frequency of 100 rad/s, the calculated Carreau data are similar to the data of the time sweep after the same time. This could be an indicator that thermal degradation does not considerably affect the Carreau model because the master curve is based on data including thermal degradation. With decreasing frequency and increasing time, the conformity decreases due to increasing external influences. Especially for low frequencies, thermal degradation can be considered as having a minor impact on the calculated Carreau values.

Overall, the comparisons show that degradation processes occur at the reference temperature of 180 °C. The long preparation time of 6 min before a rheological measurement can start has to be considered. The discussed thermal degradation processes start before the start of the actual measurement. Actual processing steps, such as extrusion and injection molding, often need shorter times in the range of only a few minutes. For this reason, there should be deeper investigations into if and how the calculated data of the complex viscosity can be used for industrial processing.

3.4. Calculated Master Curves from the Parameters at the Reference Temperature 180 °C

In the following graphs, error bars for the calculated Carreau model curves are shown for easier analysis. The error bars are based on the above-mentioned standard deviation of 12.5% since the calculated Carreau parameters are based on measurement data including errors.

First, the viscosity curves from measurement temperatures below 180 °C are investigated. Figure 5 shows the double-logarithmically plotted complex viscosity over the frequency at the temperature of 178 °C.

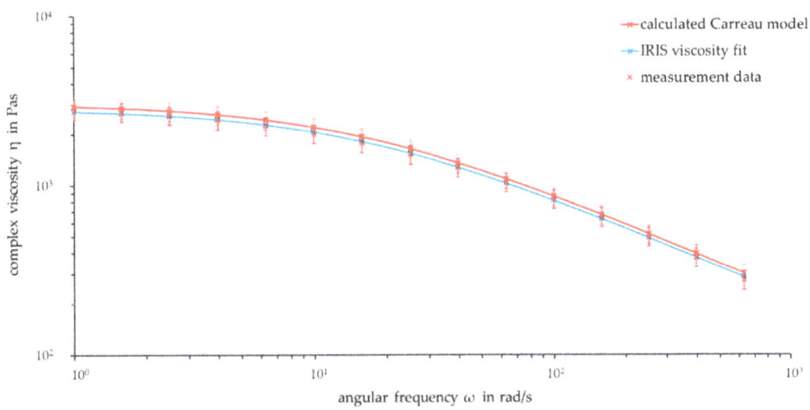

Figure 5. Frequency sweep: Calculated Carreau model, IRIS viscosity fit, and measurement data (ϑ = 178 °C, ω = 1–628 rad/s).

The calculated viscosity curve and the measured and fitted curve show a similar course, with the data of the calculated Carreau model curve being slightly higher than the measurement data. Compared to the data at 180 °C, the viscosity curves are shifted to higher viscosity values. This is due to the fact that the movement of the polymer chains slows down with a decrease in temperature, so the viscosity increases. As a result, the temperature shift factor increases for temperatures below the reference temperature.

If the measurement temperature exceeds the melting point, a stronger degradation of the PHBV chains occurs and results in lower complex viscosities. For temperatures above 182 °C, the complex viscosity curve of the calculated Carreau model is below the measurement data. It is notable that the difference between calculated and measured values increases with higher frequencies. This is interesting, because the deviation increases at lower frequencies for temperatures below 180 °C. The reason for increasing differences at higher frequencies can be found in the transition area of the viscosity curve. Especially at measurement temperatures above 185 °C, the superposition of degradation and viscosity change in the transition area begins after a shorter time and therefore at higher frequencies. Therefore, the transition area of the calculated Carreau model overlaps with a higher transition angular frequency. The Carreau model overestimates the temperature-dependent behavior of the complex viscosity at high temperatures. Higher values for the calculated viscosity were expected, because of the minor influence of degradation on the Carreau model.

The viscosity curves at 190 °C presented in Figure 6 show further details. A temperature of 190 °C is the first temperature where no overlap within the standard deviations of calculated and measured data are found and where the viscosity curves can be clearly separated. A possible reason for this is that at 190 °C, the material reaches a state in which thermal degradation has a strong influence on the PHBV chains. In general, the Carreau model underestimates the viscosity curves at temperatures above the reference temperature. On the one hand, this is explainable, because a higher complex viscosity of the Carreau model without thermal degradation was expected. On the other hand, this could be a sign that thermal degradation of PHBV chains is not as high as assumed.

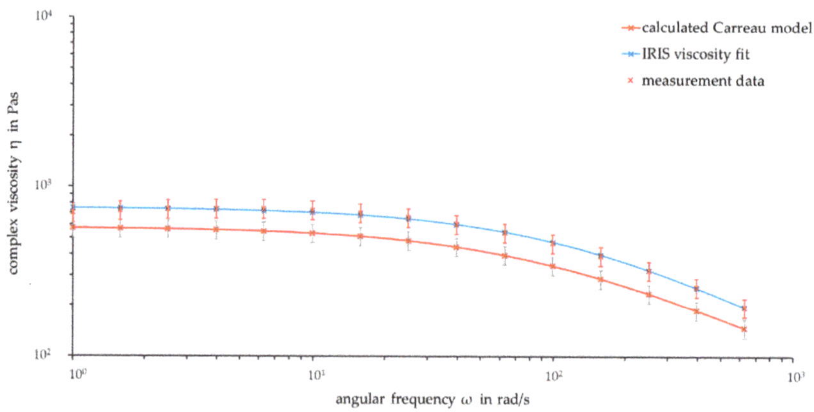

Figure 6. Frequency sweep: Calculated Carreau model, IRIS viscosity fit, and measurement data (ϑ = 190 °C, ω = 1–628 rad/s).

For temperatures below and near the reference temperature, the Carreau model overestimates the viscosity. All in all, the Carreau model calculates and illustrates the viscosity curve satisfyingly. Since the calculated viscosity curves up to a measurement temperature of 182 °C are higher than the measured data, an investigation into whether this is due to the proximity to the reference temperature or to the proximity to the melting temperature should be performed.

One possibility to further analyze the zero shear viscosity and the validity range of the viscosity curves is to extend the range of the curves from 0.1 rad/s to 628 rad/s. The analysis is only useful up to the frequencies where the calculated viscosity curve intersects with the measured viscosity curve (for high temperatures) or where the calculated viscosity curve deviates from the measured viscosity curves (for lower temperatures). Figure 7 shows the frequency sweep from 628 rad/s to 0.1 rad/s at 190 °C.

The measured viscosity curve crosses the calculated viscosity curve at 1 rad/s. For lower frequencies, the measured viscosity curve decreases while the calculated viscosity curve stays constant. Thus, the deviation between both viscosity curves increases with increasing angular frequency. At 0.25 rad/s, the error bars of the viscosity curves with the standard deviation intersect for the last time. As a result, the viscosity curve at 190 °C probably reaches zero shear viscosity between 1 rad/s and 0.25 rad/s.

The measured and calculated viscosity curves at a measurement temperature of 180 °C deviate with decreasing frequency. There is no intersection for the viscosity curves because the measured viscosity curve is lower than the calculated viscosity curve for the whole measurement. From 628 rad/s to 0.4 rad/s, there is an intersection of the error bars. Thus, the validity area at a measurement temperature of 180 °C starts at 0.4 rad/s.

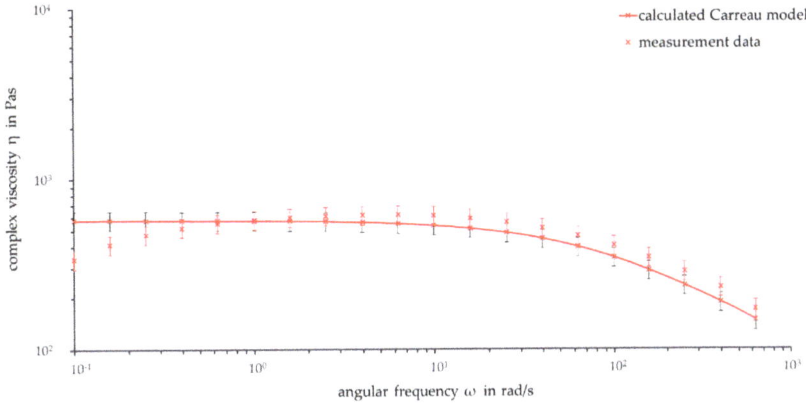

Figure 7. Frequency sweep: Calculated Carreau model and measurement data (ϑ = 190 °C, ω = 0.1–628 rad/s).

4. Conclusions

The performed rheological measurements of PHBV are in the linear viscosity area and have a standard deviation of 12.5%. The minimum rheological measurement temperature of PHBV is 178 °C. Using the IRIS software, a master curve at the reference temperature of 180 °C based on the measurement data from 178 °C to 181 °C was calculated. This master curve delivers Carreau and Arrhenius parameters that can be used to calculate the master curves for other temperatures.

Master curves for measurement temperatures below the reference temperature show higher calculated values than the measurement data. At measurement temperatures above 182 °C, the calculated viscosity data are lower than the measurement data. Thus, the Carreau model underestimates the viscosity of PHBV at higher temperatures. The error bars of the master curves until 185 °C intersect with the error bars from the measurement data, suggesting that the master curves from 178 °C to 185 °C can be used for analysis. Due to the extension of the frequency area to 0.1 rad/s, the zero shear viscosity for 190 °C can be set at 1.0 rad/s to 576.5 Pas. Analogously, the zero shear viscosity for 180 °C can be assumed at 0.4 rad/s as 1732.7 Pas.

The investigated procedure to create master curves with the Carreau model results in viable curves. To minimize the difference between calculated and measured data, other calculation models should be investigated in the future. For example, the calculation of the temperature shift factor with the Vogel-Fulcher and the WLF models should be investigated in further projects. The molar mass also has a great influence on the rheological properties of PHBV. An analogous investigation of different PHBV types would provide further highly interesting insights into the degradation behavior of PHBV.

The rheological characterization underlines the high temperature sensitivity of PHBV. A temperature shift as small as 1 °C shows visible differences in the viscosity curves. This is further confirmed by the unusually high flow activation energy E_0 = 241.8 kJ/mol and leaves only a small temperature window for processes and applications. More detailed investigations will allow better characterization of PHBV and allow for applications of the biodegradable plastic PHBV in the future.

Author Contributions: Project administration, S.L.; methodology, K.G.; data curation and investigation, A.M.; writing, A.M. and S.L.; review and editing, C.B. and K.G.; supervision, C.B. All authors have read and agreed to the published version of the manuscript.

Funding: The project on which this report is based was funded by the German Federal Ministry of Food and Agriculture under the funding code 22039518. The responsibility for the content of this publication lies with the authors.

Institutional Review Board Statement: Not applicable.

Informed Consent Statement: Not applicable.

Data Availability Statement: Not applicable.

Acknowledgments: We would like to thank our colleagues Tobias Schaible for assistance with the evaluation of the rheological measurements and Jochen Wellekötter for the linguistic revision. Furthermore, we thank the Fachagentur für Nachwachsende Rohstoffe (FNR) for funding within the framework of the "Renewable Resources" program.

Conflicts of Interest: The authors declare no conflict of interest.

References

1. Egger, M.; Sulu-Gambari, F.; Lebreton, L. First evidence of plastic fallout from the North Pacific Garbage Patch. *Sci. Rep.* **2020**, *10*, 7495. [CrossRef] [PubMed]
2. Vaishnav, A. *Microbial Polymers: Applications and Ecological Perspectives*; Springer Nature: Basingstoke, UK, 2021.
3. Zhao, H.; Cui, Z.; Wang, X.; Turng, L.-S.; Peng, X. Processing and Characterization of Solid and Microcellular Poly(Lactic Acid)/Polyhydroxybutyrate-Valerate (PLA/PHBV) Blends and PLA/PHBV/Clay Nanocomposites. *Compos. Part B Eng.* **2013**, *51*, 79–91. [CrossRef]
4. Rivera-Briso, A.L.; Serrano-Aroca, Á. Poly(3-Hydroxybutyrate-*co*-3-Hydroxyvalerate): Enhancement Strategies for Advanced Applications. *Polymers* **2018**, *10*, 732. [CrossRef] [PubMed]
5. Laycock, B.; Halley, P.; Pratt, S.; Werker, A.; Lant, P. The Chemomechanical Properties of Microbial Polyhydroxyalkanoates. *Prog. Polym. Sci.* **2013**, *38*, 536–583. [CrossRef]
6. Avella, M.; Martuscelli, E.; Raimo, M. Properties of Blends and Composites Based on Poly(3-Hydroxy)Butyrate (PHB) and Poly(3-Hydroxybutyrate-Hydroxyvalerate) (PHBV) Copolymers. *J. Mater. Sci.* **2000**, *35*, 523–545. [CrossRef]
7. Süss, R.; Leitner, V.; Paulik, C.; Putz, R.; Kamm, B. Quantifizierungsmethoden für die Produktion von PHB mittels Halomonas Halophila. *BIOspektrum* **2019**, *25*, 226–228. [CrossRef]
8. Pötter, M.; Madkour, M.H.; Mayer, F.; Steinbüchel, A. Regulation of Phasin Expression and Polyhydroxyalkanoate (PHA) Granule Formation in Ralstonia Eutropha H16. *Microbiology* **2002**, *148*, 2413–2426. [CrossRef] [PubMed]
9. Sudesh, K.; Abe, H.; Doi, Y. Synthesis, Structure and Properties of Polyhydroxyalkanoates: Biological Polyesters. *Prog. Polym. Sci.* **2000**, *25*, 1503–1555. [CrossRef]
10. European Bioplastics, Nova-Institute. Global Production Capacities of Bioplastics 2020. Available online: https://www.european-bioplastics.org/market/ (accessed on 21 January 2021).
11. Choi, H.J.; Kim, J.; Jhon, M.S. Viscoelastic Characterization of Biodegradable Poly(3-Hydroxybutyrate-Co-3-Hydroxyvalerate)1. *Polymer* **1999**, *40*, 4135–4138. [CrossRef]
12. Liu, Q.-S.; Zhu, M.-F.; Wu, W.-H.; Qin, Z.-Y. Reducing the Formation of Six-Membered Ring Ester during Thermal Degradation of Biodegradable PHBV to Enhance Its Thermal Stability. *Polym. Degrad. Stab.* **2009**, *94*, 18–24. [CrossRef]
13. Xiang, H.; Wen, X.; Miu, X.; Li, Y.; Zhou, Z.; Zhu, M. Thermal Depolymerization Mechanisms of Poly(3-Hydroxybutyrate-Co-3-Hydroxyvalerate). *Prog. Nat. Sci. Mater. Int.* **2016**, *26*, 58–64. [CrossRef]
14. Bonten, C. *Kunststofftechnik, 3. Auflage*; Hanser: München, Germany, 2020.
15. Schröder, T. *Rheologie der Kunststoffe*; Hanser: München, Germany, 2018.
16. Osswald, T.A.; Rudolph, N. *Polymer Rheology*; Hanser Publications: Cincinnati, OH, USA, 2015.
17. Morrison, F.A. *Understanding Rheology*; Topics in chemical engineering; Oxford University Press: New York, NY, USA, 2001.
18. Mours, M.; Winter, H.H. *IRIS Rheo-Hub Handbook*; Iris Development LLC: New Orleans, LA, USA, 2007.
19. Cox, W.P.; Merz, E.H. Correlation of Dynamic and Steady Flow Viscosities. *J. Polym. Sci.* **1958**, *28*, 619–622. [CrossRef]
20. McDonough, K.; Itrich, N.; Casteel, K.; Menzies, J.; Williams, T.; Krivos, K.; Price, J. Assessing the biodegradability of microparticles disposed down the drain. *Chemosphere* **2017**, *175*, 452–458. [CrossRef] [PubMed]
21. Mezger, T.G. *Das Rheologie Handbuch*; Vincentz Network: Hannover, Germany, 2019. [CrossRef]

polymers

Review

Macro and Micro Routes to High Performance Bioplastics: Bioplastic Biodegradability and Mechanical and Barrier Properties

Olivia A. Attallah [1], Marija Mojicevic [1,*], Eduardo Lanzagorta Garcia [1], Muhammad Azeem [1], Yuanyuan Chen [1], Shumayl Asmawi [2] and Margaret Brenan Fournet [1]

1 Materials Research Institute, Athlone Institute of Technology, N37 HD68 Athlone, Ireland; oadly@ait.ie (O.A.A.); e.lgarcia@research.ait.ie (E.L.G.); m.azeem@research.ait.ie (M.A.); yuanyuanchen@ait.ie (Y.C.); mfournet@ait.ie (M.B.F.)
2 Fundamental and Applied Science Department, Universiti Teknologi PETRONAS, Bandar Seri Iskandar 32610, Perak Darul Ridzuan, Malaysia; muhammad_24531@utp.edu.my
* Correspondence: mmojicevic@ait.ie

Abstract: On a score sheet for plastics, bioplastics have a medium score for combined mechanical performance and a high score for biodegradability with respect to counterpart petroleum-based plastics. Analysis quickly confirms that endeavours to increase the mechanical performance score for bioplastics would be far more achievable than delivering adequate biodegradability for the recalcitrant plastics, while preserving their impressive mechanical performances. Key architectural features of both bioplastics and petroleum-based plastics, namely, molecular weight (M_w) and crystallinity, which underpin mechanical performance, typically have an inversely dependent relationship with biodegradability. In the case of bioplastics, both macro and micro strategies with dual positive correlation on mechanical and biodegradability performance, are available to address this dilemma. Regarding the macro approach, processing using selected fillers, plasticisers and compatibilisers have been shown to enhance both targeted mechanical properties and biodegradability within bioplastics. Whereas, regarding the micro approach, a whole host of bio and chemical synthetic routes are uniquely available, to produce improved bioplastics. In this review, the main characteristics of bioplastics in terms of mechanical and barrier performances, as well as biodegradability, have been assessed—identifying both macro and micro routes promoting favourable bioplastics' production, processability and performance.

Keywords: biomaterials; biodegradation; bioplastics; mechanical performance; barrier performance; processability

Citation: Attallah, O.A.; Mojicevic, M.; Garcia, E.L.; Azeem, M.; Chen, Y.; Asmawi, S.; Brenan Fournet, M. Macro and Micro Routes to High Performance Bioplastics: Bioplastic Biodegradability and Mechanical and Barrier Properties. *Polymers* **2021**, *13*, 2155. https://doi.org/10.3390/polym13132155

Academic Editors: José Miguel Ferri, Vicent Fombuena Borràs and Miguel Fernando Aldás Carrasco

Received: 29 May 2021
Accepted: 25 June 2021
Published: 30 June 2021

1. Introduction

Pervasive plastics are leaving an indelible imprint on our planet. As high performance and energy-saving materials, plastics are ubiquitous and central to socio-economic advancement. Current mainstay plastics are processed from fossil fuel resources, with production requirements expected to double over the next 20 years. After use, these recalcitrant plastics are contributing to waste stockpiles and alarming pollution. Recycling technologies, which primarily include mechanical and thermochemical approaches, does not meet the efficiency levels required to safeguard the planet and adequately revalorise plastics as new products. The current linear economic model of resource mining, use and discarding, is now widely recognised as unsustainable. A circular approach, where resources are repurposed cyclically, akin to biological lifecycles, is essential in achieving a sustainable socio-economic ecosystem.

Nature readily operates elegant and efficient regenerative cycles for natural polymers and end of life bio-based materials. Such biodegradation and bio-regeneration processes involve microbial, enzymatic and biocatalytic activities for depolymerisation and repolymerisation. Petroleum-based plastics, with their smooth surface topographies, extensive hydrophobic chains and lack of bio-accessible organic chemical groups, are strongly bioinert and largely incompatible with bioprocessing, leading to their persistence over century timescales within land and water environments. Biomass provides a wealth of renewable and bio-waste resources for bioplastics synthesis. Many of these bio-based plastics, encompass capacities for biodegradation and bioprocessing with high performance features akin to petroleum-based plastics. The realisation of bioplastics that exhibit a complete set of mechanical and biodegradability, hold the promise of delivering material of ecologically sustainable, low carbon footprint circularity.

Bioplastics to date, however, have not achieved wide acceptability by the industry. Incompatibility with existing sorting infrastructures and high temperature mechanical recycling implemented for fossil-based plastics, along with raised production costs, are limiting factors. Technical shortcomings, such as brittleness, lower gas barrier functions and processing performances, have also played a role in keeping current market penetration levels in the region at just 2%. Combining high performance for consumer applications and continuous low carbon closed loop regeneration within plastics poses considerable challenges. At a fundamental structural level, polymeric features associated with good mechanical and fluid barrier properties are typically prohibitive to biodegradability. Petroleum-based plastics achieve the required degrees of high mechanical strength combined with flexibility and strong liquid and gas barrier properties by packing their sleek chemically structured chains into signature crystalline and amorphous regional arrangements. The tight alignment of chemically simple chains at high degrees of crystallinity also renders these plastics largely incompatible with biodegradation processes that require bioactivities, including enzymatic hydrolysis. Bioplastics, in contrast, by the very fact that they are generated from bio-based resources, are inherently more complex with more elaborate chemical structures. This provides both a means to progress their mechanical performance properties and provides amenability to bioactivity with higher levels of hydrolysable groups available for post use biodegradation and biodepolymerisation. To date, equivalent results are readily achievable, and in cases, results outperform particular mechanical properties for polylactic acid (PLA) and polyhydroxyalkanoates (PHA) bioplastics compared with conventional fossil-based thermoplastics. The potential to address performance limitations by a combination of bottom up and top-down approaches using considered chemical structure modifications and blending and composite formations, holds the promise of framing a new generation of bioplastics that encompass sustainability with performance.

In this review, the sustainability/performance triangulation between the biodegradability, mechanical and barrier properties of bioplastics is discussed. Approaches to overcoming the gap between industrially required mechanical and barrier performances and biodegradability are overviewed and related to the potential to build a new generation of high-performance sustainable plastics.

2. Bioplastic Production

Biopolymers can be obtained directly from biomass, as in the case of proteins and polysaccharides and synthetic biopolymers, such as PLA. Biodegradable polymers, including polycaprolactone (PCL), polyglycolic acid (PGA) and polybutylene succinate-co-adipate (PBSA), are primarily synthesised from petrochemicals. Microbial fermentation of biopolymers, including PHA and bacterial cellulose (BC), operates under relatively benign low energy conditions, and hence, is a highly favourable sustainable production route. Various microorganisms can accumulate PHAs as storage materials when cultivated under different nutrient and environmental conditions. This ability allows their survival under stressful conditions. The number and size of the PHA granules, the monomer composition, macromolecular structure and physico-chemical properties vary, depending on the producing microorganisms, the feedstock supplied and the operation conditions [1–3]. On the other hand, some bacteria can produce BC. This exopolysaccharide is a naturally occurring, chemically pure, free of hemicellulose, lignin and pectin, which is why BC purification is an easy process demanding low energy consumption. Nevertheless, production yields are very low, cultivation times are extensive, and thickness of layers is limited, which are major drawbacks in the conventional BC production process, affecting the range of possible applications [4]. For the industrial production of bioplastics, three limitations are very important. These include requirements of specialised growth conditions, expensive precursors, and high recovery costs. Building an increased body of knowledge on producing microbes' metabolism, biosynthetic pathways and their regulation is essential in overcoming these limitations [1,5].

3. Biosynthesised Plastics

In recent years, PHA polymers have emerged as one of the most promising biodegradable materials. Unlike commonly used fossil-based plastics or PLA, which requires the additional step of lactic acid polymerisation, PHAs are the product of bacterial metabolism and have a function of cytoplasmic inclusions. These materials, thanks to the over 150 monomer units, can have a variety of polymer properties that can compete with commonly used plastics, such as polyethene or polypropylene [6]. Monomer composition is connected to the substrate specificity of PHA synthase, hence the type of derived PHA is highly connected to the microorganism used for its production. For example, biosynthesis of short-chain length PHAs (SCL-PHAs) consisting of poly-3-hydroxybutyrate P(3HB) homopolymers is a three-step process regulated by 3-ketothiolase, acetoacetyl-CoA reductase and the SCL PHA synthase [7,8]. Bacteria, such as *Aeromonas caviae* and *Pseudomonas stutzeri*, are proven producers of PHA synthases with wide substrate specificity. Moreover, these enzymes have been identified after recombinant expression in *Ralstonia eutropha*, previously PHA negative. These PHA synthases can produce copolymers of SCL- and medium-chain length PHAs (MCL-PHA) [9]. It has been reported that combinatorial mutations in *P. aeruginosa*, *P. oleovorans*, and *P. putida* resulted in their ability to synthesise PHAs with 3-hydroxybutyrate (3HB), 3-hydroxyhexanoate (3HHx), 3-hydroxyocatanoate (3HO), 3-hydroxydecanoate (3HD), or 3-hydroxydodecanoate (3HDD) monomers [10–13]. These structures are presented in Table 1. The development of recombinant strains and using genetic engineering techniques can lead to improved mechanical and thermal characteristics of PHA materials. These properties are highly dependent on various factors during upstream and downstream processes. Modification of PHAs chemical structure, such as the introduction of functional groups or producing blends and copolymers, can also affect the quality of these materials [14]. Traditional chemical synthesis techniques have been successfully used for creating block copolymers with PHA materials. For example, block copolymers, including PHB blocks balancing with other materials (poly(6-hydroxyhexanoate), poly(3-hydroxyoctanoate), monomethoxy-terminated poly(ethylene glycol) (mPEG), and poly(ethylene glycol) (PEG)), have been reported. These structures are presented in Table 2 [8,15].

Table 1. Chemical structures of monomers described as units of PHA copolymer producing strains.

3-Hidroxyacids	Structure
butyric (3HB)	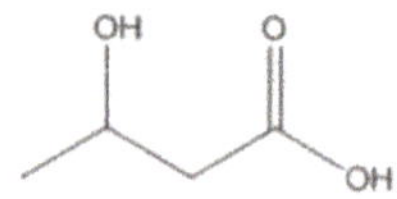
hexanoic (3HHx)	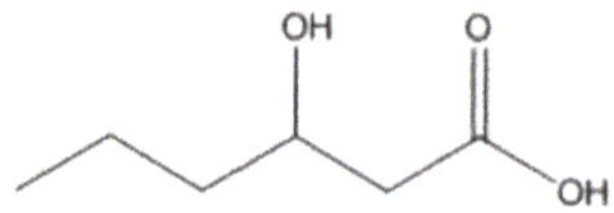
octanoic (3HO)	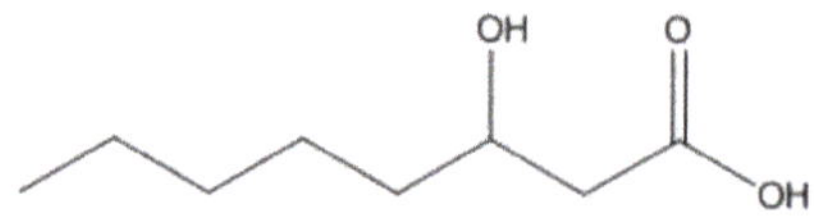
decanoic (3HD)	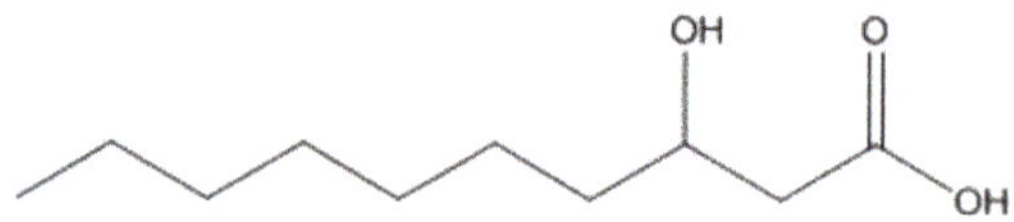
dodecanoic (3HDD)	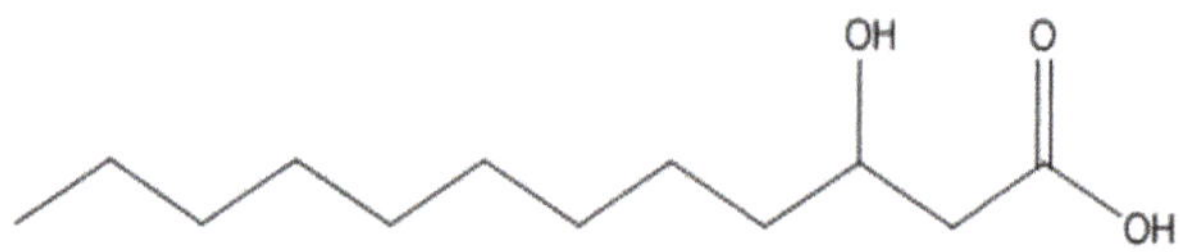

Table 2. Chemical structures of polymers commonly found as building blocks in PHA related block copolymers.

Polymer	Structure
poly(6-hydroxyhexanoate)	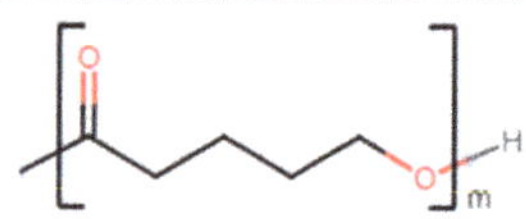
poly(3-hydroxyoctanoate)	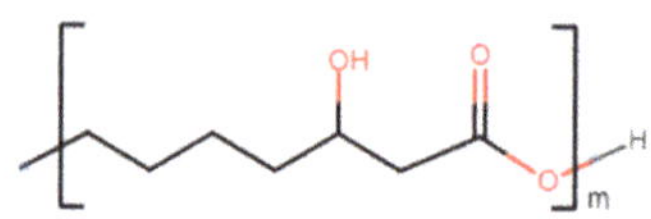
monomethoxy-terminated poly(ethylene glycol) (mPEG)	
poly(ethylene glycol)	

PHAs can also be derived from various substrates, including industrial waste streams [16], food waste [17], supplemented solid biodiesel waste, plant oils [7]. Besides these, seaweeds were found to be great feedstock for PHAs productions [18]. These materials can be produced using different strategies, including batch, fed-batch and continuous processes, and various conditions could be used for their conduction. Batch cultivations are easy and simple to operate, but production yields are very low. On the other hand, fed-batch cultivation can provide higher product and cell concentrations (no substrate inhibition) [19].

Continuous cultivation are also considered to be a viable strategy, providing required concentrations of limiting substrates, such as fatty acids and their derivatives [20]. Nevertheless, large scale application is still impractical with this type of production. Another strategy used for PHA production is solid state fermentation (SSF)—microbial cultivation on solid support made of appropriate substrates. SSF can be performed with inexpensive cultivation media, such as substrates, based on agro-industrial residues. This strategy provides disposing of waste, while valuable compounds are being produced at the same time [21]. There are few additional advantages of SSF over submerged fermentation: Easier aeration, higher substrate concentration, as well as reduced downstream processing steps. However, keeping conditions constant during the process is the biggest drawback to the market-ready production of PHAs using this strategy [22].

Bacteria from the genus *Komagataeibacter* (former *Gluconacetobacter*) synthesises another interesting, biosynthesised bioplastic—bacterial cellulose (BC). The BC production is a multi-step, strictly regulated process that involves regulatory proteins, few enzymes and catalytic complexes. The formation of 1,4-β-glucan chains is the first step in BC synthesis. This process, together with chains' assembly and crystallisation, occurs intracellularly. The second step involves extracting cellulose chains from the cells and their assembly in fibrils [4,23]. The yield and properties of the resulting material highly depend on the used bacterial strain and conditions of the conducted process, including medium composition, aeration, shaking, etc. Morphological and physical properties of the resulting material are in correlation to the cultivation broth composition. BC production can be improved using genetically modified producing strains, isolating novel strains with the ability to produce BC, and investigating process parameters and their significance. Florea et al. isolated *Komagataeibacter rhaeticus* strain able to produce a high yield of BC, while growing in low nitrogen level medium [24]. In order to increase BC production in limited oxygen conditions, genetically engineered strains have been developed [25]. Hungund et al. reported using ethyl methanesulfonate and ultraviolet radiation for *Gluconacetobacter xylinus* NCIM 2526 strain improvement that resulted in a significantly higher yield of BC (30%) [26].

Different strategies can be used to produce BC, including aerated submerged cultivation, static culture and airlift bioreactors [27]. Effects of these conditions on cellulose mechanical properties are presented in Figure 1.

Tensile strength, polymerisation degree and crystallinity index are highly influenced by BC structure. Produced in static conditions, BC forms pellicles on the surface of the medium. This gelatinous membrane can vary in thickness up to few centimetres depending on substrates' availability. Produced like this, BC has a significantly higher crystallinity index and tensile strength than the BC produced in agitated culture. Supply of air in agitated cultures will result in pellets with a higher crystallinity index. Crucial factors in BC production are the design of the reactor and proper control of conditions and process. These parameters highly affect the yield and quality of the resulting material. No matter which strategy is used, pH, oxygen supply, and temperature are essential conditions that should be carefully monitored. These parameters are mostly strain-dependent, but it was shown that the optimal pH value for cell growth and BC production is usually between 4.0 and 7.023, while the optimal temperature is in the range 28–30 °C [28].

Despite the great mechanical properties of the resulting material, static culture cannot provide uniformity of the cultivation broth; hence, cells are not equally exposed to nutrients and the thickness of the BC layer can be uneven. Additionally, productivity achieved in this strategy is very low, and it demands an extensive period of cultivation [29,30]. To improve the productivity of this process, the fed-batch strategy was developed, and a constant BC production rate was achieved for 30 days [31]. Submerged cultivations with agitation provide uniformity of nutrients, especially oxygen, resulting in higher yields in comparison to the static cultivations and making the production process cost-effective. Products of agitated fermentation depend on applied agitation speed and may include various forms of cellulose: Spheres, pellets, fibrous suspension [32]. As previously mentioned, higher productivity is the biggest advantage of submerged compared to static cultivation, but drawbacks, such as products' shape consistency and limited mechanical properties, are issued to overcome [33]. Another problem that can limit BC yield in aerated cultivation is a synthesis of gluconic acid. Due to the high agitation rates and hydrostatic stresses, the production of the secondary, protective metabolites is favoured over cellulose. Nevertheless, submerged cultivation of BC was implemented in different types of bioreactors, such as airlift, stirred tank and rotating disk. Production of this material on a large scale is still an issue, and designing new or improving existing equipment for this purpose is an important subject of research [34].

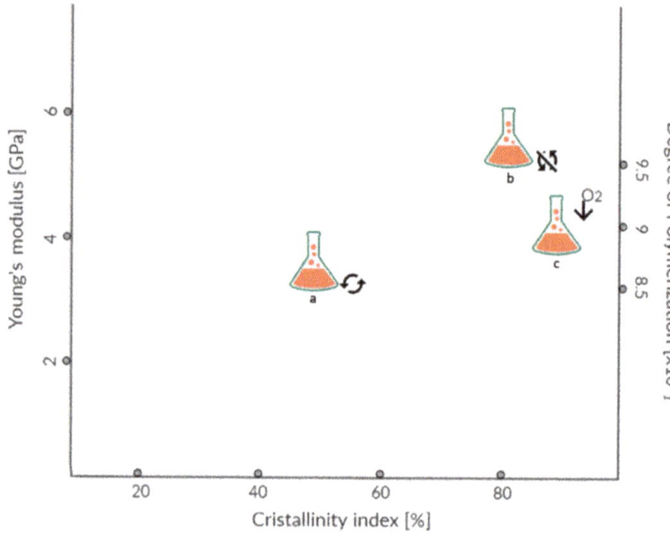

Figure 1. Young's modulus, crystallinity index and degree of polymerisation of BC depending on cultivation conditions: a—with shaking, b—without shaking, c—with additional oxygen supply.

3.1. Bioplastic Mechanical Performance

Bio-based polyesters (PLA, and PHAs) exhibit similar mechanical properties and can even exceed conventional plastic performances. Figure 2 demonstrates the Maximum Tensile Strength (MPa) and maximum Tensile Elongation (%) of bioplastics compared to petroleum-derived plastics. PLA is one of the most prominent bioplastics in terms of global consumption. It possesses several desirable properties, such as biocompatibility, biodegradability, composability and low toxicity to humans. The mechanical properties of PLA are greatly affected by the degree of PLA crystallinity. PLA derived from 93%, or more L-lactic acid can be semi-crystalline, while it is strictly amorphous when derived from 50–93% L-lactic acid. Thus, high tensile strength can be observed in films of high L-lactide content. Tensile strength and impact resistance are also influenced by the degree of crosslinking and the annealing of L-PLA, which increases the stereoregularity of the chain [35]. Comparison of mechanical properties between poly(98% L-lactide) and poly(94% L-lactide) showed a slightly greater elongation at yield for 98% than 94% L-lactide. However, poly(94% L-lactide) has an elongation at the break seven times greater than poly(98% L-lactide), indicating more plastic behaviour with 94% of L-lactide [35].

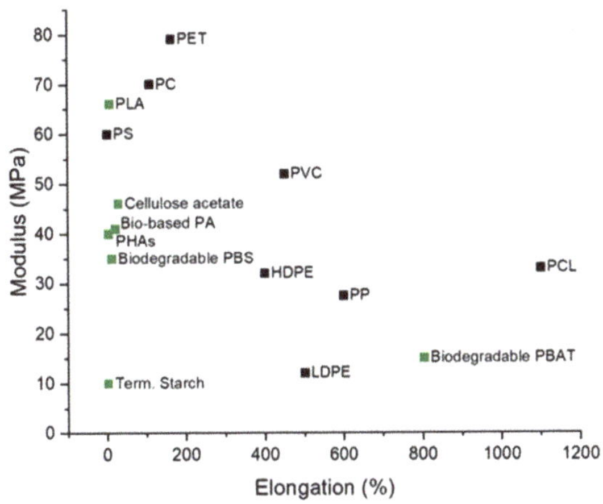

Figure 2. Maximum Tensile Strength (MPa) and Maximum Tensile Elongation (%) of bioplastics compared to petroleum-derived plastics, Data from Ref. [36].

For racemic mixtures, a study by Chen et al. demonstrated that polymerisation of 50% D-Lactide and 50% L-Lactide usually results in forming an amorphous polymer of poly (DL-lactide) [37].

As a packaging material, PLA offers high stiffness (greater than polyethene terephtalate (PET) and polystyrene (PS)), good clarity (similar to PET), relatively low processing temperatures, excellent resistance to fats and grease, and good breathability suitable for fruits and vegetable storage. Such characteristics make PLA a potential candidate to replace PS, polyethene (PE) and polypropylene (PP) in the fabrication of disposable cups, salad boxes and cold food packaging [38]. Nevertheless, PLA brittleness with less than 10% elongation at the break renders it unsuitable as a pure material for applications that require plastic deformations at higher stress levels [39]. Additionally, PLA's poor gas moisture permeability performance make it unsuitable for many beverage bottle applications [38].

On the other hand, PHAs gained considerable interest as a green alternative to petrochemically derived plastics, as they are biocompatible, biodegradable and synthesised from renewable resources [40,41]. PHB is the only polymer from the PHAs family to be produced in large quantities. This material is considered an aliphatic polyester with a linear polymer chain, composed of monomers of 3-hydroxybutyrate with a chromophoric carbonyl group. Being a member of the PHAs, PHB is also characterised by having a methyl (CH_3) as an alkyl replacing group, which provides it with a hydrophobic charge. The regularity of the polymerised PHB chain has a direct influence on its degree of crystallinity that, in turn, is influenced by the synthesis route used. Isotactic PHB, which has chiral carbon in absolute configuration R, is obtained through bacterial fermentation, while syndiotactic PHB is synthesised through a synthetic route from monomers with setting R and S. As isotactic PHB presents a more regular structure, it will allow a higher crystallinity than syndiotactic [42]. Favourable PHB properties in terms of melting point, strength, modulus and barrier properties promotes it as a substitute for PP, low density polyethene (LDPE), polyvinyl chloride (PVC) and PET in packaging applications. Differences in chemical structures between PLA, PHB and previously mentioned commonly used plastics are shown in Figure 3.

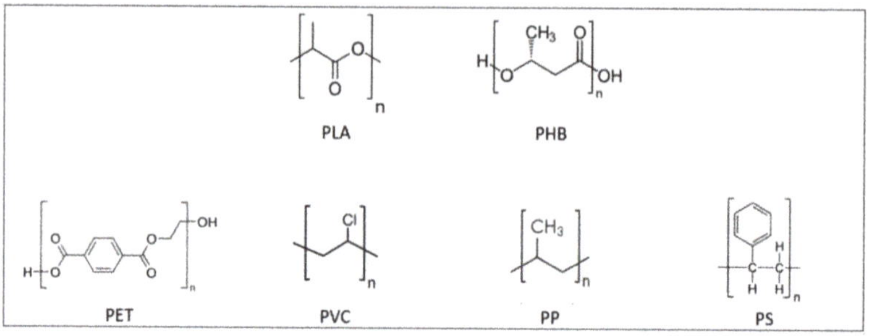

Figure 3. Chemical structures of biodegradable: Polylactic acid (PLA) and polyhydroxybutyrate PHB); and nonbiodegradable polymers: Polyethene terephatalate (PET), polyvinyl chloride (PVC), polypropylene (PP), polystyrene (PS).

Nevertheless, as a bioplastic, PHB has drawbacks, such as being brittle, hard and thermally unstable, making it challenging to use for applications like injection moulding in food industries [43,44]. As a pure material, PHB is highly crystalline (around 80%), resulting in the previously mentioned brittle nature and low elongations. The brittle nature of PHB is associated with a secondary crystallisation of the amorphous phase at ambient temperature. Another important issue is the glass temperature (T_g) of PHB. The T_g is close to room temperature resulting in secondary crystallisation taking place during storage, which, combined with a low nucleation density feature, leads to large spherulite formations which can grow over long durations leading to inter-spherulitic cracks [43]. Generally, spherulites are formed when PHB is crystallised from the melt, with band spacing between them depending on the crystallisation temperature [45]. Cracks are always present within spherulites in melt-crystallised PHB, and subsequent growth of the cracks leads to failure of the polymer. Two distinct types of crack exist in PHB spherulites, which can run either radially or circumferentially within the spherulites. Radial cracks occur more frequently in films crystallised at lower temperatures, while circumferential cracks occur when PHB is crystallised at high temperatures [46]. Another problem with PHB processing is the narrow processing window. The melting temperature of PHB is around 180 °C, therefore, processing temperature should be at least 190 °C. However, thermal degradation at this point happens rapidly, drastically reducing the acceptable residence time in the processing equipment to a few minutes only [47].

However, notwithstanding the limitations of PLA and PHAs, these bioplastic polymers have the potential to be fine-tuned to extend their application range comparable to fossil-based thermoplastics.

3.2. Bioplastic Barrier Performance

Inadequate fluid and gas barrier properties strongly impedes the utilisation of biopolymers in applications, such as the food-packaging sector. The established utilisation of PET and polyolefin family polymers in packaging applications is due to their combination of low cost, transparency, good barrier (to oxygen and water vapours), and mechanical characteristics. The oxygen permeability of PET (0.04 barrer) is much stronger than that of the recently developed biopolymers [48]. Among previously mentioned biopolymers, the most widely used—PHB stand apart having significant gas barrier properties (0.01 barrer), which are comparable to benchmark polymers, such as PET [49]. However, PHB's brittle mechanical nature precludes its suitability for food packaging applications. PLA, while having tunable mechanical properties, has oxygen permeability levels in the region of 0.26 barrer, restricting its use within several food-packaging applications [50]. As an approach to overcoming these barrier limitations, the addition of fillers to block the gas and moisture molecular pathways through the polymers, on a nano-micro scale level, is an attractive option [51].

An ideal filler should have a high surface area, aspect ratio and suitable chemical compatibility to provide enhanced mechanical and gas barrier properties at low filler content. The filler geometrical characteristics are an important factor in reducing gas permeability. The higher the aspect ratio, the greater the surface activity, which leads to an increase in mechanical properties, as well as gas barrier properties of the polymer matrix.

In terms of the orientation of fillers inside the polymer matrix, the Nielson model is commonly used. This is an ideal case where the orientation of the filler is perpendicular to the direction of diffusion and is generally not readily achievable. The modified version of the Nielson model is proposed by Bharadwaj by introducing the orientation parameter (S) [52]. If S = 0, designates perfect orientation and the Bharadwaj model will be reduced to the Nielson model, and maximum permeability reduction will be observed as indicated in Figure 4. Different models based on volume fraction and aspect ratio have also been developed to compare theoretical data with experimentally investigated gas barrier results [53].

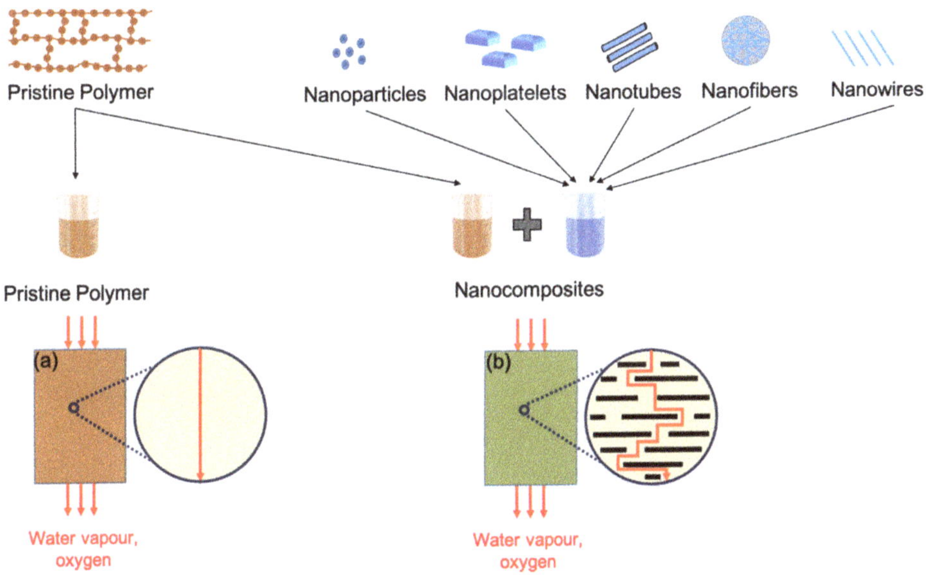

Figure 4. Effect of filler addition on gas barrier properties of nanocomposites: (**a**) Poor barrier properties in pristine polymer, due to direct diffusion pathways for gas molecules, (**b**) improved barrier properties in nanocomposites due to longer diffusion pathways.

Another aspect to be considered while studying the bioplastics' barrier properties is the dispersion and interchain compatibility of fillers within the bioplastics' matrices. Well exfoliated nano-fillers in polymer matrices give optimal reinforcement and contribute to other material performance characteristics [51]. A major issue with the dispersion of fillers in polymers is their hydrophilic nature, which causes inefficient compatibility with the hydrophobic polymer phase. Therefore, treatments are adopted to promote better interactions and good dispersion between the polymer phase and the fillers [54]. Moreover, the structural characteristics of the filler define the contribution imparted to the mechanical and gas barrier properties of polymers. The structural format of fillers dramatically impacts the gas barrier properties of their host polymers [55]. This is due to the higher crystallinity, which increases the effective path of diffusion and impedes the passage of gas molecules through the polymers rendering them suitable for packaging applications.

Figure 5 demonstrates the effect of fillers addition on the barrier properties, especially the oxygen (O_2) permeability of commonly used petroleum-based plastics and biopolymers. The data is compiled by converting oxygen permeability values from different units into a single unit (barrer). The addition of smaller amounts of fillers in biopolymers has drastically reduced the permeability of oxygen, fulfilling the criteria of ideal gas barrier material (LDPE, PET, and high density polyethene (HDPE)). Among biopolymers, PHAs showed more hindrance to the passage of gas molecules in their pristine polymers as compared to mentioned petroleum-based polymers. Functionalised graphene oxide (Gr-O) proved to be the best filler. The impressive reduction of oxygen permeability by Gr-O could be related to the strong interfacial adhesion between Gr-O and PHA polymer matrix [56].

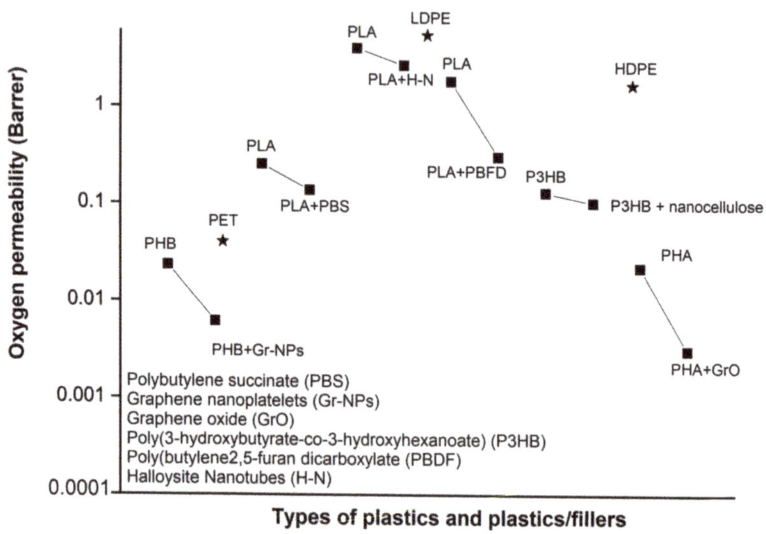

Figure 5. Effect of nanofillers on oxygen permeability of various biopolymers.

3.3. *Bioplastic Processing and Formulation*

Blending and composite formation is an established route for achieving improved technical and processing performance within polymer engineering. Blending PHAs, in particular, PHB with other polymers, offers opportunities to improve processability by lowering the processing temperature and reducing the brittle nature of these biopolymers. The physical, chemical and molecular architectural aspects dictate the enhancement of the polymer blend achievable through compounding techniques and introduced additives. Plasticiser additives can improve polymer viscosity and improve chain mobility during processing. Thermal stabilising additives can be used to eliminate premature degradation of polymers during processing, such as antioxidants, which guard against the presence of oxygen in the processing environment. Compatibiliser additives can improve miscibility between polymers by inducing flexible physical dipole-dipole interactions, or hydrogen bonding [57]. Nanocomposite additives, such as nanocrystals and nanofibres, can significantly improve the mechanical strength and gas barrier properties of the polymers if they are well dispersed in the biopolymer matrices. Natural fibres as an example of nanocomposite have been recently introduced as the main component in fibre reinforced biopolymer composites [58]. The intermolecular hydrogen bonds connecting the polymer chains of natural fibres provide a linear crystalline structure with a tensile strength reaching 15 GPa [59]. Such great strength is also accompanied by other advantages, as low cost, abundance, biodegradability, easy recyclability and fabrication of low weight composite materials [58]. All these properties made natural fibres perfect candidates as fillers in biopolymer composites and can compete with glass or carbon fibres. Accordingly, several studies were performed to evaluate the effect of incorporating natural fibres in biopolymer composites to improve the composite's mechanical and barrier properties. Among natural fibres, cellulose [60–66], hemp [67–73], kenaf [74–78] and flax [79–88] were the most studied ones. It is worth mentioning that nanocomposites of natural fibres or crystals added as fillers without plasticiser or compatibiliser, results in their poor dispersion and decrease the quality of polymer composite. Alternatively, plasticisers of hydrophilic nature, when mixed with biopolymers or their blends, tend to increase the wettability and O_2 permeability and deteriorate the barrier properties of the polymer composite.

Thus, to obtain better performance of nanocomposites with plasticisers or compatibilisers in biopolymer composites, both additives should be used together. The addition of nanocomposites with plasticisers improves the interfacial adhesion between the nanocomposites and the polymer matrix, allowing better dispersion and consequently provides a more tortuous path for gas and water and increase the barrier properties. Other approaches were introduced to improve the dispersion of the nanocomposites of natural fibres or crystals in polymer matrices. These include physical and chemical treatments of the nanocomposites before mixing with biopolymers. Bio-based coatings were also applied to natural fibres reinforced biocomposites as a means of inducing the hydrophobicity, and thus, improve the barrier properties of the biocomposite.

3.3.1. Blends and Composites

As previously mentioned, the processability and formability of PHB represent a drawback in industrial applications; blending with PLA provides a potential route to facilitating its introduction in the market, while also improving PLA properties at the same time. Several studies on PLA-PHB blends have been conducted in recent years, with the results typically showing a slightly higher Young's modulus than neat PHB and neat PLA [89]. Blends with PHB content of 50% or higher have shown lower values of tensile stress and elongation at the break in comparison to pure PLA [90]. However, above 60% PLA content has been reported to increase elongation at the break by up to 12%, with values even comparable to typical thermoplastics achieved on the addition of plasticisers [57,89,91–93]. Furthermore, the PLA-PHB 75:25 blend demonstrated higher mechanical performance than neat PLA, and greater impact resistance than the homopolymers on their own [89,94]. Jandas et al. reported the incorporation of PHB within PLA matrix in different ratios resulting in intermediate properties for the blends. The ductility of PLA increased consistently as PHB concentrations increased from 10 to 30 wt%. The maximum increase in percentage elongation was observed in the 70:30 ratio, suggesting some degree of molecular interaction between the macromolecules of PLA and PHB within the blend (Figure 6). However, tensile modulus and tensile strength were considerably decreased in the case of the blends, compared to pure PLA, as well as a corresponding decrease in stiffness. Blends prepared at 70:30 ratio were used for trial with compatibilisers and preparation of blend composites, due to the optimum elongation at the break and impact strength exhibited [57].

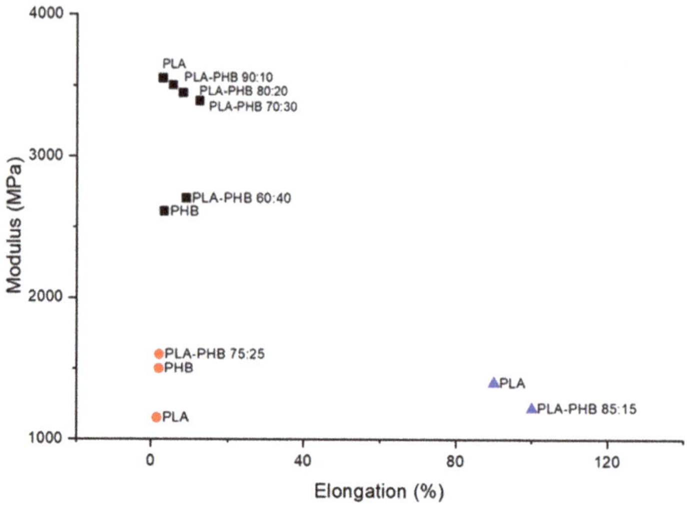

Figure 6. Young's Modulus (MPa) vs. elongation at the break (%) of PLA, PHB and their blends, reported by Jandas et al. (black squares), Data from Ref. [57], Armentano et al. (blue triangles), Data from Ref. [91] and Arrieta et al. (red circles), Data from Ref.[94].

3.3.2. Compatibilisers and Plasticisers

Facilitation of processability and improved flexibility in PLA/PHB blends is achievable using plasticisers [94]. Plasticisers are available as cost-effective, readily available materials on the market and are also generally of natural origin and include: Oxypropylated glycerin (or laprol), glycerol, glycerol triacetate, 4-nonylphenol, 4,40-dihydroxydiphenylmethane, acetyl tributyl citrate, salicylic ester, acetylsalicylic acid ester, soybean oil, epoxidised soybean oil, dibutyl phthalate, triethyl citrate, dioctyl phthalate, dioctyl sebacate, acetyl tributyl citrate, di-2-ethylhexylphthalate, tri(ethylene glycol)-bis(2-ethylhexanoate), triacetine, and fatty alcohols with or without glycerol fatty esters. Blends of PEG (2–5%) and PHB produced by solvent casting technique has been demonstrated to increase the elongation at the break by up to about four times compared with the original neat PHB. This behaviour was attributed to a plasticising effect of PEG, which acts to weaken the intermolecular forces between the adjacent polymer chains. The changes in free volume reduced the melting temperatures of the system and are also associated with an accompanying reduction in tensile strength [95].

As shown in Figure 7, compatibilisers, such as maleic anhydride, have been applied to PLA and PLA/PHB blends to impart additional flexibility and improve the blend's young's modulus by the induction of flexible physical interactions, including dipole-dipole or hydrogen bonding [57]. For industrial purposes, considerable attention needs to be paid to the selection of suitable plasticisers or compatibilisers as most tend to negatively impact other mechanical properties, such as lowering barrier properties of bioplastic blends, which can restrict their use in packaging applications [89].

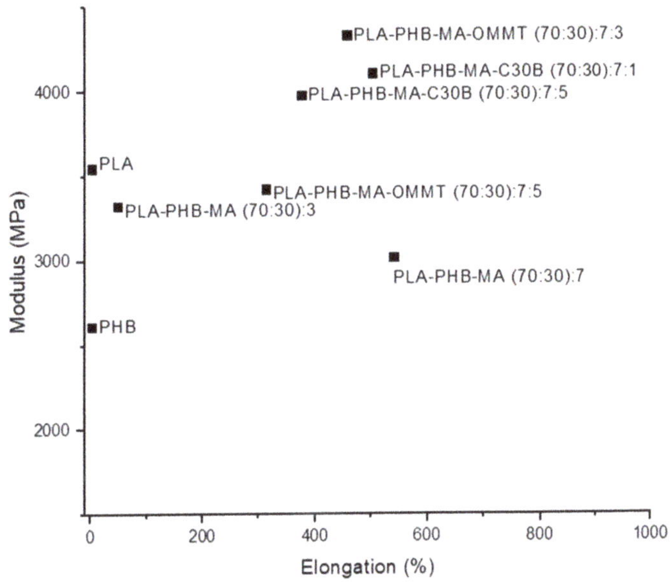

Figure 7. Young's Modulus (MPa) vs. Elongation (%) of PLA, PHB and their blends using MA as a compatibiliser, together with OMMT and C30B nanoclays, Data from Ref. [57].

3.3.3. Natural Fillers

The use of naturally sourced fibre-based fillers, with their considerable tensile strength and high sustainability, presents a key route to facilitating polymer circularity. The hydrophilic nature of bare natural fibres with no surface treatment requires an address, due to the incompatibility with hydrophobic biopolymer. Biopolymer composites reinforced with untreated natural fibres typically exhibit non-uniform fibre dispersion with interfaces promoting crack formation. These material defects can lead to premature mechanical failure of the composites [58,59,96]. These defects are attributable to the natural fibre hydrophilicity, which hinders proper mixing with the biopolymer hydrophobic matrix causing poor fibre/matrix interfacial bonding [59]. While high susceptibility of moisture absorption by the hydrophilic natural fibres can support the growth of fungi and bacteria and deteriorate the physical and mechanical properties of the bio-composites [58], this fact can be a highly useful feature that can be availed of post use for achieving biocyclability.

Different approaches are available to enhance the interfacial adhesion between the fibres and matrix, resulting in the fabrication of biocomposites with better mechanical and barrier properties. Among these solutions is the surface modification of natural fibres. Surface modification can result in increasing the sites of reaction, offering new functionality to the fibre surface and enhancing surface roughness by removing impurities [59]. Such modification will lead eventually to the improvement in mechanical properties and reduction of the water absorption of the fabricated biocomposite. The techniques applied to modify the natural surface fibres before inclusion in biopolymer matrices include physical, chemical, biological treatments and their combinations. Some examples of physical treatment techniques are calendaring, stretching, hybrid yarns production and thermal treatments; while chemical techniques include, alkali swelling, silane modifications, graft copolymerisation, and treatment with isocyanate, mercerisation [97].

A representative selection of the recently applied surface modifications for natural fibres reinforced biocomposites are given in Table 3.

Table 3. Surface treatment of natural fibres reinforced bio-composites.

Process	Natural Fibre Used	Biopolymer Matrix	Outcomes			Ref.
			Interfacial Adhesion	Mechanical Properties	Barrier Properties	
Treatment with compatibilisers; Polyglycerol polyglycidyl ether (SR-4GL), Trimethylol propane polyglycidyl ether (SR-TMP), and (Polyglycerol polypropyleneoxide (SC-P1000)	Cellulose fibres	PLA	Improved interfacial adhesion between fibres and PLA and		Inhibited degradation of the PLA matrix	[65]
STEFAC TM 8170, surfactant modification	Cellulose fibres	PLA/PHB		Enhanced mechanical performance	Improved water resistance, reduced oxygen and UV-light transmission, as well as appropriate disintegration in compost	[94]
Alkali treatment	Kenaf fibre	PHB		Reduction in the crystallinity of PHB (up to 6% reduction), making it more ductile, and improvement of the flexural modulus by up to 11%.		[75]
Silane treatment	Flax fibre	PLA	Improvement to fibre/matrix adhesion with 2% *w/w* silane content, yet further improvement of the fibre-matrix interface can be partially resolved by silane/alkali treatment combination.	Improved mechanical properties		[80]
Alkali treatment	Flax fibres	PLA		T_g values of fabricated bio-composites were lowered by 10 °C for 10% NaOH treatment and 15 °C for 30% NaOH treatment		[82]
Treatment with ethylene plasma	Flax fibres	PHB	Improved interfacial adhesion strength in the bio-composite		Improved thermal resistance	[88]

Combinations of natural materials with petroleum-based plastics is an option that is also under development. While facilitating the high mechanical performance, considerations are required when using this approach as further dilemma's may be posed regarding factors, such as continued resource depletion dependencies and degradation and biodegradation pathways, which potentially lead to increased microplastics production.

3.3.4. Bio-Coatings

The application of bio-based coatings to biocomposites and natural fibre reinforced biocomposites is a promising approach proposed to overcome the significant water uptake propensities of natural fibres and increase the moisture resistance of bioplastics for fluid barrier property application requirements. Exposure to long term environmental/hygroscopic ageing necessitates the induction of a higher level of hydrophobicity in chemically modified natural fibre reinforced bio-composites and bioplastics in general. Introducing bio-based coatings to natural fibres reinforced biocomposites, ensures the environmentally friendly and biodegradable nature of fibres. Besides, bio-based coatings are obtained from renewable resources and have superior hydrophobic characteristics [62]. For instance, polyurethane (PU) coatings were first introduced as bio-based coating resins in the 1950s [98]. PU coatings mainly provide their composites with high solvent resistance, hydrolytic stability, resistance to acid–base conditions and weather-ability [62]. Currently, most of the industrially produced PUs are petroleum-based polyols. Thus, renewable resources, such as vegetable oil [99–101], canola oil [102], soybean oil [103–105] and castor oil [106–108], are thoroughly investigated and were able to produce PU coating of competing properties to that of petroleum-based ones.

Polyfurfuryl alcohol (PFA) is another attractive type of bio-based coatings that can be used for barrier property application requirements. PFA has a low cost manufacturing process and can be obtained from natural resources as the agricultural residue of wheat, birch wood, hazelnut shells, corn, rice hulls, oat and sugar cane [62]. In addition, PFA's hydrophobicity, great heat distortion temperature and resistance to chemical erosion make it an excellent candidate as a coating material for bioplastics and natural fibres reinforced biocomposites.

Despite such potentials of PU and PFA as bio-based coatings, there is almost no work presented in the literature on the application of PU and PFA as coatings for natural fibre reinforced composites or biocomposites, except for a recent study done by Mokhothu et al. [109]. This study proposed using PU and PFA as bio-based coatings to composites containing flame-retardant treated natural fibres (flax) and phenolic resin. For three days, uncoated and coated samples were subjected to 90 °C and 90% relative humidity. Analysis was performed to the relative moisture content and mechanical properties and compared with the commercially available water-resistant product (FIRESHELL® (F1E)). Concerning the mechanical properties, PFA coated samples showed the highest modulus value (1.93 GPa) after being subjected to environmental conditioning with respect to uncoated (1.59 GPa); PU (1.05 GPa) and F1E (0.98 GPa) coated composites. Besides, the PFA and PU coated samples showed high stress at the break and a decreased elongation at the break in comparison to F1E coated ones. The moisture content of the conditioned PFA and PU coated composites was significantly reduced by 75% and 30%, respectively, when compared to uncoated and F1E coated composites [109].

4. Bioplastics Biodegradability

There is an important distinction between degradable polymers and biodegradable polymers. Degradable polymers are defined as polymers that can be depolymerised or recycled under controlled conditions and processes. According to the American Society for Testing and Materials (ASTM) definition, biodegradable polymers are polymers that can undergo decomposition into carbon dioxide, methane, water, inorganic compounds, or biomass, which can be measured by standardised tests, in a specified period, reflecting available disposal conditions (ASTM standard D6813). The mechanism of biodegradation is that the molecular weight of biodegradable polymers reduced, due to hydrolysis and oxidation, followed by breaking down into natural elements, such as water and carbon dioxide, via microorganisms. Aliphatic polyesters are the most economically viable biodegradable polymers, with PHB being among the most mechanically promising. The structural changes to the polymer molecules can be described in terms of three main categories of actions or mechanisms, namely, chain depolymerisation, random chain scission, and substituent reactions [110]. As defined by the IUPAC, depolymerisation is the process of converting a polymer into a monomer or a mixture of monomers/oligomers. Therefore, chain depolymerisation means the chain reaction is responding to the transformation of the macromolecular polymer chain into its constituent micromolecular monomers. Random scission is defined in the IUPAC Gold Book as a chemical reaction resulting in the breaking of skeletal bonds. It is also defined as a degradation mechanism that assumes a random cleavage of bonds along the macromolecular polymer chains [111]. This leads to the production of fragments that steadily decrease in length, which may eventually be small enough to allow for the removal of micromolecules. Substituent reactions refer to the kinetic reactions carried out by the constituent monomers of a polymer chain, which differ between polymers. According to Ghosh (1990), each kind of substituent has a characteristic chemical nature and reactivity [112]. However, as substituent reactions can only be observed at relatively low temperatures, substituent reactions only assume prominence when initiated and accomplished at temperatures lower than those of the breaking temperature of main chain bonds of a polymer.

Several factors affect the degradability of a polymer. In general, the surface conditions, the first-order structures, and the high order structures of a polymer play a major role in determining the rate of degradation. Surface conditions, such as hydrophilicity and surface area, directly correspond with the overall degradability of a polymer. Additionally, external environmental factors, such as humidity and temperature, also affect the overall degradability. Humidity introduces water molecules to a polymer and may result in a hydrolysis process, depending on the susceptibility or hydrophobicity of the polymer. Furthermore, the crystallinity of a polymer is also proportional to the degradability of a polymer, so that the lower the degree of crystallinity, the higher the degradability of the polymer. According to Tokiwa et al. (2009), this can be attributed to the fact that enzymes generally interact with the amorphous regions within a polymer, which are loosely packed together as compared to the crystalline regions. Moreover, the melting temperature (T_m) of polyesters greatly affects their enzymatic degradation. This is evident from the fact that aliphatic polyesters and polycarbonates with low Tm have a greater biodegradability than aliphatic polyurethanes and polyamides, which have higher Tm [113]. This is due to the large melting enthalpy change values of the latter, which can be attributed to the presence of hydrogen bonds among the polymer chains. The introduction of heat into a polymer matrix generally weakens the intermolecular bonds, resulting in an increased rate of degradation. With biodegradation specifically, the microbial species introduced to the polymer, directly correlates with the level of microbial activity, which in turn determines the rate of degradation of a biodegradable polymer [114]. The degree of microbial activity is also heavily influenced by nutrient and oxygen content in the biodegradation environment.

As described earlier, selected microorganisms can produce and storing PHAs. The ability to synthesise these molecules does not imply the capacity to also degrade them, in the case where extracellular hydrolases capable of converting polymers are also expressed [115,116]. Under nutrient-limited conditions, degradation occurs when the limitation is removed. Currently, six hundred PHA depolymerases from various microorganisms have been identified and categorised within eight families [117]. The degradation of these polymers is affected by many factors, such as type of enzyme, temperature, moisture, and nutrients composition [118]. Degradation rates of PHAs are also related to the microbial population density. It was shown that during degradation of P(3HB-co-3HV) copolymer, microbes at first attach to the polymer and then begin secreting degrading enzymes. Although PHA producing/degrading microbes usually express high specificity towards

P(3HB), many microbes have been identified with wide substrate specificity. *Xanthomonas* spp., for example, has the ability to produce enzymes for PHAs with aromatic side chain degradation and can also degrade P(3HB), P(3HO), and poly-3-hydroxy-5-phenylvalerate (P(3HPV)) [119]. The type of polymer also plays an important role in degradability. In addition to the presence of side chains, length and composition are also significant factors. Manna et al. report that homopolymers have higher degradation rates in comparison with copolymers of PHAs [120]. Other studies have demonstrated opposing results [118], which may be explained because, in these cases, the experiments were conducted in natural environments where previously mentioned factors (nutrients, moisture, temperature etc.) were non-controllable. Kusaka et al. showed that PHAs degradation ability is negatively correlated to the M_w and crystallinity [121]. The format and shape of the polymer material is also a significant factor for PHAs degradation, with thin films degrading faster than thicker films. Soil and climatic conditions are further factors that can affect the PHA degradation rate [28,118]. Boyandin et al. examined PHA films degradation response and reported that humid and the hot Vietnamese climate facilitated degradation of PHA [122].

Additives, such as fillers, are another factor that can affect the biodegradability of the bioplastic in which they are added, as demonstrated in Figures 8 and 9. There is no general guarantee that the addition of fillers will enhance or inhibit biodegradability as the effect of fillers on a polymer is mainly dependent on its chemical and physical aspects, such as size, geometry, surface area, and the surface energy of its particles [123] (Murphy, 2001). These aspects directly affect the overall degradation ability of a polymer. In general, the effect of additives on the biodegradability of a polymer is largely dependent on the properties of the additives, such as hydrophobicity and amenability to bacterial growth on the surface.

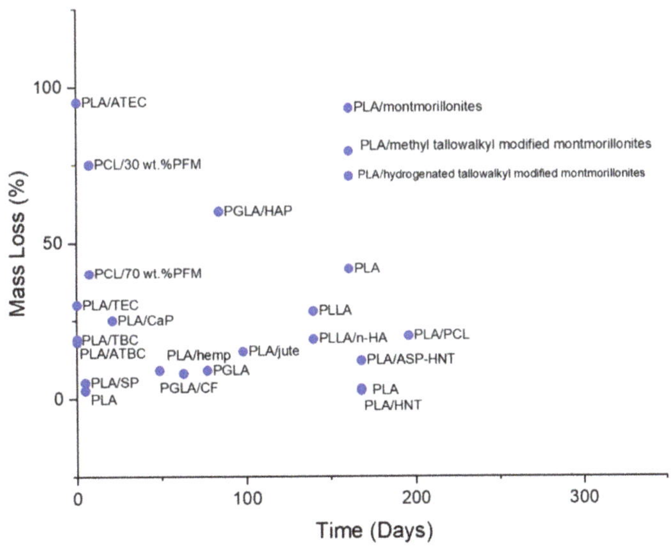

Figure 8. Mass Loss percentage of composites in various hydrolytic degradation environments.

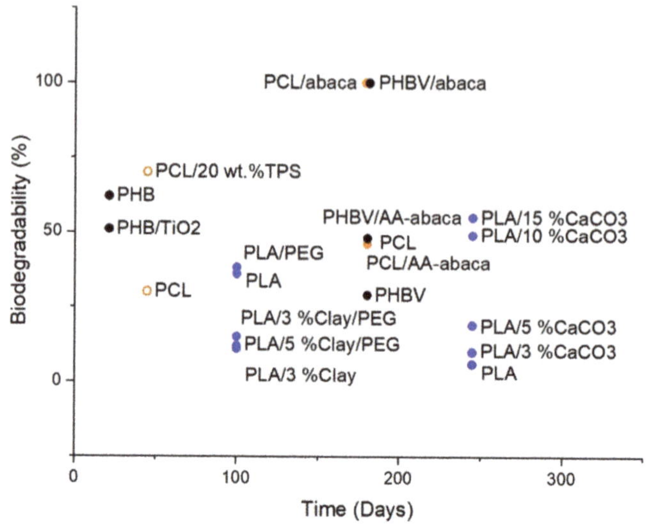

Figure 9. Biodegradability of PLA, PCL and PHB with the addition of various fillers and plasticisers in various soil burial degradation environments.

Aframehr et al. report a study on the effect of calcium carbonate ($CaCO_3$) in soil burial biodegradation where the $CaCO_3$ fillers, act to increase the biodegradability of PLA. The weight loss of $CaCO_3$ nanocomposites is approximately two times higher than other nanocomposites, with a weight loss of around 55% for PLA/15% $CaCO_3$, 49% for PLA/10% $CaCO_3$, 19% for PLA/5% $CaCO_3$, 10% for PLA/3% $CaCO_3$, and around 6% for neat PLA after a soil exposure time of 35 weeks [124]. A study by Teramoto et al. investigated the effect of treated and untreated abaca fibre filler on the biodegradability of poly(3-hydroxybutyrate-co-3-hydroxyvalerate) (PHBV) after being subjected to a soil-burial environment for a duration of 180 days. Neat PHBV exhibited the least biodegradability at only around 29% weight loss, followed by PHBV/AA-abaca at around 48% weight loss, and lastly, PHBV/untreated abaca with the highest biodegradability, which can be seen in the high degree of fragmentation after 60 days [125].

Altaee et al. conducted a study on the biodegradation of PHB and titanium oxide (PHB-TiO_2) composites in a soil burial environment with pH 7.30 and a humidity of 80% at 30 °C and found that PHB-TiO_2 exhibits a lower weight loss of only ~51% after three weeks as compared to ~62% weight loss of neat PHB through the same duration [126]. Paul et al. (2005) studied the degradation of nanocomposites of PLA with unmodified and organo-modified montmorillonites and found that montmorillonites filler enhances hydrolytic degradation. PLA with unmodified montmorillonites exhibited the greatest decrease in M_w after 23 weeks of hydrolytic degradation with a 93.1% loss in M_w. These results were followed by PLA with montmorillonites treated with bis-(2-hydroxyethyl) methyl tallowalkyl ammonium cations and PLA with montmorillonites treated with dimethyl-2-ethylhexyl (hydrogenated tallowalkyl) ammonium cations at 79.2% and 71.2% M_w loss, respectively. In comparison, unfilled PLA is found to only have a 41.6% M_w decrease compared to its initial value [127]. Chen et al. found that PLA with halloysite nanotubes (HNTs) as filler has a greater rate of hydrolytic degradation as compared to neat PLA as shown from the mass reduction of 3.1% for PLA/HNT as compared to that of neat PLA at only 2.6% in an in vitro environment in SBF at 37 °C by the end of the 24th week of degradation. It was also reported that PLA with HNTs surface treated with 3-aminopropyltriethoxysilane (ASP) has an even greater hydrolytic degradation, due to better interfacial adhesion between PLA and HNTs, which is evident from the mass reduction of 12.1% [128]. A study by Navarro et al. (2005) investigated the effect of the addition of calcium phosphate (CaP) glass to PLA on its hydrolytic degradability. During the first three weeks of the degradation, PLA/CaP composites experienced a greater weight loss than neat PLA, but an increase of weight of the PLA/CaP composite was reported after three weeks. This may be credited to forming hydrated calcium phosphate precipitate on the composite. The maximum weight loss exhibited by week 3 is 25%, and a final weight loss percentage of about 22% on week 6. In a comparison, the weight loss of neat PLA is around 1% from week 3 through week 5 [129]. Moreover, a study by Huang et al. (2013) investigated the hydrolytic degradability of poly(L-lactic acid) (PLLA)/nanohydroxyapatite (n-HA) and found that the rate of degradation of PLLA/n-HA composite was slower than neat PLLA. This is evident from the weight loss of only around 19% for PLLA/n-HA composite and about 28% for neat PLLA after 20 weeks in a PBS environment with an initial pH of around 7.4 [130]. A study by Valapa et al. (2016) investigated the hydrolytic degradation behaviour of sucrose palmitate (SP) reinforced PLA (PLA-SP) nanocomposites in acidic (pH 2), basic (pH 12), and neutral (pH 7) hydrolytic degradation environments and found that the rate of degradation is increased with the addition of sucrose palmitate. This can be seen in the mass loss percentage of about 5.1% for PLA-SP as compared to only around 2.6% mass loss percentage of neat PLA in a pH 7 degradation solution at 35 °C after 115 h [131].

Alternatively, the alterations proposed by the addition of plasticisers and compatibilisers to bioplastics directly affect their degradability since plasticisers and compatibilisers decrease the glass transition temperature (T_g) of the polymers they are blended with, as shown in Figures 8 and 9 [132]. Like fillers, there is no general guarantee that the addition of plasticisers will positively or negatively impact biodegradation as it is highly dependent on the properties of the plasticisers used. Some of the more common forms of plasticisers include citrate esters and phthalates (phthalate, isophthalate, terephthalate), the latter being biodegradable as degradation by microorganisms is considered as the most effective method of degradation for phthalates plasticisers [133]. A study by Labrecque et al. (1997) researched using triethyl citrate (TEC), tributyl citrate (TBC), acetyl triethyl citrate (ATEC) and acetyl tributyl citrate (ATBC) as plasticisers for PLA and their effect on enzyme-catalysed hydrolytic degradation. The study found that all citrate esters enhanced the degradability of PLA. At a concentration of 20%, ATEC plasticised PLA has the highest weight loss of around 95%, followed by neat PLA at around 48%, and a decreased rate of degradation in TEC, TBC, and ATBC with weight losses of around 30%, 19%, and 18%, respectively, after a degradation period of 6 h [134]. A 2009 study by Ozkoc and Kemaloglu found that the addition of PEG and clay plasticisers to PLA decreases the rate of biodegradation after exposure to the composting environment for 100 days. This is evident from the weight loss percentage values of about 15%, 12%, and 11% for PLA/3%Clay/PEG, PLA/5%Clay/PEG, and PLA/3%Clay, respectively, in comparison to that of neat PLA with a 36% weight loss. However, the weight loss percentage of PLA/PEG shows a slight increase as compared to neat PLA with a value of around 38% weight loss [135].

Careful selection and monitoring of additives and composite formation, hence, has the potential for exploitation both to enhance the target mechanical properties and simultaneously promote biodegradation.

5. Conclusions

The fundamental many-faceted aspects of biopolymer architectures afford greater versatility for configuration and exploitation compared with the main ubiquitous recalcitrant synthetic petroleum-based plastics. This is a pivotal trait that is available for harnessing to enable bioplastics to meet both the high mechanical and barrier performances of their petroleum-based counterparts together with full sustainable biodegradability and circularity. Their very bio-nature means bioplastics are inherently more elaborate at a structural level. This increases accessibility to bio-interactions for enzymatic biodegradation, biodepolymerisation and biorepolymerisation, as well as supporting routes to improving mechanical and barrier performances. While several bioplastics have already achieved certain mechanical and barrier performance criteria, which are equivalent and even exceed those of corresponding petroleum-based plastics, courses of action for resolving the remaining limitations are becoming increasingly accessible.

Here, avenues for the advancement of the performance of bioplastics with respect to mechanical and barrier properties alongside biodegradability are discussed. The key architectural features properties are M_w and crystallinity, which typically exhibit an inversely dependent relationship with mechanical performance and biodegradability for both bioplastics and petroleum-based plastics. Increasing M_w and crystallinity are generally associated with higher mechanical performance and decreased biodegradability as lower crystallinity corresponds to looser chain packing, facilitating enzyme access.

Both macro and micro strategies have the potential for dual positive correlation on the mechanical and biodegradability performances of bioplastics. Regarding the macro approach, new possibilities are afforded, such as harnessing intricate bioplastic structures and formats, which include microfibrillar frameworks and combination with advanced methods, such as in situ polymerisation. Blending and compounding with additives as selected fillers, plastisisers and compatibilisers are also being demonstrated for improved mechanical features, without decreasing biodegradability.

Whereas, regarding the micro approach, expanding PHAs families and large numbers for possible monomer units, present the potential to engineer biodegradable plastics with equivalent target petroleum plastic performances without associated environmental pollution. Further routes include metabolic pathway alteration, design of high specificity substrates, intricate copolymer and block copolymer and genetic modification to produce strains to achieve next-generation multifunctional biopolymers.

In these respects, bioplastic polymers, in contrast to Petroleum-based plastics, have not yet been tailored or even adequately explored to establish their capacities for current and future applications. Given the range and diversity of options available for bioplastics development, there are excellent prospects to extend their application range on a comparable scale to fossil-based thermoplastics and beyond. Further innovations can be expected as the knowledge and new capacities for the manipulation of biopolymers advances, and spawns outputs in related and novice disciplines. The realisation of high performance plastics, without recalcitrance, pollution or resource depletion and switching to regenerative low carbon circularity, has the potential to both safeguard and promote future prosperity for the planet and its inhabitants.

Author Contributions: O.A.A. Conceptualization, Data curation, Investigation, Writing—original draft; E.L.G. Investigation, Data curation, Formal analysis; Writing—original draft; M.A. Investigation, Data curation; S.A. Investigation, Data curation; Y.C. Data curation, Formal analysis; Writing—original draft; M.B.F. Supervision, Validation, Writing—original draft; M.M. Investigation, Visualization, Writing—original draft, Writing—review & editing. All authors have read and agreed to the published version of the manuscript.

Funding: This project has received funding from the European Union's Horizon 2020 research and innovation programme under grant agreement No 870292 (BioICEP).

Institutional Review Board Statement: Not applicable.

Informed Consent Statement: Not applicable.

Data Availability Statement: The data presented in this study are available on request from the corresponding author.

Acknowledgments: This project has received funding from the European Union's Horizon 2020 research and innovation programme under grant agreement No 870292 (BioICEP) and is supported by the National Natural Science Foundation of China (grant numbers: Institute of Microbiology, Chinese Academy of Sciences: 31961133016, Beijing Institute of Technology: 31961133015, Shandong University: 31961133014).

Conflicts of Interest: The authors declare no conflict of interest.

References

1. Madison, L.L.; Huisman, G.W. Metabolic Engineering of Poly(3-Hydroxyalkanoates): From DNA to Plastic. *Microbiol. Mol. Biol. Rev.* **1999**, *63*, 21–53. [CrossRef] [PubMed]
2. Koller, M.; Maršálek, L.; Dias, M.M.D.S.; Braunegg, G. Producing microbial polyhydroxyalkanoate (PHA) biopolyesters in a sustainable manner. *New Biotechnol.* **2017**, *37*, 24–38. [CrossRef] [PubMed]
3. Koller, M. A Review on Established and Emerging Fermentation Schemes for Microbial Production of Polyhydroxy-alkanoate (PHA) Biopolyesters. *Fermentation* **2018**, *4*, 30. [CrossRef]
4. Rajwade, J.M.; Paknikar, K.; Kumbhar, J.V. Applications of bacterial cellulose and its composites in biomedicine. *Appl. Microbiol. Biotechnol.* **2015**, *99*, 2491–2511. [CrossRef]
5. Sudesh, K.; Abe, H.; Doi, Y. Synthesis, structure and properties of polyhydroxyalkanoates: Biological polyesters. *Prog. Polym. Sci.* **2000**, *25*, 1503–1555. [CrossRef]
6. Muneer, F.; Rasul, I.; Azeem, F.; Siddique, M.H.; Zubair, M.; Nadeem, H. Microbial Polyhydroxyalkanoates (PHAs): Efficient Replacement of Synthetic Polymers. *J. Polym. Environ.* **2020**, *28*, 2301–2323. [CrossRef]
7. Ciesielski, S.; Mozejko-Ciesielska, J.; Pisutpaisal, N. Plant oils as promising substrates for polyhydroxyalkanoates pro-duction. *J. Clean. Prod.* **2015**, *106*, 408–421. [CrossRef]
8. Shuai, X.-T.; Jedlinski, Z.; Luo, Q.; Farhod, N. Synthesis of novel block copolymers of poly(3-hydroxybutyric acid) with poly(ethylene glycol) through anionic polymerisation. *Chin. J. Polym. Sci.* **2000**, *18*, 19–23.
9. Fu, J.; Sharma, P.; Spicer, V.; Krokhin, O.V.; Zhang, X.; Fristensky, B.; Wilkins, J.A.; Cicek, N.; Sparling, R.; Levin, D.B. Effects of impurities in biodiesel-derived glycerol on growth and expression of heavy metal ion homeostasis genes and gene products in Pseudomonas putida LS46. *Appl. Microbiol. Biotechnol.* **2015**, *99*, 5583–5592. [CrossRef]
10. Noda, I.; Lindsey, S.B.; Caraway, D. Nodax™ Class PHA Copolymers: Their Properties and Applications. In *Beneficial Microorganisms in Food and Nutraceuticals*; Springer Science and Business Media LLC: Berlin/Heidelberg, Germany, 2010; pp. 237–255.
11. Kunioka, M.; Doi, Y. Thermal degradation of microbial copolyesters: Poly(3-hydroxybutyrate-co-3-hydroxyvalerate) and poly(3-hydroxybutyrate-co-4-hydroxybutyrate). *Macromolecules* **1990**, *23*, 1933–1936. [CrossRef]
12. Avella, M.; Martuscelli, E.; Raimo, M. Review Properties of blends and composites based on poly(3-hydroxy)butyrate (PHB) and poly(3-hydroxybutyrate-hydroxyvalerate) (PHBV) copolymers. *J. Mater. Sci.* **2000**, *35*, 523–545. [CrossRef]
13. Qiu, Y.-Z.; Han, J.; Guo, J.-J.; Chen, G.-Q. Production of Poly(3-hydroxybutyrate-co-3-hydroxyhexanoate) from Gluconate and Glucose by Recombinant Aeromonas hydrophila and Pseudomonas putida. *Biotechnol. Lett.* **2005**, *27*, 1381–1386. [CrossRef]
14. McAdam, B.; Fournet, M.B.; McDonald, P.; Mojicevic, M. Production of Polyhydroxybutyrate (PHB) and Factors Impacting Its Chemical and Mechanical Characteristics. *Polymers* **2020**, *12*, 2908. [CrossRef]
15. Ravenelle, F.; Marchessault, R.H. One-Step Synthesis of Amphiphilic Diblock Copolymers from Bacterial Poly([R]-3-hydroxybutyric acid). *Biomacromolecules* **2002**, *3*, 1057–1064. [CrossRef]
16. Amaro, T.M.M.M.; Rosa, D.; Comi, G.; Iacumin, L. Prospects for the Use of Whey for Polyhydroxyalkanoate (PHA) Pro-duction. *Front. Microbiol.* **2019**, *10*, 992. [CrossRef]
17. Colombo, B.; Favini, F.; Scaglia, B.; Sciarria, T.P.; D'Imporzano, G.; Pognani, M.; Alekseeva, A.; Eisele, G.; Cosentino, C.; Adani, F. Enhanced polyhydroxyalkanoate (PHA) production from the organic fraction of municipal solid waste by using mixed microbial culture. *Biotechnol. Biofuels* **2017**, *10*, 1–15. [CrossRef] [PubMed]
18. Rajendran, N.; Sharanya, P.; Sneha Raj, M.; Ruth Angeeleena, B.; Rajam, C. Seaweeds can be a new source for bio-plastics. *J. Pharm. Res.* **2012**, *5*, 1476–1479.
19. Chee, J.Y.; Tan, Y.; Samian, R.; Sudesh, K. Isolation and Characterization of a Burkholderia sp. USM (JCM15050) Capable of Producing Polyhydroxyalkanoate (PHA) from Triglycerides, Fatty Acids and Glycerols. *J. Polym. Environ.* **2010**, *18*, 584–592. [CrossRef]
20. Koller, M.; Salerno, A.; Dias, M.; Reiterer, A.; Braunegg, G. Modern biotechnological polymer synthesis: A review. *Food Technol. Biotechnol.* **2010**, *48*, 255–269.
21. Pandey, A. Solid-state fermentation. *Biochem. Eng. J.* **2003**, *13*, 81–84. [CrossRef]
22. Sindhu, R.; Pandey, A.; Binod, P. Solid-state fermentation for the production of Poly(hydroxyalkanoates). *Chem. Biochem. Eng. Q.* **2015**, *29*, 173–181. [CrossRef]
23. Reiniati, I.; Hrymak, A.N.; Margaritis, A. Recent developments in the production and applications of bacterial cellulose fibers and nanocrystals. *Crit. Rev. Biotechnol.* **2017**, *37*, 510–524. [CrossRef]
24. Florea, M.; Hagemann, H.; Santosa, G.; Abbott, J.; Micklem, C.; Spencer-Milnes, X.; Garcia, L.D.A.; Paschou, D.; Lazenbatt, C.; Kong, D.; et al. Engineering control of bacterial cellulose production using a genetic toolkit and a new cellulose-producing strain. *Proc. Natl. Acad. Sci. USA* **2016**, *113*, E3431–E3440. [CrossRef] [PubMed]
25. Liu, M.; Li, S.; Xie, Y.; Jia, S.; Hou, Y.; Zou, Y.; Zhong, C. Enhanced bacterial cellulose production by Gluconacetobacter xylinus via expression of Vitreoscilla hemoglobin and oxygen tension regulation. *Appl. Microbiol. Biotechnol.* **2018**, *102*, 1155–1165. [CrossRef]
26. Hungund, B.; Gupta, S. Strain improvement of Gluconacetobacter xylinus NCIM 2526 for bacterial cellulose production. *Afr. J. Biotechnol.* **2010**, *9*, 32.
27. Neves, N.M.; Reis, R.L. (Eds.) *Biomaterials from Nature for Advanced Devices and Therapies*; Wiley: Hoboken, NJ, USA, 2016; pp. 384–399.

28. Volova, T.G.; Prudnikova, S.; Sukovatyi, A.G.; Shishatskaya, E. Production and properties of bacterial cellulose by the strain Komagataeibacter xylinus B-12068. *Appl. Microbiol. Biotechnol.* **2018**, *102*, 7417–7428. [CrossRef] [PubMed]
29. Lin, S.-P.; Calvar, I.L.; Catchmark, J.M.; Liu, J.-R.; Demirci, A.; Cheng, K.-C. Biosynthesis, production and applications of bacterial cellulose. *Cellulose* **2013**, *20*, 2191–2219. [CrossRef]
30. Shah, N.; Ul-Islam, M.; Khattak, W.A.; Park, J.K. Overview of bacterial cellulose composites: A multipurpose advanced material. *Carbohydr. Polym.* **2013**, *98*, 1585–1598. [CrossRef] [PubMed]
31. Hsieh, J.-T.; Wang, M.-J.; Lai, J.-T.; Liu, H.-S. A novel static cultivation of bacterial cellulose production by intermittent feeding strategy. *J. Taiwan Inst. Chem. Eng.* **2016**, *63*, 46–51. [CrossRef]
32. Ul-Islam, M.; Khan, S.; Ullah, M.W.; Park, J.K. Bacterial cellulose composites: Synthetic strategies and multiple applications in bio-medical and electro-conductive fields. *Biotechnol. J.* **2015**, *10*, 1847–1861. [CrossRef] [PubMed]
33. Singhsa, P.; Narain, R.; Manuspiya, H. Physical structure variations of bacterial cellulose produced by different Komaga-taeibacter xylinus strains and carbon sources in static and agitated conditions. *Cellulose* **2018**, *25*, 1571–1581. [CrossRef]
34. Parte, F.G.B.; Santoso, S.P.; Chou, C.-C.; Verma, V.; Wang, H.-T.; Ismadji, S.; Cheng, K.-C. Current progress on the production, modification, and applications of bacterial cellulose. *Crit. Rev. Biotechnol.* **2020**, *40*, 397–414. [CrossRef] [PubMed]
35. Auras, R.; Harte, B.; Selke, S. An Overview of Polylactides as Packaging Materials. *Macromol. Biosci.* **2004**, *4*, 835–864. [CrossRef]
36. Zhao, X.; Cornish, K.; Vodovotz, Y. Narrowing the Gap for Bioplastic Use in Food Packaging: An Update. *Environ. Sci. Technol.* **2020**, *54*, 4712–4732. [CrossRef]
37. Chen, S.-Q.; Lopez-Sanchez, P.; Wang, D.; Mikkelsen, D.; Gidley, M.J. Mechanical properties of bacterial cellulose synthesised by diverse strains of the genus Komagataeibacter. *Food Hydrocoll.* **2018**, *81*, 87–95. [CrossRef]
38. Jem, K.J.; Tan, B. The development and challenges of poly (lactic acid) and poly (glycolic acid). *Adv. Ind. Eng. Polym. Res.* **2020**, *3*, 60–70. [CrossRef]
39. Farah, S.; Anderson, D.G.; Langer, R. Physical and mechanical properties of PLA, and their functions in widespread applications—A comprehensive review. *Adv. Drug Deliv. Rev.* **2016**, *107*, 367–392. [CrossRef] [PubMed]
40. Visakh, P.M. CHAPTER 1. Polyhydroxyalkanoates (PHAs), their Blends, Composites and Nanocomposites: State of the Art, New Challenges and Opportunities. In *Polyhydroxyalkanoate (PHA) Based Blends, Composites and Nanocomposites*; Green Chemistry Series; Royal Society of Chemistry: Cambridge, UK; 2014; pp. 1–17. [CrossRef]
41. Goh, L.-K.; Purama, R.K.; Sudesh, K. Enhancement of Stress Tolerance in the Polyhydroxyalkanoate Producers without Mobilization of the Accumulated Granules. *Appl. Biochem. Biotechnol.* **2014**, *172*, 1585–1598. [CrossRef]
42. Dos Santos, A.J.; Valentina, L.V.O.D.; Schulz, A.A.H.; Duarte, M.A.T. From Obtaining to Degradation of PHB:Material Properties. Part I. *Ing. Cienc.* **2017**, *13*, 269–298. [CrossRef]
43. El-Hadi, A.M.; Schnabel, R.; Straube, E.; Müller, G.; Henning, S. Correlation between degree of crystallinity, morphology, glass temperature, mechanical properties and biodegradation of poly (3-hydroxyalkanoate) PHAs and their blends. *Polym. Test.* **2002**, *21*, 665–674. [CrossRef]
44. Smith, M.K.M.; Paleri, D.M.; Abdelwahab, M.; Mielewski, D.F.; Misra, M.; Mohanty, A.K. Sustainable composites from poly(3-hydroxybutyrate) (PHB) bioplastic and agave natural fibre. *Green Chem.* **2020**, *22*, 3906–3916. [CrossRef]
45. Barham, P.J.; Keller, A.; Otun, E.L.; Wills, H.H.; Holmes, P.A. Crystallization and morphology of a bacterial thermoplastic: Poly-3-hyd roxybutyrate. *J. Mater. Sci.* **1984**, *19*. [CrossRef]
46. Barham, P.J.; Keller, A. The relationship between microstructure and mode of fracture in polyhydroxybutyrate. *J. Polym. Sci. Part B Polym. Phys.* **1986**, *24*, 69–77. [CrossRef]
47. Janigová, I.; Lacıík, I.; Chodák, I. Thermal degradation of plasticized poly(3-hydroxybutyrate) investigated by DSC. *Polym. Degrad. Stab.* **2002**, *77*, 35–41. [CrossRef]
48. Shim, S.H.; Kim, K.T.; Lee, J.U.; Jo, W.H. Facile Method to Functionalize Graphene Oxide and Its Application to Poly(ethylene terephthalate)/Graphene Composite. *ACS Appl. Mater. Interfaces* **2012**, *4*, 4184–4191. [CrossRef] [PubMed]
49. Papadopoulou, E.L.; Basnett, P.; Paul, U.C.; Marras, S.; Ceseracciu, L.; Roy, I.; Athanassiou, A. Green Composites of Poly(3-hydroxybutyrate) Containing Graphene Nanoplatelets with Desirable Electrical Conductivity and Oxygen Barrier Properties. *ACS Omega* **2019**, *4*, 19746–19755. [CrossRef] [PubMed]
50. Dhar, P.; Gaur, S.S.; Soundararajan, N.; Gupta, A.; Bhasney, S.M.; Milli, M.; Kumar, A.; Katiyar, V. Reactive Extrusion of Polylactic Acid/Cellulose Nanocrystal Films for Food Packaging Applications: Influence of Filler Type on Thermomechanical, Rheological, and Barrier Properties. *Ind. Eng. Chem. Res.* **2017**, *56*, 4718–4735. [CrossRef]
51. Azeem, M.; Jan, R.; Farrukh, S.; Hussain, A. Improving gas barrier properties with boron nitride nanosheets in poly-mer-composites. *Results Phys.* **2019**, *12*, 1535–1541. [CrossRef]
52. Bharadwaj, R.K. Modeling the Barrier Properties of Polymer-Layered Silicate Nanocomposites. *Macromolecules* **2001**, *34*, 9189–9192. [CrossRef]
53. Moggridge, G.; Lape, N.K.; Yang, C.; Cussler, E. Barrier films using flakes and reactive additives. *Prog. Org. Coat.* **2003**, *46*, 231–240. [CrossRef]
54. Cui, Z.; Martinez, A.P.; Adamson, D.H. PMMA functionalized boron nitride sheets as nanofillers. *Nanoscale* **2015**, *7*, 10193–10197. [CrossRef]

55. Gitari, B.; Chang, B.P.; Misra, M.; Navabi, A.; Mohanty, A.K. A comparative study on the mechanical, thermal, and water barrier properties of PLA nanocomposite films prepared with bacterial nanocellulose and cellulose nanofibrils. *BioResources* **2019**, *14*, 1867–1889.
56. Xu, P.; Yang, W.; Niu, D.; Yu, M.; Du, M.; Dong, W.; Chen, M.; Lemstra, P.J.; Ma, P. Multifunctional and robust polyhydroxyalkanoate nanocomposites with superior gas barrier, heat resistant and inherent antibacterial performances. *Chem. Eng. J.* **2020**, *382*, 122864. [CrossRef]
57. Jandas, P.J.; Mohanty, S.; Nayak, S.K. Morphology and Thermal Properties of Renewable Resource-Based Polymer Blend Nanocomposites Influenced by a Reactive Compatibilizer. *ACS Sustain. Chem. Eng.* **2013**, *2*, 377–386. [CrossRef]
58. Kick, T.; Grethe, T.; Mahltig, B. A Natural Based Method for Hydrophobic Treatment of Natural Fiber Material. *Acta Chim. Slov.* **2017**, *64*, 373–380. [CrossRef] [PubMed]
59. Lu, N.; Oza, S.; Tajabadi, M.G. Surface Modification of Natural Fibers for Reinforcement in Polymeric Composites. In *Surface Modification of Biopolymers*; John Wiley & Sons, Inc.: Hoboken, NJ, USA, 2015; pp. 224–237.
60. Zhijiang, C.; Yi, X.; Haizheng, Y.; Jia, J.; Liu, Y. Poly(hydroxybutyrate)/cellulose acetate blend nanofiber scaffolds: Prepa-ration, characterization and cytocompatibility. *Mater. Sci. Eng. C* **2016**, *58*, 757–767. [CrossRef]
61. Panaitescu, D.M.; Nicolae, C.A.; Gabor, A.R.; Trusca, R. Thermal and mechanical properties of poly(3-hydroxybutyrate) reinforced with cellulose fibers from wood waste. *Ind. Crop. Prod.* **2020**, *145*, 112071. [CrossRef]
62. Mokhothu, T.H.; John, M.J. Review on hygroscopic aging of cellulose fibres and their biocomposites. *Carbohydr. Polym.* **2015**, *131*, 337–354. [CrossRef]
63. Joffe, R.; Pupure, L.; Berthold, F.; Varna, J. Micro-structure and Mechanical Properties in PLA Reinforced with Cellulosic Fiber Sheets Made by Wet Forming Method. In Proceedings of the 8th International Conference on Composites Testing and Model Identification (CompTest 2017), Leuven, Belgium, 5–7 April 2017.
64. Masmoudi, F.; Bessadok, A.; Dammak, M.; Jaziri, M.; Ammar, E. Biodegradable packaging materials conception based on starch and polylactic acid (PLA) reinforced with cellulose. *Environ. Sci. Pollut. Res.* **2016**, *23*, 20904–20914. [CrossRef]
65. Kyutoku, H.; Maeda, N.; Sakamoto, H.; Nishimura, H.; Yamada, K. Effect of surface treatment of cellulose fiber (CF) on durability of PLA/CF bio-composites. *Carbohydr. Polym.* **2019**, *203*, 95–102. [CrossRef]
66. Piekarska, K.; Sowinski, P.; Piorkowska, E.; Haque, M.M.U.; Pracella, M. Structure and properties of hybrid PLA nano-composites with inorganic nanofillers and cellulose fibers. *Compos. Part A Appl. Sci. Manuf.* **2016**, *82*, 34–41. [CrossRef]
67. Hu, R.; Lim, J.-K. Fabrication and Mechanical Properties of Completely Biodegradable Hemp Fiber Reinforced Polylactic Acid Composites. *J. Compos. Mater.* **2007**, *41*, 1655–1669. [CrossRef]
68. Pommet, M.; Juntaro, J.; Heng, J.; Mantalaris, A.; Lee, A.; Wilson, K.; Kalinka, G.; Shaffer, M.S.P.; Bismarck, A. Surface Modification of Natural Fibers Using Bacteria: Depositing Bacterial Cellulose onto Natural Fibers To Create Hierarchical Fiber Reinforced Nanocomposites. *Biomacromolecules* **2008**, *9*, 1643–1651. [CrossRef]
69. Xu, Z.; Yang, L.; Ni, Q.; Ruan, F.; Wang, H. Fabrication of high-performance green hemp/polylactic acid fibre composites. *J. Eng. Fibers Fabr.* **2019**, *14*. [CrossRef]
70. Michel, A.; Billington, S. Characterization of poly-hydroxybutyrate films and hemp fiber reinforced composites exposed to accelerated weathering. *Polym. Degrad. Stab.* **2012**, *97*, 870–878. [CrossRef]
71. Keller, A. Compounding and mechanical properties of biodegradable hemp fibre composites. *Compos. Sci. Technol.* **2003**, *63*, 1307–1316. [CrossRef]
72. Sawpan, M.A.; Pickering, K.L.; Fernyhough, A. Characterisation of hemp fibre reinforced Poly(Lactic Acid) composites. *Int. J. Mater. Prod. Technol.* **2009**, *36*, 229. [CrossRef]
73. Smoca, A.; Kukle, S.; Zelca, Z. Properties of Hemp Fibres Reinforced PLA Composites. *Key Eng. Mater.* **2019**, *800*, 205–209. [CrossRef]
74. Avella, M.; Buzõarovska, A.; Errico, M.E.; Gentile, G.; Grozdanov, A.; Bogoeva-Gaceva, G. Poly(3-hydroxybutyrate-co-3-hydroxyvalerate)-based biocomposites reinforced with kenaf fibers. *J. Appl. Polym. Sci.* **2007**, *104*, 3192–3200. [CrossRef]
75. Hassan, A.; Isa, M.R.M.; Ishak, Z.A.M. Improving thermal and mechanical properties of injection moulded Kenaf Fibre-reinforced Polyhydroxy-butyrate composites through fibre surface treatment. *BioResources* **2019**, *14*, 3101–3116.
76. Buzarovska, A.; Bogoeva-Gaceva, G.; Grozdanov, A.; Avella, M. Crystallization behavior of polyhydroxybutyrate in model composites with kenaf fibers. *J. Appl. Polym. Sci.* **2006**, *102*, 804–809. [CrossRef]
77. Graupner, N.; Müssig, J. A comparison of the mechanical characteristics of kenaf and lyocell fibre reinforced poly(lactic acid) (PLA) and poly(3-hydroxybutyrate) (PHB) composites. *Compos. Part A Appl. Sci. Manuf.* **2011**, *42*, 2010–2019. [CrossRef]
78. Kuciel, S.; Liber-Knec, A. Biocomposites based on PHB filled with wood or kenaf fibers. *Polimery* **2011**, *56*, 218–223. [CrossRef]
79. Bayart, M.; Gauvin, F.; Foruzanmehr, M.R.; Elkoun, S.; Robert, M. Mechanical and moisture absorption characterization of PLA composites reinforced with nano-coated flax fibers. *Fibers Polym.* **2017**, *18*, 1288–1295. [CrossRef]
80. Georgiopoulos, P.; Kontou, E.; Georgousis, G. Effect of silane treatment loading on the flexural properties of PLA/flax unidirectional composites. *Compos. Commun.* **2018**, *10*, 6–10. [CrossRef]
81. Karsli, N.G.; Aytac, A. Properties of alkali treated short flax fiber reinforced poly(lactic acid)/polycarbonate composites. *Fibers Polym.* **2014**, *15*, 2607–2612. [CrossRef]
82. Aydın, M.; Tozlu, H.; Kemaloglu, S.; Aytac, A.; Ozkoc, G. Effects of Alkali Treatment on the Properties of Short Flax Fi-ber-Poly(Lactic Acid) Eco-Composites. *J. Polym. Environ.* **2011**, *19*, 11–17. [CrossRef]

83. Shanks, R.A.; Hodzic, A.; Ridderhof, D. Composites of poly(lactic acid) with flax fibers modified by interstitial polymeri-zation. *J. Appl. Polym. Sci.* **2006**, *99*, 2305–2313. [CrossRef]
84. Foruzanmehr, M.; Vuillaume, P.; Elkoun, S.; Robert, M. Physical and mechanical properties of PLA composites reinforced by TiO^2 grafted flax fibers. *Mater. Des.* **2016**, *106*, 295–304. [CrossRef]
85. Ventura, H.; Claramunt, J.; Rodríguez-Pérez, M.; Ardanuy, M. Effects of hydrothermal aging on the water uptake and tensile properties of PHB/flax fabric biocomposites. *Polym. Degrad. Stab.* **2017**, *142*, 129–138. [CrossRef]
86. Chilali, A.; Assarar, M.; Zouari, W.; Kebir, H.; Ayad, R. Mechanical characterization and damage events of flax fabric-reinforced biopolymer composites. *Polym. Polym. Compos.* **2019**, *28*, 631–644. [CrossRef]
87. Rytlewski, P.; Stepczyńska, M.; Moraczewski, K.; Malinowski, R.; Karasiewicz, T.; Sikorska, W.; Żenkiewicz, M. Flax fibers reinforced polycaprolactone modified by triallyl isocyanurate and electron radiation. *Polym. Compos.* **2019**, *40*, 481–488. [CrossRef]
88. Lee, S.G.; Choi, S.-S.; Park, W.H.; Cho, D. Characterization of surface modified flax fibers and their biocomposites with PHB. *Macromol. Symp.* **2003**, *197*, 089–100. [CrossRef]
89. Arrieta, M.P.; Samper, M.D.; Aldas, M.; López, J. On the Use of PLA-PHB Blends for Sustainable Food Packaging Applications. *Materials* **2017**, *10*, 1008. [CrossRef]
90. Zhang, M.; Thomas, N.L. Blending polylactic acid with polyhydroxybutyrate: The effect on thermal, mechanical, and biodegradation properties. *Adv. Polym. Technol.* **2011**, *30*, 67–79. [CrossRef]
91. Armentano, I.; Fortunati, E.; Burgos, N.; Dominici, F.; Luzi, F.; Fiori, S.; Jimenez, A.; Yoon, K.; Ahn, J.; Kang, S.; et al. Processing and characterization of plasticized PLA/PHB blends for biodegradable multiphase systems. *Express Polym. Lett.* **2015**, *9*, 583–596. [CrossRef]
92. Armentano, I.; Fortunati, E.; Burgos, N.; Dominici, F.; Luzi, F.; Fiori, S.; Jiménez, A.; Yoon, K.; Ahn, J.; Kang, S.; et al. Bio-based PLA_PHB plasticized blend films: Processing and structural characterization. *LWT* **2015**, *64*, 980–988. [CrossRef]
93. Abdelwahab, M.; Flynn, A.; Chiou, B.-S.; Imam, S.; Orts, W.; Chiellini, E. Thermal, mechanical and morphological characterization of plasticized PLA–PHB blends. *Polym. Degrad. Stab.* **2012**, *97*, 1822–1828. [CrossRef]
94. Arrieta, M.P.; Samper, M.D.; Lopez, J.; Jiménez, A. Combined Effect of Poly(hydroxybutyrate) and Plasticizers on Polylactic acid Properties for Film Intended for Food Packaging. *J. Polym. Environ.* **2014**, *22*, 460–470. [CrossRef]
95. Rodrigues, J.A.F.R.; Parra, D.F.; Lugão, A.B. Crystallization on films of PHB/PEG blends. *J. Therm. Anal. Calorim.* **2005**, *79*, 379–381. [CrossRef]
96. Ali, A.; Shaker, K.; Nawab, Y.; Jabbar, M.; Hussain, T.; Militky, J.; Baheti, V. Hydrophobic treatment of natural fibers and their composites—A review. *J. Ind. Text.* **2018**, *47*, 2153–2183. [CrossRef]
97. Lalit, R.; Mayank, P.; Ankur, K. Natural Fibers and Biopolymers Characterization: A Future Potential Composite Material. *Stroj. Cas.* **2018**, *68*, 33–50. [CrossRef]
98. Awasthi, S.; Agarwal, D. Influence of cycloaliphatic compounds on the properties of polyurethane coatings. *J. Coat. Technol. Res.* **2007**, *4*, 67–73. [CrossRef]
99. Alagi, P.; Choi, Y.J.; Hong, S.C. Preparation of vegetable oil-based polyols with controlled hydroxyl functionalities for thermoplastic polyurethane. *Eur. Polym. J.* **2016**, *78*, 46–60. [CrossRef]
100. Miao, S.; Wang, P.; Su, Z.; Zhang, S. Vegetable-oil-based polymers as future polymeric biomaterials. *Acta Biomater.* **2014**, *10*, 1692–1704. [CrossRef] [PubMed]
101. Deka, H.; Karak, N. Bio-based hyperbranched polyurethanes for surface coating applications. *Prog. Org. Coat.* **2009**, *66*, 192–198. [CrossRef]
102. Kong, X.; Liu, G.; Curtis, J.M. Novel polyurethane produced from canola oil based poly(ether ester) polyols: Synthesis, characterization and properties. *Eur. Polym. J.* **2012**, *48*, 2097–2106. [CrossRef]
103. Tan, S.; Abraham, T.; Ference, D.; Macosko, C.W. Rigid polyurethane foams from a soybean oil-based Polyol. *Polymer* **2011**, *52*, 2840–2846. [CrossRef]
104. Ismail, E.A.; Motawie, A.; Sadek, E.; Ismail, E.A.; Motawie, A.; Sadek, E. Synthesis and characterization of polyurethane coatings based on soybean oil–polyester polyols. *Egypt. J. Pet.* **2011**, *20*, 1–8. [CrossRef]
105. Datta, J.; Głowińska, E. Effect of hydroxylated soybean oil and bio-based propanediol on the structure and thermal properties of synthesized bio-polyurethanes. *Ind. Crop. Prod.* **2014**, *61*, 84–91. [CrossRef]
106. Fu, C.; Zheng, Z.; Yang, Z.; Chen, Y.; Shen, L. A fully bio-based waterborne polyurethane dispersion from vegetable oils: From synthesis of precursors by thiol-ene reaction to study of final material. *Prog. Org. Coat.* **2014**, *77*, 53–60. [CrossRef]
107. Thakur, S.; Karak, N. Castor oil-based hyperbranched polyurethanes as advanced surface coating materials. *Prog. Org. Coat.* **2013**, *76*, 157–164. [CrossRef]
108. Gurunathan, T.; Mohanty, S.; Nayak, S.K. Isocyanate terminated castor oil-based polyurethane prepolymer: Synthesis and characterization. *Prog. Org. Coat.* **2015**, *80*, 39–48. [CrossRef]
109. Mokhothu, T.H.; John, M.J. Bio-based coatings for reducing water sorption in natural fibre reinforced composites. *Sci. Rep.* **2017**, *7*, 1–8. [CrossRef]
110. Van Krevelen, D.W.; Te Nijenhuis, K. *Properties of Polymers: Their Correlation with Chemical Structure; Their Numerical Estimation and Prediction from Additive Group Contributions*; Elsevier: Amsterdam, The Netherlands, 2009.
111. Sanchez, P.; A Perez-Maqueda, L.; Perejon, A.; Criado, J.M. A new model for the kinetic analysis of thermal degradation of polymers driven by random scission. *Polym. Degrad. Stab.* **2010**, *95*, 733–739. [CrossRef]

112. Ghosh, P. *Polymer Science and Technology*; Tata McGraw-Hill Education: New York, NY, USA, 1990.
113. Tokiwa, Y.; Calabia, B.P.; Ugwu, C.U.; Aiba, S. Biodegradability of Plastics. *Int. J. Mol. Sci.* **2009**, *10*, 3722–3742. [CrossRef]
114. Bernard, M. Industrial Potential of Polyhydroxyalkanoate Bioplastic: A Brief Review. *USURJ: Univ. Sask. Undergrad. Res. J.* **2014**, *1*. [CrossRef]
115. Muhammadi, S.; Afzal, M.; Hameed, S. Bacterial polyhydroxyalkanoates-eco-friendly next generation plastic: Production, biocompatibility, biodegradation, physical properties and applications. *Green Chem. Lett. Rev.* **2015**, *8*, 56–77. [CrossRef]
116. Ahmed, T.; Shahid, M.; Azeem, F.; Rasul, I.; Shah, A.A.; Noman, M.; Hameed, A.; Manzoor, N.; Manzoor, I.; Muhammad, S. Biodegradation of plastics: Current scenario and future prospects for environmental safety. *Environ. Sci. Pollut. Res.* **2018**, *25*, 7287–7298. [CrossRef] [PubMed]
117. Knoll, M.; Hamm, T.M.; Wagner, F.; Martinez, V.; Pleiss, J. The PHA Depolymerase Engineering Database: A systematic analysis tool for the diverse family of polyhydroxyalkanoate (PHA) depolymerases. *BMC Bioinform.* **2009**, *10*, 89. [CrossRef]
118. Mergaert, J.; Webb, A.; Anderson, C.; Wouters, A.; Swings, J. Microbial degradation of poly(3-hydroxybutyrate) and poly(3-hydroxybutyrate-co-3-hydroxyvalerate) in soils. *Appl. Environ. Microbiol.* **1993**, *59*, 3233–3238. [CrossRef]
119. Jendrossek, D.; Handrick, R. Microbial Degradation of Polyhydroxyalkanoates. *Annu. Rev. Microbiol.* **2002**, *56*, 403–432. [CrossRef]
120. Manna, A.; Paul, A. Degradation of microbial polyester poly(3-hydroxybutyrate) in environmental samples and in culture. *Biodegradation* **2000**, *11*, 323–329. [CrossRef] [PubMed]
121. Kusaka, S.; Iwata, T.; Doi, Y. Properties and biodegradability of ultra-high-molecular-weight poly[(R)-3-hydroxybutyrate] produced by a recombinant Escherichia coli. *Int. J. Biol. Macromol.* **1999**, *25*, 87–94. [CrossRef]
122. Boyandin, A.N.; Prudnikova, S.; Filipenko, M.L.; Khrapov, E.A.; Vasil'Ev, A.D.; Volova, T. Biodegradation of polyhydroxyalkanoates by soil microbial communities of different structures and detection of PHA degrading microorganisms. *Appl. Biochem. Microbiol.* **2011**, *48*, 28–36. [CrossRef]
123. Murphy, J. Chapter 4—Modifying Specific Properties: Mechanical Properties—Fillers. In *Additives for Plastics Handbook*, 2nd ed.; Murphy, J., Ed.; Elsevier Science: Amsterdam, The Netherlands, 2001; Volume 1, pp. 19–35.
124. Aframehr, W.M.; Molki, B.; Heidarian, P.; Behzad, T.; Sadeghi, M.; Bagheri, R. Effect of calcium carbonate nanoparticles on barrier properties and biodegradability of polylactic acid. *Fibers Polym.* **2017**, *18*, 2041–2048. [CrossRef]
125. Teramoto, N.; Urata, K.; Ozawa, K.; Shibata, M. Biodegradation of aliphatic polyester composites reinforced by abaca fiber. *Polym. Degrad. Stab.* **2004**, *86*, 401–409. [CrossRef]
126. Altaee, N.; El-Hiti, G.A.; Fahdil, A.; Sudesh, K.; Yousif, E. Biodegradation of different formulations of polyhydroxybutyrate films in soil. *SpringerPlus* **2016**, *5*, 1–12. [CrossRef]
127. Paul, M.-A.; Delcourt, C.; Alexandre, M.; Degée, P.; Monteverde, F.; Dubois, P. Polylactide/montmorillonite nanocomposites: Study of the hydrolytic degradation. *Polym. Degrad. Stab.* **2005**, *87*, 535–542. [CrossRef]
128. Chen, Y.; Murphy, A.; Scholz, D.; Geever, L.M.; Lyons, J.G.; Devine, D.M. Surface-modified halloysite nanotubes reinforced poly(lactic acid) for use in biodegradable coronary stents. *J. Appl. Polym. Sci.* **2018**, *135*, 46521. [CrossRef]
129. Navarro, M.; Ginebra, M.; Planell, J.; Barrias, C.; Barbosa, M. In vitro degradation behavior of a novel bioresorbable composite material based on PLA and a soluble CaP glass. *Acta Biomater.* **2005**, *1*, 411–419. [CrossRef]
130. Huang, J.; Xiong, J.; Liu, J.; Zhu, W.; Wang, D. Investigation of the in vitro degradation of a novel polylactide/nanohydroxyapatite composite for artificial bone. *J. Nanomater.* **2013**, *2013*, 3. [CrossRef]
131. Valapa, R.B.; Pugazhenthi, G.; Katiyar, V. Hydrolytic degradation behaviour of sucrose palmitate reinforced poly(lactic acid) nanocomposites. *Int. J. Biol. Macromol.* **2016**, *89*, 70–80. [CrossRef] [PubMed]
132. Jamarani, R.; Erythropel, H.C.; Nicell, J.A.; Leask, R.L.; Marić, M. How Green is Your Plasticizer? *Polymers* **2018**, *10*, 834. [CrossRef]
133. Boll, M.; Geiger, R.; Junghare, M.; Schink, B. Microbial degradation of phthalates: Biochemistry and environmental impli-cations. *Environ. Microbiol. Rep.* **2020**, *12*, 3–15. [CrossRef] [PubMed]
134. Labrecque, L.V.; Kumar, R.A.; Gross, R.A.; McCarthy, S.P. Citrate esters as plasticizers for poly(lactic acid). *J. Appl. Polym. Sci.* **1997**, *66*, 1507–1513. [CrossRef]
135. Ozkoc, G.; Kemaloglu, S. Morphology, biodegradability, mechanical, and thermal properties of nanocomposite films based on PLA and plasticized PLA. *J. Appl. Polym. Sci.* **2009**, *114*, 2481–2487. [CrossRef]

polymers

MDPI

Article

Nanocomposite Materials Based on TMDCs WS_2 Modified Poly(L-Lactic Acid)/Poly(Vinylidene Fluoride) Polymer Blends

Mohammed Naffakh

Escuela Técnica Superior de Ingenieros Industriales, Universidad Politécnica de Madrid (ETSII-UPM), José Gutiérrez Abascal 2, 28006 Madrid, Spain; mohammed.naffakh@upm.es

Abstract: Novel multifunctional biopolymer blend nanocomposites composed of poly(vinylidene fluoride)(PVDF) and tungsten disulfide nanotubes (INT-WS_2) that are layered transition metal dichalcogenides (TMDCs) were easily prepared by applying an economical, scalable, and versatile melt processing route. Furthermore, their synergistic effect to enhance the properties of poly(L-lactic acid) (PLLA) matrix was investigated. From morphological analysis, it was shown that the incorporation of 1D (INT)-WS_2 into the immiscible PLLA/PVDF mixtures (weight ratios: 80/20, 60/40, 40/60, and 20/80) led to an improvement in the dispersibility of the PVDF phase, a reduction in its average domain size, and consequently a larger interfacial area. In addition, the nanoparticles INT-WS_2 can act as effective nucleating agents and reinforcing fillers in PLLA/PVDF blends, and as such, greatly improve their thermal and dynamic-mechanical properties. The improvements are more pronounced in the ternary blend nanocomposites with the lowest PVDF content, likely due to a synergistic effect of both highly crystalline PVDF and 1D-TMDCs nano-additives on the matrix performance. Considering the promising properties of the developed materials, the inexpensive synthetic process, and the extraordinary properties of environmentally friendly and biocompatibe 1D-TMDCs WS_2, this work may open up opportunities to produce new PLLA/PVDF hybrid nanocomposites that show great potential for biomedical applications.

Keywords: TMDCs-WS_2; PLLA; PVDF; nanomaterials; morphology; crystallization; dynamic-mechanical properties

Citation: Naffakh, M. Nanocomposite Materials Based on TMDCs WS_2 Modified Poly(L-Lactic Acid)/Poly(Vinylidene Fluoride) Polymer Blends. *Polymers* **2021**, *13*, 2179. https://doi.org/10.3390/polym13132179

Academic Editors: José Miguel Ferri, Vicent Fombuena Borràs and Miguel Fernando Aldás Carrasco

Received: 14 June 2021
Accepted: 28 June 2021
Published: 30 June 2021

1. Introduction

The biobased, biodegradable aliphatic polyester poly(L-lactic acid) (PLLA) that is derived from natural resources, such as corn and sugar beet, is a highly versatile polymer and a promising alternative to petroleum-derived polymers in many applications as a result of its good biodegradability, renewability, reasonably good mechanical properties, and processability [1–3]. Additionally, it is known to be an excellent base polymer for biomedical applications, including drug delivery systems (DDS), sutures, and clips due to the fact that it has good biocompatibility and that its degradation products are benign to the human body. Moreover, unlike other biopolymers like, poly(hydroxyalkanoates) (PHA), poly(ethylene glycol) (PEG), and poly(ε-caprolactone) (PCL), PLLA can be processed using many different methods, such as extrusion, film casting, fiber spinning, and blow molding, due to its higher thermal processability [4]. However, in spite of its many beneficial attributes, historically, its commercial viability has been limited by poor stability during long melt molding and processing cycles, poor production efficiency, and overall high costs. Since the cost and brittleness of PLLA are quite high, it is not economically feasible to use it alone for day-to-day use as a packaging material without blending. Thus, blends of PLLA with several synthetic and biopolymers have been prepared in an effort to enhance the properties of PLLA. PLLA blends with poly(vinylidene fluoride) (PVDF), poly(butylenes succinate adipate) (PBSA), poly(ethylene glycol) (PEG), poly(methyl methacrylate) (PMMA), polypropylene (PP), polyethylene (PE), poly(ethylene oxide) (PEO), and poly(butylenes adipate-co-terephthalate) (PBAT) have been reported to improve the properties, such as toughness, modulus, impact strength, crystallization behavior, and thermal stability, compared to the neat polymer [5–9]. Notwithstanding, successful enhancement in the nucleation and crystallization behavior has been reported for PLLA by immiscible blending with PVDF via epitaxial and interface-assisted nucleation [5]. By using the classical method of fiber extrusion, without any special spinnerets, fibrous structures were obtained consisting of a polylactic acid (PLA) matrix filled with PVDF micro- and nanofibers [10]. This kind of fibers, as a hybrid system, can be successfully used for producing implants and prostheses. PVDF film was widely used in the filtration of protein because of its hydrophobicity being able to reduce surface fouling [11]. In addition, PVDF was also reported as a scaffold for cell culture because the piezoelectric properties can enhance cellular adherence and proliferation [12,13]. Previous studies also found that PLLA can facilitate the α- to β-phase transition of PVDF under eletrospinning and uniaxial stretching [14,15]. Electrospun PLA/PVDF mats exhibited higher cell proliferation for L929 fibroblasts than both PLA and PVDF mats [14]. However, to attain PLA blends with good general properties, typically, some sort of compatibilization strategy is required. For more detailed information on PLA property modification, the recent review by Zheng et al. [16] is recommended, which includes the use of copolymer, reactive polymer, nanoparticle, and low molecular weight chemical addition, as well as interfacial compatibilization, exchange reactions, and dynamic vulcanization. The conventional approach to compatibilizing polymer blends is via the use of copolymers as it is an efficient means to achieve good blend compatibility. However, commercial unavailability of specific copolymers and the fact that they must be synthesized prior to blending is one of its drawbacks [16]. Recent nanotechnology advances have been applied to PLLA-based polymers, resulting in improved chemical, mechanical, and biological properties. Advanced nanocomposite materials were obtained by filling the polymer matrix with both synthetic and natural nanoparticles [17–20]. The addition of CNTs to polymer composite structures with natural fiber has opened a new era of polymer composites for various structural applications. As polymer matrix reinforcements, different types of CNTs with specific and unique functional groups interact with hydroxyl groups in natural fiber cellulose chains, thus modifying the natural fiber surface [17]. The emerging halloysite-based bionanocomposites are used in applications such as biomedicine, packaging, corrosion protection, and restoration of cultural heritages [19]. In this way, Lisuzzo et al. demonstrated that the chitosan coating of halloysite nanotubes driven by electrostatic interactions can be considered a suitable strategy to obtain drug delivery systems with tunable properties [20]. Of particular interest is the use of layered transition metal dicalcogenide (TMDC) nanostructures, such as tungsten disulfide (WS_2) and molybdenum disulfide (MoS_2), which are broadband semiconductors with multidimensional structural anisotropy, 0D (IF), 1D (INT), and 2D [21–26]. These environmentally friendly and biocompatible TMDCs nanoparticles also have demonstratable processing, performance, design, and cost advantages over nanoclays, CNTs, etc. when manufacturing advanced biopolymer nanocomposites poly(3-hydroxybutyrate) (PHB), Bio-PNC INT-WS_2, poly(propylene fumarate) (PPF), poly(ether ether ketone) (PEEK), PLLA, etc. [27–30]. Furthermore, the cytotoxicity of INT-WS_2 is comparable to regular environmental particulate matter and much lower than other nanoparticles, like silica or carbon black [31,32]. Moreover, blending nanoparticles with an immiscible mixture (PLLA-blend) expands the possible routes for compatibilization, and unlike polymeric compatibilizers, nanoparticles are not specific to the nature of the immiscible components of the mixture and are easily incorporated by mixing. Further, the addition of nanoparticles can significantly improve the materials' properties, combining the attributes of the base polymer blend with the characteristics of the nanoparticle [16,33].

The object of the current research is to demonstrate the advantage of using INT-WS_2 as a suitable nano-reinforcement to improve the performance of promising PLLA/PVDF polymer blends. New melt-processable nanocomposites were prepared via a scalable, versatile, and cheap procedure, without the addition of compatibilizers or modifiers. In particular, the effect of INT-WS_2 on the morphology, thermal, processability, and mechanical properties of the resulting PLLA/PVDF/INT-WS_2 nanocomposites is considered.

2. Experimental Section

2.1. Materials and Processing

Poly(L-lactic acid) (PLLA) and poly(vinylidene fluoride) (PVDF) were purchased from Goodfellow Ltd. (Huntingdon, UK). Multiwall WS_2 1D nanotubes (INT-WS_2) with diameters of 30–150 nm and lengths of 1–20 nm were obtained from NanoMaterials Ltd. (Yavne, Israel) (see Section 3.1). Both the blends and nanocomposites were prepared following the same procedure: each mixture of PVDF and PLLA, with or without INT-WS_2, was dispersed in a small volume of ethanol (HPLC grade, Sigma-Aldrich Química SL, Madrid, Spain) and homogenized by mechanical stirring and bath ultrasonication for approximately 15 min. Subsequently, the dispersion was partially dried in vacuum at 60 °C under a pressure of about 70 mbar for 24 h. The PLLA/PVDF blends are designated as 80/20, 60/40, 40/60, and 20/80, where the numbers indicate the weight percentages of PLLA and PVDF, respectively. In the case of PLLA/PVDF/INT-WS_2 nanocomposites, the INT-WS_2 fraction was 0.5 wt.% of the total composite weight and the ratio of PLLA and PVDF was the same as in the binary blends (80/20-INT(79.6/19.9/0.5), 60/40-INT(59.7/39.8/0.5), 40/60-INT (39.8/59.7/0.5), and 20/80-INT(19.9/79.6/0.5)). In a previous study [30], the crystallization behavior and mechanical properties of PLLA filled with different amounts of INT–WS_2 (0.1 wt.%, 0.5 wt.%, and 1.0 wt.%) were investigated, and it was found that 0.5 wt.% loading led to the highest property improvements. Therefore, 0.5 wt.% INT–WS_2 was chosen as an optimum concentration to prepare the ternary PLLA/PVDF/INT–WS_2 nanocomposites. For the sake of comparison, reference samples of PLLA/INT-WS_2 (0.5 wt.%) (PLLA-INT) and PVDF/INT-WS_2 (0.5 wt.%) (PVDF-INT) nanocomposites were also prepared in the same way. The melt-mixing of the resulting dispersions (~6 g) was performed using a micro-extruder (Thermo-Haake Minilab system) operating at 190 °C with a rotor speed of 100 rpm for 10 min. Then, the samples were pressed into films of 0.5 mm thickness in a hot press system using two heating/cooling plates.

2.2. Characterization Studies

The morphology of the samples was characterized using an ultra-high field-emission scanning microscopy (FESEM) (SU8000, Hitachi Co., Tokyo, Japan). Cryogenically fractured surfaces from film specimens were coated with a ~5 nm Au/Pd layer to avoid charging during electron irradiation.

Wide-angle x-ray diffraction (WAXS) analysis were carried out using a Bruker D8 Advance diffractometer (Bruker AXS GmbH, Karlsruhe, Germany) and a Ni-filtered CuKα radiation source. Diffractograms were recorded at a 0.2 °/s scan speed and a resolution of 40 points/degree over the 2θ region of 5 to 35 °C. Diffractograms were also recorded at a scanning rate of approximately 7 °C/min to understand the dynamic crystallization and melting behavior of the samples. This was performed by initially holding them at a temperature of 190 °C for 5 min to erase any thermal history, and then cooling from 190 to 30 °C, followed by reheating them over the same temperature range and rate.

Thermogravimetric analysis (TGA) was performed using a TA Instruments Q50 thermobalance (Waters Cromatografía, S.A., Cerdanyola del Vallès, Spain) under nitrogen gas (flow rate = 60 mL/min) at 10 °C/min, over a temperature range of 100–800 °C.

Differential scanning calorimetry (DSC) was performed on a Perkin Elmer DSC7/7700 differential scanning calorimeter (Perkin-Elmer España SL, Madrid, Spain), calibrated with indium (T_m = 156.6 °C, ΔH_m = 28.45 kJ/kg) and zinc (T_m = 419.47 °C, ΔH_m = 108.37 kJ/kg) under the flow of nitrogen gas (25 mL/min). The samples were first heated to 220 °C and held at the same temperature for 5 min to erase the thermal history. Then, the crystallization of the samples was carried out by cooling from 220 to 40 °C, followed by heating cycles at 10 °C/min over the interval of temperatures between 40 and 220 °C. The crystallization/melting enthalpy of PLLA in the blend nanocomposites was determined by considering the weight fraction of PLLA in the nanocomposites.

Dynamic mechanical analysis (DMA) was performed on rectangular shaped samples using a Mettler DMA 861 device (Mettler-Toledo, Greifensee, Switzerland), at three frequencies of 0.1, 1, and 10 Hz in the tensile mode. An 8 N oscillating dynamic force using an amplitude of 17 μm at fixed frequency was adopted. The relaxation spectra were recorded over the temperature range −100 °C to 150 °C, at a heating rate of 3 °C/min.

3. Results

3.1. Morphology

Tailoring phase morphology of immiscible blends through the addition of nanoparticles is a universally accepted strategy of forming compatible nanocomposites polymer blends, and typically results in improved upon physical properties. The reason for this is that the added nanoparticles typically locate at the interface of the polymer domains and act as interfacial modifiers, strengthening the interfacial adhesion. Nanoparticles can induce the formation of fine dispersed phase particles from coalescence during melt processing, thus stabilizing the fine morphology and therefore maintaining the properties of the blend.

The cryogenically fractured surface morphologies using SEM of the PLLA/PVDF blends can be seen in Figure 1. For the 80/20 PLLA/PVDF mixture, a distinct two-phase morphology was noted, with the PVDF phase dispersed evenly within the PLLA matrix. Furthermore, the mean diameter of the domains augmented with increasing PVDF content. When the PLLA content < 60 wt.%, the phase morphology was reversed and the PLLA phase became dispersed in the PVDF. For the ternary hybrid PLLA/PVDF/INT nanocomposites and individual PLLA/INT and PVDF/INT nanocomposites (data are not shown), it was found that the INTs were uniformly dispersed at the nanoscale without evidence of aggregates or agglomerates (see arrows pointing to individual INT-WS_2 tubes in the images), verifying the effectiveness of the melt extrusion conditions (Figure 2). The INT-WS_2 nanoparticles also improved the compatibility of the two phases, as demonstrated by an important reduction in the size of the dispersed PVDF domain. This can be attributed to the formed morphological structure in which INT-WS_2 was mainly dispersed in the PLLA matrix and at the PLLA/PVDF interface.

To achieve similar morphology, other authors employed more elaborate methodologies. For example, carboxyl functionalized multi-walled carbon nanotubes (MWCNTs) were used by Wu et al. to improve the compatibility of immiscible PLA/PCL mixtures [34]. In the case of PLA and PBS, Chen et al. employed double functionalized organoclay (TFC) to increase phase compatibility [35]. They also found that the concentration of TFC had a significant effect on the blend morphology, though when the TFC content < 0.5 wt.%, the PBS domain did not change in size and it was almost exclusively found within the PLA regions. Another nanoparticle that has been used to make immiscible PLA mixtures compatible is silica, as reported by Odent et al. in a mixture of PLA with a gummy copolyester based on 3-caprolactone, P[CL-*co*-LA] [36]. It was found that P[CL-*co*-LA] with spherical nodules dispersed regularly in the PLA matrix in a blank mixture containing 10% by weight of P[CL-*co*-LA], while surface treated (5 wt.% hexamethyldisilazane) spherical nodules disappeared. On the other hand, the compatibility of hydroxyl functionalized and poly(3-caprolactone)-*b*-poly(L-lactide) diblock copolymer grafted polyhedral oligomeric silsesquioxane (POSS-OH and POSS-PCL-*b*-PLLA, respectively) on PLA/PCL blends were analyzed by Monticelli et al. [37]. In both cases, the adhesion between PLA and PCL increased, and with POSS-PCL-*b*-PLLA, a nearly homogeneous microstructure was formed.

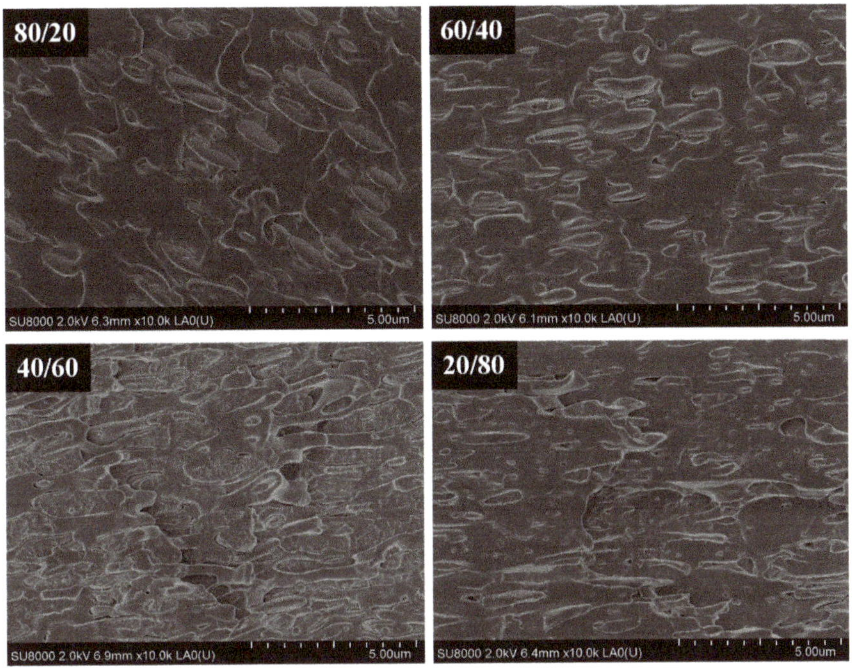

Figure 1. High-resolution SEM image for PLLA/INT-WS_2, PVDF/INT-WS_2, binary PLLA/PVDF, and ternary PLLA/PVDF/INT-WS_2 hybrid nanocomposites.

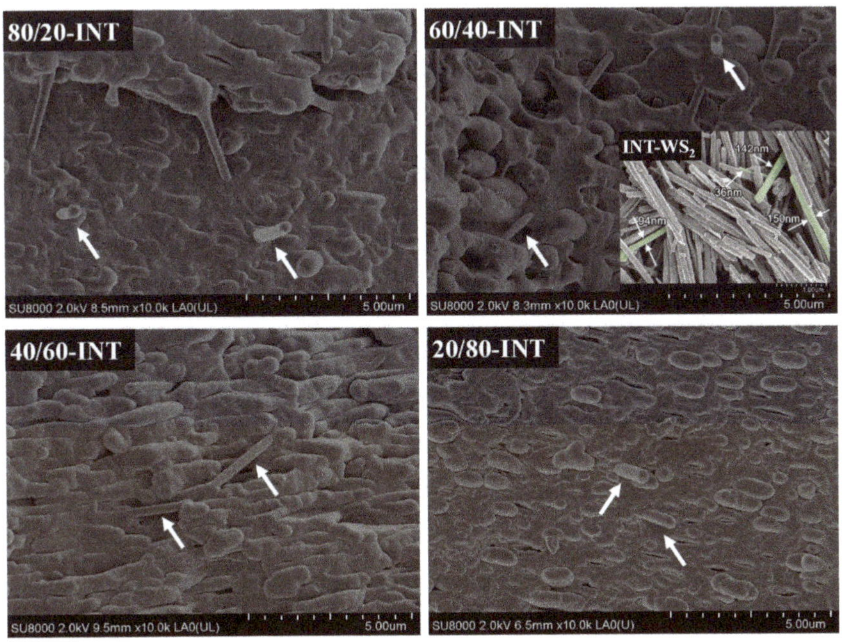

Figure 2. High-resolution SEM image for ternary PLLA/PVDF/INT-WS_2 hybrid nanocomposites; the inset shows the nanotube dimensions (thicknesses) of as received INT-WS_2.

3.2. Thermal Stability

The thermal stability conditions the processing limits of this type of system, and as such, the study of the effect of blend composition on the degradation temperatures is of utmost importance. Figure 3 shows the variation of the integral and differential thermogravimetric curves, TGA and DTG, of PLLA/PVDF blends as a function of the blend composition. The initial weight loss of PLLA started at around 326 °C, and in the case of PVDF, at a temperature about 100 °C higher. The differential curves show a single peak for each polymer which implies that the degradation processes occurred in a single step. The values of the characteristic degradation temperatures T_i (temperature for 2% weight loss), T_{10} (temperature for 10% weight loss), T_{max} (temperature corresponding to the maximum rate of weight loss), and R_{max} (rate of maximum decomposition) are summarized in Table 1.

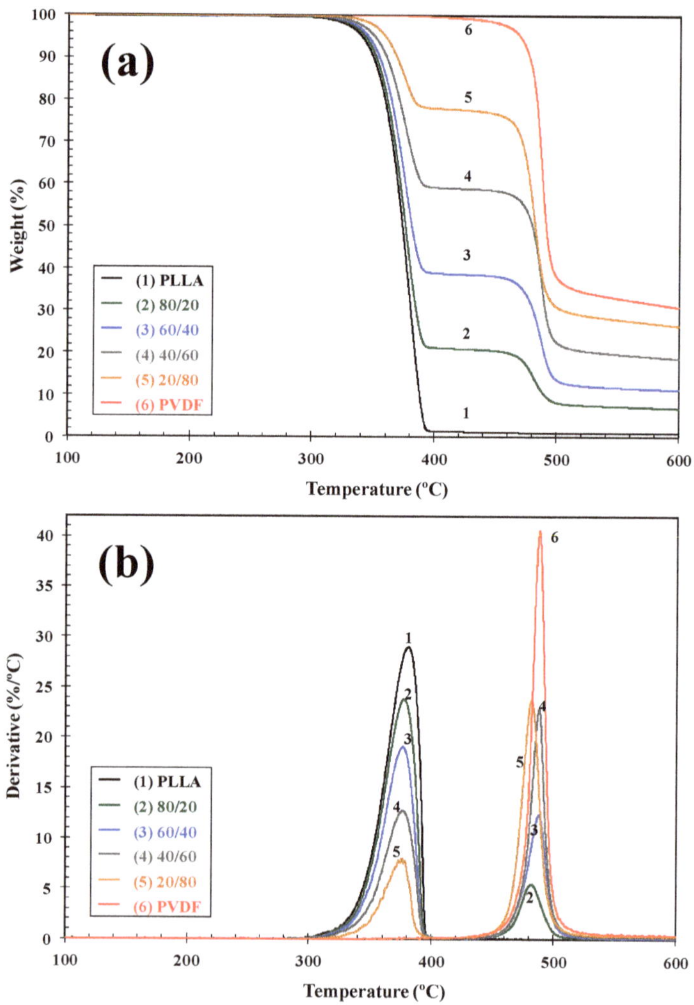

Figure 3. (**a**) Thermogravimetric (TGA) and (**b**) derivative thermogravimetric (DTG) curves of PLLA/PVDF blends.

Table 1. TGA parameters of different PLLA/PVDF/INT-WS_2 blend nanocomposites based on PLLA, PVDF, and INT-WS_2.

Material	T_i (°C)	T_{10} (°C)	T_{max} (°C)	R_{max} (%/°C)
PLLA	325.6	347.0	381.0	29.01
80/20	327.4	351.9	378.2/480.7	23.85/05.34
60/40	332.8	353.7	376.4/488.0	19.00/12.21
40/60	338.3	360.0	377.3/488.0	12.75/22.97
20/80	346.5	368.2	375.5/481.6	8.02/23.73
PVDF	420.9	453.5	478.4	40.34
PLLA-INT	325.8	349.0	377.4	29.39
80/20-INT	327.4	351.9	378.0/489.2	22.61/05.58
60/40-INT	330.2	352.0	377.1/485.3	18.39/09.34
40/60-INT	336.4	358.3	374.8/482.0	13.27/14.96
20/80-INT	344.8	367.1	371.9/476.5	6.95/23.86
PVDF-INT	396.6	437.0	454.5	17.68

T_i: temperature for 2% weight loss; T_{10}: temperature for 10% weight loss; T_{max}: temperature corresponding to the maximum rate of weight loss; and R_{max}: rate of maximum decomposition.

It is clear that the presence of PVDF influenced the degradation behavior of PLLA and vice versa. The addition of PVDF to PLLA hardly increased the thermal stability of the PLLA biopolymer (T_i and T_{10}), and the thermal degradation of both components took place via different mechanisms, since two well-defined decomposition stages can be observed. Although the change in T_{max} values was small (Figure 4a), the change in the corresponding values of maximum rate of decomposition (R_{max}) is representative of the variation in the composition of the blends, Figure 4b. In particular, the R_{max} of PLLA/PVDF blends clearly fell dramatically with respect to the values observed of neat PLLA and PVDF. Analogously, the addition of 0.5 wt.% INT-WS_2 to the PLLA/PVDF blends slightly affected the thermal stability (Table 1), without apparent perturbation of the mechanisms of degradation of the blend components, as can be deduced from the presence of the degradation maxima corresponding to PVDF and PLLA (Figure 5).

3.3. Crystallization Behavior

DSC was used to study the crystallization behavior of the biopolymers and it was found that PVDF, despite having a similar melting point to PLLA, crystallizes at a faster rate and has a higher crystallization temperature [5]. It can be understood therefore that the PVDF will separate from the melt and crystallize first during cooling. Any modification to the PLLA/PVDF domain interface would affect this process of phase separation, crystallization rate, and hence, crystal morphology.

Dynamic DSC cooling scans at a rate of 10 °C/min of the prepared PLLA/PVDF blends (Figure 6a) show that as the ratio of PVDF increased, the PLLA crystallization exotherms shifted to higher temperatures and had a higher enthalpy. It should be noted that the crystallization exotherm of pure PLLA with a crystallization enthalpy of only 8.0 J/g can be difficult to detect. For comparison, the samples containing only 0.5 wt.% INT-WS_2 are shown in Figure 6b, where it is seen that the crystallization exotherms shifted toward higher temperatures and the enthalpy increased, and in the case of neat PLLA, up to 41.6 J/g. To be able to compare this change in T_c vs. PLLA concentration for the PLLA/PVDF blends and those containing INT-WS_2 nanoparticles, Figure 7a is presented. Regarding the blends, a strong increase in the crystallization temperature of PLLA was found upon increasing PVDF content, from 92.2 °C for neat PLLA to 133.1 °C for raw PVDF. A noticeable increase up to 111.3 °C was already found with the incorporation of only 20 wt.% PVDF, but no significant change in T_c was observed with a further increase of PVDF content. The T_c of PLLA increased dramatically after blending with PVDF, indicating that PVDF accelerated the melt-crystallization of PLLA. In the same way, the presence of INT-WS_2 caused an increase in the crystallization temperature, both in the neat polymers and in the blends, the rise being higher than 25 °C for the neat PLLA and about 2 °C for the neat PVDF (see Table 2). With regard to the nanocomposites, this aspect was also observed, though it seems to be less dependent on the concentration of the polymers. As such, it would suggest that the nanofillers provoked nucleation in both polymeric components, with the effect being more pronounced for PLLA. In contrast, 2D-WS_2 nanosheets have been shown to slow down the crystallization rate of PLLA [25]. Such differences suggest that the nanoparticle shape plays a fundamental role in PLLA crystallization. This discrepancy is likely related to several factors, including the nanofiller geometry, its surface energy, roughness, and crystalline structure as well as on the filler ability to form the critical nucleus [17,22,23]. It should be noticed that, in the case of the 60/40-INT nanocomposite, a double crystallization exotherm was found (Figure 6b), with T_c values of 114.1 °C and 134.5 °C, which is likely related to the presence of two distinct macrophases, one containing the majority of the 0.5 wt.% INT-WS_2 and the other encompassing very little. This tendency was also observed for the 40/60-INT and 20/80-INT nanocomposite samples.

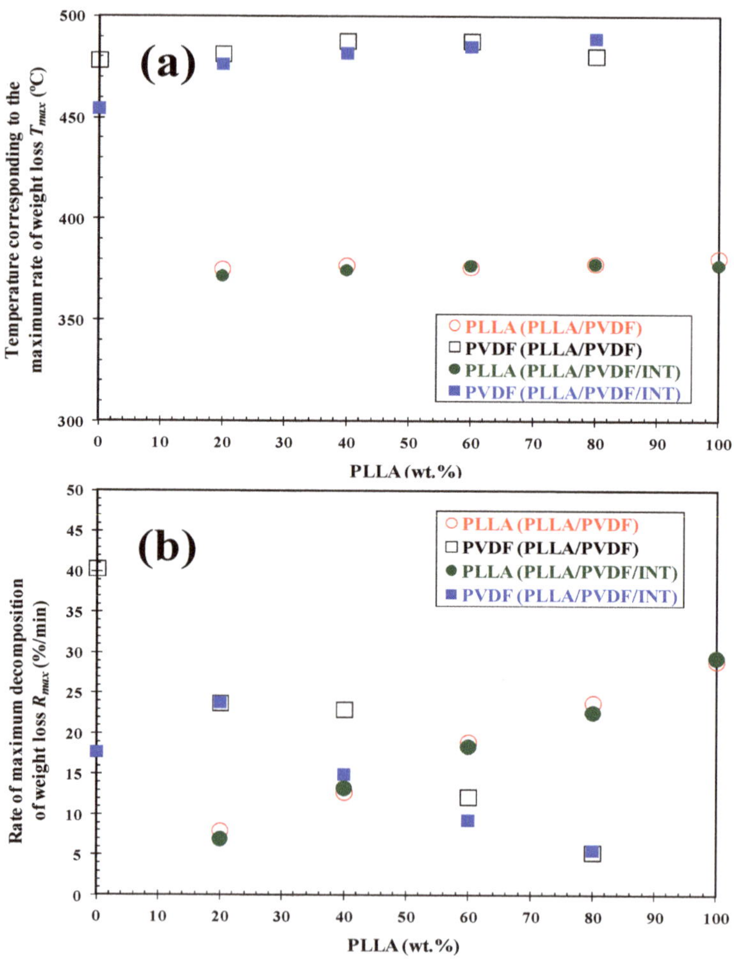

Figure 4. (**a**) Variation of the temperature and (**b**) rate of maximum decomposition (T_{max}/R_{max}) of PLLA and PVDF in the binary PLLA/PVDF and ternary PLLA/PVDF/INT-WS2 hybrid nanocomposites with composition.

the Dutch coast, sand nourishment is considered a strategic approach and a state trend for coastal protection. By 2000, the area of sand was increased from 6.4 to 12 million m^3/yr., adding to the coastline 432 km, which enabled the coastal zone to stay in equilibrium with sea level rise [63].

It is interesting to shed some light on the problem of coastal sand dune quarrying. Natural sand dunes and backshores along the coast of the Mediterranean region act as natural defense measures, reducing erosion and shoreline retreat [64]. Removal of these dunes is a risky course [28]. Most of the dunes along the Nile Delta, between Baltim and Damietta, have been removed for urbanization [65], including the construction of recreational resorts and new communities [66,67]. Such anthropogenic intervention, while claiming to be sustainable development and investment in the tourism industry, is actually a major threat to the environment and the stability of the coast. Due to the lack of awareness of ICZM, several strategic projects were built in inconvenient coastal areas. Focusing on human demands for entertainment and investments such as tourism-accommodating beaches and recreational facilities in face of strategic industries imply risks and pollution. Such imbalance causes deterioration of highly vulnerable systems, compromising both ecosystem integrity and health, as well as tourism itself [68].

Finally, a discussion about the possible inundation scenarios of the northern Nile Delta coast seems very interesting. Recent research has postulated two scenarios of 0.5 m and 1 m SLR of future inundated land areas and the affected populations. The extreme scenario of SLR to 1 m will affect about 3900 km^2 of cropland, 1280 km^2 of vegetation, 205 km^2 of wetland, 146 km^2 of urban areas at the northern borders, and cause more than 6 million people to lose their houses [69]. However, a more optimistic scenario is based on the measurements of a SLR gauge and referred to a value ranges between 6.7–6.9 mm/yr. [70]. An interpolation process has been performed for the SLR rates up to year 2050 (a time period of 35 yeas) has been calculated at 0.24 m. In addition to the average subsidence rate (calculated at Damietta Promontory to be 2.6 mm/yr., which appears consistent with the global positioning system rate of 3.5 mm/yrm. Accordingly, the areas vulnerable to inundation by year 2050 will be 38.40 km^2, 3.80 km^2, 5.20 km^2, and 2.60 km^2 for the urban, agricultural lands, fishing farms, and bare areas, respectively [70].

Tenths of papers have been devoted to the effect of SLR and land subsidence since 1988, when Stanley and his colleagues [9–11] published their pioneering studies on the Nile Delta subsidence scenarios, with suggested inundated areas. Since then several scenarios have been published, estimating future inundation of hundreds of kilometres and the migration of millions of people from their first homes to other areas to the south [11–13]. A 35 years period has passed (1988 to 2022) during which a SLR equal to 0.33 m. is supposed to inundate 1/3 of the suggested areas, a case that has not yet occurred. The present situation undoubtfully forced us to conclude that such a gradual increase in SLR is a secondary player and the predicted 2050 [70] scenario will be similar to that of Stanley [9–11] unless a new strong and rapid player, such as Tsunami [71], enters the game. Therefore, the anthropogenic factors are strongly involved. Such as compaction of the Holocene sediments (8000 to 2500) [13], and urban-induced loading at major cities (12 to 20 mm/yr.). The major cities such as Cairo, Tanta, Mahala, Mansoura, Damietta, and Port Said, show estimated subsidence are around 6.4 ± 0.4 mm/year, 4.0 ± 0.6 mm/year, 4.8 ± 1.0 mm/year, 10.0 ± 1.2 mm/yr., 10.3 ± 1.6 mm/yr., and 4.9 ± 1.6 mm/yr. respectively [72,73]. In addition, the role of groundwater overexploitation (16 to 20 mm/yr.) that recorded at newly reclaimed lands and most agricultural lands (like Menoufia governorate) [74], and high subsidence rates (up to 0.7 mm/year) over onshore gas fields, particularly the Abu Madi [74]. The localization of subsidence signals on InSAR and GPS imited at the big cities [75], and about 15.56% of the coastal cities are affected by subsidence [76], the subsidences related to anthropogenic activities seem to be more effective rather than SLR [76].

4. Conclusions

Owing to its historical values, economic resources of oil and gas and dense population, the Nile Delta is a traditional global model for delta research. Several scenarios of inundation due to SLR are emerging, however; inspection of several areas of the coast confirms the role of anthropogenic interventions in coastal erosion problems rather than SLR and land subsidence. In unawareness with the ICZM several economic projects, resorts, and profession communities were constructed in inconvenient coastal areas, which were later subjected to erosion. These led the state to implement a plan for coastal protection. The current study used high-resolution satellite images to detect the shoreline changes at the Damietta Promontory and to assess the efficiency of the implemented protection structures in securing sustainability of the coast against SLR and climate changes.

- The current study confirms the efficiency of detached breakwaters in protecting the coast, with annual sand nourishment preferred in order to secure wide and gently sloped beaches convenient for swimming. However, the dimensions of the breaker bodies, gaps, and shadow areas must be taken into considerations.
- Seawalls, on the other hand, are effective in protecting the coast-facing non-available beaches. The case presented was at the promontory facing the Manzala triangle, where the seawall may offer stability of the coast against most storms, which allows better investments for the islands along the Manzala triangle in a model similar in nature to that of Venice, Italy.
- The seawall at the Damietta Promontory should be extended to the El-Diba Inlet, the ancient mouth of the Nile's Mendisian branch. Additionally, we encourage the presence of detached breakwaters to offer water circulation within the Manzala triangle and restore beaches for the new resort communities.
- Sand nourishment is the preferred way for protection; this depends on the presence of a suitable source of sand of the right grain size.
- Due to the absence of continuous maintenance, several areas of collapses are observed and parts of the breaker body reached an elevation of less than 1 m a.s.l. Such a situation represents a great risk and continuous maintenance is a critical task to keep the elevation of the breakers not less than 6 m a.s.l. in order to impede tsunamic waves.
- The groins failed in protecting the coast in cases where there are multidirectional currents or rip currents perpendicular to the coast due to the resultant sand undermining, resulting in collapse as seen at Ras El-Bar.
- It is still too early for the Y groins to be evaluated; however, these create a new pattern of erosion between groins and accretion along the groins' bodies that will transform the groins to tombolos.
- The protection structures must be built to cover an entire cell or sector, not just a small segment. A segment of coast between two bodies of water or outlying headlands is referred to as a segment in this definition.
- Building the protection structures in an opposite direction to currents is not recommended. Such a situation creates unusual waves and currents that may threaten the works.
- The current study confirms the role of anthropogenic intervention rather than SLR by comparing the 35 year time period from 1988 postulated by Stanley and another 35 year period of a new scenario for SLR through to 2050. In fact, nothing has changed along the coast since 1988, except the construction of protecting structures. The historical changes are linked to the entrapment of the discharge rather than action of SLR in addition to some anthropogenic activities along the coast near big cities. However, a sudden tsunami is not ruled out.

Author Contributions: Conceptualization, Investigation, Writing original draft, H.M.E.-A.; Remote Sensing Investigation and Software Processing M.M.N.T. All authors have read and agreed to the published version of the manuscript.

Funding: This research received no external funding.

Institutional Review Board Statement: Not applicable.

Informed Consent Statement: Not applicable.

Data Availability Statement: The data are available on request from the corresponding author.

Acknowledgments: I would like to thank the unidentified reviewers who provided assistance throughout the review process. Many thanks to the guest editors who worked tirelessly to get this issue published. I would like to thank Mansoura University, Egypt for scientific support through funding the publishing fees.

Conflicts of Interest: The authors declare no conflict of interest.

References

1. Moawad, M.B. Recent landscape changes at Damietta Promontory, Nile Delta-Egypt. *Egypt. J. Remote Sens. Space Sci.* **2021**, *24*, 525–535. [CrossRef]
2. Coutellier, V.; Stanley, D.J. Late Quaternary Stratigraphy and Paleogeography of the Eastern Nile Delta Shelf. *Mar. Geol.* **1987**, *77*, 257–275. [CrossRef]
3. Sestini, G. Nile Delta: A Review of Depositional Environments and Geological History. *Geol. Soc.* **1989**, *41*, 99–127. [CrossRef]
4. El-Asmar, H.M.; Taha, M.M.N.; El-Sorogy, A.S. Morphodynamic Changes as an Impact of Human Intervention at the Ras El-Bar-Damietta Harbor Coast, NW Damietta Promontory, Nile Delta, Egypt. *J. Afr. Earth Sci.* **2016**, *124*, 323–339. [CrossRef]
5. Frihy, O.E.; Khafagy, A.A. Climate and Human Induced Changes in Relation to Shoreline Migration Trends at the Nile Delta Promontories, Egypt. *CATENA* **1991**, *18*, 197–211. [CrossRef]
6. Masria, A.; Nadaoka, K.; Negm, A.; Iskender, M. Detection of Shoreline and Land Cover Changes around Rosetta Promontory, Egypt, Based on Remote Sensing Analysis. *Land* **2015**, *4*, 216–230. [CrossRef]
7. Frihy, O.E.; Debes, E.A.; El Sayed, W.R. Processes Reshaping the Nile delta Promontories of Egypt: Pre- and Post-Protection. *Geomorphology* **2003**, *53*, 263–279. [CrossRef]
8. Frihy, O.E.; Lawrence, D. Evolution of the Modern Nile Delta Promontories: Development of Accretional Features during Shoreline Retreat. *Environ. Geol.* **2004**, *46*, 914–931. [CrossRef]
9. Stanley, D.J. Recent Subsidence and Northeast Tilting of the Nile Delta, Egypt. *Mar. Geol.* **1990**, *94*, 147–154. [CrossRef]
10. Stanley, D.J.; Warne, A.G. Nile Delta: Recent Geological Evolution and Human Impact. *Science* **1993**, *260*, 628–634. [CrossRef]
11. Stanley, J.D.; Clemente, P.L. Increased land Subsidence and Sea-Level Rise are Submerging Egypt's Nile Delta Coastal Margin. *GSA Today* **2016**, *27*, 4–11. [CrossRef]
12. Marriner, N.; Flaux, C.; Morhangem, C.; Kaniewski, D. Nile Delta's sinking past: Quantifiable Links with Holocene Compaction and Climate-Driven Changes in Sediment Supply. *Geology* **2012**, *40*, 1083–1086. [CrossRef]
13. Becker, R.H.; Sultan, M. Land Subsidence in the Nile Delta: Inferences from Radar Interferometry. *Holocene* **2009**, *19*, 949–954. [CrossRef]
14. El-Asmar, H.M. Short term coastal changes along Damietta-Port Said coast Northeast of the Nile Delta. *Egypt. J. Coast. Res.* **2002**, *18*, 433–441.
15. Abaza, J.; Mostafa, M.; El-Fekki, A. The Forgotten Fishermen of Ezbet El-Borg. *Daily News*, 14 April 2015; 1/9–9/9.
16. Aly, M.H.; Giardino, J.R.; Klein, A.G.; Zebker, H.A. InSAR study of shoreline change along the Damietta Promontory, Egypt. *J. Coast. Res.* **2012**, *28*, 1263–1269. [CrossRef]
17. ASRT. *Sedimentation in Damietta Harbor*; Academy of Scientific Research and Technology Final Report; ASRT: Cairo, Egypt, 1988.
18. Tetra Tech. Shoreline Master Plan for the Nile Delta Coast. *Prog. Rep.* **1984**, *1*, 143.
19. Sogreah, M. Effects on the construction of the Port of Damietta on the evolution of the littoral drift. *Consult. Rep.* **1982**, *9*, 35/1202.
20. AboZed, B.I. Effects of Waves and Currents on the Siltation Problem of Damietta harbor, Nile Delta coast, Egypt. *Mediterr. Mar. Sci.* **2007**, *8*, 33–47. [CrossRef]
21. El-Asmar, H.M. Impact of protection structures on Physical and Sedimentary Parameters along the Damietta coastal area, Nile Delta, Egypt. *J. Sedimentol. Egypt.* **1995**, *3*, 111–124.
22. Ezzeldin, M.M.; Rageh, O.S.; Saad, M.E. Assessment impact of the Damietta Harbor (Egypt) and its deep navigation channel on adjacent shorelines. *J. Integr. Coast. Zone Manag.* **2020**, *20*, 265–281. [CrossRef]
23. El-Asmar, H.M.; White, K. Changes in coastal sediment transport processes due to construction of New Damietta Harbor, Nile Delta, Egypt. *Coast. Eng.* **2002**, *46*, 127–138. [CrossRef]
24. Deabes, E.A. Sedimentation Processes at the Navigation Channel of the Liquefied Natural Gas (LNG) Port, Nile Delta, Egypt. *Int. J. Geosci.* **2010**, *1*, 14–20. [CrossRef]

25. Stein, A.; Ge, Y.; Fabris-Rotelli, I. Introduction to the Special Issue "Uncertainty in Remote Sensing Image Analysis". *Remote Sens.* **2018**, *10*, 1975. [CrossRef]
26. Wang, W.; Hall-Beyer, M.; Wu, C.; Fang, W.; Nsengiyumva, W. Uncertainty Problems in Image Change Detection. *Sustainability* **2020**, *12*, 274. [CrossRef]
27. Wu, W.; Mhaimeed, A.S.; Al-Shafie, W.M.; Ziadat, F.; Dhehibi, B.; Nangia, V.; De Pauw, E. Mapping soil salinity changes using remote sensing in Central Iraq. *Geoderma Reg.* **2014**, *2*, 21–31. [CrossRef]
28. Taha, M.M.N. Monitorig Changing Position of the Shoreline along Burullus–Baltim, Nile Delta, Egypt, and the Role of Coastal Sand dunes as a Natural Defense Against Erosion in a Geoheritage Site: Remote sensing application. *Egypt. J. Geol.* **2020**, *64*, 111–121.
29. Do, A.T.; Vries, S.D.; Stive, M.J. The Estimation and Evaluation of Shoreline Locations, Shoreline-Change Rates, and Coastal Volume Changes Derived from Landsat Images. *J. Coast. Res.* **2019**, *35*, 56–71. [CrossRef]
30. Available online: http://earthexplorer.usgs.gov/ (accessed on 21 August 2022).
31. Available online: https://livingatlas.arcgis.com/wayback/#active=47471&ext=-115.34940,36.03763,-115.24760,36.09036 (accessed on 21 August 2022).
32. Thieler, E.R.; Himmelstoss, E.A.; Zichichi, J.L.; Ergul, A. *The Digital Shoreline Analysis System (DSAS) Version 4.0—An ArcGIS Extension for Calculating Shoreline Change*; Open-File Report, 2008-1278; U.S. Geological Survey: Reston, VA, USA, 2009. [CrossRef]
33. Dewidar, K.; Bayoumi, S. Forecasting Shoreline Changes along the Egyptian Nile Delta Coast using Landsat Image Series and Geographic Information System. *Environ. Monit. Assess.* **2021**, *193*, 429. [CrossRef]
34. Fountoulis, I.; Vassilakis, E.; Mavroulis, S.; Alexopoulos, J.; Dilalos, S.; Erkeki, A. Synergy of Tectonic Geomorphology, Applied Geophysics and Remote Sensing Techniques Reveals New Data for Active Extensional Tectonism in NW Peloponnese (Greece). *Geomorphology* **2015**, *237*, 52–64. [CrossRef]
35. Thieler, E.R.; Hammar-Klose, E.S. *National Assessment of Coastal Vulnerability to Sea-Level Rise, U.S. Atlantic Coast*; U.S. Geological Survey: Reston, VA, USA, 1999; Volume 1, pp. 99–593.
36. Elkafrawy, S.B.; Basheer, M.A.; Mohamed, H.M.; Naguib, D.M. Applications of remote sensing and GIS techniques to evaluate the effectiveness of coastal structures along Burullus headland-Eastern Nile Delta, Egypt. *Egypt. J. Remote Sens. Space Sci.* **2021**, *24*, 247–254. [CrossRef]
37. Basheer, A.K.K.; Chandra, A.P. Assessment and Prediction of Shoreline Change using Multi-temporal Satellite Data and Geostatistics: A Case Study on the Eastern Coast of India. *J. Water Clim. Chang.* **2022**, *13*, 147. [CrossRef]
38. E1-Fishawi, N.M.; Badr, A.A. Volumetric changes of nearshore sediments between Rosettaand Damietta, Egypt. *INQUA Comm. Quat. Shorel. Newsl.* **1989**, *11*, 39–42.
39. Fanos, A.M.; Khafagy, A.A.; El Kady, M.M. Variations of the Nile River discharge and sediment regime during the periodfrom 1800 to 2000 and its effects on the Nile Delta coast. In Proceedings of the 8th International Symposium on River Sedimentation, Cairo, Egypt, 3–5 November 2001.
40. Frihy, O.E. Sea-level rise and shoreline retreat of the Nile Delta Promontories, Egypt. *Nat. Hazards* **1992**, *5*, 65–81. [CrossRef]
41. El-Asmar, H.M.; El-Kafrawy, S.; Taha, M.M.N. Monitoring coastal changes along Damietta promontory and the barrier beach toward Port Said east of the Nile Delta. Egypt. *J. Coast. Res.* **2014**, *30*, 993–1005. [CrossRef]
42. Ali, E.; El-Magd, I.A. Impact of Human Interventions and Coastal Processes along the Nile Delta Coast, Egypt during the Past Twenty-Five Years. *Egypt. J. Aquat. Res.* **2016**, *42*, 1–10. [CrossRef]
43. Dewidar, K.; Frihy, O.E. Automated Techniques for Quantification of Beach Change Rates using Landsat Series along the North-Eastern Nile Delta, Egypt. *J. Oceanog. Mar. Sci.* **2010**, *1*, 28–39.
44. Abou Samra, R.M.; Ali, R.R. Applying DSAS Tool to Detect Coastal Changes along the Nile Delta Egypt. *Egypt. J. Remote Sesing Space Sci.* **2021**, *24*, 463–470. [CrossRef]
45. Frihy, O.E.; Komar, P.D. Long-term Shoreline Changes and the Concentration of Heavy Minerals in Beach Sands of the Nile Delta, Egypt. *Mar. Geol.* **1993**, *115*, 253–261. [CrossRef]
46. El-Banna, M. Erosion and Accretion Rates and their Associated Sediment Characters along Ras El Bar Coast, Northeast Nile Delta, Egypt. *Environ. Geol.* **2007**, *52*, 41–49. [CrossRef]
47. Fanos, A.M.; Khafagy. A.A.; Dean, R.G. Protective Works on the Nile Delta Coast. *J. Coast. Res.* **1995**, *11*, 516–528.
48. Elnabwy, M.T.; Emad Elbeltagi, E.; El Banna, M.M.; Elshikh, M.M.Y.; Motawa, I.; Kaloop, M. An Approach Based on Landsat Images for Shoreline Monitoring to Support Integrated Coastal Management—A Case Study, Ezbet Elborg, Nile Delta, Egypt. *ISPRS Int. J. Geo-Inf.* **2020**, *9*, 199. [CrossRef]
49. Ghoneim, E.; Mashaly, J.; Gamble, D.; Halls, J.; Abu Bakr, M. Nile Delta exhibited a spatial reversal in the rates of shoreline retreat on the Rosetta promontory comparing pre- and post-beach protection. *Geomorphology* **2015**, *228*, 1–14. [CrossRef]
50. White, K.; El-Asmar, H.M. Monitoring Changing Position of Coastlines using Thematic Mapper Imagery, an example from the Nile Delta. *Geomorphology* **1999**, *29*, 93–105. [CrossRef]
51. Masria, A.; Negm, A.; Iskender, M.; Savedra, O.C. Numircal Modeling of Rosetta River Mouth, Egypt. *Int. Water Technol. J. IWTJ* **2014**, *4*, 190.
52. Deabes, E.A.M. Applying ArcGIS to Estimate the Rates of Shoreline and Back-Shore Area Changes along the Nile Delta Coast, Egypt. *Int. J. Geosci.* **2017**, *8*, 332–348. [CrossRef]

53. El-Gamal, A.; Balbaa, S.H.; Rashed, M.A.; Mansour, A.S. Three Decades Monitoring of Shoreline Change Pattern of Damietta Promontory, Nile Delta, Egypt. *Aquat. Sci. Technol.* **2020**, *8*, 1–14. [CrossRef]
54. Khalifa, A.M.; Soliman, M.R.; Yassin, A.A. Assessment of a combination between hard structures and sand nourishment eastern of Damietta harbor using numerical modeling. *Alex. Eng. J.* **2017**, *56*, 545–555. [CrossRef]
55. Masria, A.; Nadaoka, K.; Kuriyama, Y.; Negm, A.; Iskender, M.; Savedra, O.C. Neashore and Beach nourishments Effects on the Stability of the Rosette Promontory, Egypt. In Proceedings of the Eighteenth International Water Technology Conference, IWTC18 2015, Sharm El-Sheikh, Egypt, 12–14 March 2015.
56. Masria, A.; Iskender, M.; Negm, A.; Savedra, O.C. The Effect of potential Discharges on the Stability of the Rosetta Promontory, Egypt. *J. Geogr. Tech.* **2014**, *2*, 6976.
57. Negm, A.; Masria, A.; Iskender, M.; Savedra, O.C. Towards a Sustainable Stability of Coastal Zone at Rosetta Promontory/Mouth, Egypt. *Oceanography* **2015**, *3*, 132. [CrossRef]
58. Pranzini, E.; Anfuso, G.; Cinelli, I.; Marco Piccardi, M.; Vitale, G. Shore Protection Structures Increase and Evolution on the Northern Tuscany Coast (Italy): Influence of Tourism Industry. *Water* **2018**, *10*, 1647. [CrossRef]
59. Saı¨di, H.; Souissi, R.; Zargouni, F. Environmental Impact of Detached Breakwaters on the Mediterranean Coastline of Soliman (NE of Tunisia) end. *Fis. Acc. Lincei* **2012**, *23*, 339–347. [CrossRef]
60. Zidan, A.R.; Rageh, O.S.; Sarhan, T.E.; Esmail, M. Effect of Breakwaters on Wave Energy Dissipation: Case Study, Ras El-Bar beach, Egypt. *Int. Water Technol. J.* **2012**, *2*, 268–283.
61. Saengsupavanich, C. Detached Breakwaters: Communities' Preferences for Sustainable Coastal Protection. *J. Environ. Manag.* **2013**, *115*, 106–113. [CrossRef] [PubMed]
62. Bitan, M.; Zviely, D. Sand Beach nourishment: Experience from the Mediterranean Coast of Israel. *J. Mar. Sci. Eng.* **2020**, *8*, 273. [CrossRef]
63. Brand, E.; Ramaeker, G.; Lodder, Q. Dutch Experience with Sand Nourishments for Dynamic Coastline Conservation- An Operational Overview. *Ocean Coast. Manag.* **2022**, *217*, 106008. [CrossRef]
64. Enríquez, A.R.; Marcos, M.; Falqués, A.; Roelvink, D. Assessing Beach and Dune Erosion and Vulnerability under Sea Level Rise: A case Study in the Mediterranean Sea. *Front. Mar. Sci.* **2019**, *6*, 4. [CrossRef]
65. Iskader, M. Stability of the Northern coast of Egypt under the effect of urbanization and climate change. *Water Sci.* **2021**, *35*, 1–10. [CrossRef]
66. El-Banna, M. Nature and human impact on Nile Delta coastal sand dunes, Egypt. *Environ. Geol.* **2004**, *45*, 690–695. [CrossRef]
67. El-Banna, M. Vulnerability and Fate of a Coastal Sand Dune Complex, Rosetta-Idku, northwestern Nile Delta, Egypt. *Environ. Geol.* **2008**, *54*, 1291–1299. [CrossRef]
68. Della Bella, A.; Fantinato, E.; Scarton, F.; Buffa, G. Mediterranean Developed Coasts: What Future for the Foredune Restoration? *J. Coast. Conserv.* **2021**, *25*, 49. [CrossRef]
69. Hasan, E.; Khan, S.I.; Yang, H. Investigation of Potential Sea Lvel Rise Impact on the Nile Delta, Egypt using Digital Elevation Models. *Environ. Monit. Assess.* **2015**, *187*, 649. [CrossRef] [PubMed]
70. El-Quilish, M.; El-Ashquer, M.; Dawod, G.; El Fiky, G. Development of an Inundation Model for the Northern Coastal Zone of the Nile Delta Region, Egypt Using High-Resolution DEM. *Arab. J. Sci. Eng.* **2022**. [CrossRef]
71. Polonia, A.; Vaiani, S.C.; Lange, G.J. Did the A.D. 365 Crete Earthquake/Tsunami Trigger Synchronous Giant Turbidity Currents in the Mediterranean Sea? *Geology* **2016**, *44*, 191–194. [CrossRef]
72. Rateb, A.; Abotalib, A.Z. Inferencing the Land Subsidence in the Nile Delta using Sentinel-1 Satellites and GPS between 2015 and 2019. *Sci. Total Environ.* **2020**, *729*, 138868. [CrossRef] [PubMed]
73. El-Bastawesy, M.; Cherif, O.H.; Sultan, M. The Geomorphological Evidences of Subsidence in the Nile Delta: Analysis of High Resolution Topographic DEM and Multi-temporal Satellite Images. *J. Afr. Earth Sci.* **2017**, *136*, 252–261. [CrossRef]
74. Gebremichael, E.; Sultan, M.; Becker, R.; El-Bastawesy, M.; Cherif, O.H.; Emil, M. Assessing Land Deformation and Sea Encroachment in the Nile Delta: A Radar Interferometric and Inundation Modeling Approach. *J. Geophys. Res. Solid Earth* **2018**, *123*, 3208–3224. [CrossRef]
75. Saleh, M.; Becker, M. New estimation of Nile Delta Subsidence Rates from InSAR and GPS Analysis. *Environ. Earth Sci.* **2019**, *78*, 6. [CrossRef]
76. Hassaan, M.A.; Abdrabo, M.A. Vulnerability of the Nile Delta Coastal Areas to Inundation by Sea Level Rise. *Environ. Monit. Assess.* **2013**, *185*, 6607–6616. [CrossRef]

sustainability

MDPI

Article

Characterization of Hydrologic Sand and Dust Storm Sources in the Middle East

Ramin Papi [1,2], Sara Attarchi [1,*], Ali Darvishi Boloorani [1,*] and Najmeh Neysani Samany [1]

[1] Department of Remote Sensing and GIS, Faculty of Geography, University of Tehran, Azin Alley. 50, Vesal Str, Tehran 11369, Iran

[2] National Cartographic Center (NCC), Meraj Alley, Azadi Square, Tehran 11369, Iran

* Correspondence: satarchi@ut.ac.ir (S.A.); ali.darvishi@ut.ac.ir (A.D.B.)

Abstract: Due to diverse hydroclimatic conditions and human interventions, the Middle East hosts a variety of active sources of sand and dust storms (SDS). Discrimination of different types of SDS sources is the most important factor for adopting optimal mitigation measures to combat SDS. This study employed a binary mask-based modeling framework to identify Middle East SDS sources. Accordingly, using time series of remotely sensed data of land surface and atmospheric aerosol parameters, SDS sources covering an area of 1 million Km^2 were identified with an overall accuracy of 82.6%. Considering the type of land use and spatial-temporal changes in water bodies, SDS sources were categorized into seven types in terms of origin. Desert sources have the largest share (>79%), whereas hydrologic sources accounted for about 8.4%. The results showed that water bodies had a declining trend after 2000. The occurrence of two severe drought periods in 2000–2001 and 2007–2012 led to a 52% decrease in water bodies and a 14–37% increase in SDS emission compared to the pre-2000 period. The latter drought period also led to a sharp decrease in groundwater resources across the region. Our results revealed that natural circumstances and drought actively contribute to the depletion of water resources that led to the formation of SDS sources in the Middle East, while the role of anthropogenic factors is predominant in the case of hydrologic SDS sources.

Keywords: sand and dust storms; hydrology; surface and ground water changes; spatial-temporal analysis; remote sensing; Middle East

Citation: Papi, R.; Attarchi, S.; Darvishi Boloorani, A.; Neysani Samany, N. Characterization of Hydrologic Sand and Dust Storm Sources in the Middle East. *Sustainability* **2022**, *14*, 15352. https://doi.org/10.3390/su142215352

Academic Editors: Mohamed El-Alfy, Ahmed El Kenawy, Petra-Manuela Schuwerack and Zhongfeng Xu

Received: 4 August 2022
Accepted: 7 November 2022
Published: 18 November 2022

1. Introduction

Middle East (ME) is the most affected region by climate change with a significant contribution (15–20%) to global SDS emission [1]. Accordingly, as a prominent feature of this region, SDS has affected climatic parameters [2,3], human health [4,5], socio-economic factors [6], and terrestrial and marine ecosystems [7]. ME is the second most SDS-affected region in the world after North Africa. Considering its location within the subtropical high-pressure belt of the Northern Hemisphere and its expansive deserts [8], ME is experiencing SDS events in hot and dry seasons. However, the maximum activity period varies in different regions of ME. SDS events generally occur in the northwestern parts of the region in winter and spring. In addition, the most intense events form in the southwestern parts of the region in the summer [9].

SDS events are driven by a mixture of natural and anthropogenic factors [10]. Natural factors such as topographic diversity, climatic variability, and several seasonal effective air masses have created a highly complex situation in this region [11]. With mean annual precipitation of less than 200 mm, most parts of the Middle East suffer from dry and semi-arid climates. Therefore, water scarcity has been a lasting problem in the region with serious consequences for the stability and development of the region's economy [12]. The spatial-temporal patterns of precipitation and temperature vary to a great extent across the region with over 1800 mm along the Caspian Sea in Northern Iran and zero annual precipitation in some desert areas in Central Iran and Saudi Arabia [11]. In addition to climatic conditions, global warming, climate change, and drought occurrence have intensified the spatial-temporal frequency of SDS events in the region [13,14]. Mismanagement of water, soil, and vegetation resources in different parts of the ME, especially in Iran, Iraq, Syria, and Turkey are recounted as anthropogenic factors affecting the SDS occurrence and intensity over the past few decades [13,15].

Numerous studies have explored the identification of the ME SDS sources, mostly identifying local scale sources in, e.g., Khuzestan [16] and Razavi Khorasan [17] provinces in Southwestern and Northeastern Iran, respectively. Other studies have been conducted on the catchment scale, for instance, the Euphrates River Basin [18], the Central Iran Basin [15], and the Tigris and Euphrates Rivers Basin [19,20]. Rashki et al. 2021 [21] identified and characterized the SDS sources in Iran on a national scale. Some studies such as [22–25] investigated ME SDS sources on a regional scale. As reported in the literature and by [10], the highest density of SDS sources is in the northern and northwestern parts of Iraq between the Tigris and Euphrates rivers and along the Syrian–Iraqi border. Considering the location of the mountains and the direction of the dominant regional wind (Shamal wind) [26], these sources affect Iraq as well as Iran and Kuwait, in the spring and summer. These sources are generally located in lands with extensive desert cover, low population density, and sparse agriculture along rivers. Iraq, Saudi Arabia, and the Persian Gulf margins, respectively, experience the greatest number of SDS events [10].

SDS occurrence is essentially a function of geographic and hydroclimatic conditions. Therefore, SDS sources located in different geographical regions may have different origins and types. Furthermore, various environmental parameters such as wind speed, evapotranspiration, precipitation, soil physicochemical properties (e.g., soil moisture, texture, and organic matter), vegetation cover, and topography are known as effective SDS drivers [19]. These drivers are interactively involved in the formation of SDS. For instance, water resources affect vegetation cover and are recognized as an essential factor in SDS emission [27]. Water bodies (e.g., rivers, wetlands, and permanent and seasonal lakes) accumulate fine-grained sediments, which are highly susceptible to wind erosion. These erodible lands are so-called "hydrological SDS sources". As confirmed by the literature, changes in water resources (e.g., Urmia Lake [28,29], Hamoun Lakes [30,31], and Iraq's lakes and wetlands [13]) have played a key role in the Middle East SDS emission. Therefore, it can be argued that the hydrological cycle parameters (e.g., precipitation, surface runoff, evapotranspiration, soil moisture, river flows, and surface water and groundwater resources [32]) have a determining role in the formation of SDS sources, which calls for further investigations.

A wide range of natural and anthropogenic factors contribute to the formation of SDS sources in the ME [22]. In addition to the identification of SDS sources, discriminating between types of sources is an essential way to adopt mitigation measures. Discrimination and characterization of SDS sources can provide useful information for better understanding and management of this environmental hazard. In this regard, at the basin scale, recently [15] identified SDS sources in Central Iran using remote sensing data and characterized them in terms of their origin. SDS is naturally a highly dynamic phenomenon driven by the large-scale interactions of environmental parameters of Earth's ecosystems, which highlights the importance of identifying SDS sources at the regional scale. In [22], the authors identified the ME SDS sources and categorized them by their origin on a regional scale using atmospheric aerosol indices. This approach relied on the thresholding of atmospheric aerosol indices, which is not accurate enough and so the location of the identified sources is not entirely reliable.

Remote sensing methods are equipped with special capabilities for spatial-temporal modeling of SDS from the local [15] to global scales [33,34]. Previous studies have developed different approaches for SDS source identification using remote sensing data. They can be in three categories, namely aerosol loading-based [35], spectral indices-based [36], and susceptibility mapping [19]. Some innovative approaches have recently been developed, including the multi-step binary masking approach proposed by [15]. This approach benefits from not only simplicity but also provides more acceptable results considering the inclusion of the main SDS drivers and atmospheric aerosol indices in the modeling process.

Considering the effects of water resources on SDS activity in the Middle East, the main goal of this study is to identify hydrologic SDS sources. To achieve this goal, SDS sources are identified and then categorized based on their origin. Then, the long-term changes in surface and ground water in the hydrologic SDS sources are analyzed.

2. Materials and Methods

Methodology

As illustrated in Figure 1, the developed methodology to characterize hydrological SDS sources of the Middle East using remote sensing data can be summarized in three main steps. In the first step, SDS sources were identified by using a combination of atmospheric aerosol indices and land surface parameters (e.g., vegetation cover, wind speed, soil moisture, and soil sediment thickness) based on a multi-step masking approach. The second step involved categorizing the identified SDS sources in terms of their origin and type based on different land cover classes. In the final step, focusing on hydrologic SDS sources, spatial-temporal patterns of changes in surface and ground water resources were analyzed.

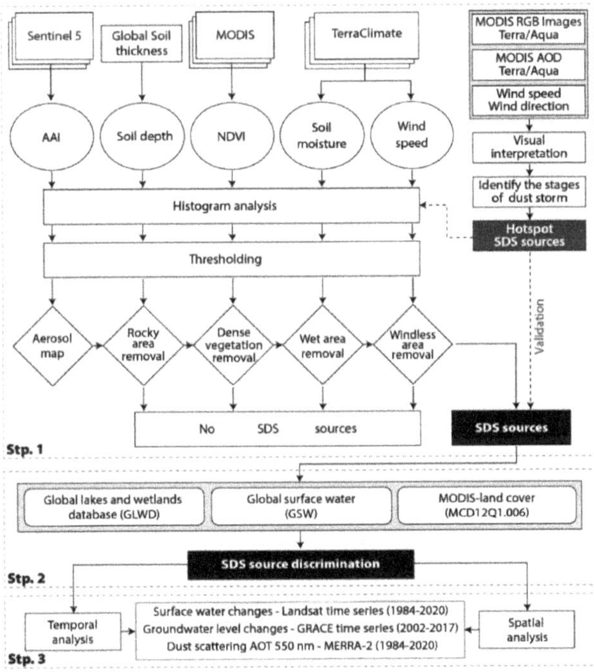

Figure 1. Stepwise approach to characterize hydrologic SDS sources of the Middle East.

- *Identifying SDS sources (Stp. 1)*

This study used a multi-step binary masking procedure [15] to identify SDS sources in the Middle East. In general, SDS sources are associated with long-term dense atmospheric aerosol content. However, a high aerosol content does not necessarily correspond to an SDS source. As [15] acknowledged, taking advantage of the spatial-temporal patterns of effective environmental parameters (i.e., SDS drivers) in combination with satellite atmospheric aerosol indices in a modeling framework can be useful to improve the accuracy of identified SDS sources. Because in this way, the long-term conditions of the atmosphere and the Earth's surface are simultaneously considered to identify SDS emission sources. Accordingly, the preliminary SDS sources map was produced using atmospheric aerosol indices. Then, the accurate SDS sources map was generated by accounting for other effective SDS drivers through binary masks.

Sentinel 5 satellite was launched on 13 October 2017, by ESA to perform atmospheric measurements to be used for air quality, ozone, and UV radiation, and climate monitoring and forecasting. By thresholding, the mean absorbing aerosol index (AAI) of the Sentinel 5 (OFFL/L3_AER_AI) in the period from 11 July 2018 to 1 June 2022, the preliminary SDS sources map was generated. AAI is an aerosol index based on wavelength-dependent changes in the UV radiation for a specific wavelength pair (354 and 388 nm) during Rayleigh scattering. It is calculated from the difference between the observed top of the atmosphere reflectance and the modeled reflectance (the theoretical reflectance calculated from an atmosphere with only Rayleigh scattering). Positive values of AAI indicate the existence of UV-absorbing aerosols such as dust and smoke in the atmosphere. Moreover, ozone absorption is extremely low in the wavelengths used to produce the AAI product. Unlike aerosol optical depth (AOD) measurements that are highly affected by cloud cover [37], the AAI can be calculated even in cloudy conditions [38]. The mean AAI map shows the overall spatial patterns of atmospheric aerosol over a four-year period in the study area. An appropriate threshold must be defined to separate SDS from other aerosols when attempting to identify SDS sources using atmospheric aerosol indices [39], since only high atmosphere aerosol content in a region could not be an accurate indicator for identifying the emission sources. Therefore, surface SDS drivers must be considered to identify accurate SDS sources.

To identify SDS sources, it could be assumed that future sources will be created by the same past drivers and conditions [17]. Various environmental parameters play a role in the formation of SDS sources [40]. There is no specific rule for choosing these drivers [17]. In this study, the most important drivers including vegetation status, wind speed, soil moisture, and soil sediment thickness, which are mainly employed in the literature, were considered. Accordingly, MODIS 16-day normalized difference vegetation index (NDVI) (1 km spatial resolution) (MOD13A2.v006), TerraClimate monthly average 10 m wind speed and monthly soil moisture (4 km) [41], and global 1 km gridded thickness of soil [42] were used as SDS drivers. The 4-years mean maps (11 July 2018 to 1 June 2022) were obtained for all parameters except for soil thickness. There are four main stages in SDS formation from an emission source: start, maximum lifting, cutoff, and dispersion. Satellite imagery can be used to identify SDS sources with high accuracy in the initial stages (start and maximum lifting) via visual interpretation. By creating a color composite image (RGB), SDS events and their emission sources can be discriminated from other phenomena in the satellite image through image interpretation keys, including shape, size, pattern, tone, texture, etc. Based on the approach proposed by [19], a total of 1014 SDS hotspots were identified via visual interpretation of MODIS-Terra/Aqua sub-daily true-color composite images (RGB_{143}) from 2018 to 2022 (Figure 2). Here, 70% of SDS hotspots were used as training samples for histogram analysis and determining appropriate thresholds for generating preliminary SDS sources maps and binary masks from SDS drivers. Histogram slicing and thresholding were performed based on the histograms of SDS drivers in hotspots. The appropriate thresholds were determined at the 99% confidence interval, and values outside this interval were assumed to be errors or outliers. Accordingly, the maps of SDS drivers were separated into two binary classes 0 (no SDS) and 1 (SDS). In the next step, the bitmaps were masked from the preliminary SDS sources to obtain the accurate SDS sources map, which was validated using the remaining 30% of SDS hotspots.

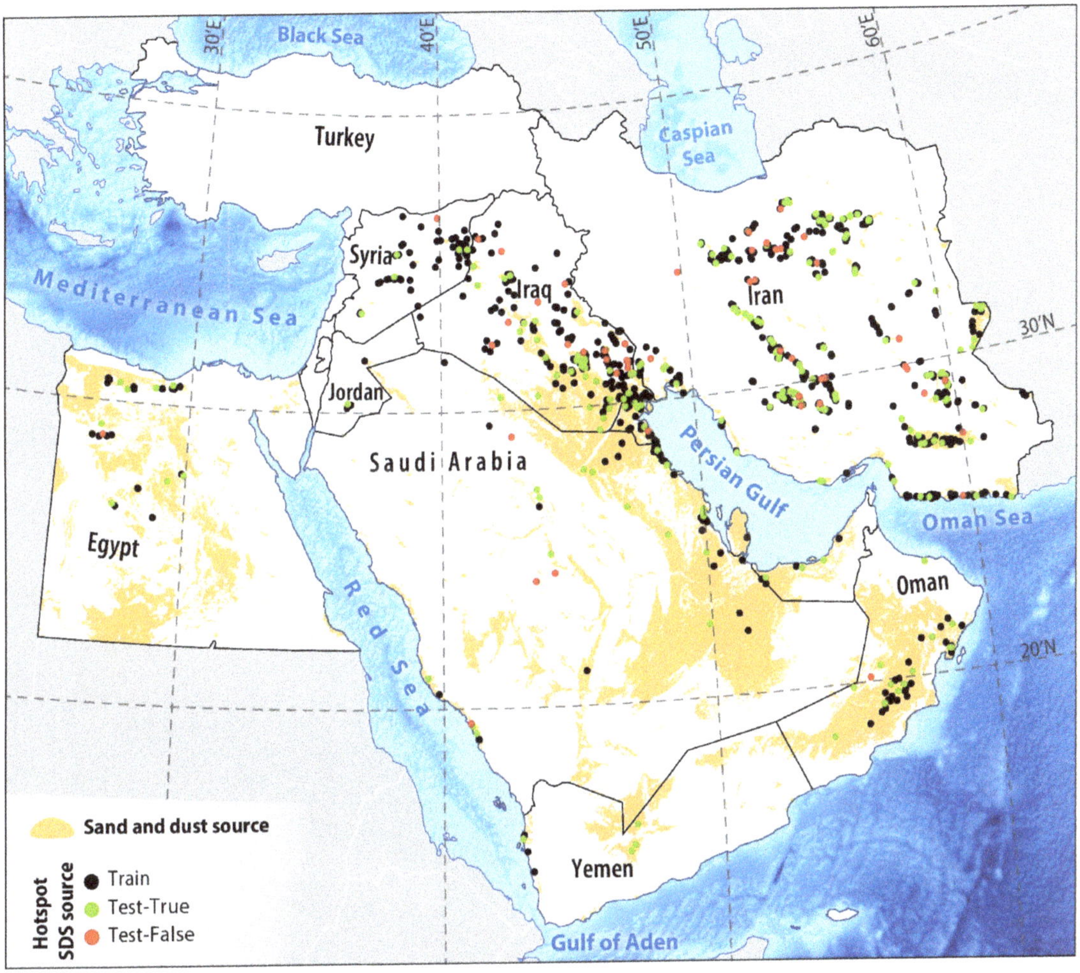

Figure 2. SDS sources of Middle East. The colored solid dots are SDS hotspots identified using visual interpretation of MODIS-RGB images used for histogram analysis (70%: black dots) and accuracy assessment (30%: green and red dots).

- *SDS Source Discrimination (Stp. 2)*

The global lakes and wetlands database (GLWD), global surface water (GSW), and MODIS-land cover 2020 map (MCD12Q1.006) were used to attribute the identified SDS sources to the land cover type and surface conditions. GLWD is obtained from an aggregation of the best maps, data, and information available for global-scale lakes and wetlands using GIS functions [43]. GSW was produced using the time series of Landsat 5, 7, and 8 images from 16 March 1984 to 31 December 2020. Each pixel is classified into water and no-water classes using an expert system. The monthly GSW data available from 1984–2020 are applicable to change detection purposes [44]. The change detection map demonstrates the spatial-temporal pattern of changes in surface water. This makes it possible to identify intermittent lakes and wetlands and discriminate the corresponding SDS sources.

The identified sources were discriminated in terms of origin by overlaying the SDS sources map on different land cover classes. Ephemeral water bodies (wetlands, lakes, and rivers) contribute to SDS emissions [13]. Therefore, GLWD and GSW products were used to discriminate hydrologic SDS sources. Here, hydrologic SDS sources correspond to the margins and beds of permanent, seasonal, and dried lakes, wetlands, and rivers. Coastal SDS sources were discriminated by proximity analysis based on shorelines.

- *Surface and ground water changes (Stp. 3)*

Surface and ground water resources in hydrologic SDS sources were analyzed through Landsat and Gravity Recovery and Climate Experiment (GRACE) satellite data. The normalized difference water index (NDWI) [45] based on Landsat 5, 7, and 8 data from 1984 to 2022 was used to extract surface water and assess their spatial-temporal changes. In this index, values greater than zero are assumed as water [45].

Monitoring of the groundwater changes based on GRACE data was performed through three products from different data processing centers, namely CSR (UT Center for Space Research), GFZ (GeoForschungsZentrum Potsdam), and JPL (NASA Jet Propulsion Laboratory) [46]. These centers are part of the GRACE Ground System and produce the level 2 data used in this dataset. The values of these three products are slightly different. Therefore, the average of these products from April 2002 to July 2017 was used to detect the spatial-temporal changes in the underground water level.

3. Results

3.1. Identification and Attribution of SDS Sources

Histogram analysis based on reference SDS hotspots (Figure 2) was performed and the thresholds of 0, 0.15, 2.7 m/s, 5 mm, and 3 m at a 95% confidence level were determined for AAI, NDVI, wind speed, soil moisture, and soil thickness, respectively. Based on the thresholding of the average AAI map, a preliminary SDS sources map was produced for the Middle East. The binary masks of SDS drivers were then multiplied in the preliminary SDS sources map to obtain the accurate SDS sources (Figure 2). By using the reference SDS hotspots, the overall accuracy of the SDS sources map was obtained at 82.6%, which shows the reliability of the results. The green dots (Figure 2) are SDS hotspots that correspond correctly to the identified SDS sources. In contrast, the red dots are incorrectly identified as SDS sources. Despite the acceptable accuracy, it is worth noting that uncertainty is an inevitable part of modeling process. However, the precise amount for each independent parameter is unknown. Nevertheless, the accuracy assessment shows the overall uncertainty (17.4%) imposed on the model by the input parameters. Finally, the SDS sources were categorized into seven groups in terms of origin and type (Figure 3).

Table 1 shows the area of identified SDS sources by type and country. The largest and smallest sources corresponded to the desert (~79.1%) and coastal wetland (~0.9%), respectively. Saudi Arabia hosted the largest SDS sources (~41%) in the Middle East.

According to Figure 3, hydrologic SDS sources include permanent lake, intermittent lake/wetland, and freshwater marsh & floodplain classes. Coastal wetlands are also considered SDS sources with a hydrological origin. They are not highly affected by human activities and climatic events (e.g., drought) since they can benefit from seawater, unlike other hydrologic sources. Therefore, coastal wetlands were considered independently in this study. In terms of area, in total, hydrologic SDS sources cover 8.4% of the ME, of which the largest area (~5.9%) is corresponding to intermittent lake/wetland (Table 1). Intermittent lake/wetland SDS sources were identified in all countries listed in Table 1 with the largest area belonging to Iran (~41%). Meanwhile, permanent lake SDS sources were observed only in Iran (~68%), Iraq (~26%), and Syria (~6%). Freshwater marsh & floodplain sources were observed only in Iraq (~77%) and Iran (~23%). Overall, Iran (41%), Iraq (21%), Oman (14%), and Saudi Arabia (13%) have the largest area of hydrologic SDS sources (Table 1). Sources in Oman and Saudi Arabia correspond to only the intermittent lake/wetland class.

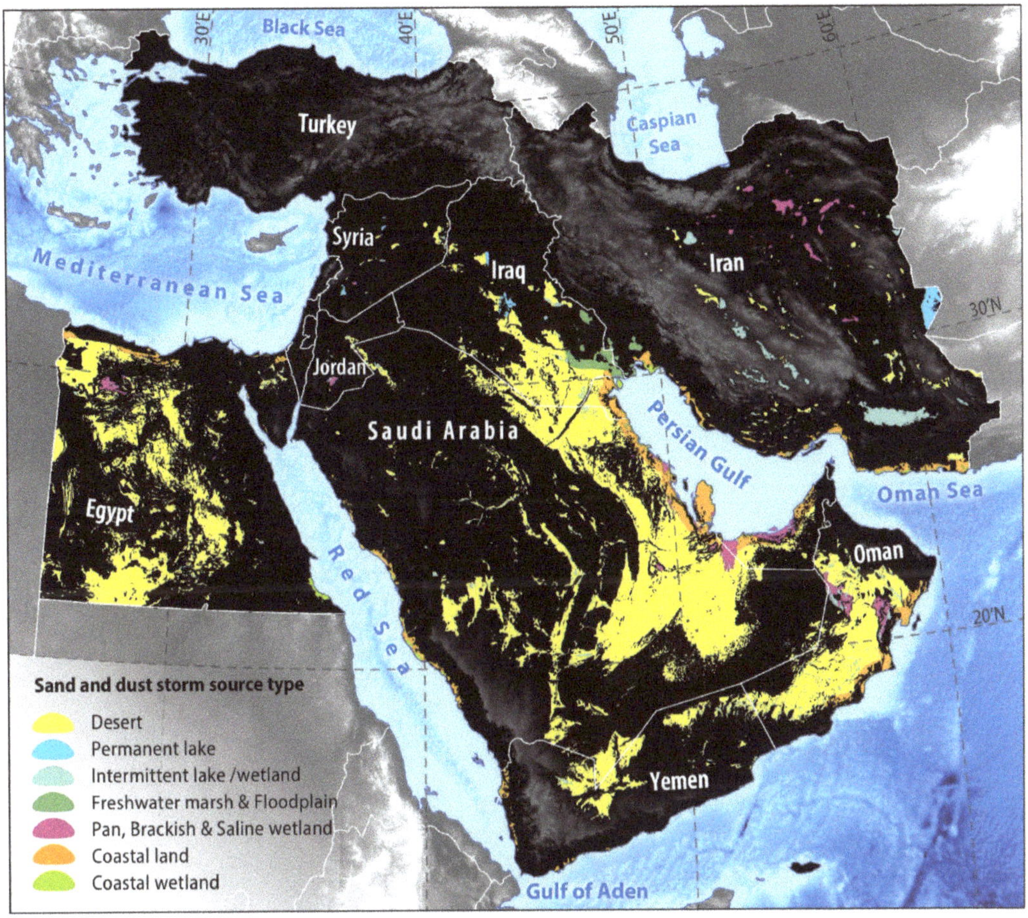

Figure 3. SDS source discrimination in the Middle East in terms of origin.

Table 1. The area (km^2) of different types of SDS sources in the Middle Eastern countries.

Country	Desert	Permanent Lake	Freshwater Marsh & Floodplain	Coastal Wetland	Pan, Brackish & Saline Wetland	Intermittent Lake/Wetland	Coastal Land	Total	%
Bahrain	-	-	-	12	-	31	43	86	<0.1
Egypt	201,503	-	-	1938	3068	2755	6476	215,739	19.9
Iran	14,934	7362	3685	2972	9555	26,392	16,000	80,899	7.5
Iraq	74,858	2834	12,282	-	-	3974	695	94,643	8.8
Jordan	936	-	-	-	999	440	30	2406	0.2
Kuwait	5972	-	-	743	-	395	6042	13,152	1.2
Oman	108,898	-	-	415	10,977	13,192	12,572	146,053	13.5
Qatar	-	-	-	148	372	241	8417	9179	0.9
Saudi Arabia	396,711	-	-	2720	7528	12,117	24,562	443,639	41
Syria	3039	698	-	-	157	522	-	4416	0.4
UAE	10,980	-	-	-	7137	805	6843	25,765	2.4
Yemen	38,386	-	-	448	-	3531	3434	45,800	4.2
Total	856,215	10,894	15,966	9397	39,793	64,397	85,114	1,081,776	100
%	79.1	1.0	1.5	0.9	3.7	5.9	7.9	100	

3.2. Temporal Analysis of Surface and Ground Water Changes in Hydrologic SDS Sources

The temporal pattern of water area changes in the ME lakes and wetlands was extracted using the long-term data archive (1984–2022) of Landsat 5, 7, and 8 satellites (Figure 4). The temporal pattern of SDS in the same period was obtained using the reanalysis MERRA-2 dust scattering AOT 550 nm, monthly 0.5 × 0.625 deg product (Figure 4).

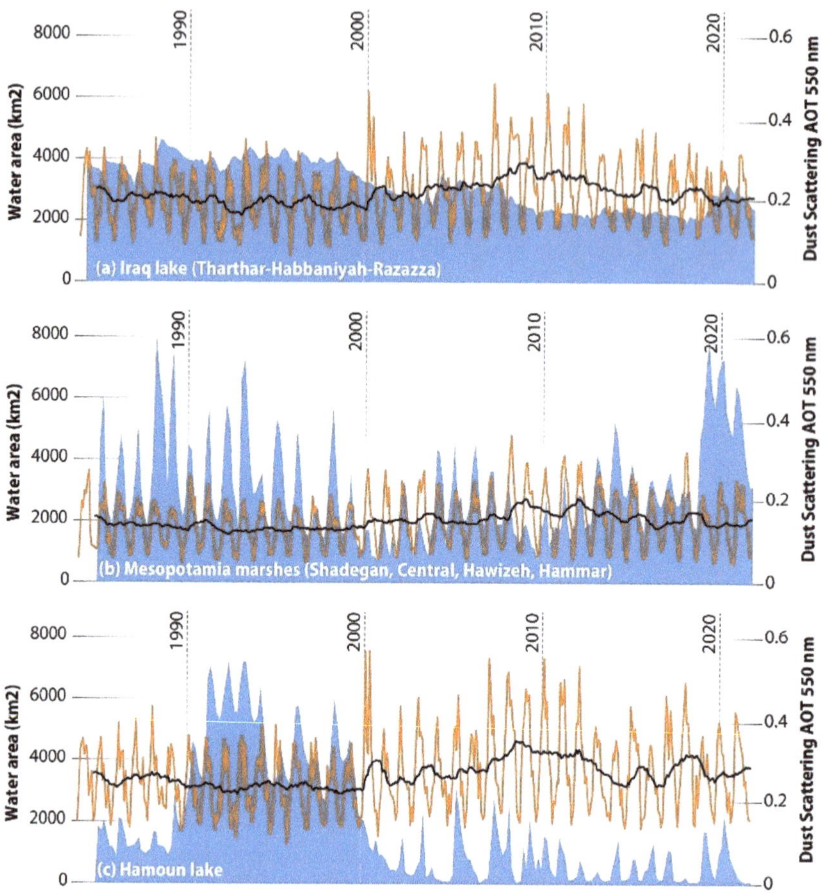

Figure 4. The temporal pattern of changes in water bodies and SDS levels in the Middle East. Blue graph shows the surface water area changes and the brown and black graphs are dust scattering and its moving average, respectively.

As shown in Figure 4a,c, the lakes generally had an increasing trend before 2000 and a decreasing trend after 2000. The first drop in the lakes area occurred in the 2000–2001 period. The second drop can be seen from 2007 onwards, which continued nearly until 2012. Disregarding the seasonal changes, Mesopotamian wetlands exhibited an almost steady trend in the pre-2000 period (Figure 4b). Still, a decreasing trend can be seen since the 1990s, with the highest reduction observed in 2001. From 2003 to 2007, there has been a moderate increasing trend, which again reversed after this period. The second sharp drop occurred in 2008–2009 period and continued until 2012. A sharp and sudden increase can be seen after 2019 in Iran and Iraq due to the increase in rainfall and floods [47].

Dust scattering AOT temporal pattern shows a smaller change in the pre-2000 compared to the post-2000 period. In terms of seasonal patterns, extreme peaks can be seen after

2000, especially in spring and summer (brown graph in Figure 4). Regardless of seasonal changes, as indicated by the dust-scattering AOT moving average (black graph in Figure 4), SDS has increased by an average of 18.4% since 2000.

The correlation between changes in water bodies and SDS was calculated at −0.72, −0.64, and −0.39 for Iraqi lakes, Mesopotamian wetlands, and Hamoun Lake, respectively. Negative values indicate an inverse correlation between water body changes and SDS. The correlation is high for Iraqi lakes and Mesopotamian wetlands while it is low for Hamoun Lake. The correlation analysis showed an inverse but not very strong relationship between the area of water bodies and SDS emission.

Figure 5 shows the Mann–Kendall test for trend analysis of surface water area and dust scattering AOT. Where the z-value equal to −3.15 indicates that surface water has experienced a decreasing trend, while 3.18 for dust scattering AOT shows an increasing trend. These findings are approved by the Sen's estimate lines of both graphs.

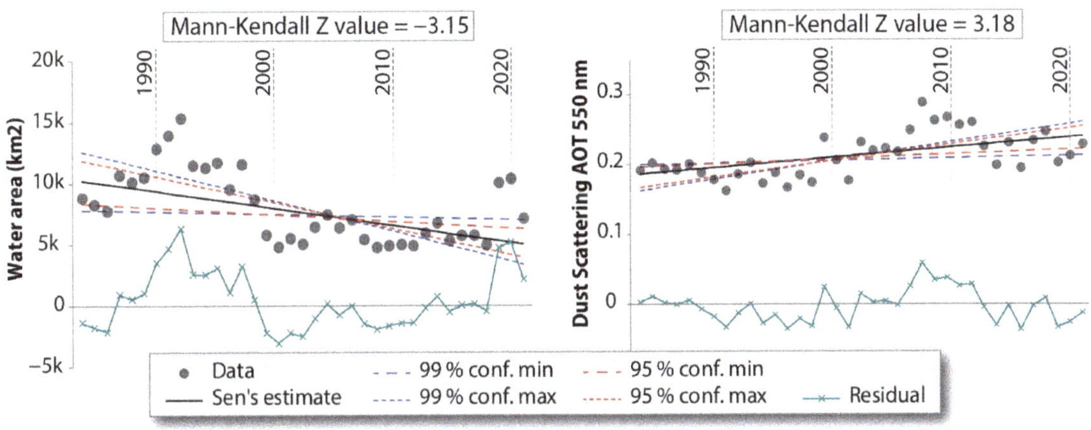

Figure 5. Mann–Kendall test for trend analysis of surface water area and dust emission AOT.

Figure 6 shows the temporal pattern of monthly changes in the groundwater level using GRACE data for SDS sources. The temporal signals of groundwater level changes in SDS sources indicate a high correlation. Accordingly, all SDS sources, except for freshwater marsh & floodplain, had an almost identical temporal change pattern, but their levels of change are different. Overall, a clear decreasing trend can be seen in the groundwater level of the ME SDS sources since 2008. The highest negative changes occurred in hydrologic SDS sources, particularly in freshwater marsh & floodplains. The least changes were observed in desert sources.

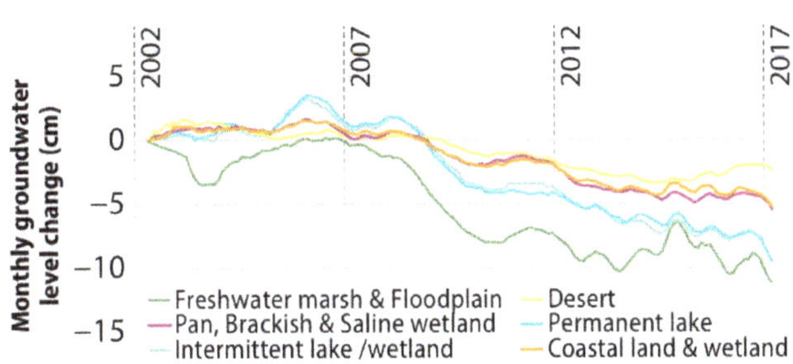

Figure 6. Temporal pattern of groundwater level changes in the Middle East SDS sources obtained from GRACE monthly data.

3.3. Spatial Analysis of Surface and Ground Water Changes in Hydrologic SDS Sources

The spatial pattern of changes in the surface water area (including lakes, wetlands, and rivers) was acquired for two periods before 2000 (1984–1999) and after 2000 (2000–2020) using GSW data (Figure 7a). The increasing and decreasing changes are illustrated in blue and red, respectively. As shown in Figure 7b, the highest changes in surface water occurred in Iran, Iraq, Turkey, and Egypt, respectively. In Iran, negative changes were roughly three times more than positive changes. For Iraq, negative changes were almost twice the positive changes. For all countries except Egypt, the decrease in surface water was greater than the increase. Negative changes in surface water in Iran were about 30% more than in Iraq, twice as much as in Turkey, and three times as much as in Egypt. The highest density of surface water changes was detected in the hydrologic SDS sources of Iraq, especially the southwestern areas and Mesopotamian wetlands (Figure 7a).

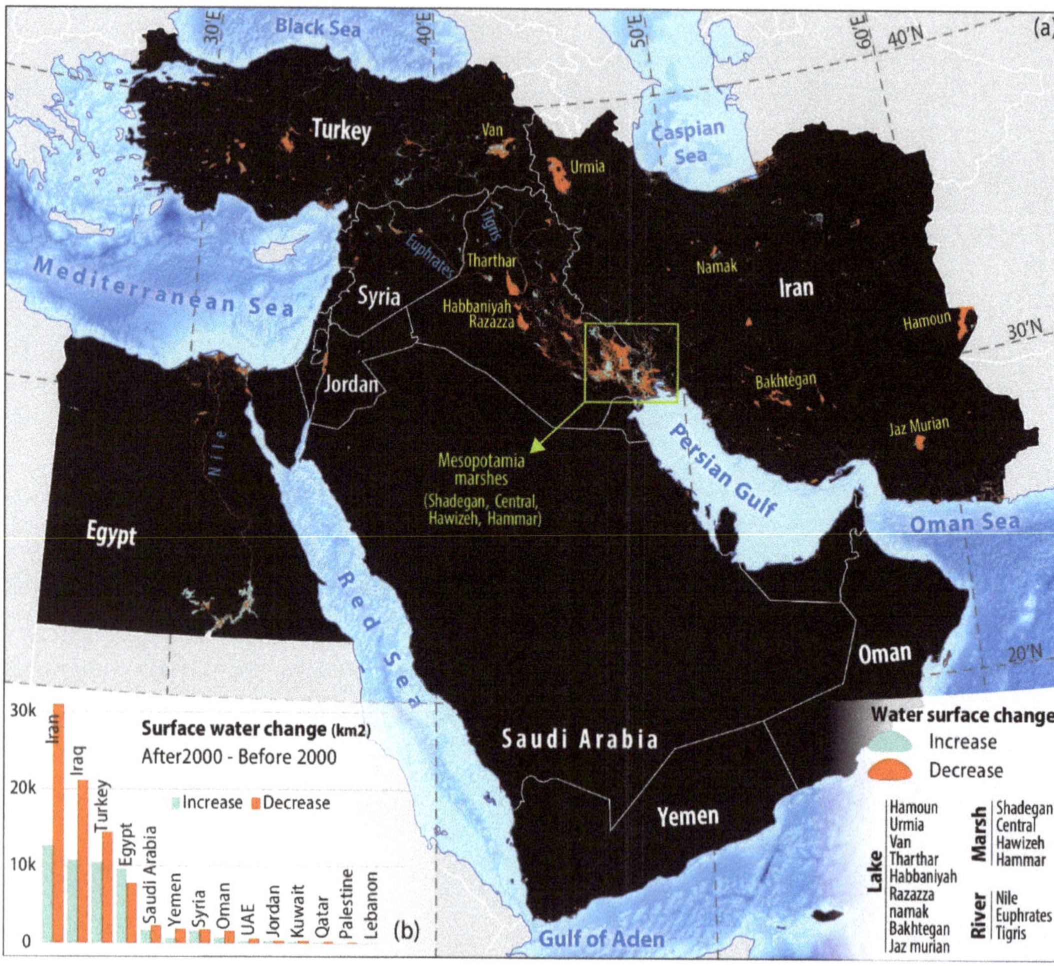

Figure 7. Changes in the surface water area of the Middle Eastern hydrologic SDS sources based on GSW data (**a**) and surface water area changes by country (**b**).

The spatial pattern of total groundwater level changes in the Middle East was extracted from April 2002 to July 2017 (Figure 8a). The variation of groundwater level in this period is between −10 and +3 m. The highest decrease and increase in groundwater level were observed for the northwestern regions of Iran and the western regions of Turkey, respectively. In hydrologic SDS sources, the groundwater level has mainly decreased. On the other hand, the spatial pattern of the groundwater level shows far fewer changes in desert SDS sources. Groundwater level changes by country are presented in Figure 8b. On average, the highest drop in the groundwater level occurred in Iraq, Iran, and Syria, respectively. Additionally, the largest area of hydrologic SDS sources can be found in Iraq and Iran. In addition, permanent lakes and freshwater marsh & floodplain SDS sources are located only in these three countries. The lowest decrease in groundwater level occurred in Oman. Oman has the largest area of hydrologic SDS sources after Iran and Iraq (Table 1).

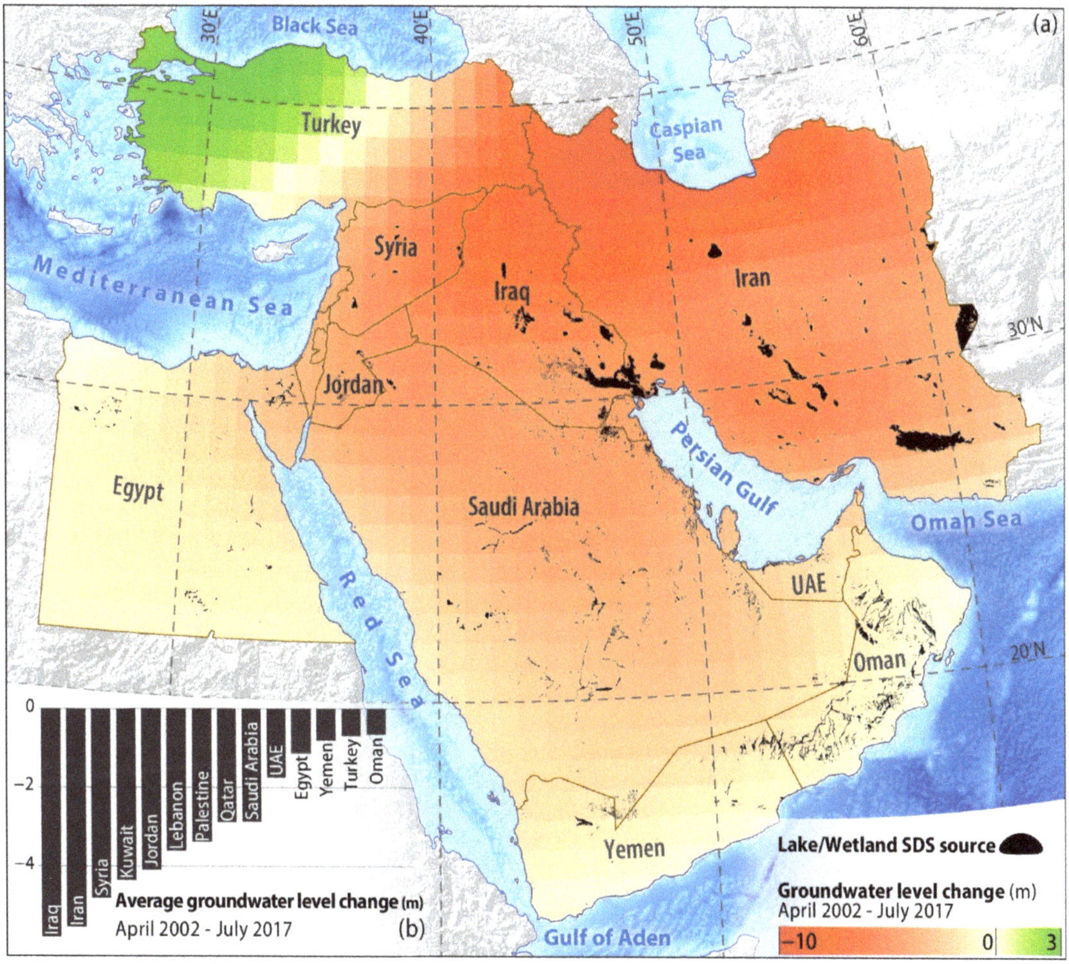

Figure 8. Groundwater level changes in the Middle East from April 2002 to July 2017, obtained from GRACE monthly data (**a**) and average changes by country (**b**).

4. Discussion

4.1. Characterization of Middle Eastern SDS Sources

The occurrence of SDS in different regions and from different emission sources exerts various impacts on the environment. Therefore, combating SDS in different sources requires specific mitigation measures [48]. Accordingly, an essential step for the proper management of SDS sources is identifying the source type. In desert sources, for example, common mitigation measures include stabilizing soil surface by mulching, planting climate-compatible shrubs and trees, using windbreaks, creating mechanical and biological barriers (e.g., dead vegetation), and employing other soil and plant resource management methods [48]. On the other hand, hydrologic SDS sources, including lakes, wetlands, and rivers, demand water resource management [27,49]. Water resource management in countries with shared basins such as Iran, Iraq, Turkey, and Syria could be undertaken through joint efforts.

The accurate identification of SDS sources is considered a challenging task. In the last few decades, various approaches have been used to identify and discriminate the SDS sources in ME. Most of these approaches fall under the atmospheric aerosol loading-based category (e.g., [21,22,35]). In these approaches, a region with a high concentration of atmospheric aerosols is generally considered an SDS source. The aerosol loading-based approaches often detect SDS events regardless where they are emitted from. Due to the highly dynamic nature of SDS, detected events may originate from more distant regions. Accordingly, using only atmospheric aerosol indices such as AOD and AAI may lead to errors in modeling and significantly reduce the accuracy of identifying SDS sources. To tackle this problem, by considering the multi-year average of thresholded atmospheric aerosol indices and SDS drivers through a supervised approach (by taking advantage of the SDS hotspots identified by visual interpretation), it is possible to identify SDS sources with satisfactory accuracy. However, compared to previous studies that have identified SDS sources only based on aerosol indices, the present approach gives more reliable results.

Most of the Middle East is located within the global desert belt given its low annual precipitation [50]. Here, as the results showed, over 79% of the identified SDS sources are in the desert class. SDS sources of Saudi Arabia, Egypt, Oman, and Yemen are mainly sandy deserts. Those in Iran, Iraq (especially Mesopotamian wetlands), and Syria are generally composed of fine-grained alluvial sediments. The larger size of the sources does not essentially mean greater activity and SDS emission. Accordingly, despite the high area, desert SDS sources in Saudi Arabia, Egypt, Oman, and Yemen they have low activity in terms of SDS frequency and intensity. Previous studies indicate that the highest SDS activity occurs in the northern and northwestern parts of Iraq and along the Syrian–Iraqi border [25]. This issue is dependent on the source type, particle diameter, time, duration, and intensity of erosive winds. Fine-grained particles (e.g., SDS sources in the alluvial plains of Iran, Iraq, and Syria) are lightweight and are easily detached from the Earth's surface and lifted into the upper layers of the atmosphere, so they can travel over long distances [51]. These fine-grained SDS particles are easier to detect in satellite images due to their durability and the longer movement of the plume in the atmosphere.

The spatial-temporal analysis of water bodies showed that that both surface water (Figures 4 and 6) and groundwater (Figures 5 and 8) have decreased in the Middle East, which reveals the instability of the water balance conditions in the region. The results show that the water balance in hydrologic SDS sources is more negative than in other sources. In this regard, both surface water and groundwater resources have experienced severe reductions. Regarding coastal SDS sources, it is worth mentioning that these sources are influenced by the sea and its physicochemical reactions. Although they are considered hydrologic sources, reduction in their groundwater levels is not evident since they are recharged by the sea. Desert SDS sources have experienced much less groundwater depletion than other sources, which may be attributed to limited agricultural and low groundwater exploitation in these areas.

4.2. Influence of Drought on SDS Sources in Middle East

The long-term analysis of surface water changes in hydrologic SDS sources showed a relatively sharp decrease in the 2000–2001 and 2007–2012 periods compared to pre-2000. Further examination revealed that these two periods correspond almost to two very severe drought periods hitting the ME in the last 50 years [47,52]. As shown in Table 2, surface water resources in the first and second drought periods have decreased by 51% and 52.9%, respectively, compared to the pre-2000. Hamoun Lakes have experienced a very severe decrease (over 72%) in both drought periods compared to other water bodies. Iraqi lakes and Hamoun Lake experienced a sharper decrease in the second period compared to the first. In contrast, the Mesopotamia wetlands show a lower decline in the second period.

Table 2. Surface water changes and dust scattering AOT from hydrologic SDS sources during drought periods.

	Name	Pre-2000	First Drought	Change in First Drought Compared to Post-2000 (%)	Second Drought	Change in Second Drought Compared to Pre-2000 (%)
Water area (km^2)	Iraq lake	3942	3037	−23.0	2496	−36.7
	Marshes	3343	1275	−61.9	1699	−49.2
	Hamoun	3535	990	−72.0	899	−74.6
	Total	10821	5301	−51.0	5094	−52.9
AOT	Iraq lake	0.19	0.22	12.0	0.26	33.3
	Marshes	0.13	0.15	18.1	0.18	40.6
	Hamoun	0.24	0.27	13.5	0.33	37.9
	Mean	0.19	0.21	14.5	0.26	37.3

In the first and second periods, dust scattering AOT in water bodies increased by 14.5% and 37.3%, respectively, compared to the pre-2000 (Table 2), which the greatest increase occurred in Mesopotamian wetlands in both episodes. This increase in SDS from hydrological sources mainly occurred in spring and summer. This can be attributed to the predominance of seasonal wet conditions and regional wind regimes (e.g., Shamal Winds). Interestingly, SDS occurrence is a response to drought-induced climatic conditions (Figure 6), since the increase in SDS does not exactly coincide with the decrease in water bodies, and, generally, a rather irregular time lag is observed between the two. This is essentially because there is a delay between the reduction of water bodies and their complete desiccation and the erodibility of their substrate sediments. The duration of this time lag is a function of the hydroclimatic conditions and the soil's physicochemical properties. Therefore, based on correlation analysis, there is no straightforward statistical relationship between water area changes and SDS activity. It can be argued that, in general, the reduction of surface water in most cases (especially during drought episodes) has been accompanied by an increase in SDS activity and vice versa. Another possible reason for the lack of a high correlation between surface water changes and SDS activity is the effect of human activities.

There was no information available about groundwater changes before 2002. Still, the effect of the second drought period on the reduction of ground water levels is clearly visible in Figure 8. The continuation of the decreasing groundwater trends after 2012 is likely due to insufficient aquifer recharge resulting from low precipitation and overexploitation. It can be argued that periodic droughts have directly affected the water balance and hydrological resources of the Middle East, leading to increased SDS activity.

A severe flood occurred in 2019 in the central and eastern regions of the Middle East, including Iran and Iraq [53], which caused the temporary restoration of wetlands and lakes in these regions (Figure 6). However, after that, the area of water bodies has been decreasing. The connection between surface water reduction and SDS occurrence means that the continuation of these conditions will intensify SDS activity in the Middle East in the future.

4.3. *Influence of Human Interventions on Middle Eastern SDS Sources*

Human intervention and exploitive intentions have always been artificial agents manipulating the environment. The population of the ME countries is estimated at 460 million in 2022, which is a significant growth compared to the previous decades. The rapid population growth in the past few decades has intensified the demand for water, food, and energy [12,54]. This has in turn increased and intensified water diversion projects (e.g., the construction of dams, pumping stations, irrigation canals, or other man-made structures that modifies the natural flow of a waterway) in the region to provide drinking water, expand agricultural lands, meet industrial uses, and hydropower energy [13]. In addition, the non-cooperation of countries, war, and insecurity in Iraq, Syria, and Yemen [55,56] have also contributed to the mismanagement of water resources and SDS in the Middle East.

Turkey's GAP projects on the Tigris and Euphrates Rivers and Egypt's dam construction on the Nile River are among the most intrusive anthropogenic activities affecting water resources in the Middle East. However, SDS sources in Egypt are mostly non-hydrologic (Figure 3) and this project has not so far affected SDS activity to a great extent. Still, the formation of new hydrologic SDS sources, particularly on the margins of the Nile in the future is not far-fetched. However, GAP dam construction projects have already reduced the environmental water right of wetlands and led to the drying of some lakes downstream. The same has happened in Iran due to the mismanagement of water resources in basins located in Central Iran [15]. Accordingly, as shown in Figure 3, some SDS sources in Southern and Central Iraq and Central and Eastern Iran have hydrologic origins. It can be concluded that human intervention has played a key role in the formation of hydrologic SDS sources and a limited role in coastal and desert sources.

Despite the occurrence of severe drought episodes in the region, there have been cases of increase in surface water resources in countries such as Egypt and Turkey (Figure 4). The analysis of the past 40-year trends does not show a significant increase in their precipitation. Therefore, this increase has been mainly due to human activities such as water diversion projects on the Nile River in Egypt and the Tigris and Euphrates Rivers in Turkey. Nevertheless, these projects have reduced the amount of downstream water flow and have disrupted the natural hydrological balance of their respective basins and sub-basins. On the other hand, the creation of artificial surface water resources may trigger the formation of new SDS sources in the future, as happened in the case of the artificial lakes of Tharthar, Habbaniyah, and Razzaza in Iraq. Therefore, the increase in surface water resources does not necessarily ensure SDS control in long term. Streams and rivers carry and deposit fine-grained sediments in lakes, wetlands, and floodplains. The permanent or seasonal desiccation of these water bodies accompanied by erosive winds prompts the formation of new SDS sources. This accentuates the critical importance of protecting natural and artificial surface water sources from instabilities caused by hydroclimatic conditions and human intervention to combat SDS.

5. Conclusions

This study investigated the Middle East as one of the global sand and dust storms (SDS) sources focusing on hydrological parameters. First, the SDS sources were identified using remote sensing data and a multi-step binary mask approach with an overall accuracy of 82.6%. The sources were categorized into seven classes based on land cover type, which included desert, permanent lake, freshwater marsh & floodplain, coastal wetland, pan, brackish & saline wetland, intermittent lake/wetland, and coastal land.

According to the findings, desert sources (over 79%) have the highest area of SDS sources. Hydrologic SDS sources including the permanent lake, intermittent lake/wetland, and freshwater marsh & floodplain accounted for 8.4% of SDS sources in the Middle East. The largest area of hydrologic SDS sources belonged to Iran (41%), Iraq (21%), Oman (14%), and Saudi Arabia (13%).

Our results revealed that Iran, Iraq, Turkey, and Egypt showed the highest amount of decreasing and increasing surface water changes, respectively. The highest density of surface water change was observed in those of the Mesopotamian wetlands in Southern Iraq. The highest decrease in groundwater was observed in Iran and Iraq, which includes the highest areas of hydrologic SDS sources.

The correlation analysis showed an inverse but not very strong relationship between surface water changes and SDS activities. The temporal pattern of changes in surface waters, including wetlands and lakes, showed a declining trend in the post-2000 compared to the pre-2000 period. Similarly, SDS activities have increased by an average of 18.4% since 2000. In the 2000–2001 and 2007–2012 periods, severe droughts corresponded to the shrinking of surface water up to 51% and 52.9%, respectively. SDS activities in surface water in the first and second periods increased by an average of 14.5% and 37.3%, respectively. Overall, it can be argued that periodic droughts have negatively affected the water resources in the Middle East, leading to an increase in SDS activity, especially from hydrologic SDS sources.

Despite the occurrence of severe drought episodes in the region, human interventions such as water diversion projects have reduced the water flow to downstream regions, which caused the SDS activities. For instance, the Mesopotamian wetlands of Iraq have experienced the highest density of hydrologic SDS sources, which is the consequence of anthropogenic activities upstream.

In general, it can be concluded that the main causes of SDS sources formation in the Middle East consist of natural factors such as the predominance of drought. Specifically, the hydrologic SDS sources are suffering from mismanagement of water resources, which greatly contributes to SDS activities.

The most important challenge of this approach was the identifying SDS hotspots using visual interpretation. Because these hotspots, as a training dataset, strongly affect the accuracy of the binary mask-based model. To perform this task, the interpreter must have sufficient knowledge about SDS and the study area. In addition, in general, visual interpretation is a time-consuming and expensive process, especially on a large scale. However, this approach can be extended to other areas, and in this way, global SDS sources can be discriminated.

Author Contributions: Conceptualization, R.P., S.A., A.D.B. and N.N.S.; data curation, R.P.; formal analysis, R.P.; investigation, R.P.; methodology, R.P.; project administration, S.A., A.D.B. and N.N.S.; software, R.P.; supervision, S.A., A.D.B. and N.N.S.; visualization, R.P.; writing—original draft, R.P.; writing—review and editing, R.P., S.A., A.D.B. and N.N.S. All authors have read and agreed to the published version of the manuscript.

Funding: The publication fees of this article have been supported by Mansoura University.

Institutional Review Board Statement: Not applicable.

Informed Consent Statement: Not applicable.

Data Availability Statement: Data are available on request due to their robustness and restrictions for sharing publicly.

Acknowledgments: The authors would like to express their appreciation to Masoud Soleimani, Mohsen Bakhtiari, and Saham Mirzaei for their supports in reviewing the paper.

Conflicts of Interest: The authors declare no conflict of interest.

References

1. Zittis, G.; Almazroui, M.; Alpert, P.; Ciais, P.; Cramer, W.; Dahdal, Y.; Fnais, M.; Francis, D.; Hadjinicolaou, P.; Howari, F.; et al. Climate Change and Weather Extremes in the Eastern Mediterranean and Middle East. *Rev. Geophys.* **2022**, *60*, e2021RG000762. [CrossRef]
2. Boloorani, A.D.; Najafi, M.S.; Soleimani, M.; Papi, R.; Torabi, O. Influence of Hamoun Lakes' dry conditions on dust emission and radiative forcing over Sistan plain, Iran. *Atmospheric Res.* **2022**, *272*, 106152. [CrossRef]
3. Francis, D.; Chaboureau, J.P.; Nelli, N.; Cuesta, J.; Alshamsi, N.; Temimi, M.; Pauluis, O.; Xue, L. Summertime dust storms over the Arabian Peninsula and impacts on radiation, circulation, cloud devel-opment and rain. *Atmos. Res.* **2021**, *250*, 105364. [CrossRef]
4. Khaniabadi, Y.O.; Daryanoosh, S.M.; Amrane, A.; Polosa, R.; Hopke, P.K.; Goudarzi, G.; Mohammadi, M.J.; Sicard, P.; Armin, H. Impact of Middle Eastern Dust storms on human health. *Atmos. Pollut. Res.* **2017**, *8*, 606–613. [CrossRef]
5. Soleimani, Z.; Teymouri, P.; Boloorani, A.D.; Mesdaghinia, A.; Middleton, N.; Griffin, D.W. An overview of bioaerosol load and health impacts associated with dust storms: A focus on the Middle East. *Atmos. Environ.* **2020**, *223*, 117187. [CrossRef]
6. Middleton, N.; Kashani, S.S.; Attarchi, S.; Rahnama, M.; Mosalman, S.T. Synoptic Causes and Socio-Economic Consequences of a Severe Dust Storm in the Middle East. *Atmosphere* **2021**, *12*, 1435. [CrossRef]
7. Jish Prakash, P.; Stenchikov, G.; Kalenderski, S.; Osipov, S.; Bangalath, H. The impact of dust storms on the Arabian Pen-insula and the Red Sea. *Atmos. Chem. Phys.* **2015**, *15*, 199–222. [CrossRef]

8. Rezazadeh, M.; Irannejad, P.; Shao, Y. Climatology of the Middle East dust events. *Aeolian Res.* **2013**, *10*, 103–109. [CrossRef]
9. Kutiel, H.; Furman, H.K.H. Dust Storms in the Middle East: Sources of Origin and Their Temporal Characteristics. *Indoor Built Environ.* **2003**, *12*, 419–426. [CrossRef]
10. World Bank. *Sand and Dust Storms in the Middle East and North Africa Region: Sources, Costs, and Solutions*; World Bank: Washington, DC, USA, 2019.
11. Heydarizad, M.; Minaei, M.; Ichiyanagi, K.; Sorí, R. The effects of local and regional parameters on the δ18O and δ2H values of precipitation and surface water resources in the Middle East. *J. Hydrol.* **2021**, *600*, 126485. [CrossRef]
12. Bozorg-Haddad, O.; Zolghadr-Asli, B.; Sarzaeim, P.; Aboutalebi, M.; Chu, X.; Loáiciga, H.A. Evaluation of water shortage crisis in the Middle East and possible remedies. *J. Water Supply: Res. Technol.* **2020**, *69*, 85–98. [CrossRef]
13. Boloorani, A.D.; Papi, R.; Soleimani, M.; Karami, L.; Amiri, F.; Samany, N.N. Water bodies changes in Tigris and Euphrates basin has impacted dust storms phenomena. *Aeolian Res.* **2021**, *50*, 100698. [CrossRef]
14. Hamzeh, N.H.; Karami, S.; Opp, C.; Fattahi, E.; Jean-François, V. Spatial and temporal variability in dust storms in the Middle East, 2002–2018: Three case studies in July 2009. *Arab. J. Geosci.* **2021**, *14*, 538. [CrossRef]
15. Papi, R.; Kakroodi, A.; Soleimani, M.; Karami, L.; Amiri, F.; Alavipanah, S.K. Identifying sand and dust storm sources using spatial-temporal analysis of remote sensing data in Central Iran. *Ecol. Inform.* **2022**, *70*, 101724. [CrossRef]
16. Gholami, H.; Mohamadifar, A.; Collins, A.L. Spatial mapping of the provenance of storm dust: Application of data mining and ensemble modelling. *Atmos. Res.* **2020**, *233*, 104716. [CrossRef]
17. Boroughani, M.; Pourhashemi, S.; Hashemi, H.; Salehi, M.; Amirahmadi, A.; Asadi, M.A.Z.; Berndtsson, R. Application of remote sensing techniques and machine learning algorithms in dust source detection and dust source susceptibility mapping. *Ecol. Inform.* **2020**, *56*, 101059. [CrossRef]
18. Papi, R.; Argany, M.; Moradipour, S.; Soleimani, M. Modeling the potential of Sand and Dust Storm sources formation using time series of remote sensing data, fuzzy logic and artificial neural network (A Case study of Euphrates basin). *Eng. J. Geo-spatial Inf. Technol.* **2021**, *8*, 61–82. [CrossRef]
19. Boloorani, A.D.; Samany, N.N.; Papi, R.; Soleimani, M. Dust source susceptibility mapping in Tigris and Euphrates basin using remotely sensed imagery. *Catena* **2022**, *209*, 105795. [CrossRef]
20. Boloorani, A.D.; Kazemi, Y.; Sadeghi, A.; Shorabeh, S.N.; Argany, M. Identification of dust sources using long term satellite and climatic data: A case study of Tigris and Euphrates basin. *Atmos. Environ.* **2020**, *224*, 117299. [CrossRef]
21. Rashki, A.; Middleton, N.J.; Goudie, A.S. Dust storms in Iran–Distribution, causes, frequencies and impacts. *Aeolian Res.* **2021**, *48*, 100655. [CrossRef]
22. Ginoux, P.; Prospero, J.M.; Gill, T.E.; Hsu, N.C.; Zhao, M. Global-scale attribution of anthropogenic and natural dust sources and their emission rates based on MODIS Deep Blue aerosol products. *Rev. Geophys.* **2012**, *50*. [CrossRef]
23. Cao, H.; Amiraslani, F.; Liu, J.; Zhou, N. Identification of dust storm source areas in West Asia using multiple environmental datasets. *Sci. Total Environ.* **2015**, *502*, 224–235. [CrossRef] [PubMed]
24. Moridnejad, A.; Karimi, N.; Ariya, P.A. A new inventory for middle east dust source points. *Environ. Monit. Assess.* **2015**, *187*, 1–11. [CrossRef]
25. Moridnejad, A.; Karimi, N.; Ariya, P.A. Newly desertified regions in Iraq and its surrounding areas: Significant novel sources of global dust particles. *J. Arid Environ.* **2015**, *116*, 1–10. [CrossRef]
26. Parajuli, S.P.; Yang, Z.; Kocurek, G. Mapping erodibility in dust source regions based on geomorphology, meteorology, and remote sensing. *J. Geophys. Res. Earth Surf.* **2014**, *119*, 1977–1994. [CrossRef]
27. Hamidi, M. The key role of water resources management in the Middle East dust events. *Catena* **2020**, *187*, 104337. [CrossRef]
28. Boroughani, M.; Hashemi, H.; Hosseini, S.H.; Pourhashemi, S.; Berndtsson, R. Desiccating Lake Urmia: A New Dust Source of Regional Importance. *IEEE Geosci. Remote Sens. Lett.* **2019**, *17*, 1483–1487. [CrossRef]
29. Ghale, Y.A.G.; Tayanc, M.; Unal, A. Dried bottom of Urmia Lake as a new source of dust in the northwestern Iran: Understanding the impacts on local and regional air quality. *Atmospheric Environ.* **2021**, *262*, 118635. [CrossRef]
30. Rashki, A.; Kaskaoutis, D.; Goudie, A.; Kahn, R. Dryness of ephemeral lakes and consequences for dust activity: The case of the Hamoun drainage basin, southeastern Iran. *Sci. Total Environ.* **2013**, *463–464*, 552–564. [CrossRef]
31. Miri, A.; Maleki, S.; Middleton, N. An investigation into climatic and terrestrial drivers of dust storms in the Sistan region of Iran in the early twenty-first century. *Sci. Total Environ.* **2021**, *757*, 143952. [CrossRef]
32. Thakur, P.K.; Nikam, B.R.; Garg, V.; Aggarwal, S.P.; Chouksey, A.; Dhote, P.R.; Ghosh, S. Hydrological Parameters Estimation Using Remote Sensing and GIS for Indian Region: A Review. *Proc. Natl. Acad. Sci. India Sect. A Phys. Sci.* **2017**, *87*, 641–659. [CrossRef]
33. O'Loingsigh, T.; Mitchell, R.; Campbell, S.; Drake, N.; McTainsh, G.; Tapper, N.; Dunkerley, D. Correction of dust event frequency from MODIS Quick-Look imagery using in-situ aerosol measurements over the Lake Eyre Basin, Australia. *Remote Sens. Environ.* **2015**, *169*, 222–231. [CrossRef]
34. Baddock, M.C.; Bryant, R.G.; Acosta, M.D.; Gill, T.E. Understanding dust sources through remote sensing: Making a case for CubeSats. *J. Arid Environ.* **2021**, *184*, 104335. [CrossRef]
35. Prospero, J.M.; Ginoux, P.; Torres, O.; Nicholson, S.E.; Gill, T.E. Environmental characterization of global sources of at-mospheric soil dust identified with the Nimbus 7 Total Ozone Mapping Spectrometer (TOMS) absorbing aerosol product. *Rev. Geophys.* **2002**, *40*, 1–2. [CrossRef]

36. Karimi, N.; Moridnejad, A.; Golian, S.; Samani, J.M.V.; Karimi, D.; Javadi, S. Comparison of dust source identification techniques over land in the Middle East region using MODIS data. *Can. J. Remote Sens.* **2012**, *38*, 586–599. [CrossRef]
37. Soleimani, M.; Argany, M.; Papi, R.; Amiri, F. Satellite aerosol optical depth prediction using data mining of climate parameters. *Phys. Geogr. Res. Q.* **2021**, *53*, 319–333. [CrossRef]
38. Stein Zweers, D.C. TROPOMI ATBD of the UV aerosol index. S5P-KNMI-L2-0008-RP. 1.0. Available online: http://www.tropomi.eu/sites/default/files/files (accessed on 1 August 2022).
39. Nabavi, S.O.; Haimberger, L.; Samimi, C. Climatology of dust distribution over West Asia from homogenized remote sensing data. *Aeolian Res.* **2016**, *21*, 93–107. [CrossRef]
40. Shepherd, G.; Terradellas, E.; Baklanov, A.; Kang, U.; Sprigg, W.; Nickovic, S.; Joowan, C. *Global Assessment of Sand and Dust Storms*; WMO: Geneva, Switzerland, 2016.
41. Abatzoglou, J.T.; Dobrowski, S.Z.; Parks, S.A.; Hegewisch, K.C. TerraClimate, a high-resolution global dataset of monthly climate and climatic water balance from 1958–2015. *Sci. Data* **2018**, *5*, 170191. [CrossRef]
42. Pelletier, J.D.; Broxton, P.D.; Hazenberg, P.; Zeng, X.; Troch, P.A.; Niu, G.; Williams, Z.C.; Brunke, M.A.; Gochis, D. *Global 1-km Gridded Thickness of Soil, Regolith, and Sedimentary Deposit Layers*; ORNL DAAC: Oak Ridge, TN, USA, 2016.
43. Lehner, B.; Döll, P. Development and validation of a global database of lakes, reservoirs and wetlands. *J. Hydrol.* **2004**, *296*, 1–22. [CrossRef]
44. Pekel, J.-F.; Cottam, A.; Gorelick, N.; Belward, A.S. High-resolution mapping of global surface water and its long-term changes. *Nature* **2016**, *540*, 418–422. [CrossRef]
45. Gao, B.-C. Normalized Difference Water Index for Remote Sensing of Vegetation Liquid Water from Space. In Proceedings of the SPIE'S 1995 Symposium on OE/Aerospace Sensing and Dual Use Photonics, Orlando, FL, USA, 17–21 April 1995; pp. 225–236. [CrossRef]
46. Darvishi Boloorani, A.; Soleimani, M.; Papi, R.; Alavipanah, S.K.; Al-Quraishi, A.M.F. Zoning Areas Susceptible to Land Subsidence in Tigris and Euphrates Basins. *Eng. Technol. J.* **2019**, *37*, 265–272. [CrossRef]
47. Barlow, M.; Zaitchik, B.; Paz, S.; Black, E.; Evans, J.; Hoell, A. A Review of Drought in the Middle East and Southwest Asia. *J. Clim.* **2016**, *29*, 8547–8574. [CrossRef]
48. Middleton, N.; Kang, U. Sand and Dust Storms: Impact Mitigation. *Sustainability* **2017**, *9*, 1053. [CrossRef]
49. Al-Ansari, N.; Adamo, N.; Sissakian, V. Water shortages and its environmental consequences within Tigris and Euphrates Rivers. *J. Earth Sci. Geotech. Eng.* **2019**, *9*, 27–56.
50. Batanouny, K.H. Climatic Aridity in the Deserts of the Middle East. In *Plants in the Deserts of the Middle East*; Springer: Berlin/Heidelberg, Germany, 2001; pp. 11–24. [CrossRef]
51. Bolin, B.; Aspling, G.; Persson, C. Residence time of atmospheric pollutants as dependent on source characteristics, at-mospheric diffussion processes and sink mechanisms. *Tellus* **1974**, *26*, 185–195. [CrossRef]
52. Trigo, R.M.; Gouveia, C.; Barriopedro, D. The intense 2007–2009 drought in the Fertile Crescent: Impacts and associated atmospheric circulation. *Agric. For. Meteorol.* **2010**, *150*, 1245–1257. [CrossRef]
53. Yadollahie, M. The Flood in Iran: A Consequence of the Global Warming? *Int. J. Occup. Environ. Med.* **2019**, *10*, 54–56. [CrossRef]
54. Roudi-Fahimi, F.; Creel, L.; De Souza, R.-M. *Finding the Balance: Population and Water Scarcity in the Middle East and North Africa*; Population Reference Bureau: Washington, DC, USA, 2002.
55. Allan, J.A. Hydro-Peace in the Middle East. *SAIS Rev.* **2002**, *22*, 255–272. [CrossRef]
56. Amery, H.A. Water wars in the Middle East: A looming threat. *Geogr. J.* **2002**, *168*, 313–323. [CrossRef]

sustainability MDPI

Article

Climate Change during the Third Millennium—The Gulf Cooperation Council Countries

Mohamed E. Hereher [1,2]

[1] Department of Environmental Sciences, Faculty of Science, Damietta University, New Damietta 34517, Egypt; mhereher@gmail.com
[2] Geography Department, College of Arts and Social Sciences, Sultan Qaboos University, Muscat 123, Oman

Abstract: The Gulf Cooperation Council (GCC) is a union occurring in the arid subtropical belt of the world. Contemporary climate change is a serious environmental issue at the regional and global levels. The main purpose of this study was to investigate the changes and trends in the regional climate in the GCC countries in terms of the land surface temperatures and surface anomalies, precipitation, and sea surface temperatures over the first two decades of this millennium. Research data exclusively relied on satellite remote sensing. Spatial, statistical, and cartographic analyses were performed to extract, manipulate, and display thematic maps reflecting the changes and trends of the regional climate. The results showed that notable climate changes were obvious and frequent throughout the GCC countries, with symptoms influencing the atmosphere, lithosphere, and the hydrosphere of the region. However, a prominent gradient in the severity of climate change occurred from north to south. Remarkably, serious impacts were observed in Kuwait and eastern Saudi Arabia, while the least effects were recorded in Oman. The study denotes the competence of remote sensing for monitoring regional climate change.

Keywords: greenhouse gases; geospatial analysis; remote sensing; Arabian Peninsula

Citation: Hereher, M.E. Climate Change during the Third Millennium—The Gulf Cooperation Council Countries. *Sustainability* **2022**, *14*, 14181. https://doi.org/10.3390/su142114181

Academic Editor: Swadhin Behera

Received: 21 April 2022
Accepted: 27 May 2022
Published: 31 October 2022

1. Introduction

The global CO_2 emissions increased from 6 billion tons in 1950 to 22 billion tons in 1990 and eventually approached 34 billion tons in 2020 [1]. As it contains more than 60% of the world's population, Asia is the largest contributor of CO_2 in the world, with about 53% of global emissions. There is a direct correlation between the high rates of greenhouse gas (GHG) emissions (mostly CO_2 and methane) from burning fossil fuels (oil, gas, and coal) and global warming [2]. This issue of climate change has received universal interest in the last decades. International endeavors pursue the goal of keeping global warming from approaching 2 °C by encouraging the countries that emit the most GHGs to restrict the increase in global temperatures below 1.5 °C [3]. However, the rates of GHG emission could increase the global temperatures by 6 °C by the end of this century [4]. Recently, the United Nations Climate Change Conference (COP26), which was held in November 2021 in Glasgow, Scotland, emphasized again the responsibility of all countries to limit the world temperature increase by 1.5 °C above the pre-industrial revolution levels. The world witnesses not only enormous emissions of GHGs but many other air pollutants as well. Hence, global warming is just one facet of the problem. Air pollution is considered the most significant environmental health issue in the world. It ranks among the top five risks for attributable deaths globally [5] and causes about 4.2 million premature deaths every year [6]. Air pollution consists of particulate matter and gaseous constituents, e.g., sulfur dioxide (SO_2), nitrogen dioxide (NO_2), ozone, particulates, and carbon monoxide (CO).

The Gulf Cooperation Council (GCC) is a regional geographical union established in 1981 between six Arab countries (Kuwait, Saudi Arabia, Bahrain, Qatar, United Arab Emirates, and Oman) (Figure 1). These oil-producing countries are all experiencing remarkable socio-economic developments depending on their tremendous fossil fuel production

(Table 1). They hold approximately 40% of the world's known oil and about 24% of the world's known gas reserves [7]. These states have also experienced accelerated population growth accompanied by extensive energy utilization. Energy consumption in the GCC countries is the highest in the world and is as much as 2.5 times the consumption of the European Union [8]. There is a direct relationship between energy consumption and CO_2 emissions as well as between economic growth and energy consumption in the GCC countries [9]. All the GCC countries are listed as being among the largest per capita CO_2 emitters in the world, as they emit 3.08% of global CO_2, with the maximum being from Saudi Arabia (1.8%) in 2020. Major sources of GHGs and air pollutants in the GCC countries are numerous. However, the burning of fossil fuels for energy, transportation, petrochemical industries, and oil refineries are at the top. Unfortunately, regional studies on GHGs and air pollution studies are generally few. However, there are some local investigations that demonstrate the rates of GHG emissions and their projections for the near future. For example, CO_2 emissions have tripled in Oman between 2000 and 2014 due to the expansion of the industrial sector [10]. Ref. [11] showed that the CO_2 emissions from the transportation sector of Saudi Arabia increased from 2000 through 2010 to 2020 to a total of 56 million tons, 98 million tons, and 163 million tons, respectively. In the United Arab Emirates (UAE), the major emissions (90%) of GHGs are produced from the energy sector [12].

The GCC region is under a hot desert climate (BWh), according to the Koppen climate classification. Summers are generally hot to extremely hot, and air temperatures usually approach 40 °C during the summer months. However, winters are generally mild with temperatures seldom falling below 10 °C, particularly in coastal regions. Rainfall in the area is generally scant, except for in Oman, where Al-Hajar Mountain receives up to 300 mm/year. The relative humidity is high (up to 85%) along the coastal zones. The GCC countries are usually exposed to frequent dust storms accompanying the *Al-Shamal* (northern in Arabic) winds, and these storms can significantly reduce visibility and cause problems regarding the air and traffic [13]. The region is experiencing fast population growth. The majority of the region's population resides in Saudi Arabia, UAE, Kuwait, and Oman. The total population of the GCC countries exceeds 50 million [14].

The impacts of climate change on the GCC region were also addressed in some previous studies, with a particular focus on air temperatures, precipitation, and the Arabian Gulf environment. For example, Ref. [15] observed a tendency toward increasing drought intensity and frequency between 1981 and 2000 over the Arabian Peninsula due to climate change. Ref. [13] reported a significant increase in air temperatures between 1996 and 2010. Ref. [16] observed that between 2003 and 2018, the annual increase in the gulf water surface temperatures exceeded the global ocean warming that occurred since 1900. In addition, Ref. [17] reported a correlation between the warming of seawater and the locations of bleached coral reefs in the Arabian Gulf. Moreover, Ref. [18] observed significant nighttime warming caused by the urban heat islands of the major cities in the region, particularly in Dubai (1.9 °C/decade). The warming of the climate is generally associated with drought, an increase in the frequency of heatwaves, the deterioration of terrestrial biodiversity and marine ecosystems, and sea-level rise [19–21]. To the best of the author's knowledge, there has been little major research on regional climate change, and most of the published work was at the country level. In addition, the majority of the previous studies relied on data from existing ground meteorological stations, which are relatively little in space compared to the huge area of the region. The present study entirely depends on remotely sensed data, which covers all the spatial gaps and boundaries of the region. In addition, there is a necessity for regional climate change studies in order to determine appropriate management procedures and to prioritize the available natural resources in the region with particular focus upon the renewable resources [22–26]. Therefore, the present study attempts to fill the gap in the literature and provides a comprehensive regional and synoptic overview on the status of climate change over the past two decades using remotely sensed data. The goal is to focus on the decadal changes in the land surface temperatures, precipitation, and sea surface temperatures, as revealed by satellite images. Therefore, the current research aims to answer the following questions: 1—what are the temporal regional changes in the emission of GHGs and air pollution? 2—are there any changes in land surface temperatures, precipitation, and sea surface temperatures? And 3—what are the decadal rates of these changes? The time frame of most studied parameters extends to two decades, while some others extend to 40 years.

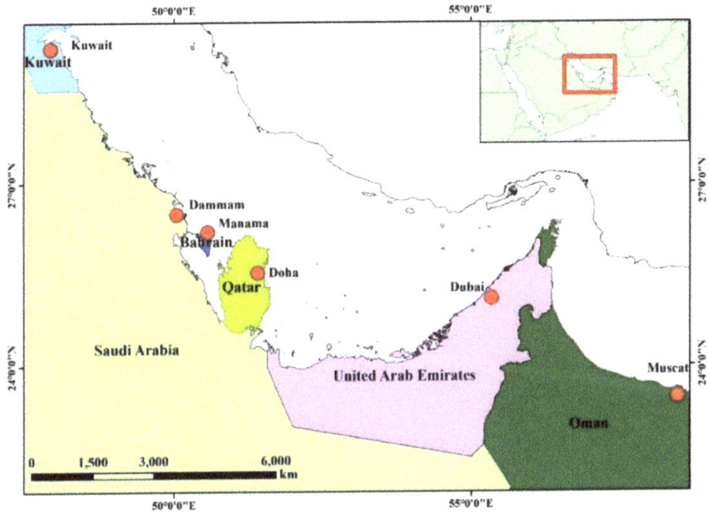

Figure 1. The study region of the GCC countries (the red square in the upper map) and the major cities along the coast of the Arabian Gulf and the Sea of Oman.

Table 1. Synopsis of the GCC countries (Source: https://www.gccstat.org/en; accessed on 18 February 2022).

Country	Area (km^2)	Population in 2017 (Million)	Oil Production in 2017 (Million Barresls/Day)	Natural Gas-Marketed Production in 2019 (Million m^3)	Electricity Production in 2019 (GWH)
Kuwait	17,818	4.2	2.70	13,952	75,082
KSA	2,149,690	32.6	9.96	117,000	342,622
Bahrain	778	1.5	0.19	17,149	17,854
Qatar	11,600	2.7	0.60	176,000	49,873
UAE	83,600	9.3	2.96	55,097	138,454
Oman	309,500	4.5	0.97	36,462	37,513

2. Materials and Methods

2.1. Atmospheric Data

In this investigation, the goal was to probe into the geographical changes of atmospheric constituents, considering those that are the most related to global warming and climate change in this important fossil fuel stock region. The atmospheric data include two GHGs: carbon dioxide (CO_2) and methane (CH_4), as well as three air pollutants: sulfur dioxide (SO_2), nitrogen dioxide (NO_2), and black carbon (BC). The data utilized are gridded (monthly/daily records) time series raster images covering the entirety of the GCC countries along the Arabian Gulf (Table 2). All these data were acquired by means of remote sensing. The CO_2 and CH_4 datasets cover the periods 2003–2016 and 2003–2020, respectively. On the other hand, the dataset of BC and SO_2 extends to 40 years (1980–2020), while NO_2 data cover the period 2005–2020. The data were acquired for a window covering the entirety of the GCC countries overlooking the Arabian Gulf. All the data were downloaded from the Geospatial Interactive Online Visualization and Analysis Infrastructure (Giovanni) (https://giovanni.gsfc.nasa.gov/giovanni/, (accessed on 11 November 2021)) portal; Version (V.4.36). Giovanni is a web-based platform developed by the Goddard Earth Sciences Data and Information Services Center (GES DISC) that provides a simple and instinctive way to visualize, analyze, and access the earth science remotely sensed data acquired from many sources and satellites, such as TRMM, TOMS, OMI, AIRS, MERRA, MLS, MODIS, etc. [27]. These data were provided as daily or monthly records shown as timely averaged maps in different spatial resolutions ranging from 0.25 × 0.25 to 2 × 2.5 degrees in GeoTiff format and at the Geographical Coordinate System (GCS).

Table 2. Characteristics of the analyzed data (Source: https://giovanni.gsfc.nasa.gov/giovanni/; accessed on 18 February 2022).

Parameter	Source	Spatial Resolution (Degree)	Temporal Resolution	Temporal Coverage	Units of Measurements
CO_2	AIRS *	2 × 2.5	Monthly	2003–2016	Parts per million (ppm)
CH_4	AIRS	1	Monthly	2003–2020	Parts per billion per volume (ppbv)
SO_2	MERRA-2 **	0.5 × 0.6	Monthly	1980–2020	Kg/m^3
NO_2	OM I ***	0.25	Daily	2005–2020	$Molecules/cm^2$
Black carbon	MERRA-2	0.5 × 0.6	Monthly	1980–2020	Kg/m^3
Nighttime Temp. anomaly	AIRS	1	Monthly	2003–2016	Degrees Celsius

* Atmospheric Infrared Sounder. ** The Modern-Era Retrospective analysis for Research and Applications. *** Ozone Monitoring Instrument.

2.2. Land Surface Temperatures and Precipitation Data

The Land Surface Temperature (LST) data were acquired from the Moderate Resolution Imaging Spectroradiometer (MODIS) instrument onboard the *Aqua* satellite. In this study, LST data were selected from the MYD11B3.V6 product, which provides average monthly land surface temperature in a pixel size of 0.05° × 0.05° (5600 m × 5600 m) within a 1200 by 1200 km (km) tile. This product provides both daytime (at 13:30) and nighttime (01:30) land surface temperatures. MODIS data were found to well agree with ground LSTs, as they were validated in different environments, including lakes, vegetation, and soil sites in clear sky conditions by the radiance-based approach to have a mean bias within 1 K [28]. The LST data were obtained from the NASA's Land Processes Distributed Active Archive Center (LP DAAC) (https://lpdaac.usgs.gov, (accessed on 1 December 2021)) from the tiles H22V06 and H23V06 for the period from January 2003–December 2020. In this investigation, the daytime LST images were extracted from the MYD11B3 product on a monthly basis, and the digital numbers were converted to degrees Celsius at the pixel level. The two tiles were mosaicked together in order to compile one continuous image, and a subset covering the study area was then produced. In addition, the nighttime anomalies of surface air temperatures were also investigated. Surface nighttime temperature anomalies data were also acquired from the Giovanni web portal for the period from 2003–2016. The data were acquired from the Atmospheric Infrared Sounder (AIRS) satellite for the main cities along the Arabian Gulf. Regarding precipitation, gridded data were acquired from the World Climate (https://www.worldclim.org, (accessed on 19 February 2022)) web portal. This platform provides a monthly raster database of the global monthly precipitation obtained via the reanalysis of 9000 to 60,000 ground weather stations worldwide for the period from 1960–2018 via interpolation techniques. The global cross-validation correlations were 0.86 for precipitation, revealing high confidence estimations. Details on acquiring, processing, and validating these climate data are shown in Ref. [29]. In the present study, the monthly precipitation dataset acquired between 2000 and 2018 was downloaded at 2.5° × 2.5° (~20 km^2) for the GCC region in GeoTiff format. The precipitation profile for the study area was generated, from which the mean annual precipitation and the decadal trend were statistically determined.

2.3. Sea Surface Temperatures Data

The GCC countries overlook the western side of the Arabian Gulf. The oceanographic characteristics of this semi-enclosed inland sea are affected by the regional climate. Therefore, climate change should influence not only the lithosphere but the hydrosphere as well. The sea surface temperatures (SST) of the gulf water were extracted by thermal remote sensing satellites. In this context, MODIS data are highly beneficial for mapping SST gradients and have high confidence results, as they significantly correlate to the in situ SST measurements [30]. They have been processed and calibrated by the NASA technical laboratories using the most recent updates and algorithms. In this study, level-3 (9 km) global monthly SST data for the period of 2003–2018 were downloaded from the NASA Ocean Color web portal (http://oceancolor.gsfc.nasa.gov, (accessed on 19 May 2019)). The monthly images were stacked together, the entire time series profile was produced, and the geographical subset of the gulf was extracted.

2.4. Data Analysis and Visualization

In the beginning, all the acquired data were re-projected to the Geographical Coordinate System. However, the pixel size of all images in each parameter was kept without resampling, as each element was independently treated. For each parameter, images of the individual monthly records were stacked together, forming one file of multi-date profiles (from the first to the last date) for this parameter. This raster dataset was then converted to vector data in a tabular format (the American Standard Code for Information Interchange, ASCII) for consequent analysis. For each parameter, the converted ASCII table shows the coordinates and the record of each point for every date in the time series. This table was eventually used for statistical analysis. The monthly and annual average values for each point for each parameter were computed from the monthly records to yield a mean layer for that parameter. In addition, the rate of change at every point of that parameter was obtained from the slope of the linear regression relationship of the time series records. For example, the LST for a given pixel at a given date is calculated as follows:

$$Y = \alpha X + \beta \tag{1}$$

where X, Y, α, and β are the time, LST, slope, and intercept, respectively. The modeled slope of the regression (α) was defined as the temperature rate of change (°C/decade). In the same manner, the decadal trend for every studied parameter was determined at the pixel level. The tables of the mean and trend for each parameter were displayed in ArcMap for interpolation analysis. The Inverse Distance Weighted (IDW) algorithm is the interpolator used to produce the surface of distribution maps of the mean and rate of change throughout the study region. This technique, which considers the spatial density of data points when producing the continuous surface, is widely used in environmental and climatic studies [31,32]. Thematic maps for the mean and decadal trends of each parameter, including the atmospheric, terrestrial, and sea surface temperature changes, were prepared and visualized. In addition to the regional mean and trend maps, all the GHGs and air pollutants were investigated to reveal the monthly profile during the study period of each parameter. Then the monthly distribution pattern was analyzed for the main cities overlooking the Arabian Gulf (Kuwait City, Dammam, Manama, Doha, Dubai, and Muscat), and a window of 50 × 50 km was selected for each city. The monthly profiles for GHG/air pollutant parameters for these cities were displayed in order to visualize the temporal variations of these inputs to the atmosphere.

3. Results

3.1. Greenhouse Gases and Air Pollution

All the GCC countries are experiencing notable high emissions of greenhouse gases [33]. There has been a significant increase in the CO_2 gas concentration throughout the entire region, which increased from 375 ppm in 2003 to 402 ppm in 2016. Figure 2a shows that the mean CO_2 emissions were remarkably high in Kuwait, KSA, Bahrain, Qatar, and UAE. This high concentration reflects the dependence of these countries upon crude oil for their energy, industry, and transport sectors. Globally, two-thirds of carbon dioxide (CO_2) originated from the energy sector [26]. The decadal trend of CO_2 gas emissions (Figure 2b) revealed maxim emission rates in Kuwait, KSA, and UAE (up to 21.6 ppm/decade). Saudi Arabia alone contributed 1.8% of the global CO_2 emissions in 2020 [26]. The lower rates of emissions were clearly observed in areas that were farther away from the gulf in the Arabian desert. At the monthly level, it was observed that the concentrations of CO_2 are higher during spring and summer, with the maximum values occurring during May and June. The lowest concentrations were recorded in January and October (Figure 2c). The highest monthly CO_2 record ever was observed in Dubai in June.

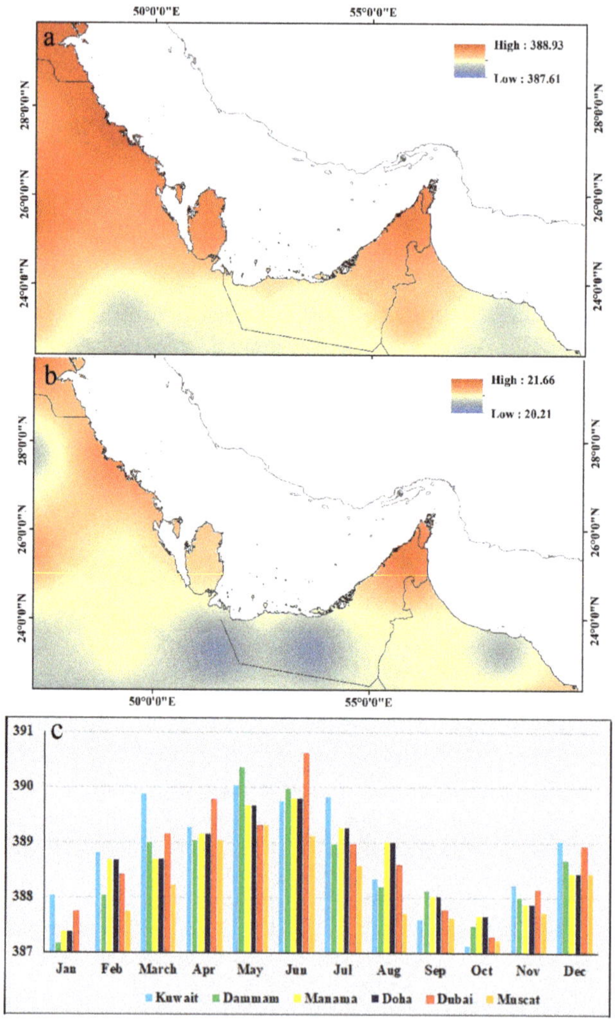

Figure 2. (**a**) is the distribution of the mean CO_2 gas emissions in ppm over the GCC countries between 2003 and 2016, (**b**) represents the trend (ppm/decade), which are maximal in Kuwait, KSA, and UAE, and (**c**) shows the average monthly concentration of CO_2 in the six main cities. The source of the raw data is: https://giovanni.gsfc.nasa.gov/giovanni/; accessed on 18 February 2022.

Methane (CH_4) is considered a major greenhouse gas that is responsible for global warming. In contrast to CO_2 gas, methane has a short lifetime (~12 years); however, it absorbs much energy in the troposphere compared to CO_2 [34]. The energy sector, which depends on oil and gas reserves, is the primary source of atmospheric methane. The distribution of the mean monthly concentration of CH_4 emissions between 2003 and 2020 (Figure 3a) revealed that many emissions were clear in UAE (1862 ppbv), Kuwait (1860 ppbv), and Saudi Arabia (1854 ppbv). In addition, there has been a notable increase in the decadal rate of emissions in the GCC countries, with maximal rates in Oman (456 ppbv/decade) and UAE (437 ppbv/decade). This is evident in the increase in CH_4 emissions in Oman (mainly from the energy sector), which increased from 288 Gg in 2000 to 825 Gg in 2015 [35]. The peak emission was observed during August, whereas the lowest record was observed during February (Figure 3c).

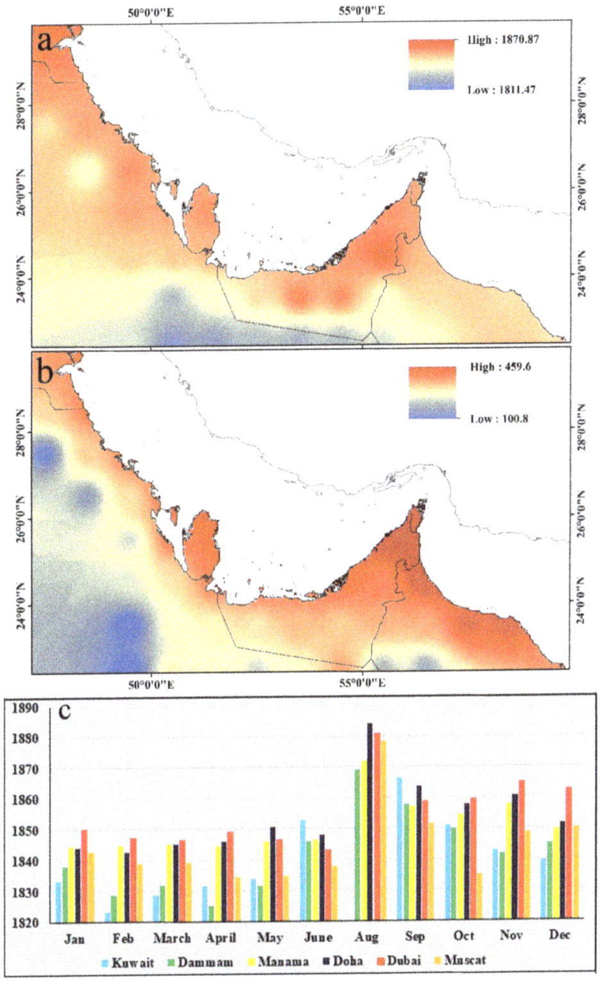

Figure 3. (**a**) is the distribution of the mean monthly CH_4 gas emissions in ppbv over the GCC countries between 2003 and 2020, (**b**) represents the trend (ppbv/decade), which are maximal in Oman and UAE, and (**c**) shows the average monthly concentration of CH_4 in the six main cities. The source of the raw data is: https://giovanni.gsfc.nasa.gov/giovanni/; accessed on 18 February 2022.

The three air pollutants analyzed in this study are SO_2, NO_2, and black carbon. SO_2 concentrations and trends are shown in Figure 4. The primary sources of SO_2 are oil refineries, desalination, and power plants using diesel [14,36]. The mean monthly concentrations of SO_2 for the period from 1980–2020 revealed maximum values in both the Dammam region of Saudi Arabia (15 μg/m^3) and Kuwait (13 μg/m^3). The decadal trends concurred with the monthly averages, where maximal trends occurred in these two locations. Specifically, KSA and Kuwait are among the top 25 countries producing SO_2 in the world in 2018 [37]. In addition, Ref. [38] reported that SO_2 emissions from power plants in Kuwait are greatly exceeding the annual national standards (80 μg/m^3). At the monthly level (Figure 4c) and in contrast to the CO_2 emissions, winter and fall witnessed maximum SO_2 emissions, while the minimum was recorded during June and July.

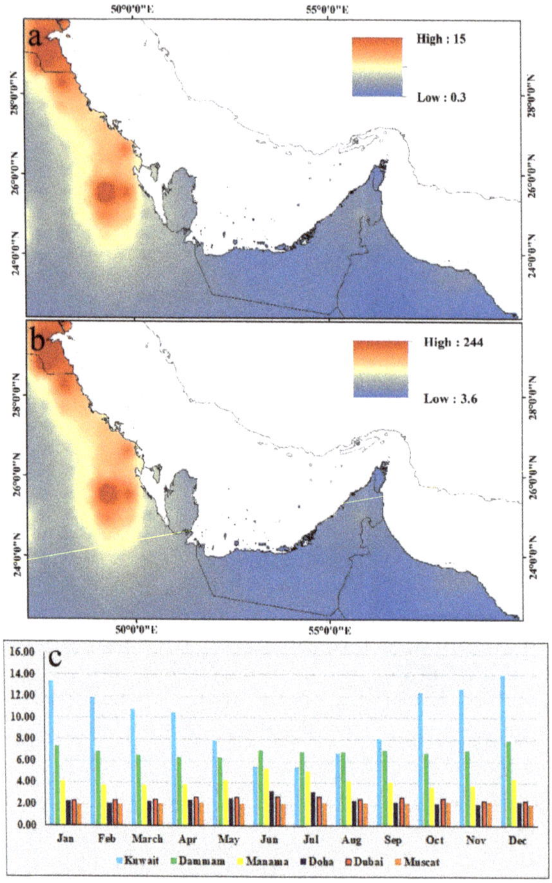

Figure 4. (**a**) is the distribution of the mean monthly SO_2 emissions in μgm/m^3 over the GCC countries between 1980 and 2020, (**b**) represents the trend (μgm/m^3 per decade), which are maximal in Dammam and Kuwait, and (**c**) shows the average monthly concentration of SO_2 in the six main cities. The source of the raw data is: https://giovanni.gsfc.nasa.gov/giovanni/; accessed on 18 February 2022.

The distribution of NO_2 is correlated mainly to the road traffic and to the indoor use of stoves, heaters, and fossil-fuel-burning appliances [39]. Therefore, it is obvious that the high mean concentrations of this pollutant are apparent in all the metropolitan agglomerations of the GCC countries, such as Kuwait City, Dammam, Manama, Doha, and Dubai (Figure 5a) during the period from 2005–2020. The maximum monthly concentrations were observed in Abu Dhabi (9.2 Molecules/cm^2) and Kuwait cities (8.9 Molecules/cm^2). Interestingly, Ref. [40] reported that the GCC countries with high NO_2 emissions are UAE, KSA, and Kuwait. In addition, higher NO_2 from traffic was also reported in Kuwait City [41] and Abu Dhabi, UAE [40]. However, the decadal rate of NO_2 revealed recent inputs from other regions, such as Sohar in northern Oman (Figure 5b). This is confirmed by Ref. [42], who reported that the traffic near Sohar City in Oman yields double the average concentration of NO_2. In contrast, some cities, such as Dammam, experienced a decrease in the decadal trend. It is also clear that all the GCC cities have comparable monthly emissions, with relatively high emissions during the winter months (Figure 5c).

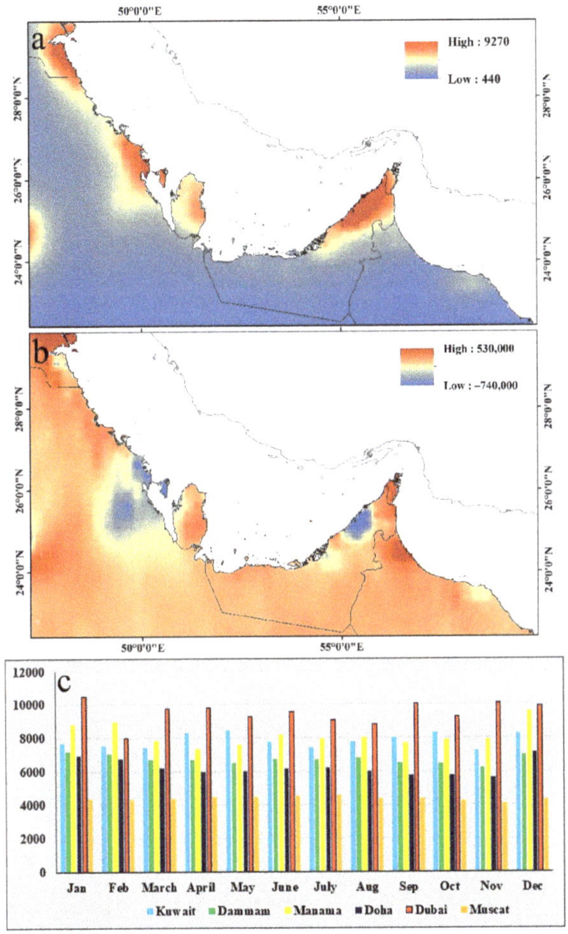

Figure 5. (**a**) is the distribution of the mean monthly NO_2 gas emissions in trillion molecules/cm^2 throughout the GCC countries between 2005 and 2020, (**b**) represents the trend (trillion molecules/cm^2 per decade), which is maximal in Sohar (Oman), and (**c**) shows the average monthly concentration of NO_2 in the six main cities. The source of the raw data is: https://giovanni.gsfc.nasa.gov/giovanni/; accessed on 18 February 2022.

Black carbon (BC) is considered to be a fine carbonaceous particulate matter (PM2.5) and occurs in the exhaust of the diesel-engine vehicles due to the incomplete combustion of fossil fuel. It is also produced from non-fossil fuels, such as biomass combustion [43]. It is a significant source of climate change. Specifically, BC is the second largest anthropogenic contributor to global warming after CO_2 [44]. This pollutant was clearly observed in the GCC countries between 1980 and 2020, particularly in UAE and Kuwait. The maximum monthly averages were observed in Dubai (0.73 $\mu g/m^3$) and Kuwait (0.57 $\mu g/m^3$). The decadal trends were also remarkably the highest in both Kuwait and UAE (Figure 6a,b). The monthly distribution pattern was shown to be high during the late spring and early summer months, while the minimum values were recorded in January (Figure 6c). The peak occurrence of BC was encountered in Dubai, either at the monthly level, the annual mean, or the decadal trend.

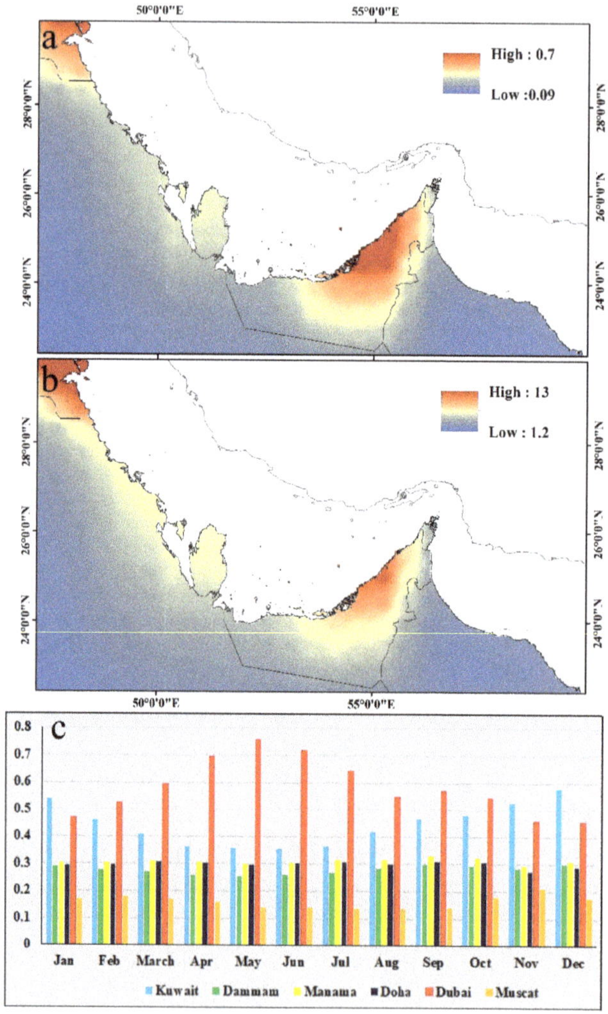

Figure 6. (**a**) is the distribution of the mean monthly BC emissions in $\mu gm/m^3$ throughout the GCC countries between 1980 and 2020, (**b**) represents the trend ($\mu gm/m^3$ per decade), which is maximal in Kuwait and Dubai, and (**c**) shows the average monthly concentration of BC in the six main cities. The source of the raw data is: https://giovanni.gsfc.nasa.gov/giovanni/; accessed on 18 February 2022.

3.2. Precipitation, Land Surface Temperatures and Anomalies of Surface Temperatures

The annual rainfall along the GCC countries for the period 2000–2018 significantly varies from one country to another. It is discernable that the mountainous ranges in Oman and UAE receive the maximum rainfall (>300 mm/year) by the orographic and monsoonal rains from the Indian Ocean [13]. The remaining parts receive less precipitation (~50 mm/year), except for northeastern Saudi Arabia, which receives relatively higher annual rainfall (up to 150 mm) (Figure 7a). The majority of rainfall occurs during winter; however, heavy torrential rainstorms may hit the region during fall and spring. The statistical analysis of the precipitation trends for this period (2000–2018) indicated a clear decrease (9.7 mm/decade) in precipitation, mostly over Kuwait and some parts of the KSA. On the other hand, Al-Hajar Mountain in northern Oman and the northern part of UAE are witnessing a rise in precipitation rates (up to 12.6 mm/decade). This is obvious due to the recent increase in the frequency of torrential rainstorms over Al-Hajar Mountain in Oman.

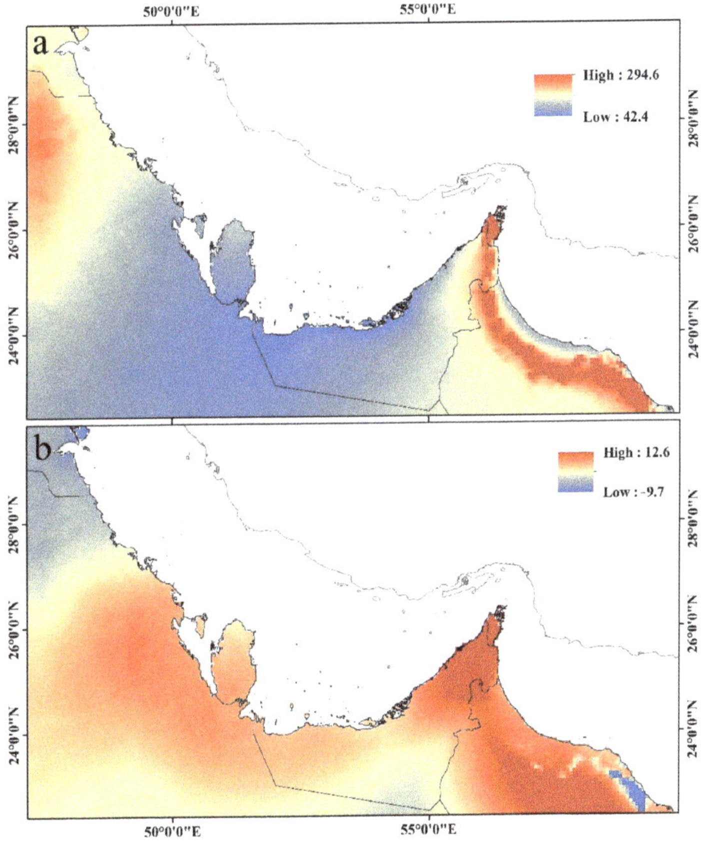

Figure 7. (**a**) is the distribution of the mean annual precipitation in mm throughout the GCC countries between 2000 and 2018, and (**b**) is the trend (mm/decade), which is positive (wetting) over Oman and negative (drying) over Kuwait. The source of the raw data is: https://www.worldclim.org; accessed on 18 February 2022.

The mean values of the land surface temperatures (LST) for the period from 2003–2020 showed higher records for the southern GCC countries, e.g., UAU and Oman, where the mean LSTs approached 50 °C for the plains and sandy terrains of these countries. The highlands in Oman and UAE have conspicuously lower LST than surrounding plateaus

(Figure 8a). The northern section of the GCC courtiers, including Kuwait, KSA, Bahrain, and Qatar, exhibited relatively lower mean LST, with a difference approaching 10–15 °C, as this region experiences colder winters than UAE and Oman due to the continental and latitudinal effects. The statistical LST trend for the study period (2003–2020) revealed a contradictory pattern to the mean values. Certainly, it is very clear that Kuwait, Bahrain, Qatar, the eastern province of KSA, and the middle part of the UAE have been exposed to a decadal LST warming (Figure 8b). Specifically, this warming has reached up to 2.0 °C/decade in Kuwait, 1.7 °C in Bahrain, and 1.5 in Dammam and Abu Dhabi (UAE). On the other hand, the inland desert of KSA and the highlands of Oman have experienced a decadal cooling in LST by about 0.1 to 0.8 °C/decade. The regional change of the LST for the entire region was toward warming by 0.62 °C/decade.

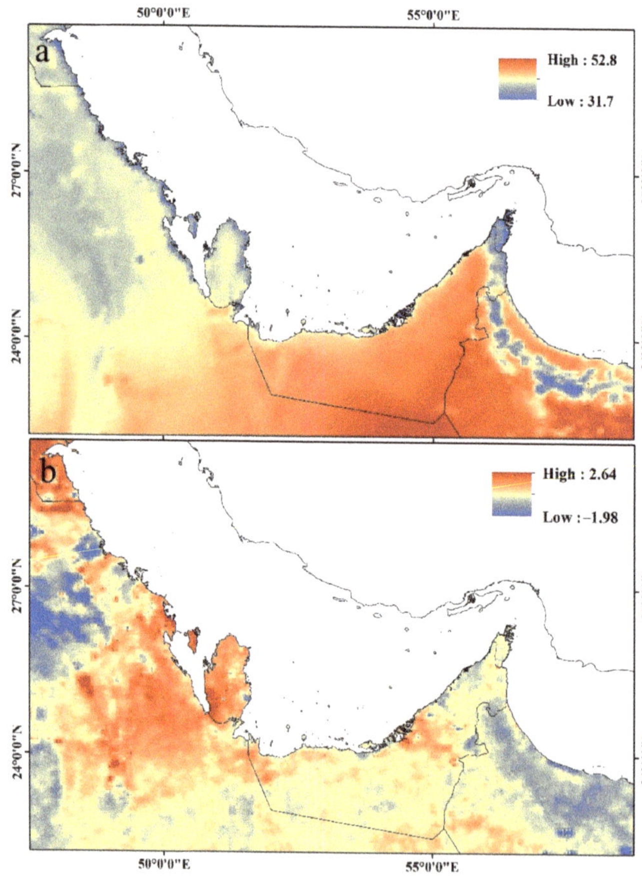

Figure 8. (**a**) is the distribution of the mean annual LST in degree Celsius over the GCC countries (2003–2020), and (**b**) is the trend (°C/decade), which reveals warming over Kuwait, Dammam, Bahrain, and Qatar and cooling over Oman and parts of KSA. The source of the data is: https://lpdaac.usgs.gov; accessed on 18 February 2022.

The average anomaly of nighttime surface temperature, as acquired from the AIRS satellite, showed that there has been an increase in the anomalies of extraordinary temperatures, particularly toward the warming of nights. However, it was observed that some of the GCC countries were exposed to several colder nights during the period from 2003–2016. Minimal nighttime anomalies occurred in 2005, 2008, and 2011, mostly in Kuwait and

Dubai, and reached 0.6 °C below average. Nevertheless, during the years 2010, 2015, and 2016, all the GCC countries experienced extreme warmer nights that were above the annual average. This is conspicuous, with maximal anomalies approaching 0.99 °C in Kuwait, Doha, and Manama (Figure 9a). Ref. [45] attributes the anomalous increase in temperatures during 2015/2016 to the El Niño/La Niña Southern Oscillation (ENSO) event. At the monthly level, it was observed that surface temperature anomalies are frequent during summer and early fall (from July to October), and the maximal values occurred in Kuwait and Muscat. However, during March, there was a decrease in temperatures that went below the average, with a minimum value in Manama. The remaining months exhibited values close to the average records. Generally, extraordinary nighttime heatwaves were frequent in Kuwait, Manama, Doha, and Dammam between 2003 and 2016, with minimal occurrences in Muscat.

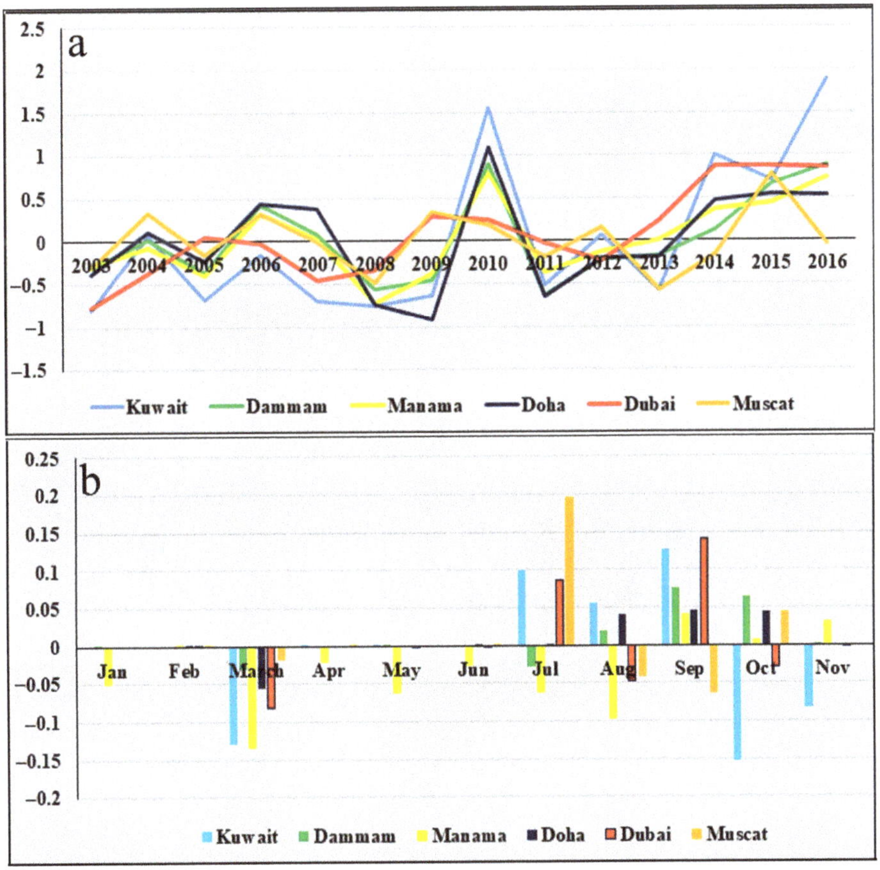

Figure 9. (**a**) is the average nighttime surface temperature anomalies in degree Celsius for the main cities along the gulf between 2003 and 2016, and (**b**) is the monthly distribution of nighttime surface temperature anomalies. The source of the raw data is: https://giovanni.gsfc.nasa.gov/giovanni/; accessed on 18 February 2022.

3.3. Sea Surface Temperatures

The Arabian Gulf, which is a shallow inland sea with a mean depth of about 40 m, reveals considerable variations in its sea surface temperatures. This is attributed mainly to the latitudinal location of this semi-enclosed, inland water body. In addition, the gulf environment is significantly influenced by evaporation, circulation, and lateral exchange with the Arabian Sea [46]. There is a thermal gradient in the mean LST of the gulf from south to north. The mean annual SST for the period from 2003–2018 was maximal along the coasts of UAE and Oman (~28 °C) in the south, while the coldest waters occurred near the offshores of Kuwait (~24 °C) in the north (Figure 10a). The decadal trends of SST agree with the regional LST decadal trend, where the northern section is warming up at maximal rates. This is obvious in areas where there has been a warming of 1.64 °C/decade, such as the coasts of Kuwait and KSA. The least warming was recorded near the entrance of the gulf at the Strait of Hurmuz (~0.2 °C/decade). Considerable warming (0.6–1.0 °C/decade) was also observed near the offshores of Qatar, Bahrain, and UEA (Figure 10b). The overall decadal rate of the entire gulf is 0.7 °C/decade. The low warming trend in the southern part of the gulf might be attributed to the connection with the Indian Ocean and the exchange of energy with the deep, colder water body along the Sea of Oman.

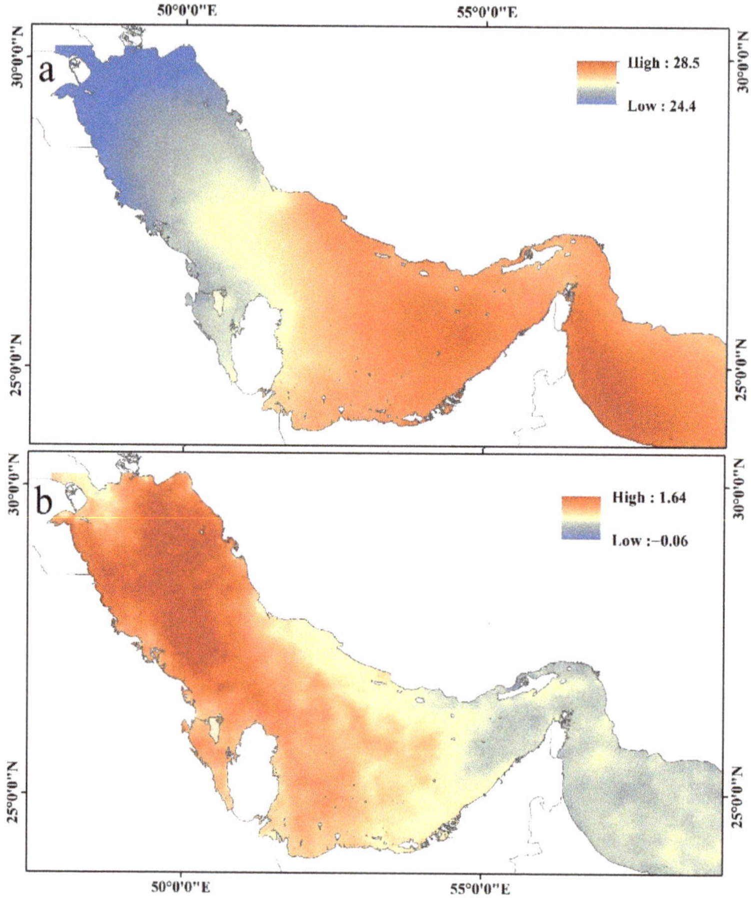

Figure 10. (**a**) is the distribution of the mean annual SST degree Celsius over the Arabian Gulf (2003–2018), and (**b**) is the trend (°C/decade), which reveals high warming near Kuwait, Dammam, Bahrain, and Qatar. The source of the raw data is: http://oceancolor.gsfc.nasa.gov; accessed on 18 February 2022.

4. Discussion and Conclusions

The GCC union includes some of the countries with the most excessive energy consumption and is considered a global hub for oil and gas reserves. Lately, fossil-fuel related activities have bloomed in the region, resulting in substantial and immense emissions of greenhouse gases and air pollutants. The GCC countries have a unique socio-economic setting that relies on huge oil reserves. These countries account for about a quarter of the world's crude oil production, mostly stemming from Saudi Arabia, the UAE, and Kuwait, and these three countries were among the ten largest producers of crude oil in the world in 2018 [47]. In addition, the majority of the GCC countries can be found in the top ten worldwide ranks of energy consumption per capita [48]. Moreover, the low energy prices have attracted foreign companies to establish investments, which has not only influenced the economy but also increased the energy demand [49]. In addition, the number of motor vehicles in GCC countries was 221% more than the average value in Asia, Oceania, and the Middle East in 2015 [50]. The excessive use of energy, specifically fossil fuels, imposes considerable pressure upon these non-renewable resources. As such, the current study proves that there are considerable emissions of GHGs and air pollution. Looking at the emissions of the two greenhouse gases (CO_2 and CH_4), it is clear that there has been a substantial increase in both gases during the past two decades. CO_2 increased from 375 ppm to more than 400 ppm. In this context, Ref. [13] mentioned that the CO_2 emissions in 2012 alone were four times the average per capita in the world. However, this increase in CO_2 emission has been documented to be ubiquitous and remarkable worldwide. The present study indicates that the highest rates of emissions were encountered in KSA, UAE, and Kuwait, where energy sources rely mainly on fossil fuels [51]. In Saudi Arabia, for example, the demand for crude oil for the energy sector increased by 340% between 2006 and 2011 [52]. Simultaneously, CH_4 emissions also showed increased rates in the third millennium. For a country like Oman, where the decadal rate of emission is the highest of the GCC countries, CH_4 emissions from the oil and gas sectors are the primary contributor to the country's total GHG emissions [35]. It is also remarkable that GHGs were at a maximum during the summer months and at a minimum during winter. Regionally, the GCC countries contributed 2.4% of GHG emissions during a time when only 0.6% of the world's population was alive [53]. In addition to these GHGs, significant air pollutants also showed higher outputs into the atmosphere in GCC countries, particularly Kuwait, UAE, and KSA. In contrary to GHGs, SO_2 and NO_2 emissions are maximal during winter, while BC emissions are maximal during summer. The present study demonstrates that GCC countries reveal higher emissions of SO_2, NO_2, and BC. Many concentrations of these air pollutants are quite observable in the industrial regions of GCC countries. Consequent impacts involve the deterioration of air quality and the blooming of air-borne diseases. The health impacts of air pollution are noteworthy in GCC countries due to the higher relative humidity in the region [54]. Some studies attribute the high health risk and mortality in Kuwait to the increased levels of air pollution [55,56], while Ref. [57] claims that air pollution is responsible for 9% of mortality in Saudi Arabia. The lowest emissions of the studied air pollutants were observed in Oman.

Decadal changes in regional precipitation, land surface temperatures, and sea surface temperatures were apparent from the remotely sensed data. Statistical analysis revealed a decline in precipitation in the northern part, including the entire State of Kuwait and some parts of Saudi Arabia. This observation is confirmed by Ref. [58], who reported declining rates of rainfall in this region of the Arabian Peninsula. These results also agree with the projected climate change scenarios in terms of the aridity and drought of the Arabian Peninsula, where wetter conditions will prevail south of 25 °N and drier conditions will occur north of 25 °N [59]. Interestingly, the present study showed a relative wetting in some parts of Oman (south of 25 °N). It is worth noting that there was a substantial increase in the surface temperature anomalies during summer nights. This concept was previously established in Saudi Arabia by Ref. [60], who observed that the number of abnormal warm nights increased by 21 days during the decades between 1979 and 2019.

Concurrently, remotely sensed land surface temperature analysis showed considerable warming, particularly along the northern section. The present study demonstrates that GCC countries revealed a net warming of the land by 0.6 °C/decade over the past twenty years, with maximal warming (2 °C/decade) in Kuwait. It is important to note that the rate of global temperatures increased globally by 1.1 °C in the last 150 years, with higher rates after 1970 [61]. Remarkably, Ref. [58], in their analysis regarding the changes in the regional air temperatures using data acquired from 25 meteorological stations for the period from 1980–2010, reported an increase in the air temperatures of the Arabian Peninsula by 0.63 °C/decade. In addition, Ref. [62], using data from 44 stations, reported that the Arabian Peninsula witnessed a warming trend of 0.55/decade for the period from 1980–2008. Climate change has been also observed to influence the Arabian Gulf's surface water temperatures. Although the northern section of the gulf is the coldest, it experiences the most warming of the entire water body. The study indicates that the warming of this part of the gulf approaches 0.7 °C/decade. This was also confirmed by Ref. [63], who reported the comparable warming (0.6/decade) of the Kuwait Bay for the period from 1985–2002. The southern section of the Gulf witnessed the least warming during the study period due to the connection with the Indian Ocean through the Sea of Oman. Overall, it was observed that there is a dipole pattern in the severity of climate change elements (LST, precipitation, and SST) in the GCC countries. The northern section of the GCC countries, particularly Kuwait and northern KSA, revealed the highest degree of climate change in terms of decreasing rates of precipitation, increasing LST, warm nighttime anomalies, and a warming of the sea surface temperatures of the gulf water. On the other hand, the southern side of the GCC countries, mostly Oman and parts of UAE, exhibited lower degrees of climate change, mostly a localized warming in LST.

The present study could help to prioritize the management of the natural resources of the GCC region. However, the truthfulness of the results is definitely related to the quality of the data used. Limitations of this research include the coarse resolution of the data, the short period of coverage for some parameters, and the absence of ground-based measurements for validation. The launch of the recent atmosphere-monitoring satellites, such as the Sentinel-5 Precursor in 2017, will provide detailed information about the atmosphere and its constituents via fine spatial resolution data. Also, as the study period in the current investigation is still limited to draw a solid trend of climate change and there should be longer time-series data to yield reliable results, observations of temperature and precipitation changes agree with localized previous studies, which utilized a longer time span. Furthermore, future research will focus on the integration between space-based and ground-based measurements for more trustworthy outcomes.

Funding: This research received no external funding.

Institutional Review Board Statement: Not applicable.

Informed Consent Statement: Not applicable.

Data Availability Statement: The data are available on request from the corresponding author.

Conflicts of Interest: The author declares no conflict of interest.

References

1. Ritchie, H.; Roser, M. CO_2 and Greenhouse Gas Emissions. 2020. Available online: https://ourworldindata.org/co2-and-other-greenhouse-gas-emissions (accessed on 4 April 2022).
2. Intergovernmental Panel on Climate Change (IPCC). *Climate Change 2013: The Physical Science Basis*; Working Group I Contribution to the IPCC Fifth Assessment Report; Cambridge University Press: Cambridge, UK, 2013.
3. Nyambuu, U.; Semmler, W. Climate change and the transition to a low carbon economy—Carbon targets and the carbon budget. *Econ. Model.* **2020**, *84*, 367–376. [CrossRef]
4. Department of Energy and Climate Change (DECC). *The UK Low Carbon Transition Plan: National Strategy for Climate and Energy*; Department of Energy and Climate Change: London, UK, 2009.
5. Borroni, E.; Pesatori, A.; Bollati, V.; Buoli, M.; Carugno, M. Air pollution exposure and depression: A comprehensive updated systematic review and meta-analysis. *Environ. Pollut.* **2022**, *292*, 118245. [CrossRef]

6. WHO. *Ambient Air Pollution: A Global Assessment of Exposure and Burden of Disease*; World Health Organization: Geneva, Switzerland, 2016.
7. *Statistical Review of World Energy, June 2008*; BP: London, UK, 2008. Available online: https://www.bp.com/en/global/corporate/energy-economics/statistical-review-of-world-energy.html (accessed on 15 January 2022).
8. Asif, M. Growth and sustainability trends in the buildings sector in the GCC region with particular reference to the KSA and UAE. *Renew. Sustain. Energy Rev.* **2016**, *55*, 1267–1273. [CrossRef]
9. Salahuddin, M.; Gow, J. Economic growth, energy consumption and CO_2 emissions in Gulf Cooperation Council countries. *Energy* **2014**, *73*, 44–58. [CrossRef]
10. Charabi, Y.; Al Nasiri, N.; Al Awadhi, T.; Choudri, B.; Al Bimani, A. GHG emissions from the transport sector in Oman: Trends and potential decarbonization pathways. *Energy Strategy Rev.* **2020**, *32*, 100548. [CrossRef]
11. Rahman, S.; Khondaker, A.; Hasan, M.; Reza, I. Greenhouse gas emissions from road transportation in Saudi Arabia—A challenging frontier. *Renew. Sustain. Energy Rev.* **2017**, *69*, 812–821. [CrossRef]
12. Khondaker, A.N.; Hassan, M.; Rahman, S.; Malik, K.; Shafiullah, M.; Muhyedeen, M. Greenhouse gas emissions from energy sector in the United Arab Emirates—An overview. *Renew. Sustain. Energy Rev.* **2016**, *59*, 1317–1325. [CrossRef]
13. Al-Maamary, H.M.; Kazem, H.A.; Chaichan, M.T. Climate change: The game changer in the Gulf Cooperation Council Region. *Renew. Sustain. Energy Rev.* **2017**, *76*, 555–576. [CrossRef]
14. Omidvarborna, H.; Baawain, M.; Al-Mamun, A. Ambient air quality and exposure assessment study of the Gulf Cooperation Council countries: A critical review. *Sci. Total Environ.* **2018**, *636*, 437–448. [CrossRef] [PubMed]
15. Kim, D.-W.; Byun, H.-R. Future pattern of Asian drought under global warming scenario. *Arch. Meteorol. Geophys. Bioclimatol. Ser. B* **2009**, *98*, 137–150. [CrossRef]
16. Scafetta, N. Climate change and its causes, a discussion about some key issues. *arXiv* **2010**, arXiv:1003.1554.
17. Hereher, M. Assessment of climate change impacts on sea surface temperatures and sea level rise—The Arabian Gulf. *Climate* **2020**, *8*, 50. [CrossRef]
18. Al-Fazari, A.; El-Kenawy, A.; Al-Nasiri, N.; Hereher, M. *Monitoring Urban Heat Islands in Selected Cities of the Gulf Region Based on Nighttime MODIS LST Data (2003–2018). Urban Heat Island (UHI) Mitigation: Hot and Humid Regions*; Springer: Berlin/Heidelberg, Germany, 2020; Chapter 12; pp. 249–276.
19. IPCC. *Global Warming of 1.5 °C*; IPCC Special Report on the impacts of global warming of 1.5 °C above pre-industrial levels and related global greenhouse gas emission pathways, in the context of strengthening the global response to the threat of climate change, sustainable development, and efforts to eradicate poverty; Masson-Delmotte, V., Zhai, P., Pörtner, H.-O., Roberts, D., Skea, J., Shukla, P.R., Pirani, A., Moufouma-Okia, W.C., Péan, R., Pidcock, S., et al., Eds.; The Intergovermnetal Panel on Climate Change: Geneva, Switzerland, 2018; Available online: https://www.ipcc.ch/sr15 (accessed on 20 February 2022).
20. Martins, F.; Benassi, R.; Torres, R.; Neto, F. Impacts of 1.5 °C and 2 °C global warming on Eucalyptus plantations in South America. *Sci. Total Environ.* **2022**, *825*, 153820. [CrossRef] [PubMed]
21. Fotso-Nguemo, T.; Vondou, D.; Diallo, I.; Diedhiou, A.; Weber, T.; Tanessong, R.; Nghonda, J.; Yepdo, Z. Potential impact of 1.5, 2 and 3 °C global warming levels on heat and discomfort indices changes over Central Africa. *Sci. Total Environ.* **2021**, *804*, 150099. [CrossRef]
22. Hereher, M.; El Kenawy, A. Exploring the potential of solar, tidal, and wind energy resources in Oman using an integrated climatic-socioeconomic approach. *Renew. Energy* **2020**, *161*, 662–675. [CrossRef]
23. Elrahmani, A.; Hannun, J.; Eljack, F.; Kazi, M.-K. Status of renewable energy in the GCC region and future opportunities. *Curr. Opin. Chem. Eng.* **2020**, *31*, 100664. [CrossRef]
24. Ünal, F.; Temir, G.; Köten, H. Energy, exergy and exergoeconomic analysis of solar-assisted vertical ground source heat pump system for heating season. *J. Mech. Sci. Technol.* **2018**, *32*, 3929–3942. [CrossRef]
25. Cura, D.; Yilmaz, M.; Koten, H.; Senthilraja, S.; Awad, M.M. Evaluation of the technical and economic aspects of solar photovoltaic plants under different climate conditions and feed-in tariff. *Sustain. Cities Soc.* **2022**, *80*, 103804. [CrossRef]
26. Alharbi, F.; Csala, D. Gulf Cooperation Council Countries' Climate Change Mitigation Challenges and Exploration of Solar and Wind Energy Resource Potential. *Appl. Sci.* **2021**, *11*, 2648. [CrossRef]
27. Flores, R.; Kaya, N.; Eser, O.; Saltan, S. The effect of mineral dust transport on PM 10 concentrations and physical properties in Istanbul during 2007–2014. *Atmos. Res.* **2017**, *197*, 342–355. [CrossRef]
28. Wan, Z. New refinements and validation of the MODIS Land-Surface Temperature/Emissivity products. *Remote Sens. Environ.* **2008**, *112*, 59–74. [CrossRef]
29. Fick, S.E.; Hijmans, R.J. WorldClim 2: New 1-km spatial resolution climate surfaces for global land areas. *Int. J. Climatol.* **2017**, *37*, 4302–4315. [CrossRef]
30. Carlson, D.F.; Yarbro, L.A.; Scolaro, S.; Poniatowski, M.; McGee-Absten, V.; Carlson, P.R., Jr. Sea surface temperatures and seagrass mortality in Florida Bay: Spatial and temporal patterns discerned from MODIS and AVHRR data. *Remote. Sens. Environ.* **2018**, *208*, 171–188. [CrossRef]
31. Apaydin, H.; Sonmez, F.; Yildirim, Y. Spatial interpolation techniques for climate data in the GAP region in Turkey. *Clim. Chang.* **2004**, *28*, 31–40. [CrossRef]
32. Ozelkan, E.; Bagis, S.; Ozelkan, E.; Ustundag, B.; Yucel, M.; Ormeci, C. Spatial interpolation of climatic variables using land surface temperature and modified inverse distance weighting. *Int. J. Remote Sens.* **2015**, *36*, 1000–1025. [CrossRef]

33. Mahmood, H. The effects of natural gas and oil consumption on CO2 emissions in GCC countries: Asymmetry analysis. *Environ. Sci. Pollut. Res.* **2022**, 1–17. Available online: https://pubmed.ncbi.nlm.nih.gov/35359204/ (accessed on 15 January 2022). [CrossRef]
34. Farahat, A. The Impact of the 2020 Oil Production Fluctuations on Methane Emissions over the Gulf Cooperation Council (GCC) Countries: A Satellite Approach. *Atmosphere* **2022**, *13*, 11. [CrossRef]
35. Charabi, Y. Digging deeper into cutting methane emissions from the oil and gas industry in the era of volatile prices. *Mitig. Adapt. Strateg. Glob. Chang.* **2021**, *26*, 6. [CrossRef]
36. Al-Jahdali, M.O.; Bin Bisher, A.S. Sulfur Dioxide (SO_2) Accumulation in Soil and Plant's Leaves around an Oil Refinery: A Case Study from Saudi Arabia. *Am. J. Environ. Sci.* **2008**, *4*, 84–88. [CrossRef]
37. Dahiya, S.; Myllyvirta, L.; Global SO2 Emission Hotspot Database. Greenpeace Environment Trust. 2019. Available online: https://www.greenpeace.org (accessed on 29 March 2022).
38. Al-Rashidi, M.; Nassehi, V.; Wakeman, R. Investigation of the efficiency of existing air pollution monitoring sites in the state of Kuwait. *Environ. Pollut.* **2005**, *138*, 219–229. [CrossRef]
39. Hereher, M.; Eissa, R.; Alqasemi, A.; El Kenawy, A.M. Assessment of air pollution at Greater Cairo in relation to the spatial variability of surface urban heat island. *Environ. Sci. Pollut. Res.* **2021**, *29*, 21412–21425. [CrossRef]
40. Lelieveld, J.; Hoor, P.; Jöckel, P.; Pozzer, A.; Hadjinicolaou, P.; Cammas, J.P.; Beirle, S. Severe ozone air pollution in the Persian Gulf region. *Atmos. Chem. Phys.* **2009**, *9*, 1393–1406. [CrossRef]
41. Khan, A.; Al-Salem, S. Seasonal variation effect on airborne pollutants in an urban area of the state of Kuwait. *J. Environ. Res. Dev.* **2007**, *1*, 215–218.
42. Chaichan, M.T.; Al-Asadi, K.A. Environmental impact assessment of some vehicles pollutants in Sohar, Oman. *Int. J. Sci. Eng. Res.* **2015**, *6*, 493–496.
43. Wyche, K.P.; Cordell, R.L.; Smallbone, K.L.; Lyons, P.; Hama, S.M.L.; Monks, P.S.; Staelens, J.; Hofman, J.; Stroobants, C.; Roekens, E.; et al. The spatio-temporal evolution of black carbon in the North-West European 'air pollution hotspot'. *Atmos. Environ.* **2020**, *243*, 117874. [CrossRef]
44. Comer, B. Including Estimates of Black Carbon Emissions in the Fourth IMO GHG Study. The International Council of Clean Transportation (ICCT). 2019. Available online: https://wwwcdn.imo.org/localresources/en/OurWork/Environment/Documents/CSC-Bryan%20Comer.pdf (accessed on 31 March 2021).
45. Varotsos, C.A.; Cracknell, A.P.; Efstathiou, M.N. The global signature of the El Niño/La Niña Southern Oscillation. *Int. J. Remote Sens.* **2018**, *39*, 5965–5977. [CrossRef]
46. Xue, P.; Eltahir, E. Estimation of the Heat and Water Budgets of the Persian (Arabian) Gulf Using a Regional Climate Model. *J. Clim.* **2015**, *28*, 5041–5062. [CrossRef]
47. BP. Statistical Review of World Energy. 2019. Available online: https://www.bp.com/en/global/corporate/news-and-insights/press-releases/bp-statistical-review-of-world-energy-2019.html (accessed on 5 April 2022).
48. Iriani, M.; Trabelsi, M. The economic impact of phasing out energy consumption subsidies in GCC countries. *J. Bus. Econ. Manag.* **2016**, *87*, 35–49. [CrossRef]
49. Bhutto, A.W.; Bazmi, A.A.; Zahedi, G.; Klemeš, J.J. A review of progress in renewable energy implementation in the Gulf Cooperation Council countries. *J. Clean. Prod.* **2014**, *71*, 168–180. [CrossRef]
50. OICA. International Organization of Motor Vehicle Manufacturers, Vehicle in Use by Country and Type 2005–2015. Available online: https://www.oica.net/wp-content/uploads//Total_in-use-All-Vehicles.pdf (accessed on 9 April 2022).
51. Statistica Research Department. Volume of Greenhouse Gas Emissions in the Gulf Cooperation Council in 2014, by Country. 2017. Available online: https://www.statista.com/statistics/814843/gcc-volume-of-ghg-emissions-by-country/ (accessed on 5 April 2022).
52. IEA. World Energy Outlook 2010, Paris Cedex, France. 2011. Available online: https://www.iea.org/reports/world-energy-outlook-2010 (accessed on 1 March 2022).
53. Reiche, D. Energy Policies of Gulf Cooperation Council (GCC) Countries—Possibilities and Limitations of Ecological Modernization in Rentier States. *Energ. Policy* **2010**, *38*, 2395–2403. [CrossRef]
54. Amoatey, P.; Omidvarborna, H.; Baawain, S.; Al-Mamun, A. Indoor air pollution and exposure assessment of the gulf cooperation council countries: A critical review. *Enviro. Int.* **2018**, *121*, 491–506. [CrossRef]
55. Achilleos, S.; Al-Ozairi, E.; Alahmad, B.; Garshick, E.; Eophytou, A.; Bouhamra, W.; Yassin, M.; Koutrakis, P. Acute effects of air pollution on mortality: A 17-year analysis in Kuwait. *Environ. Int.* **2019**, *126*, 476–483. [CrossRef]
56. Al-Hemoud, A.; Gasana, J.; Al-Dabbous, A.; Alajeel, A.; Al-Shatti, A.; Behbehani, W.; Malak, M. Exposure levels of air pollution (PM2.5) and associated health risk in Kuwait. *Environ. Res.* **2019**, *179*, 108730. [CrossRef] [PubMed]
57. Rojas-Rueda, D.; Alsufyani, W.; Herbst, C.; AlBalawi, S.; Alsukait, R.; Alomran, M. Ambient particulate matter burden of disease in the Kingdom of Saudi Arabia. *Environ. Res.* **2021**, *197*, 111036. [CrossRef]
58. Almazroui, M.; Islam, N.; Saeed, S.; Saeed, F.; Ismail, M. Future Changes in Climate over the Arabian Peninsula based on CMIP6 Multimodel Simulations. *Earth Syst. Environ.* **2020**, *4*, 611–630. [CrossRef]
59. Waha, K.; Krummenauer, L.; Adams, S.; Aich, V.; Baarsch, F.; Coumou, D.; Schleussner, C.F. Climate change impacts in the Middle East and Northern Africa (MENA) region and their implications for vulnerable population groups. *Reg. Environ. Change* **2017**, *17*, 1623–1638. [CrossRef]

60. Almazroui, M. Changes in Temperature Trends and Extremes over Saudi Arabia for the Period 1978–2019. *Adv. Meteorol.* **2020**, *2020*, 1–21. [CrossRef]
61. Intergovernmental Panel on Climate Change (IPCC). Summary for Policymakers. In *Climate Change 2021: The Physical Science Basis*; Contribution of Working Group I to the Sixth Assessment Report of the Intergovernmental Panel on Climate Change; Masson-Delmotte, V., Zhai, P., Pirani, A., Connors, S.L., Péan, C., Berger, S., Caud, N., Chen, Y., Goldfarb, L., Gomis, M., et al., Eds.; Cambridge University Press: Cambridge, UK, 2021; *in press*.
62. AlSarmi, S.; Washington, R. Recent observed climate change over the Arabian Peninsula. *J. Geophys. Res. Atmos.* **2011**, *116*, 1–15. [CrossRef]
63. Al-Rashidi, T.B.; El-Gamily, H.I.; Amos, C.L.; Rakha, K.A. Sea surface temperature trends in Kuwait bay, Arabian Gulf. *Nat. Hazards* **2009**, *50*, 73–82. [CrossRef]

sustainability

Article

Mathematical Modelling of Biogas Production in a Controlled Landfill: Characterization, Valorization Study and Energy Potential

Mabrouki Jamal

Laboratory of Spectroscopy, Molecular Modeling, Materials, Nanomaterial, Water and Environment, CERNE2D, Faculty of Science, Mohammed V University in Rabat, Avenue Ibn Battouta, Agdal, Rabat BP1014, Morocco; jamal.mabrouki@um5r.ac.ma

Abstract: Methane potential is the volume of methane gas produced during anaerobic degradation in the presence of the bacteria of an initially inserted sample. This paper presents a degradation study of the green and industrial fermentable waste sheltered by the landfill of Mohammedia in which the biogas deposit and the associated recoverable energy at the end of exploitation is estimated and the power of the gas engine of the proposed cogeneration unit is calculated. The Total potential biogas production value of the household waste of the city of Mohammedia is much higher than that of the American and French household waste recommended by the US EPA and French ADEME. This calls into question the adaptability of the modeling tools for biogas production to Moroccan waste. The four modeling equations for landfill will be evaluated. The results show that the ADEME model proved to be more descriptive and better adapted to this case.

Keywords: modeling; biogas; methane; landfill of Mohammedia; upgrading

Citation: Jamal, M. Mathematical Modelling of Biogas Production in a Controlled Landfill: Characterization, Valorization Study and Energy Potential. *Sustainability* **2022**, *14*, 15490. https://doi.org/10.3390/su142315490

Academic Editors: Ahmed El Kenawy, Petra-Manuela Schuwerack, Zhongfeng Xu and Mohamed El-Alfy

Received: 30 September 2022
Accepted: 11 November 2022
Published: 22 November 2022

1. Statement of Novelty

The recuperation of biogas from the Mohammedia site landfill was calculated utilizing four demonstrating conditions, and the present models included only family waste, with methanogenic potential estimations of 100 m^3 to 170 m^3 of CH_4/ton of waste for the American models and 50 out of 100 for the French ADEME model. To adjust to the Moroccan setting, and especially to the instance of Mohammedia, we extrapolated a lot of information on waste from various regions and enterprises so as to get values depicting the methanogenic potential specific to the various substrates. This permitted us to gauge the biogas deposit indicated by the given operating horizon.

2. Introduction

Landfill is an easy to implement and relatively inexpensive waste disposal technique. Without proper management, however, it can lead to a variety of hygienic, health and environmental problems. Only a landfill that has been stabilized, and is therefore without further development, can be defined as no longer being harmful to the environment. When it comes to renewable energies, wind turbines, solar collectors and hydropower are most often mentioned. However, there are other solutions, such as energy production from biomass: wood, biofuels or biogas [1]. The population of developing countries is growing, leading to an increase in the needs of the poor and the production of waste and effluents. Waste recycling contributes to poverty alleviation and environmental sanitation [2].

Mechanization is an anaerobic digestion process that generally achieves double the energy yield of the original process. The objective of energy recovery by methane (CH_4) is the recovery and stabilisation of organic waste with a view to material recovery by its partial restitution to the ground [3,4]. With ever-increasing and more diversified consumption all over the world, waste production is constantly increasing in quantity and quality, thus creating enormous risks to the environment and both the safety and health of local populations [3]. Landfilling remains the predominant method of disposal of household and similar waste in Africa, particularly in Morocco, in part because of its simplicity, but also because of its lower cost compared with other methods, such as incineration.

The Mohammedia control landfill receives a significant amount of waste with high methanogenic potential every day, such as household waste (61% organic matter), green waste, poultry droppings and tannery waste. By anaerobic decomposition, this mixture generates a good quality biogas (CH_4 55.6%, CO_2 32%, H_2S 600 ppm, O_2 1%) which reminds us of its value in other applications [3]. This study was carried out in order to quantify the biogas deposit at the Mohammedia site using four modelling equations, these models using only household waste, with methanogenic potential values in the order of 100 m^3 to 170 m^3 of CH_4/tonne of waste for the American models [5] and 50 of 100 for the French ADEME model [6], corresponding to the specificities of the waste and the regions where these tools were developed [7]. In order to adapt them to the Moroccan context, in particular to the case of Mohammedia, we extrapolated a heap of data on waste from different municipalities and industries in order to obtain values describing the methanogen potential specific to the different substrates. This allowed us to estimate the biogas deposit according to the given operating horizon.

3. Description of the Studied Zone

The Mohammedia interprovincial control landfill is located in the municipality of Ben Yakhlef, on the shoreline. It is west of Chaaba el Hamra, a tributary of the west bank of the Nfifikh river, about 270 m south of the Dayat Al Hila security perimeter (X = 32440, Y = 338979) and occupies an area of 47 hectares.

The zone is moderately hilly and ends at the edge of the west bank of a talweg (Chaaba El Hamra) perpendicular to the west bank of Oued Nfifikh. From upstream to downstream, the site has a height difference of 27 m. Its proximity to the ocean gives this region a temperate and humid climate (80% humidity) with a mild winter and a summer cooled by the ocean breezes. The average temperature is 23 °C and the annual precipitation level is 400 mm, in addition to a daily evapotranspiration potential of 5–6 mm/12 h. Eleven rural and urban municipalities (including Mohammedia, Ain Harrouda, Bouznika, Ben Sliman, El Mansouria, Ech-Challalat, Ben Yakhlaf, Sidi Mousa ben ali and Sidi Mousa El Majdoub) are served by the so-called landfill centre, which started in 2012 and is scheduled to close in 2032. The project area is divided between the landfill area and other landfill accessories, as described in the plan below, which shows the biogas collection network of crates 1 and 2, already in operation [8].

4. Materials and Methods

4.1. Experimental Design

In order to measure the amount of biogas produced by the waste studied, an anaerobic digestion device and a device for determining the volume of biogas generated by water displacement were established in the laboratory. A mass of 20 g of each sample was crushed and mixed with 100 mL of water and incubated for 40 days in a bioreactor placed in a water bath at a constant temperature (35 °C), promoting bio-mechanisation (Figure 1) [9].

4.2. Modeling Equations

4.2.1. EPA Model

The US EPA (Environmental Protection Agency) has also carried out a study; this led to a model based on data collected on site. This Model is based on a first-order Equation (1) with a decreasing generation rate of biogas over time [10,11]:

$$\mathbf{Qt} = 2 \times \mathbf{L_0} \times \mathbf{R}\left(\mathbf{e}^{(-\mathbf{KC})} - \mathbf{e}^{(-\mathbf{Kt})} \right) \tag{1}$$

Qt: quantity of biogas generated over time t (m^3/year);
L_0: total potential biogas production (m^3 CH_4/t of waste);

K: kinetic constant for biogas generation (year^{-1});
T: time elapsed since storage began (year);
R: average rate of waste accepted during the site's operating period (t/year);
C: time since site closure (C = 0 year for active sites) (year).

$$L_0 = \mathbf{FCM.\ COD.\ CODF.A.\frac{16}{12}\ .\ 1000} \tag{2}$$

FCM: correction factor of the CH_4, expressed as a percentage;
COD: degradable organic carbon, expressed as t of C/t of waste;
CODF: concealed COD fraction;
A: fraction of CH_4 in biogas;
16/12: stoichiometry coefficient.

$$\mathbf{COD = 0.4E + 0.17F + 0.15G + 0.3H} \tag{3}$$

E: fraction of waste consisting of paper and textiles;
F: fraction of waste consisting of garden and/or park waste;
G: fraction of waste consisting of food waste;
H: fraction of waste consisting of wood and/or straw.

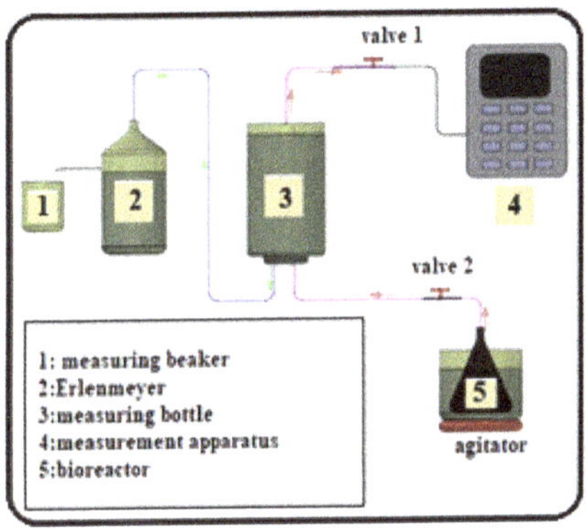

Figure 1. Device for measuring methanogenic potential by the displacement of water [9].

4.2.2. LANDGEM Model

The LANDGEM model is based on a first-order degradation equation that is estimated over several years. Indeed, for a mass of waste accepted in year i (Mi), methane production follows a decreasing exponential law. For several years, production is evaluated every tenth of a year. The equation used to estimate the total amount of methane produced in a TEC is [12,13]:

$$Q(CH_4) = \sum_{i=1}^{n} \sum_{j=0.1} K\,L_0\,\frac{M_i}{10}\,e^{(-Kt_{ij})} \tag{4}$$

i: time increment of 1 year;
j: cutting the year into tenths.

4.2.3. ADEME Model

ADEME estimates the methane emissions from the TECs by calculating the quantity of methane produced (uncaptured methane and captured methane) using this expressions [14,15]

$$Q(CH4) = \sum L_0 \sum_{i=1}^{3} A_i P_i K_i e^{(-K_i(t-x))} \tag{5}$$

$$L_0 = 0.934.C_0(0.014.T + 0.28)\ en\frac{m^3}{t} \tag{6}$$

i: the subdivision into three categories of waste;
$\mathbf{P_i}$: the fraction of waste with degradation constant i;
$\mathbf{C_0}$: biodegradable organic carbon;
T: degradation temperature 30 °C;
$\mathbf{A_i}$: factor of the mass of waste accepted in year i;
x: year of landfilling of waste.
The three degradation constants K depend on the biodegradability of the waste:
$\mathbf{K_1}$ = 0.5 in order to degrade 15% of waste (easily biodegradable);
$\mathbf{K_2}$ = 0.10 in order to degrade 55% of waste (moderately biodegradable);
$\mathbf{K_3}$ = 0.04 in order to degrade 30% of waste (poorly biodegradable);
The degradation kinetics are assumed to be the same regardless of the composition of the waste [16].

4.2.4. Scholl Canyon Model

The Scholl Canyon Model is a first-order decomposition model. It allows the calculation of CH_4 resulting from the decomposition of waste, taking into account the fact that this waste decomposes over many years. It is expressed in Equation (6) [17]:

$$Qt = \sum(K\,L_0\ M_x\ (\ e^{(-K(\ t-x))})\ ) \tag{7}$$

Qt: quantity of methane produced during the year in question (T) (kg of CH_4/year);
x: year of entry of the waste;
$\mathbf{M_x}$: amount of waste landfilled during the year × (Mt);
$\mathbf{L_0}$: methane production potential (kg of CH_4/t of waste);
T: considered year.

4.3. Sampling and Analysis

The samplings were made in Tedlar bags and glass ampules. The H_2 and N_2 were analyzed by chromatography on a molecular sieve using a detector with thermal conductivity. The CH_4 and CO_2 were measured by porous polymer analysis with a thermal conductivity detector (TCD). The C_2 to C_5 were analyzed by chromatography on a porous polymer with a flame ionization detector (FID). The CO was analyzed using the non-dispersive infrared technique.

5. Results and Discussion

5.1. Tonnage of Waste

The Mohammedia controlled landfill receives on average 500 t/d of DMA that constitutes 74% of the total tonnage of various types (Table 1): household waste (OM); green waste; mixtures of household waste, soil and gravel; and common industrial waste considered to be AMD.

Table 1. Percentage by type of waste.

Designation	Value (t/d)	%
Household garbage	387	70%
Green waste	5	1%
Household garbage, soil and rubble	100	18%
Non-hazardous industrial waste	64	11%

Among the wastes with high methanogenic potential destined for landfill are DM, green waste and two types of industrial waste, namely: waste from the Mohammedia tannery and poultry droppings brought in by the Delicate-meat company. The two graphs that follow show the tonnage of this waste since the landfill opened (Figures 2 and 3).

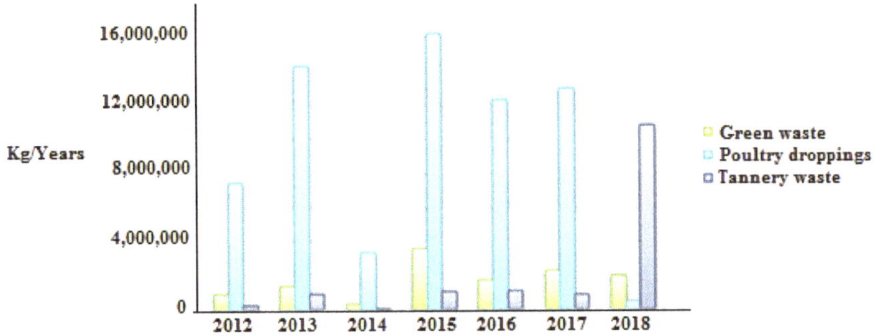

Figure 2. Tonnage of green waste, poultry droppings and tannery waste.

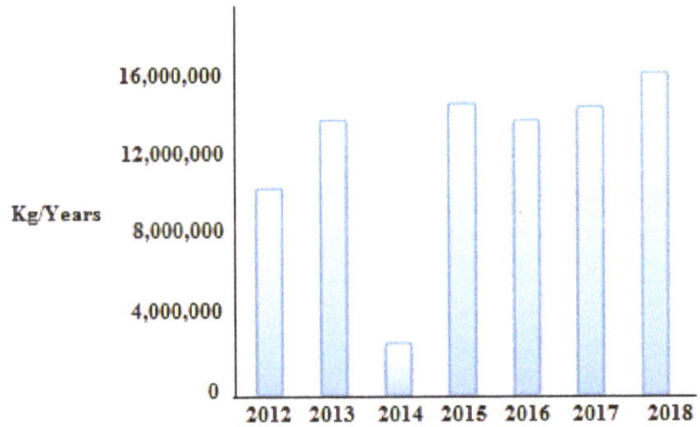

Figure 3. Annual tonnage of household waste.

5.2. Waste Characterization

The studies on the characterization of household and similar waste in the city of Mohammedia conducted by A. Ouatmane in 2018 and A. El Maguiri et al. in 2016 [7–9] report that the fraction is <80 mm, which represents fermentable organic matter in the order of 61%. However, the >80 mm fraction can be divided into categories and sub-categories, as shown in Figure 4.

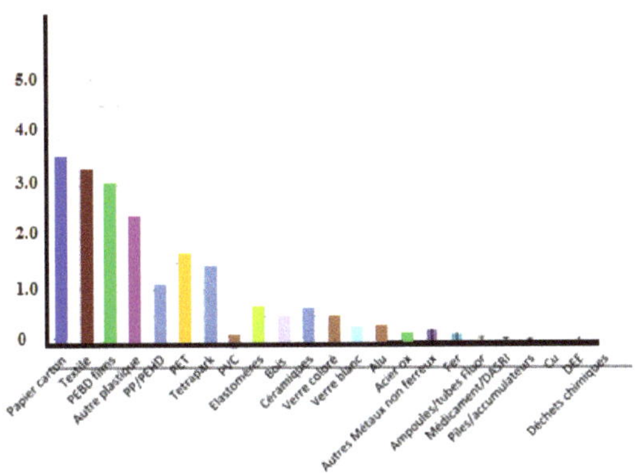

Figure 4. Average of the mass percentages of the fractions from the household and similar waste in the city of Mohammedia.

Bi-monthly sampling of the same landfill waste pile showed (Table 2) a clear evolution in the physical and chemical parameters during the landfilling of the waste, in particular a significant decrease in the organic carbon content due to mineralisation. The total nitrogen content showed a significant increase during fermentation. This variation corresponds in fact to a relative enrichment in nitrogen of the residual dry matter of the compost, the total amount of nitrogen actually decreasing as illustrated. The mineral nitrogen contents are always low, and their evolution is typical of what is found in compost heaps with a low level of ammoniac nitrogen and traces of nitrate, which then forms nitrate at the end of maturation. This maturation results in a lowering of the C/N ratio from 32, indicating a stabilisation of the organic compounds. Similarly, the equivalent humidity of the product falls, this last point being related to the concomitant rise in pH. Table 2 shows the results of the physicochemical analyses carried out during the DMA characterization.

Table 2. Results of physicochemical analyses of the waste fraction <80 mm.

Description	Unit	Value
Density: fraction < 80 mm	t/m^3	1.09
Humidity	%	38.02
Organic matter	g/100 g MS	69.93
Total organic carbon	g/100 g MS	40.26
Nitrogen	g/100 g MS	1.25
Report C/N	-	32.21
PCI fraction < 80 mm	Kcal/Kg	1002
PCI fraction > 80 mm	Kcal/Kg	2071

All the results are presented in Table 2. Total organic carbon corresponds to approximately 40% of the dry matter of the composts analyzed. Given the very heterogeneous composition of these materials and the diversity of their origins, we can assume that the variations observed were moderate (Table 3).

Table 3. TOC value in g/100 g [3,4].

Type of Waste	TOC (g/100 g)	C/N
Household waste	40.26 [2]	32.21
Green waste	27 [3]	55
Poultry droppings	13.59 [4]	3.68
Tannery waste	14 [4]	3

The C/N ratio is around 32.21 at the beginning of the first phase of the process. Subsequently, a decrease in this ratio is noted, which becomes equal to 28 at the end of the first phase. This reduction is explained by the active transformation of the carbon into carbon dioxide, accompanied by a decrease in the content of organic acids in the waste mass. The purpose of this input is to amplify the microbial activity to prepare for the start of the next stage.

5.3. Methanogenic Potential

The potential for biogas generation by anaerobic decomposition of the various fermentable wastes sheltered by the TEC is a critical parameter for modelling biogas production. The US EPA recommends L0 values ranging from 170 m^3 of CH_4 per tonne of waste for arid areas to 96 m^3 for wetlands [17,18], though these values take into consideration the composition and physicochemical properties of household waste in the United States, which are certainly different from those in Morocco (Table 4).

Table 4. Table of COD and L0 calculation results.

	COT (g/100 g)	COD	FCM	CS	F (%)	L0 M^3/t
Household waste	40.26	0.099	1	1.333	55.6	563.77
Green waste	27	0.37	1	1.333	65 [18]	1649.77
Poultry droppings	13.59	0.542	1	1.333	60 [18]	1124.23
Tannery waste	14	0.3	1	1.333	60 [18]	640.24

Equation (2) includes in its expression three key elements (TOC, COD and F) that define the methanogenic potential of waste. The first expresses the carbon content, an essential element in the formation of methane. The term COD (3) relates the composition of waste, and knowing the composition and physical and chemical properties of its waste enables the calculation of L0 specific to a region.

The estimate of L0 using Equation (6) of the ADEME model gives a value of 26.3 m^3/t. Compared with the values recommended by the EPA and LANDGEM, this is extremely small and does not reflect the methanogenic potential of Mohammedia household waste (Figure 5), whose fermentable organic matter fraction is around 61% [6], which is probably significant compared with the % MO of waste in France and the US.

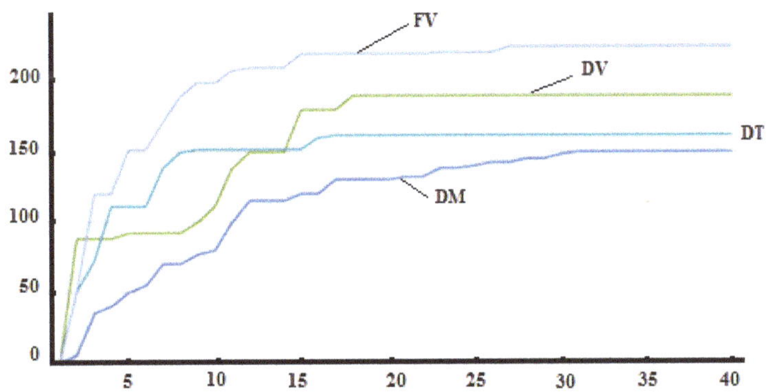

Figure 5. Biogas production kinetics for the four types of waste in ml.

After 40 days of fermentation of the different substrates at a temperature of 35 °C, the graph of the biogas production kinetics, which is strongly related to temperature and C/N ratio [7], shows that poultry droppings produce the largest volume of biogas (266 mL), which is certainly due to the abundance of lipids in the substrate. Then come in decreasing order green waste, tannery waste and household waste, with respective volumes of 189 mL, 160 mL and 150 mL.

These experimental results do not coincide at all with the empirically calculated values of L0 since this series of experiments is limited in time to 40 days, while the methanogenic potential calculation equation given by the US EPA takes into account the total consumption of the substrate.

In addition, the monitoring of the quality of biogas generated for all samples shows that the oxygen content increases from 19%/V to 7%/V during the first week, with carbon dioxide production averaging 5%/V. However, methane only appeared during the last week in insignificant quantities ranging from 1% to 2% by volume.

5.4. Biogas Production Modelling via the Four Models

The different modelling equations mentioned in this work utilize three terms (tonnage, K and L0) that define the volume of biogas/methane generated in a time interval.

The methane generation constant (K) represents the decomposition rate. This depends mainly on waste and precipitation on site. High levels of K indicate a higher level of gas production over time [7]. As with L0, the US EPA has set values of K ranging from 0.02 year^{-1} to 0.7 year^{-1} for arid and humid areas respectively, as well as a conventional value of 0.05 year^{-1} [19].

The French model developed by ADEME, on the other hand, uses three degradation constants K according to the biodegradability of the waste:

$\mathbf{K_1}$ = 0.5 (easily biodegradable);

$\mathbf{K_2}$ = 0.10 (moderately biodegradable);

$\mathbf{K_3}$ = 0.04 (poorly biodegradable).

The values of the methane generation constant adopted by ADEME are somewhat higher than those of the American models. However, due to the high humidity content of the waste and the climatological conditions characterizing the study area, the ADEME model is the most suitable for modelling biogas production for waste from the city of Mohammedia (Figure 6). In addition, it offers the possibility of modelling methane production for waste of a different nature, which is the case in this study, by summing the L0i and Ki for n substrates (Equation (5)).

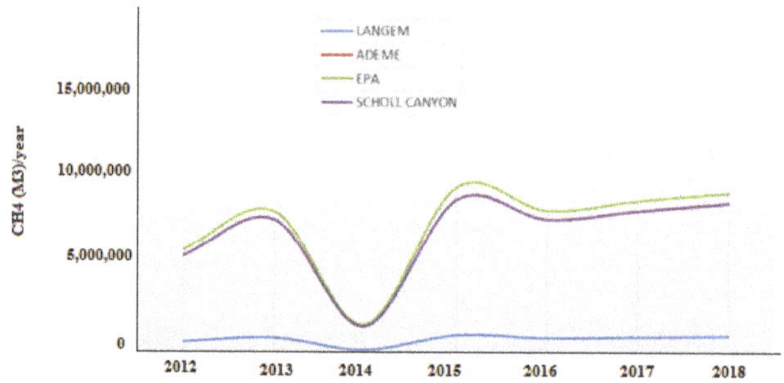

Figure 6. Modeling results of CH_4 production via the four models.

In this study, we assigned values of K to the different types of waste studied (0.1 for household waste, 0.5 for green waste and 0.04 for poultry droppings and tannery waste) according to their degree of biodegradability by comparing the carbon/nitrogen ratios.

The modelling results show only a slight variation between three of the models, the EPA, SCHOLL CANYON and ADEME, but the LANDGEM model showed a significant difference compared with the others, which is why production is evaluated on a tenth of a year basis [20].

Details of the calculations are provided in Appendix A.

5.5. Estimate of the Biogas Field

The estimation of the tonnage of household waste from the different urban and rural municipalities between 2018 and 2032, according to their respective waste production ratios of 0.76 and 0.3 Kg/inhab/d [19], begins with the calculation of the population evolution and the ratio using Equations (8)–(10) (Appendix B).

$$\mathbf{EP = Ai \times \left(1 - \frac{rate\ of\ increase}{100}\right)^{Ai-A0}} \quad (8)$$

$$\mathbf{ER = 0.76(ou\ 0.3) \times (1 + 1.36/100)^{Ai-A0}} \quad (9)$$

$$\mathbf{T = (population \times ratio \times 365\ /1000)} \quad (10)$$

EP: population evolution;

ER: ratio evolution;

T: annual tonnage;

Ai: population i.

In view of the difficulty of estimating the tonnage of green and industrial waste, the average percentages of the apparent tonnage recorded between 2012 and 2018 of each substrate and their percentages of contribution to methane production were used to predict the tonnage of fermentable waste until the end of operation, including the volume of biogas related to it (Figure 7).

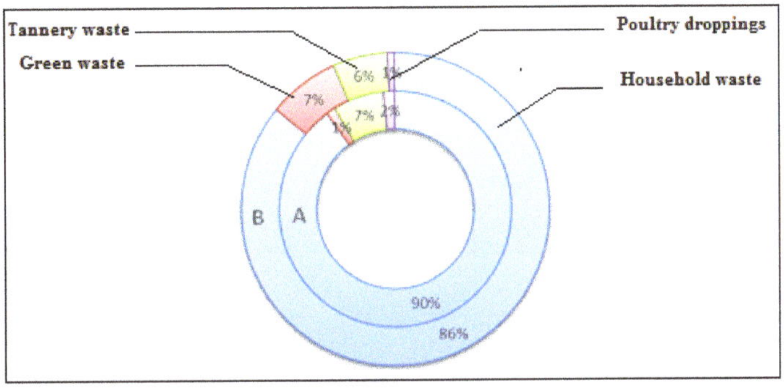

Figure 7. (A) % tonnage; (B) % contribution to CH_4 production of the four substrates.

The calculation of greenhouse gas emissions is performed using Equation (11) below:

$$\mathbf{GH\,G_p} = \mathbf{21}\left(\mathbf{0.016} \times \frac{\mathbf{Q_p}}{\mathbf{22.4}}\right) \tag{11}$$

GHG_p: equivalent CO_2 emissions (t CO_2/year);
Q_p: quantity of methane produced (m^3/year);
21: ratio of CH_4 to 1 CO_2.

Table 5 below shows the modelling results of biogas production from anaerobic decomposition for four types of waste studied using the ADEME model.

Table 5. Calculation results for the annual production of biogas, CH_4 and CO_2 equivalent.

Year	Tonnage DM	Tonnage (DM, DV, FV, DT)	Biogas	CH4	GHGp
Unit	MKg/Years	MKg/Years	MM^3/Years	MM^3/Years	KT CO_2/Year
2012	103.88	114.27	10.64	5.92	88.75
2013	138.26	152.09	14.60	8.12	121.81
2014	26.35	28.99	2.863	1.59	23.88
2015	146.72	161.40	16.81	9.34	140.19
2016	138.23	152.05	14.68	8.16	122.44
2017	144.72	159.19	15.62	8.67	130.34
2018	161.96	178.15	16.48	9.16	137.47
2019	101.27	111.40	9.29	5.17	77.49
2020	101.60	111.76	9.32	5.18	77.74
2021	101.93	112.12	9.35	5.20	78.00
2022	102.26	112.49	9.38	5.21	78.25
2023	102.59	112.85	9.41	5.23	78.50
2024	102.93	113.22	9.44	5.25	78.76
2025	103.26	113.59	9.47	5.27	79.01
2026	103.60	113.96	9.50	5.28	79.27
2027	103.93	114.33	9.53	5.32	79.53
2028	103.93	114.33	9.54	5.31	79.53
2029	104.61	115.07	9.60	5.34	80.05
2030	104.95	115.45	9.63	5.35	80.31
2031	105.29	115.82	9.66	5.37	80.57
Total	**230.79**	**253.87**	**224.55**	**124.85**	**1872.73**

5.6. Potential for Energy Recovery by Cogeneration

The cogeneration unit consists of a gas engine or turbine, an alternator and optional heat recovery circuits. The gas is burned in the engine and then the mechanical energy of the engine is transformed into electricity via the alternator (Figure 8).

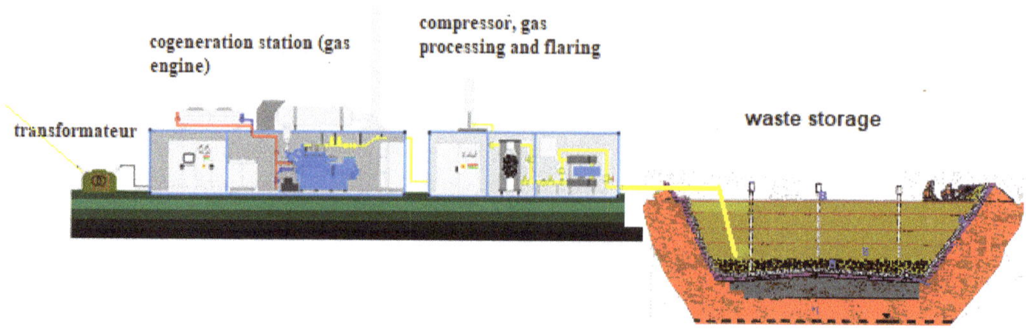

Figure 8. Overview of the cogeneration station [20].

Well before cogeneration, the biogas purification step is necessary. This requires the presence of hydrogen sulphide at levels exceeding 900 ppm, which is the case for the controlled landfill in Fez, where the H_2S content in the biogas is around 1200 ppm compared with 600 ppm for the Mohammedia landfill (Table 6).

Table 6. Biogas energy balance.

Year	MCH_4 m^3/Year	Total Energy GWh	Recoverable Energy GWh	Energy MWh
2012	5.91	58.81	55.87	6.38
2013	8.12	80.72	76.68	8.75
2014	1.60	15.82	15.03	1.77
2015	9.34	92.90	88.25	10.07
2016	8.16	81.13	77.08	8.79
2017	8.69	86.37	82.05	9.37
2018	9.16	91.09	86.54	9.88
2019	5.17	51.35	48.78	5.57
2020	5.19	51.51	48.94	5.59
2021	5.20	51.68	49.10	5.60
2022	5.22	51.85	49.26	5.62
2023	5.23	52.02	49.42	5.64
2024	5.25	52.19	49.58	5.66
2025	5.28	52.36	49.74	5.68
2026	5.28	52.53	49.90	5.70
2027	5.30	52.70	50.07	5.71
2028	5.30	52.70	50.07	5.71
2029	5.34	53.04	50.39	5.75
2030	5.35	53.21	50.56	5.771
2031	5.37	53.39	50.72	5.79
2032	5.39	53.56	50.88	5.80
totaux	124.85	1240.99	1178.94	134.58

The total annual energy produced from biogas is the product of the volume of methane multiplied by its lower calorific value, which is 9.94 kWh/m^3 under normal temperature and pressure conditions [21–28].

$$\mathbf{E_{totale} = PCI_{CH4} \times V_{CH4}\ [KWh]} \quad (12)$$

We allow for 5% energy loss in order to be sure that the engine is more supercharged than underfuelled [22]. The energy recoverable by the motor is therefore as follows:

$$\mathbf{E_{Valorisable} = 0.95 \times E_{totale}\ [KWh]} \quad (13)$$

$$\mathbf{E_{t=1h} = \frac{E_{Valorisable}}{365 \times 24}\ [KW]} \quad (14)$$

Taking into account an Et = 1 h average of 7288 KW, the gas engine would need to be designed to operate between 50% and 100% of its rated load, with an optimal efficiency around 75%. We are therefore looking for an engine with a power of about 9715 kW to be close to this optimum [12,21–28].

6. Conclusions

The projected depletion of fossil energy resources associated with the environmental issue of global warming has intensified the interest in renewable energies and the possible options they can offer. In this research study, it was possible to determine the value of the methanogenic potential of household waste and other substrates of the landfill of the city Mohammedia, which were significantly higher than the values usually used in modelling efforts due to the high proportion of organic waste. From an analysis of the total stability of the mathematical model of the process equilibrium, we constructed a criterion that, based on the inputs of the process and the parameters of the model, determines whether the mode of operation represents a risk to the sustainability of the process. The volume of methane that would be generated after twenty years of operation is of the order of 124,848,407 m^3, thus producing 1,178,943,512 KWh of recoverable energy, a very large deposit that justifies investment.

Funding: I thank the University of Mansoura (Egypt) for supporting the publication costs of this article within the framework of COP27.

Data Availability Statement: No data are available.

Acknowledgments: The corresponding author would like to thank the Center for Water, Natural Resources, Environment and Sustainable Development for its support and participation in carrying out research at the Centre. This research was carried out as part of the doctorate of the Faculty of Sciences of Rabat. ECOMED's support in providing the required data is greatly appreciated.

Conflicts of Interest: The authors declare no conflict of interest regarding the publication of this manuscript.

Abbreviations

DMA	household and similar waste
CET	Technical Landfill Centre
DM	household waste
DV	green waste
FV	poultry droppings
DT	tannery waste
MO	organic matter
K	methane generation constant
L_0	methane production potential

Appendix A

Table A1. Results the Calculations of Modeling By LANDGEM.

	Modeling By LANDGEM (KM^3 of CH4 per T of Waste)			
	DM	DV	FV	DT
2012	579.84	45.13	33.54	0.59
2013	771.72	66.62	64.39	2.25
2014	147.083	16.13	14.97	0.058
2015	818.96	17.76	72.91	26.26
2016	771.53	86.42	55.36	27.21
2017	807.76	11.328	58.50	21.24
2018	903.98	96.65	20.52	27.67

Table A2. Results the Calculations of Modeling By SCHOLL CANION.

	Modeling By SCHOLL CANION (MM^3 of CH_4 per T of Waste)			
	DM	DV	FV	DT
2012	5.28	0.29	0.32	0.005
2013	7.03	0.42	0.62	0.0216
2014	1.34	0.10	0.14	0.0005
2015	7.46	0.11	0.70	0.0252
2016	7.03	0.55	0.53	0.0261
2017	7.37	0.72	0.56	0.0204
2018	8.24	0.61	0.02	0.2661

Table A3. Results the Calculations of Modeling By EPA.

	Modeling By EPA (MM3 of CH4 per T of Waste)			
	DM	DV	FV	DT
2012	5.57	0.37	0.330	0.0058
2013	7.41	0.55	0.633	0.0221
2014	1.41	0.13	0.147	0.0006
2015	7.87	0.15	0.718	0.0258
2016	7.41	0.71	0.544	0.0269
2017	7.76	0.94	0.576	00209
2018	8.69	0.79	0.020	0.2723

Table A4. Results the Calculations of Modelling by ADEME.

	Modelling by ADEME (MM^3 of CH4 per T of Waste)			
	DM	DV	FV	DT
2012	5.30	0.29	0.32	0.0060
2013	7.05	0.42	0.62	0.0220
2014	1.34	0.10	0.14	0.0005
2015	7.48	0.11	0.70	0.0253
2016	7.05	0.55	0.53	0.0262
2017	7.38	0.72	0.56	0.0205
2018	8.261	0.62	0.20	0.2669

Appendix B

Table A5. Population evolution.

Population Evolution									
			Urban Communes				Rural Communities		
Communes	Mohammedia	Ain Harouda	Bouznika	Bensliman	El Mansouria	Ech-Challalat	Ben yakhlef	Sidi Moussa Ben Ali	Sidi Moussa El Majdoub
Initial population	187,708	41,853	27,028	46,478	12,955	40,311	18,233	9,368	12412
Rate of increase	1.01	1.04	1.03	1.02	1.04	1.03	1.1	1.02	1.05
2005	185,812.149	41,417.7288	26,749.6116	46,003.9244	12,820.268	39,895.7967	18,032.437	9272.4464	12,281.674
2006	183,935.446	40,986.9844	26,474.0906	45,534.6844	12,686.9372	39,429.0356	17,862.9438	9172.30482	12,412
2007	182,077.698	40,560.7198	26,201.4075	45,070.2306	12,554.9931	39,078.1758	17,637.9053	9084.2532	12,025.1129
2008	180,238.714	40,138.8883	25,931.533	44,610.5142	12,424.4211	38,675.6706	17,443.8884	8991.59382	11,898.84922
2009	178,418.303	39,721.4439	25,664.4382	44,155.487	12,295.2072	38,277.3112	17,252.0056	8899.87956	11,773.9113
2010	176,616.278	39,308.3408	25,400.0945	43,705.101	12,167.337	37,883.0549	17,062.2335	8809.10079	11,650.28523
2011	174,832.453	38,899.5341	25,138.4735	43,259.309	12,040.7967	37,492.8594	16,874.549	8719.24796	11,527.95723
2012	173,066.646	38,494.9789	24,879.5472	42,818.064	11,915.5724	37,106.683	16,688.9289	8630.31163	11,406.91368
2013	171,318.673	38,094.6312	24,623.2879	42,381.3198	11,791.6505	36,724.4842	16,505.3507	8542.28245	11,287.14109
2014	169,588.354	37,698.447	24,369.668	41,949.303	11,669.173	36,346.222	16,323.7918	8455.15117	11,168.62611
2015	167,875.512	37,306.3831	24,118.6604	41,521.1502	11,547.6595	35,971.8559	16,144.2301	8368.90863	11,051.35553
2016	166,179.969	36,918.3968	23,870.2382	41,097.6345	11,427.5639	35,601.3458	15,966.6436	8283.54576	10,935.3163
2017	164,501.551	36,534.4454	23,624.3748	40,678.4386	11,308.7172	35,234.6519	15,791.0105	8199.05359	10,820.49548
2018	162,840.086	36,154.4872	23,381.0437	40,263.5185	11,191.1065	34,871.735	15,617.3094	8115.42325	10,706.88028
2019	161,195.401	35,778.4805	23,140.219	39,852.8307	11,074.719	34,512.5561	15,445.519	8032.64593	10,594.45803
2020	159,567.327	35,406.3843	22,901.8747	39,446.3318	10,959.5419	34,157.0768	15,275.6183	7950.71294	10,483.21623
2021	157,955.697	35,038.1579	22,665.9854	39,043.9792	10,845.5627	33,805.2589	15,107.5865	7869.61567	10,373.14246
2022	156,360.345	34,673.7611	22,432.5258	38,645.7306	10,732.7689	33,457.0647	14,941.403	7789.34559	10,264.22446
2023	154,781.105	34,313.154	22,201.4707	38,251.5442	10,621.1481	33,112.457	14,777.0476	7709.89426	10,156.4501
2024	153,217.816	33,956.2972	21,972.7956	37,861.3784	10,510.6881	32,771.3987	14,614.5001	7631.25334	10,049.80738
2025	151,670.316	33,603.1517	21,746.4758	37,475.1923	10,401.377	32,433.8533	14,453.7406	7553.41456	9944.284399
2026	150,138.446	33,253.6789	21,522.4871	37,092.9454	10,293.2026	32,099.7846	14,294.7494	7476.36973	9839.869413
2027	148,622.048	32,907.8407	21,300.8055	36,714.5973	10,186.1533	31,769.1568	14,137.5072	7400.11076	9736.550784
2028	147,120.965	32,565.5991	21,081.4072	36,340.1084	10,080.2173	31,441.9345	13,981.9946	7324.62963	9634.317001
2029	145,635.04	32,226.917	20,864.269	35,969.439	9975.3831	31,118.083	13,828.193	7249.9184	9533.156672
2030	144,164.13	31,891.757	20,649.367	35,602.551	9871.6391	30,797.566	13,676.083	7175.9692	9433.058527
2031	142,708.071	31,560.0827	20,436.6782	35,239.405	9768.97405	30,480.3514	13,525.6456	7102.77435	9334.011413
2032	141,266.72	31,231.8578	20,226.1805	34,879.9631	9667.37672	30,166.4037	13,376.8635	7030.32606	9236.004293

Table A6. Evolution of the Ratio.

Evolution of the Ratio									
			Urban Communes				Rural Communities		
Communes	Mohammedia	Ain Harouda	Bouznika	Bensliman	El Mansouria	Ech-Challalat	Ben yakhlef	Sidi Moussa Ben Ali	Sidi Moussa El Majdoub
2005	0.770336	0.770336	0.770336	0.770336	0.770336	0.30408	0.30408	0.30408	0.30408
2006	0.78081257	0.78081257	0.78081257	0.78081257	0.78081257	0.30821549	0.30821549	0.30821549	0.30821549
2007	0.79143162	0.79143162	0.79143162	0.79143162	0.79143162	0.31240722	0.31240722	0.31240722	0.31240722
2008	0.80219509	0.80219509	0.80219509	0.80219509	0.80219509	0.31665596	0.31665596	0.31665596	0.31665596
2009	0.81310494	0.81310494	0.81310494	0.81310494	0.81310494	0.32096248	0.32096248	0.32096248	0.32096248
2010	0.82416317	0.82416317	0.82416317	0.82416317	0.82416317	0.82416317	0.82416317	0.82416317	0.82416317
2011	0.83537179	0.83537179	0.83537179	0.83537179	0.83537179	0.83537179	0.83537179	0.83537179	0.83537179
2012	0.84673285	0.84673285	0.84673285	0.84673285	0.84673285	0.33423665	0.33423665	0.33423665	0.33423665
2013	0.85824841	0.85824841	0.85824841	0.85824841	0.85824841	0.33878227	0.33878227	0.33878227	0.33878227
2014	0.86992059	0.86992059	0.86992059	0.86992059	0.86992059	0.34338971	0.34338971	0.34338971	0.34338971
2015	0.88175151	0.88175151	0.88175151	0.88175151	0.88175151	0.34805981	0.34805981	0.34805981	0.34805981
2016	0.89374333	0.89374333	0.89374333	0.89374333	0.89374333	0.35279342	0.35279342	0.35279342	0.35279342
2017	0.90589824	0.90589824	0.90589824	0.90589824	0.90589824	0.35759141	0.35759141	0.35759141	0.35759141
2018	0.91821846	0.91821846	0.91821846	0.91821846	0.91821846	0.36245465	0.36245465	0.36245465	0.36245465
2019	0.93070623	0.93070623	0.93070623	0.93070623	0.93070623	0.36738404	0.36738404	0.36738404	0.36738404
2020	0.94336383	0.94336383	0.94336383	0.94336383	0.94336383	0.37238046	0.37238046	0.37238046	0.37238046
2021	0.95619358	0.95619358	0.95619358	0.95619358	0.95619358	0.37744483	0.37744483	0.37744483	0.37744483
2022	0.96919781	0.96919781	0.96919781	0.96919781	0.96919781	0.38257808	0.38257808	0.38257808	0.38257808
2023	0.9823789	0.9823789	0.9823789	0.9823789	0.9823789	0.38778115	0.38778115	0.38778115	0.38778115
2024	0.99573926	0.99573926	0.99573926	0.99573926	0.99573926	0.39305497	0.39305497	0.39305497	0.39305497
2025	1.00928131	1.00928131	1.00928131	1.00928131	1.00928131	1.00928131	1.00928131	1.00928131	1.00928131
2026	1.02300754	1.02300754	1.02300754	1.02300754	1.02300754	0.40381876	0.40381876	0.40381876	0.40381876
2027	1.03692044	1.03692044	1.03692044	1.03692044	1.03692044	0,4093107	0.4093107	0.4093107	0.4093107
2028	1.05102256	1.05102256	1.05102256	1.05102256	1.05102256	0.41487733	0.41487733	0.41487733	0.41487733
2029	1.06531646	1.06531646	1.06531646	1.06531646	1.06531646	0.42051966	0.42051966	0.42051966	0.42051966
2030	1.07980477	1.07980477	1.07980477	1.07980477	1.07980477	0.42623872	0.42623872	0.42623872	0.42623872
2031	1.09449011	1.09449011	1.09449011	1.09449011	1.09449011	0.43203557	0.43203557	0.43203557	0.43203557
2032	1.10937518	1.10937518	1.10937518	1.10937518	1.10937518	0.43791125	0.43791125	0.43791125	0.43791125

References

1. Bouhadiba, B.; Mezouri, F.; Kehila, Y.; Matejka, G. For an integrated Management for Municipal solid wast in Algeria.Systemic and Methodological Approaches. *Int. Rev. Chem. Eng. Rapid Commun.* **2010**, *2*, 426–429.
2. Gu, Y.; Zhou, G.; Wu, Y.; Xu, M.; Chang, T.; Gong, Y.; Zuo, T. Environmental performance analysis on resource multiple-life-cycle recycling system: Evidence from waste pet bottles in China. *Resour. Conserv. Recycl.* **2020**, *158*, 104821. [CrossRef]
3. Gautam, P.; Kumar, S.; Lokhandwala, S. Energy-aware intelligence in megacities. In *Current Developments in Biotechnology and Bioengineering*; Elsevier: Amsterdam, The Netherlands, 2019; pp. 211–238.

4. Joseph, O.; Rouez, M.; Métivier-Pignon, H.; Bayard, R.; Emmanuel, E.; Gourdon, R. Adsorption of heavy metals on to sugar cane bagasse: Improvement of adsorption capacities due to anaerobic degradation of the biosorbent. *Environ. Technol.* **2009**, *30*, 1371–1379. [CrossRef] [PubMed]
5. Mabrouki, J.; Moufti, A.; Bencheikh, I.; Azoulay, K.; El Hamdouni, Y.; El Hajjaji, S. Optimization of the Coagulant Flocculation Process for Treatment of Leachate of the Controlled Discharge of the City Mohammedia (Morocco). In Proceedings of the International Conference on Advanced Intelligent Systems for Sustainable Development, Tangier, Morocco, 8–11 July 2019; Springer: Cham, Switzerland, 200; pp. 200–212.
6. Finon, D. La pénétration à grande échelle des ENR dans les marchés électriques. La perte de repère des évaluations économiques. *Rev. De L'energie* **2016**, *633*, 366–390. (In French)
7. Mabrouki, J.; Moufti, A.; Bencheikh, I.; Azoulay, K.; El Hamdouni, Y.; El Hajjaji, S. Optimization of the Coagulant Flocculation Process for Treatment of Leachate of the Controlled Discharge of the City Mohammedia (Morocco). In *Advances in Intelligent Systems and Computing, Proceedings of the International Conference on Advanced Intelligent Systems for Sustainable Development (AI2SD-2018), Tangier, Morocco, 12–14 July 2018*; Ezziyyani, M., Ed.; Springer: Berlin/Heidelberg, Germany, 2018; pp. 200–212.
8. Andriani, D.; Atmaja, T.D. The potentials of landfill gas production: A review on municipal solid waste management in Indonesia. *J. Mater. Cycles Waste Manag.* **2019**, *21*, 1572–1586. [CrossRef]
9. Mabrouki, J.; El Yadini, A.; Bencheikh, I.; Azoulay, K.; Moufti, A.; El Hajjaji, S. Hydrogeological and hydrochemical study of underground waters of the tablecloth in the vicinity of the controlled city dump Mohammedia (Morocco). In *Advances in Intelligent Systems and Computing, Proceedings of the International Conference on Advanced Intelligent Systems for Sustainable Development (AI2SD-2018), Tangier, Morocco, 12–14 July 2018*; Ezziyyani, M., Ed.; Springer: Berlin/Heidelberg, Germany, 2018; pp. 22–33.
10. El Maguiri, A.; Fawaz, N.; Abouri, M.; Idrissi, L.; Taleb, A.; Souabi, S.; Vincent, R. Caractérisation physique et valorisation des déchets ménagers produits par la ville de Mohammedia, Maroc. *Rev. Nat. Technol. C. Sci. L'environ.* **2016**, *14*, 16–25.
11. Elkadi, A.; Maatouk, M.; Raissouni, M.; Chafik, T.; Mouhssine, A. Caractérisation des déchets ménagers et assimilés de la ville de Tanger/Characterization of household and assimilated waste in the city of Tangier. *Int. J. Innov. Appl. Stud.* **2016**, *18*, 512.
12. Mabrouki, J.; Bencheikh, I.; Azoulay, K.; Es-soufy, M.; el Hajjaji, S. Smart Monitoring System for the Long-Term Control of Aerobic Leachate Treatment: Dumping Case Mohammedia (Morocco). In *Big Data and Networks Technologies, Proceedings of the 3rd International Conference on Big Data and Networks Technologies (BDNT 2019), Leuven, Belgium, 29 April–2 May 2019*; Springer: Berlin/Heidelberg, Germany, 2019; pp. 220–230.
13. El Asri, O.; Afilal, M.E.; Laiche, H.; Elfarh, A. Evaluation of physicochemical, microbiological, and energetic characteristics of four agricultural wastes for use in the production of green energy in Moroccan farms. *Chem. Biol. Technol. Agric.* **2020**, *7*, 1–11.
14. Mabrouki, J.; Benbouzid, M.; Dhiba, D.; El Hajjaji, S. Simulation of wastewater treatment processes with Bioreactor Membrane Reactor (MBR) treatment versus conventional the adsorbent layer-based filtration system (LAFS). *Int. J. Environ. Anal. Chem.* **2020**, 1–11. [CrossRef]
15. Porowska, D. Review of Research Methods for Assessing the Activity of a Municipal Landfill Based on the Landfill Gas Analysis. *Period. Polytech. Chem. Eng.* **2021**, *65*, 167–176. [CrossRef]
16. Mabrouki, J.; Fattah, G.; Al-Jadabi, N.; Abrouki, Y.; Dhiba, D.; Azrour, M.; Hajjaji, S.E. Study, simulation and modulation of solar thermal domestic hot water production systems. *Model. Earth Syst. Environ.* **2022**, *8*, 2853–2862. [CrossRef]
17. Ortiz-Oliveros, H.B.; Flores-Espinosa, R.M. Design of a mobile dissolved air flotation system with high rate for the treatment of liquid radioactive waste. *Process. Saf. Environ. Prot.* **2020**, *144*, 23–31. [CrossRef]
18. Spokas, K.; Bogner, J.; Chanton, J.P.; Morcet, M.; Aran, C.; Graff, C.; Hebe, I. Methane mass balance at three landfill sites: What is the efficiency of capture by gas collection systems? *Waste Manag.* **2006**, *26*, 516–525. [CrossRef] [PubMed]
19. Hijazi, O.; Munro, S.; Zerhusen, B.; Effenberger, M. Review of life cycle assessment for biogas production in Europe. *Renew. Sustain. Energy Rev.* **2016**, *54*, 1291–1300. [CrossRef]
20. Scharff, H.; Jacobs, J. Applying guidance for methane emission estimation for landfills. *Waste Manag.* **2006**, *26*, 417–429. [CrossRef]
21. Mouhssine, A.; Brigui, J. Organic waste characterization in Tangier City and evaluation of its potential biogas. *Int. J. Innov. Appl. Stud.* **2017**, *20*, 1042–1052.
22. Mabrouki, J.; Azrour, M.; Hajjaji, S.E. Use of internet of things for monitoring and evaluating water's quality: A comparative study. *Int. J. Cloud Comput.* **2021**, *10*, 633–644. [CrossRef]
23. El Baz, F.; Arjdal, S.; Bakrim, M.; Mediouni, T.; Faska, N. Methanisation of Agadir urban solid waste: Theoretical evaluation of the energy production potential. In *Materials Today, Proceedings of the International Congress: Applied Materials for the Environment (CIMAE-2018), Agadir, Morocco, 5–7 December 2018*; Dahham, O.S., Zulkepli, N.N., Eds.; Elsevier: Amsterdam, The Netherlands, 2019; pp. 97–99.
24. Wu, Z.L.; Lin, Z.; Sun, Z.Y.; Gou, M.; Xia, Z.Y.; Tang, Y.Q. A comparative study of mesophilic and thermophilic anaerobic digestion of municipal sludge with high-solids content: Reactor performance and microbial community. *Bioresour. Technol.* **2020**, *302*, 122851. [CrossRef]
25. Trianni, A.; Cagno, E.; Accordini, D. Energy efficiency measures in electric motors systems: A novel classification highlighting specific implications in their adoption. *Appl. Energy* **2019**, *252*, 113481. [CrossRef]
26. Blewitt, J. Deschooling society? A lifelong learning network for sustainable communities, urban regeneration and environmental technologies. *Sustainability* **2010**, *2*, 3465–3478. [CrossRef]

27. Suther, T.; Fung, A.; Koksal, M.; Zabihian, F. Macro level modeling of a tubular solid oxide fuel cell. *Sustainability* **2010**, *2*, 3549–3560. [CrossRef]
28. Richardson, R.B. Ecosystem services and food security: Economic perspectives on environmental sustainability. *Sustainability* **2010**, *2*, 3520–3548. [CrossRef]

sustainability

MDPI

Review

Biochar as a Soil Amendment for Restraining Greenhouse Gases Emission and Improving Soil Carbon Sink: Current Situation and Ways Forward

Ahmed Mosa [1,*], Mostafa M. Mansour [1], Enas Soliman [1], Ayman El-Ghamry [1], Mohamed El Alfy [2] and Ahmed M. El Kenawy [3]

[1] Department of Soils, Faculty of Agriculture, Mansoura University, Mansoura 35516, Egypt
[2] Department of Geology, Faculty of Science, Mansoura University, Mansoura 35516, Egypt
[3] Department of Geography, Mansoura University, Mansoura 35516, Egypt
* Correspondence: ahmedmosa@mans.edu.eg

Abstract: The global exponential rise in greenhouse gas (GHG) emissions over the last few decades has triggered an urgent need to contextualize low-cost and evergreen technologies for restraining GHG production and enhancing soil carbon sink. GHGs can be mitigated via incorporating biochar into soil matrix to sequestrate the mineralized carbon in a stable form upon organic matter decomposition in soil. However, the efficiency of using biochar to offset GHG emissions from soil and terrestrial ecosystems is still debatable. Moreover, in the literature, biochar shows high functionality in restraining GHG emissions in short-term laboratory studies, but it shows minimal or negative impacts in field-scale experiments, leading to conflicting results. This paper synthesizes information on the ability of biochar to mitigate carbon dioxide (CO_2), nitrous oxide (N_2O), and methane (CH_4) emissions from soil and organic biomass, with an emphasis on cropland soils. The feedstock type, pyrolysis temperature, and application rate factors showed significant effects on controlling the effectiveness of biochar in restraining GHG emissions. Our study demonstrates that biochar, taken as a whole, can be seen as a powerful and easy-to-use tool for halting the rising tide of greenhouse gas emissions. Nonetheless, future research should focus on (i) identifying other indirect factors related to soil physicochemical characters (such as soil pH/EH and $CaCO_3$ contents) that may control the functionality of biochar, (ii) fabricating aged biochars with low carbon and nitrogen footprints, and (iii) functionalizing biologically activated biochars to suppress CO_2, CH_4, and N_2O emissions. Overall, our paradoxical findings highlight the urgent need to functionalize modern biochars with a high capacity to abate GHG emissions via locking up their release from soil into the carbonaceous lattice of biochar.

Keywords: biochar; croplands and rangelands; carbon sequestration; organic manures

Citation: Mosa, A.; Mansour, M.M.; Soliman, E.; El-Ghamry, A.; El Alfy, M.; El Kenawy, A.M. Biochar as a Soil Amendment for Restraining Greenhouse Gases Emission and Improving Soil Carbon Sink: Current Situation and Ways Forward. *Sustainability* **2023**, *15*, 1206. https://doi.org/10.3390/su15021206

Academic Editor: Bo Jiang

Received: 30 October 2022
Revised: 2 January 2023
Accepted: 4 January 2023
Published: 9 January 2023

1. Introduction

The exponential increase in greenhouse gas (GHG) emissions following anthropogenic activities in the last few decades has caused a pronounced rise in climatic changes with consequent environmental crises in global warming, drought, salinity, biodiversity, and diseases [1]. Human agricultural practices account for about 13.5% of global GHG emissions, including 80.4 petagrams of CO_2 per year (11 times the current rate of fossil fuel combustion), and about 63% of the world's non–CO_2 GHG emissions, including 84% of global N_2O and 54% of global CH_4 emissions [2]. Other indirect emissions, such as those from machinery and transportation, are also produced by common agricultural practices. The greenhouse effect, which raises the earth's temperature, is caused when the produced GHGs trap infrared (IR) radiation that is emitted from the earth's surface [3]. It was reported that seven countries (China, USA, India, Australia, Brazil, Canada, and Chile) contribute to producing more than 50% of the world's total soil emissions, and about

49% of the agricultural–related emissions [4]. Since irrigated agriculture is the dominant cultivation system in arid and semi-arid regions, large amounts of GHGs are emitted due to different agricultural practices including manure application, rice cultivation, enteric fermentation, burning crop residues, manure storage in the open air, and using energy for operating irrigated pumps [5]. Egypt's GHG emissions (as an example for arid countries) increased rapidly to more than 133% between 1990 and 2012, with a total emission of around 318 million tonnes eq. CO_2 [6].

Although agriculture led to substantial increases in GHGs, some agricultural practices could have significant impacts on reducing these emissions. Agriculture can make considerable contributions to mitigating GHG emissions in the atmosphere by (i) increasing soil organic carbon sinks, (ii) reducing the carbon and nitrogen footprint of soil organic amendments, (iii) recycling crop residues into value-added products instead of burning, and (iv) reducing GHG emissions generated during organic matter decomposition [7–9]. Pyrolysis has received attention recently as an effective method of treating organic waste and agricultural byproducts. Thermal processing of agricultural crop residues lowers waste volume and transportation costs while producing value-added products. Biochar has become a focal point of multidisciplinary study over the last decade as a solution to various worldwide challenges due to its high functionality, non-sophisticated processing, and renewability potential. Biochar is a charcoal-like substance made from pyrolyzed biomass intended for utilization as a soil improver. It has been credited with multiple benefits, including its abilities to improve the fertility and water-holding capacity of soil, protect water quality, capture greenhouse gases, generate carbon neutral energy, and increase agricultural output, as well as its contributions to carbon sequestration and the removal of pathogens [10–13]. The appropriate pyrolysis technology for biochar production should be considered based on the targeted field of application. Slow pyrolysis is the most common technique for biochar production, with a pyrolysis temperature ranging between 300 and 700 °C at a long residence time (300 to 7200 s) and low heating rate (0.1 to 1 °C/s) [14]. Fast pyrolysis is operated at high pyrolysis temperature within the range of 500–1200 °C at a high heating rate (10–200 °C/s) [15]. Flash pyrolysis, however, features a higher pyrolysis temperature (>900 °C) and heating rate (>1000 °C/s) [16]. Vacuum pyrolysis is another technique, in which biomass is converted under sub-atmospheric pressure (pyrolysis temperature, heating rate, and pressure are within the ranges of 300–700 °C, 0.1–1 °C/s, and 0.01–0.20 MPa, respectively) [17]. Hydro-pyrolysis is another technology, which involves using a high-pressure hydrogen atmospheric condition inside the pyrolysis reactor (pyrolysis temperature = 350–600 °C, heating rate = 10–300 °C/s, pressure = 10–17 MPa, and residence time > 60 s) [18]. Unlike other pyrolysis techniques, the heating energy in microwave pyrolysis penetrates the carbonaceous biomass and causes a vibration in their internal molecules [19].

The high functionality of biochar, including its physical properties (porosity, large surface area, and high water-holding capacity) as well as its chemical properties (abundance of oxygen-containing functional groups, surface charge, and pH-modulating effect), suggests its potential utilization in reducing GHG emissions [20]. However, a great deal of uncertainty remains surrounding the competitiveness of biochar with traditional soil amendments (e.g., compost and farmyard manures) given its low nutrient content and high pH value [21]. Additionally, the efficacy of pristine biochar for restraining GHG emissions under field-scale applications is still questionable. The wide range of variation between biochars depends upon the multiplicity of factors that underlie thermochemical conversion of organic biomass [22–24]. The novelty of this review relies on assessing the key factors that might impede the functionality of biochar in restraining GHG emissions. In addition, this review highlights the urgent need to develop fit-for-purpose forms of functionalized biochars tailored to improving soil carbon sink and restraining GHG emissions from soil matrix.

Several key factors control the properties of biochar and its effectiveness in reducing GHG emissions, including (i) feedstock type (e.g., agricultural wastes, sludge and manures, algal biomass, and crustacean shell wastes), (ii) pyrolysis type (slow/fast pyrolysis, flash pyrolysis, microwave pyrolysis, vacuum pyrolysis, and hydro-pyrolysis), (iii) thermal processing protocol (e.g., heating rate, pressure and carries gas, residence time, and reactor design), and (iv) soil application rates and methods (e.g., broadcasting, in-furrow, or mixture with soil amendments) [25]. Consequently, biochar functionalization has emerged as a new trend, providing a roadmap for enhancing the competitiveness of biochar and its sustainable soil application for reducing GHG emissions.

In this review, we aimed to (i) summarize the recent research undertaken to remove CO_2, N_2O, and CH_4 emissions from soil and terrestrial ecosystems; (ii) review the anomalies and similarities among several investigations of factors affecting the efficiency of biochar in restraining GHG emissions; (iii) examine research done to lessen greenhouse gas emissions from the composting process; and (iv) highlight recent attempts undertaken to functionalize modern biochars with high capacity to achieve neutrality of carbon/nitrogen and net zero emissions. This review will help the academic/research community, as well as decision-making entities and environmental agencies, in establishing a decision-making framework for the large-scale application of biochar to mitigate GHG emissions. Overall, this review synthesizes data from different dimensions of biochar utilization for achieving carbon neutrality and reducing GHG emissions from agricultural ecosystems.

2. Bibliographic Data Collection

Books, book chapters, research articles, review articles, and proceedings were all scoured for this review. All selected sources were written in English and published within the last decade (2010 onwards). All of these articles came straight from reputable sources (e.g., Scopus, Web of Science, ProQuest, EBSCO, and JSTOR). Figure 1 shows a sample of the growing body of literature in the Scopus database on the topic of using biochar to mitigate agricultural GHG emissions. In our review, the International Biochar Initiative (IBI), the United States Department of Agriculture (USDA), and the European Environment Agency all contributed to the credibility of our review by providing official reports, statistics, and proceedings. Use of Get-Data Graph Digitizer (ver. 2.22, Russian Federation) allowed us to convert the data visualizations into corresponding numerical values. In this review, biochar + carbon dioxide emissions, biochar + nitrogen oxide emissions, biochar + chlorofluorocarbon emissions, biochar + soil + carbon dioxide emissions, biochar + soil + nitrogen oxide emissions, and biochar + soil + chlorofluorocarbon emissions were the initial search keywords. In addition, a number of meta-analyses were reported from a variety of published articles to establish a solid assessment of the current state of the potential use of biochar for limiting GHG emissions and the prospects for this direction.

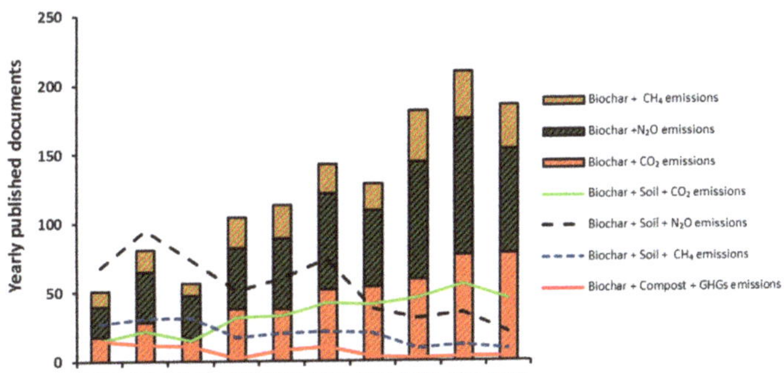

Figure 1. Number of documents published yearly in the Scopus database based on a query that employed the following keywords: biochar + CO_2 emissions, biochar + N_2O emissions, biochar + CH_4 emissions, biochar + soil + CO_2 emissions, biochar + soil + N_2O emissions, and biochar + soil + CH_4 emissions.

3. Effect of Biochar Application on Reducing GHG Emissions from Soil and Terrestrial Ecosystems

The research on biochar has attained significant momentum over the last decade. However, research related to field-scale applications has rarely been approached, highlighting the need for systematic efforts to fabricate end-use-appropriate biochars. The efficiency of biochar in reducing GHG emissions as compared with other soil amendments is summarized in Table 1. Endowed with unique functionality, biochar has been introduced as a promising adsorbent able to abate GHG emissions. Attempts have been undertaken to deploy biochar and its derivatives for mitigating CO_2, N_2O, and CH_4 emissions (Table 2). A comprehensive discussion of the mechanisms involved in CO_2, N_2O, and CH_4 emissions by biochar is provided below based on recent published results.

3.1. Effect of Biochar Application on Reducing CO_2 Emissions

Biochar exhibits a high potential for soil carbon sequestration given its high and resistant carbon content (particularly when its oxygen:carbon ratio is less than 0.2), prevalence of aromatic structures, and abundance of active functional groups, which improves its recalcitrant nature against decomposition in soil [33]. The carbon footprint of biochar ranges between 0.04 tCO_2eq (net emissions) and 1.67 tCO_2eq per t of biomass (net reduction) based on the feedstock type, system boundaries, and modalities of life cycle assessment studies [34]. Upon its soil application, biochar tends to change soil physical characteristics (e.g., water-holding capacity and bulk density) given its low skeletal density and higher surface area compared with particles of soil matrix [35,36]. The key role of biochar in modulating the pore structure parameters of soil (porosity, pore size distribution, connectivity, anisotropy, and fractal dimension) was a significant mechanism involved in restraining CO_2 emissions [37]. According to literature data, the overall average ability of biochar to reduce CO_2 emissions is −0.43%. This poor efficiency highlights the critical need to functionalize contemporary biochars with greater sorption capacity (Figure 2).

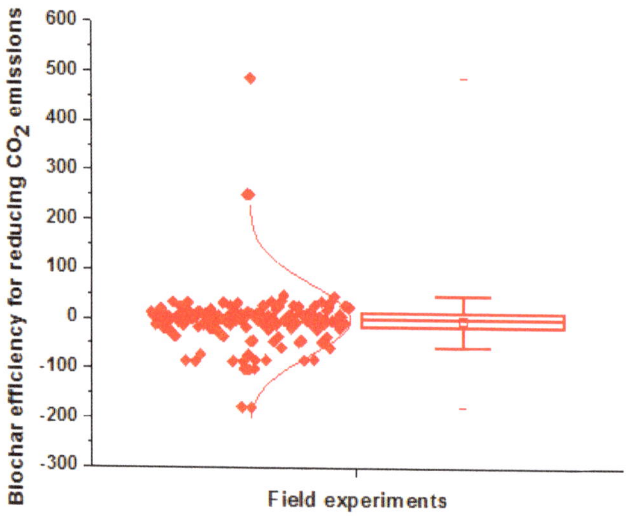

Figure 2. Biochar efficiency in reducing CO_2 emissions (%). Data are extracted from 38 field investigations comprising 165 individual observations. Box chart is illustrated by the mean (dot), median (centerline), lower and upper quartiles (the lower and upper borders of the box, respectively), and whiskers-error bars (the minimum and maximum observations).

Table 1. A comparision between biochar and other soil amendments in reducing GHG emissions.

Country	Climatic Conditions	Soil Properties			Soil Amendments	Application Rate (Mg ha^{-1})	Cultivated Crop	Years of Study	Yield (%) Compared with Control	GHGs Emission Rate Compared with Control (%)			Reference
		Texture	pH	OC (g kg^{-1})						CO_2	N_2O	CH_4	
South Korea	-	Clay loam	5.8	22.98	Fly ash	2.0	Rice	1	7.33	-	3.53	−3.68	[26]
					Silicate slag				21.75	-	5.74	33.74	
					Phosphogypsum				17.02	-	12.58	31.90	
					Revolving furnace slag				20.57	-	7.73	26.99	
					Blast furnace slag				13.00	-	8.83	9.82	
Japan	-	Sandy loam	6.1	21.32	Biochar	2.0	Rice	1	10.47	-	31.83	−13.99	
					Silicate slag				25.58	-	17.65	14.68	
					Phosphogypsum				23.26	-	14.88	20.14	
					Biochar + azolla-cyanobacteria	2.0 + 5.0			27.91	-	26.30	7.85	
					Silicate slag + azolla-cyanobacteria				37.21	-	15.22	22.39	
					Phosphogypsum + azolla-cyanobacteria				30.23	-	11.07	25.60	
Bangladesh	-	Clay loam	5.9	10.35	Biochar	2.0	Rice	1	15.85	-	20.00	−9.49	
					Silicate slag				28.05	-	14.18	18.35	
					Phosphogypsum				24.39	-	9.87	27.22	
					Biochar + azolla-cyanobacteria	2.0 + 5.0			25.61	-	25.06	9.49	
					Silicate slag + azolla-cyanobacteria				40.24	-	12.15	26.58	
					Phosphogypsum + azolla-cyanobacteria				34.15	-	7.34	29.11	
China	Warm temperate and semi-humid monsoon	Clay	7.4	5.79	Wheat straw	7.5	Soybean	1	32.79	-	−37.37	-	[27]
					Pig manure	15			24.59	-	−49.49	-	
					Cattle manure	30			52.46	-	−61.61	-	
					Wheat straw	7.5		2	11.57	-	−20.55	-	
					Pig manure	15			34.77	-	−31.51	-	
					Cattle manure	30			64.81	-	−31.51	-	
					Wheat straw	7.5	Wheat	1	7.25	-	−5.74	-	
					Pig manure	15			3.63	-	−30.33	-	
					Cattle manure	30			8.69	-	−62.30	-	
					Wheat straw	7.5		2	10.32	-	−44.21	-	
					Pig manure	15			12.20	-	−52.63	-	
					Cattle manure	30			16.87	-	−67.37	-	
Southeastern China	-	Silt loam	6.5	18.10	Steel slag	8.0	Early rice	1	0.86	4.08	27.56	10.45	[28]
					Biochar rice straw				6.05	9.87	15.75	−7.37	
					Steel slag + biochar	8.0 + 8.0			9.29	18.71	20.47	34.10	
					Steel slag	8.0	Late rice	1	1.49	12.25	27.40	14.89	
					Biochar rice straw				3.57	25.36	15.07	43.60	
					Steel slag + biochar	8.0 + 8.0			6.98	21.32	20.55	33.43	

Table 1. *Cont.*

Country	Climatic Conditions	Soil Properties			Soil Amendments	Application Rate (Mg ha^{-1})	Cultivated Crop	Years of Study	Yield (%) Compared with Control	GHGs Emission Rate Compared with Control (%)			Reference
		Texture	pH	OC (g kg^{-1})						CO_2	N_2O	CH_4	
India	Sub-tropical, semi-arid	Loam	8.1	5.90	Azolla Blue-green algae (BGA) Azolla + BGA Hyphomicrobium facile (A) Burkholderia sp. (B) Methylobacteruim oryzae (C) A + B + C	- - - - - - -	Rice	2	14.33 6.39 9.97 4.86 −0.84 1.71 4.60	- - - - - - -	8.73 12.04 40.74 1.85 −1.59 −3.83 1.06	9.62 7.07 13.27 4.95 4.10 19.91 −10.41	[29]
China	Sub-tropical monsoon	Clay	8.6	11.77	Humic acid Gypsum Humic acid + gypsum	0.6 0.6 0.6 + 0.6	Rice	1	18.37 2.30 10.45	- - -	−3.77 9.43 −26.42	−6.20 19.36 27.25	[30]
China	Temperate continental monsoon	Loam	6.8	8.81	Humic acid + controlled-release fertilizer	-	Maize	1 2	4.72 11.10	−2.47 −2.94	−40.94 −40.40	- -	[31]
Australia	Humid sub-tropical	Clay	7.8	15.00	Chicken manure + conventional N application rate Composted chicken manure +conventional N application rate Chicken manure + reduced N application rate Composted chicken manure + reduced N application rate	-	Green beans + sorghum + broccoli + lettuce	1	−4.55 0.00 −6.82 2.27	14.16 11.80 11.98 −1.81	−23.53 8.82 −41.18 23.53	- - - -	[32]

Table 2. Effect of biochar application on mitigating GHG emissions under field conditions.

Country	Soil Properties				Biochar Feedstock	Pyrolysis temp. (°C)	Application Rate (Mg ha^{-1})	Cultivated Crop	Years of Study	Yield (%) Compared with Control	GHG emissions Rate (%) Compared with Control			Reference
	Texture	pH	EC ds m^{-1}	OC g kg^{-1}							CO_2	N_2O	CH_4	
Australia	Clay loam	4.6	-	4.5	Cattle feedlot waste	550	10.0	Ryegrass	3	-	0.27	−13.67	-	[38]
China	loam	-	-	-	Wheat	550	20 40	Corn-wheat	2	- -	0.13 −5.40	20.79 31.54	- -	[26]
USA	Silt loam	7.7	0.4	18	Wood	500	22.4	Corn Silage	1 2 3	28.1 32.37 41.62	6.25	25.67	13.61	[39]

Table 2. *Cont.*

Country	Soil Properties				Biochar Feedstock	Pyrolysis temp. (°C)	Application Rate (Mg ha^{-1})	Cultivated Crop	Years of Study	Yield (%) Compared with Control	GHG emissions Rate (%) Compared with Control			Reference
	Texture	pH	EC ds m^{-1}	OC g kg^{-1}							CO_2	N_2O	CH_4	
China	-	-	-	-	Wheat straw	450	20.0 40.0	Corn	1 2	-	−4.17 −7.99	34.34 33.62	- -	[40]
Japan	- -	8.45 8.30	- -	6.3 14.1	Bamboo	700–800	20.0	Kabocha squash Bok choy Kabocha squash Bok choy	1 1 1 1	- −173.68 3.76 −38.02	−346.4 −347.0 −108.45 −146.83	−42.85 −29.50 −163.59 −0.56	- - - -	[41]
Australia	Loamy sand Sandy loam	7.1 6.4	0.11 0.04	2.6 1.8	Wood	500	10.0 20.0 10.0 20.0	Grass	2	- - - -	6.25 −9.38 −18.24 17.57	19.50 1.58 9.18 19.77	- - - 40.00	[42]
China	-	6.04	-	20.1	Wheat straw	350–550	10 20 40	Rice	1	10.0 25.1 26.3	−1.05 15.81 23.91	7.14 30.71 48.57	−14.00 −11.33 −30.67	[43]
China	-	8.1	-	49.6	Corn straw	550	26.0 64.0 128.0	Cabbage + carrot	1	- - -	−20.12 −59.98 −87.21	11.19 −18.33 5.48	- - -	[44]
Germany	-	7.1	-	-	Beech wood	400	60.0	Corn	3	-	42.86	62.96	-	[45]
USA	Clay loam	6.1	0.17	20.1	Pinewood Corn stover Switchgrass	- - -	10.0 10.0 10.0	Corn + soybean Corn + soybean Corn + soybean	1 2 3 1 2 3 1 2 3	- - - - - - - - -	−13.04 9.16 1.33 −6.60 11.90 −13.65 −19.90 5.64 −17.94	20.72 7.58 32.41 19.55 9.41 11.11 15.67 5.84 28.93	- 20.76 - - 20.13 - - 2.39 -	[46]
South korea	Silt loam	5.18	0.50	17.8	South korea	Barley straw	400	Chinese cabbage	1	-	-	60.60	-	[47]
China	Sandy loam	6.5	-	18.1	China	Rice Straw	600	Rice	Early paddy Late paddy	5.7 3.6	9.87 25.36	15.75 43.60	−7.37 15.07	[28]
China	-	7.4	-	12.7	Rice straw	600	20.0 40.0	Rice	1 2	9.4 15.9 24.0 36.3	- - - -	−65.46 11.64 58.01 43.12	29.67 6.35 29.87 15.58	[48]
China	Sandy loam	8.5	0.32	14.5	Corn straw	400	15.0 30.0 45.0 15.0 30.0 45.0	Corn	1 2	6.29 7.34 1.77 1.45 7.57 3.15	24.66 17.57 22.13 19.27 25.85 40.60	71.13 - - 43.71 46.69 38.74	- - - - - -	[49]
Canada	silty clay loam	5.6	-	60.9	Wheat straw	450	20	Barley	1 2	20.37	6.22 3.56	33.18 34.32	- -	[50]
China	-	4.5	-	21.4	-	-	20.0	Tea	1 2	0.0 0.0	- -	14.36 7.10	- -	[51]

In temperate forests, proper management strategies (including biochar application) should be undertaken since about 70% of the global carbon fluxes are generated from respiration [52]. This modulating effect on the respiration of temperate forests might maximize the carbon pool of the soil matrix. In this regard, bamboo leaf biochar application at rates of 5 and 10 Mg ha^{-1} increased carbon stock in moso bamboo forest by about 486 and 253%, respectively [53]. In the same regard, Ge et al. [54] recommended the combined application of biochar and nitrogen fertilizers for maintaining soil fertility and reducing CO_2 emissions from moso bamboo forest. The recalcitrant nature of biochar, its high water holding capacity, and its high potential to form soil aggregates with labile organic components may support its impact on maximizing soil carbon sequestration [53]. It is suggested that the safeguard effect of biochar against CO_2 emissions is associated with the predominance of soil bacterial species dealing with the tricarboxylic acid cycle [55]. Moreover, the key role of biochar in stimulating soil catalase, sucrose, urease, and β–glucosidase activities might support the protective effect of biochar against CO_2 emission from soil [56]. The stimulation of CO_2-fixing bacteria (e.g., *Chloroflexi*) following soil addition of biochar was also reported as a mechanism responsible for reducing CO_2 emissions [57].

On the other hand, cultivated land has almost lost 30–75% of its antecedent organic carbon pool as a result of soil respiration [58]. Therefore, effective soil management is highly recommended to maintain its soil carbon pool and offset the CO_2 emissions from cultivated land. To sustain the soil carbon pool in soil, organic additives application (e.g., farmyard manures) has been widely used as a common technique to sustain the productivity of soil. However, the high mineralization rate of organic carbon in soil (particularly under arid conditions) increases the potential of high CO_2 emissions from soil. Therefore, scientists recommend the application of organic manure in its pyrolyzed form given its carbon-negative nature. For instance, pig manure biochar application to sandy loam soil (northeast Toledo, Spain) decreased carbon mineralization compared to the original pig manure form [59]. In Zheng et al.'s work, biochar application over four years in sandy loam soil located in a semi-arid region (Henan Province, China) led to an increase in soil organic carbon (up to 22.1%), as well as the recalcitrant organic carbon fraction (up to 32.3%) [60]. Moreover, biochar application altered the bacterial community structure by motivating the abundance of *Chloroflexi* phylum (a bacterial community with low carbon mineralization and high carbon fixation potential).

In addition, biochar application was effective in reducing CO_2 emissions from saline soils in arid regions. Specifically, 400 °C corn stalks biochar application to coastal saline soil increased grain yield by about 28% without a significant effect on GHGs (CO_2, CH_4, and N_2O); however, the raw corn stalks increased N_2O emission by 18% [61]. Conversely, other reports point to a neutral or stimulating effect of biochar in increasing CO_2 emission. For instance, the contribution of biochar additives (*Conocarpus erectus* L. at 400 °C) in mitigating CO_2 emissions from sandy–calcareous soil in Saudi Arabia was negligible relative to the original biomass form [62]. The stimulating effect of biochar application increasing CO_2 emissions could be due to the enhancement of dissolved organic carbon content, as well as the enrichement of *Proteobacteria* (copiotrophic bacteria) and inhibition of *Acidobacteriota* (oligotrophic bacteria) [63]. Besides, the role of biochar in reducing soil bulk density might provide a hospitable environment for microorganism respiration [56].

3.2. *Effect of Biochar Application on Reducing N_2O Emissions*

Nitrous oxide (N_2O) is an atmospheric GHG derived mainly from agricultural practices (60%) that has greater global warming potential (300 folds) relative to CO_2 [64]. Scientists have noted a steady increase of N_2O emissions of about 12% over the last 75 years (290 vs. 330 ppbv) [65]. Due to the progressive transition toward intensive agriculture, the excessive utilization of mineral fertilizers has led to the increase of N_2O emissions by about 80% over the past century [66]. It is estimated that nearly 70% of global N_2O emissions are generated from soil nitrogen transformation processes, which are associated with synthetic nitrogen inputs and soil tillage, including nitrification (conversion of NH_4

to NO_3^-) and denitrification (conversion of NO_3^- to N_2) [67]. Interestingly, compost has been recognized as an important source for generating both of CO_2 and N_2O. The high CO_2 and N_2O footprints of compost and other organic fertilizers cause substantial losses from the inherent carbon (0.1–10%) and nitrogen (2.0–3.0%) [68]. For example, livestock manure compost is produced annually by about 1.2×10^6 metric tons of N_2O [69].

Biochar has been suggested as a non-sophisticated solution for maximizing the stability of organic fertilizers and reducing N_2O emissions given its high porosity and high number of active functional groups. Data extracted from the literature point to a relatively low efficiency of biochar (24.64%) in mitigating N_2O emissions (Figure 3).

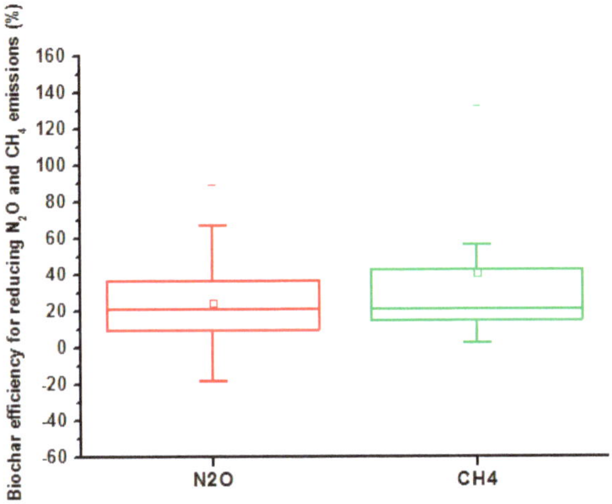

Figure 3. Biochar efficiency in reducing N_2O and CH_4 emissions (%). Data are extracted from 19 investigations comprising 66 individual observations. Box chart is illustrated by the mean (dot), median (centerline), lower and upper quartiles (the lower and upper borders of the box, respectively), and whiskers-error bars (the minimum and maximum observations).

Authors have introduced numerous mechanisms that are responsible for alleviating N_2O emissions. Song et al. suggested that biochar reduced N_2O emissions directly by reducing the gross nitrification/denitrification rates of soil, and indirectly by reducing the content of soil-available N (ammonium, nitrate, and soluble organic nitrogen) and the activities of urease and protease [70]. Ji et al. [71] demonstrated that the application of biochar as a soil amendment was responsible for minimizing fungal abundance, thereby reducing N_2O emissions by about 28%. The promotion of nitrifying bacteria and the inhibition of denitrifying bacteria by biochar application were also highlighted as mechanisms responsible for mitigating N_2O emissions [72]. In addition, the inhibitory effect of biochar on bacterial-related nitrification/denitrification and N-cycle bacterial genes was further recorded in the rhizospheric layer of soil [73]. Another investigation pointed to the modulating effect of soil bulk density, nitrate reductase, nitrite reductase, and hydroxylamine reductase activities related to the denitrification process [74].

The ability of biochar to suppress N_2O emissions has a long-lasting effect, even seven years following its soil application [75]. In a ^{15}N-tracer incubation experiment, the protective effect of raw wood biochar (600 °C) on an alkaline soil (pH = 8.57) was due to the inhibition of the mineralization rate (gross autotrophic/heterotrophic nitrification and mineralization) and increasing the gross immobilization rates in soil [76]. The ability of biochar to modulate soil properties (e.g., aeration, pH/E_H, and organic carbon) has a subsequent effect on regulating soil nitrogen transformations and reducing N_2O emissions [77].

Biochar soil application along with mineral nitrogen fertilizers might have a protective effect in reducing cumulative N_2O emissions. In a microplot experiment, biochar application (12 Mg ha^{-1} of maize straw @ 450 °C) with ^{15}N-labeled urea to sandy loam soil for three years maintained the retention of mineral nitrogen in the rhizosphere through reducing N_2O emissions and NO_3^- leaching [78]. Moreover, the combined application of biochar with nitrification inhibitors shows a high potential to reduce N_2O emissions. In this regard, the application of manure biochar and nitrification inhibitor (nitrapyrin) reduced N_2O emissions from urea by about 45.2% in a 60-day laboratory incubation experiment [79].

In contrast, other reports have suggested an adverse effect of biochar on increasing N_2O emissions. For example, application of peanut shell biochar (pyrolyzed at 550 °C) along with nitrogen fertilizers to a grazing grassland (sandy loam in texture) in Queensland, Australia, increased N_2O emissions compared to the unamended treatment (without biochar). This is mainly due to the increment of soil pH, which caused an abundance of narG, nirS, and AOB genes in the soil [80]. In this regard, the volatile matter of biochar has a key role in controlling the capacity of biochar for reducing N_2O emissions since volatile matter content acts as a decomposable source of organic carbon for the denitrifying organisms [81]. Fresh biochar application was also responsible for increasing N_2O emissions given the stimulation of the AOB-*amoA* gene abundance through autotrophic nitrification and denitrification [82].

3.3. Effect of Biochar Application on Reducing CH_4 Emissions from Rice Basins

Rice is the second largest cereal crop grown worldwide, representing the staple diet of two-thirds of the human population. A large amount (about 95%) of the world's rice yield is produced in developing countries, since it is considered an important source of employment and high income in rural areas. Among all agricultural ecosystems, paddy rice basins are one of the major sources of CH_4 emissions. The annual CH_4 emissions from paddy rice basins range between 31 and 112 Tg per year, which contributes about 5–19% of total greenhouse gas emissions [83]. Due to intensive wet rice farming all over the world, a tremendous increase in CH_4 emissions have been recorded by about 1.2 Tg/decade between 1961 and 2016 [84]. CH_4 is generated during the decomposition of organic matter by the aid of methanogenic archaea (methanogens) [85]. Although the contribution of CH_4 to total GHG emissions is not significant, it has a 25-fold higher global warming potential than CO_2 [86]. In addition, under waterlogging paddy ecosystems, the denitrification process is always very active and tends to convert nitrate to nitrous oxide [87].

The contribution of biochar to the reduction of CH_4 emissions has been highlighted in the literature. Analysis of recent literature showed moderate removal efficiency (40.49%) of CH_4 by biochar application to rice basins (Figure 3). The motivating effect of biochar application on activities of Acetyl-CoA synthase and β-glucosidase involved in carbon fixation reduced coenzyme activities related to methanogens [88]. Sriphirom et al. illustrated that 500 °C *Rhizophora apiculata* biochar application up to 4% led to a reduction in CH_4 emissions (9–21%) due to the abundance of electron donors (organic acids) and acceptors (NO_3^-, SO_4^{2-}, and Fe^{3+}), which accelerate redumethanogenesis reduction [89]. The beneficial effect of biochar in improving soil aeration and the readiness of O_2 supplies may inhibit methanogenesis [90]. In Wang et al.'s study, application of straw-derived biochar at 24 and 48 Mg ha^{-1} mitigated CH_4 emissions by 20–51% through inhibiting the abundances of some methanogen populations (e.g., *Methanosaeta* and *Methanoregula*) [84]. Further study demonstrated that amending paddy soil with 550 °C biochar derived from *Rosa anemoniflora* branches inhibited the emission of CH_4 by motivating an abundance of mcrA and a high ratio of pmoA/mcrA [91]. The large surface area of biochar might also favor the electron transfer between bacteria and Fe minerals, thereby motivating the domination of Fe-reducing bacteria that discourage methanogens and inhibit CH_4 emissions [92]. The porous nature of biochar might promote CH_4 oxidation after adsorption into the abundant pores [93]. The biochar aging process might also facilitate the interaction with soil organic matter, Fe/Al oxides, and silicon, thereby forming coating layers on the internal and external surfaces of biochar [94].

The long-lasting effect of biochar application showed a high effectiveness for reducing CH_4 emissions. In the study of Nan et al., biochar derived from rice straw reduced the emissions of CH_4 over three successive years by about 43, 31, and 30%, and increased rice productivity by 8.0%, 1.6%, and 7.3%, respectively [95]. This long–term retardation of CH_4 emissions highlights the safeguarding effect of biochar as a stable and suitable microenvironment for carbon sequestration in rice basins. According to Wang et al., biochar application (24–48 Mg ha^{-1}) into rice paddies inhibited CH_4 emissions by 20–51% over four years of rice cropping due to the aerating effect of biochar, which enhanced the abundance of methanotrophic bacteria and decreased the abundance ratio of methanogens to methanotrophs [96].

In contrast, some reports have pointed to an increment of cumulative CH_4 emissions due to the abundance of methanogenic and methanotrophic genes in soil following biochar application [97]. The safeguarding effect of biochar against CH_4 emissions depends upon the pyrolysis technique and its temperature. Biochar showed higher efficacy for methane oxidation as compared with hydrochar, with higher pyrolysis temperature being superior. According to Liu et al., hydrochar application suppressed the growth of *Bacillus*, *Methylocystis*, and *Methylobacter*; however, biochar motivated an abundance of methane-oxidizing bacteria (methanotrophs) such as *Methylobacter* and *Methylocystis* [98].

4. Factors Affecting the Efficiency of Biochar in Mitigating GHG Emissions

4.1. Effect of Feedstock Type

The efficiency of biochar in reducing GHGs from soil depends mainly upon the inherent components of the feedstock. From recent published data (147 observations comprising 34 investigations), it is found that the average values of biochar efficiency in mitigating GHGs were 9.81, 6.68, and 20.37% with feedstocks of agricultural residuals, woody materials, and manures, respectively (Figure 4).

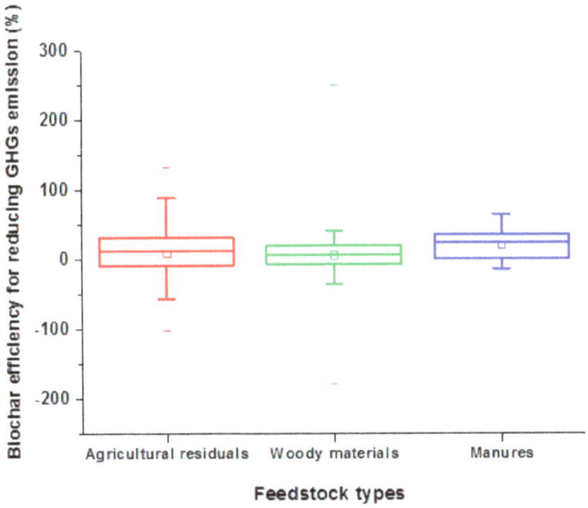

Figure 4. Effect of feedstock type on biochar efficiency in reducing GHG emissions from soil. Data are extracted from 34 investigations comprising 147 individual observations. Box chart is illustrated by the mean (dot), median (centerline), lower and upper quartiles (the lower and upper borders of the box, respectively), and whiskers-error bars (the minimum and maximum observations).

In other global meta-analyses, the magnitudes of N_2O reduction index of biochars were feedstock-dependent: bamboo (31.9%) > field crop straw (27.1%) > manures (27.0%) > hardwood (18.1%) > field crop husks (0.47%) [99]. Contradictory results, however, were reported by other investigations since the high nitrogen content in manure-derived biochars favors its higher N_2O emission footprint relative to other woody/herbaceous-derived biochars [100]. Another study showed that plant-derived biochars showed higher values of aromatic carbon with high stability and resistance against microbial decay than other manure-derived biochars [101]. In a further study, there were no significant differences between manure and sawdust biochar (2.4 kg m^{-2}) for inhibiting emissions of N_2O and CH_4 from soil [102]. It was also noticed that soils amended with biochars with a small C:N ratio exhibited higher CO_2 efflux than those amended with other biochars with a large C:N ratio [103].

4.2. Effect of Pyrolysis Temperature

Pyrolysis temperature plays a pivotal role in regulating GHG emissions following biochar soil application. The slow pyrolysis technique involves using low temperature, low heating rates, and high residence times to generate a high yield of high-quality biochar [104]. However, other pyrolysis types (fast and flash pyrolysis) generate low biochar yield with low surface functionality [105]. Specifically, the low rate of heating (24 °C min^{-1}) can form biochars with high aromaticity relative to the heating rate (62 °C min^{-1}) [106]. In terms of biochar stability, a higher mineralization rate was noticed with biochars produced under shorter residence time of pyrolysis compared to slow-pyrolyzed biochars, which exhibit less mobile organic matter and high resistance against microbial degradation [107]. The carrier gas (e.g., N_2, CO_2, or Ar) pointedly impact the yield, active functional groups, and volatile organic carbon content of biochar [108]. Moreover, the reactor design significantly affects the physicochemical properties of biochar. In this regard, a microwave pyrolysis reactor produces highly stable biochar relative to fixed/fluidized bed, rotating cone, screw feeder/auger, and vacuum pyrolizers [109]

Data extracted from recent literature illustrated that pyrolysis temperature was crucial for the performance of biochar in mitigating GHG emissions (Figure 5). Average values of inhibition efficiency of biochar-amended soils relative to the unamended ones are–60.71, 2.82, 25.42, and 7.86% with pyrolysis temperatures of 300–399, 400–499, 500–599, and 600–699 °C, respectively. Therefore, the utilization of low-pyrolyzed biochar in mitigating GHG emissions is not recommended given its high carbon footprint.

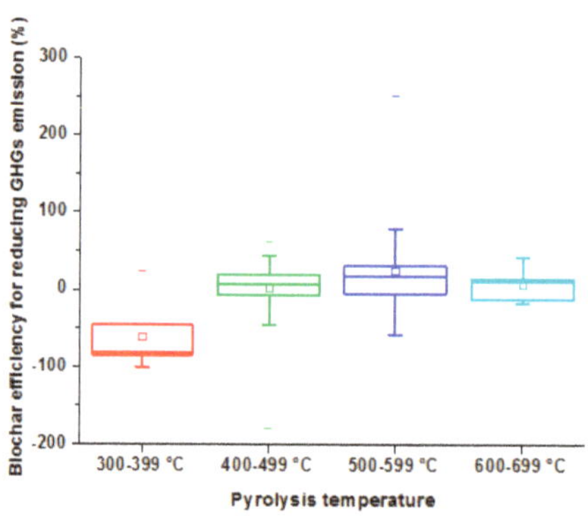

Figure 5. Effect of pyrolysis temperature on biochar efficiency in reducing GHG emissions from soil. Data are extracted from 26 investigations comprising 82 individual observations. Box chart is illustrated by the mean (dot), median (centerline), lower and upper quartiles (the lower and upper borders of the box, respectively), and whiskers-error bars (the minimum and maximum observations).

Based on a global meta-analysis, biochars produced at higher pyrolysis temperatures (>500 °C) exhibited higher potentials in reducing GHG emissions due to: (i) higher values of specific surface area, ash content, and polycondensed moieties; (ii) lower values of dissolved organic carbon, aliphatic compounds, and total surface charge; and (iii) suppressing the activity of soil microorganisms [110]. To summarize, low pyrolysis temperatures generate biochars with high volatile matter contents, low aromaticity, and high O:C ratio that are less stable than those generated at high pyrolysis temperatures [111,112]. In the study of Yang et al., 300 °C maize straw biochar increased CO_2 emissions over the control treatment by about 46% owing to the increment of dissolved organic matter following the stimulation of copiotrophic bacteria (*Proteobacteria*) and the inhibition of oligotrophic bacteria (*Acidobacteriota*) [84]. In their study, they also reported that increasing pyrolysis temperature up to 450 and 600 °C reduced CO_2 emissions by about 10.5 and 14.0%, respectively, due to a subsequent decline in dissolved organic matter following biochar application. In Spain, biochar application (pig manure at 300 °C) to sandy loam soil resulted in a positive impact on dehydrogenase enzyme activity; however, the pyrolysis temperature of 500 °C did not show positive impacts on the activities of soil enzymes [59]. This modulating effect of biochar on soil enzymes was significantly correlated with CO_2 emissions by soil. According to Al-Rabaiai et al., high amounts of water-soluble organic compounds in biochars derived at low pyrolysis temperature might have a priming effect on stimulating microbial activities and soil respiration [85]. In contrast, spent-mushroom-substrate-derived biochar applied to moso bamboo forest soil at the rate of 50 g kg^{-1} caused a considerable increase in CO_2 emissions by about 73, 43, and 16.6% with pyrolysis temperatures of 300, 450, and 600 °C, respectively [113].

Likewise, the performance of biochar in mitigating N_2O emissions declined sharply at low pyrolysis temperatures owing to the smaller surface area and the low aromaticity of the produced biochar [114]. In addition, the extremely high pyrolysis temperature (900 °C) led to an increase in N_2O emissions from soil amended with walnut shell biochar due to favoring the denitrification process [115]. It was also reported that a high pyrolysis temperature of biochar (700–900 °C) can increase the cumulative emission of CH_4 fluxes, given the formation of biochars with highly condensed aromatic graphite structures, motivating electronic transfer ability of methanogens and the inherent salts in biochar additives [116,117]. In an incubation experiment, CH_4 generation from a paddy soil following straw biochar application was ranked as follows: biochar at 300 °C > biochar at 500 °C > biochar at 700 °C [118].

4.3. *Effect of Application Rate*

Restraining GHG emissions by biochar is an application-dependent strategy. Biochar often applied through broadcasting and mixing with soil matrix by tillage methods [58]. However, this application method is responsible for wind loss of biochar by about 25% of the applied amount [119]. In addition, biochar is frequently applied to a trench as furrow application after crop planting with a lower amount of application and minimal soil disturbance [120]. On the other hand, biochar could be indirectly applied to the soil by mixing with several soil amendments (e.g., lime, compost, and manure) [121,122].

A subsequent initial flush of CO_2 is emitted from soil following biochar application, which declines sharply with the recalcitrant aged biochar [123]. The inhibitory effect of biochar application to CO_2 emissions can be associated with the sorption of rhizodeposits and enzymes onto biochar, which lead to a reduction in carbon-degrading microbial activity in soil [124]. Numerous attempts have been made to specify the optimum rate of biochar soil application; however, a great deal of uncertainty remains surrounding the appropriate rate for each soil type. The extracted data from 31 investigations comprising 146 individual observations illustrate the average removal efficiency values of GHGs from biochar-amended soils relative to the unamended ones. Biochar efficiency in reducing GHG emissions was, on average, 15.91, 6.13, −2.13, and −4.45% with application rates of 1–10, 11–20, 21–40, and >40 Mg ha^{-1} (Figure 6). According to the obtained results, there seems to be a consensus that the low rate of biochar (up to 10 Mg ha^{-1}) is more preferable to achieve net carbon neutrality. However, a high application rate of biochar might increase the cumulative GHG emissions.

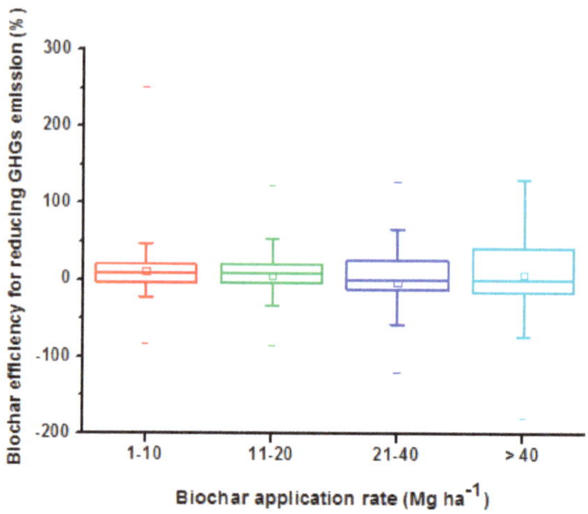

Figure 6. Effect of biochar application rate on efficiency in reducing GHG emissions. Data are extracted from 26 investigations comprising 82 individual observations. Box chart is illustrated by the mean (dot), median (centerline), lower and upper quartiles (the lower and upper borders of the box, respectively), and whiskers-error bars (the minimum and maximum observations).

Former studies have monitored the effect of the biochar application rate on the efficacy of regulating GHG emissions. Under upland rice production, the highest application rate of 350 °C rice-husk-derived biochar (25 Mg ha^{-1}) showed the highest CO_2 emissions (3.06 CO_2–C g/m^2) relative to the one-fifth application rate (2.78 CO_2–C g/m^2) [125]. Similarly, a low level of biochar (5 Mg ha^{-1} of 500 °C bamboo branches) could be more effective than the high application dosage (20 Mg ha^{-1}) in reducing CO_2 emissions and improving soil carbon sequestration in forest soils [126]. Another investigation revealed that increasing the application rate of bamboo leaf biochar did not show substantial alterations in CO_2 emission [127].

As mentioned earlier, the optimum application rate of biochar might change according to the soil type. For instance, soil biochar application (20 Mg ha^{-1} 650 °C) to deciduous mixed forest led to a substantial decline in CO_2 emissions; however, this emission rate was not significantly changed in a long-term fertilized apple orchard [128]. Other reports suggested raising the application rate of biochar to offset the cumulative GHG emissions. In the study of Shen et al., increasing the application rate of 450 °C maize straw biochar from 10 up to 30 Mg ha^{-1} was associated with a progressive reduction in cumulative CO_2 emissions (from 3.9 to 11.8%) due to the sorption of labile carbon onto internal and external surfaces of biochar, thus suppressing the rate of soil respiration [129].

The application rate also affects the performance of biochar in regulating GHG emissions from different soil layers. For instance, cumulative N_2O emissions from the topsoil (0–5 cm) following the application of pruning waste biochar (@ 600 °C) at rates of 2 and 10% declined N_2O emissions by 12.5% and 26.3%, respectively. However, the safeguard effect of pruning waste biochar in reducing N_2O emissions from a soil layer of 0–10 and the rhizospheric layer was only observed with the rate of 10% (15.1 and 13.8%, respectively) [130]. Biochar application as film-mulching has recently been investigated on cropping systems grown under drip irrigation. In view of that, corn-residue-derived biochar (produced under pyrolysis at 400–500 °C) was applied as a film mulch to drip-irrigated corn grown in sandy loam soil in Inner Mongolia, China. Results showed that increasing the application rate from 15 up to 45 Mg ha^{-1} as a film mulch system was associated with significant reductions in GHGs over two growing seasons: CO_2 (19 –33%), CH_4 (124–132%), and N_2O (55–79%) [49].

5. Effect of Biochar on Reducing GHG Emissions during Composting

Composting is a process in which organic wastes are transformed via complicated biochemical reactions into recalcitrant organic products (humic substances in particular) that can serve as fertilizers to sustain soil fertility and productivity [131]. However, the composting process is responsible for emitting substantial amounts of GHGs that raised concerns from ecological point of view. In particular, CH_4 accounts for about 6% of the total carbon loss during the composting process, and the emission of N_2O accounts for approximately 3.8% of the total nitrogen losses [132]. Furthermore, the release of ammonia (NH_3) during composting not only harms the ecosphere but also declines the agricultural revenues from compost additives. Related research investigations revealed that animal husbandry is the main source of agricultural non-CO_2 emissions, accounting for about 37 and 65% of CH_4 and N_2O, respectively (12% of the anthropogenic GHG emissions globally) [133]. The emission of GHGs is more pronounced in the initial phase of the composting process (the thermophilic phase), in which the temperature reaches about 70 °C; however, this emission declines dramatically in the ultimate maturation phase (the mesophilic phase), when the temperature drops to 40–50 °C [134].

The applicability of biochar additives to reducing GHG emissions during composting is illustrated in Table 3. The high functionality of biochar has attracted research attention for utilization as a supplemental additive during the composting process to accelerate the startup of decomposition, shorten the period of composting, and reduce the amounts of GHG emissions [135]. Furthermore, biochar might reduce the mobility of water-soluble organic substances and avoid their losses during the composting process [136]. In view of this, biochar application during composting might inhibit the abundance of the nirK gene in denitrifying bacteria, which causes a significant reduction of N_2O emissions during composting [137]. In the study of Guo et al., bamboo charcoal application during composting led to minimizing NH_3 emissions following the active nitrification by *Nitromonas* [138]. In another study to evaluate the effect of 5% biochar application during pig manure composting, it was reported that biochar could shorten the composting period and reduce emissions of CO_2, CH_4, N_2O, and NH_3 by about 35.9, 15.4, 19.9, and 18.8%, respectively [139]. However, in another study, chicken manure biochar application (up to 10%) during chicken manure composting declined the release of GHGs: N_2O (19.0–27.4%), CH_4 (9.3–55.9%), and NH_3 (24.2–56.9%) [132]. Likewise, the co-application of bamboo biochar with poultry manure during composting (up to 10% w:w) reduced CO_2 and NH_3 losses by about 542–149% and 48–11%, respectively [140].

The composting of sewage sludge is a green technology for reducing the negative impacts associated with its ecological hazards [141]. However, a great deal of uncertainty arises surrounding its high GHG footprint [142]. To address this environmental constraint, Awasthi et al. [143] added rice straw biochar at a high rate (8–18%) during sewage sludge composting and recorded remarkable declines in GHG emissions: CH_4 (93–95%), NH_3 (58–65%), and N_2O (95–97%). In yet another investigation, bamboo biochar application along with bacterial agents during sewage sludge composting mitigated CH_4 and N_2O emissions (45.7% and 3.7%, respectively) due to the beneficial effect of biochar on filling the space between compost particles, thereby minimizing the potential heat losses and motivating microbial activity and consequent heat production [144]. Vermicomposting is a benign and modern eco-friendly technique for addressing the vast accumulation of organic wastes, although the incorporation of earthworms in these organic fertilizers showed higher emissions of N_2O relative to other traditional composts [145]. According to Wu et al., the incorporation of 500 °C rice straw biochar with vermicompost significantly reduced cumulative N_2O emissions (~19%) [146].

The application of biochar as a ruminant diet showed a significant impact on improving the digestion performance of animals, reducing enteric CH_4 and increasing the value of post-excretion biochar–manure mixture. In a recent study, the effect of pine-based biochar application on cattle diet increased stockpile/compost aromaticity with a high content of more humic-like organic matter precursors [147].

Table 3. Effect of biochar application on mitigating GHG emissions during composting.

Feedstock	Biochar Characteristics			Composting Materials	Compost Characteristics		GHG Emissions Reduction Effect (% Compared with Control)				References
	Pyrolysis Temperature (°C)	Biochar Application (% *w*/*w*)	Particle Size (mm)		Bulk Density (g cm^{-3})	C/N Ratio	CO_2	CH_4	N_2O	NH_3	
Hardwood+ softwood	500–700	27.4	≤16	Hen manure + barley straw	0.49	17.4–13.7	21.5–22.9	77.9–83.6	35.3–43.0	35.3–43.0	[148]
Woody material	-	4	-	Sewage sludge + woodchips	-	22–25	-	-	-	8.5–9.2	[149]
Holm oak	650	10	-	Green waste + municipal solid waste	-	27.5–16.2	52.9	95.1	14.2	-	[150]
Green waste + poultry litter	550	10	-	Poultry litter + sugarcane straw	-	-	-	77.8–83.3	68.2–74.9	54.9 –60.2	[151]
Wheat straw	500–600	2–18	2–5	Sewage sludge + wheat straw	0.50	25.0	-	92.8–95.3	95.1–97.3	58.0–65.2	[143]
Bamboo	-	5	2–3	Pig manure + sawdust	0.50	-	-	54.4	36.1	12.4	[152]
Chicken manure + wheat straw	550–600	2–10	-	Chicken manure	0.50	16.8–14.2	-	24.4–63.4	6.8–16.9	22.9–50.5	[132]
Bamboo	-	2–10	-	Poultry manure + wheat straw	0.50	24.3–18.9	5.5–72.6	12.5–72.9	12.4–81.6	19.0–77.4	[140]
Cornstalk	450–500	10.0	≤2.0	Hen manure, sawdust	0.42	-	-	-	-	12.4	[153]
Waste wood pellets	520	10 15	-	Chicken mortalities	-	20.3 20.0	- -	- -	- -	40.0 56.8	[154]

6. Designer/Functionalized Biochar for Efficient Retardation of GHG Emissions from Soil and Terrestrial Ecosystems

The scientific community is interested in maximizing the functionality of biochar so that it can be tailored to a variety of agro-environmental applications [11–13,155]. In a meta-analysis study, acidic oxidation was the most efficient method for enhancing the physicochemical properties of biochar (specific surface area, micro-pores, oxygen-containing functional groups, and cation exchange capacity) relative to other oxidation methods (physical, alkaline, metal oxide, and natural oxidation methods) [156]. Recently, designer biochar functionalization (specialization) has attracted increasing attention in recent research to lessen GHG emissions from agricultural ecosystems. Several reports highlighted the activation of pristine biochar (raw biochar) via physical and chemical modification methods to improve the functionality of biochar (increase its surface area and porous structure) for the sorption of GHGs. However, most of these reports are still under lab-scale experimentations. In addition, most of these reports are focused on capturing CO_2 emissions. Data extracted from published reports (97 values) showed that the average values of CO_2 sorption with pristine, physically activated, and chemically activated biochars were 37.8, 56.5, and 59.4 mg g^{-1}, respectively (Figure 7).

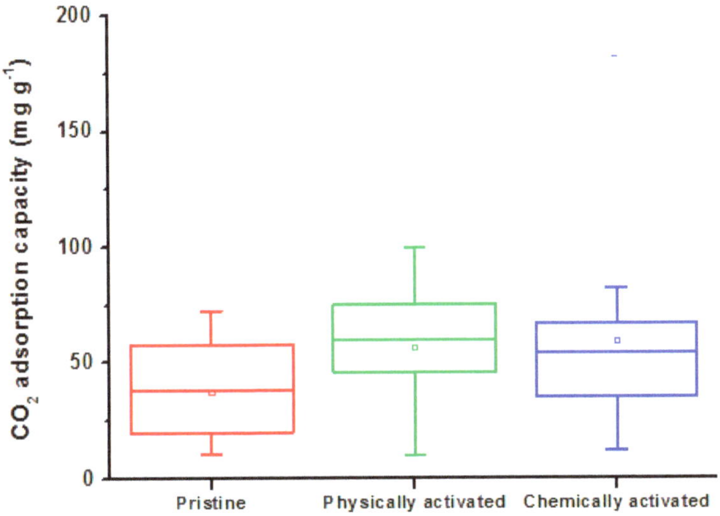

Figure 7. Effect of activation methods on sorption capacity of CO_2 (mg g^{-1}) with pristine and designer biochars. Data are extracted from 97 individual observations. Box chart is illustrated by the mean (dot), median (centerline), lower and upper quartiles (the lower and upper borders of the box, respectively), and whiskers-error bars (the minimum and maximum observations).

A wide range of researchers have highlighted physical modification as the preferable activation method to maximize the functionality of biochar relative to chemical modifications methods [157,158]. Proponents of physical modification point to the high risk of deteriorating the carbonaceous lattice following chemical modifications and the potential to block the pore structures of the biochar matrix, which might reduce gases sorptivity [159]. Steam and high-temperature gas activation have been attempted to improve the porosity of biochar; however, the NH_3 activation not only improved the pore structure of activated biochar, but also introduced active functional groups onto the carbonaceous lattice [160]. Physical treatment with NH_3 can also increase the alkaline nature of biochar and the base–acid interaction between CO_2 and the originated nitrogen-containing functional groups [161]. Furthermore, NH_3 treatment grafted pyrrolic–N groups onto the biochar matrix, which facilitated the H–bonding interaction between CO_2 molecules, and the proton of pyrrolic–N [162].

Opponents of physical activation methods, however, point toward the higher energy consumption, longer activation time, and higher activation temperature. Several chemical activation methods have been reviewed for the functionalization of biochar for capturing GHGs. Among them, alkali and acid-modified biochars have shown high sorptivity with low cost and simple processing. Alkali treatment of wood pellet biochar maximized the capture capacity of CO_2 by about five folds relative to untreated biochar (50.73 vs. 10.45 mg g^{-1}) due to improving surface area, porosity, and abundance of active functional groups on the biochar matrix [163]. High CO_2 capture (160 mg g^{-1}) was further recorded by 350 °C pine cone biochar activated with KOH [164]. In yet another study, high capacities of CO_2 retardation (136.7–182.0 mg g^{-1}) were recorded by KOH-activated biochars derived from pine sawdust and sewage sludge mixture as compared with plain biochar (35.5–42.9 mg g^{-1}) given the formation of tunable porous features in biochar matrix [165].

In their studies on rice plant, Shin and coworkers [166] reported that activated biochar (alkali-treated rice hull pellets)-doped mineral fertilizer (40% N) reduced cumulative CO_2 and N_2O emissions by about 10 and 0.003 kg ha^{-1}, respectively, compared to the control treatment, with a negligible effect on the emission of CH_4. In another experiment, 700 °C rice husk biochar was acid modified with H_3PO_4 and further combined with nano-zero-valent iron (nZVI) to enhance its sorption capacity for GHGs. The functionalized biochar form reduced CO_2 and N_2O emissions; however, CH_4 emission showed a noticeable increase [167].

The doping of heteroatoms (e.g., Mg, N, S, and Ni) into the carbonaceous lattice of biochar exhibited promising values for capturing GHGs due to the electron-withdrawal effect. In view of this, Mg-doped rice straw biochar application (at 9%) to a dryland soil in Hunan Province, China, showed a minimal effect on CH_4 emission but reduced cumulative emissions of CO_2 (9%) and N_2O (32%) as compared to the control treatment [168]. Nitrogen-doped biochar (mixture of rice straw and waste wood pyrolyzed at 600 °C and applied at 0–8 Mg ha^{-1}) was further studied in rice cropping soils under a short-term study [169]. Compared with the control treatment, nitrogen-doped biochar application increased CO_2 emissions and reduced emissions of CH_4 with the application rate of 8 Mg ha^{-1}.

Iron species receive high attention for biochar specialization in several agro-environmental applications. As a result of its effect on denitrification functional genes (nirk, narG, nirS, and nosZ), methanogenesis (mcrA), and methanotrophs (pmoA), modified biochar (derived from reed, walnut, saw dust, and sludge) supported by nano-zero-valent iron (nZVI) was able to reduce N_2O, CO_2, and CH_4 emissions [170]. Biochar (600 °C *Camellia oleifera* fruit shell) modified with $Fe(NO_3)_3$/KOH showed a high efficacy in N_2O retardation with an increment of about 8.6% over pristine biochar [171]. Developing enriched biochar derivatives with higher functionality (organo-mineral complexes, surface area, exchanging capacity of ions and dissolved organic carbons) has been proposed to enhance the sequestration of GHGs. The carbon footprint of either conventional rice straw biochar or enriched biochar with lime, clay, ash, and manure was compared under paddy field conditions [172]. They indicated that the difference in the carbon footprint between biochar types is mainly associated with variations in CH_4 emissions among plain and functionalized biochars.

7. Conclusions and Future Prospects

In recent years, lowering the rise in emissions of greenhouse gases has become one of the issues that requires global attention. In this work, we performed a review of the recent literature on biochar, as a tool to lessen the impact of emissions and mitigate their negative consequences. Overall, this review assesses the different methods of biochar production and their effectiveness. This review primarily summarizes the biochemical processes occurring in the charosphere. In addition, recent developments in our understanding of how to activate biochar for maximum effectiveness in achieving carbon neutrality goals are covered. To conclude:

- Biochar and its derivatives have been shown to reduce emissions of CO_2, N_2O, and CH_4 from soil and organic manures. Recent field-scale studies have found that biochar has the potential to reduce emissions of CO_2, N_2O, and CH_4 on the order of −0.430, 24.681, and 40.486%, respectively.
- While biochar showed promising results in reducing greenhouse gas emissions when tested in controlled laboratory settings, studies conducted on a larger scale have shown either no effect or even a negative one on efforts to lower GHG emissions.
- Notably, biochar's effectiveness in cutting greenhouse gas emissions is proportional to its application rate, pyrolysis temperature, and the type of feedstock used to make it.
- To produce effective amendments with a high capacity for restraining GHG production and enhancing soil carbon sink, it is recommended that manure-derived biochars be pyrolyzed between 500 and 600 °C, and applied at a rate of less than 10 Mg ha^{-1}.
- Overall, biochar can be seen as a highly effective and relatively simple tool for reversing the upward trend in greenhouse gas emissions.
- Since carbon and nitrogen transformation processes are microbially dependent, future research should be directed toward (i) investigating other indirect factors related to soil physicochemical characters (such as soil pH/Eh, colloidal and $CaCO_3$ contents) that may control the functionality of biochar, (ii) fabricating aged biochars with low carbon and nitrogen footprints, and (iii) functionalizing biologically activated biochars to suppress CO_2, CH_4, and N_2O emissions.

Author Contributions: Conceptualization, A.M. and M.M.M.; methodology, A.M. and E.S.; software, A.E.-G. and M.E.A.; validation, A.M., M.M.M. and A.M.E.K.; formal analysis, A.M. and E.S.; investigation, A.M. and M.M.M.; resources, M.E.A. and A.M.E.K.; data curation, A.M. and E.S.; writing—original draft preparation, A.M., M.M.M. and E.S.; writing—review and editing, A.M., M.E.A. and A.M.E.K.; visualization, A.E.-G. and E.S.; supervision, A.M. and A.E.-G.; project administration, A.M. and A.M.E.K.; funding acquisition, A.M.E.K. and M.E.A. All authors have read and agreed to the published version of the manuscript.

Funding: This research received no external funding.

Institutional Review Board Statement: We choose to exclude this statement because our study did not involve humans or animals.

Informed Consent Statement: We choose to exclude this statement because our study did not involve humans or animals.

Data Availability Statement: Not applicable.

Acknowledgments: The authors would like to acknowledge Mansoura University for supporting the publication fees of this manuscript.

Conflicts of Interest: The authors declare no conflict of interest.

References

1. Thakur, I.S.; Kumar, M.; Varjani, S.J.; Wu, Y.; Gnansounou, E.; Ravindran, S. Sequestration and utilization of carbon dioxide by chemical and biological methods for biofuels and biomaterials by chemoautotrophs: Opportunities and challenges. *Bioresour. Technol.* **2018**, *256*, 478–490. [CrossRef] [PubMed]
2. Chen, M.; Cui, Y.; Jiang, S.; Forsell, N. Toward carbon neutrality before 2060: Trajectory and technical mitigation potential of non-CO_2 greenhouse gas emissions from Chinese agriculture. *J. Clean. Prod.* **2022**, *368*, 133186. [CrossRef]
3. IPCC. *Climate Change 2013: The Physical Science Basis. Contribution of Working Group I to the Fifth Assessment Report of the Intergovernmental Panel on Climate Change*; Cambridge University Press: Cambridge, UK, 2013; p. 1535.
4. Maraseni, T.N.; Qu, J. An international comparison of agricultural nitrous oxide emissions. *J. Clean. Prod.* **2016**, *135*, 1256–1266. [CrossRef]
5. Filho, W.L.; Setti, A.F.F.; Azeiteiro, U.M.; Lokupitiya, E.; Donkor, F.K.; Etim, N.N.; Matandirotya, N.; Olooto, F.M.; Sharifi, A.; Nagy, G.J.; et al. An overview of the interactions between food production and climate change. *Sci. Total Environ.* **2022**, *838*, 156438. [CrossRef] [PubMed]
6. Moussa, R.R. Reducing carbon emissions in Egyptian roads through improving the streets quality. *Environ. Dev. Sustain.* **2022**, 1–22. [CrossRef]

7. Alshaal, T.; El-Ramady, H.; Al-Saeedi, A.H.; Shalaby, T.; Elsakhawy, T.; Omara, A.E.D.; Gad, A.; Hamad, E.; El-Ghamry, A.; Mosa, A. The rhizosphere and plant nutrition under climate change. In *Essential Plant Nutrients*; Springer: Berlin/Heidelberg, Germany, 2017; pp. 275–308.
8. Mosa, A.; Taha, A.A.; Elsaeid, M. In-situ and ex-situ remediation of potentially toxic elements by humic acid extracted from different feedstocks: Experimental observations on a contaminated soil subjected to long-term irrigation with sewage effluents. *Environ. Technol. Innov.* **2021**, *23*, 101599. [CrossRef]
9. Frank, S.; Havlík, P.; Soussana, J.-F.; Levesque, A.; Valin, H.; Wollenberg, E.; Kleinwechter, U.; Fricko, O.; Gusti, M.; Herrero, M. Reducing greenhouse gas emissions in agriculture without compromising food security? *Environ. Res. Lett.* **2017**, *12*, 105004. [CrossRef]
10. Laird, D.A.; Brown, R.C.; Amonette, J.E.; Lehmann, J. Review of the pyrolysis platform for coproducing bio-oil and biochar. *Biofuels Bioprod. Biorefining* **2009**, *3*, 547–562. [CrossRef]
11. El-Naggar, A.; Mosa, A.; Ahmed, N.; Niazi, N.K.; Yousaf, B.; Sarkar, B.; Rinklebe, J.; Cai, Y.; Chang, S.X. Modified and pristine biochars for remediation of chromium contamination in soil and aquatic systems. *Chemosphere* **2022**, *303*, 134942. [CrossRef]
12. Shaheen, S.M.; Mosa, A.; El-Naggar, A.; Faysal Hossain, M.; Abdelrahman, H.; Khan Niazi, N.; Shahid, M.; Zhang, T.; Tsang, Y.F.; Trakal, L.; et al. Manganese oxide-modified biochar: Production, characterization and applications for the removal of pollutants from aqueous environments—A review. *Bioresour. Technol.* **2022**, *346*, 126581. [CrossRef]
13. El-Naggar, A.; Mosa, A.; Amin, A.E.-E.; Yang, X.; Yousaf, B.; El-Naggar, A.H.; Cai, Y.; Chang, S.X. Chapter 18—Biochar for remediation of alkaline soils contaminated with toxic elements. In *Biochar in Agriculture for Achieving Sustainable Development Goals*; Tsang, D.C.W., Ok, Y.S., Eds.; Academic Press: Cambridge, MA, USA, 2022; pp. 223–240. [CrossRef]
14. Tsang, D.C.W.; Ok, Y.S. *Biochar in Agriculture for Achieving Sustainable Development Goals*; Elsevier: Amsterdam, The Netherlands, 2022.
15. Danesh, P.; Niaparast, P.; Ghorbannezhad, P.; Ali, I. Biochar Production: Recent Developments, Applications, and Challenges. *Fuel* **2022**. [CrossRef]
16. Li, Y.; Gupta, R.; Zhang, Q.; You, S. Review of biochar production via crop residue pyrolysis: Development and perspectives. *Bioresour. Technol.* **2023**, *369*, 128423. [CrossRef] [PubMed]
17. Gabhane, J.W.; Bhange, V.P.; Patil, P.D.; Bankar, S.T.; Kumar, S. Recent trends in biochar production methods and its application as a soil health conditioner: A review. *SN Appl. Sci.* **2020**, *2*, 1–21. [CrossRef]
18. Oh, S.; Lee, J.; Lam, S.S.; Kwon, E.E.; Ha, J.-M.; Tsang, D.C.W.; Ok, Y.S.; Chen, W.-H.; Park, Y.-K. Fast hydropyrolysis of biomass Conversion: A comparative review. *Bioresour. Technol.* **2021**, *342*, 126067. [CrossRef] [PubMed]
19. Ethaib, S.; Omar, R.; Kamal, S.M.M.; Awang Biak, D.R.; Zubaidi, S.L. Microwave-assisted pyrolysis of biomass waste: A mini review. *Processes* **2020**, *8*, 1190. [CrossRef]
20. Wang, L.; Ok, Y.S.; Tsang, D.C.W.; Alessi, D.S.; Rinklebe, J.; Mašek, O.; Bolan, N.S.; Hou, D. Biochar composites: Emerging trends, field successes and sustainability implications. *Soil Use Manag.* **2022**, *38*, 14–38. [CrossRef]
21. Shaheen, S.M.; Mosa, A.; Jeyasundar, A.; Soundari, P.G.; Hassan, N.E.E.; Yang, X.; Antoniadis, V.; Li, R.; Wang, J.; Zhang, T. Pros and Cons of Biochar to Soil Potentially Toxic Element Mobilization and Phytoavailability: Environmental Implications. *Earth Syst. Environ.* **2022**, 1–25. [CrossRef]
22. Mosa, A.; El-Ghamry, A.; Tolba, M. Functionalized biochar derived from heavy metal rich feedstock: Phosphate recovery and reusing the exhausted biochar as an enriched soil amendment. *Chemosphere* **2018**, *198*, 351–363. [CrossRef]
23. Zou, R.; Qian, M.; Wang, C.; Mateo, W.; Wang, Y.; Dai, L.; Lin, X.; Zhao, Y.; Huo, E.; Wang, L. Biochar: From by-products of agro-industrial lignocellulosic waste to tailored carbon-based catalysts for biomass thermochemical conversions. *Chem. Eng. J.* **2022**, *441*, 135972. [CrossRef]
24. Lin, J.-C.; Mariuzza, D.; Volpe, M.; Fiori, L.; Ceylan, S.; Goldfarb, J.L. Integrated thermochemical conversion process for valorizing mixed agricultural and dairy waste to nutrient-enriched biochars and biofuels. *Bioresour. Technol.* **2021**, *328*, 124765. [CrossRef]
25. Abhishek, K.; Srivastava, A.; Vimal, V.; Gupta, A.K.; Bhujbal, S.K.; Biswas, J.K.; Singh, L.; Ghosh, P.; Pandey, A.; Sharma, P. Biochar application for greenhouse gas mitigation, contaminants immobilization and soil fertility enhancement: A state-of-the-art review. *Sci. Total Environ.* **2022**, *853*, 158562. [CrossRef] [PubMed]
26. Liu, W.-J.; Zeng, F.-X.; Jiang, H.; Zhang, X.-S. Preparation of high adsorption capacity bio-chars from waste biomass. *Bioresour. Technol.* **2011**, *102*, 8247–8252. [CrossRef] [PubMed]
27. Zhang, X.; Qian, H.; Hua, K.; Chen, H.; Deng, A.; Song, Z.; Zhang, J.; Raheem, A.; Danso, F.; Wang, D. Organic amendments increase crop yield while mitigating greenhouse gas emissions from the perspective of carbon fees in a soybean-wheat system. *Agric. Ecosyst. Environ.* **2022**, *325*, 107736. [CrossRef]
28. Wang, C.; Wang, W.; Sardans, J.; Singla, A.; Zeng, C.; Lai, D.Y.F.; Peñuelas, J. Effects of steel slag and biochar amendments on CO_2, CH_4, and N_2O flux, and rice productivity in a subtropical Chinese paddy field. *Environ. Geochem. Health* **2019**, *41*, 1419–1431. [CrossRef]
29. Malyan, S.K.; Bhatia, A.; Tomer, R.; Harit, R.C.; Jain, N.; Bhowmik, A.; Kaushik, R. Mitigation of yield-scaled greenhouse gas emissions from irrigated rice through Azolla, Blue-green algae, and plant growth–promoting bacteria. *Environ. Sci. Pollut. Res.* **2021**, *28*, 51425–51439. [CrossRef]
30. Sun, L.; Ma, Y.; Liu, Y.; Li, J.; Deng, J.; Rao, X.; Zhang, Y. The combined effects of nitrogen fertilizer, humic acid, and gypsum on yield-scaled greenhouse gas emissions from a coastal saline rice field. *Environ. Sci. Pollut. Res.* **2019**, *26*, 19502–19511. [CrossRef]

31. Guo, Y.; Ma, Z.; Ren, B.; Zhao, B.; Liu, P.; Zhang, J. Effects of Humic Acid Added to Controlled-Release Fertilizer on Summer Maize Yield, Nitrogen Use Efficiency and Greenhouse Gas Emission. *Agriculture* **2022**, *12*, 448. [CrossRef]
32. De Rosa, D.; Rowlings, D.W.; Biala, J.; Scheer, C.; Basso, B.; Grace, P.R. N_2O and CO_2 emissions following repeated application of organic and mineral N fertiliser from a vegetable crop rotation. *Sci. Total Environ.* **2018**, *637*, 813–824. [CrossRef]
33. Sohi, S.P.; Krull, E.; Lopez-Capel, E.; Bol, R. A review of biochar and its use and function in soil. *Adv. Agron.* **2010**, *105*, 47–82.
34. Leppäkoski, L.; Marttila, M.P.; Uusitalo, V.; Levänen, J.; Halonen, V.; Mikkilä, M.H. Assessing the carbon footprint of biochar from willow grown on marginal lands in Finland. *Sustainability* **2021**, *13*, 10097. [CrossRef]
35. Brewer, C.E.; Chuang, V.J.; Masiello, C.A.; Gonnermann, H.; Gao, X.; Dugan, B.; Driver, L.E.; Panzacchi, P.; Zygourakis, K.; Davies, C.A. New approaches to measuring biochar density and porosity. *Biomass Bioenergy* **2014**, *66*, 176–185. [CrossRef]
36. Lim, T.J.; Spokas, K.A.; Feyereisen, G.; Novak, J.M. Predicting the impact of biochar additions on soil hydraulic properties. *Chemosphere* **2016**, *142*, 136–144. [CrossRef]
37. Fan, R.; Zhang, B.; Li, J.; Zhang, Z.; Liang, A. Straw-derived biochar mitigates CO_2 emission through changes in soil pore structure in a wheat-rice rotation system. *Chemosphere* **2020**, *243*, 125329. [CrossRef] [PubMed]
38. Scheer, C.; Grace, P.R.; Rowlings, D.W.; Kimber, S.; van Zwieten, L. Effect of biochar amendment on the soil-atmosphere exchange of greenhouse gases from an intensive subtropical pasture in northern New South Wales, Australia. *Plant Soil* **2011**, *345*, 47–58. [CrossRef]
39. Lentz, R.D.; Ippolito, J.A.; Spokas, K.A. Biochar and manure effects on net nitrogen mineralization and greenhouse gas emissions from calcareous soil under corn. *Soil Sci. Soc. Am. J.* **2014**, *78*, 1641–1655. [CrossRef]
40. Zheng, M.; Chen, H.; Li, D.; Zhu, X.; Zhang, W.; Fu, S.; Mo, J. Biological nitrogen fixation and its response to nitrogen input in two mature tropical plantations with and without legume trees. *Biol. Fertil. Soils* **2016**, *52*, 665–674. [CrossRef]
41. Watanabe, A.; Ikeya, K.; Kanazaki, N.; Makabe, S.; Sugiura, Y.; Shibata, A. Five crop seasons' records of greenhouse gas fluxes from upland fields with repetitive applications of biochar and cattle manure. *J. Environ. Manag.* **2014**, *144*, 168–175. [CrossRef]
42. Keith, A.; Singh, B.; Dijkstra, F.A.; van Ogtrop, F. Biochar field study: Greenhouse gas emissions, productivity, and nutrients in two soils. *Agron. J.* **2016**, *108*, 1805–1815. [CrossRef]
43. Zhang, A.; Bian, R.; Li, L.; Wang, X.; Zhao, Y.; Hussain, Q.; Pan, G. Enhanced rice production but greatly reduced carbon emission following biochar amendment in a metal-polluted rice paddy. *Environ. Sci. Pollut. Res.* **2015**, *22*, 18977–18986. [CrossRef]
44. Fang, B.; Lee, X.; Zhang, J.; Li, Y.; Zhang, L.; Cheng, J.; Wang, B.; Cheng, H. Impacts of straw biochar additions on agricultural soil quality and greenhouse gas fluxes in karst area, Southwest China. *Soil Sci. Plant Nutr.* **2016**, *62*, 526–533. [CrossRef]
45. Hagemann, N.; Harter, J.; Kaldamukova, R.; Guzman-Bustamante, I.; Ruser, R.; Graeff, S.; Kappler, A.; Behrens, S. Does soil aging affect the N_2O mitigation potential of biochar? A combined microcosm and field study. *GCB Bioenergy* **2017**, *9*, 953–964. [CrossRef]
46. Abagandura, G.O.; Chintala, R.; Sandhu, S.S.; Kumar, S.; Schumacher, T.E. Effects of biochar and manure applications on soil carbon dioxide, methane, and nitrous oxide fluxes from two different soils. *J. Environ. Qual.* **2019**, *48*, 1664–1674. [CrossRef]
47. Kang, S.-W.; Kim, S.-H.; Park, J.-H.; Seo, D.-C.; Ok, Y.S.; Cho, J.-S. Effect of biochar derived from barley straw on soil physicochemical properties, crop growth, and nitrous oxide emission in an upland field in South Korea. *Environ. Sci. Pollut. Res.* **2018**, *25*, 25813–25821. [CrossRef]
48. Yang, S.; Xiao, Y.n.; Sun, X.; Ding, J.; Jiang, Z.; Xu, J. Biochar improved rice yield and mitigated CH_4 and N_2O emissions from paddy field under controlled irrigation in the Taihu Lake Region of China. *Atmos. Environ.* **2019**, *200*, 69–77. [CrossRef]
49. Yang, W.; Feng, G.; Miles, D.; Gao, L.; Jia, Y.; Li, C.; Qu, Z. Impact of biochar on greenhouse gas emissions and soil carbon sequestration in corn grown under drip irrigation with mulching. *Sci. Total Environ.* **2020**, *729*, 138752. [CrossRef]
50. Duan, M.; Wu, F.; Jia, Z.; Wang, S.; Cai, Y.; Chang, S.X. Wheat straw and its biochar differently affect soil properties and field-based greenhouse gas emission in a Chernozemic soil. *Biol. Fertil. Soils* **2020**, *56*, 1023–1036. [CrossRef]
51. Han, Z.; Lin, H.; Xu, P.; Li, Z.; Wang, J.; Zou, J. Impact of organic fertilizer substitution and biochar amendment on net greenhouse gas budget in a tea plantation. *Agric. Ecosyst. Environ.* **2022**, *326*, 107779. [CrossRef]
52. Hawthorne, I.; Johnson, M.S.; Jassal, R.S.; Black, T.A.; Grant, N.J.; Smukler, S.M. Application of biochar and nitrogen influences fluxes of CO_2, CH_4 and N_2O in a forest soil. *J. Environ. Manag.* **2017**, *192*, 203–214. [CrossRef]
53. Xu, L.; Fang, H.; Deng, X.; Ying, J.; Lv, W.; Shi, Y.; Zhou, G.; Zhou, Y. Biochar application increased ecosystem carbon sequestration capacity in a Moso bamboo forest. *For. Ecol. Manag.* **2020**, *475*, 118447. [CrossRef]
54. Ge, X.; Cao, Y.; Zhou, B.; Xiao, W.; Tian, X.; Li, M.-H. Combined application of biochar and N increased temperature sensitivity of soil respiration but still decreased the soil CO_2 emissions in moso bamboo plantations. *Sci. Total Environ.* **2020**, *730*, 139003. [CrossRef]
55. Li, X.; Yao, S.; Wang, Z.; Jiang, X.; Song, Y.; Chang, S.X. Polyethylene microplastic and biochar interactively affect the global warming potential of soil greenhouse gas emissions. *Environ. Pollut.* **2022**, 120433. [CrossRef] [PubMed]
56. Wang, Q.; Yuan, J.; Yang, X.; Han, X.; Lan, Y.; Cao, D.; Sun, Q.; Cui, X.; Meng, J.; Chen, W. Responses of soil respiration and C sequestration efficiency to biochar amendment in maize field of Northeast China. *Soil Tillage Res.* **2022**, *223*, 105442. [CrossRef]
57. Wang, L.; Gao, C.; Yang, K.; Sheng, Y.; Xu, J.; Zhao, Y.; Lou, J.; Sun, R.; Zhu, L. Effects of biochar aging in the soil on its mechanical property and performance for soil CO_2 and N_2O emissions. *Sci. Total Environ.* **2021**, *782*, 146824. [CrossRef]

58. Song, B.; Almatrafi, E.; Tan, X.; Luo, S.; Xiong, W.; Zhou, C.; Qin, M.; Liu, Y.; Cheng, M.; Zeng, G.; et al. Biochar-based agricultural soil management: An application-dependent strategy for contributing to carbon neutrality. *Renew. Sustain. Energy Rev.* **2022**, *164*, 112529. [CrossRef]
59. Gascó, G.; Paz-Ferreiro, J.; Cely, P.; Plaza, C.; Méndez, A. Influence of pig manure and its biochar on soil CO_2 emissions and soil enzymes. *Ecol. Eng.* **2016**, *95*, 19–24. [CrossRef]
60. Zheng, H.; Liu, D.; Liao, X.; Miao, Y.; Li, Y.; Li, J.; Yuan, J.; Chen, Z.; Ding, W. Field-aged biochar enhances soil organic carbon by increasing recalcitrant organic carbon fractions and making microbial communities more conducive to carbon sequestration. *Agric. Ecosyst. Environ.* **2022**, *340*, 108177. [CrossRef]
61. Lin, X.W.; Xie, Z.B.; Zheng, J.Y.; Liu, Q.; Bei, Q.C.; Zhu, J.G. Effects of biochar application on greenhouse gas emissions, carbon sequestration and crop growth in coastal saline soil. *Eur. J. Soil Sci.* **2015**, *66*, 329–338. [CrossRef]
62. El-Mahrouky, M.; El-Naggar, A.H.; Usman, A.R.; Al-Wabel, M. Dynamics of CO_2 Emission and Biochemical Properties of a Sandy Calcareous Soil Amended with Conocarpus Waste and Biochar. *Pedosphere* **2015**, *25*, 46–56. [CrossRef]
63. Yang, Y.; Sun, K.; Liu, J.; Chen, Y.; Han, L. Changes in soil properties and CO_2 emissions after biochar addition: Role of pyrolysis temperature and aging. *Sci. Total Environ.* **2022**, *839*, 156333. [CrossRef]
64. Van Kessel, C.; Venterea, R.; Six, J.; Adviento-Borbe, M.A.; Linquist, B.; van Groenigen, K.J. Climate, duration, and N placement determine N_2O emissions in reduced tillage systems: A meta-analysis. *Glob. Chang. Biol.* **2013**, *19*, 33–44. [CrossRef]
65. Thompson, R.L.; Lassaletta, L.; Patra, P.K.; Wilson, C.; Wells, K.C.; Gressent, A.; Koffi, E.N.; Chipperfield, M.P.; Winiwarter, W.; Davidson, E.A. Acceleration of global N_2O emissions seen from two decades of atmospheric inversion. *Nat. Clim. Chang.* **2019**, *9*, 993–998. [CrossRef]
66. Tian, H.; Yang, J.; Xu, R.; Lu, C.; Canadell, J.G.; Davidson, E.A.; Jackson, R.B.; Arneth, A.; Chang, J.; Ciais, P. Global soil nitrous oxide emissions since the preindustrial era estimated by an ensemble of terrestrial biosphere models: Magnitude, attribution, and uncertainty. *Glob. Chang. Biol.* **2019**, *25*, 640–659. [CrossRef] [PubMed]
67. Mateo-Marín, N.; Quílez, D.; Guillén, M.; Isla, R. Feasibility of stabilised nitrogen fertilisers decreasing greenhouse gas emissions under optimal management in sprinkler irrigated conditions. *Agric. Ecosyst. Environ.* **2020**, *290*, 106725. [CrossRef]
68. Ge, J.; Huang, G.; Huang, J.; Zeng, J.; Han, L. Particle-scale modeling of methane emission during pig manure/wheat straw aerobic composting. *Environ. Sci. Technol.* **2016**, *50*, 4374–4383. [CrossRef] [PubMed]
69. Wang, C.; Lu, H.; Dong, D.; Deng, H.; Strong, P.J.; Wang, H.; Wu, W. Insight into the effects of biochar on manure composting: Evidence supporting the relationship between N_2O emission and denitrifying community. *Environ. Sci. Technol.* **2013**, *47*, 7341–7349. [CrossRef]
70. Song, Y.; Li, Y.; Cai, Y.; Fu, S.; Luo, Y.; Wang, H.; Liang, C.; Lin, Z.; Hu, S.; Li, Y.; et al. Biochar decreases soil N_2O emissions in Moso bamboo plantations through decreasing labile N concentrations, N-cycling enzyme activities and nitrification/denitrification rates. *Geoderma* **2019**, *348*, 135–145. [CrossRef]
71. Ji, C.; Han, Z.; Zheng, F.; Wu, S.; Wang, J.; Wang, J.; Zhang, H.; Zhang, Y.; Liu, S.; Li, S.; et al. Biochar reduced soil nitrous oxide emissions through suppressing fungal denitrification and affecting fungal community assembly in a subtropical tea plantation. *Agric. Ecosyst. Environ.* **2022**, *326*, 107784. [CrossRef]
72. Tang, Z.; Liu, X.; Li, G.; Liu, X. Mechanism of biochar on nitrification and denitrification to N_2O emissions based on isotope characteristic values. *Environ. Res.* **2022**, *212*, 113219. [CrossRef]
73. Zhong, L.; Li, G.; Qing, J.; Li, J.; Xue, J.; Yan, B.; Chen, G.; Kang, X.; Rui, Y. Biochar can reduce N_2O production potential from rhizosphere of fertilized agricultural soils by suppressing bacterial denitrification. *Eur. J. Soil Biol.* **2022**, *109*, 103391. [CrossRef]
74. Yang, X.; Sun, Q.; Yuan, J.; Fu, S.; Lan, Y.; Jiang, X.; Meng, J.; Han, X.; Chen, W. Successive corn stover and biochar applications mitigate N_2O emissions by altering soil physicochemical properties and N-cycling-related enzyme activities: A five-year field study in Northeast China. *Agric. Ecosyst. Environ.* **2022**, *340*, 108183. [CrossRef]
75. Kalu, S.; Simojoki, A.; Karhu, K.; Tammeorg, P. Long-term effects of softwood biochar on soil physical properties, greenhouse gas emissions and crop nutrient uptake in two contrasting boreal soils. *Agric. Ecosyst. Environ.* **2021**, *316*, 107454. [CrossRef]
76. Liao, X.; Mao, S.; Chen, Y.; Zhang, J.; Müller, C.; Malghani, S. Combined effects of biochar and biogas slurry on soil nitrogen transformation rates and N_2O emission in a subtropical poplar plantation. *Sci. Total Environ.* **2022**, *848*, 157766. [CrossRef] [PubMed]
77. Pokharel, P.; Qi, L.; Chang, S.X. Manure-based biochar decreases heterotrophic respiration and increases gross nitrification rates in rhizosphere soil. *Soil Biol. Biochem.* **2021**, *154*, 108147. [CrossRef]
78. Liao, X.; Liu, D.; Niu, Y.; Chen, Z.; He, T.; Ding, W. Effect of field-aged biochar on fertilizer N retention and N_2O emissions: A field microplot experiment with 15N-labeled urea. *Sci. Total Environ.* **2021**, *773*, 145645. [CrossRef] [PubMed]
79. Cai, Y.; Akiyama, H. Effects of inhibitors and biochar on nitrous oxide emissions, nitrate leaching, and plant nitrogen uptake from urine patches of grazing animals on grasslands: A meta-analysis. *Soil Sci. Plant Nutr.* **2017**, *63*, 405–414. [CrossRef]
80. Lan, Z.M.; Chen, C.R.; Rezaei Rashti, M.; Yang, H.; Zhang, D.K. Linking feedstock and application rate of biochars to N_2O emission in a sandy loam soil: Potential mechanisms. *Geoderma* **2019**, *337*, 880–892. [CrossRef]
81. Ameloot, N.; De Neve, S.; Jegajeevagan, K.; Yildiz, G.; Buchan, D.; Funkuin, Y.N.; Prins, W.; Bouckaert, L.; Sleutel, S. Short-term CO_2 and N_2O emissions and microbial properties of biochar amended sandy loam soils. *Soil Biol. Biochem.* **2013**, *57*, 401–410. [CrossRef]

82. Zhang, X.; Zhang, J.; Song, M.; Dong, Y.; Xiong, Z. N_2O and NO production and functional microbes responding to biochar aging process in an intensified vegetable soil. *Environ. Pollut.* **2022**, *307*, 119491. [CrossRef]
83. Bernstein, L.; Bosch, P.; Canziani, O.; Chen, Z.; Christ, R.; Riahi, K. *IPCC, 2007: Climate Change 2007: Synthesis Report*; IPCC: Geneva, Switzerland, 2008.
84. Wang, C.; Shen, J.; Liu, J.; Qin, H.; Yuan, Q.; Fan, F.; Hu, Y.; Wang, J.; Wei, W.; Li, Y.; et al. Microbial mechanisms in the reduction of CH_4 emission from double rice cropping system amended by biochar: A four-year study. *Soil Biol. Biochem.* **2019**, *135*, 251–263. [CrossRef]
85. Sun, B.; Bai, Z.; Li, Y.; Li, R.; Song, M.; Xu, S.; Zhang, H.; Zhuang, X. Emission mitigation of CH_4 and N_2O during semi-permeable membrane covered hyperthermophilic aerobic composting of livestock manure. *J. Clean. Prod.* **2022**, *379*, 134850. [CrossRef]
86. Solomon, S.; Qin, D.; Manning, M.; Averyt, K.; Marquis, M. *Climate Change 2007-the Physical Science Basis: Working Group I Contribution to the Fourth Assessment Report of the IPCC*; Cambridge University Press: Cambridge, UK, 2007; Volume 4.
87. Zhang, A.; Bian, R.; Pan, G.; Cui, L.; Hussain, Q.; Li, L.; Zheng, J.; Zheng, J.; Zhang, X.; Han, X. Effects of biochar amendment on soil quality, crop yield and greenhouse gas emission in a Chinese rice paddy: A field study of 2 consecutive rice growing cycles. *Field Crops Res.* **2012**, *127*, 153–160. [CrossRef]
88. Lin, X.; Wang, N.; Li, F.; Yan, B.; Pan, J.; Jiang, S.; Peng, H.; Chen, A.; Wu, G.; Zhang, J.; et al. Evaluation of the synergistic effects of biochar and biogas residue on CO_2 and CH_4 emission, functional genes, and enzyme activity during straw composting. *Bioresour. Technol.* **2022**, *360*, 127608. [CrossRef] [PubMed]
89. Sriphirom, P.; Towprayoon, S.; Yagi, K.; Rossopa, B.; Chidthaisong, A. Changes in methane production and oxidation in rice paddy soils induced by biochar addition. *Appl. Soil Ecol.* **2022**, *179*, 104585. [CrossRef]
90. Kim, J.; Yoo, G.; Kim, D.; Ding, W.; Kang, H. Combined application of biochar and slow-release fertilizer reduces methane emission but enhances rice yield by different mechanisms. *Appl. Soil Ecol.* **2017**, *117–118*, 57–62. [CrossRef]
91. Ji, B.; Chen, J.; Li, W.; Mei, J.; Yang, Y.; Chang, J. Greenhouse gas emissions from constructed wetlands are mitigated by biochar substrates and distinctly affected by tidal flow and intermittent aeration modes. *Environ. Pollut.* **2021**, *271*, 116328. [CrossRef]
92. Wang, N.; Chang, Z.-Z.; Xue, X.-M.; Yu, J.-G.; Shi, X.-X.; Ma, L.Q.; Li, H.-B. Biochar decreases nitrogen oxide and enhances methane emissions via altering microbial community composition of anaerobic paddy soil. *Sci. Total Environ.* **2017**, *581*, 689–696. [CrossRef]
93. Liu, Y.; Yang, M.; Wu, Y.; Wang, H.; Chen, Y.; Wu, W. Reducing CH_4 and CO_2 emissions from waterlogged paddy soil with biochar. *J. Soils Sediments* **2011**, *11*, 930–939. [CrossRef]
94. Joseph, S.; Graber, E.R.; Chia, C.; Munroe, P.; Donne, S.; Thomas, T.; Nielsen, S.; Marjo, C.; Rutlidge, H.; Pan, G.-X. Shifting paradigms: Development of high-efficiency biochar fertilizers based on nano-structures and soluble components. *Carbon Manag.* **2013**, *4*, 323–343. [CrossRef]
95. Nan, Q.; Wang, C.; Yi, Q.; Zhang, L.; Ping, F.; Thies, J.E.; Wu, W. Biochar amendment pyrolysed with rice straw increases rice production and mitigates methane emission over successive three years. *Waste Manag.* **2020**, *118*, 1–8. [CrossRef]
96. Wang, C.; Liu, J.; Shen, J.; Chen, D.; Li, Y.; Jiang, B.; Wu, J. Effects of biochar amendment on net greenhouse gas emissions and soil fertility in a double rice cropping system: A 4-year field experiment. *Agric. Ecosyst. Environ.* **2018**, *262*, 83–96. [CrossRef]
97. Qi, L.; Pokharel, P.; Chang, S.X.; Zhou, P.; Niu, H.; He, X.; Wang, Z.; Gao, M. Biochar application increased methane emission, soil carbon storage and net ecosystem carbon budget in a 2-year vegetable–rice rotation. *Agric. Ecosyst. Environ.* **2020**, *292*, 106831. [CrossRef]
98. Liu, F.; Ji, M.; Xiao, L.; Wang, X.; Diao, Y.; Dan, Y.; Wang, H.; Sang, W.; Zhang, Y. Organics composition and microbial analysis reveal the different roles of biochar and hydrochar in affecting methane oxidation from paddy soil. *Sci. Total Environ.* **2022**, *843*, 157036. [CrossRef] [PubMed]
99. Lee, S.-I.; Park, H.-J.; Jeong, Y.-J.; Seo, B.-S.; Kwak, J.-H.; Yang, H.I.; Xu, X.; Tang, S.; Cheng, W.; Lim, S.-S.; et al. Biochar-induced reduction of N_2O emission from East Asian soils under aerobic conditions: Review and data analysis. *Environ. Pollut.* **2021**, *291*, 118154. [CrossRef] [PubMed]
100. Nguyen, T.T.N.; Xu, C.-Y.; Tahmasbian, I.; Che, R.; Xu, Z.; Zhou, X.; Wallace, H.M.; Bai, S.H. Effects of biochar on soil available inorganic nitrogen: A review and meta-analysis. *Geoderma* **2017**, *288*, 79–96. [CrossRef]
101. Domingues, R.R.; Trugilho, P.F.; Silva, C.A.; Melo, I.C.N.A.d.; Melo, L.C.A.; Magriotis, Z.M.; Sanchez-Monedero, M.A. Properties of biochar derived from wood and high-nutrient biomasses with the aim of agronomic and environmental benefits. *PLoS ONE* **2017**, *12*, e0176884. [CrossRef]
102. Lin, Z.; Liu, Q.; Liu, G.; Cowie, A.L.; Bei, Q.; Liu, B.; Wang, X.; Ma, J.; Zhu, J.; Xie, Z. Effects of Different Biochars on Pinus elliottii Growth, N Use Efficiency, Soil N_2O and CH_4 Emissions and C Storage in a Subtropical Area of China. *Pedosphere* **2017**, *27*, 248–261. [CrossRef]
103. Rittl, T.F.; Butterbach-Bahl, K.; Basile, C.M.; Pereira, L.A.; Alms, V.; Dannenmann, M.; Couto, E.G.; Cerri, C.E.P. Greenhouse gas emissions from soil amended with agricultural residue biochars: Effects of feedstock type, production temperature and soil moisture. *Biomass Bioenergy* **2018**, *117*, 1–9. [CrossRef]
104. Al-Rumaihi, A.; Shahbaz, M.; McKay, G.; Mackey, H.; Al-Ansari, T. A review of pyrolysis technologies and feedstock: A blending approach for plastic and biomass towards optimum biochar yield. *Renew. Sustain. Energy Rev.* **2022**, *167*, 112715. [CrossRef]
105. Amenaghawon, A.N.; Anyalewechi, C.L.; Okieimen, C.O.; Kusuma, H.S. Biomass pyrolysis technologies for value-added products: A state-of-the-art review. *Environ. Dev. Sustain.* **2021**, *23*, 14324–14378. [CrossRef]

106. Yang, W.; Shang, J.; Sharma, P.; Li, B.; Liu, K.; Flury, M. Colloidal stability and aggregation kinetics of biochar colloids: Effects of pyrolysis temperature, cation type, and humic acid concentrations. *Sci. Total Environ.* **2019**, *658*, 1306–1315. [CrossRef] [PubMed]
107. Nguyen, D.H.; Scheer, C.; Rowlings, D.W.; Grace, P.R. Rice husk biochar and crop residue amendment in subtropical cropping soils: Effect on biomass production, nitrogen use efficiency and greenhouse gas emissions. *Biol. Fertil. Soils* **2016**, *52*, 261–270. [CrossRef]
108. Zhao, T.; Yao, Y.; Li, D.; Wu, F.; Zhang, C.; Gao, B. Facile low-temperature one-step synthesis of pomelo peel biochar under air atmosphere and its adsorption behaviors for Ag (I) and Pb (II). *Sci. Total Environ.* **2018**, *640*, 73–79. [CrossRef] [PubMed]
109. Lewandowski, W.M.; Januszewicz, K.; Kosakowski, W. Efficiency and proportions of waste tyre pyrolysis products depending on the reactor type—A review. *J. Anal. Appl. Pyrolysis* **2019**, *140*, 25–53. [CrossRef]
110. Zhang, Q.; Xiao, J.; Xue, J.; Zhang, L. Quantifying the effects of biochar application on greenhouse gas emissions from agricultural soils: A global meta-analysis. *Sustainability* **2020**, *12*, 3436. [CrossRef]
111. McBeath, A.V.; Smernik, R.J.; Krull, E.S.; Lehmann, J. The influence of feedstock and production temperature on biochar carbon chemistry: A solid-state 13C NMR study. *Biomass Bioenergy* **2014**, *60*, 121–129. [CrossRef]
112. Shaheen, S.M.; Mosa, A.; Abdelrahman, H.; Niazi, N.K.; Antoniadis, V.; Shahid, M.; Song, H.; Kwon, E.E.; Rinklebe, J. Removal of toxic elements from aqueous environments using nano zero-valent iron-and iron oxide-modified biochar: A review. *Biochar* **2022**, *4*, 1–21. [CrossRef]
113. Deng, B.; Shi, Y.; Zhang, L.; Fang, H.; Gao, Y.; Luo, L.; Feng, W.; Hu, X.; Wan, S.; Huang, W.; et al. Effects of spent mushroom substrate-derived biochar on soil CO_2 and N_2O emissions depend on pyrolysis temperature. *Chemosphere* **2020**, *246*, 125608. [CrossRef]
114. Wang, Z.; Zheng, H.; Luo, Y.; Deng, X.; Herbert, S.; Xing, B. Characterization and influence of biochars on nitrous oxide emission from agricultural soil. *Environ. Pollut.* **2013**, *174*, 289–296. [CrossRef]
115. Mukome, F.N.D.; Six, J.; Parikh, S.J. The effects of walnut shell and wood feedstock biochar amendments on greenhouse gas emissions from a fertile soil. *Geoderma* **2013**, *200–201*, 90–98. [CrossRef]
116. Nan, Q.; Xin, L.; Qin, Y.; Waqas, M.; Wu, W. Exploring long-term effects of biochar on mitigating methane emissions from paddy soil: A review. *Biochar* **2021**, *3*, 125–134. [CrossRef]
117. Chiappero, M.; Norouzi, O.; Hu, M.; Demichelis, F.; Berruti, F.; di Maria, F.; Mašek, O.; Fiore, S. Review of biochar role as additive in anaerobic digestion processes. *Renew. Sustain. Energy Rev.* **2020**, *131*, 110037. [CrossRef]
118. Cai, F.; Feng, Z.; Zhu, L. Effects of biochar on CH_4 emission with straw application on paddy soil. *J. Soils Sediments* **2018**, *18*, 599–609. [CrossRef]
119. Husk, B.; Major, J. Commercial Scale Agricultural Biochar Field Trial in Québec, Canada over Two Years: Effects of Biochar on Soil Fertility, Biology and Crop Productivity and Quality. 2010. Available online: http://www.blue-leaf.ca/main-en/report_a3.php (accessed on 3 March 2011).
120. Rahman, M.A.; Jahiruddin, M.; Kader, M.A.; Islam, M.R.; Solaiman, Z.M. Sugarcane bagasse biochar increases soil carbon sequestration and yields of maize and groundnut in charland ecosystem. *Arch. Agron. Soil Sci.* **2022**, *68*, 1338–1351. [CrossRef]
121. Novak, J.M.; Ippolito, J.A.; Watts, D.W.; Sigua, G.C.; Ducey, T.F.; Johnson, M.G. Biochar compost blends facilitate switchgrass growth in mine soils by reducing Cd and Zn bioavailability. *Biochar* **2019**, *1*, 97–114. [CrossRef] [PubMed]
122. An, X.; Wu, Z.; Shi, W.; Qi, H.; Zhang, L.; Xu, X.; Yu, B. Biochar for simultaneously enhancing the slow-release performance of fertilizers and minimizing the pollution of pesticides. *J. Hazard. Mater.* **2021**, *407*, 124865. [CrossRef] [PubMed]
123. Zimmerman, A.R.; Gao, B.; Ahn, M.-Y. Positive and negative carbon mineralization priming effects among a variety of biochar-amended soils. *Soil Biol. Biochem.* **2011**, *43*, 1169–1179. [CrossRef]
124. Van Zwieten, L.; Singh, B.P.; Tavakkoli, E.; Joseph, S.; Macdonald, L.M.; Rose, T.J.; Rose, M.T.; Kimber, S.W.L.; Morris, S.; Cozzolino, D. Biochar built soil carbon over a decade by stabilizing rhizodeposits. *Nat. Clim. Chang.* **2017**, *7*, 371–376.
125. Olaniyan, J.O.; Isimikalu, T.O.; Raji, B.A.; Affinnih, K.O.; Alasinrin, S.Y.; Ajala, O.N. An investigation of the effect of biochar application rates on CO_2 emissions in soils under upland rice production in southern Guinea Savannah of Nigeria. *Heliyon* **2020**, *6*, e05578. [CrossRef]
126. Ge, X.; Cao, Y.; Zhou, B.; Wang, X.; Yang, Z.; Li, M.-H. Biochar addition increases subsurface soil microbial biomass but has limited effects on soil CO_2 emissions in subtropical moso bamboo plantations. *Appl. Soil Ecol.* **2019**, *142*, 155–165. [CrossRef]
127. Lu, X.; Li, Y.; Wang, H.; Singh, B.P.; Hu, S.; Luo, Y.; Li, J.; Xiao, Y.; Cai, X.; Li, Y. Responses of soil greenhouse gas emissions to different application rates of biochar in a subtropical Chinese chestnut plantation. *Agric. For. Meteorol.* **2019**, *271*, 168–179. [CrossRef]
128. Walkiewicz, A.; Kalinichenko, K.; Kubaczyński, A.; Brzezińska, M.; Bieganowski, A. Usage of biochar for mitigation of CO_2 emission and enhancement of CH_4 consumption in forest and orchard Haplic Luvisol (Siltic) soils. *Appl. Soil Ecol.* **2020**, *156*, 103711. [CrossRef]
129. Shen, Y.; Zhu, L.; Cheng, H.; Yue, S.; Li, S. Effects of biochar application on CO_2 emissions from a cultivated soil under semiarid climate conditions in Northwest China. *Sustainability* **2017**, *9*, 1482. [CrossRef]
130. Oo, A.Z.; Sudo, S.; Win, K.T.; Shibata, A.; Gonai, T. Influence of pruning waste biochar and oyster shell on N_2O and CO_2 emissions from Japanese pear orchard soil. *Heliyon* **2018**, *4*, e00568. [CrossRef] [PubMed]
131. Mosa, A.A.; Taha, A.; Elsaeid, M. Agro-environmental applications of humic substances: A critical review. *Egypt. J. Soil Sci.* **2020**, *60*, 211–229. [CrossRef]

132. Chen, H.; Awasthi, S.K.; Liu, T.; Duan, Y.; Ren, X.; Zhang, Z.; Pandey, A.; Awasthi, M.K. Effects of microbial culture and chicken manure biochar on compost maturity and greenhouse gas emissions during chicken manure composting. *J. Hazard. Mater.* **2020**, *389*, 121908. [CrossRef]
133. Millennium Ecosystem Assessment. *Synthesis Report*; Millennium Ecosystem Assessment: Washington, DC, USA, 2005.
134. Mahapatra, S.; Ali, M.H.; Samal, K. Assessment of compost maturity-stability indices and recent development of composting bin. *Energy Nexus* **2022**, *6*, 100062. [CrossRef]
135. Chung, W.J.; Chang, S.W.; Chaudhary, D.K.; Shin, J.; Kim, H.; Karmegam, N.; Govarthanan, M.; Chandrasekaran, M.; Ravindran, B. Effect of biochar amendment on compost quality, gaseous emissions and pathogen reduction during in-vessel composting of chicken manure. *Chemosphere* **2021**, *283*, 131129. [CrossRef]
136. Zhang, L.; Sun, X. Changes in physical, chemical, and microbiological properties during the two-stage co-composting of green waste with spent mushroom compost and biochar. *Bioresour. Technol.* **2014**, *171*, 274–284. [CrossRef]
137. Li, S.; Song, L.; Jin, Y.; Liu, S.; Shen, Q.; Zou, J. Linking N_2O emission from biochar-amended composting process to the abundance of denitrify (nirK and nosZ) bacteria community. *AMB Express* **2016**, *6*, 1–9. [CrossRef]
138. Guo, H.; Gu, J.; Wang, X.; Yu, J.; Nasir, M.; Zhang, K.; Sun, W. Microbial driven reduction of N_2O and NH_3 emissions during composting: Effects of bamboo charcoal and bamboo vinegar. *J. Hazard. Mater.* **2020**, *390*, 121292. [CrossRef]
139. Wang, X.; Liu, X.; Wang, Z.; Sun, G.; Li, J. Greenhouse gas reduction and nitrogen conservation during manure composting by combining biochar with wood vinegar. *J. Environ. Manag.* **2022**, *324*, 116349. [CrossRef] [PubMed]
140. Awasthi, M.K.; Duan, Y.; Awasthi, S.K.; Liu, T.; Zhang, Z. Influence of bamboo biochar on mitigating greenhouse gas emissions and nitrogen loss during poultry manure composting. *Bioresour. Technol.* **2020**, *303*, 122952. [CrossRef] [PubMed]
141. Ghorbani, M.; Konvalina, P.; Walkiewicz, A.; Neugschwandtner, R.W.; Kopecký, M.; Zamanian, K.; Chen, W.-H.; Bucur, D. Feasibility of Biochar Derived from Sewage Sludge to Promote Sustainable Agriculture and Mitigate GHG Emissions—A Review. *Int. J. Environ. Res. Public Health* **2022**, *19*, 12983. [CrossRef] [PubMed]
142. Jiang, J.; Kang, K.; Wang, C.; Sun, X.; Dang, S.; Wang, N.; Wang, Y.; Zhang, C.; Yan, G.; Li, Y. Evaluation of total greenhouse gas emissions during sewage sludge composting by the different dicyandiamide added forms: Mixing, surface broadcasting, and their combination. *Waste Manag.* **2018**, *81*, 94–103. [CrossRef]
143. Awasthi, M.K.; Wang, M.; Chen, H.; Wang, Q.; Zhao, J.; Ren, X.; Li, D.-s.; Awasthi, S.K.; Shen, F.; Li, R. Heterogeneity of biochar amendment to improve the carbon and nitrogen sequestration through reduce the greenhouse gases emissions during sewage sludge composting. *Bioresour. Technol.* **2017**, *224*, 428–438. [CrossRef] [PubMed]
144. Xue, S.; Zhou, L.; Zhong, M.; Kumar Awasthi, M.; Mao, H. Bacterial agents affected bacterial community structure to mitigate greenhouse gas emissions during sewage sludge composting. *Bioresour. Technol.* **2021**, *337*, 125397. [CrossRef] [PubMed]
145. Lv, B.; Cui, Y.; Wei, H.; Chen, Q.; Zhang, D. Elucidating the role of earthworms in N_2O emission and production pathway during vermicomposting of sewage sludge and rice straw. *J. Hazard. Mater.* **2020**, *400*, 123215. [CrossRef]
146. Wu, D.; Feng, Y.; Xue, L.; Liu, M.; Yang, B.; Hu, F.; Yang, L. Biochar Combined with Vermicompost Increases Crop Production While Reducing Ammonia and Nitrous Oxide Emissions from a Paddy Soil. *Pedosphere* **2019**, *29*, 82–94. [CrossRef]
147. Romero, C.M.; Redman, A.-A.P.H.; Owens, J.; Terry, S.A.; Ribeiro, G.O.; Gorzelak, M.A.; Oldenburg, T.B.P.; Hazendonk, P.; Larney, F.J.; Hao, X.; et al. Effects of feeding a pine-based biochar to beef cattle on subsequent manure nutrients, organic matter composition and greenhouse gas emissions. *Sci. Total Environ.* **2022**, *812*, 152267. [CrossRef]
148. Chowdhury, M.A.; de Neergaard, A.; Jensen, L.S. Potential of aeration flow rate and bio-char addition to reduce greenhouse gas and ammonia emissions during manure composting. *Chemosphere* **2014**, *97*, 16–25. [CrossRef]
149. Malińska, K.; Zabochnicka-Świątek, M.; Dach, J. Effects of biochar amendment on ammonia emission during composting of sewage sludge. *Ecol. Eng.* **2014**, *71*, 474–478. [CrossRef]
150. Vandecasteele, B.; Sinicco, T.; D'Hose, T.; Nest, T.V.; Mondini, C. Biochar amendment before or after composting affects compost quality and N losses, but not P plant uptake. *J. Environ. Manag.* **2016**, *168*, 200–209. [CrossRef] [PubMed]
151. Agyarko-Mintah, E.; Cowie, A.; Singh, B.P.; Joseph, S.; van Zwieten, L.; Cowie, A.; Harden, S.; Smillie, R. Biochar increases nitrogen retention and lowers greenhouse gas emissions when added to composting poultry litter. *Waste Manag.* **2017**, *61*, 138–149. [CrossRef] [PubMed]
152. Mao, H.; Lv, Z.; Sun, H.; Li, R.; Zhai, B.; Wang, Z.; Awasthi, M.K.; Wang, Q.; Zhou, L. Improvement of biochar and bacterial powder addition on gaseous emission and bacterial community in pig manure compost. *Bioresour. Technol.* **2018**, *258*, 195–202. [CrossRef] [PubMed]
153. Zhang, H.; Marchant-Forde, J.N.; Zhang, X.; Wang, Y. Effect of cornstalk biochar immobilized bacteria on ammonia reduction in laying hen manure composting. *Molecules* **2020**, *25*, 1560. [CrossRef]
154. Wang, Y.; Akdeniz, N.; Yi, S. Biochar-amended poultry mortality composting to increase compost temperatures, reduce ammonia emissions, and decrease leachate's chemical oxygen demand. *Agric. Ecosyst. Environ.* **2021**, *315*, 107451. [CrossRef]
155. El-Naggar, A.; Ahmed, N.; Mosa, A.; Niazi, N.K.; Yousaf, B.; Sharma, A.; Sarkar, B.; Cai, Y.; Chang, S.X. Nickel in soil and water: Sources, biogeochemistry, and remediation using biochar. *J. Hazard. Mater.* **2021**, *419*, 126421. [CrossRef]
156. Ghorbani, M.; Konvalina, P.; Kopecký, M.; Kolář, L. A meta-analysis on the impacts of different oxidation methods on the surface area properties of biochar. *Land Degrad. Dev.* **2022**. [CrossRef]
157. Liu, W.-J.; Jiang, H.; Yu, H.-Q. Development of biochar-based functional materials: Toward a sustainable platform carbon material. *Chem. Rev.* **2015**, *115*, 12251–12285. [CrossRef]

158. Zhang, X.; Zhang, S.; Yang, H.; Feng, Y.; Chen, Y.; Wang, X.; Chen, H. Nitrogen enriched biochar modified by high temperature CO_2–ammonia treatment: Characterization and adsorption of CO_2. *Chem. Eng. J.* **2014**, *257*, 20–27. [CrossRef]
159. Lian, F.; Cui, G.; Liu, Z.; Duo, L.; Zhang, G.; Xing, B. One-step synthesis of a novel N-doped microporous biochar derived from crop straws with high dye adsorption capacity. *J. Environ. Manag.* **2016**, *176*, 61–68. [CrossRef] [PubMed]
160. Zhang, X.; Wu, J.; Yang, H.; Shao, J.; Wang, X.; Chen, Y.; Zhang, S.; Chen, H. Preparation of nitrogen-doped microporous modified biochar by high temperature CO_2–NH_3 treatment for CO_2 adsorption: Effects of temperature. *RSC Adv.* **2016**, *6*, 98157–98166. [CrossRef]
161. Xing, W.; Liu, C.; Zhou, Z.; Zhang, L.; Zhou, J.; Zhuo, S.; Yan, Z.; Gao, H.; Wang, G.; Qiao, S.Z. Superior CO_2 uptake of N-doped activated carbon through hydrogen-bonding interaction. *Energy Environ. Sci.* **2012**, *5*, 7323–7327. [CrossRef]
162. Ma, X.; Li, L.; Wang, S.; Lu, M.; Li, H.; Ma, W.; Keener, T.C. Ammonia-treated porous carbon derived from ZIF-8 for enhanced CO_2 adsorption. *Appl. Surf. Sci.* **2016**, *369*, 390–397. [CrossRef]
163. Zhang, C.; Sun, S.; Xu, S.; Wu, C. CO_2 capture over steam and KOH activated biochar: Effect of relative humidity. *Biomass Bioenergy* **2022**, *166*, 106608. [CrossRef]
164. Kaya, N.; Uzun, Z.Y. Investigation of effectiveness of pine cone biochar activated with KOH for methyl orange adsorption and CO_2 capture. *Biomass Convers. Biorefinery* **2021**, *11*, 1067–1083. [CrossRef]
165. Li, K.; Zhang, D.; Niu, X.; Guo, H.; Yu, Y.; Tang, Z.; Lin, Z.; Fu, M. Insights into CO_2 adsorption on KOH-activated biochars derived from the mixed sewage sludge and pine sawdust. *Sci. Total Environ.* **2022**, *826*, 154133. [CrossRef]
166. Shin, J.; Park, D.; Hong, S.; Jeong, C.; Kim, H.; Chung, W. Influence of activated biochar pellet fertilizer application on greenhouse gas emissions and carbon sequestration in rice (*Oryza sativa* L.) production. *Environ. Pollut.* **2021**, *285*, 117457. [CrossRef]
167. Liu, Z.; Tang, J.; Ren, X.; Schaeffer, S.M. Effects of phosphorus modified nZVI-biochar composite on emission of greenhouse gases and changes of microbial community in soil. *Environ. Pollut.* **2021**, *274*, 116483. [CrossRef]
168. Khan, M.N.; Li, D.; Shah, A.; Huang, J.; Zhang, L.; Núñez-Delgado, A.; Han, T.; Du, J.; Ali, S.; Sial, T.A.; et al. The impact of pristine and modified rice straw biochar on the emission of greenhouse gases from a red acidic soil. *Environ. Res.* **2022**, *208*, 112676. [CrossRef]
169. Yang, X.; Wang, W.; Chen, X.; Sardans, J.; Wang, C.; Vancov, T.; Fang, Y.; Wang, S.; Yuan, X.; Llusià, J.; et al. Effects of N-enriched biochar on ecosystem greenhouse gas emissions, rice yield, and bacterial community diversity in subtropical rice paddy soils. *Eur. J. Soil Biol.* **2022**, *113*, 103440. [CrossRef]
170. Kong, F.; Wang, J.; Hou, W.; Cui, Y.; Yu, L.; Zhang, Y.; Wang, S. Influence of modified biochar supported sulfidation of nano-zero-valent-iron (S-nZVI/BC) on nitrate removal and greenhouse gas emission in constructed wetland. *J. Environ. Sci.* **2023**, *125*, 568–581. [CrossRef] [PubMed]
171. Gao, Y.; Wang, B.; Luo, L.; Deng, B.; Shad, N.; Hu, D.; Aly, H.M.; Zhang, L. Effects of hydroxyapatite and modified biochar derived from Camellia oleifera fruit shell on soil Cd contamination and N_2O emissions. *Ind. Crops Prod.* **2022**, *177*, 114476. [CrossRef]
172. Mohammadi, A.; Cowie, A.; Mai, T.L.A.; de la Rosa, R.A.; Brandao, M.; Kristiansen, P.; Joseph, S. Quantifying the greenhouse gas reduction benefits of utilising straw biochar and enriched biochar. *Energy Procedia* **2016**, *97*, 254–261. [CrossRef]

sustainability

Article

Cogeneration Systems Performance Analysis as a Sustainable Clean Energy and Water Source Based on Energy Hubs Using the Archimedes Optimization Algorithm

Magda I. El-Afifi [1,2], Magdi M. Saadawi [1] and Abdelfattah A. Eladl [1,*]

1 Electrical Engineering Department, Faculty of Engineering, Mansoura University, Mansoura 35516, Egypt
2 Nile Higher Institute of Engineering and Technology, Mansoura 35511, Egypt
* Correspondence: eladle7@mans.edu.eg

Abstract: Different energy requirements of the residential sector are varied, such as electricity, heating, cooling, water, etc., and these necessities are met by multi-energy systems using various energy sources and converters. In this paper, an optimal day-ahead operation of a large residential demand sector is presented based on the energy hub (EH) model with combined heat and power (CHP) as a cogeneration system. The purpose of the optimization is to maximize social welfare (SW) and minimize environmental emissions subjected to numerous technical constraints. To explore the effectiveness of the proposed model, real cases were studied and results were analyzed. Moreover, to evaluate the efficiency of the proposed methodology, the Archimedes optimization algorithm (AOA) is implemented for optimizing the EH system. The performance of the AOA is compared with the genetic algorithm, and the results depict that the AOA is better in terms of convergence speed and global search ability. Implementation of the proposed framework shows that the total SW is increased by 27.44% and environmental emissions are reduced by 18.36% compared to the base case without the EH. Additionally, there is 512.26 MWh and 149.4 m^3 as a surplus in the electricity and water that are sold to every network, respectively.

Keywords: energy hub; cogeneration systems; Archimedes optimization algorithm; sustainable; emissions; water management

Citation: El-Afifi, M.I.; Saadawi, M.M.; Eladl, A.A. Cogeneration Systems Performance Analysis as a Sustainable Clean Energy and Water Source Based on Energy Hubs Using the Archimedes Optimization Algorithm. *Sustainability* **2022**, *14*, 14766. https://doi.org/10.3390/su142214766

Academic Editors: Mohamed El-Alfy, Ahmed El Kenawy, Petra-Manuela Schuwerack and Zhongfeng Xu

Received: 25 August 2022
Accepted: 7 November 2022
Published: 9 November 2022

1. Introduction

Currently, urban areas comprise more than half of the world's population, and this number is expected to increase to nearly 5 billion by 2030 [1]. The increase in the world population has prompted researchers to conduct more studies on energy. To achieve this conclusion, researchers face serious economic and public issues that include the development of low carbon, pollution reduction, energy efficiency, shared energy supply, fresh water supply, and more. Figure 1 summarizes the percentage increase in world water withdrawals and consumption, population, and carbon dioxide emissions from 1900 to 2040 [2]. It can be seen that carbon dioxide emissions are predicted to reach a significant amount by 2040. Moreover, water withdrawal and consumption have also increased by more than 1000% [2]. On the other hand, the world was shocked in late December 2019 by the sudden outbreak of COVID-19, which impeded progress toward sustainable energy [3] and caused many major damages to the global economy [4]. As these circumstances persist, post-COVID-19 energy planning should be factored into an appropriate level of investment cost as the global economy continues to recover.

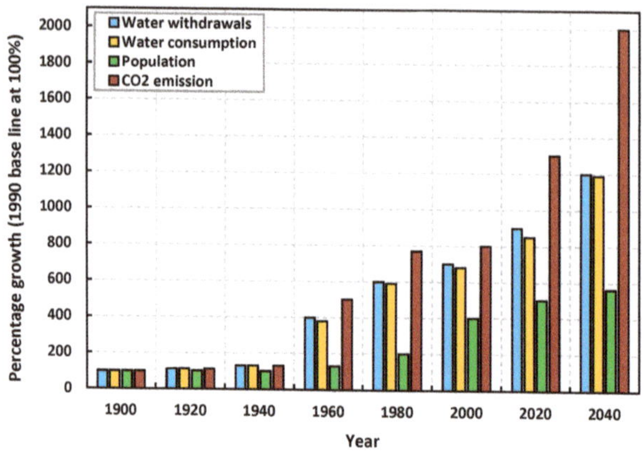

Figure 1. Water, population and CO_2 emission percentage growth rate from 1900 to 2040 (1900 baseline at 100%) [2].

One such suitable candidate is a cogeneration or combined heat and power (CHP) system. The operation mechanism of the CHP units depends on the generation of electricity and heat simultaneously, which increases its efficiency by 80–90%, compared to the traditional stand-alone generation, which has a typical efficiency of 35–55% [5]. Approximately 10–15% of the power transmission losses in the network can be eliminated by using CHP generators [6]. Applying the CHP units reduces overall energy production costs and greenhouse gas emissions because they use less fuel. Many energy systems are integrated through the CHP units such as electricity, gas, and heating systems, which are known as multi-energy systems. An energy hub (EH) concept is introduced for multi-energy systems modeling.

Nowadays, several works of literature have been presented investigating the scheduling of EH. Some different work has been carried out recently on EH planning in the stochastic situation. A new energy management technique for sustainable urban energy systems through EH is presented in [7]. This technique consists of two levels of integrated system modeling based mainly on CHP and renewable energy sources (RESs) but without taking into account these resources' uncertainty. Residential, commercial, and industrial hubs with similar internal configurations and equipment are proposed in [8]. CHP is a major component of these hubs to feed both electrical and thermal requirements. Work in [9] investigated the optimal scheduling of an EH integrated with CHP and solar energy for two days considering uncertainties via various scenarios. In [10], a technical economic evaluation of EH based on CHP for a medical complex was presented. In this work, an improvement of EH model efficiency and the economic benefits of using CHP were discussed. The risk-constrained probabilistic schedule of an integrated EH with a CHP module has been optimized considering the demand-side resources of a multi-objective model in [11]. Additionally, a multi-objective scheduling model for a small network based on CHP with the integration of an energy storage system (ESS) was presented in [12]. In [13], the optimum size of the EH system was presented using a mid-size CHP unit, where the reliability of supply was evaluated. A robust operating model for an EH with multicarrier including electric vehicles and CHP was presented in [14]. An EH model with CHP was designed for a residential building in [15].

On the other hand, some studies investigated the role of electrical heat pumps (EHPs) in hub schemes. In [16], the performance of residential EH integrated with EHP was discussed. The optimal model of bidding strategy for EH to participate in the target energy market which includes the EHP system was proposed in [17]. The authors investigated the flexibility of an EH operation including CHP and EHP in [18]. A day-ahead optimized scheduling model for an EH including EHP was presented and discussed [19]. Mixed integer linear programming (MILP) was implemented to model the optimization problem in the general algebraic modeling language (GAMS) program for an EH with CHP and ESS to reduce annual costs and emissions without considering RESs [20]. The effect of CHP units feeding many loads to achieve less fuel consumption and less pollution was studied [21]. In [22], an optimal operating strategy for CHP was proposed to reduce the total cost while also satisfying the thermoelectric coupling reduction. An optimal operation technique has been proposed in buildings supplied with CHP units integrated with RESs [23]. The particle swarm algorithm (PSO) algorithm was applied for the optimum planning of multi-source EH systems including RES to feed electrical and thermal loads [24]. An optimal multi-objective method was proposed to optimize a hybrid combined cooling, heating, and power system in order to reduce the total cost, and carbon dioxide emissions and increase the system flexibility [25]. A multi-objective technique based on a genetic algorithm (GA) was proposed in [26] to optimize the performance of an integrated natural gas combined, cooling, heating, and power unit and a ground source heat pump from the energy, economy, and emission perspectives without considering uncertainties of RESs.

Providing clean and reliable fresh water sources for various uses of human societies has always been a major global problem, especially in water-deficient regions. Recently, water desalination (WD) based on reverse osmosis technology has received more attention compared to other technologies because of its economic performance [27]. An optimal day-ahead operation of a microgrid based on coastal EH to minimize operational and environmental costs was presented [2]. Authors in [1] proposed a new model of a smart island in the power system composed of different energy resources including smart EH equipped with a desalination unit to supply electricity, thermal, and water demand. Authors in [28] studied the operation of multi-carrier energy systems integrated with CHP, wind turbines (WTs), gas-fired power plants, heat buffer tanks, pumped-storage systems, and gas storage technologies to meet daily electric power, gas, heating, and water loads. In [29], the optimal management of a water grid constructed for a pilot agro-industrial district, based on greenhouse cultivation, was analyzed. The proposed algorithm was based on the EHs approach and took into account economic terms and the optimal use of the available resources. EH scheduling for day-ahead has been studied in the presence of WT and electrical, thermal, and pico-hydel energy storage [30]. In [31], a proposed EH consisting of electricity, heat, and WD production equipment was optimized based on the GA to reduce costs and pollutants and increase the exergy efficiency.

In light of the previous studies, the current study contributes to presenting an optimization-based model to address the planning and operation of the EH system. Different loads, unpredictable wind power, photovoltaic (PV) power, and ESSs are taken into account. The optimal integration among various types of resources, environmental requirements, and economic criteria will be explored. In addition, the optimization problem will be solved based on the Archimedes optimization algorithm (AOA) to maximize total social welfare (SW) by reducing the total system operation cost and carbon dioxide emissions. A comparison between the most recent applied research and the proposed method is shown in Table 1.

Table 1. Comparison between the proposed system and some related literature.

Ref.	Year	PV	WT	CHP	EHP	GB	WD	ES	HS	WS	Demand	CO #	EA &	UA ¥	Grid-Connected	SA ¤
[1]	2020	×	×	✓	×	✓	✓	✓	×	✓	E-H-W *	✓	×	✓	✓	GA
[2]	2021	✓	✓	×	×	×	✓	×	✓	✓	E-H-W	✓	✓	✓	✓	GAMS
[12]	2017	×	×	✓	×	✓	×	✓	✓	×	E-H +	✓	×	✓	✓	MILP
[16]	2018	×	✓	✓	✓	✓	×	✓	×	×	E-H	✓	×	✓	✓	MILP
[19]	2021	×	✓	✓	✓	✓	×	×	×	×	E-H	✓	✓	✓	✓	MINLP
[20]	2021	×	×	✓	×	✓	×	✓	✓	×	E-H	✓	×	×	✓	MILP-GAMS
[21]	2020	✓	✓	✓	×	✓	×	✓	✓	×	E-H-C ǂ	✓	✓	✓	✓	QPSO
[22]	2019	×	✓	✓	×	×	×	×	✓	×	E-H	✓	×	✓	✓	-
[23]	2020	✓	✓	×	×	×	×	✓	×	×	E-H-C	×	×	×	✓	-
[24]	2020	×	×	✓	×	×	×	×	×	×	E-H	✓	✓	×	✓	PSO

Table 1. *Cont.*

Ref.	Year	PV	WT	CHP	EHP	GB	WD	ES	HS	WS	Demand	CO #	EA &	UA ¥	Grid-Connected	SA ¤
[25]	2020	✓	×	×	✓	×	×	✓	×	✓	E-H-C	✓	✓	×	✓	NSGA
[26]	2019	✓	×	×	✓	×	×	✓	×	×	E-H-C	✓	✓	×	✓	NSGA-II
Proposed		✓	✓	✓	✓	✓	✓	✓	✓	✓	E-H-W	✓	✓	✓	✓	AOA

GB: Gas boiler, ES: Electrical storage, HS: Heating storage, WS: Water storage, E-H-W *: Electrical-Heating-Water, E-H +: Electrical-Heating, E-H-C ‡: Electrical-Heating-Cooling, CO #: Cost analysis, EA &: Environmental analysis, UA ¥: Uncertainty analysis, SA ¤: Solution algorithm MINLP: Mixed integer nonlinear programming, QPSO: Quantum Particle Swarm Optimization, NSGA: Non-dominated sorting genetic algorithm.

The major contributions of this paper can be briefed as:

- Proposing a methodology to address integrations among EH components and energy networks, while satisfying different system constraints. Economic and environmental objective functions are considered to address the configuration and optimal operation of the EH system. Moreover, seeking to address reliability enhancement for electrical, thermal, and drinking water as vital requirements.
- Applying the AOA and GA to optimally solve the optimization problem to maximize the total SW and minimize emissions. Moreover, analyzing the performance of CHP units and verifying their effectiveness with EHs.
- Verifying the effectiveness of the proposed method in a short-term situation; a day is examined within one hour.

2. Energy Hub (EH) Modelling

Figure 2 shows an architecture of a proposed EH. The proposed EH consists of three integrated networks. The main energy sources of the hub are electricity, natural gas, and water whereas the output side supplies electricity, heat, and freshwater loads. The integration among these networks occurs through the EH components, such as CHP, EHP, boiler, and WD. Additionally, electrical, thermal, and water storage units were used to maximize utilization during periods of low load and to fill the deficit at peak loads. RESs (PV and WT) are also used in the EH to minimize total operating costs and emissions. The EH input and output are coupled by mathematical relationships which are defined in (1) and (2) as follows [32]:

$$[L] = [C][P] \tag{1}$$

$$[L] = \begin{bmatrix} L_1 \\ L_2 \\ \vdots \\ L_M \end{bmatrix}_{m\times 1}, \ [C] = \begin{bmatrix} C_{11} & \cdots & C_{1n} \\ \vdots & \ddots & \vdots \\ C_{m1} & \cdots & C_{mn} \end{bmatrix}_{m\times n}, \ [P] = \begin{bmatrix} P_1 \\ P_2 \\ \vdots \\ P_n \end{bmatrix}_{n\times 1} \tag{2}$$

where L, C, and P are output loads, coupling matrix, and input energy carriers, respectively, which are illustrated in detail in the following subsections.

Nowadays, many companies all over the world apply the EH concept and distribute water, gas, electric power, and RES. The first real EH was presented in Switzerland by the Regionalwerke AG Baden company [33]. The hub was composed of wood chips, a cogeneration plant, and methane gasification. The objective was to obtain natural gas and heat from burning wood chips available in the company's supply area. Natural gas was pumped either directly into the natural gas system or converted into electricity through the cogeneration unit and integrated into the electrical grid. In either case, the wasted heat was injected into the local heating network. Nitrogen and steam were used for the gasification process which was also provided at the hub entrance in addition to the above-mentioned energy carriers.

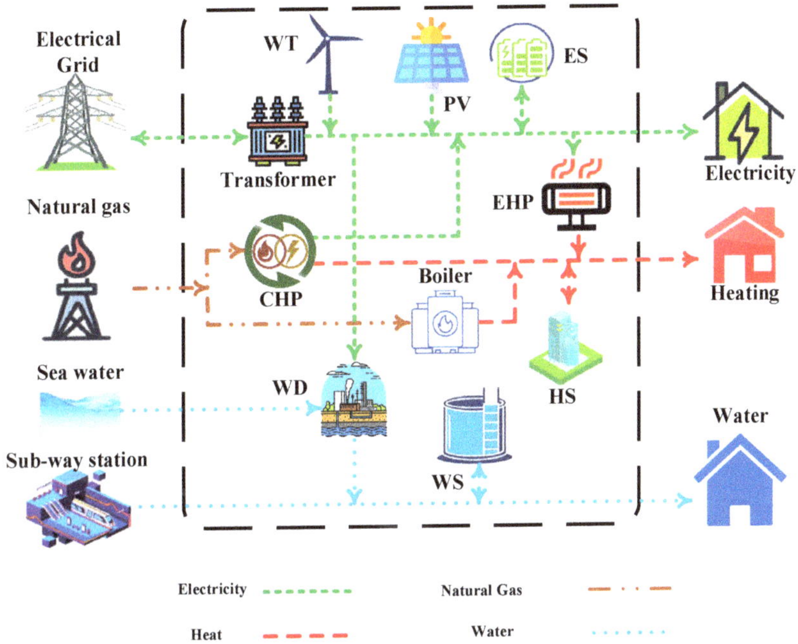

Figure 2. The architecture of studied EH.

2.1. PV Unit Output Power Modelling

Integrating PV modules into the electrical grid is a great solution to meet electrical demand and reduce carbon dioxide emissions. On the other hand, the development of PV modules faces a major challenge which is their unpredictable nature [34]. The power produced from PV arrays $(P_{PV}(t))$ is calculated by:

$$P_{PV}(t) = \frac{G(t)}{1000} P_{PVr} \ \eta_{PV} \tag{3}$$

where $G(t)$ is the solar radiation at time t (kW/m^2), P_{PVr} is the PV cell rated power (MW), and η_{PV} is PV cell efficiency.

2.2. WT Output Power Modelling

The generation of electrical energy from wind is one of the most promising types of generation in the world. The flow of air is converted into electrical energy without a negative impact on the environment [34]. Despite the different geographical conditions, seasonal changes, and other factors, the proposed model in this work of wind power is applicable under any operating conditions. Additionally, the power generation of WTs is highly uncertain and is formulated as a function of wind speed as represented in this work. The power generated $(P_W(t))$ of a WT is calculated by (4):

$$P_W(t) = \begin{cases} 0 & 0 \leq v(t) \leq v_i \ \& \ v_o \leq v(t) \\ P_{Wr} \times \frac{(v(t)-v_i)}{(v_r-v_i)} & v_i \leq v(t) \leq v_r \\ P_{Wr} & v_r \leq v(t) \leq v_o \end{cases} \tag{4}$$

where $v(t)$, v_i, v_o, and v_r are the actual wind speed at time t, cut-in speed, cut-out speed, and rated speed (m/s), respectively. P_{Wr} is the rated value of wind power (MW).

2.3. ESSs Model

ESSs are used to feed loads in case of low power generation. ESSs were modeled similarly to the authors presented in [35].

2.4. CHP Model

Both heat and power are generated from the CHP units. Each CHP unit has a thermal-electrical characteristic. One of the most used and feasible thermoelectric operation regions is shown in Figure 3. The complete model of CHP is illustrated in [11,36].

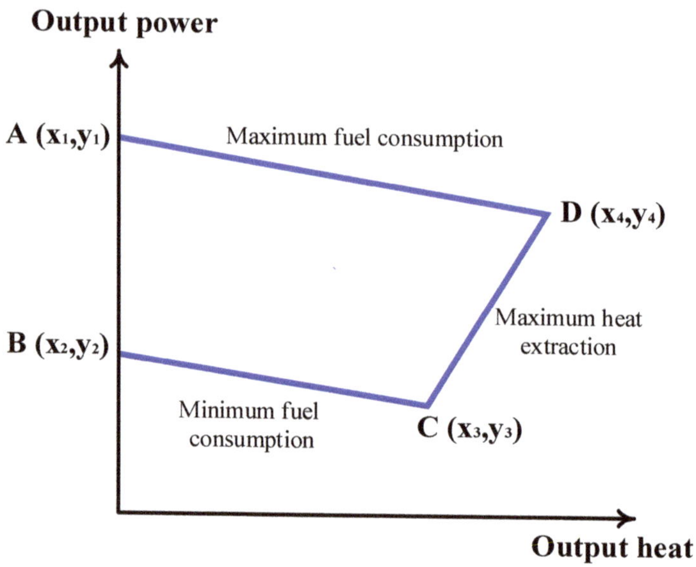

Figure 3. The thermal-electrical characteristic of the CHP unit.

2.5. EHP Model

EHP is powered by electric energy and extracted the heat from the cold ambient air and transfers it to heat the water in the heating system. The electrical power ($P_{EHP}(t)$) taken from EH by EHP is [37]:

$$P_{EHP}(t) = \frac{Q_{EHP}(t)}{COP} \tag{5}$$

where COP is the EHP coefficient of performance, and $Q_{EHP}(t)$ is EHP thermal energy produced (MWth) at time t.

2.6. GB Model

In the proposed EH system, GB is a standby unit for supplying thermal demand requirements when the heat produced by the CHP and EHP, and the available ESS capacity, is insufficient. The heat produced by the GB unit is given as [21].

$$Q_{GB}(t) = \gamma_{GB}\ P_{GB}(t) \tag{6}$$

where $Q_{GB}(t)$ is EHP thermal energy produced (MWth) at time t, γ_{GB} is the GB coefficient of performance, and $P_{GB}(t)$ is the quantity of natural gas absorbed by GB at time t (MW).

2.7. WD Unit Modelling

In this paper, one of the components of a water supply system is a WD unit, which can be established anywhere in the system (near the sea or groundwater). This paper focused on reducing costs and emissions without addressing the optimal location of the WD units (which can be considered in future work). Additionally, the WD unit is considered an aggregate device that imports electrical energy from EH and exports fresh water. Therefore, the detailed thermodynamic formula for the water grid is not taken into account. The WD unit is supplied with electric power to produce fresh water from seawater. The relationship between the produced water ($W_{WD}(t)$) and the consumed electricity ($P_{WD}(t)$) can be calculated as [2]:

$$P_{WD}(t) = \frac{W_{WD}(t)}{\eta_{WD}} \tag{7}$$

where η_{WD} is the performance coefficient of WD (m^3/MW).

3. Uncertainty Analysis of RESs

The high uncertainty in the energy produced from RESs makes it important to model the stochastic behavior of these resources. So, the planning and operation of an EH are managed, taking into account the unpredictable nature of these units. In this section, the uncertainties in RESs power generation are addressed. In order to account for differences in wind speed data, the Weibull distribution is used to continuously track changes in wind speed through a scale factor k and a form factor c. The Weibull distribution probability density function (pdf) is defined as [32]:

$$f_W(P_{W_{av}}) = \begin{cases} \left(\frac{klv_i}{c}\right)\left(\frac{(1+\rho l)v_i}{c}\right)^{k-1} exp\left(\left(\frac{(1+\rho l)v_i}{c}\right)^k\right) & 0 \le P_{W_{av}} \le P_{W_r} \\ 1 - exp\left[-\left(\frac{v_i}{c}\right)^k\right] + exp\left[-\left(\frac{v_o}{c}\right)^k\right] & P_{W_{av}} = 0 \\ exp\left[-\left(\frac{v_r}{c}\right)^k\right] - exp\left[-\left(\frac{v_o}{c}\right)^k\right] & P_{W_{av}} = P_{W_r} \end{cases} \tag{8}$$

where $P_{W_{av}}$ is the total available wind power (MW), $\rho = \frac{P_{W_{av}}}{P_{W_r}}$ and $l = \frac{v_r - v_i}{v_i}$.

To model the uncertainties in the PV modules' generated power, the beta distribution function is used [32]:

$$f_{PV}(P_{PV_{av}}) = \begin{cases} G^{\alpha-1}(1-G)^{\beta-1}\frac{\Gamma(\alpha+\beta)}{\Gamma(\alpha)+\Gamma(\beta)} & 0 \le G \le 1, \alpha, \beta \ge 0 \\ 0 & otherwise \end{cases} \tag{9}$$

where $P_{PV_{av}}$ is the total available PV power (MW), α and β are beta pdf parameters.

4. Thermal Generation Emissions

In most electric power grids, the bulk of the electric power generated is by thermal power plants which produce high carbon dioxide emissions. Hence, the number of emissions (E_G) can be expressed as [38]:

$$E_G = a_E\, P_G^2 + b_E\, P_G + d_E + \gamma_E\, exp(\delta_E\, P_G) \tag{10}$$

where a_E, b_E, d_E,γ_E, and δ_E are emission coefficients of the thermal generators and P_G is the power of the thermal generator (MW).

5. Problem Formulation and Methodology

The primary objective of this paper is to provide the electricity, heating, and water requirements of EH at the highest SW by minimizing total system operating costs and lowering CO_2 emissions while taking into account various system constraints.

5.1. Objective Function

This work aims to obtain the optimum values of power, heat, and water output from available generators subject to coordinated constraints. The total operating cost can be expressed as a summation of individual operating costs.

This paper deals with a system that includes a conventional power unit, a CHP unit, and an HOU. Reliable models describing the off-design process (partial load relationships between power production and fuel consumption) per unit as well as reliable optimization methods are required in order to improve the optimization problem. Depending on the nature of these relationships, the appropriate method should be applied to solve the optimization problem. However, in practice, second-order or linear convex input/output relationships are applied. The cost function per unit is obtained when the input/output curve is multiplied by the cost of fuel burned per unit. Assuming convex input/output curves for conventional power, CHP and HOU, their cost functions will also be convex [39] as follows:

- Conventional power generators' operating cost can be represented by a quadratic form as [36]:

$$C_G(P_G) = a_G P_G^2 + b_G P_G + c_G \quad (11)$$

where C_G is the conventional generator-produced power cost ($), and $a_G, b_G,$ and c_G are the conventional generator cost coefficients.

- CHP generator operation cost can be represented by a quadratic form as [32]:

$$C_{CHP}(P_{CHP}, Q_{CHP}) = a_{Co} + b_{Co} P_{CHP} + c_{Co} P_{CHP}^2 + d_{Co} Q_{CHP} + e_{Co} Q_{CHP}^2 + f_{Co} P_{CHP} Q_{CHP} \quad (12)$$

where C_{CHP} is the CHP generator produced energy cost ($), $a_{Co}, b_{Co}, c_{Co}, d_{Co}, e_{Co},$ and f_{Co} are the CHP generation cost coefficients, and P_{CHP}, Q_{CHP} are the electrical and heating outputs of the CHP unit.

- The operating cost of a heat-only unit (HOU) can be represented by a quadratic form as [36]:

$$C_{HOU}(Q_{HOU}) = V_{HOU} Q_{HOU}^2 + M_{HOU} Q_{HOU} + R_{HOU} \quad (13)$$

where C_{HOU} is the produced heat cost ($), and $V_{HOU}, M_{HOU},$ and R_{HOU} are the heat generation cost coefficients of an HOU.

- Because of the uncertainty of the available RESs at any given time, the factors for overestimating and underestimating the available RESs must be included in the model. The overestimation factor can be easily explained in that if a certain amount of RES power is assumed and that power is not available at the assumed time, then the power must be purchased from an alternative source or the loads must be disposed of. In the case of an underestimation penalty, if the available RES power is more than was assumed, then that power will be wasted, and it is reasonable for the system operator to pay a cost to the RES power product for the wastage of available capacity. Surplus RES power is usually sold to neighboring utilities, or by rapid redistribution. The output of non-RES generators is correspondingly reduced. Only if this cannot be achieved should the phantom load resistors be connected to "waste" the excess power. Obviously, these actions can be modeled by a simple minimization penalty cost function as [36]:

$$C_W(P_W) = d_W f_W(P_{W_{av}}) P_W + c_{p.W}(P_{W_{av}} - P_W) + c_{r.W}(P_W - P_{W_{av}}) \quad (14)$$

$$C_{PV}(P_{PV}) = h_{PV} f_{PV}(P_{PV_{av}})(P_{PV}) + c_{p.PV}(P_{PV_{av}} - P_{PV}) + c_{r.PV}(P_{PV} - P_{PV_{av}}) \quad (15)$$

where C_W and C_{PV} are the total cost of WT and PV generators ($), d_W and h_{PV} are the cost coefficient of WT and PV generators ($/MW), respectively, $f_W(P_W)$ and $f_{PV}(P_{PV})$ are the Weibull *pdf* and beta *pdf* of WT and PV generator, respectively, $c_{p.W}$ and $c_{p.PV}$ are the cost coefficient of WT and PV generators because of over-generation ($/MW), P_W and P_{PV} are the scheduled output of WT and PV generators, and $c_{r.W}$ and $c_{r.PV}$ are the cost coefficient of WT and PV generators because of under-generation ($/MW). The cost coefficients of WT and PV generation are calculated as follows [36]:

$$c_{p.W}(P_{W_{av}} - P_W) = c_{p.W} \int_{P_W}^{P_{W_r}} (P_{W_{av}} - P_W) f_W(P_{W_{av}}) dP_W \tag{16}$$

$$c_{r.W}(P_W - P_{W_{av}}) = c_{r.W} \int_0^{P_W} (P_W - P_{W_{av}}) f_W(P_{W_{av}}) dP_W \tag{17}$$

$$c_{p.PV}(P_{PV_{av}} - P_{PV}) = c_{p.PV} \int_{P_{PV}}^{P_{PV_r}} (P_{PV_{av}} - P_{PV}) f_{PV}(P_{PV_{av}}) dP_{PV} \tag{18}$$

$$c_{r.PV}(P_{PV} - P_{PV_{av}}) = c_{r.PV} \int_0^{P_{PV}} (P_{PV} - P_{PV_{av}}) f_{PV}(P_{PV_{av}}) dP_{PV} \tag{19}$$

- The operating cost of charge/discharge of ESSs can be represented by different models but this paper deals with a simple linear function, as the ESSs should be considered as a load when being charged and be considered as a generation source when discharging to the network [40]:

$$C_{ESSi}(P_{ESSi}) = C_{dsi} P_{dsi} - C_{chi}\, P_{chi} \tag{20}$$

where C_{ESSi} is the ESS power cost ($), C_{dsi} and C_{chi} are the discharging and charging cost of the ith ESSs ($/MW), respectively, and P_{dsi} and P_{chi} are the discharging and charging power of the ith ESSs (MW), respectively.

The operation cost of GB, EHP, WD, and WG is supposed to be linear. Where the output energy of each unit is linearly related to the quantity of fuel entering the unit and the cost coefficients.

- The operating cost of EHP can be represented by a linear function as [37]:

$$C_{EHP}(Q_{EHP}) = a_{EHP}\, P_{EHP} \tag{21}$$

where C_{EHP} is the power absorbed cost ($) and a_{EHP} is the EHP unit cost coefficient ($/MW).

- The operating cost of GB can be represented by a linear function as [41]:

$$C_{GB}(P_{GB}) = a_{GB} P_{GB} \tag{22}$$

where C_{GB} is the natural gas absorbed cost ($) and a_{GB} is GB unit cost coefficient ($/MW).

- The operating cost of a WD unit can be represented by a linear function as [42]:

$$C_{WD}(P_{WD}) = K_{WD}\, P_{WD} \tag{23}$$

where C_{WD} is the power absorbed cost ($) and K_{WD} is the WD unit cost coefficient ($/MW).

- The operating cost of the water grid (WG) can be represented in a linear form as [35]:

$$C_{WG}(W_{WG}) = K_{WG}\, W_{WG} \tag{24}$$

where C_{WG} is the total cost of water produced from WG ($), K_{WG} is the cost coefficient of the WG ($/m^3), and W_{WG} is the volume of the water produced from the WG (m^3).

To calculate the total system operation cost during period T, (10)–(24) are integrated into one formula as follows:

$$Cost = \sum_{t=1}^{T} C_G(P_G) + \sum_{t=1}^{T} C_{CHP}(P_{CHP}, H_{CHP}) + \sum_{t=1}^{T} C_W(P_W) + \sum_{t=1}^{T} C_{PV}(P_{PV}) + \sum_{t=1}^{T} \sum_{i=1}^{N_{ESSs}} C_{ESSi}(P_{ESSi}) + \sum_{t=1}^{T} C_{EHP}(Q_{EHP}) + \sum_{t=1}^{T} C_{HOU}(Q_{HOU}) + \sum_{t=1}^{T} C_{GB}(P_{GB}) + \sum_{t=1}^{T} C_{WD}(P_{WD}) + \sum_{t=1}^{T} C_{WG}(W_{WG}) \quad (25)$$

In this work, a single objective optimization problem is solved to find the minimum cost. So, a penalty factor h (\$/kg) is used to convert emissions value into cost [38]. The optimum solutions are then achieved while minimizing operation cost, minimizing emission cost, and meeting the load requirements and system operation constraints. The maximum penalty rate factor is defined as the ratio between the highest predictable value of operating cost ($cost^{max}$) and the highest estimated value for emissions (E_G^{max}), which is given by:

$$h = cost^{max} / E_G^{max} \quad (26)$$

Then, the corresponding cost of emissions (C_E) can be calculated by total emissions (E_G) as follows [38]:

$$C_E = h \ E_G \quad (27)$$

The revenue (R_D) achieved from selling the energy to end users can be expressed as:

$$R_D = \lambda_e \ P_{ED} + \lambda_h \ Q_{HD} + \lambda_W \ W_{WR} \quad (28)$$

where λ_e is electrical energy consumed cost (\$/MW), λ_h is heat energy consumed cost (\$/MWth), λ_w is water consumed cost (\$/m^3), and P_{ED}, Q_{HD}, and W_{WR} are the electrical, heat, and water demand, respectively.

Social welfare is given by:

$$SW = R_D - Cost - C_E \quad (29)$$

The overall objective function (F) is defined as:

$$Max \rightarrow F = SW \quad (30)$$

This objective is constrained by the following:

1. Electrical power balance

The sum of power generated by the EH components in addition to electricity supplied by the electrical grid should be equal to the total electrical requirements, EHP, WD, and electrical losses of the system. So, the electric power balance constraint is defined as:

$$P_G + P_W + P_{PV} + P_{CHP} \pm P_{ES} = P_{EHP} + P_{WD} + P_{ED} + P_{loss} \quad (31)$$

where P_{ES} is the ES power (MW) and P_{loss} is the value of total power losses (MW) which is illustrated in [32].

2. Heating power balance

The sum of heat produced by CHP, GB, and EHP should be enough to supply heating demand in the EH. Therefore, the heat power balance can be met as follows:

$$Q_{HOU} + Q_{CHP} + Q_{GB} + Q_{EHP} \pm Q_{HS} = Q_{HD} + Q_{loss} \quad (32)$$

where Q_{HS} is the power of HS (MWth) and Q_{loss} is the heat loss (MWth) which is illustrated in detail in [32].

3. Water balance

The total generated water from the WD unit, WS, and WG should supply water demand in the EH. This constraint can be met by water balance as follows:

$$W_{WD} + W_{WG} \pm W_{WS} = W_{WR} \quad (33)$$

where W_{WS} is the volume of water storage (m^3).

4. Line flow and bus voltage limits:

$$\left.\begin{array}{c}\left|S_{flow,i}\right| \leq S_{flow,i}^{max}\\ V_i^{min} \leq V_i \leq V_i^{max}\end{array}\right\} \quad (34)$$

where $S_{flow,i}$ is the flow of apparent power in the *i*th line (MVA), $S_{flow,i}^{max}$ is the maximum apparent power flow limit in the ith line (MVA), and V_i^{max}, V_i^{min} are the maximum and minimum voltage limits of the ith bus (p.u).

5. Ramp rate limits for thermal generator, CHP, EHP, GB, WD, WG, and HOU:

$$\left.\begin{array}{c}P_{i,t-1} - P_{i,t} < DR_i\\ P_{i,t} - P_{i,t-1} < UR_i\end{array}\right\} \quad (35)$$

where DR_i, UR_i are the down-ramp rate and up-ramp rate limit of the ith unit, $P_{i,t-1}$ is the power of the *i*th unit at time $t-1$, and $P_{i,t}$ is the power of the ith unit at time t.

6. Real operating power limits for wind, PV, CHPs, EHP, and WD units:

$$\left.\begin{array}{c}0 \leq P_W \leq P_{W_r}\\ 0 \leq P_{PV} \leq P_{PV(K_{t\ max})}\\ P_{CHP}^{min} \leq P_{CHP} \leq P_{CHP}^{max}\\ 0 \leq P_{EHP} \leq P_{EHP}^{max}\\ 0 \leq P_{WD} \leq P_{WD}^{max}\end{array}\right\} \quad (36)$$

where $P_{PV(K_{t\ max})}$ is PV output power at maximum solar radiation (MW), P_{CHP}^{max}, P_{CHP}^{min} are the maximum and minimum power outputs of a cogeneration generator (MW), P_{EHP}^{max} is the maximum installed capacity of the EHP unit (MW), and P_{WD}^{max} is the maximum installed capacity of the WD unit (MW).

7. Heat limits for CHPs, HOUs, EHP, and GB:

$$\left.\begin{array}{c}Q_{CHP}^{min} \leq Q_{CHP} \leq Q_{CHP}^{max}\\ Q_{HOU}^{min} \leq Q_{HOU} \leq Q_{HOU}^{max}\\ 0 \leq Q_{EHP} \leq Q_{EHP}^{max}\\ 0 \leq Q_{GB} \leq Q_{GB}^{max}\end{array}\right\} \quad (37)$$

where Q_{CHP}^{max}, Q_{CHP}^{min} are the maximum and minimum heat outputs of the CHP generator (MWth), Q_{HOU}^{max}, Q_{HOU}^{min} are the maximum and minimum heat outputs of HOU (MWth), Q_{EHP}^{max} is the maximum installed capacity of the EHP unit (MWth), and Q_{GB}^{max} is the maximum installed capacity of a GB unit (MWth).

8. Charging/discharging limits of ESSs:

$$P_{ESS_i}^{min} \leq P_{ESSC_i}, \qquad P_{ESSD_i} \leq P_{ESS_i}^{max} \quad (38)$$

where $P_{ESS_i}^{max}$,$P_{ESS_i}^{min}$ are the maximum and minimum power limits of ESSs (MW).

9. Initial and final energy in ESSs:

$$E_{ESS_i}^{min} \leq E_{ESS_i}^{ini} + P_{ESSC_i}\Delta t - P_{ESSD_i}\Delta t \leq E_{ESS_i}^{max} \quad (39)$$

where $E_{ESS_i}^{max}$, $E_{ESS_i}^{min}$ are the maximum and minimum limit energy of ESSs (MWh), and $E_{ESS_i}^{ini}$ is the initial energy of the ith ESSs (MWh).

5.2. Optimization Algorithm

The AOA is a novel optimization algorithm that has been proposed to address real-world problems [43]. This optimizer derives the idea of his work from Archimedes' principles states, and it depends on the behavior of the force exerted when an object is partially or completely immersed in a liquid. Similar to most swarm optimizers and metaheuristic algorithms, the AOA proposes population-based solutions. In this case, the proposed solutions are represented by immersed objects. The optimization process starts with proposing an initial set of random particles (objects/locations/solutions). Each particle has its size, density, and acceleration which are updated in an iterative process. Theoretically, the AOA includes many explorations and exploitations processes because it is a global optimization algorithm. Figure 4 introduces the pseudo-code of the AOA; including population initialization, population assessment, and parameter updating [44].

Archimedes Optimization Algorithm (AOA) : Pseudo code

- **Inputs:**
  $N, T_{max}, C_1, C_2, C_3, and\ C_4$
- **Output:**
  Object with best fitness value
- **Steps:**

1. Initialize objects population with random positions, densities and volumes using:

$$O_i = lb_i + rand \times (ub_i - lb_i); i = 1, 2, \ldots, N$$
$$den_i = rand$$
$$vol_i = rand$$
$$acc_i = lb_i + rand \times (ub_i - lb_i)$$

2. Evaluate initial population and select the one with the best fitness value
3. Set iteration counter $t = 1$
4. While $t \leq t_{max}$ do

For each object i do

5. Update density and volume of each object by

$$den_i^{t+1} = den_i^t + rand \times (den_{best} - den_i^t)$$
$$vol_i^{t+1} = vol_i^t + rand \times (vol_{best} - vol_i^t)$$

6. Update transfer and density decreasing factors by

$$= exp\left(\frac{t - t_{max}}{t_{max}}\right)$$
$$d^{t+1} = exp\left(\frac{t - t_{max}}{t_{max}}\right) - \left(\frac{t}{t_{max}}\right)$$

if $TF \leq 0.5$ then ▶Exploration phase

7. Update acceleration and normalization acceleration by

$$den_i^{t+1} = \frac{den_{mr} + vol_{mr} \times acc_{mr}}{den_i^{t+1} \times vol_i^{t+1}}$$
$$acc_{i-norm}^{t+1} = u \times \frac{acc_i^{t+1} - \min(acc)}{\max(acc) - \min(acc)} - 1$$

8. Update position by

$$x_i^{t+1} = x_i^t + C_1 \times rand \times acc_{i-norm}^{t+1} \times d \times (x_{rand} - x_i^t)$$

else ▶Exploration phase

9. Update acceleration and normalization acceleration by

$$den_i^{t+1} = \frac{den_{best} + vol_{best} \times acc_{best}}{den_i^{t+1} \times vol_i^{t+1}}$$
$$acc_{i-norm}^{t+1} = u \times \frac{acc_i^{t+1} - \min(acc)}{\max(acc) - \min(acc)} - 1$$

10. Update direction flag F by

$$F = \begin{cases} +1\ if\ P \leq 0.5 \\ -1\ if\ P > 0.5 \end{cases}$$
$$P = 2 \times rand - C_4$$
$$x_i^{t+1} = x_{best}^t + F \times C_2 \times rand \times acc_{i-norm}^{t+1} \times d \times (T \times x_{best} - x_i^t)$$

end if
end for

11. Evaluate each object and select the one with the best fitness value
12. Set $t = t + 1$

end while
return object with the best fitness value

Algorithm parameters	
N	Population size
T_{max}	Maximum iterations
C_1, C_2, C_3, C_4	Control variables
O_i	i^{th} object in a population
lb_i, ub_i	Lower and upper bounds of the search-space
vol_i, den_i, acc_i	Volume, density and acceleration for each ith object
$rand$	**D** dimensional vector randomly generates number between [0, 1]
$vol_{best}, den_{best}, acc_{best}$	Volume, density and acceleration associated with the best object found
TF	Transfer operator
t	Iteration number
d	Density decreasing factor
$vol_{mr}, den_{mr}, acc_{mr}$	Volume, density and acceleration of random material
x_i	Position for each ith object

Figure 4. AOA Pseudo code.

5.3. Proposed Stochastic Planning Structure

The primary objective of stochastic planning is the total SW and emissions reduction since all demand requirements are satisfied. In this paper, the AOA algorithm was used to improve the EH system performance. The suggested methodology for EH planning taking into account electrical, thermal, and water energy balancing is illustrated in Figure 5. Heat network, electrical grid, water network specifications, economic data, weather data, power generation of RESs, and various types of demands are the inputs to the proposed system. Further, the electrical, thermal, and water energy balances are analyzed in the planning procedure.

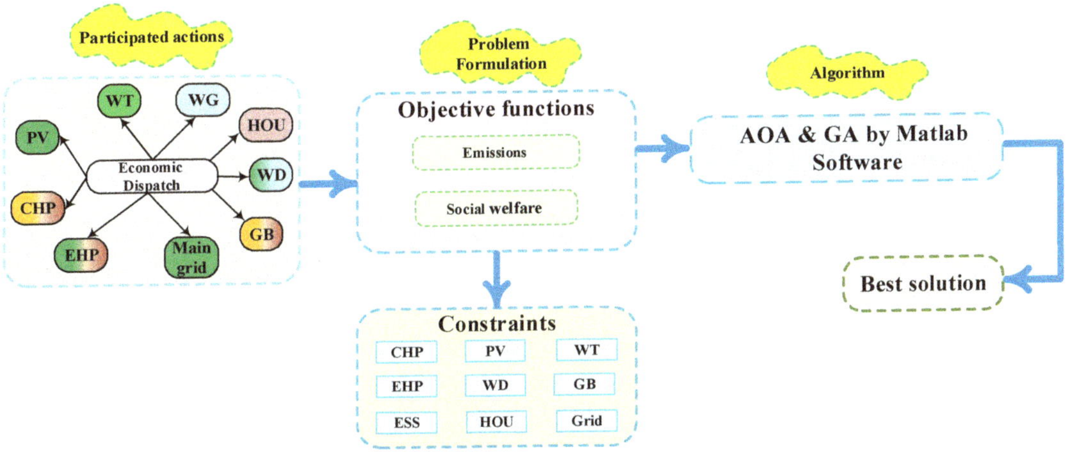

Figure 5. Proposed algorithm.

5.4. EH Methodology to Satisfy Electrical, Thermal, and Water Demand

In this study, electrical, thermal, and water demands are considered. The overall methodology for supplying all electrical, thermal, and water loads is discussed. In addition, priority should be given to the various components and networks of EH to supply aggregate demand. The methodology is illustrated in the following subsections.

5.4.1. A proposed Methodology to Satisfy Electrical Requirements

In the proposed EH, electrical loads are supplied by PV modules and WT. If the energy generated by RESs is more than the electrical demand, the surplus generated energy will be stored in the ES. Otherwise, if the energy produced by RESs is not sufficient to meet all the electrical loads, then the energy stored in the ES supplies the remaining part of the electrical loads. In all cases, if the energy produced from the RESs, and ES is not enough to supply demand requirements, the deficit energy is taken from CHP. If the energy produced from all EH units is not enough for demand, the required energy will be purchased from the electrical grid. On the other hand, whenever ES is fully charged and the energy produced from RESs and CHP is more than the electrical loads, the excess energy is sold to the electricity grid. Figure 6a illustrates the proposed methodology.

5.4.2. A Proposed Methodology to Satisfy Thermal Requirements

The GB, CHP, and EHP are considered heat generators in the proposed EH configuration. If the thermal energy generated by the CHP is greater than the heating requirements, the additional thermal energy is stored in HS. Otherwise, when the energy produced from CHP is not sufficient to supply the heating demand, the energy stored in the HS is used to feed the remainder of the load. In all cases, if the energy produced from the CHP, and HS, is not enough to supply demand requirements, the deficit energy is taken from EHP, GB, and HOU, respectively. Figure 6b shows the proposed heating demand-saving strategy.

5.4.3. A Proposed Methodology to Satisfy Water Requirements

In the proposed EH system, the WD unit is considered the main water source. If the water generated by the WD unit is greater than the water demand for the loads, the additional water is stored in WS. In cases where the energy produced from the WD unit and WS is not enough to supply demand requirements, the deficit water is taken from the public WG. On the other hand, when WS is charged and the water produced from the WD unit is more than the water demand, the excess water is sold to the WG. Figure 6c shows the proposed water demand-saving strategy.

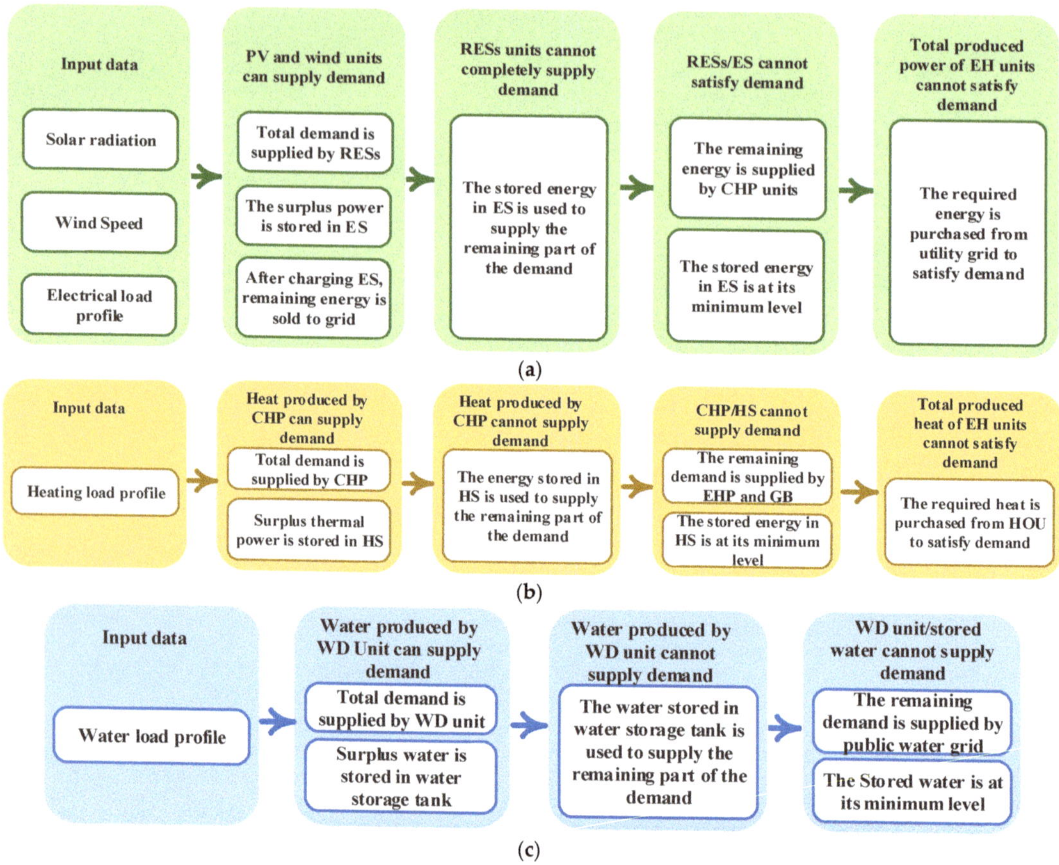

Figure 6. Proposed methodologies to satisfy load requirements (**a**) electrical (**b**) heat (**c**) water.

6. Simulation Results and Discussion

6.1. Simulation Setup

The methodology proposed is applied to an EH represented by a modified IEEE 5-bus electrical network over one day during a one-hour time interval. The modified IEEE 5-bus network, heat system, and water system are shown in Figure 7. Thermal, electrical, and water demands are assumed in the proposed EH as the curves shown in Figure 8 [19]. The electrical demand is divided into two parts, a fixed part (base load) and an elastic part (20% of the base demand). A percentage of the elastic part is fed only when the EH has a surplus of electrical power generation depending on the electricity price and amount of surplus power. Daily wind speed and solar radiation are shown in Figure 9 [32]. The data on the EH technologies are given in Appendix A. The optimization problem was solved for many case studies with various operating conditions and the results were analyzed. These simulations were carried out using MATLAB 2017a 64-bit version on a PC with an Intel (R) Core (TM) i5-8250U CPU 1.60 GHz, RAM 12 GB system, and 1 TB of storage.

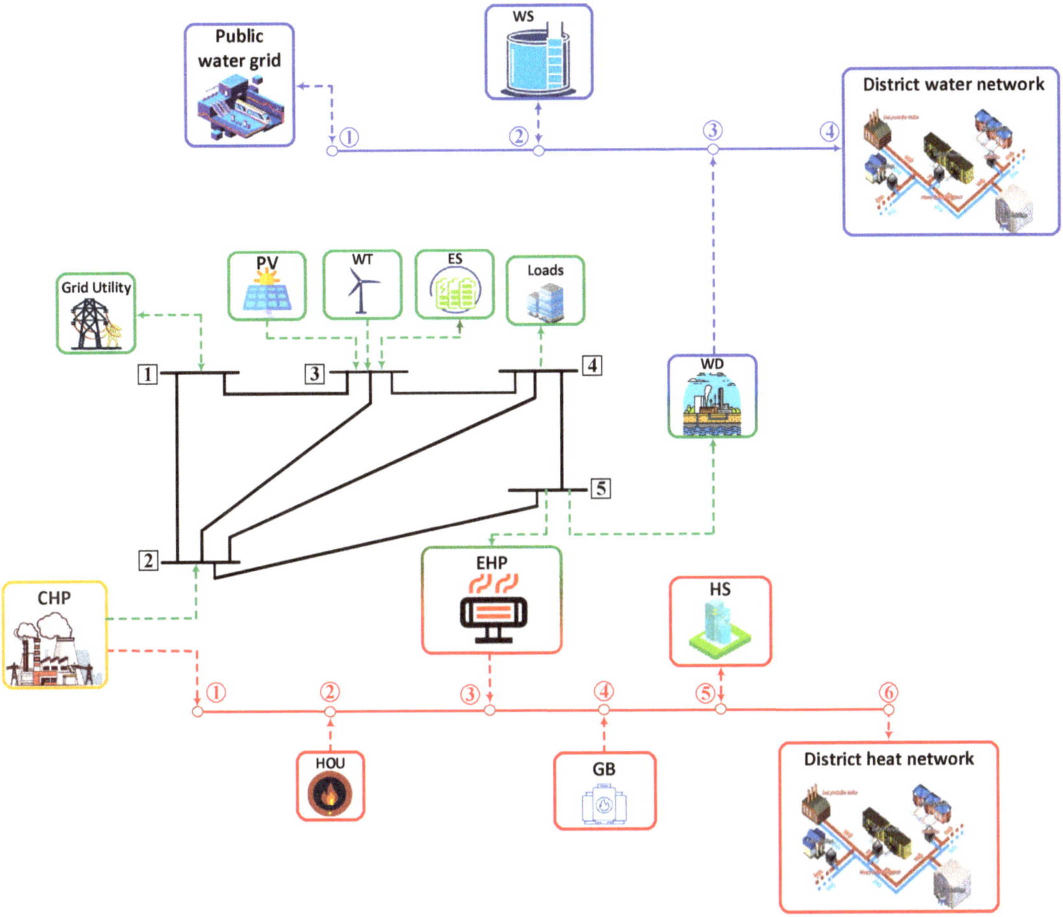

Figure 7. The layout of the proposed EH system.

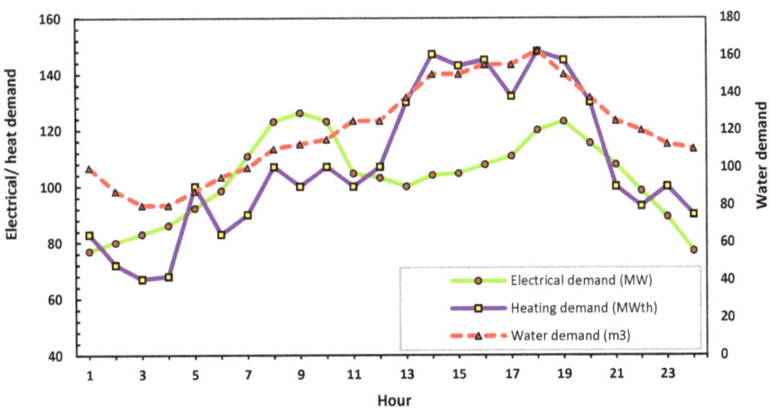

Figure 8. Energy carriers' demand for the day under study.

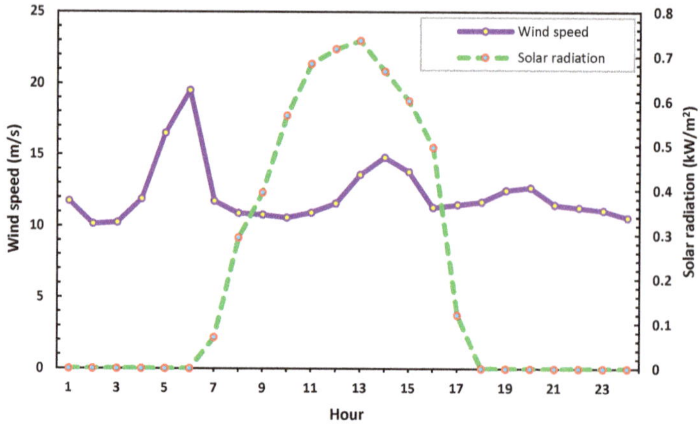

Figure 9. Wind speed and solar radiation.

6.2. Case Studies

Four cases are studied to explain the performance of CHP, WD, GB, and EHP, as shown in Table 2. In each case, many indicators are calculated, such as total SW, emissions, losses, and sold power/water to the electrical/water network are calculated (see Table 3).

Table 2. Case studies in this work.

Cases	Electricity/Heat/ Water Network	CHP	EHP	GB	RESs	WD	ESSs
Case 1 (base case)	✓	×	×	×	×	×	×
Case 2	✓	✓	×	×	×	×	×
Case 3	✓	✓	✓	✓	×	✓	✓
Case 4	✓	✓	✓	✓	✓	✓	✓

Table 3. Different EH configurations' daily operation results.

Case	SW ($)	Emissions (kg)	Energy Loss		Electrical Energy Requirement (MWh)	Power Sold to the Grid (MWh)	Water Sold to the Network (m^3)
			Electrical (MWh)	Heat (MWhth)			
Case 1 (base case)	275,467.99	5638.27	3.05	108.27	2466.40	-	-
Case 2	292,150.20	5075.92	2.71	128.20	2484.00	23.20	-
Case 3	336,786.09	5848.00	2.72	131.40	3151.24	90.44	-
Case 4	379,648.00	4603.00	4.12	131.04	4269.82	512.26	149.4

By analyzing the data in the previous table, it can be observed that:

- In case 1 (base case), there is no EH and the loads are fed directly from electricity, HOU, and WG, respectively. This case was studied to explore the impact of EH on total SW, emissions, and losses. As reported in Table 3, total SW, emissions, and electrical and heat losses are 275,467.99 USD, 5638.27 kg, 3.05 MWh, and 108.27 MWhth, respectively. Additionally, there is no electrical power or water sold to the electricity grid and water network, respectively. The total electrical demand, in this case, is 2466.4 MWh (the base load curve).
- In case 2, CHP was integrated with the EH to show the effect of this unit on the performance parameters. The main source to supply heating demand requirements is the CHP unit during the day because of the low price of natural gas as shown in Figure 10a,b. With the integration of the CHP unit, SW increased by 5.71% and emissions decrease by 9.97%, respectively, compared with the base case. Additionally, the elastic part of the electrical demand supplied by the hub increased by 0.7% and the hub sold 23.2 MWh to the electricity grid.

- In case 3, EHP, GB, WD units, and ESSs are added to the CHP in the previous case. This case results in an increase of 3151.24 MWh in the electrical demand because of adding the EHP and WD units. So, the total SW increased to 336,786.09 USD. On the other hand, total emissions increased a little compared to the base case. Figure 11 presents the results for this case.
- In the last case (case 4), CHP, RESs, ESSs, EHP, WD, and GB are operating at the same time. This configuration enables the EH not only to meet demand requirements but also to sell electricity and water to the electrical and water networks during light load (11 AM to 5 PM in the electrical system and 1 AM to 10 AM in the water network) as shown in Figure 12a–c. So, compared with the base case, all performance parameters are improved, and total SW and emissions are 379,648 USD and 4603 kg. In addition, the total electrical demand supplied by the hub increased to 4269.82 and the hub sold 512.26 MWh to the electricity grid and 149.4 m^3 to the water network.

Figure 13 shows the available power for both wind turbines and PV panels compared to the scheduled values. It has been found that this power changes throughout the day depending on the availability of wind and PV production for the plants and the electricity prices. With the increase in demand for electricity, electricity prices are relatively high. Thus, wind turbines and PV panels are at high risk when delivering a large amount of power generation at that time because they may contain high imbalanced charges in event of non-fulfillment of the scheduled power. Meanwhile, they can also provide high power generation when demand is low, hence the electricity prices are becoming lower.

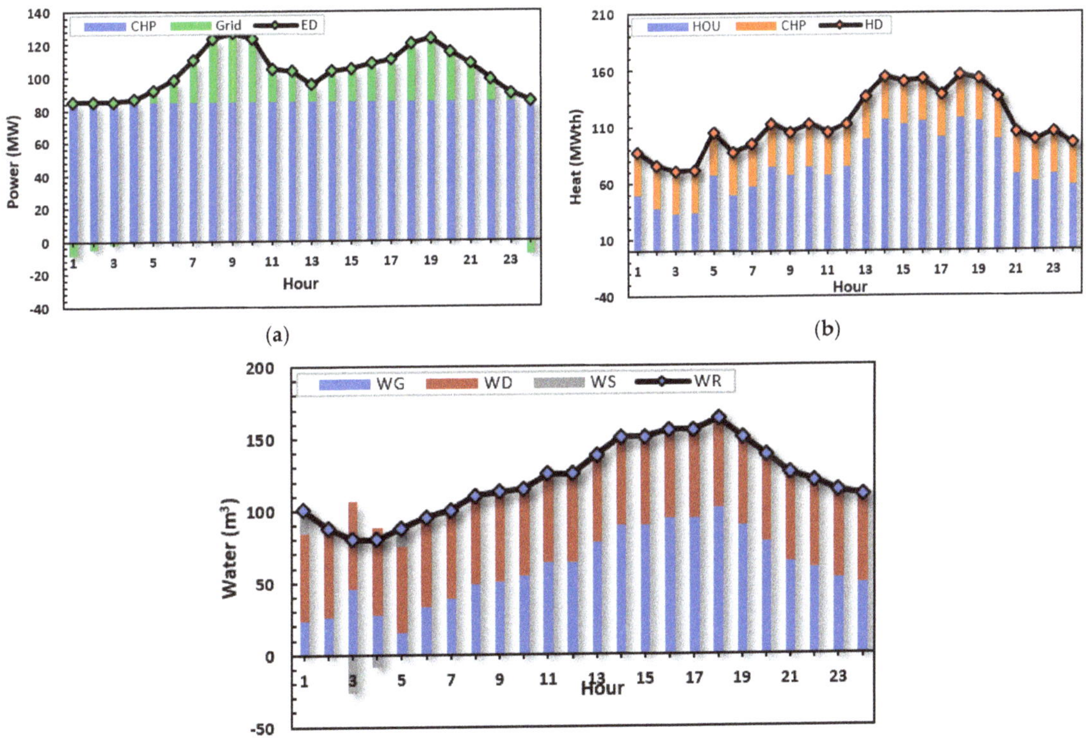

(a) (b)

(c)

Figure 10. Generation outputs for case 2 (**a**) electricity (**b**) heat (**c**) water.

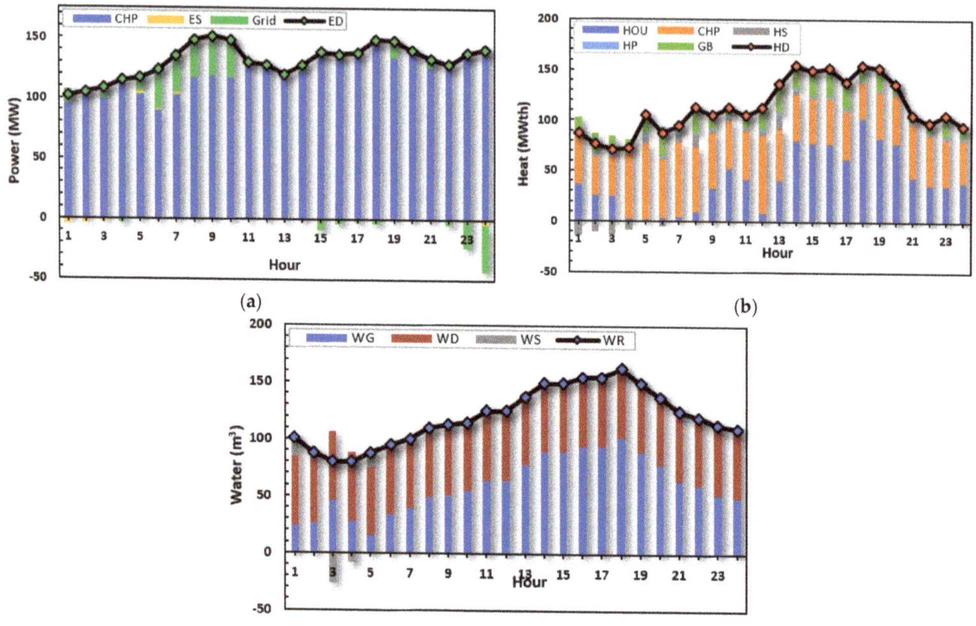

Figure 11. Generation outputs for case 3 (**a**) electricity (**b**) heat (**c**) water.

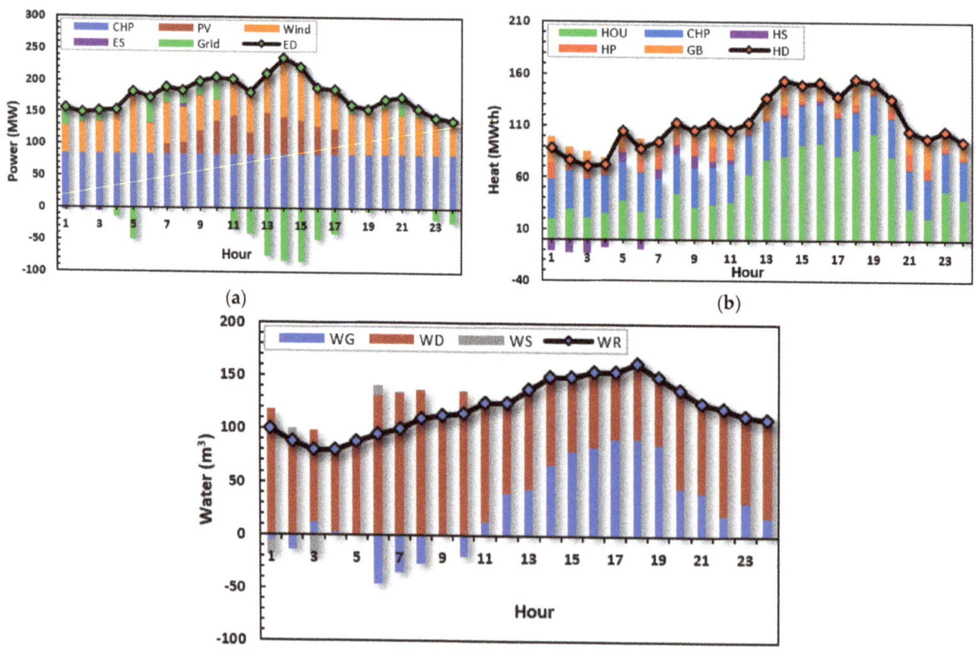

Figure 12. Generation outputs for case 4 (**a**) electricity (**b**) heat (**c**) water.

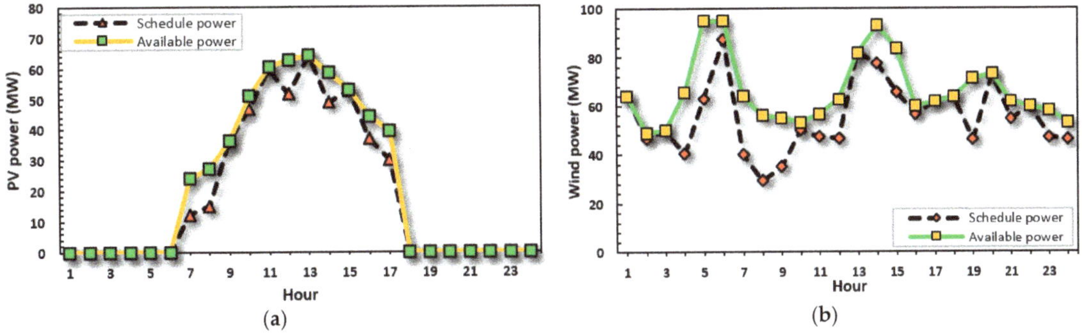

Figure 13. Comparison of available and scheduled power for: (**a**) PV power plant. (**b**) wind farm.

6.3. Proposed Algorithm Validation

To verify the efficiency of the AOA, the results of case 4 (the best case) solved by the AOA are compared with that obtained by GA. Table 4 shows the characteristics of each algorithm. Although GAs is often criticized for being too slow, the studied problem is a planning problem that only needs to be solved offline. That is why the simulation time is not of great importance as the capability to converge. It was noticed that the convergence curve of the AOA is quite smooth and with no oscillations, as shown in Figure 14. Additionally, the AOA needs less time than GA.

Table 4. AOA and GA characteristics.

AOA Characteristics	
Population Size	50
C_1	2
C_2	6
C_3	2
C_4	0.5
Number of iterations	100
GA Characteristics	
Population size	50
Number of iterations	100

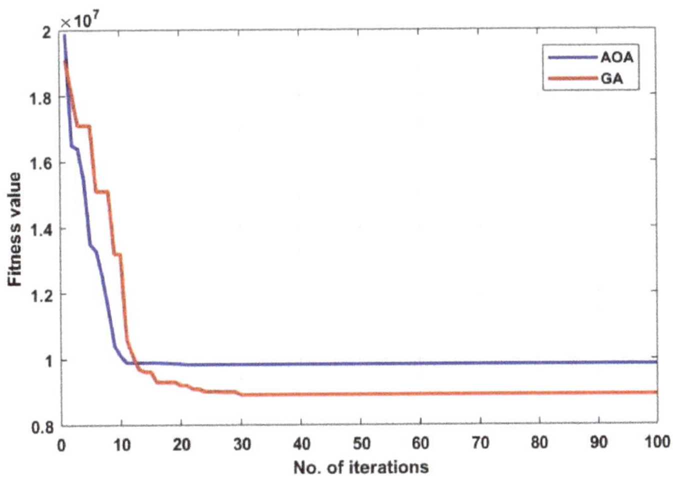

Figure 14. Fitness function convergence curves.

Tables 5 and 6 present the hourly optimal values of EH variables for case 4 by applying the AOA and GA, respectively. It is clear from these tables that the power/water taken from the electrical/water network in the case of the AOA is less than GA, which is reflected in the sold capacity of the network, as shown in Figure 15a,b. Additionally, the power taken from the CHP unit increases in the case of GA. On the other hand, the power taken from RESs increases in the case of the AOA which leads to a decrease in the total emissions, as shown in Figure 16 and Table 7, a decrease in the total losses, as shown in Figure 17a,b, and an increase in the total electrical demand, as shown in Figure 18.

Table 5. Hourly optimal values generated from the EH system of case 4 solved by AOA.

Hr.	Grid (MW)	PV (MW)	Wind (MW)	CHP-E (MW)	ES (MW)	HOU (MWth)	CHP-H (MWth)	EHP (MWth)	GB (MWth)	HS (MWth)	WG (m^3)	WD (m^3)	WS (m^3)
1	8.69	0.00	43.84	85.10	−3.52	20.31	37.53	15.93	25.19	−11.23	−5.59	118.33	−12.23
2	−7.68	0.00	48.92	85.00	−3.95	29.00	37.51	15.93	6.37	−12.88	−13.83	88.25	12.58
3	−16.15	0.00	49.87	85.01	−5.81	21.03	37.51	15.99	10.37	−14.06	11.80	86.62	−18.92
4	−12.82	0.00	65.55	85.01	2.60	24.84	37.50	10.79	6.11	−7.48	2.96	76.90	−0.36
5	−45.59	0.00	95.00	85.00	1.98	37.48	37.50	4.54	16.62	8.93	−1.55	86.92	1.63
6	18.01	0.00	47.75	85.01	0.26	26.81	37.51	12.44	20.47	−9.55	−46.37	131.52	9.34
7	3.20	15.98	64.12	85.00	0.03	21.12	37.50	13.85	12.61	9.31	−34.85	133.29	2.06
8	14.17	17.57	56.05	85.00	4.39	44.70	37.50	6.62	14.92	8.55	−26.89	135.94	0.45
9	4.44	36.21	55.09	85.02	0.00	31.21	37.50	14.21	9.97	11.79	0.77	111.25	0.48
10	36.60	50.92	34.33	85.03	0.00	33.21	37.52	14.63	22.49	4.51	−20.37	135.18	0.69
11	−14.67	60.45	56.52	85.02	0.00	35.59	37.50	11.03	17.03	3.90	12.57	111.93	0.00
12	−52.25	33.73	62.70	85.07	0.00	63.19	37.51	3.86	6.50	1.26	39.22	85.18	0.10
13	−58.47	64.17	61.34	85.04	0.00	77.26	37.51	3.61	17.98	0.17	43.18	94.82	0.00
14	−38.49	57.30	93.09	85.01	0.00	80.59	37.50	8.73	24.20	3.16	66.21	83.79	0.00
15	−40.87	53.02	83.59	85.14	0.00	91.68	37.52	3.88	14.81	2.06	79.19	70.81	0.00
16	−16.40	44.29	59.84	85.03	0.00	93.49	37.51	3.56	14.43	3.03	83.61	70.89	0.00
17	10.10	39.38	61.74	85.00	0.00	80.94	37.53	5.97	14.04	0.00	91.29	63.21	0.00
18	12.12	0.00	63.65	85.01	0.00	87.14	37.50	7.55	23.10	0.00	91.55	70.45	0.00
19	31.11	0.00	71.25	85.02	0.00	102.84	37.52	2.77	8.85	0.00	85.64	64.36	0.00
20	28.41	0.00	73.13	85.01	0.00	80.31	37.50	9.79	8.62	0.00	44.10	93.90	0.00
21	17.68	0.00	61.75	85.00	0.00	30.60	37.51	15.99	21.01	0.00	40.87	83.63	0.00
22	38.54	0.00	59.85	85.00	0.00	21.92	37.52	10.06	28.60	0.00	19.50	100.50	0.00
23	2.86	0.00	57.95	85.00	0.00	48.03	37.54	3.86	15.82	0.00	31.93	80.57	0.00
24	−4.01	0.00	53.20	85.01	−0.24	39.86	37.51	3.41	14.03	0.00	17.42	92.08	0.00
Tot.	255.93	473.01	1480.11	2040.55	−4.26	1223.16	900.26	218.99	374.15	1.47	761.81	2270.3	−4.18

Table 6. Hourly optimal values generated from the EH system of case 4 solved by GA.

Hr.	Grid (MW)	PV (MW)	Wind (MW)	CHP-E (MW)	ES (MW)	HOU (MWth)	CHP-H (MWth)	EHP (MWth)	GB (MWth)	HS (MWth)	WG (m^3)	WD (m^3)	WS (m^3)
1	−21.73	0.00	64.13	85.00	−6.00	40.98	37.50	2.00	16.24	−8.93	6.27	119.98	−25.75
2	−14.61	0.00	46.78	86.25	−1.21	8.75	38.95	3.03	27.15	−1.47	−20.18	89.84	17.34
3	−15.61	0.00	49.88	89.65	−0.24	16.37	37.50	4.51	24.93	−12.01	−8.90	88.40	0.00
4	−4.29	0.00	40.32	85.04	6.00	31.79	38.12	4.97	5.00	−8.00	18.71	85.55	−24.76
5	−22.34	0.00	63.14	85.26	0.00	36.89	41.01	2.07	23.69	1.81	−1.79	87.53	1.26
6	−28.69	0.00	87.24	85.74	0.00	44.12	38.45	5.71	5.00	−5.84	−1.25	95.75	0.00
7	24.79	12.23	39.95	85.00	0.00	36.21	37.50	4.76	5.00	10.92	−19.69	120.19	0.00
8	44.97	15.06	29.42	85.00	0.00	53.16	38.17	3.17	18.11	0.00	−34.33	131.25	12.58
9	36.84	36.23	35.04	85.36	0.00	31.62	37.50	9.37	17.71	8.80	−32.00	132.22	12.28
10	5.00	46.71	49.85	89.67	0.00	32.63	38.71	9.28	18.72	12.91	−20.35	135.85	0.00
11	−17.91	59.81	47.22	85.00	0.00	51.15	37.50	11.23	5.00	0.00	−0.44	124.94	0.00
12	−34.19	51.63	46.75	87.02	0.00	45.82	38.25	3.65	25.02	0.00	6.69	117.81	0.00
13	−89.16	64.18	81.70	88.01	0.00	81.19	43.49	6.65	5.00	0.00	38.28	99.72	0.00
14	−45.57	48.86	77.34	85.00	0.00	87.96	37.50	10.76	17.85	0.00	42.39	107.61	0.00
15	−47.54	52.72	65.86	85.28	0.00	88.09	37.89	3.62	20.52	0.00	21.35	128.65	0.00
16	−24.46	36.80	56.46	85.00	0.00	94.72	37.50	2.00	17.96	0.00	30.90	123.60	0.00
17	−9.85	29.96	61.75	85.00	0.00	61.80	37.50	10.00	29.41	0.00	59.18	95.32	0.00
18	15.42	0.00	63.63	85.23	0.00	75.21	49.37	4.40	26.68	0.00	61.12	100.88	0.00
19	51.59	0.00	46.20	85.00	0.00	71.96	50.41	9.50	20.46	0.00	41.40	108.60	0.00
20	4.35	0.00	72.89	88.23	0.00	87.47	37.95	2.77	8.17	0.00	6.69	131.31	0.00
21	17.57	0.00	54.48	94.34	0.00	44.30	41.41	9.46	9.92	0.00	19.27	105.23	0.00
22	−13.50	0.00	59.64	89.10	0.00	35.40	41.86	3.50	17.33	0.00	34.92	85.08	0.00
23	−1.23	0.00	47.02	85.15	0.00	51.81	39.13	6.42	7.71	0.00	35.89	76.61	0.00
24	−16.60	0.00	46.15	86.22	0.00	35.79	37.50	2.00	19.66	0.00	6.59	102.91	0.00
Tot.	200.53	454.18	1332.81	2075.53	−1.45	1245.20	950.67	134.85	392.25	−1.81	429.65	2594.82	−7.05

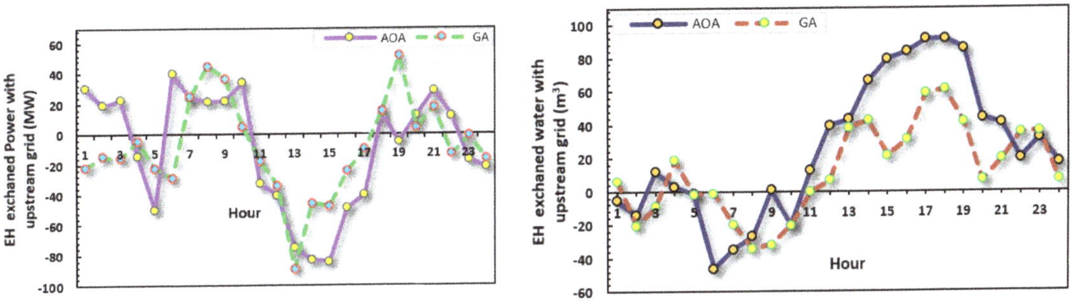

(a) (b)

Figure 15. EH exchanged energy (**a**) EH exchanged power with the upstream grid (**b**) EH exchanged water with the water network.

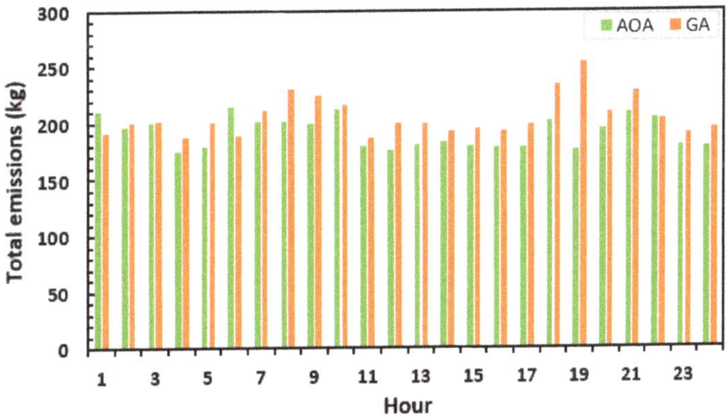

Figure 16. EH emissions.

Table 7. The results for daily operation in case 4.

Algorithm	SW (USD)	Emission (kg)	Energy Loss		Electrical Energy Demand (MWh)	Power Sold to the Grid (MWh)	Water Sold to the Network (m^3)	Fitness Function
			Electrical (MWh)	Heat (MWhth)				
GA	364,203.14	4936.9	4.82	134.18	4061.62	407.28	138.89	8.9286×10^6
AOA	379,648.53	4603.0	4.12	131.04	4269.82	512.26	149.40	9.8439×10^6
Improvement (%)	4.06%	6.76%	14.52%	2.34%	4.87%	20.49%	7.03%	10.25%

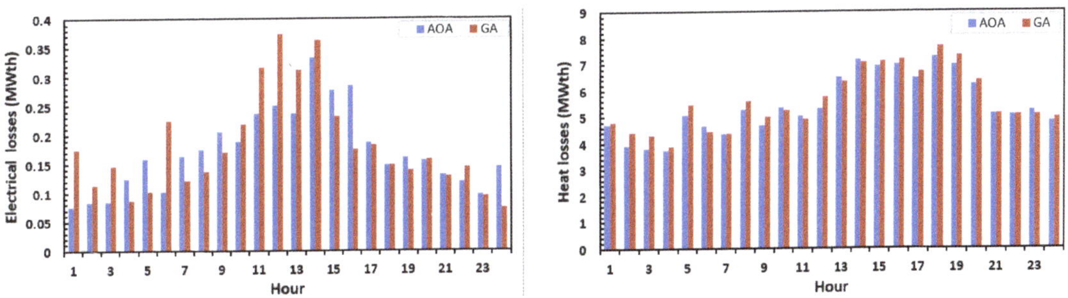

(a) (b)

Figure 17. EH losses (**a**) electrical (**b**) heat.

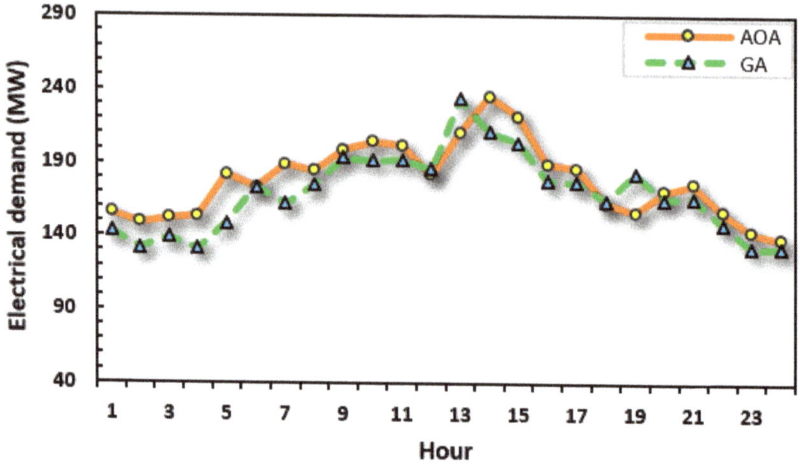

Figure 18. EH electrical demand.

Table 7 summarized the main results obtained from the two algorithms. It can be observed that all operating parameters of the EH system are improved when the AOA is used to solve the optimization problem. Total emissions and electrical/heat losses are reduced by 6.76%, 14.52%, and 2.34%, respectively, and total electrical demand, total SW, electrical power sold to the grid, and amount of water sold to the water network are increased by 4.06%, 20.49%, and 7.03%, respectively.

7. Conclusions

An optimum operation and configuration for an EH integrated with CHP was developed in this paper. The proposed configuration of the EH is composed of different types of energy sources, generation, and ESSs to feed different demands. The main objective was to maximize the total SW by reducing total operating costs and reducing total emissions value. The emissions of the EH were transformed into a penalty cost function via the emissions coefficient. Many case studies were proposed to satisfy the demands of the EH. The optimal operation problem was solved using the AOA algorithm to determine the energy generated from each source and satisfy objective functions and operational constraints. The optimization problem was implemented in the MATLAB environment. Four cases with different configurations were studied to analyze the performance of CHP with the EH. According to the results, the best performance in terms of configuration, operation, and emissions was obtained with the hub including a combination of CHP, PV, WT, EHP, GB, WD, and ESSs. The AOA was validated by solving the same problem of case 4 by GA. The results proved that the performance of the AOA is better than GA in terms of total operating cost and emissions. The numerical results illustrated the following points:

With the integration of the CHP unit, SW increased by 5.71% and emissions decreased by 9.97%, respectively, compared with the base case. Additionally, the total electrical demand supplied by the hub increased by 0.7% and the hub sold 23.2 MWh to the electricity grid.

- In case 4, all performance parameters were improved; the total SW and emissions are 379,648 USD and 4603 kg. Additionally, the total electrical demand supplied by the hub increased to 4269.82 MWh; the hub sold 512.26 MWh to the electricity grid and 149.4 m^3 to the water network.
- All operating parameters of the system (case 4) were improved by applying the AOA to solve the optimization problem. Total emissions and electrical/heat losses were reduced by 6.76%, 14.52%, and 2.34%, respectively, and total electrical demand, total SW, electrical power sold to the grid, and amount of water sold to the water network were increased by 4.06%, 20.49%, and 7.03%, respectively, compared to case 4 solved with GA.
- The number of emissions was converted to cost by using a penalty factor h. The aim was achieved and emissions were reduced, which in turn reduced all associated social and environmental aspects.

In future work, the problem must be represented as a long-term economic model with detailed modeling of the water distribution system, as well as taking into account the losses in the water system and all types of emissions.

Author Contributions: M.I.E.-A.: Conceptualization, Methodology, and Validation; Resources, Visualization, and Investigation; M.M.S.: Data curation, Writing, and Original draft preparation; A.A.E.: Supervision, Writing, Reviewing, and Editing the Manuscript. All authors have read and agreed to the published version of the manuscript.

Funding: Publication fees of this article have been covered by Mansoura University, Egypt.

Institutional Review Board Statement: Not applicable.

Informed Consent Statement: Not applicable.

Data Availability Statement: Not applicable.

Acknowledgments: The authors would like to thank the financial support provided by funding 60 research proposals for the master's and Ph.D. students' projects, which have received funding from the research fund account of Mansoura University. The publication fees of this article have been supported by Mansoura University.

Conflicts of Interest: The authors declare no conflict of interest.

Nomenclature

A. Abbreviations	
AOA	Archimedes optimization algorithm
CHP	Combined heat and power
ED	Electric demand
EH	Energy hub
EHP	Electric heat pump
ES	Electric storage
ESS	Energy storage system
GA	Genetic algorithm
GAMS	General algebraic modeling language
GB	Gas boiler
HD	Heat demand
HOU	Heat only unit
HS	Heat storage
MILP	Mixed integer linear programming
MINLP	Mixed-integer nonlinear programming
NSGA-II	Non-dominated sorting genetic algorithm-II
C. Variables	
PV	Photovoltaic
QPSO	Quantum Particle Swarm Optimization
RES	Renewable energy resource
SW	Social welfare
WD	Water desalination
WG	Water grid
WR	Water demand
WS	Water storage
WT	Wind turbine
B. Input values	
$a_E, b_E, d_E, \gamma_E, \delta_E$	Emission coefficients of the thermal generators
a_G, b_G, c_G	Conventional generator cost coefficients
$a_{Co}, b_{Co}, c_{Co}, d_{Co}, e_{Co}, f_{Co}$	CHP generation cost coefficients
a_{EHP}	EHP unit cost coefficient ($/MW)
a_{GB}	GB unit cost coefficient ($/MW)
c	Form factor
$c_{p.W}, c_{p.PV}$	Over-estimation cost coefficient of WT and PV generators ($/MW)
$c_{r.W}, c_{r.PV}$	Under-estimation cost coefficient of WT and PV generators ($/MW)
C_{dsi}, C_{chi}	Discharging and charging cost of the ith ESSs ($/MW)
COP	EHP coefficient of performance
d_W, h_{PV}	Operating cost of WT and PV generators ($/MW)
DR_i, UR_i	Down-ramp rate and up-ramp rate limit of the ith unit
$E_{ESS_i}^{max}$, $E_{ESS_i}^{min}$	Maximum and minimum limit energy of ESSs (MWh)
$E_{ESS_i}^{ini}$	Initial energy of the ith ESSs (MWh)

P_{CHP}^{max}, P_{CHP}^{min}	Maximum and minimum power outputs of cogeneration generator (MW)
P_{EHP}^{max}	Maximum installed capacity of the EHP unit (MW)
P_{WD}^{max}	Maximum installed capacity of the WD unit (MW)
Q_{CHP}^{max}, Q_{CHP}^{min}	Maximum and minimum heat outputs of the CHP generator (MWth)
Q_{HOU}^{max}, Q_{HOU}^{min}	Maximum and minimum heat outputs of HOU (MWth)
Q_{EHP}^{max}	Maximum installed capacity of the EHP unit (MWth)
Q_{GB}^{max}	Maximum installed capacity of GB unit (MWth)
$S_{flow,i}^{max}$	Maximum apparent power flow limit in the ith line (MVA)
V_i^{max}, V_i^{min}	Maximum and minimum voltage limits of the ith bus (p.u)
V_{HOU}, M_{HOU}, R_{HOU}	Heat generation cost coefficients of HOU
v_i, v_o, v_r	Cut-in speed, cut-out speed, and rated speed (m/s)
γ_{GB}	GB coefficient of performance
η_{WD}	Performance coefficient of WD (m^3/MW)
η_{PV}	PV cell efficiency
α, β	Beta *pdf* parameters
$\lambda_e, \lambda_h, \lambda_w$	Electrical, heat, and water consumed cost
$f_W(P_W)$, $f_{PV}(P_{PV})$	Weibull *pdf* and beta *pdf* of WT and PV generator
$cost^{max}$	Highest predictable value of operating cost ($)
C_{ESSi}	ESS power cost ($)
C_{EHP}	Power absorbed by EHP cost ($)
C_{GB}	Natural gas absorbed by GB cost ($)
C_{HOU}	Produced heat by HOU cost ($)
C_{WD}	Power absorbed by WD cost ($)
C_{WG}	Total cost of water produced from WG ($)
C_E	Corresponding cost of emissions ($)
C_G	Total cost of power produced from grid ($)
C_{CHP}	Total cost of energy produced from CHP ($)
C_W, C_{PV}	Total cost of WT and PV generators ($)
E_G	Total value of emissions (kg)
E_G^{max}	Highest estimated value for emissions (kg)
$G(t)$	Solar radiation at time t (kW/m^2)
$P_{GB}(t)$	Quantity of natural gas absorbed by GB at time t (MW)
$P_{EHP}(t)$	EHP electrical energy produced (MW) at time t
$P_{WD}(t)$	Consumed electricity by WD (MW)
P_G	Power of the thermal generator (MW)
P_{CHP}, Q_{CHP}	Electrical and heating outputs of the CHP (MW/MWth)
P_W, P_{PV}	Scheduled output of WT and PV generator (MW)
P_{dsi}, P_{chi}	Discharging and charging power of the ith ESSs (MW)
P_{ES}	ES power (MW)
P_{loss}	Total power losses (MW)
$P_{i,t-1}$	Power of the ith unit at time $t-1$

$\hbar$	Penalty rate factor ($/kg)	$P_{i,t}$	Power of the ith unit at time t
k	Scale factor	Q_{HS}	Power of HS (MWth)
K_{WD}	WD unit cost coefficient ($/MW)	Q_{loss}	Heat loss (MWth)
K_{WG}	Cost coefficient of the WG ($/m^3)	$Q_{EHP}(t)$	EHP thermal energy produced at time t (MWth)
L, C, P	Output loads, coupling matrix, input energy carriers	$Q_{GB}(t)$	EHP thermal energy produced at time t (MWth)
P_{PV_r}, P_{W_r}	PV and WT rated power (MW)	$S_{flow,i}$	Flow of apparent power in i^{th} line (MVA)
$P_{PV_{av}}$, $P_{W_{av}}$	Available PV and WT power (MW)	$v(t)$	Actual wind speed at time t (m/s)
P_{ED}, Q_{HD}, W_{WR}	Electrical, heat, and water demand	W_{WG}	Volume of the water produced from the WG (m^3)
$P^{max}_{ESS_i}$, $P^{min}_{ESS_i}$	Maximum and minimum power limits of ESSs (MW)	$W_{WD}(t)$	Produced water by WD (m^3)
$P_{PV(K_{t\ max})}$	PV output power at maximum solar radiation (MW)	W_{WS}	Volume of water storage (m^3)

Appendix A

	Parameter	Unit	Value		Parameter	Unit	Value
CHP	a_{CO_i}	$	1250	Grid	a_G	$	550
	b_{CO_i}	$/MW	14.5		b_G	$/MW	8.1
	c_{CO_i}	$/MW^2	0.0345		c_G	$/MW^2	0.00028
	d_{CO_i}	$/MWth	4.2	HOU	R_{HOU}	$	100
	e_{CO_i}	$/MWth^2	0.03		M_{HOU}	$/MWth	0.1
	f_{CO_i}	$/MW.MWth	0.031		V_{HOU}	$/MWth^2	0.001
EHP	COP	-	2.5	WD	η_{WD}	m^3/MW	3.03
	P^{max}_{EHP}	MW	40		P^{max}_{WD}	MW	60
	a_{EHP}	$/MW	3.25		K_{WD}	$/MW	2.66
HS	η_{ch}	-	1	WG	K_{WG}	$/m^3	4
	η_{dis}	-	1	WS	η_{ch}	-	0.9
	E_{min}	MWhth	8		η_{dis}	-	0.9
	E_{max}	MWhth	60		Q_{min}	m^3/h	8
	E_{ini}	MWhth	12		Q_{max}	m^3/h	40
ES	η_{ch}	-	0.9		Q_{ini}	m^3/h	10
	η_{dis}	-	0.9				
	E_{min}	MWh	3.3				
	E_{max}	MWh	30				
	E_{ini}	MWh	3.3				

References

1. Mohamed, M.A.; Almalaq, A.; Mahrous Awwad, E.; El-Meligy, M.A.; Sharaf, M.; Ali, Z.M. An Effective Energy Management Approach within a Smart Island Considering Water-Energy Hub. *IEEE Trans. Ind. Appl.* **2020**, *9994*, 1–8. [CrossRef]
2. Jalili, M.; Sedighizadeh, M.; Sheikhi Fini, A. Optimal operation of the coastal energy hub considering seawater desalination and compressed air energy storage system. *Therm. Sci. Eng. Prog.* **2021**, *25*, 101020. [CrossRef]
3. Klemeš, J.J.; Van Fan, Y.; Jiang, P. COVID-19 pandemic facilitating energy transition opportunities. *Int. J. Energy Res.* **2021**, *45*, 3457–3463. [CrossRef] [PubMed]
4. Buheji, M.; da Costa Cunha, K.; Beka, G.; Mavrić, B.; Leandro do Carmo de Souza, Y.; Souza da Costa Silva, S.; Hanafi, M.; Chetia Yein, T. The Extent of COVID-19 Pandemic Socio-Economic Impact on Global Poverty. A Global Integrative Multidisciplinary Review. *Am. J. Econ.* **2020**, *10*, 213–224. [CrossRef]
5. Bilgen, S.; Keleş, S.; Sarikaya, I.; Kaygusuz, K. A perspective for potential and technology of bioenergy in Turkey: Present case and future view. *Renew. Sustain. Energy Rev.* **2015**, *48*, 228–239. [CrossRef]
6. Bhatia, S.C. *Advanced Renewable Energy Systems*; CRC Press: Boca Raton, FL, USA, 2014.
7. Sani, M.M.; Sani, H.M.; Fowler, M.; Elkamel, A.; Noorpoor, A.; Ghasemi, A. Optimal energy hub development to supply heating, cooling, electricity and freshwater for a coastal urban area taking into account economic and environmental factors. *Energy* **2022**, *238*, 121743. [CrossRef]
8. Mansouri, S.A.; Javadi, M.S.; Ahmarinejad, A.; Nematbakhsh, E.; Zare, A.; Catalao, J.P. A coordinated energy management framework for industrial, residential and commercial energy hubs considering demand response programs. *Sustain. Energy Technol. Assess.* **2021**, *47*, 101376. [CrossRef]
9. Xu, X.; Hu, W.; Liu, W.; Du, Y.; Huang, R.; Huang, Q.; Chen, Z. Look-ahead risk-constrained scheduling for an energy hub integrated with renewable energy. *Appl. Energy* **2021**, *297*, 117109. [CrossRef]
10. Karkhaneh, J.; Allahvirdizadeh, Y.; Shayanfar, H.; Galvani, S. Risk-constrained probabilistic optimal scheduling of FCPP-CHP based energy hub considering demand-side resources. *Int. J. Hydrogen Energy* **2020**, *45*, 16751–16772. [CrossRef]
11. Shahinzadeh, H.; Moradi, J.; Gharehpetian, G.B.; Abedi, M.; Hosseinian, S.H. Multi-Objective Scheduling of CHP-Based Microgrids with Cooperation of Thermal and Electrical Storage Units in Restructured Environment. In Proceedings of the 2018 Smart Grid Conference (SGC), Sanandaj, Iran, 28–29 November 2018; pp. 1–10. [CrossRef]

12. Moradi, S.; Ghaffarpour, R.; Ranjbar, A.M.; Mozaffari, B. Optimal integrated sizing and planning of hubs with midsize/large CHP units considering reliability of supply. *Energy Convers. Manag.* **2017**, *148*, 974–992. [CrossRef]
13. Zafarani, H.; Taher, S.A.; Shahidehpour, M. Robust operation of a multicarrier energy system considering EVs and CHP units. *Energy* **2020**, *192*, 116703. [CrossRef]
14. Rastegar, M.; Fotuhi-Firuzabad, M.; Lehtonen, M. Home load management in a residential energy hub. *Electr. Power Syst. Res.* **2015**, *119*, 322–328. [CrossRef]
15. Xie, S.; Wang, X.; Qu, C.; Wang, X.; Guo, J. Impacts of different wind speed simulation methods on conditional reliability indices. *Int. Trans. Electr. Energy Syst.* **2015**, *25*, 359–373. [CrossRef]
16. Davatgaran, V.; Saniei, S.; Mortazavi, S.S. Optimal bidding strategy for an energy hub in energy market. *Energy* **2018**, *148*, 482–493. [CrossRef]
17. Zhang, L.; Zhu, Y. Modeling of CHP-EHP coupled energy station considering load side flexibility. In Proceedings of the IEEE International Conference on Energy Internet, ICEI 2019, Nanjing, China, 21–31 May 2019; pp. 71–74. [CrossRef]
18. Mirhedayati, A.S.; Shahinzadeh, H.; Nafisi, H.; Gharehpetian, G.B.; Benbouzid, M.; Shaneh, M. CHPs and EHPs Effectiveness Evaluation in a Residential Multi-Carrier Energy Hub. In Proceedings of the 2021 25th Electrical Power Distribution Conference (EPDC), Karaj, Iran, 18–19 May 2021; pp. 42–47. [CrossRef]
19. Nosratabadi, S.M.; Jahandide, M.; Guerrero, J.M. Robust scenario-based concept for stochastic energy management of an energy hub contains intelligent parking lot considering convexity principle of CHP nonlinear model with triple operational zones. *Sustain. Cities Soc.* **2020**, *68*, 102795. [CrossRef]
20. Tay, Z.X.; Lim, J.S.; Alwi, S.R.W.; Manan, Z.A. Optimal Planning for the Cogeneration Energy System using Energy Hub Model. *Chem. Eng. Trans.* **2021**, *88*, 349–354. [CrossRef]
21. Shahrabi, E.; Hakimi, S.M.; Hasankhani, A.; Derakhshan, G.; Abdi, B. Developing optimal energy management of energy hub in the presence of stochastic renewable energy resources. *Sustain. Energy Grids Netw.* **2021**, *26*, 100428. [CrossRef]
22. Teng, Y.; Sun, P.; Leng, O.; Chen, Z.; Zhou, G. Optimal Operation Strategy for Combined Heat and Power System Based on Solid Electric Thermal Storage Boiler and Thermal Inertia. *IEEE Access* **2019**, *7*, 180761–180770. [CrossRef]
23. Zhou, Y.; Cao, S. Quantification of energy flexibility of residential net-zero-energy buildings involved with dynamic operations of hybrid energy storages and diversified energy conversion strategies. *Sustain. Energy Grids Netw.* **2020**, *21*, 100304. [CrossRef]
24. Zhang, H.; Cao, Q.; Gao, H.; Wang, P.; Zhang, W.; Yousefi, N. Optimum design of a multi-form energy hub by applying particle swarm optimization. *J. Clean. Prod.* **2020**, *260*, 121079. [CrossRef]
25. Wang, J.; Liu, Y.; Ren, F.; Lu, S. Multi-objective optimization and selection of hybrid combined cooling, heating and power systems considering operational flexibility. *Energy* **2020**, *197*, 117313. [CrossRef]
26. Ren, F.; Wang, J.; Zhu, S.; Chen, Y. Multi-objective optimization of combined cooling, heating and power system integrated with solar and geothermal energies. *Energy Convers. Manag.* **2019**, *197*, 111866. [CrossRef]
27. Shahzad, M.W.; Burhan, M.; Ng, K.C. Pushing desalination recovery to the maximum limit: Membrane and thermal processes integration. *Desalination* **2017**, *416*, 54–64. [CrossRef]
28. Nazari-Heris, M.; Mohammadi-Ivatloo, B.; Asadi, S. Optimal operation of multi-carrier energy networks with gas, power, heating, and water energy sources considering different energy storage technologies. *J. Energy Storage* **2020**, *31*, 101574. [CrossRef]
29. Ramos-Teodoro, J.; Gil, J.D.; Roca, L.; Rodríguez, F.; Berenguel, M. Optimal water management in agro-industrial districts: An energy hub's case study in the southeast of Spain. *Processes* **2021**, *9*, 333. [CrossRef]
30. Mokaramian, E.; Shayeghi, H.; Sedaghati, F.; Safari, A. Four-Objective Optimal Scheduling of Energy Hub Using a Novel Energy Storage, Considering Reliability and Risk Indices. *J. Energy Storage* **2021**, *40*, 102731. [CrossRef]
31. Mostafavi Sani, M.; Noorpoor, A.; Shafie-Pour Motlagh, M. Optimal model development of energy hub to supply water, heating and electrical demands of a cement factory. *Energy* **2019**, *177*, 574–592. [CrossRef]
32. Eladl, A.A.; El-Afifi, M.I.; Saeed, M.A.; El-Saadawi, M.M. Optimal operation of energy hubs integrated with renewable energy sources and storage devices considering CO_2 emissions. *Int. J. Electr. Power Energy Syst.* **2019**, *117*, 105719. [CrossRef]
33. Geidl, M. Integrated Modeling and Optimization of Multi-Carrier Energy Systems. Ph.D. Thesis, ETH Zurich, Zurich, Switzerland, 2007.
34. Hasankhani, A.; Hakimi, S.H. Stochastic energy management of smart microgrid with intermittent renewable energy resources in electricity market. *Energy* **2021**, *219*, 119668. [CrossRef]
35. Mokaramian, E.; Shayeghi, H.; Sedaghati, F.; Safari, A.; Alhelou, H.H. A CVaR-Robust-based multi-objective optimization model for energy hub considering uncertainty and E-fuel energy storage in energy and reserve markets. *IEEE Access* **2021**, *9*, 109447–109464. [CrossRef]
36. Eladl, A.A.; ElDesouky, A.A. Optimal economic dispatch for multi heat-electric energy source power system. *Int. J. Electr. Power Energy Syst.* **2018**, *110*, 21–35. [CrossRef]
37. Neves, D.; Silva, C.A. Modeling the impact of integrating solar thermal systems and heat pumps for domestic hot water in electric systems—The case study of Corvo Island. *Renew. Energy* **2014**, *72*, 113–124. [CrossRef]
38. Damodaran, S.K.; Kumar, T.K.S. Economic and emission generation scheduling of thermal power plant incorporating wind energy. In Proceedings of the IEEE Region 10 International Conference TENCON, Penang, Malaysia, 5–8 November 2017; pp. 1487–1492. [CrossRef]

39. Hetzer, J.; David, C.Y.; Bhattarai, K. An economic dispatch model incorporating wind power. *IEEE Trans. Energy Convers.* **2008**, *23*, 603–611. [CrossRef]
40. Wang, Z.; Zhong, J.; Chen, D.; Lu, Y.; Men, K. A multi-period optimal power flow model including battery energy storage. *IEEE Power Energy Soc. Gen. Meet.* 2013. [CrossRef]
41. Najafi, A.; Falaghi, H.; Contreras, J.; Ramezani, M. Medium-term energy hub management subject to electricity price and wind uncertainty. *Appl. Energy* **2016**, *168*, 418–433. [CrossRef]
42. Zhang, G.; Hu, W.; Cao, D.; Liu, W.; Huang, R.; Huang, Q.; Blaabjerg, F. Data-driven optimal energy management for a wind-solar-diesel-battery-reverse osmosis hybrid energy system using a deep reinforcement learning approach. *Energy Convers. Manag.* **2020**, *227*, 113608. [CrossRef]
43. Hashim, F.A.; Hussain, K.; Houssein, E.H.; Mabrouk, M.S.; Al-Atabany, W. Archimedes optimization algorithm: A new metaheuristic algorithm for solving optimization problems. *Appl. Intell.* **2021**, *51*, 1531–1551. [CrossRef]
44. Houssein, E.H.; Helmy, B.E.D.; Rezk, H.; Nassef, A.M. An enhanced Archimedes optimization algorithm based on Local escaping operator and Orthogonal learning for PEM fuel cell parameter identification. *Eng. Appl. Artif. Intell.* **2021**, *103*, 104309. [CrossRef]

sustainability

Article

Experimental Investigation on Performance Enhancement of Parabolic Trough Concentrator with Helical Rotating Shaft Insert

Mohamed Allam [1,2], Mohamed Tawfik [2,*], Maher Bekheit [2] and Emad El-Negiry [2]

1 Misr International Technological University MITU, Cairo 11725, Egypt
2 Mechanical Power Engineering Department, Faculty of Engineering, Mansoura University, Mansoura 35516, Egypt
* Correspondence: m_tawfik@mans.edu.eg

Abstract: The parabolic trough collector provides an extensive range of solar heating and electricity production applications in solar power plants. The receiver tube of the parabolic trough collector has a vital role in enhancing its performance by using different inserts inside it. In the present work, outdoor experimental tests were conducted to study the performance of a small-scale parabolic trough collector equipped with a centrally placed rotating helical shaft. Three cases were studied: a parabolic trough collector without helical shaft insert, a parabolic trough collector with stationary helical shaft insert, and a parabolic trough collector with a rotating helical shaft insert. The experiments are performed for different shaft rotational speeds (4, 11, and 21 RPM) and various flow rates (0.5, 1, 1.5, 2, and 2.5 LPM) of water as a heat transfer fluid. The fluid flow and heat transfer parameters (friction factor, Reynolds number, Nusselt number, and thermal enhancement factor) and performance parameters (thermal, overall, and exergetic efficiencies) are studied. The results indicated that the helical shaft insert had increased the required pumping power for the same flow rate. However, the parabolic trough collector thermal performance has enhanced with the shaft rotational speed. For all cases, the parabolic trough collector efficiency increases with the flow rate of the heat transfer fluid, but the percentage enhancement in efficiency decreases. Using a shaft rotational speed of 21 RPM and heat transfer fluid flow rates of 0.5 LPM leads to maximum thermal efficiency enhancement and a maximum friction factor ratio of 46.47% and 7.7 times, respectively, compared to plain tube. A comparison based on the same pumping power (thermal enhancement factor) shows that the maximum enhancement occurs at a flow rate of 1 LPM, and the efficiency enhancement is about 37% at a shaft rotational speed of 21 RPM. From an economic point of view, using a rotating helical shaft produces the lower annual cost of useful heat per kWh.

Keywords: solar concentrator; thermal; inserts; helical rotating shaft; active methods

Citation: Allam, M.; Tawfik, M.; Bekheit, M.; El-Negiry, E. Experimental Investigation on Performance Enhancement of Parabolic Trough Concentrator with Helical Rotating Shaft Insert. *Sustainability* **2022**, *14*, 14667. https://doi.org/10.3390/su142214667

Academic Editors: Pablo García Triviño, Ahmed El Kenawy, Petra-Manuela Schuwerack, Zhongfeng Xu and Mohamed El-Alfy

Received: 16 July 2022
Accepted: 15 August 2022
Published: 8 November 2022

1. Introduction

Fossil fuel consumption still significantly contributes to energy demand worldwide, so pollution and global warming have become critical problems [1]. At the same time, from 2003 to 2030, worldwide energy consumption is expected to grow by 71% [2]. Therefore, researchers try to address this issue using renewable energy sources, which may offer a solution. Solar thermal power is one of the valuable alternatives for reducing fossil fuel use, reducing greenhouse gases, and satisfying human needs. The concentrated solar power (CSP) technology is classified as line focus or point focus based on the receiving mechanism. The line focus systems include linear Fresnel reflectors and the parabolic trough, while the point focus systems include the parabolic dish, solar tower, and beam-down central receivers [3,4].

The most often applied technique in solar plants to produce electricity and in industrial and thermal processes is the parabolic trough collector (PTC) [5,6]. The concentrator and the receiver tube are significant components of the solar field. The reflector surface of the concentrator reflects solar rays onto the receiving tube, which converts solar power into heat transmitted to the heat transfer fluid (HTF), such as air, water, or oil [7]. The receiver comprises a selective-coated metallic tube to enhance the absorption of solar irradiance. In addition, the entire tube is enveloped with a vacuumed glass tube with an anti-reflective coating to minimize heat loss by convection [8]. The most critical component of the PTC is the receiver tube, whose performance directly reflects the PTC thermal efficiency. The higher surface temperature of the receiver tube is the major problem that leads to uneven temperature distribution over the receiver. Thus, thermal stresses and losses of the receiver tube increase [9]. Therefore, the thermal efficiency decreases. Because of this, researchers have investigated ways to boost the heat transfer rate within the receiver to improve the PTC thermal performance.

Three techniques enhance heat transfer: passive, active, and compound [10]. The first technique is the passive method that does not consume energy, and the heat transfer rate can be augmented by inserting various inserts inside the receiver tube with a particular engineering design or modifying the entire receiver tube surface. The second technique is the active method, such as surface vibration and magnetic fields, which requires external energy to change the path of HTF to improve the heat transfer rate. Finally, the third technique is the compound method, which combines active and passive methods [11,12]. Low cost and simplicity are why different inserts inside the receiver tube make it the most common ordinary technique to improve heat transfer rate [13].

Using inserts inside the receiver tube augments the velocity fluctuation by forcing the flow beside the tube wall to mix within the flow at core regions [7], enhancing the heat transfer coefficient. Such enhancement reduces thermal losses of the receiver tube and temperature gradient, leading to enhanced thermal efficiency [11], while the significant penalty increases the cost of pumping power through the pressure drop increase [14]. Twisted tape, wire coils, porous inserts, fins, cylindrical/rods, rings, and other insert configurations represent the most common inserts [15]. Twisted tapes are the most often used insert to increase the swirling of the HTF flow to boost the thermal performance [16] and have been investigated in studies [17–24], as introduced in the following paragraphs.

Waghole and Shrivastva [17] assessed the influence of twisted tape within a PTC receiver with twisted ratios of 0.577, 1, and 1.732. Water and silver nanofluid were selected as HTFs under Re numbers (500–6000). Their results showed a rise in thermal efficiency, friction factor, and Nu number of 135–205%, 1–1.75 times, and 1.25–2.1 times, respectively, higher than expected by the conventional receiver. Furthermore, the water friction factor and heat transfer coefficient are lower than silver nanofluid of 0–0.1% volume concentration.

Nakhchi et al. [19] experimentally tested the thermal performance within a pipe with twisted tape inserts with double-cut. Water is chosen as the HTF under the Re numbers range (5000–15,000). The friction factors and Nusselt numbers were increased up to 489% and 177.4%, respectively, for pipes fitted with twisted tapes with a cut ratio of 0.90, compared to the typical tube.

Arunachala [20] experimentally analyzed the PTC performance when equipped with twisted tape and using SYLTHERM 800 as an HTF under Re numbers range of 544–1742 and the twisting ratio of 5.2, 4.1, and 2.7; then, they compared the results with an analytical model. Results showed that, at a minimum twisted ratio of 2.7, the maximum thermal efficiency of 40% is achieved, compared to 19.1% for a plain tube, while the thermo-hydraulic efficiency was decreased to 39.7% because of the increase in pressure drop.

Hosseinalipour et al. [21] numerically investigated decreasing temperature differences within a PTC receiver using a twisted tape insert. The investigation considered superheated steam as HTF with a Re number range (10^4–9.79×10^5) and used different twist ratios (2.5, 5, and 10) and various clearance ratios (0, 0.2, 0.3, 0.4, and 1). The results revealed friction factors and Nu number increases of 1.8–4.1 and 1.0–1.7, respectively. Moreover, the temperature differential over the circumference was reduced by 10–45%.

Elton and Arunachala [22] developed a correlation for the Nu number with twisted tape presence and absence. In addition, they used Soltrace to simulate non-uniform solar radiation. THERMINOL VP-1 was considered HTF and a range of Re numbers varying between (2300–2500). They found that the error in the developed correlations for standard and modified tubes lies within a range (±19% and ±20%), respectively.

Ghadirijafarbeigloo et al. [23] numerically investigated twisted louvered tapes and compared the findings with typically twisted tape inserts at various twisting ratios of 2.67, 4, and 5.33. BEHRAN thermal oil was applied as an HTF. In their study, the Re number range was (5000–25,000). They observed that twisted louvered tape had better thermal performance than standard twisted tape. A 210% and 150% increase in friction factor and Nu number was also found when the twisted louvered tape was used.

Song et al. [24] numerically studied using a helical insert and its impact on the PTC receiver thermal performance. First, they employed the Monte Carlo ray-tracing (MCRT) method to model the heat flux distribution over the receiver wall. Next, DOWTHERM-A was considered an HTF within a range of Re number (10,000–75,000) and inlet temperature (373–640 K). They found that at rates of mass flow of 0.11 and 0.6 kg/s, the heat losses were reduced by six and three times, fitting the helical insert. Moreover, a helical screw insert led to a pressure drop increase of 23 times.

Shashank and Sarviya [18] experimentally and numerically investigated the influence of double and single helical inserts with the various twisted ratios of 1.5, 2.5, and 3 inside tubes. They selected water as an HTF with different Re numbers (4000–16,000). They observed that the double-helical insert's Nu number and friction factor were higher than the single helical insert. Moreover, at a low twisted ratio, the maximum enhancement of Nu number was 112% for a double helical insert compared to an empty tube with an average friction factor of 9.1. Finally, the optimum value of entropy production number was achieved for double helical insert compared to single helical insert.

Another method to improve the thermal performance is rotating twisted tapes within the tube. This method was analyzed experimentally by Abdullah et al. [25]. The different twisted tape ratios and variable operating rotational speeds, under constant heat flux with air as HTF and three Re numbers 8000, 10,000, and 13,000. They observed a thermal enhancement for tubes with twisted rotational tape than tubes with fixed insert. Furthermore, the higher twisted tape rotational speed and lower twisted ratio demonstrated improved thermal performance with a higher friction factor.

Cong et al. [26] experimentally studied the flow characteristic and thermal performance of a tube with a stationary twisted tape and a rotating one at 4 RPM with TiO_2-water nanofluid as an HTF. In addition, different Re numbers from 600 to 7000 and different mass fractions of nanoparticles of 0.1%, 0.3%, and 0.5% were studied. They observed that using the applied nanofluid with rotating twisted tape improved the thermal performance by 53.1–101.6%. Furthermore, the exergy efficiency of the stationary twisted tape was increased compared to the plain tube case under similar pumping power. On the other hand, it can be reduced by using the combination of nanofluids and rotating twisted tape.

Pavan et al. [27] investigated experimentally using rotating twisted tape of twisted ratio 2.37 within a tube. Under constant heat flux affecting the tube, water-CuO nanofluid of 0.2% volume concentration was used as HTF for different twisted tape rotational speeds and flow rates. It was concluded that, at a higher rotational speed of 300 RPM, the highest increase in heat transfer coefficient was 347% while using rotating twisted tape and nanofluid together, compared with plain tube, but the friction factor was increased 3.3 times. Other researchers used different types of inserts or compared them to increase the heat transfer performance.

Promvonge [28] experimentally studied thermal performance enhancement in tubes supplied with twisted tapes and wire coils. The applied HTF was the air with Re numbers (3000–18,000) under constant heat flux. The experiment included wire coils with different pitch ratios and twisted tapes with various twist ratios. They observed a double rise in heat transfer when using twisted tapes with wire coils compared to twisted tape or wire coils. They also observed that using wire coils and twisted tapes with a lower twist ratio and pitch could enhance thermal performance.

Varun et al. [29] compared the PTC performance with a wire matrix and twisted tape. The variable solar irradiance was simulated using the SOLTRACE software program. Two twisted tapes were presented with twisted ratios of 3.37 and 7.86, while three different configurations of wire matrices were presented. The selected HTF was SYNTHETIC oil under the Re number range of 1200–8200 and varying irradiation from 450 to 800 W/m^2. The experimental work showed that the augmentation in thermal efficiency for both twisted tape of 3.37 and medium dense wire matrix were 15.5% and 55.5%, respectively, with a comparable rise in pressure drop of 242% and 3602%, respectively. Regarding performance evaluation criterion analysis, the twisted tape of 3.37 is higher, whereas, in thermo-hydraulic analysis, the medium dense wire matrix is higher.

Other inserts were used to enhance heat transfer; Chakraborty et al. [30] utilized a helical coil within the receiver of a PTC. They simulated the cases using the ANSYS software for pure water mass flow rates (0.016–0.033 kg/s). The exergy and thermal efficiency were increased by 2.4–3.3% and 10–16%, respectively, when helical coils were inserted within the receiver, despite the increased pumping power. Zhao et al. [31] experimentally studied using pin fines within the PTC receiver tube, with air as an HTF for Re number range of 40,000–60,000. It can be concluded from the obtained results that PTC fitted with pin fines enhanced both exergy and energy efficiency by 2.55–4.29% and 10.4–14.5% in comparison with the plain receiver, while the pressure losses were increased about 10–20 times.

The use of metal foam within a PTC receiver was examined experimentally by Jamal-Abad et al. [32] using water as HTF at various Re numbers ranging from 750 to 2000. Results illustrated that using the receiver with metal foam resulted in Nu number and thermal efficiency increase by 1.01–1.80 and 1.03 to 1.08, respectively. Furthermore, the friction factor was intensified. The overall loss coefficient was reduced, which boosted the PTC thermal efficiency. The current study investigates the effect of rotating helical shaft within a PTC receiver using outdoor experimental work with water as HTF. The experimental work is performed for different shaft rotational speeds and flow rates. According to the literature, using a rotating helical shaft inside the PTC receiver as an enhanced insert has not been investigated before. Three cases are compared and studied: a PTC without insert, a PTC with the stationary helical shaft, and a PTC with the rotating helical shaft, to determine the optimum case with better heat transfer performance with a low penalty on pressure drop. The primary criteria for evaluating the examined cases depend on fluid flow, heat transfer, and performance parameters. Finally, an economic analysis is carried out to specify the benefits of the proposed systems from an economic point of view.

2. Materials and Methods

2.1. Experimental Test Rig

The experimental test rig was designed mainly to study the PTC performance with a helical shaft rotating centrally inside the receiver. An experimental test rig of PTC is shown in Figure 1, while its schematic diagram is illustrated in Figure 2. The main elements of the test rig comprise a PTC, a solar tracking mechanism, and measuring instruments.

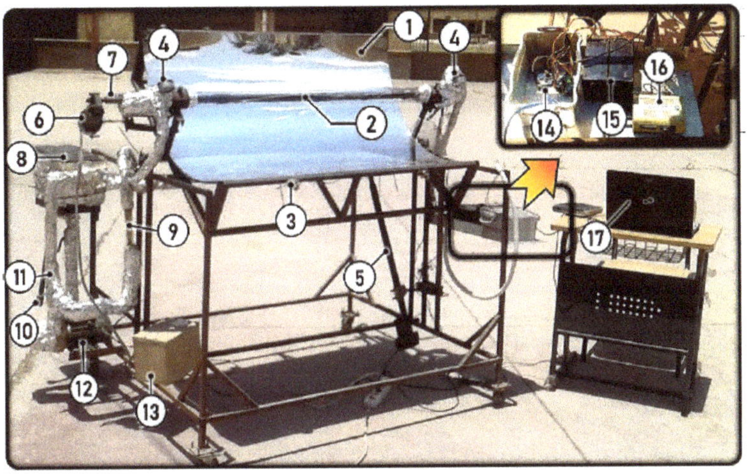

(1) Acrylic reflector sheet	(2) Receiver tube	(3) LDR sensor	(4) Pressure gauge
(5) Linear actuator	(6) DC motor gearbox	(7) Coupling	(8) Storage water tank
(9) Flowmeter	(10) Drain tube	(11) Bypass valve	(12) Circulation pump
(13) DC motor control circuit	(14) ARDUINO circuit	(15) Batteries	(16) Data logger
(17) Laptop			

Figure 1. Experimental test rig of PTC.

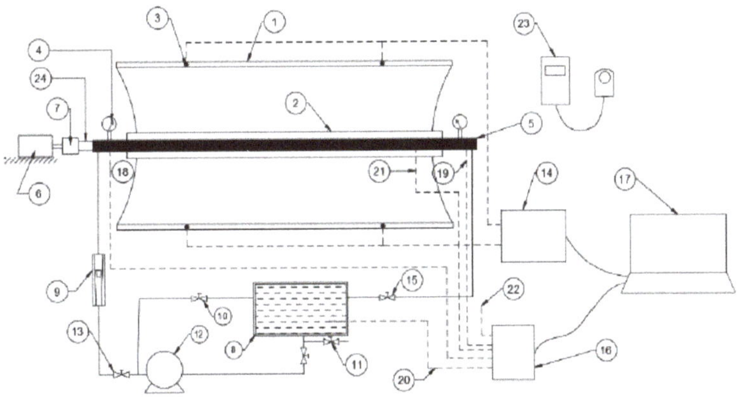

(1) Acrylic reflector sheet	(2) Glass tube	(3) LDR sensor	(4) Pressure gauge
(5) Copper tube	(6) DC motor gearbox	(7) Coupling	(8) Storage water tank
(9) Flowmeter	(10) Bypass valve	(11) Drain valve	(12) Circulation pump
(13) Flow control valve	(14) ARDUINO circuit	(15) Exit valve	(16) Data logger
(17) Laptop	(18) Inlet temperature sensor	(19) Outlet temperature sensor	(20) Inside tank temperature sensor
(21) Temperature sensors (T1-T16)	(22) Ambient temperature sensor	(23) Solar power meter	(24) Helical shaft

Figure 2. The experimental test rig schematic diagram.

2.1.1. Parabolic Trough Solar Collector

The PTC comprises two main components, reflector and receiver.

The Reflector

A 2-mm-thick acrylic mirror sheet was used as a reflector, which can resist outdoor conditions, including heat, cold, and dampness, besides their lightweight and easy to fabricate. The solar-weighted spectral reflectance was tested using a JASCO V-570 Spectrophotometer. The results are plotted in Figure 3, which indicates an average solar weighted solar reflectance of 0.84. The reflector was designed in a parabola shape using two parameters of aperture width (W) and focal distance (F), as shown in Figure 4. The reflector sheet was fixed onto a 1.5-mm-thick steel sheet attached to three steel ribs to increase resistance against the wind load.

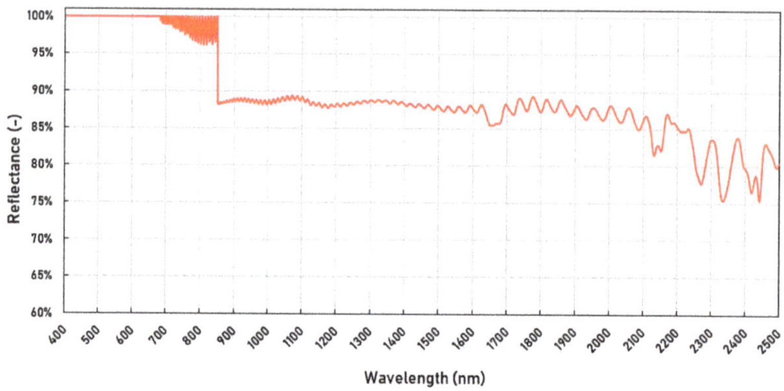

Figure 3. The spectral reflectance of the acrylic mirror sheet.

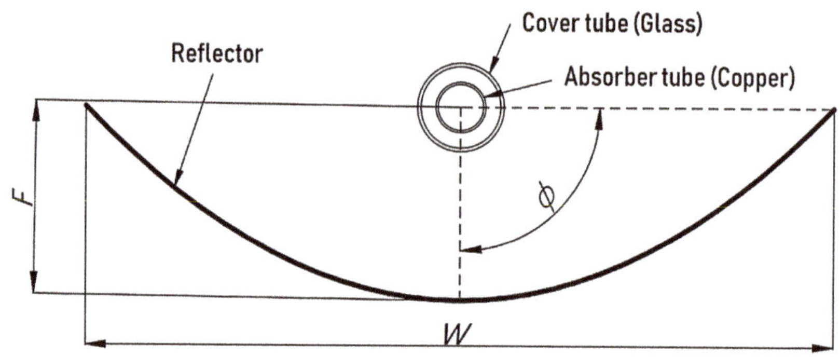

Figure 4. The characteristic dimensions of the tested PTC.

The width of the aperture, W, of the parabola and its focal length, F, are determined using the following Equations [33,34]:

$$W = \frac{2\,Stan(0.5\phi)}{[sec(0.5\phi) + tan(0.5\phi) + ln(sec(0.5\phi) + tan(0.5\phi))]} \tag{1}$$

$$F = \frac{W}{4\,tan(0.5\phi)} \tag{2}$$

where ϕ is the rim angle, and S is the acrylic mirror width (arc length). The following equations were used to assess the correctness of the parabola drawing [35]:

$$y = \frac{x^2}{4F} \tag{3}$$

The concentration ratio, *C*, representing the concentrator aperture area to the receiver areas, is given by [36,37]:

$$C = \frac{W}{\pi D_o} \tag{4}$$

Specifications of the reflector are provided in Table 1

Table 1. Specification of the reflector.

Parameter	Value
Aperture width, W	0.871 m
The parabolic trough length, L	1.200 m
Aperture area, A_a	1.050 m^2
Rim angle, ϕ	90°
Focal line distance, F	0.218 m
wide of mirror acrylic sheet, S	1.000 m
Reflector sheet thickness, t	0.002 m
Reflectance of the mirror, ρ	0.840
The concentration ratio, C	11.70

The Receiver

The receiver is the most critical PTC component, directly affecting PTC performance. It was a black-painted red copper tube to enhance solar radiation absorbtance. It was enveloped by a borosilicate glass cylinder to prevent heat loss by convection, with a transmittance of 0.90 [38]. Specification of the receiver tube is given in Table 2. A representation of the receiver tube is demonstrated in Figure 5.

Table 2. Specification of the receiver tube.

Parameter	Value
Receiver absorptivity, α [39]	0.92
Copper tube length, L_c	1440 mm
Copper tube inside diameter, D_i	26.4 mm
Copper tube outside diameter, D_o	28.6 mm
Glass cover inside diameter, D_{gi}	54 mm
Glass cover outside diameter, D_{go}	60 mm
Glass tube length, L_g	1100 mm

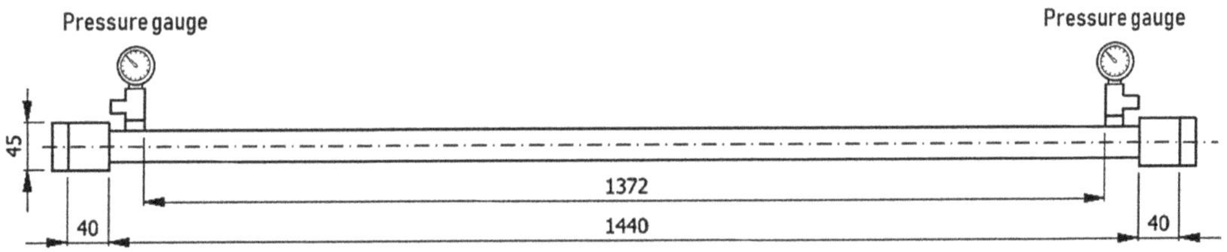

Figure 5. The receiver tube schematic diagram, dimensions in mm.

The Helical Shaft

Steel was selected as shaft material for its high strength, elasticity, and ability to withstand machining. Dimensions of the helical shaft are shown in Figure 6. The helical shaft was immersed in a nickel-chrome basin to prevent corrosion during experimental work. The two sides of the shaft were fixed onto an oil seal ring that bears a high temperature of 250 °C to prevent HTF leakage. Specification of the helical shaft is given in Table 3. DC motor gearbox of 24 V and 50 RPM was used to rotate the shaft at variable speeds of 4, 11, and 21 RPM, and it was connected to the shaft through a coupling and was fixed to the PTC structure.

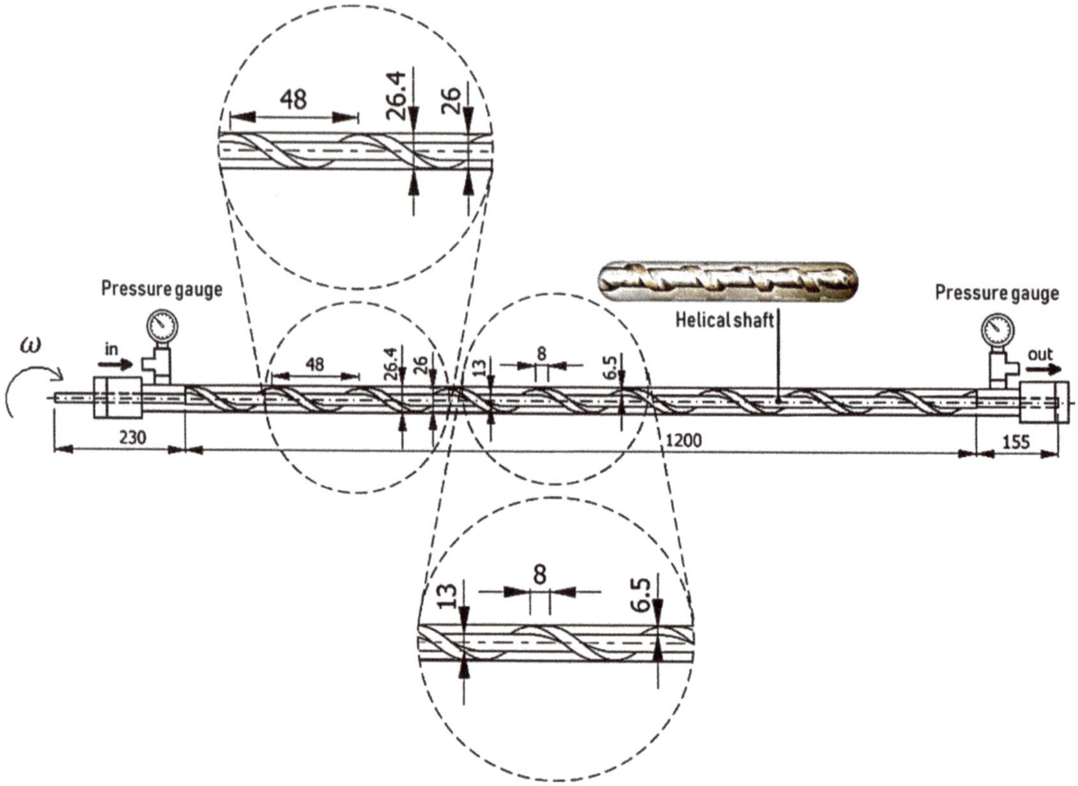

Figure 6. Longitudinal section of the helical shaft inside the receiver tube, dimensions in mm.

Table 3. Specification of the helical shaft.

Parameter	Value
The total length of the helical shaft, L_h	1585 mm
length of the helical section, l_h	1200 mm
Diameter of the helical shaft, D_h	26 mm
Pitch length of the helical shaft, p_h	48 mm
The thickness of the tooth, t_h	8 mm
Height of the tooth. H_h	6.5 mm

2.1.2. The Sun Tracking Mechanism

The PTC tracks the sun using a single-axis tracking approach where the tracking axis can be positioned in the East-West or the North-South directions. Since the North-South sun-tracking axis provides more solar energy yearly than the East-West sun-tracking axis [40], the collector was oriented to rotate about a horizontal North-South axis. The tracking mechanism consists of four light-dependent resistors (LDRs), detecting the solar irradiation and instructing the linear actuators through an Arduino circuit to track the sun.

2.1.3. The Measuring Instruments

The test rig was equipped with the following measuring instruments:

- Four K-type thermocouples with an accuracy of (±0.1 °C) were calibrated and employed to detect the HTF and ambient temperatures. Two were located at the receiver tube entrance and exit, one in the HTF tank and one in the ambient atmosphere.

- The average receiver surface temperature was determined by placing sixteen calibrated thermocouples on the copper tube's outer surface in circular and longitudinal directions, as illustrated in Figure 7. The thermocouples were connected to a data logger recording the measured values with a range of (−129–1372 °C).
- A digital portable solar power meter was used to measure the beam solar radiation at normal incidence through a scale of (1–1999 W/m^2) with an accuracy of (±10 W/m^2).
- A rotameter was employed to measure the volume flow rate with a scale of (0.5–4 LPM) and accuracy (±4% full scale) installed at the pump downstream.
- Since the HTF pressure is inversely proportional to the flow rate, three couples of pressure gauges were used to measure the pressure with an accuracy of (±1% full scale). Two pressure gauges were located at the receiver tube inlet and outlet to evaluate the pressure drop across the receiver under technical data and operating conditions listed in Table 4. The third couple of pressure gauge monitored the HTF pressure over the outdoor tests to maintain the HTF temperature below its saturation point to maintain the HTF in the liquid state.
- A digital multimeter (40 mA–10 A) was used to measure the DC motor power consumption with an accuracy of (±0.8% + 3).
- A digital tachometer measured the helical shaft rotating speed with a scale of (2–20,000 RPM) and accuracy (±0.05% + 1 digits).

Table 4. Specification of pressure gauges used and operating conditions.

Gauge Number	Range (Pa)	Resolution (Pa)	Range of Flow Rates (LPM) *
1	2500	50	0.5–1.0 (indoor tests)
2	16,000	100	1.5–2.5 (indoor tests)
3	250,000	10,000	0.5–2.5 (outdoor tests)

* The range of flow rates where the gauge is used to determine the pressure drop over the receiver.

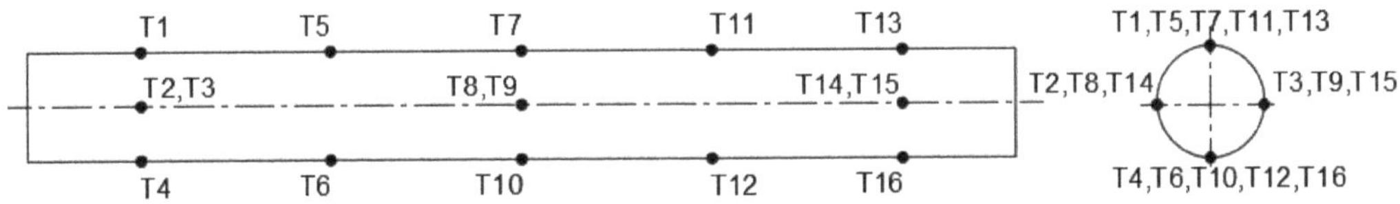

Figure 7. Thermocouple distribution over the copper tube surface.

2.1.4. System Operation

The outdoor experiments were conducted at Integrated Technical Education Institute, Cairo, Egypt. The latitude and the longitude of Cairo city are 30°02′ N and 31°14′ E, respectively. Using a linear actuator, the PTC axis was oriented to the North-South. The solar radiation HTF temperatures at the tank, ambient, test section inlet, and outlet were continuously detected and recorded. Over twenty-five clear days, within June, July, and August, the measured data were recorded every 15 min from 10:00 to 16:30. The irradiation over all experiments ranged from 600 to 1200 W/m^2. A 225 W pump circulated the HTF between the tank and the test section. After the HTF temperature increased in the receiver tube, the hot HTF was returned to the tank and then to the tube in a closed-loop cycle. The HTF flow rate was varied to obtain five variable flow rates of 0.5, 1, 1.5, 2, and 2.5 LPM using ball valves located at the inlet and exit of the HTF stream. DC motor rotational speeds (4, 11, and 21 RPM) were used.

2.2. Governing Equations

2.2.1. Friction Coefficient Evaluation

For Plain Receiver Tube

The average velocity (U_{av}) and Reynolds number (Re) are given as [6]:

$$U_{av} = \frac{\dot{V}}{\frac{\pi}{4}{D_i}^2} \tag{5}$$

$$\mathrm{Re} = \frac{U_{av}\ D_i}{v} \tag{6}$$

For the Receiver with Helical Shaft Insert

U_{av} and Re are determined based on the measured flow rate according to the equivalent hydraulic diameter, D_{eq}. Where D_{eq} is evaluated experimentally by measuring the HTF volume, V, required to fill the entire length of the copper tube, L_c with the helical shaft insert as follows [41]:

$$V = \pi/4\ D_{eq^2}\ \times\ L_c \tag{7}$$

$$D_{eq} = (4V/\pi L_c)^{0.5} \tag{8}$$

The Re number of the flow within a tube is estimated as follows [42]:

$$\mathrm{Re} = \frac{U_{av}\ D_{eq}}{v} \tag{9}$$

where v is the kinematic coefficient of viscosity of water and U_{av} is the average velocity in the receiver tube with helical shaft insert; determined from as:

$$U_{av} = \frac{\dot{V}}{\frac{\pi}{4}{D_{eq}}^2} + Np \tag{10}$$

where $\dot{V}$ is the volumetric flow rate of the HTF, N is the number of revolutions per minute, and p is the helical shaft pitch.

The Darcy–Weisbach friction factor is described as [18,30]:

$$f = \frac{\Delta P}{0.5\ \rho_f\ {U_{av}}^2\ \frac{l}{D_{eq}}} \tag{11}$$

where ΔP is pressure drop difference between pipe inlet and exit, that is determined experimentally, ρ_f is the density of HTF at the average bulk temperature $\left[T_{av} = 0.5(T_{fo} + T_{fi})\right]$ and l is the test section length (1.372 m).

2.2.2. Nusselt Number Evaluation

The average inner wall temperature of the receiver is [20,22]:

$$T_{wi}{}^{\sim} = T_{wo}{}^{\sim} - \frac{Q_u\ \ln\left(D_o/D_i\right)}{2\pi k_c l} \tag{12}$$

$$T_{wo}{}^{\sim} = \frac{\sum T_{wo}}{16} \tag{13}$$

where k_c, $T_{wo}{}^{\sim}$ are the thermal conductivity of the copper and average outer wall temperature of the receiver, respectively.

Log mean temperature difference:

$$LMTD = (\Delta T_1 - \Delta T_2)/ln\ (\Delta T_1/\Delta T_2) \tag{14}$$

where:

$$\Delta T_1 = T_{wi}^{\sim} - T_{fi} \tag{15}$$

$$\Delta T_2 = T_{wi}^{\sim} - T_{fo} \tag{16}$$

The average heat transfer coefficient for HTF was calculated by [43]:

$$h = \frac{Q_u}{A_i\ LMTD} \tag{17}$$

where Q_u is the heat absorbed by HTF. A_i is the inner surface area of the receiver, which can be expressed as:

$$A_i = \pi D_i l \tag{18}$$

The average Nusselt number is determined as follows [18]:

$$Nu = \frac{hD_{eq}}{k} \tag{19}$$

2.3. *Performance Evaluation*

2.3.1. Thermal Enhancement Factor

The thermal enhancement factor (TEF) is the ratio of the augmented surface heat transfer coefficient (inserted receiver tube), h, to that of a smooth surface (plain receiver tube), h_o, at the same pumping power. It is given as [11]:

$$TEF = \left[\frac{h}{h_o}\right]_{pp} = \left[\frac{Nu}{Nu_o}\right]_{pp} = \frac{(Nu/Nu_o)}{(f/f_o)^{(1/3)}} \tag{20}$$

where Nu and Nu_o are for the receiver with insert and the receiver without insert, respectively. The same is true for the friction factor (f and f_o).

2.3.2. Thermal Efficiency

The thermal efficiency is the heat absorbed by HTF (Q_u) to the total incident direct irradiance to the collector aperture. The instantaneous thermal efficiency can be determined as follows [36]:

$$\eta_{th,i} = \frac{Q_{u,i}}{H_{bn,i}\ A_a} \tag{21}$$

$$'Q_{u,i} = \dot{m}_f C_p \left(T_{fo} - T_{fi}\right)_i \tag{22}$$

The average daily thermal efficiency is:

$$\eta_{th} = \frac{Q_u}{Q_s} = \frac{\sum Q_{u,i}}{\sum H_{bn,i} A_a} \tag{23}$$

$$'Q_u = \sum Q_{u,i} \tag{24}$$

$$Q_s = \sum H_{bn,i} A_a \tag{25}$$

where $i = 1, 2, \ldots, n$ and n is the number of time intervals in the day.

T_{fi}, T_{fo}, H_{bn} and A_a are the inlet and outlet temperatures of the HTF, normal beam solar radiation at the collector plane, and aperture area, respectively.

2.3.3. Overall Efficiency

The average daily overall efficiency, η_{ov}, evaluates the net heat production, which can be represented as [44]:

$$\eta_{ov} = \frac{Q_u - \frac{P_p}{\eta_{el}} - P_{mo}}{Q_s} \tag{26}$$

where η_{el} and η_P are electrical and pump efficiency, 0.327 and 0.80, respectively [45–47]. While P_P and P_m are the pumping power and the power of the DC rotating motor, respectively. The pumping power is a critical parameter as using its value increases as a side effect of applying an insert turbulator within the thermal receiver. It can be calculated as follows [44]:

$$P_P = \frac{\Delta P\ V}{\eta_P} \tag{27}$$

2.3.4. Exergtic Efficiency

The exergtic efficiency of the PTC, η_{ex}, is a critical parameter that indicates the maximum work production extracted from the PTC, which is given below [48]:

$$\eta_{ex} = \frac{E_u}{E_s} \tag{28}$$

where, E_u and E_s are the useful output exergy of the HTF and the available solar exergy, respectively. E_u is obtained by [49]:

$$E_u = Q_u\left(1 - \frac{T_{am}}{T_m}\right) - \frac{T_{am}}{T_m}W_T \tag{29}$$

where, W_T is the total of power consumed by the pump and the DC motor. The power consumption of the DC motor is equal to 26.40, 13.68, and 7.44 at speeds of 4, 11, and 21 RPM, respectively. The available solar exergy is obtained from [50,51]:

$$E_s = Q_s\left(1 - \frac{4}{3}\cdot\frac{T_{am}}{T_{sun}} + \frac{1}{3}\cdot\left[\frac{T_{am}}{T_{sun}}\right]^4\right) \tag{30}$$

The mean temperature of fluid, T_m, can be written as [52]:

$$T_m = \frac{\left(T_{fo} - T_{fi}\right)}{ln\left(T_{fo}/T_{fi}\right)} \tag{31}$$

T_{am} is the ambient temperature and T_{sun} is the sun temperature, and it is equal to 5770 K.

The uncertainty analysis in experimental measuring parameters was presented by [53,54], which could be determined using a root–sum–square combination of each of the individual inputs. For example, the uncertainty of a function X can be determined as follows:

$$w_X = \sqrt{\left(\frac{\partial X}{\partial x_1}w_1\right)^2 + \left(\frac{\partial X}{\partial x_2}w_2\right)^2 + \cdots\left(\frac{\partial X}{\partial x_n}w_n\right)^2} \tag{32}$$

where w_X is the uncertainty of the value X, w is the uncertainty of the measured parameter, and X is a given function of the independent variables $x_1, x_2, x_3, \ldots, x_n$. Thus, $X = X(x_1, x_2, x_3, \ldots, x_n)$.

And then, the relative error λ_R is defined as:

$$\lambda_R = \frac{w_X}{X} \tag{33}$$

The relative error of friction factor, Nu number, thermal efficiency, and overall efficiency are 9.29%, 5.10%, 4.92%, and 5.12%, respectively.

3. Results and Discussion

The experimental runs were performed for three cases: (1) plain receiver tube (without insert), (2) receiver with stationary helical shaft (SHS) insert, and (3) receiver with helical shaft (RHS) insert rotating with different rotational speeds of 4 RPM (RHS1), 11 RPM

(RHS2), and 21 RPM (RHS3). All cases were tested at HTF flow rates of 0.5, 1.0, 1.5, 2.0, and 2.5 LPM. The data were continuously recorded every 15 min. A nomenclature summary of the tested cases is listed in Table 5.

Table 5. Nomenclature of tested cases; denotations and conditions.

Case	Denotation	Insert Rotating Speed (RPM)
WIR	Without-insert receiver (Plain tube)	–
SHS	Stationary helical shaft receiver	0
RHS1	Rotating helical shaft receiver—Case 1	4
RHS2	Rotating helical shaft receiver—Case 2	11
RHS3	Rotating helical shaft receiver—Case 3	21

All cases are tested at HTF flow rates of 0.5, 1.0, 1.5, 2.0, and 2.5 LPM.

The variation of Re with flow rate is presented in Figure 8. It can be noted that the Re rises with increasing the flow rate for all cases due to the axial velocity increase with the flow rate, while at the same flow rate, the axial velocity increased in the RHS cases more than in the SHS case as the axial velocity increased with the helical shaft rotation.

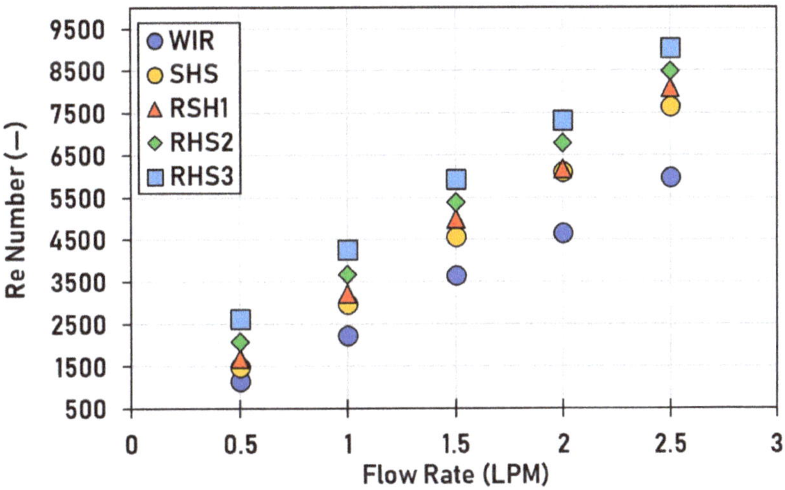

Figure 8. Reynolds number against flow rate in all tested cases.

Figure 9 shows the average heat transfer coefficient for all the examined cases of the PTC, which clearly shows that increasing the value of the average heat transfer coefficient with increasing the flow rate, which may be explained because more flow disturbance is created along the tube walls when the flow rate increases. Intense swirls enhance the fluid mixing between the core and near the wall regions and consequently enhance the average heat transfer coefficient by increasing the flow rate. In the cases of SHS and RHS and at a constant HTF flow rate, the helical shaft increases the intensity of swirl flow and resulting in higher contact surface area between the receiver and HTF, leading to an increase in the useful heat gain and consequently enhancing the average heat transfer coefficient with introducing SHS and RHS.

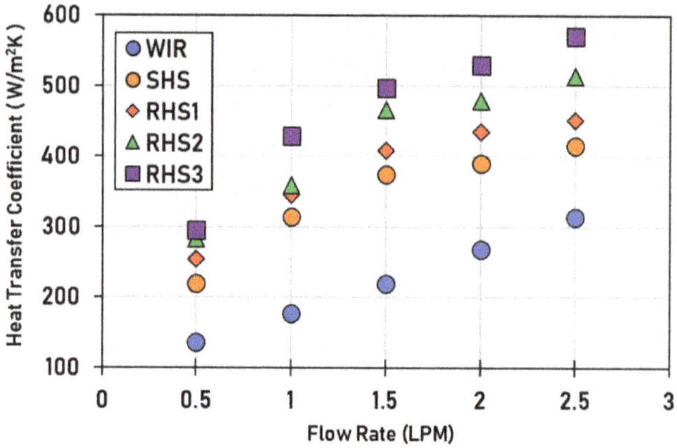

Figure 9. Average heat transfer coefficient against flow rate.

The variation of the Nu number corresponding to the average heat transfer coefficient in Figure 9 is shown in Figure 10a. The Nu number ratio with flow rate variation is illustrated in Figure 10b. The higher Nu number ratio is achieved at 1 LPM compared to the other flow rates, and it achieves a maximum of 2.42 for the case of RSHS3. Furthermore, the Nu number ratio decreases with the rise of flow rate from 1.5 LPM to 2.5 LPM for all cases, while at 0.5 LPM, the Nu number ratio reaches a higher value of 2.17 for the case of RSHS3 and reaches a lower value of 1.62 for the case of SHS.

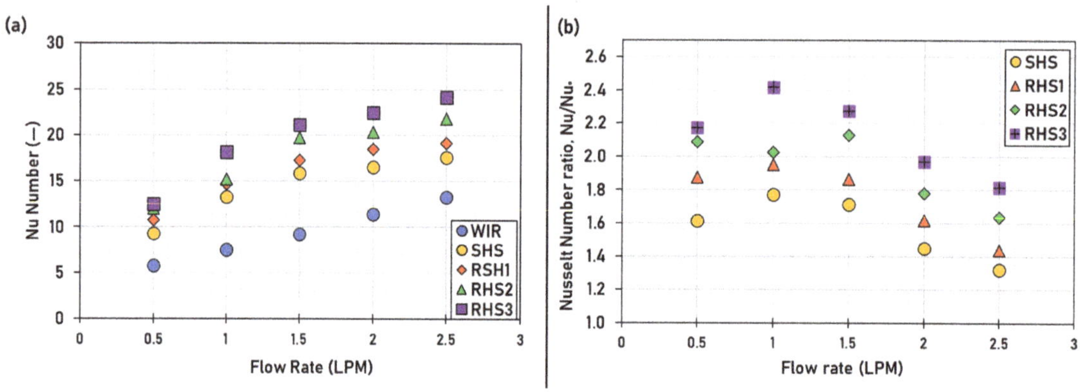

Figure 10. Nusselt number relations ratio against flow rate. (**a**) Nusselt number against flow rate. (**b**) Nusselt number ratio against flow rate.

The thermal efficiencies of all tested cases are shown in Figure 11. It can be noted that the instantaneous thermal efficiency varies with the experiment time over the day due to the rapid incident solar radiation variation. However, the thermal efficiency slightly varies when inserted the helical shaft inside the receiver tube since the helical shaft can compensate for the incident solar radiation variations over the daytime. Moreover, Figure 11a to Figure 11e indicate that increasing the helical shaft insert's rotation speed results in higher and more stabilized instantaneous thermal efficiency over the daytime. This result may be attributed to improved mixing, more intense secondary flow, and higher thermal boundary layer disruption due to shaft rotation. As a consequence, the convective heat transfer is improved.

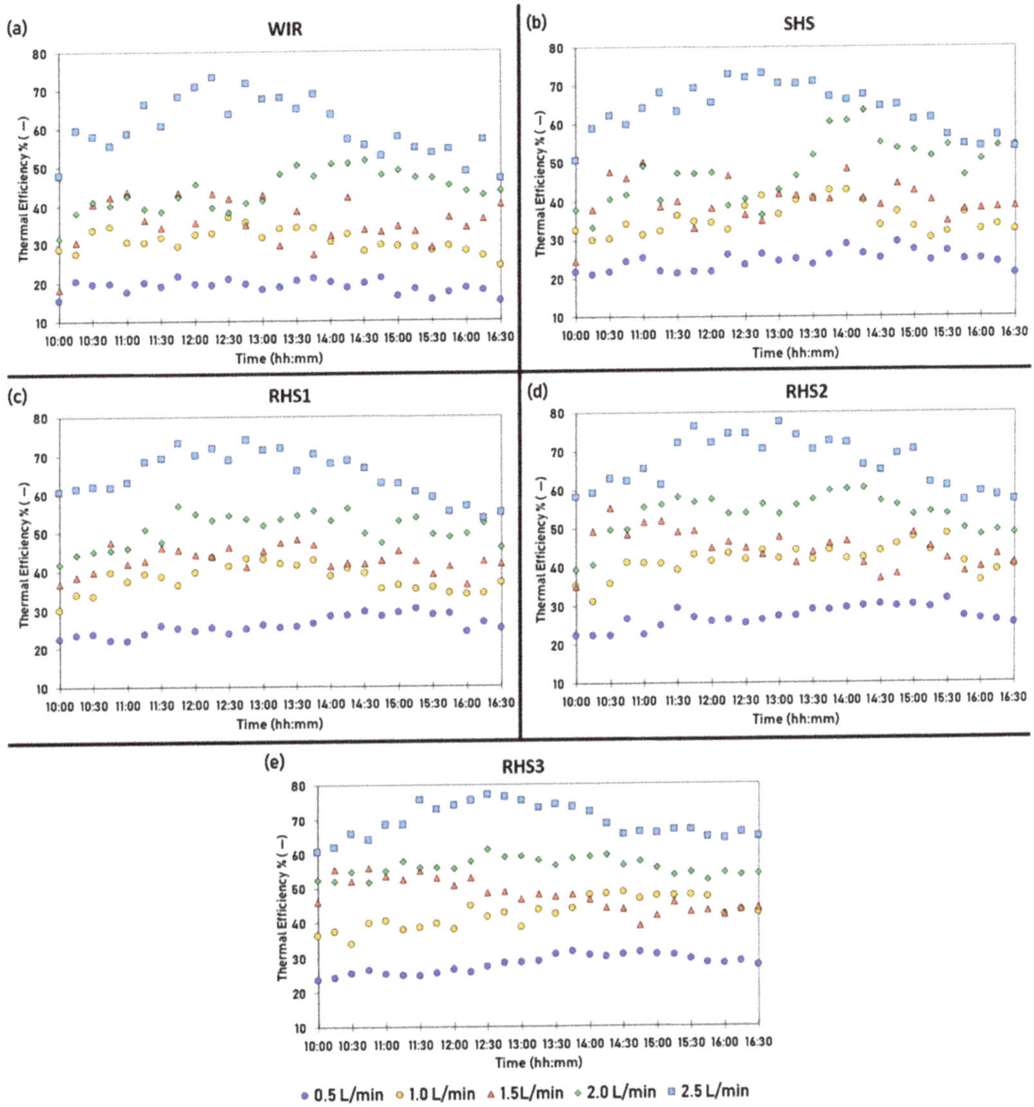

Figure 11. The instantaneous thermal efficiency against time in all tested cases.

Figure 12 shows temperature differences at a different flow rate for all cases. As the flow rate decreases from 2.5 to 0.5 LPM, the temperature difference increases, as stated in [55]. The reason is that, at the low flow rate, the HTF has a long time to increase its temperature along the receiver tube, so the temperature difference has a higher value. On the contrary, the HTF does not have enough time to increase its temperature at a high flow rate, so the temperature difference has a lower value.

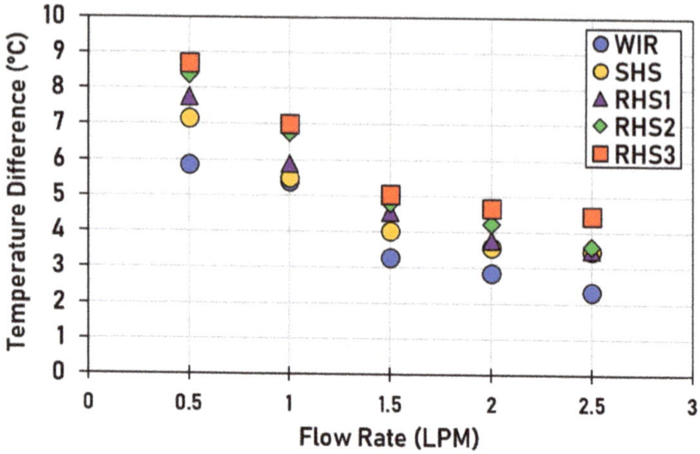

Figure 12. Temperature difference against flow rate.

Compared to a plain tube, the temperature differential is enhanced by using an insert within the receiver at a constant HTF flow rate. Furthermore, due to expanding the contact surface area for HTF with the shaft, the convective heat transfer rate is enhanced, and higher outlet temperature is observed, consequently increasing temperature difference. In addition, when the helical shaft rotates, it causes more intense secondary flow, more disturbance of the thermal boundary layer, and improved mixing. Consequently, the convective heat transfer rate and outlet temperature increase the temperature difference.

In addition, from Figure 13a, it can be observed that the increase in the HTF flow rate boosts the instantaneous thermal efficiency. These observed results are not limited to the instant thermal efficiency. However, they extend to average daily thermal efficiencies, as illustrated in Figure 13a. The thermal efficiency increase with increasing the flow rate is compatible with what was stated in [52,56] due to higher confrontation between the HTF particles and higher turbulence and, as a result, augmented thermal performance and efficiency. In the cases of RHS1, RHS2, and RHS3, the rotating helical shaft amplifies the intensity of swirl flow and enhances thermal efficiency compared to SHS. The efficiency of PTC for WIR, SHS, RHS1, RHS2, and RHS3 varies from 19% to 60.3%, 24.51% to 63.82%, 25.9% to 65.02%, 27.1% to 66.9%, and 27.9% to 69.25% with flow rate ranging from 0.5 to 2.5 LPM.

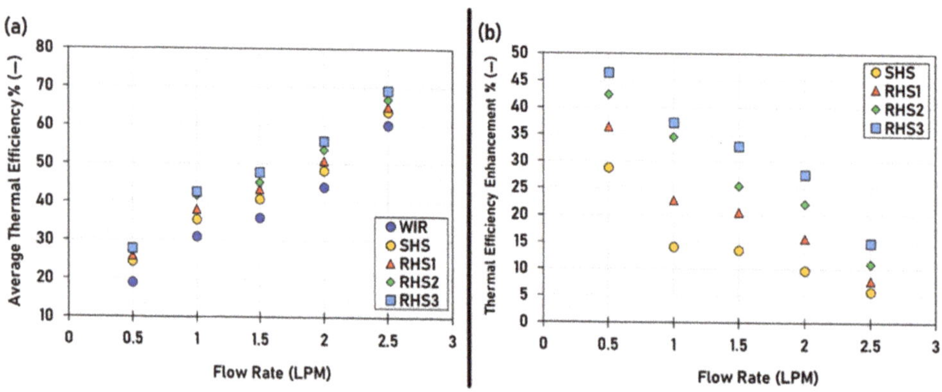

Figure 13. Thermal efficiency against flow rate. (**a**) Average daily efficiency vs. flow rate. (**b**) Efficiency enhancement % vs. flow rate.

However, the thermal efficiency enhancement, as plotted in Figure 13b, is observed to decrease with the rise of flow rate as the rate of growth of thermal efficiency for the plain tube with increasing flow rates is higher than the rate of growth in thermal efficiency for the receiver tube fitted with stationary or rotating helical shaft. As a result, the enhancement in thermal efficiency is reduced with increasing flow rates. Moreover, the enhancement is higher when the shaft rotates at a higher rotational speed than the stationary one. Using RHS3 leads to the maximum enhancement, ranging from 14.82 to 46.47% while using SHS leads to enhancement from 5.82% up to 28.78% compared to the plain tube.

The friction factor against flow rates for all tested cases is evaluated and plotted as illustrated in Figure 14a. As noted, the friction factor reduces as the flow rate increases, as indicated in [52]. Moreover, it increases in the cases of SHS, RHS1, RHS2, and RHS3 compared to the plain tube. This is because the shaft causes increases in swirl flow intensity, residence time, and contact surface area for SHS. In contrast, the long time inside the receiver tube increases the contact between the HTF particles and the receiver tube wall and, as a result, increases the friction factor and pressure drop. Moreover, for the cases of RHS1, RHS2, and RHS3, increasing the rotational speed magnifies the intensity of swirl flow and turbulence. As a result, it makes the HTF move over the longer paths, increasing the friction factor compared to RSHS. Consequently, the maximum friction factor is 1.3, and it is reached at a lower flow rate of 0.5 LPM and a higher rotational speed of 21RPM (RHS3), while the lower friction factor is 0.156 and reached at a higher flow rate of 2.5 LPM and stationary insert (SHS). Therefore, the friction factor ratio reduces with the growth in flow rate, while for the RHS3 and SHS cases, the friction factor ranges from 4.1 to 7.7 times and 1.9 to 4.1 times, respectively, compared to the WIR case, as seen in Figure 14b.

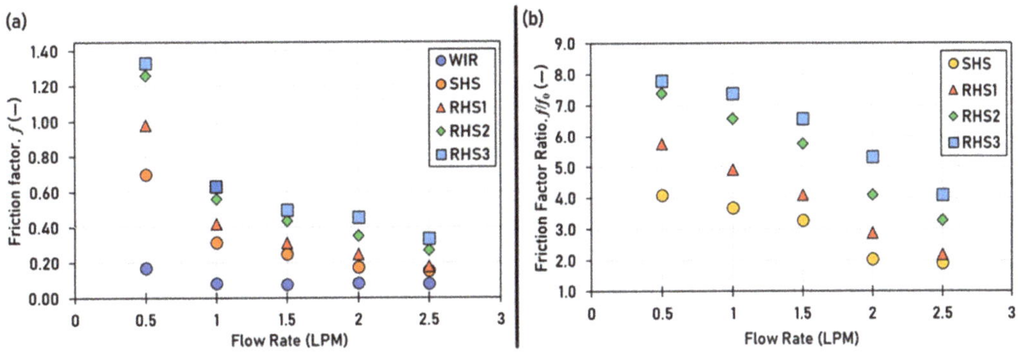

Figure 14. Friction factor against flow rate. (**a**) Friction factor vs. flow rate. (**b**) Friction factor ratio vs. flow rate.

The pumping power increases by increasing the flow and using the helical shaft insert, as illustrated in Figure 15a and stated in [18]. Moreover, using a shaft insert leads to higher pumping power in the cases of SHS, RHS1, RHS2, and RHS3 compared with the WIR. Raising the flow rate from 0.5 to 2.5 LPM increases the pressure drop from 250 to 1400 Pa, 350 to 1600 Pa, 450 to 2400 Pa, and 500 to 3000 Pa in the cases of SHS, RHS1, RHS2, and RHS3, respectively, and consequently increasing pumping power. However, the increase in pumping power, as plotted in Figure 15b, is observed to reduce with the flow rate rise as the pumping power is greater for the plain tube when increasing the flow rates compared to the rate of growth in pumping power for the receiver tube fitted with stationary or rotating helical shaft. As a result, the percentage rise in pumping power is decreased with rising flow rates. Moreover, the increase is higher when the shaft rotates at a higher rotational speed than the stationary one. Using RHS3 leads to the maximum increase, ranging from 404 to 850%, while using SHS increases 130 to 400% compared to the plain tube.

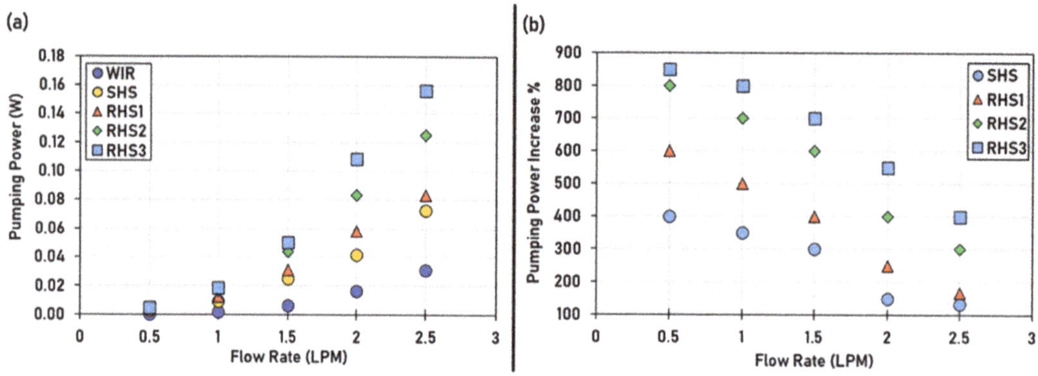

Figure 15. Pumping power relations ratio against flow rate. (**a**) Pumping power against flow rate (**b**) % increase in pumping power against flow rate.

Considering the effect of power consumed in the pump and the DC motor rotating the helical shaft, the overall efficiency can be used to evaluate the system; as shown in Figure 16a. It indicates an improvement of thermal efficiency for all cases with the flow rate increase due to the slight rise in the total power consumed compared to the higher growth in beneficial heat gain. So, the overall efficiency trend is the same for thermal, and the overall efficiencies enhancement is the same for the thermal efficiency enhancement, as shown in Figure 16b.

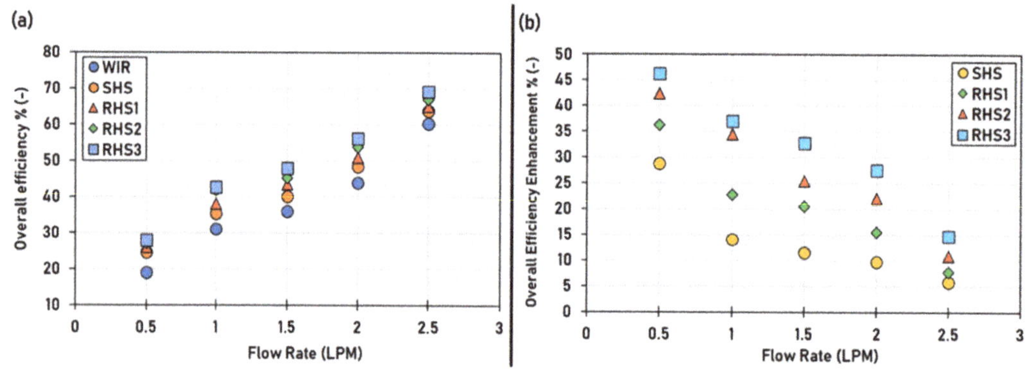

Figure 16. Efficiency relations against flow rate. (**a**) Overall efficiency against flow rate. (**b**) % enhancement of overall efficiency.

Figure 17 shows the variations of TEF against flow rate. TEF for all cases is higher than unity, while the heat transfer enhancement is higher than the pressure drop rise. The maximum TEF is 1.24, achieved at 1 LPM and in the case of RHS3.

The exergy efficiency with various flow rates in all tested cases is shown in Figure 18. Based on results plotted in Figure 18, raising the flow rate from 0.5 to 2.5 LPM increases the useful heat gain and results in exergy efficiency enhancement from 11.11% to 36.35% in the plain tube case stated in [50]. Inserting the helical shaft increases exergy efficiency from 14.16, 14.52, 15.14, and 15.23% to 37.81, 39.21, 38.98, and 40.41% for the cases of SHS, RHS1, RHS2, and RHS3, respectively, with raising the flow rate from 0.5 to 2.5 LPM. This result is attributed to the rising logarithmic mean temperature of the fluid, T_m. With boosting the flow rate that leads to both decreasing in the second term $(T_0/T_m W_T)$ and increasing the useful heat gain in the first term and consequently increasing the exergy efficiency. Increasing the total power consumption at the same flow rate due to rotating the helical shaft in the cases of RHS1, RHS2, and RHS3 does not prevent enhancing the exergy efficiency because the rate of increasing the useful heat gain is higher than increasing the total power consumption that leads to enhancement in exergy efficiency.

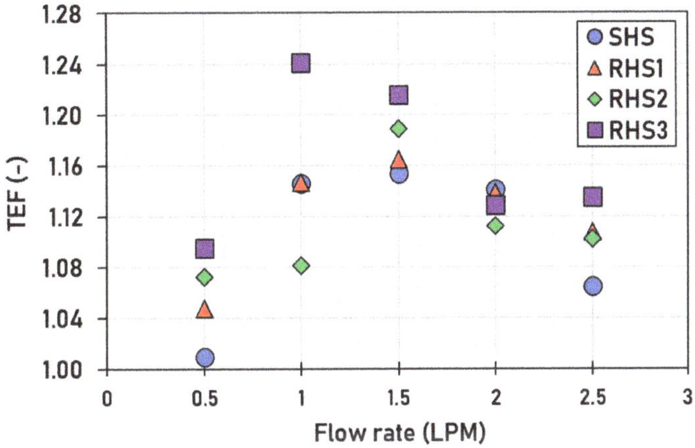

Figure 17. Variation of TEF against flow rate.

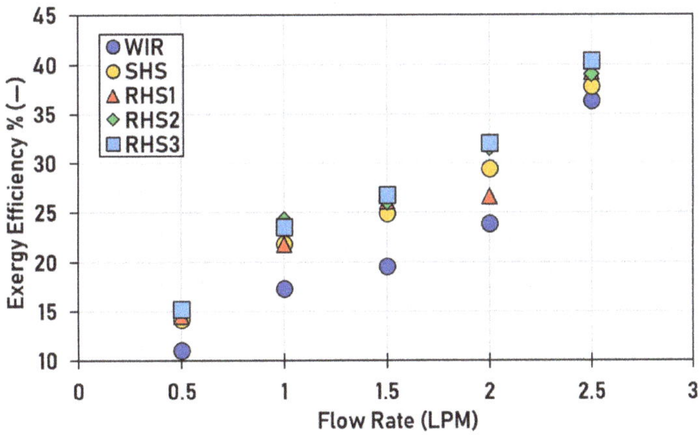

Figure 18. Exergetic efficiency against flow rate.

A comparison between the present work (static and rotating helical shaft) and different types of inserts in terms of thermal efficiency ratio, Nu number ratio, and friction ratio is provided in Table 6. The highest efficiency ratio is 2.33 for wire matrix and twisted tape [29]. On the other hand, they achieved a higher friction ratio of 36.02.

The filling porous [57] achieved a maximum Nu number ratio of 17.24 and a maximum friction factor ratio of 412.62. According to the reviewed literature, using twisted tape and fins can lead to higher thermal enhancement with a lower impact on the friction factor. The experimental study of the static helical shaft has a higher efficiency ratio of 1.28 and a lower friction ratio of 4.10 compared with a numerical study of twisted tape of 1.07 and a higher friction factor ratio of 6.15. While using the rotating helical shaft in the current study led to a maximum efficiency ratio of 1.46 and a maximum Nu number ratio of 2.41 compared to fins and twisted tape, but as a penalty of maximum friction factor ratio of 7.79.

Table 6. Comparison between present work and various inserts inside PTC.

Study Type	Type of Insert	η/η_o	Nu/Nu_o	f/f_o	Ref.
Exp.	Copper foam	1.03–1.08	1.01–1.80	95–250	[32]
Exp.	Wire matrix and twisted tape	1.35–2.33	1.67–2.60	36.02	[29]
Num.	Toroidal rings	1.00–1.037	1.00–2.33	1.00–14.62	[58]
Num.	Perforated plate	1.005–1.018	1.8–2.0	13.34–13.46	[59]
Num.	Filling porous	1.07	3.49–17.24	17.60–412.62	[57]
Num.	Twisted tape	1.005–1.015	1.70–1.77	4.17–6.15	[59]
Num.	Internal longitudinal fins	1.37–1.41	1.96–2.19	2.03–2.12	[60]
Exp.	Static helical shaft	1.05–1.28	1.32–1.77	1.91–4.10	Present work
Exp.	Rotating helical shaft	1.06–1.46	1.43–2.41	2.18–7.79	Present work

4. Economic Analysis

The utility of using the inserts inside PTC can be evaluated based on its economic viability. Fixed cost (FC) and annual variable cost (VC) are considered in order to estimate the total cost of PTC, while the initial cost is estimated by considering all components of building a system, and the operating cost is estimated by considering the effect of power consumed in the pump and DC motor. The cost data of components of PTC for three WIR, SHS, and RHS cases are listed in Table 7 according to the Egyptian market. The economic analysis is carried out at maximum TEF, achieved at 1 LPM and RHS3.

Table 7. The cost data of various components of PTC.

Sr.	Details of Component	Quantity (Units)	Unit Cost (USD/Unit)	Total Cost (USD)
1	Acrylic reflector sheet	1	66.7	66.7
2	Copper tube	1	16.7	16.7
3	Glass tube	1	11.1	11.1
4	Linear actuator	1	17.8	17.8
5	Pipes and accessories	1	31.4	31.4
6	valves	5	2.2	11.1
7	Storage water tank	1	38.9	38.9
8	Silicon	5	0.8	4.2
9	Insulation material	1	16.7	16.7
10	Circulation pump	1	19.4	19.4
11	ARDUINO circuit	1	33.3	33.3
12	Batteries	2	38.9	77.8
13	The metallic structure of PTC	1	61.1	61.1
14	Manufacture cost	1	55.6	55.6
A	**Total fixed cost**	**WIR**		**462**
15	Insert tape (steel)	1	27.8	27.8
16	Manufacture cost	1	61.1	61.1
B	**Total fixed cost**	**SHS**		**551**
17	Oil seal ring	2	2.2	4.4
18	DC motor gearbox and control circuit	1	25.0	25.0
19	Coupling	1	2.5	2.5
C	**Total fixed cost**	**RHS**		**583**

Bold rows (A, B, and C) represent the summation of total fixed costs of each case (WIR, SHS, and RHS), respectively.

Total Cost Estimation of PTC

The annual cost of useful heat per kWh (ACUH) can be calculated as follows [54]:

$$\text{ACUH} = \text{TC}/\text{TUH} \tag{34}$$

where, TC is the total cost and can be determined from Equation (35), while TUH is the total useful heat during the life span and is expressed by Equation (36).

$$TC = FC + VC \tag{35}$$

$$TUH = n \times \text{Yearly productivity} \tag{36}$$

where, n is the expected PTC lifetime, assumed to be 10 years. The yearly productivity can be calculated as follows:

$$\text{Yearly productivity} = (Q_u/\text{day}) \times \text{operating days} \tag{37}$$

where, the operating days were assumed to be 340 sunny days per year and (Q_u/day) is the useful heat per day for one cycle can be calculated as follows:

$$(Q_u/\text{day}) = Q_u \times \text{operating hours} \tag{38}$$

where, the operating hours were 7 h (from 9:30 to 16:30).

The economics of different cases of PTC based on estimated ACUH are presented in Table 8.

Table 8. The economic analysis for PTC with WIR, SHS, and RHS.

	WIR	SHS	RHS
Fixed cost (FC), USD	462	551	583
Annual variable cost (VC), USD	0.0004	0.0014	5.810
(Q_u/day), kWh	2.5	2.8	3.4
TUH, kWh	8414.3	9405.8	11,543.0
Total cost (TC), USD	462	551	588.4
ACUH, USD	0.055	0.059	0.051

The economic analysis in Table 8 shows that ACUH for PTC with WIR, SHS, and RHS are USD 0.055, USD 0.059, and USD 0.051, respectively. PTC with RHS has minimum ACUH, while SHS with PTC uses maximum ACUH. Therefore, using RHS is beneficial from an economic point of view compared to WIR and SHS, and using SHS without rotation has a poor impact on ACUH.

5. Conclusions

The influence of inserting a stationary and rotating helical shaft within the receiver on PTC performance was investigated in the present study. Three cases were investigated: A PTC without helical shaft insert, a PTC with stationary helical shaft insert, and a PTC with rotating helical shaft insert. The experiments were performed for different shaft rotational speeds of (4, 11, and 21 RPM) and various flow rates (0.5, 1, 1.5, 2, and 2.5 LPM) of water as an HTF. Moreover, the fluid flow and thermal parameters (friction factor, Re number, Nu number, and TEF) and performance parameters (thermal, overall, and exergetic efficiencies) were investigated. The following are the key findings of this investigation:

- The helical shaft insert has increased the required pumping power for the same flow rate.
- The pumping power rises with the shaft rotating speed, particularly at the high HTF flow rate.
- For the same flow rate, the helical shaft insert has enhanced the thermal performance (thermal, overall, and exergetic efficiencies) of the PTC.
- The thermal performance enhancement increases with growing the shaft rotating speed and the HTF flow rate.
- Compared to WIR, the friction factor ratios for RHS3 are higher than SHS, which are in the ranges of 4.1 to 7.79 times and 1.91 to 4.10 times, respectively. Furthermore,

Nu number ratios for RHS3 are higher than SHS in the ranges of 1.81 to 2.41 and 1.32 to 1.77.
- The maximum enhancement in thermal efficiency is 46.47%, achieved at a minimum flow rate of 0.5 LPM and RHS3.
- The maximum exergy efficiency is 40.41%, achieved at a higher flow rate of 2.5 LPM and RHS3.
- A comparison between the investigated three cases at the same pumping power using criteria of TEF shows that the maximum enhancement in thermal performance occurs at an HTF flow rate of 1 LPM, and the percentage enhancement in thermal efficiency is about 37% at shaft rotational speed of 21 RPM.
- The cost analysis shows that ACUH for RHS (USD 0.051) is lower than WIR and SHS (USD 0.059) at maximum TEF.

It can be concluded that using static and rotating helical shaft insert inside the receiver of PTC may be used for solar heating processes and power generation applications to enhance the performance of the PTC with a low penalty on pumping power compared to the other inserts. In the future, the use of nanofluids or hybrid nanofluids with rotating helical shafts is recommended, in order to investigate their effects on the thermal and hydraulic performance of PTC. Furthermore, a PV system coupled with PTC to supply the DC motor with power should be planned to reduce power consumption.

Author Contributions: Conceptualization, M.T., M.B. and E.E.-N.; methodology, M.A. and M.T.; data analysis, M.A. and M.T.; investigation, M.A., M.T., M.B. and E.E.-N.; writing—original draft preparation, M.A.; writing—review and editing, M.A. and M.T.; visualization, M.T.; supervision, M.T., M.B. and E.E.-N. All authors have read and agreed to the published version of the manuscript.

Funding: The Article Processing Charge (APC) was funded by Mansoura University, Egypt.

Informed Consent Statement: Not applicable.

Acknowledgments: The publication fees of this article have been supported by Mansoura University.

Conflicts of Interest: The authors declare no conflict of interest.

Nomenclature

Symbols

C	concentration ratio (-)
C_p	specific heat capacity (J/kg K)
D	diameter (m)
E	exergy rate (W)
F	focal length (m)
f	friction factor (-)
H_{bn}	normal beam solar radiation (W/m^2)
H	height (m)
h	heat transfer coefficient, W/m^{2}K
L	copper tube length (m)
l	test section length (m)
$\dot{m}$	mass flow rate (kg/s)
N	revolutions per minute (RPM)
P	power (W)
p	pitch (m)
Q	heat rate (W)
S	reflector width (m)
T	temperature (°C)
t	thickness (m)
U	velocity (m/s)
V	volume (m^3)
$\dot{V}$	volumetric flow rate (m^3/s)
W	aperture width (m)

Subscripts

a	aperture
av	average
am	ambient
c	copper tube
el	electrical
eq	equivilant
ex	exergetic
f	fluid
g	glass tube
h	helical shaft
i	inner or inlet
m	logarithmic mean
mo	motor
o	outer or outlet
ov	overall
p	pump
PP	constant pumping power
s	solar
T	total
u	useful
w	wall

Abbreviations

HTF	heat transfer fluid

w	uncertainty (-)	LMTD	Log mean temperature difference
Greek Letters		LPM	liter per minute
α	absorptance (-)	PTC	parabolic trough collector
γ	intercept factor (-)	RHS	rotating helical shaft receiver
Δ	difference (-)	SHS	stationary helical shaft receiver
η	efficiency (-)	TEF	thermal enhancement factor
λ	relative error (-)	WIR	without-insert receiver
μ	dynamic viscosity (kg/m.s)	ACUH	annual cost of useful heat per kWh
v	kinematic coefficient of viscosity (m^2/s)	TC	total cost
ρ_f	density of HTF (kg/m^3)	TUH	total useful heat during the life span
ρ	reflectance (-)	FC	fixed cost
τ	transmittance (-)	VC	annual variable cost
ϕ	rim angle (°)	***Dimensionless Numbers***	
ω	angular velocity (rad/s)	Nu	Nusselt number
		Re	Reynolds number

References

1. Bilal, A.; Khan, M.N.; Zubair, M.; Bellos, E. Commercial Parabolic Trough CSP Plants: Research Trends and Technological Advancements. *Sol. Energy* **2020**, *211*, 1422–1458. [CrossRef]
2. Sarbu, I.; Sebarchievici, C. Introduction. In *Solar Heating and Cooling Systems*; Elsevier: Cambridge, MA, USA, 2017; pp. 1–11. ISBN 9780128116623.
3. Shouman, E.R.; Khattab, N.M. Future Economic of Concentrating Solar Power (CSP) for Electricity Generation in Egypt. *Renew. Sustain. Energy Rev.* **2015**, *41*, 1119–1127. [CrossRef]
4. Tawfik, M. A Review of Directly Irradiated Solid Particle Receivers: Technologies and Influencing Parameters. *Renew. Sustain. Energy Rev.* **2022**, *167*, 112682. [CrossRef]
5. Fernández-García, A.; Zarza, E.; Valenzuela, L.; Pérez, M. Parabolic-Trough Solar Collectors and Their Applications. *Renew. Sustain. Energy Rev.* **2010**, *14*, 1695–1721. [CrossRef]
6. Anabela, E.; Panduro, C.; Finotti, F.; Largiller, G.; Yngve, K. A Review of the Use of Nanofluids as Heat-Transfer Fluids in Parabolic-Trough Collectors. *Appl. Therm. Eng.* **2022**, *211*, 118346. [CrossRef]
7. Abed, N.; Afgan, I. An Extensive Review of Various Technologies for Enhancing the Thermal and Optical Performances of Parabolic Trough Collectors. *Int. J. Energy Res.* **2020**, *44*, 5117–5164. [CrossRef]
8. Kreith, F.; Goswami, D.Y. *Handbook of Energy Efficiency and Renewable Energy*; Taylor & Francis Group: New York, NY, USA, 2007.
9. Peng, H.; Li, M.; Liang, X. Thermal-Hydraulic and Thermodynamic Performance of Parabolic Trough Solar Receiver Partially Filled with Gradient Metal Foam. *Energy* **2020**, *211*, 119046. [CrossRef]
10. Parlamıs, H.; Ozden, E.; Sami, M. Experimental Performance Analysis of a Parabolic Trough Solar Air Collector with Helical-Screw Tape Insert: A Comparative Study. *Sustain. Energy Technol. Assess.* **2021**, *47*, 101562. [CrossRef]
11. Sharma, M.; Jilte, R. A Review on Passive Methods for Thermal Performance Enhancement in Parabolic Trough Solar Collectors. *Int. J. Energy Res.* **2021**, *45*, 4932–4966. [CrossRef]
12. Kumar, C.N.; Murugesan, P. Review on Twisted Tapes Heat Transfer Enhancement. *Int. J. Sci. Eng. Res.* **2012**, *3*, 1–9.
13. Mousa, M.H.; Miljkovic, N.; Nawaz, K. Review of Heat Transfer Enhancement Techniques for Single Phase Flows. *Renew. Sustain. Energy Rev.* **2021**, *137*, 110566. [CrossRef]
14. Hasanpour, A.; Farhadi, M.; Sedighi, K. A Review Study on Twisted Tape Inserts on Turbulent Flow Heat Exchangers: The Overall Enhancement Ratio Criteria. *Int. Commun. Heat Mass Transf.* **2014**, *55*, 53–56. [CrossRef]
15. Allam, M.; Tawfik, M.; Bekheit, M.; El-Negiry, E. Heat Transfer Enhancement in Parabolic Trough Receivers Using Inserts: A Review. *Sustain. Energy Technol. Assess.* **2021**, *48*, 101671. [CrossRef]
16. Garg, M.O.; Nautiyal, H.; Khurana, S.; Shukla, M.K. Heat Transfer Augmentation Using Twisted Tape Inserts: A Review. *Renew. Sustain. Energy Rev.* **2016**, *63*, 193–225. [CrossRef]
17. Waghole, D.R.; Warkhedkar, R.M.; Kulkarni, V.S.; Shrivastva, R.K. Experimental Investigations on Heat Transfer and Friction Factor of Silver Nanofliud in Absorber/Receiver of Parabolic Trough Collector with Twisted Tape Inserts. *Energy Procedia* **2014**, *45*, 558–567. [CrossRef]
18. Chaurasia, S.R.; Sarviya, R. Numerical and Experimental Thermal Performance with Entropy Generation Analysis on Tube with Helical Screw Tape Inserts at Number of Strips in Turbulent Flow. *Proc. Inst. Mech. Eng. Part C J. Mech. Eng. Sci.* **2021**, *235*, 1057–1070. [CrossRef]
19. Nakhchi, M.E.; Hatami, M.; Rahmati, M. Experimental Investigation of Heat Transfer Enhancement of a Heat Exchanger Tube Equipped with Double-Cut Twisted Tapes. *Appl. Therm. Eng.* **2020**, *180*, 115863. [CrossRef]
20. Arunachala, U.C. Experimental Study with Analytical Validation of Energy Parameters in Parabolic Trough Collector with Twisted Tape Insert. *J. Sol. Energy Eng.* **2020**, *142*, 1–12. [CrossRef]

21. Hosseinalipour, S.M.; Rostami, A.; Shahriari, G. Numerical Study of Circumferential Temperature Difference Reduction at the Absorber Tube of Parabolic Trough Direct Steam Generation Collector by Inserting a Twisted Tape in Superheated Region. *Case Stud. Therm. Eng.* **2020**, *21*, 100720. [CrossRef]
22. Elton, D.N.; Arunachala, U.C. Twisted Tape Based Heat Transfer Enhancement in Parabolic Trough Concentrator—An Experimental Study. *IOP Conf. Ser. Mater. Sci. Eng.* **2018**, *376*, 012034. [CrossRef]
23. Ghadirijafarbeigloo, S.; Zamzamian, A.H.; Yaghoubi, M. 3-D Numerical Simulation of Heat Transfer and Turbulent Flow in a Receiver Tube of Solar Parabolic Trough Concentrator with Louvered Twisted-Tape Inserts. *Energy Procedia* **2014**, *49*, 373–380. [CrossRef]
24. Song, X.; Dong, G.; Gao, F.; Diao, X.; Zheng, L.; Zhou, F. A Numerical Study of Parabolic Trough Receiver with Nonuniform Heat Flux and Helical Screw-Tape Inserts. *Energy* **2014**, *77*, 771–782. [CrossRef]
25. Mohammed, A.A.; Mohammed, A.A.; Falleh, M.A. Heat Transfer Augmentation in Tube Fitted with Rotating Twisted Tape Insert. *J. Mech. Eng. Res. Dev.* **2020**, *43*, 308–316.
26. Qi, C.; Wang, G.; Yan, Y.; Mei, S.; Luo, T. Effect of Rotating Twisted Tape on Thermo Hydraulic Performances of Nano Fluids in Heat Exchanger Systems. *Energy Convers. Manag.* **2018**, *166*, 744–757. [CrossRef]
27. Pavan, K.N.; Chethan, K.S.; Suresh, R.B. Investigation on Heat Transfer Augmentation in Tubes with Rotating Twisted Tape Insert Using Water and Copper Oxide Nanofluid as Heat Transfer Medium. *J. Crit. Rev.* **2020**, *7*, 1595–1600.
28. Promvonge, P. Thermal Augmentation in Circular Tube with Twisted Tape and Wire Coil Turbulators. *Energy Convers. Manag.* **2008**, *49*, 2949–2955. [CrossRef]
29. Varun, K.; Arunachala, U.C.; Elton, D.N. Trade-off between Wire Matrix and Twisted Tape: SOLTRACE® Based Indoor Study of Parabolic Trough Collector. *Renew. Energy* **2020**, *156*, 478–492. [CrossRef]
30. Chakraborty, O.; Das, B.; Gupta, R. Impact of Helical Coil Insert in the Absorber Tube of Parabolic Trough Collector. *Smart Innov. Syst. Technol.* **2021**, *206*, 177–187. [CrossRef]
31. Zhao, Z.; Bai, F.; Zhang, X.; Wang, Z. Experimental Study of Pin Finned Receiver Tubes for a Parabolic Trough Solar Air Collector. *Sol. Energy* **2020**, *207*, 91–102. [CrossRef]
32. Jamal-Abad, M.T.; Saedodin, S.; Aminy, M. Experimental Investigation on a Solar Parabolic Trough Collector for Absorber Tube Filled with Porous Media. *Renew. Energy* **2017**, *107*, 156–163. [CrossRef]
33. Anand, S.C.; Rai, D.A.K.; Sachan, V. Comparative Study of Parabolic Trough Concentrators. *Int. J. Mech. Eng. Technol.* **2014**, *5*, 65–73.
34. Subramaniyan, C.; Subramani, J.; Kalidasan, B.; Anbuselvan, N.; Yuvaraj, T.; Prabaharan, N.; Senjyu, T. Investigation on the Optical Design and Performance of a Single-Axis-Tracking Solar Parabolic Trough Collector with a Secondary Reflector. *Sustainability* **2021**, *13*, 9918. [CrossRef]
35. Arasu, A.V.; Sornakumar, T. Design, Manufacture and Testing of Fiberglass Reinforced Parabola Trough for Parabolic Trough Solar Collectors. *Sol. Energy* **2007**, *81*, 1273–1279. [CrossRef]
36. Kalogirou, S.A. *Solar Energy Engineering: Processes and Systems*, 2nd ed.; Elsevier Inc.: Amsterdam, The Netherlands, 2014.
37. Büker, M.S.; Parlamış, H.; Alwetaishi, M.; Benjeddou, O. Experimental Investigation on the Dehumidification Performance of a Parabolic Trough Solar Air Collector Assisted Rotary Desiccant System. *Case Stud. Therm. Eng.* **2022**, *34*, 102077. [CrossRef]
38. Goodfellow Ceramic & Glass Division. *Properties of Borosilicate Glass*; Goodfellow Ceramic & Glass Division: Pittsburgh, PA, USA, 2009.
39. Reddy, K.S.; Ravi Kumar, K.; Ajay, C.S. Experimental Investigation of Porous Disc Enhanced Receiver for Solar Parabolic Trough Collector. *Renew. Energy* **2015**, *77*, 308–319. [CrossRef]
40. Romero-Alvarez, M.; Zarza, E. Handbook of Energy Efficiency and Renewable Energy. *Choice Rev. Online* **2008**, *45*, 45-2629. [CrossRef]
41. Ibrahim, E.Z. Augmentation of Laminar Flow and Heat Transfer in Flat Tubes by Means of Helical Screw-Tape Inserts. *Energy Convers. Manag.* **2011**, *52*, 250–257. [CrossRef]
42. Mwesigye, A.; Bello-Ochende, T.; Meyer, J.P. Heat Transfer and Thermodynamic Performance of a Parabolic Trough Receiver with Centrally Placed Perforated Plate Inserts. *Appl. Energy* **2014**, *136*, 989–1003. [CrossRef]
43. Eiamsa-ard, S.; Thianpong, C.; Eiamsa-ard, P. Turbulent Heat Transfer Enhancement by Counter/Co-Swirling Flow in a Tube Fitted with Twin Twisted Tapes. *Exp. Therm. Fluid Sci.* **2010**, *34*, 53–62. [CrossRef]
44. Chakraborty, O.; Das, B.; Gupta, R.; Debbarma, S. Heat Transfer Enhancement Analysis of Parabolic Trough Collector with Straight and Helical Absorber Tube. *Therm. Sci. Eng. Prog.* **2020**, *20*, 100718. [CrossRef]
45. Xiao, H.; Liu, P.; Liu, Z.; Liu, W. Performance Analyses in Parabolic Trough Collectors by Inserting Novel Inclined Curved-Twisted Baffles. *Renew. Energy* **2021**, *165*, 14–27. [CrossRef]
46. Kumar, B.N.; Reddy, K.S. Numerical Investigations on Metal Foam Inserted Solar Parabolic Trough DSG Absorber Tube for Mitigating Thermal Gradients and Enhancing Heat Transfer. *Appl. Therm. Eng.* **2020**, *178*, 115511. [CrossRef]
47. Bellos, E.; Tzivanidis, C. Investigation of a Star Flow Insert in a Parabolic Trough Solar Collector. *Appl. Energy* **2018**, *224*, 86–102. [CrossRef]
48. Bellos, E.; Tzivanidis, C.; Antonopoulos, K.A.; Gkinis, G. Thermal Enhancement of Solar Parabolic Trough Collectors by Using Nano Fl Uids and Converging-Diverging Absorber Tube. *Renew. Energy* **2016**, *94*, 213–222. [CrossRef]

49. Bozorg, M.V.; Hossein Doranehgard, M.; Hong, K.; Xiong, Q. CFD Study of Heat Transfer and Fluid Flow in a Parabolic Trough Solar Receiver with Internal Annular Porous Structure and Synthetic Oil–Al_2O_3 Nanofluid. *Renew. Energy* **2020**, *145*, 2598–2614. [CrossRef]
50. Vahidinia, F.; Khorasanizadeh, H.; Aghaei, A. Comparative Energy, Exergy and CO_2 Emission Evaluations of a LS-2 Parabolic Trough Solar Collector Using Al_2O_3/SiO_2-Syltherm 800 Hybrid Nanofluid. *Energy Convers. Manag.* **2021**, *245*, 114596. [CrossRef]
51. Bellos, E.; Tzivanidis, C.; Daniil, I.; Antonopoulos, K.A. The Impact of Internal Longitudinal Fins in Parabolic Trough Collectors Operating with Gases. *Energy Convers. Manag.* **2017**, *135*, 35–54. [CrossRef]
52. Akbarzadeh, S.; Valipour, M.S. Energy and Exergy Analysis of a Parabolic Trough Collector Using Helically Corrugated Absorber Tube. *Renew. Energy* **2020**, *155*, 735–747. [CrossRef]
53. Holman, J.P. *Experimental Methods for Engineers*, 8th ed.; McGraw-Hill Series in Mechanical Engineering: New York, NY, USA, 2011; Volume s1-VIII.
54. Kabeel, A.E.; Sharshir, S.W.; Abdelaziz, G.B.; Halim, M.A.; Swidan, A. Improving Performance of Tubular Solar Still by Controlling the Water Depth and Cover Cooling. *J. Clean. Prod.* **2019**, *233*, 848–856. [CrossRef]
55. Kumar, D.; Kumar, S. Thermal Performance of the Solar Parabolic Trough Collector at Different Flow Rates: An Experimental Study. *Int. J. Ambient Energy* **2018**, *39*, 93–102. [CrossRef]
56. Bellos, E.; Tzivanidis, C. Enhancing the Performance of a Parabolic Trough Collector with Combined Thermal and Optical Techniques. *Appl. Therm. Eng.* **2020**, *164*, 114496. [CrossRef]
57. Zheng, Z.; Xu, Y.; He, Y. Thermal Analysis of a Solar Parabolic Trough Receiver Tube with Porous Insert Optimized by Coupling Genetic Algorithm and CFD. *Sci. China Technol. Sci.* **2016**, *59*, 1475–1485. [CrossRef]
58. Ahmed, K.A.; Natarajan, E. Thermal Performance Enhancement in a Parabolic Trough Receiver Tube with Internal Toroidal Rings: A Numerical Investigation. *Appl. Therm. Eng.* **2019**, *162*, 114224. [CrossRef]
59. Bellos, E.; Tzivanidis, C. Enhancing the Performance of Evacuated and Non-Evacuated Parabolic Trough Collectors Using Twisted Tape Inserts, Perforated Plate Inserts and Internally Finned Absorber. *Energies* **2018**, *11*, 1129. [CrossRef]
60. Laaraba, A.; Mebarki, G. Enhancing Thermal Performance of a Parabolic Trough Collector with Inserting Longitudinal Fins in the Down Half of the Receiver Tube. *J. Therm. Sci.* **2020**, *29*, 1309–1321. [CrossRef]

sustainability

Article

Novel Biosynthesis of Graphene-Supported Zero-Valent Iron Nanohybrid for Efficient Decolorization of Acid and Basic Dyes

Mahmoud Samy [1], Marwa Elkady [2,3,*], Ayman Kamal [4,5], Noha Elessawy [6], Sahar Zaki [4] and Marwa Eltarahony [4,*]

1 Department of Public Works Engineering, Faculty of Engineering, Mansoura University, Mansoura 35516, Egypt
2 Chemical and Petrochemical Engineering Department, Egypt-Japan University of Science and Technology, New Borg El-Arab City, Alexandria 21934, Egypt
3 Fabrication Technology Department, Advanced Technology and New Materials Research Institute (ATNMRI), City of Scientific Research and Technological Applications, New Borg El-Arab City, Alexandria 21934, Egypt
4 Environmental Biotechnology Department, Genetic Engineering and Biotechnology Research Institute (GEBRI), City of Scientific Research and Technological Applications (SRTA-City), Alexandria 21934, Egypt
5 Faculty of Agriculture, Alexandria University, Alexandria 21544, Egypt
6 Computer Based Engineering Applications Department, Informatics Research Institute IRI, City of Scientific Research & Technological Applications (SRTA-City), Alexandria 21934, Egypt
* Correspondence: marwa.elkady@ejust.edu.eg (M.E.); m_eltarahony@yahoo.com (M.E.); Tel.: +20-100-978-2536 (Marwa Eltarahony)

Citation: Samy, M.; Elkady, M.; Kamal, A.; Elessawy, N.; Zaki, S.; Eltarahony, M. Novel Biosynthesis of Graphene-Supported Zero-Valent Iron Nanohybrid for Efficient Decolorization of Acid and Basic Dyes. *Sustainability* **2022**, *14*, 14188. https://doi.org/10.3390/su142114188

Academic Editors: Mohamed El-Alfy, Ahmed El Kenawy, Petra-Manuela Schuwerack and Zhongfeng Xu

Received: 25 August 2022
Accepted: 21 September 2022
Published: 31 October 2022

Abstract: Herein, respiratory nitrate reductases (NAR) were utilized in the biosynthesis of zero-valent iron (ZVI) graphene nanocomposite as a simultaneous reducing and capping agent, for the first time, to efficiently adsorb methylene blue (MB) and direct red-81 (DR-81). Under anaerobic conditions, the greenly synthesized graphene was incubated with iron precursor in the presence of crude-NAR enzyme for 48 h to obtain the ZVI graphene composite followed by characterizing this composite using physiochemical analyses. Scanning and transmission electron microscopy, energy dispersive X-ray spectroscopy and X-ray diffraction techniques assured the chemical composition and the interaction between ZVI and graphene. The influences of operating conditions such as contact time, pH and adsorbent dose on the adsorption efficacy were explored in the case of ZVI graphene, graphene and ZVI. ZVI graphene nanocomposite displayed the highest removal efficiency of MB and DR-81 compared to graphene and ZVI-NPs. The removal percentages of DR-81 and MB by ZVI graphene nanocomposites were 88.3 ± 2.66% and 87.6 ± 2.1%, respectively, at pH 7, adsorbent dose 20 mg/50 mL, initial MB or DR-81 concentration of 10 mg/L and shaking speed of 150 rpm. A pseudo first-order model could describe the adsorption kinetics, and the adsorption mechanism was discussed. The promising results of the current study support the potential of the recruitment of ZVI graphene nanocomposites in eliminating various pollutants from industrial effluents on a larger scale. Further, the prepared nanohybrid can be used in other applications such as photocatalysis, Fenton and persulfate activation processes.

Keywords: decolorization; direct red-81; graphene; methylene blue; nitrate reductases; operating parameters; zero-valent iron

1. Introduction

The expansion of industrial activities such as cosmetics, leather, ink, textiles and printing contributes to the release of heavily contaminated effluents containing dyes to water streams [1,2]. Azo dyes constitute almost 50% of the global production (700,000 ton/year) and during the dyeing process, about 20% of the dyes can be discharged to water sources [3]. The presence of dyes in water bodies inhibits light penetration and photosynthesis and reduces the dissolved oxygen ratios resulting in threats and risks to humans and animals [4].

Therefore, it is imperative to effectively treat effluents containing dyes before discharge to the environment. Many physical, chemical and biological treatment methods have been proposed for the removal of dyes [5–7]. The azo dyes are highly resistant to being degraded by microorganisms and they are toxic to bacteria which make the biological treatment ineffective [8]. Similarly, the removal of dyes by physical–chemical treatment processes such as adsorption, photocatalysis, ozonation and membrane filtration has been investigated [9–12]. However, these methods are associated with some drawbacks such as high costs and the production of toxic by-products [13].

The adsorption process is a promising, inexpensive, simple and efficient treatment process for the removal of industrial effluents (e.g., dyes) compared to previously mentioned techniques [14]. Activated carbon is a commonly employed adsorbent for the removal of bio-resistant pollutants (dyes) from aqueous solutions [15]. However, activated carbon challenges the high cost and requirement of frequent regeneration which obstruct the reusability and full-scale application [16]. Hence, there is an increasing need for the preparation of effective adsorbents with low costs. Zero-valent iron (ZVI) has recently pulled wide heed due to its low cost, high reactivity, green impact, high surface area as well as its capability to reduce oxidized contaminants [12,17]. ZVI is commonly prepared via the liquid-phase reduction in iron salts using strong reducing agents (e.g., sodium borohydride). In spite of the simplicity, short time and high reactivity of the liquid reduction method, this method suffers from shortcomings such as high costs, toxicity of the reducing agent and poor stability of prepared nanoparticles [18]. Moreover, the release of nanomaterials prepared from harmful chemicals after the treatment process has a negative impact on the aquatic and terrestrial environments [19]. Therefore, the efforts have been directed to prepare nanoparticles using ecofriendly materials (e.g., bacteria, plants, algae). In this study, we employed respiratory nitrate reductase (NAR) enzyme as a green reducing and stabilizing agent to replace toxic reducing agents needed in the preparation of ZVI nanoparticles. Due to the toxicity of metal ions to bacteria, bacteria start to protect themselves and then, metal ions can bioaccumulate or biomineralize into the cells after the reduction of metal ions to a lower state [20]. The biosynthesis of ZVI can reduce the cost, protect the environment from toxic chemicals and produce stable nanoparticles.

Another problem facing ZVI is the fast aggregation and formation of the iron oxide layer which reduces the reactivity and reducibility of ZVI [12,21]. To overcome this problem, pure ZVI can be supported with different materials such as silica, kaolinite, starch and graphene [22–24]. Several studies in recent years recorded the efficiency of graphene and its oxide, in particular as a nanosorbent, in different remediation applications of polluted environments by the dint of its high theoretical surface area (~2620 m^2/g), high electrical conductivity and stability [5]. In this work, a ZVI graphene nanocomposite was prepared by reducing iron salt in the presence of graphene, which was produced from plastic waste, by the catalysis of NAR enzyme. The use of graphene can reduce the agglomeration and the oxidation of ZVI, facilitate the electron transfer from ZVI surface to the pollutant and improve the dispersibility of the nanoparticles [25]. Moreover, the conversion of plastic waste (polyethylene terephthalic) to graphene participates in the management of environmental problems related to the release of plastic wastes to the environment and reduction in costs related to the disposal of plastic waste [26]. Liu et al. (2014) prepared ZVI supported by graphene for the removal of phosphorus from aqueous solutions [25]. They stated that the prepared composite attained a high removal of phosphorus.

In this study, the novel procedures for the biological synthesis of ZVI graphene nanocomposite were provided. The synthesized ZVI graphene nanocomposite was utilized for the removal of methylene blue (MB) and direct red-81 (DR-81). The effects of operating parameters such as contact time, pH and adsorbent dosage were investigated. The removal mechanism and adsorption kinetics of MB and DR-81 were explored.

2. Materials and Methods

2.1. Materials

Plastic bottle wastes were collected from the garbage of City of Scientific Research and Technological Applications in Borg Al Arab, Alexandria. Sodium citrate ($Na_3C_6H_5O_7$, 98%), hydrochloric acid (HCl, 97%), sodium chloride (NaCl, 97%), magnesium sulfate heptahydrate ($MgSO_4{\cdot}7H_2O$, 99%), ferrous chloride tetrahydrate ($FeCl_2{\cdot}4H_2O$, 95%), sodium molybdate ($Na_2MoO_4{\cdot}2H_2O$, 99%), copper sulfate ($CuSO_4{\cdot}5H_2O$, 95%), cobalt chloride hexahydrate ($CoCl_2{\cdot}6H_2O$, 98%), manganese chloride tetrahydrate ($MnCl_2{\cdot}4H_2O$, 97%), zinc sulfate ($ZnSO_4$, 93%), potassium phosphate (KH_2PO_4, 95%), dipotassium hydrogen phosphate ($K_2HPO_4{\cdot}7H_2O$, 97%) and potassium nitrate (KNO_3, 98%) were procured from Sigma-Aldrich and used without modifications.

2.2. Biosynthesis of ZVI Graphene Nanocomposite

Initially, graphene and ZVI-NPs were prepared following the procedures described in detail by El Essawy et al. (2017) and Zaki et al. (2019) [27,28]. For the preparation of ZVI graphene nanocomposite, respiratory membrane-bound nitrate reductase enzyme (NAR) was recruited as reducing and stabilizing agent simultaneously in an in situ synthesis process. NAR enzyme as well as electron shuttling molecules commence nitrate reduction and shuttle electrons to iron ions. Then, the oxidation states of metal ions can change via redox reaction until reaching zero-valent state and forming ZVI-NPs [28,29]. In the presence of graphene, the enzymatically generated ZVI-NPs bind to graphene particles, forming ZVI graphene nanocomposites. Firstly, about 10^8 CFU/mL (0.5 McFarland) of bionanofactory strain *P. mirabilis* 10B was precultured anaerobically in the optimized broth that enhances and induces the growth of NAR enzyme, of the following ingredients (g/L): sodium citrate 7.5, NaCl 0.5, $MgSO_4{\cdot}7H_2O$ 0.12, $FeCl_2{\cdot}4H_2O$ 0.24, $Na_2MoO_4{\cdot}2H_2O$ 0.015, $CuSO_4{\cdot}5H_2O$ 0.06, $CoCl_2{\cdot}6H_2O$ 0.04, $MnCl_2{\cdot}4H_2O$ 0.03, $ZnSO_4$ 0.31, KH_2PO_4 3.0, $K_2HPO_4{\cdot}7H_2O$ 1.0 and KNO_3 5.0 at pH 7.0. The inoculated broth was incubated anaerobically at 30 °C for 48 h in anaerobic jar. After incubation, a considerable bacterial biomass was harvested by centrifugation at 10,000 rpm for 20 min. The harvested cells were washed several times by sterile distilled water and the, suspended in 80 mM potassium phosphate buffer (pH 7.0) and exposed to mild osmotically shock disruption followed by centrifugation at 5000 rpm for 5 min to eliminate unbroken cells and cell debris. Thereafter, the obtained spheroplast was subjected to centrifugation at 10,000 rpm, 4 °C for 60 min. The precipitated pellets were resuspended in 80 mM potassium phosphate buffer (pH 7.0) and utilized as a complete membrane-bound $\alpha\beta\gamma$ complex of NAR. Finally, 0.2 gm of graphene was added to crude-NAR suspension that was supplemented by 1.5 mM of Fe $(NO_3)_3{\cdot}9H_2O$. The overall mixture was incubated under anaerobic conditions for 48 h at 30 °C. The obtained ZVI graphene nanocomposite was collected, washed and dried at 80 °C for 2 h for subsequent characterization and application.

2.3. Characterization of ZVI Graphene Nanocomposite

The crystallographic information of ZVI graphene nanocomposite was investigated by an X-ray diffraction (XRD, Shimadzu-7000, USA) using Cu Kα radiation (λ = 1.54056 Å), voltage of 40 KV, current of 30 mA and scan speed of 10 °/min. The incorporation of ZVI in carbon matrix of graphene was affirmed via the specification of elemental composition using energy dispersive X-ray spectroscopy (EDX) analyzer and elemental mapping combined with transmission electron microscope (TEM, JEOL JSM 6360LA, Japan). Additionally, TEM and scanning electron microscopy (SEM) (JEOL JEM-1230) with an accelerating voltage of 200 kV were employed to study the morphology of the synthesized composite. Raman spectra (Jaso, Japan) were recorded to specify the molecular structure of the synthesized nanoparticles. Raman spectra were set in the 4000–400 cm^{-1} region with a step size of 1 cm^{-1}. Fourier transform infrared spectroscopy (Shimadzu, FTIR-8400S) was performed to specify the functional groups using KBr pellets in the 4000–400 cm^{-1} region with a step size of 1 cm^{-1}. Surface area was estimated using Belsorp-max automated apparatus (BEL Japan) and the sample was degassed at 200 °C for 3 h. The data of nitrogen (adsorption–desorption isotherms) were recorded at 77.53 K.

2.4. Experimental Procedures

A stock solution with a concentration of 1000 mg/L was prepared for direct red-81 (DR-81) and methylene blue (MB) by dissolving 1 g of the dye in 1 L of distilled water. The adsorption batch experiments were conducted in 100-mililter beaker with 50 mL of the dye solution and the solution was shaken at a speed of 150 rpm. The effects of different parameters such as contact time (15–120) min, adsorbent dose (20, 40, 60, 100 mg/50 mL), and pH (3, 7, 9, 11) were studied at an initial dye concentration of 10 mg/L. Samples were withdrawn and then centrifuged prior to the measurements by UV–Visible spectrophotometer (Labomed model, Inc., Los Angeles, CA, USA) at wavelengths of 500 and 664 nm for DR-81 and MB, respectively. The comparison between the performance of pure ZVI-NPs, graphene and ZVI graphene nanocomposite was performed at pH 7, initial dye concentration of 10 mg/L, adsorbent dose of 20 mg/50 mL and shaking speed of 150 rpm. The adsorption kinetics were investigated using pseudo first- and second-order models and intraparticle diffusion on a time range (0–90 min) at pH 7, adsorbent dose 20 mg/50 mL, initial MB or DR-81 concentration of 10 mg/L and shaking speed of 150 rpm. The equations and discussion of these models were provided in detail in our previous work [9]. In reusability study, the particles were compiled after each run and left at room temperature to dry, thence they can be used in next cycles.

3. Results and Discussion

3.1. Characterization of ZVI Graphene Nanocomposite

In the present study, the NAR enzyme was successfully utilized as a catalyzing and stabilizing agent to synthesize ZVI graphene nanohybrid. Generally, the biological approaches of nanomaterial synthesis are advantageous over other traditional physicochemical means. They are characterized by their cost-effectiveness, non-toxicity, biosafety and ecofriendliness. They reckon on the utilization of biological molecules in the reduction of bulk parent materials into their nanoform; neglecting by such way the usage of hazardous reductants, high-temperature or energy-intensive processing [28,29]. It is noteworthy to mention the characteristic role of nitrate reductases (NRs), which belong to oxidoreductases groups. They are mainly catalyzing organic material oxidation and play crucial function in pollution remediation and eventually participate in the nitrogen cycle. Recently, NRs were reported in various investigations addressing nanostructure synthesis in green biological methods [29,30]. The ZVI graphene nanocomposite was analyzed by various characterization techniques. SEM and TEM images of pure graphene were provided in the Supplementary Materials (Figure S1) to show the changes in the morphology in the case of the composite. Figure 1a,b shows TEM and SEM images of ZVI graphene biocomposites with particle size ranged from 2.3 to 7 nm, indicating the incorporation of ZVI-NPs in the graphene matrix. However, the magnetic properties of ZVI-NPs resulted in the aggregation of ZVI-NPs in chain-like forms.

Additionally, the EDX pattern in Figure 1c confirmed the presence of both carbon and iron at their characteristic positions (0.277 and 6.4 keV, respectively) with atomic percentages of 50.8% and 36.5%, respectively. Other peaks at 0.39 and 2.3 keV were ascribed to nitrogen and sulfur with atomic ratios of 9.2 and 3.5%, respectively. Interestingly, the existence of nitrogen and sulfur might be owing to the bounded bacterial biomolecules during the synthesis process, such as amino groups ($-NH_3^+$) and sulfhydryl (SH) groups, which could eventually provide the nanocomposite with self-functionalizing and stabilizing properties. However, a Cu peak was also detected due to the copper grid that was used to hold the samples. The XRD pattern in Figure 1d demonstrates the XRD pattern of bare graphene showing only the diffraction planes of graphene. The XRD pattern in Figure 1e affirmed the interaction between the graphene and ZVI-NPs. The peak at 24.67° is attributed to the (002) diffraction plane of graphene [3]. The sharp peak at 44.4° is attributed to the

(110) diffraction plane of ZVI-NPs, whereas the peaks at 31.7°, 35.56°, 51.82° and 59.92° are attributed to the iron oxides formed on the ZVI surface due to the oxidation that might be occurred during washing, drying and processing of the nanocomposite [2,31].

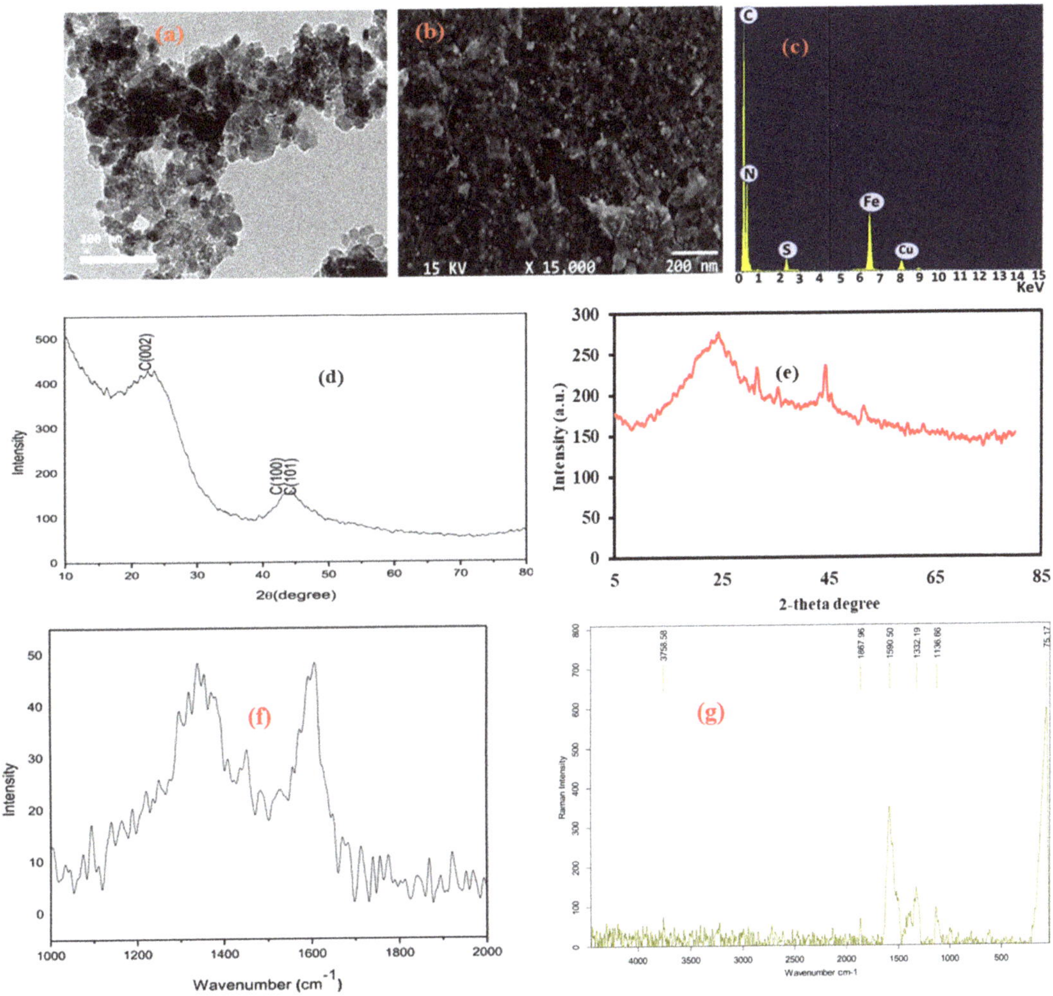

Figure 1. Physicochemical properties (**a**) TEM image, (**b**) SEM, (**c**) EDX pattern of ZVI graphene, (**d**) XRD pattern of graphene, (**e**) XRD of ZVI graphene, (**f**) Raman spectra of graphene and (**g**) Raman spectra of ZVI graphene [27].

Raman spectra of pristine graphene are given in Figure 1f, showing only the D and G bands of graphene. Figure 1g shows the Raman spectra of ZVI graphene. The peaks at 1332.19 cm^{-1} and 1590.5 cm^{-1} are ascribed to the D and G bands of graphene [25]. The peaks at wavenumbers lower than 1000 cm^{-1} (around 75 cm^{-1}, 550 cm^{-1} and 750 cm^{-1}) are attributed to the Fe-O bond, confirming the presence of iron oxides on the surface of ZVI-NPs [31]. Supplementary Materials Figure S2a,b shows the adsorption–desorption isotherms and FTIR spectra of graphene. The surface area was 721.7 m^2/g. The bands in FTIR spectra at 3447, 1636, 1219 and 1105 cm^{-1} are indexed to O-H, C=O, O-H, C-O-C and C-O bonds, respectively [27]. Additionally, the band at 1600 cm^{-1} is imputed to the

C==C bond [27]. Figure S2c,d demonstrates the FTIR spectra and adsorption–desorption isotherms of ZVI graphene. The presence of the hydroxyl group (O-H) was confirmed through the bands at 1610 and 3420 cm^{-1}. The band at 1350 cm^{-1} is assigned to the COO^{-} group [32]. Moreover, the band at 1020 cm^{-1} is attributed to the C-O bond. The bands at 476 and 628 cm^{-1} are imputed to the Fe-O bond. The surface area of ZVI graphene was estimated (55 m^2/g). Figure S2e,f,g shows the elemental mapping of graphene and ZVI graphene [33]. In the case of graphene, only carbon was detected. Whereas carbon, iron and oxygen were detected in the elemental mapping of the composite.

3.2. Application of ZVI Graphene Nanocomposite in Cationic and Anionic Dye Removal

As observed in Figure 2a,b, the alterations in the UV–vis spectra (from 200 to 800 nm) of DR-81 and MB were monitored upon treatment with ZVI graphene nanocomposite, ZVI and graphene. The spectrum patterns of DR-81 and MB (controls) exhibited the main absorption peaks at 500 and 664 nm for DR-81 and MB, respectively, due to the azo bond, which almost disappeared in the supernatant of the treated samples especially in the case of ZVI graphene. Such vanishing of both peaks reflected the successful removal of both dyes. Our results are coincident with those investigated by Samy et al. (2020) [34]. However, the adsorption process was confirmed by SEM in Figure 2c,d. The pores in the case of ZVI graphene nanocomposite approximately disappeared due to the filling of pores by the dye molecules confirming the high adsorption performance of the dyes by the biosynthesized composite. These things considered, new peaks were formed due to the formation of chemical bonds between the iron oxide layer and dye molecules, as revealed by Raman spectroscopy (Figure 2e,f).

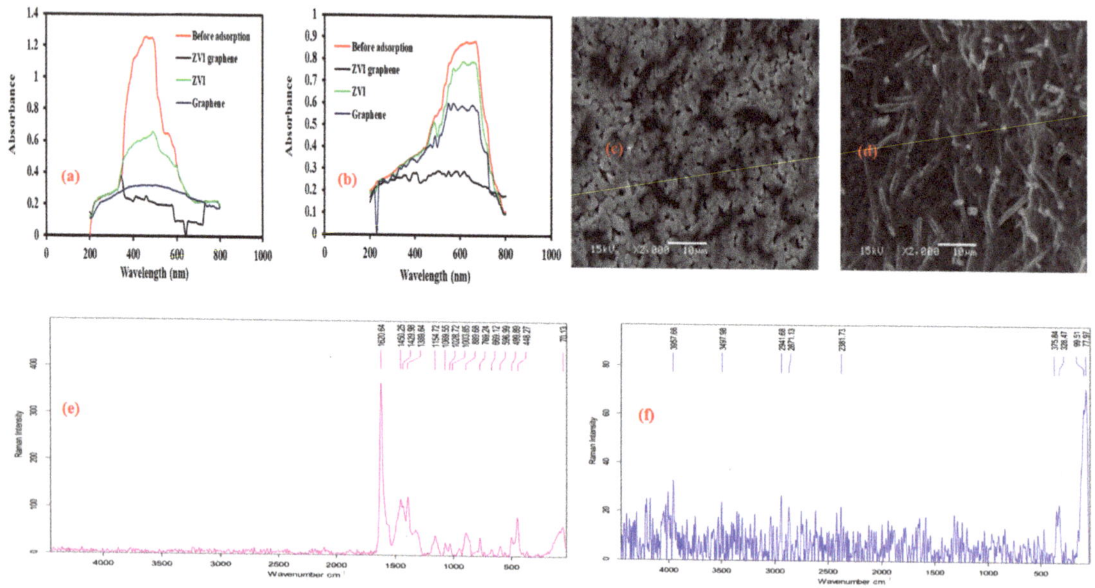

Figure 2. Absorption spectra of (**a**) DR-81 and (**b**) MB before and after adsorption over ZVI graphene, ZVI and graphene; SEM image of ZVI graphene after adsorption of (**c**) DR-81 and (**d**) MB; Raman spectra of ZVI graphene after adsorption of (**e**) DR-81 and (**f**) MB.

Additionally, XRD and FTIR of ZVI graphene after adsorption of DR-81 and MB are given in Supplementary Materials Figure S3a,b. The intensity and transmittance decreased after the adsorption of DR-81 and MB. The change in the intensity and transmittance might be due to the bonds formed between the adsorbent's surface and adsorbate. EDX of ZVI graphene after adsorption in Figure S3c,d showed the introduction of Cl and Na in the case of MB and DR-81, reaffirming the bonding between the adsorbent's surface and the adsorbate.

3.3. Effect of Contact Time

A comparison between the adsorption performance of ZVI-NPs, graphene and ZVI graphene nanocomposite was conducted to specify the required time for achieving equilibrium and choose the best material, as shown in Figure 3a,b. The experiments were conducted at pH 7, adsorbent dose 20 mg/50 mL, initial MB or DR-81 concentration of 10 mg/L and shaking speed of 150 rpm. The results showed that 90 min was the equilibrium time in the cases of MB and DR-81. The prolonged equilibrium time indicated the porous structure and high surface area of the prepared composite. Gajera et al. (2022) reported an equilibrium time of 2 h [35]. Moreover, the ZVI graphene nanocomposite showed higher adsorption performance towards MB and DR-81 compared to ZVI-NPs and graphene. The removal efficiencies of DR-81 were 61.4 ± 1.6%, 49.8 ± 1.7% and 88.3 ± 2.66% in the cases of graphene, ZVI-NPs and ZVI graphene nanocomposite, respectively, whereas the removal efficiencies of MB were 68.8 ± 3.1%, 53.3 ± 2.39% and 87.6 ± 2.1% after 90 min with the same order. The extension of the reaction time did not greatly improve the removal percentage because the pores became saturated and the adsorbents' surfaces reached an equilibrium state [25]. The biogenic ZVI graphene nanocomposite attained a removal efficiency of 91.8% and 91.3% for DR-81 and MB, respectively, after 120 min. The synergistic effects of graphene and ZVI-NPs were the major reason for such high removal ratios. The graphene participated in the enhancement of ZVI-NPs dispersion and reduction in aggregation [36]. Graphene exhibited higher performance than ZVI-NPs due to its high surface area, while pure ZVI-NPs might agglomerate which decreased the surface area, thereby reducing the adsorption efficiency.

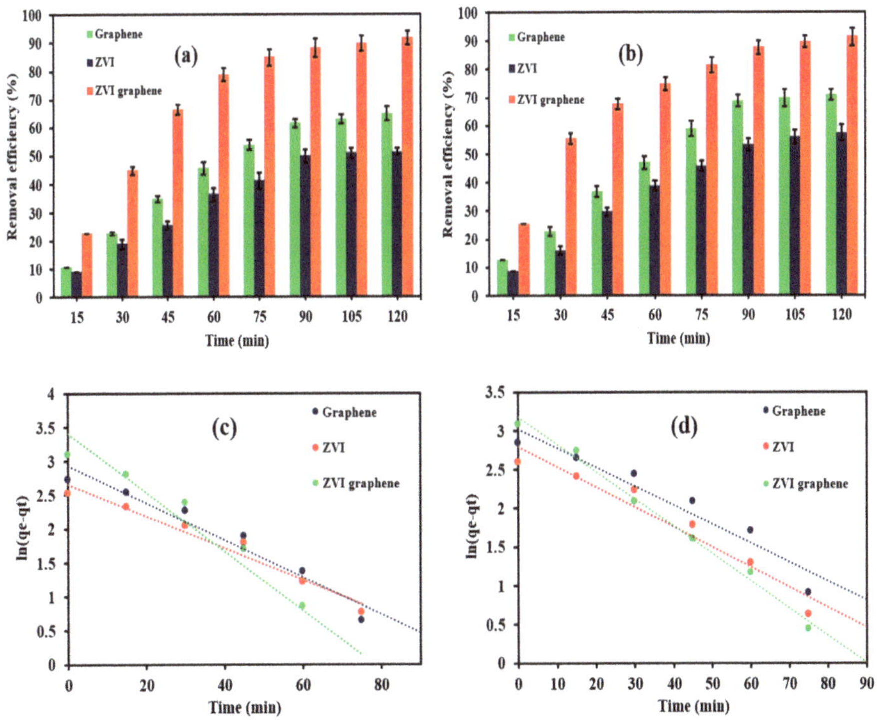

Figure 3. Effect of contact time on the adsorption efficiency of (**a**) DR-81 and (**b**) MB, and pseudo first-order kinetic model in the case of (**c**) DR-81 and (**d**) MB.

The rate constants of the pseudo first-order model were estimated in the case of graphene, ZVI and ZVI graphene via the linear plot of ln(q_e-q_t) versus time, as shown in Figure 3c,d. The high coefficient of determination values and the slight difference between experimental adsorption capacity and estimated adsorption capacity from the first-order model at equilibrium affirmed the suitability of the first-order model to describe adsorption kinetics. The rate constants were 0.0273, 0.0236 and 0.0434 min^{-1} for graphene, ZVI and ZVI graphene, respectively, in the case of DR-81, whereas the rate constants were 0.026, 0.0245 and 0.0352 min^{-1} for the synthesized materials with the same order in the case of MB. The high-rate constant in the case of the composite compared to graphene and ZVI was in agreement with the results obtained from the comparison between the three materials. Second-order and intraparticle diffusion models were also provided in the Supplementary Materials (Figure S4). Table S1 shows the constant rate (K_1 and K_2), coefficient of determination (R^2) and adsorption capacity at equilibrium (q_e) in the cases of first- and second-order models as well as the constants and R^2 of the intraparticle diffusion model. According to R^2 values in Table S1, the second-order model could not describe the adsorption process. Additionally, obtained q_e from the second-order model is far from the experimental value. The intraparticle diffusion model displayed the multilinearity of the adsorption process.

3.4. *Effect of Adsorbent Dose*

Figure 4a,b shows the effect of adsorbent dose on the removal efficiency of DR-81 and MB at pH 7, shaking speed at 150 rpm and initial dye concentration of 10 mg/L after a contact time of 90 min. The increase in the graphene dose from 20 mg/50 mL to 60 mg/50 mL resulted in the raising of the removal efficiency from 61.4 ± 1.6% to 88.9 ± 1.9% in the case of DR-81 and from 68.8 ± 3.1% to 89.8 ± 2.4% in the case of MB. However, the increase in graphene dose above 60 mg/50 mL did not show any amelioration in the removal efficiency in both dyes. Regarding ZVI, the increase in the ZVI dose above 60 mg/50 mL showed slight improvement in the adsorption efficiency. In the case of ZVI graphene, upon increasing the dose from 20 mg/50 mL to 40 mg/50 mL, the enhancement of the removal efficiency from 88.3 ± 2.66% to 96.88 ± 1.9% in the case of DR-81 and from 87.6 ± 2.1% to 98.66 ± 1.9% in the case of MB was noticed. The raising of ZVI graphene to 100 mg/50 mL just increased the removal efficiency to 99.77 ± 3.1% and 99.6 ± 2.6% in the cases of DR-81 and MB, respectively. In general, the raising of adsorbent dose increases the number of binding sites which contributes to the amelioration of adsorption performance. Howbeit, doses above the optimum value might decrease or cause a slight improvement of the removal efficiency because of the possible agglomeration of nanoparticles. The agglomeration of nanoparticles could decrease the surface area and, consequently, reduce the active adsorption sites [37]. Mensah et al. (2022) reported the same observation and showed that the increase in the adsorbent dose above 0.2 mg/mL did not boost the adsorption ratio [3].

The stability of the prepared composite was examined under five repetitive runs (450 min, 90 min for each run) for the adsorption of DR-81 and MB, as shown in Supplementary Materials Figure S5. The slight decrease in adsorption percentage in the sequential runs assured the stability of the composite. Howbeit, the decrease in the adsorption ratio might be due to the loss of adsorbent during sampling.

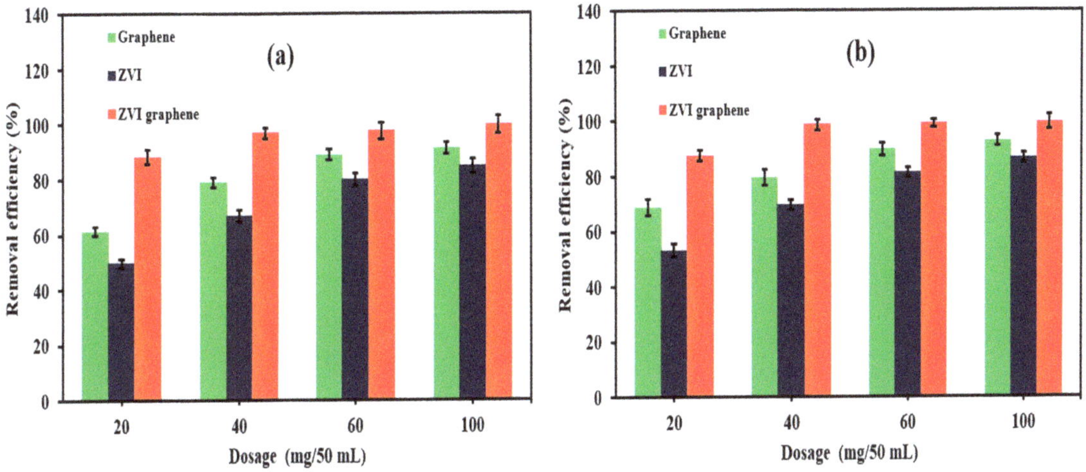

Figure 4. Effect of adsorbent dose on the removal efficiency of (**a**) DR-81 and (**b**) MB.

3.5. *Effect of pH*

The effect of pH was studied by varying pH from 3 to 11 (using 1 M of HCl or NaOH) at initial dye concentration of 10 mg/L, adsorbent dose of 40 mg/50 mL and shaking speed of 150 rpm, as shown in Figure 5. Point of zero charge (PZC) of ZVI is nearly eight, as reported by Sun et al. (2015) [38]. At pH < PZC, the adsorbent's surface charge becomes positive, and DR-81 is an ionic dye. Therefore, the removal efficiency was improved in acidic and neutral conditions due to the attractive forces between the adsorbent's surface and DR-81 molecules. The elimination efficiency increased from 96.88 $\pm$ 1.67% to 99.4 $\pm$ 1.9% by decreasing pH from 7 to 3 using the ZVI graphene nanocomposite. On the other hand, the removal efficacy went down in alkaline conditions due to the repulsive forces between negatively charged adsorbent's surface and anionic dye molecules. The removal efficiencies were 85.6 $\pm$ 2.3% and 78.9 $\pm$ 2.67% at pH 9 and 11, respectively, using the ZVI graphene nanocomposite. In the case of MB, the removal efficiencies decreased at pH 3 and it was 81.4 $\pm$ 1.9% due to the repulsive forces between positively charged sorbent's surface and cationic dye. However, at neutral pH, the removal efficiency was high. The increase in pH from 3 to 7 resulted in the raising of hydroxyl ions which could be adsorbed on the adsorbent's surface and, consequently, MB as a cationic dye could be adsorbed on the surface [39]. At high pH values, the removal efficiencies decreased to 88.7 $\pm$ 1.4% and 80.1 $\pm$ 1.1% at pH 9 and 11, respectively, in spite of the attractive forces between the adsorbent and adsorbate. At alkaline conditions, ZVI could be easily corroded and, consequently, iron hydroxides would be formed and precipitated [39]. The formed iron hydroxides could block the active sites and inhibit the electron transfer between ZVI and target pollutants which resulted in the decrease in removal efficiency [40]. Hamdy et al. (2018) reported the decrease in MB removal efficiency in alkaline condition due to the block of active sites of ZVI by the corrosion products [41]. In the case of the adsorption of MB and DR-81 over ZVI at pH > 7, the adsorption efficiency decreased due to the accelerated corrosion of ZVI, resulting in the formation of iron hydroxides that could block the binding sites and suppress the electron transfer [40]. At pH $\leq$ 7, the adsorption ratio went up in the case of DR-81 due to the attraction between DR-81 molecules and ZVI surface, whereas the adsorption percentage declined in the case of MB due to the repulsive forces [39]. The adsorption at pH 3 was lower than that of pH 7 due to the competition between MB molecules and Cl^- on ZVI surface [42]. At pH $\leq$ 7, the graphene surface charge is positive, whereas it was negative at pH > 7, as reported by Yang et al. (2022) [43]. So, the adsorption efficiency improved in the case of DR-81 at pH < 7 due to the attraction forces, while the adsorption ratio of DR-81 went down at pH > 7 due to the repulsive forces. In the case of MB, the adsorption proportion decreased at pH < 7 due to the repulsive forces and the adsorption went up at pH > 7 owing to the attractive forces. In the case of MB at pH 7, the adsorbed hydroxyl ions on the adsorbent's surface might increase the attractive forces, thus improving the degradation ratio.

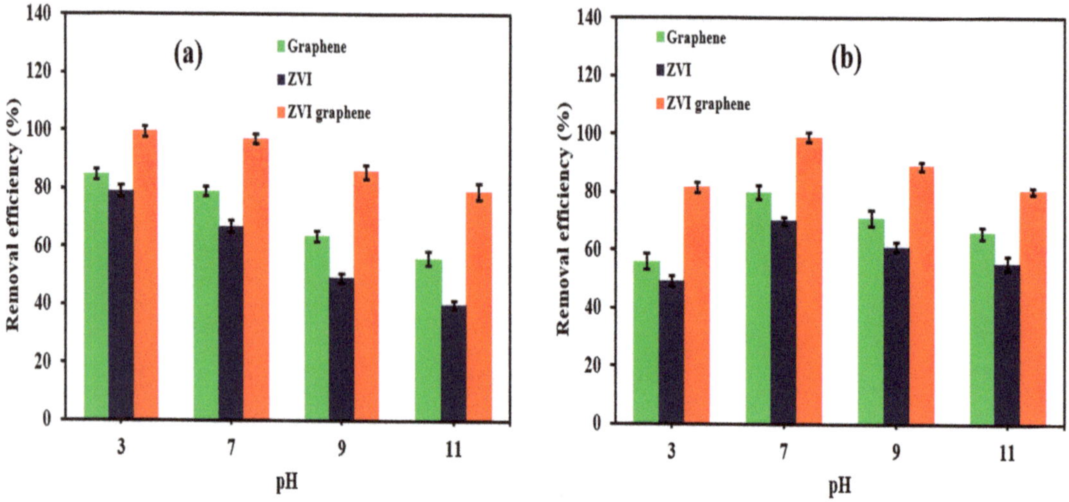

Figure 5. Effect of pH on the removal efficiency of (**a**) DR and (**b**) MB.

3.6. Adsorption Mechanism

Figure 6 demonstrates the adsorption mechanism of DR-81 and MB on ZVI graphene nanocomposite's surface. The adsorption of MB on the adsorbent's surface might be owing to the hydrogen bonding between OH and COOH groups of the adsorbent and NH group of MB [44]. In the case of DR-81, the adsorption took place via hydrogen bonding between SO_3^-, NH and N=N of DR-81 and the hydroxyl group in the adsorbent [45]. Additionally, the hydrogen bonding could happen between COOH of the adsorbent and the NH group of DR-81. The adsorption could take place due to the electrostatic interaction between the COO^- group (deprotonated form of COOH) of the adsorbent and NH^+ of MB. Moreover, the adsorption might occur due to the electrostatic interaction between Na^+ and the deprotonated form of COOH (COO^-) in the case of DR-81 [46]. Further, the adsorption of MB or DR-81 might take place due to the π-π stacking between aromatic groups of MB or DR-81 and oxygen of the hydroxyl group in the prepared adsorbent [45,47]. Additionally, the adsorption could take place due to the Yoshida hydrogen bonding between benzene ring of DR-81 or MB and H in the hydroxyl group [45,48]. The decolorization of DR-81 and MB might also take place due to the reduction by atomic hydrogen (H*) formed owing to the reaction between electrons and H^+ on the adsorbent's surface [49]. Furthermore, the formed iron hydroxides could adsorb DR-81 and MB. The adsorption of DR-81 or MB on ZVI graphene is governed by hydrogen bonding, electrostatic interaction (attractive and repulsive forces), Yoshida hydrogen bonding and π-π stacking. So, the increase or the decrease in the adsorption percentage can be affected by the attractive or repulsive forces between the adsorbent and adsorbate, as explained in Section 3.5. The adsorption in the case of graphene could take place via the same mechanisms (hydrogen bonding, electrostatic interaction (attractive and repulsive forces), Yoshida hydrogen bonding and π-π stacking based on its functional groups, as previously described in the literature) [3]. In the case of ZVI, adsorption could occur via hydrogen bonding, electrostatic interaction (attractive and repulsive forces), Yoshida hydrogen bonding and π-π stacking [49]. Additionally, the removal of dyes via ZVI might take place by reduction and precipitation [50].

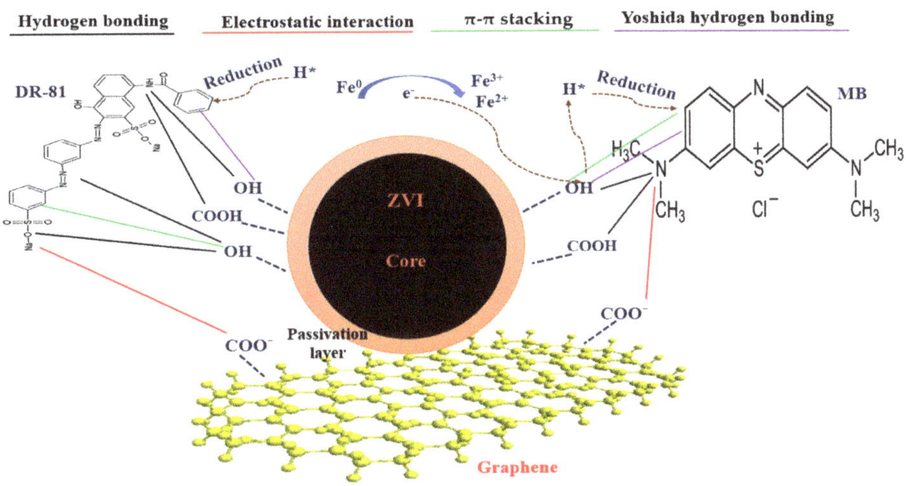

Figure 6. Adsorption mechanism of DR-81 and MB over ZVI graphene nanocomposite.

4. Conclusions

The excellent interaction between the graphene prepared from plastic waste and the biosynthesized ZVI under the catalysis of bacterial NAR enzyme was confirmed by different analytical techniques such as TEM, SEM, EDX, XRD and Raman spectra. The ZVI graphene nanocomposite achieved the highest removal efficiency compared to ZVI-NPs and graphene. The ZVI graphene nanocomposite attained a removal efficiency of 88.3 $\pm$ 2.66% and 87.6 $\pm$ 2.1% in the case of DR-81 and MB, respectively, after an equilibrium contact time of 90 min. The increase in adsorbent dose from 20 to 40 mg/50 mL resulted in the increase in the removal efficiency from 88.3 $\pm$ 2.66% to 96.88 $\pm$ 1.9% in the case of DR-81, compared to 87.6 $\pm$ 2.1% and 98.66 $\pm$ 1.9% in the case of MB. High removal efficiency of DR-81 was attained at acidic and neutral conditions, whereas the removal performance of MB was improved at neutral conditions using the ZVI graphene nanocomposite. The ZVI graphene had the highest adsorption rate in the case of MB and DR-81. The adsorption of MB or DR-81 on ZVI graphene surface could be due to hydrogen bonding, electrostatic interaction, Yoshida hydrogen bonding and π-π stacking. Such biogenic nanocomposite can contribute to lowering the toxicity and the costs of chemicals employed in the synthesis of ZVI-NPs and can overcome the hazards related to the plastic wastes. Additionally, the utilization of plastic wastes in the synthesis of graphene can contribute to the efficient management, reduction in greenhouse gases resulted from the burning or/and landfilling of plastic waste and scalable production of graphene. The high performance, biocompatibility and inexpensiveness of ZVI graphene nanocomposites can encourage decision makers to the full-scale treatment of industrial effluents by the biologically fabricated ZVI graphene nanocomposite. Further, we plan and advise other researchers to employ this efficient nanohybrid in other applications such as photocatalysis, Fenton reaction or persulfate activation system expecting high performance.

Supplementary Materials: The following supporting information can be downloaded at: https://www.mdpi.com/article/10.3390/su142114188/s1.

Author Contributions: Writing—original draft, M.S., M.E. (Marwa Elkady) and M.E. (Marwa Eltarahony); validation, M.S., S.Z. and A.K.; investigation, M.S., N.E. and M.E. (Marwa Eltarahony); formal analysis, M.S., N.E. and M.E. (Marwa Elkady); writing—review and editing, M.S., M.E. (Marwa Elkady) and M.E. (Marwa Eltarahony); methodology, M.E. (Marwa Eltarahony) and A.K.; supervision, M.E. (Marwa Eltarahony), S.Z. and M.E. (Marwa Elkady). All authors have read and agreed to the published version of the manuscript.

Funding: This research received no external funding.

Institutional Review Board Statement: Not applicable.

Informed Consent Statement: Not applicable.

Data Availability Statement: All data are provided.

Acknowledgments: The authors are grateful to Mansoura University for publication fee support.

Conflicts of Interest: The authors declare no conflict of interest.

References

1. Samy, M.; Ibrahim, M.G.; Gar Alalm, M.; Fujii, M. Application of CNTs/LaVO 4 on Photocatalytic Degradation of Methylene Blue in Different Contact Modes. In Proceedings of the 2020 Advances in Science and Engineering Technology International Conferences (ASET), Dubai, United Arab Emirates, 4 February–9 April 2020.
2. Mensah, K.; Samy, M.; Ezz, H.; Elkady, M.; Shokry, H. Utilization of Iron Waste from Steel Industries in Persulfate Activation for Effective Degradation of Dye Solutions. *J. Environ. Manag.* **2022**, *314*, 115108. [CrossRef]
3. Mensah, K.; Mahmoud, H.; Fujii, M.; Shokry, H. Novel Nano-Ferromagnetic Activated Graphene Adsorbent Extracted from Waste for Dye Decolonization. *J. Water Process Eng.* **2022**, *45*, 102512. [CrossRef]
4. Wang, G.; Liao, Y.; Stejskal, J. Fabrication of Polyaniline/Poly(Vinyl Alcohol)/Montmorillonite Hybrid Aerogels toward Efficient Adsorption of Organic Dye Pollutants. *J. Hazard. Mater.* **2022**, *435*, 129004. [CrossRef]
5. Samy, M.; Ibrahim, M.G.; Gar Alalm, M.; Fujii, M.; Diab, K.E.; ElKady, M. Innovative Photocatalytic Reactor for the Degradation of Chlorpyrifos Using a Coated Composite of ZrV2O7 and Graphene Nano-Platelets. *Chem. Eng. J.* **2020**, *395*, 124974. [CrossRef]
6. Samy, M.; Ibrahim, M.G.; Fujii, M.; Diab, K.E.; Elkady, M.; Alalm, G. CNTs/MOF-808 Painted Plates for Extended Treatment of Pharmaceutical and Agrochemical Wastewaters in a Novel Photocatalytic Reactor. *Chem. Eng. J.* **2021**, *406*, 7. [CrossRef]
7. Samy, M.; Alalm, M.G.; Mossad, M. Utilization of Iron Sludge Resulted from Electro-Coagulation in Heterogeneous Photo-Fenton Process. *Water Pract. Technol.* **2020**, *15*, 1228–1237. [CrossRef]
8. Venkatesan, S.; Chen, Y.; Teng, H.; Lee, Y. Enhanced Adsorption on TiO_2 Photoelectrodes of Dye-Sensitized Solar Cells by Electrochemical Methods Dye. *J. Alloys Compd.* **2022**, *903*, 163959. [CrossRef]
9. Mahanna, H.; Azab, M. Adsorption of Reactive Red 195 Dye from Industrial Wastewater by Dried Soybean Leaves Modified with Acetic Acid. *Desalin. Water Treat.* **2020**, *178*, 312–321. [CrossRef]
10. Samy, M.; Gar Alalm, M.; Fujii, M.; Ibrahim, M.G. Doping of Ni in MIL-125(Ti) for Enhanced Photocatalytic Degradation of Carbofuran: Reusability of Coated Plates and Effect of Different Water Matrices. *J. Water Process Eng.* **2021**, *44*, 102449. [CrossRef]
11. Elsayed, I.; Madduri, S.; El-giar, E.M.; Barbary, E. Effective Removal of Anionic Dyes from Aqueous Solutions by Novel Polyethylenimine-Ozone Oxidized Hydrochar (PEI-OzHC) Adsorbent. *Arab. J. Chem.* **2022**, *15*, 103757. [CrossRef]
12. Jankowska, K.; Su, Z.; Zdarta, J.; Jesionowski, T.; Pinelo, M. Synergistic Action of Laccase Treatment and Membrane Filtration during Removal of Azo Dyes in an Enzymatic Membrane Reactor Upgraded with Electrospun Fibers. *J. Hazard. Mater.* **2022**, *435*, 129071. [CrossRef]
13. Muthukumar, P.; Sowmiya, E.; Arunkumar, G.; Pannipara, M.; Al-sehemi, A.G.; Philip, S. Highly Enhanced Dye Adsorption of MoO_3 Nanoplates Fabricated by Hydrothermal-Calcination Approach in Presence of Chitosan and Thiourea. *Chemosphere* **2022**, *291*, 132926. [CrossRef]
14. Chen, C.; Tseng, W.J. Preparation of TiN-WN Composite Particles for Selective Adsorption of Methylene Blue Dyes in Water. *Adv. Powder Technol.* **2022**, *33*, 103423. [CrossRef]
15. Zhang, R.; Zeng, L.; Wang, F.; Li, X.; Li, Z. Influence of Pore Volume and Surface Area on Benzene Adsorption Capacity of Activated Carbons in Indoor Environments. *Build. Environ.* **2022**, *216*, 109011. [CrossRef]
16. Chung, J.; Sharma, N.; Kim, M.; Yun, K. Activated Carbon Derived from Sucrose and Melamine as Low-Cost Adsorbent with Fast Adsorption Rate for Removal of Methylene Blue in Wastewaters. *J. Water Process Eng.* **2022**, *47*, 102763. [CrossRef]
17. Ren, J.; Zheng, L.; Su, Y.; Meng, P.; Zhou, Q.; Zeng, H. Competitive Adsorption of Cd(II), Pb(II) and Cu(II) Ions from Acid Mine Drainage with Zero-Valent Iron/Phosphoric Titanium Dioxide: XPS Qualitative Analyses and DFT Quantitative Calculations. *Chem. Eng. J.* **2022**, *445*, 136778. [CrossRef]
18. Puthukkara, A.R.P.; Jose, S.T.; Lal, D.S. Plant mediated synthesis of zero valent iron nanoparticles and its application in water treatment. *J. Environ. Chem. Eng.* **2021**, *9*, 104569. [CrossRef]
19. Bundschuh, M.; Filser, J.; Lüderwald, S.; Mckee, M.S.; Metreveli, G.; Schaumann, G.E.; Schulz, R.; Wagner, S. Nanoparticles in the Environment: Where Do We Come from, Where Do We Go To? *Environ. Sci. Eur.* **2018**, *30*, 6. [CrossRef]
20. Iravani, S. Bacteria in Nanoparticle Synthesis: Current Status. *Int. Sch. Res. Not.* **2014**, *2014*, 59316.
21. He, Y.; Fang, T.; Wang, J.; Liu, X.; Yan, Z.; Lin, H.; Li, F.; Guo, G. Insight into the Stabilization Mechanism and Long-Term Effect on As, Cd, and Pb in Soil Using Zeolite-Supported Nanoscale Zero-Valent Iron. *J. Clean. Prod.* **2022**, *355*, 131634. [CrossRef]

22. Ma, Q.; Teng, W.; Sun, Y.; Chen, Y.; Xue, Y.; Chen, X.; Zhang, C.; Zhang, H.; Fan, J.; Qiu, Y.; et al. Multi-Component Removal of Pb (II), Cd (II), and As (V) over Core-Shell Structured Nanoscale Zero-Valent Iron @ Mesoporous Hydrated Silica. *Sci. Total Environ.* **2022**, *827*, 154329. [CrossRef] [PubMed]
23. Kumari, B.; Dutta, S. Integrating Starch Encapsulated Nanoscale Zero-Valent Iron for Better Chromium Removal Performance. *J. Water Process Eng.* **2020**, *37*, 101370. [CrossRef]
24. Kakavandi, B.; Takdastan, A.; Pourfadakari, S. Heterogeneous Catalytic Degradation of Organic Compounds Using Nanoscale Zero-Valent Iron Supported on Kaolinite: Mechanism, Kinetic and Feasibility Studies. *J. Taiwan Inst. Chem. Eng.* **2019**, *96*, 329–340. [CrossRef]
25. Liu, F.; Yang, J.; Zuo, J.; Ma, D.; Gan, L.; Xie, B.; Wang, P.; Yang, B. Graphene-Supported Nanoscale Zero-Valent Iron: Removal of Phosphorus from Aqueous Solution and Mechanistic Study ScienceDirect Graphene-Supported Nanoscale Zero-Valent Iron: Removal of Phosphorus from Aqueous Solution and Mechanistic Study. *J. Environ. Sci.* **2014**, *26*, 1751–1762. [CrossRef] [PubMed]
26. Pet, P.; Waste, B. A Novel One Step Synthesis for Carbon Based Nanomaterials from a Novel One-Step Synthesis for Carbon-Based Nanomaterials from Polyethylene Terephthalate (PET) Bottles Waste. *J. Air Waste Manag. Assoc.* **2016**, *67*, 358–370. [CrossRef]
27. El, N.A.; Ali, S.M.; Farag, H.A.; Konsowa, A.H. Green Synthesis of Graphene from Recycled PET Bottle Wastes for Use in the Adsorption of Dyes in Aqueous Solution. *Ecotoxicol. Environ. Saf.* **2017**, *145*, 57–68. [CrossRef]
28. Zaki, S.A.; Eltarahony, M.M.; Abd-el-haleem, D.A. Disinfection of Water and Wastewater by Biosynthesized Magnetite and Zerovalent Iron Nanoparticles via NAP-NAR Enzymes of Proteus Mirabilis 10B. *Environ. Sci. Pollut. Res.* **2019**, *26*, 23661–23678. [CrossRef]
29. Eltarahony, M.; Zaki, S.; Elkady, M.; Abd-el-haleem, D. Biosynthesis, Characterization of Some Combined Nanoparticles, and Its Biocide Potency against a Broad Spectrum of Pathogens. *J. Nanomater.* **2018**, *2018*, 5263814. [CrossRef]
30. Golla, N.; Pradesh, A. Antibacterial and antiviral properties. *Int. J. Pharma Sci. Res.* **2019**, *10*, 1223–1228. [CrossRef]
31. Adel, A.; Alalm, M.G.; El-Etriby, H.K.; Boffito, D.C. Optimization and Mechanism Insights into the Sulfamethazine Degradation by Bimetallic ZVI/Cu Nanoparticles Coupled with H_2O_2. *J. Environ. Chem. Eng.* **2020**, *8*, 104341. [CrossRef]
32. Wu, M.; Ma, Y.; Wan, J.; Wang, Y.; Guan, Z.; Yan, Z. Investigation of Factors Affecting the Physicochemical Properties and Degradation Performance of NZVI@mesoSiO_2 Nanocomposites. *J. Mater. Sci.* **2019**, *54*, 7483–7502. [CrossRef]
33. Ratanaphain, C.; Viboonratanasri, D.; Prompinit, P.; Krajangpan, S.; Khan, E.; Punyapalakul, P. Reactivity Characterization of SiO_2-Coated Nano Zero-Valent Iron for Iodoacetamide Degradation: The Effects of SiO_2 Thickness, and the Roles of Dehalogenation, Hydrolysis and Adsorption. *Chemosphere* **2022**, *286*, 131816. [CrossRef]
34. Samy, M.; Ibrahim, M.G.; Alalm, M.G.; Fujii, M. Modeling and Optimization of Photocatalytic Degradation of Methylene Blue Using Lanthanum Vanadate. *Mater. Sci. Forum* **2020**, *1008*, 97–103.
35. Gajera, R.; Patel, R.V.; Yadav, A.; Labhasetwar, P.K. Adsorption of Cationic and Anionic Dyes on Photocatalytic Flyash/TiO_2 Modified Chitosan Biopolymer Composite. *J. Water Process Eng.* **2022**, *49*, 102993. [CrossRef]
36. Xing, R.; He, J.; Hao, P.; Zhou, W. Graphene oxide-supported nanoscale zero-valent iron composites for the removal of atrazine from aqueous solution. *Colloids Surfaces A Physicochem. Eng. Asp.* **2020**, *589*, 124466. [CrossRef]
37. Jing, Q.; Qiao, S.; Xiao, W.; Tong, L.; Ren, Z. Efficient Removal of 2,4-DCP by Nano Zero-Valent Iron-Reduced Graphene Oxide: Statistical Modeling and Process Optimization Using RSM-BBD Approach. *Adsorpt. Sci. Technol.* **2021**, *2021*, 7130581. [CrossRef]
38. Sun, X.; Kurokawa, T.; Suzuki, M.; Takagi, M.; Kawase, Y. Removal of Cationic Dye Methylene Blue by Zero-Valent Iron: Effects of PH and Dissolved Oxygen on Removal Mechanisms. *J. Environ. Sci. Health Part A* **2015**, *50*, 1057–1071. [CrossRef]
39. Elkady, M.F.; Hassan, H.S. Invention of Hollow Zirconium Tungesto-Vanadate at Nanotube Morphological Structure for Radionuclides and Heavy Metal Pollutants Decontamination from Aqueous Solutions. *Nanoscale Res. Lett.* **2015**, *10*, 1–6. [CrossRef]
40. Khan, A.; Muthu, S.; Park, J.; Lee, W.; Chon, C.; Sung, J.; Lee, G. Azo Dye Decolorization by ZVI under Circum-Neutral PH Conditions and the Characterization of ZVI Corrosion Products. *J. Ind. Eng. Chem.* **2016**, *47*, 86–93. [CrossRef]
41. Hamdy, A.; Mostafa, M.K.; Nasr, M. Zero-Valent Iron Nanoparticles for Methylene Blue Removal from Aqueous Solutions and Textile Wastewater Treatment, with Cost Estimation from Aqueous Solutions and Textile Wastewater Treatment, with Cost Estimation. *Water Sci. Technol.* **2018**, *78*, 367–378. [CrossRef]
42. Samy, M.; Mossad, M.; El-Etriby, H.K. Synthesized Nano Titanium for Methylene Blue Removal under Various Operational Conditions. *Desalin. Water Treat.* **2019**, *165*, 374–381. [CrossRef]
43. Yang, J.; Shojaei, S.; Shojaei, S. Removal of Drug and Dye from Aqueous Solutions by Graphene Oxide: Adsorption Studies and Chemometrics Methods. *NPJ Clean Water* **2022**, *5*, 5. [CrossRef]
44. Thi, T.; Nguyet, P.; Trinh, X.; Minh, D.; Hoang, T. Preparing Three-Dimensional Graphene Aerogels by Chemical Reducing Method: Investigation of Synthesis Condition and Optimization of Adsorption Capacity of Organic Dye. *Surf. Interfaces* **2021**, *23*, 101023.
45. Shahinpour, A.; Tanhaei, B.; Ayati, A.; Beiki, H.; Sillanpää, M. Binary Dyes Adsorption onto Novel Designed Magnetic Clay-Biopolymer Hydrogel Involves Characterization and Adsorption Performance: Kinetic, Equilibrium, Thermodynamic, and Adsorption Mechanism. *J. Mol. Liq.* **2022**, *366*, 120303. [CrossRef]

46. Galdames, A.; Ruiz-rubio, L.; Orueta, M.; Miguel, S. Zero-Valent Iron Nanoparticles for Soil and Groundwater Remediation. *Int. J. Environ. Res. Public Health* **2020**, *17*, 5817. [CrossRef]
47. Pasinszki, T.; Krebsz, M. Synthesis and Application of Zero-Valent Iron Nanoparticles in Water Treatment, Environmental Remediation, Catalysis, and Their Biological Effects. *Nanomaterials* **2020**, *10*, 917. [CrossRef]
48. Figueiredo, T.; Mayara, P.; Roberto, V.; Nunes, B.; Siqueira, C.; Picone, F.; Prediger, P. Instantaneous Adsorption and Synergic Effect in Simultaneous Removal of Complex Dyes through Nanocellulose/Graphene Oxide Nanocomposites: Batch, Fixed-Bed Experiments and Mechanism. *Environ. Nanotechnol. Monit. Manag.* **2021**, *16*, 100584. [CrossRef]
49. Eltaweil, A.S.; El-tawil, A.M.; El-monaem, E.M.A.; El-subruiti, G.M. Zero Valent Iron Nanoparticle-Loaded Nanobentonite Intercalated Carboxymethyl Chitosan for Efficient Removal of Both Anionic and Cationic Dyes. *ACS Omega* **2021**, *6*, 6348–6360. [CrossRef]
50. Boparai, H.K.; Joseph, M.; O'Carroll, D.M. Kinetics and Thermodynamics of Cadmium Ion Removal by Adsorption onto Nano Zerovalent Iron Particles. *J. Hazard. Mater.* **2011**, *186*, 458–465. [CrossRef]

sustainability

MDPI

Article

A Micro-Metal Inserts Based Microchannel Heat Sink for Thermal Management of Densely Packed Semiconductor Systems

Essam M. Abo-Zahhad [1,2], Chaouki Ghenai [1,3], Ali Radwan [3,4], Osama Abdelrehim [4], Mohamed S. Salem [4], Mohamed R. Elmarghany [4], Asmaa Khater [4] and Mahmoud A. Shouman [4,*]

[1] Renewable Energy and Energy Efficiency Research Group, Sustainable Energy and Power Systems Research Centre, Research Institute for Sciences and Engineering (RISE), University of Sharjah, Sharjah P.O. Box 27272, United Arab Emirates
[2] Mechanical Power Engineering Department, Faculty of Energy Engineering, Aswan University, Aswan 81528, Egypt
[3] Department of Sustainable and Renewable Energy Engineering, College of Engineering, University of Sharjah, Sharjah P.O. Box 27272, United Arab Emirates
[4] Mechanical Power Engineering Department, Faculty of Engineering, Mansoura University, Mansoura 35516, Egypt
* Correspondence: m_shouman@mans.edu.eg

Citation: Abo-Zahhad, E.M.; Ghenai, C.; Radwan, A.; Abdelrehim, O.; Salem, M.S.; Elmarghany, M.R.; Khater, A.; Shouman, M.A. A Micro-Metal Inserts Based Microchannel Heat Sink for Thermal Management of Densely Packed Semiconductor Systems. *Sustainability* **2022**, *14*, 14182. https://doi.org/10.3390/su142114182

Academic Editor: Wei Wu

Received: 6 September 2022
Accepted: 14 October 2022
Published: 31 October 2022

Abstract: The thermal management of high-heat-density devices is essential for reliable operation. In this work, a novel procedure is proposed and investigated for the efficient thermal management of such devices. The proposed procedure introduces different arrangements of metal inserts within a cooling channel heat sink. The objective of those inserts is to form boundary layers to prevent any hot spots from appearing within the flow and increase temperature uniformity. Five different arrangements are introduced and numerically investigated using the commercial software package ANSYS FLUENT 2021R1. The model was validated against previous findings and showed a good agreement with errors of less than 5.5%. The model was then used to study the heat transfer characteristics of the proposed cases compared to traditional straight channels under the same operating conditions. All the proposed arrangements displayed better heat transfer characteristics than the traditional configuration within the studied range. They also exhibited lower temperature nonuniformities, implying better temperature distribution. The temperature contours over the heat source top surface and the flow streamlines are also introduced. Among all the proposed arrangements cases, a microchannel with micro metal insert located at the top wall along with a second row of inserts covering two-thirds of the bottom wall is studied. This case achieved the best heat transfer characteristics and highest temperature uniformity, making it a viable candidate for high power density devices' thermal management.

Keywords: high-heat devices; electronics cooling; thermal management; CFD; heat transfer enhancement; microchannels; inserts

1. Introduction

Recently, there has been a considerable increase in compact, high-performance electronic devices in almost every aspect, including, but not limited to, manufacturing, education, communication, transportation, and entertainment. This has led to increasing demand for high-power micro-electronic chips [1,2]. To maintain the reliability and the high performance of said microchips, their operating temperatures should be kept below 85 °C. Therefore, controlling the microchips' temperatures has become one of the main challenges threatening the widespread electronics industry. This thermal management challenge arises due to two factors. The first one is the high-power density of these chips, which generates high heat flux rates exceeding 1000 W/cm^2 in some local spots [3,4]. The second factor is the demand for a compact design of these chips, which reduces the available space for heat removal. These two factors eliminated the ability to use conventional thermal management techniques. Therefore, finding novel thermal management techniques that can fulfill the required heat dissipation rates has become an intriguing issue that has concerned many researchers. As part of these devoted efforts, researchers have employed nanofluids [5,6], phase change materials [7,8], and liquid metal [1] as cooling fluids.

Utilizing microchannel heat sinks is considered an efficient tool that can be used to cool down electronic components at a high heat rate. The first microchannel heat sink was presented by Tuckerman and Pease, who presented a simple design comprising several parallel channels [9] capable of dissipating up to 790 W/cm^2 from the surface of the chip with a relatively low pumping power. Since then, many researchers have tried to enhance the thermal performance of the microchannel heat sink with variable degrees of success.

The effect of using different channel configurations on heat transfer performance was previously investigated [10,11]. It was found that rectangular cross-section microchannels exhibit better heat transfer performance compared to trapezoidal and triangle cross-section microchannels. It was also concluded that the pressure loss is directly proportional to the channel aspect ratio. Hung et al. [12] numerically examined the effect of enlarging the outlet of the microchannel heat sink. The pressure drop was found to be reduced by the increase in the enlargement ratio. On the other hand, Nusselt number, temperature control, and heat transfer efficiency increased.

The impact of adding ribs to the side walls of straight microchannels on the heat transfer performance was also studied. Different cross sections of ribs were tested, including rectangular, semi-circular, and triangular, with different angles [13–15]. Using ribs improved the heat transfer performance at lower values of Re. On the other hand, a pressure drop increase was observed. Adding cavities to the straight channel reduced the pressure drop and the heat transfer performance. Thus, a combination between cavities and ribs could enhance heat transfer performance while maintaining an acceptable pressure drop value [16–18].

The use of wavy microchannels to reduce the temperature of the electronic components was also investigated [19–21]. It was noticed that a secondary flow was generated when the flow passed through the wavy channels, which resulted in an enhancement in the fluid mixing and an improved heat transfer performance [22]. A numerical investigation led by Mohammed et al. [23] estimated the optimal amplitude of the wavy channel to be between 0.063 and 0.22 μm. However, increasing the amplitude resulted in poor performance compared to conventional straight channels. The pressure drop and the friction factor were found to be increased with the increase of the amplitude. The effect of adding secondary branches to the wavy microchannels was numerically and experimentally studied by Chiam et al. [24]. The advantages of the secondary branches could be found at lower values of Re (<100), while increasing Re over that range resulted in a higher pressure drop compared to conventional wavy channels. This was not the case when secondary branches were added to wavy channels at relatively small amplitudes (<0.075 μm), with high transfer performance and lower pressure drop values.

Chuan et al. [25] and Gong et al. [26] utilized porous media within the walls of the microchannels. It was reported that porous walls reduced the viscous stresses and thus the pressure drop. The pressure drop was reduced by 48% when porous media was used compared to solid walls [25]. Also, the heat transfer performance was enhanced when porous walls were used.

The effect of adding a secondary channel to the flow path on the heat transfer performance of the microchannels was numerically studied [27,28]. Using secondary channels along with ribs reduced the pressure drop by up to 50%. The heat transfer was also enhanced by increasing the flow mixing between the channels [27]. Multi-objective algorithm optimization was used to find the best heat sink geometry with secondary flow channel design [28]. The optimization aimed to minimize the pressure drop and maximize the heat transfer rate. The optimized microchannel reduced the pumping power and the thermal resistance by 23% and 29%, respectively.

As can be seen from the previous research, various efforts were applied to enhance heat transfer within microchannels. However, there is always room for more innovative ideas. The current work investigates the effect of inserting a structured micro-sized metal mesh into the streamline on the heat transfer performance. The proposed concept depends on inducing the formation of a boundary layer by attaching metal inserts with different placements to avoid any hot spots within the device. The design is based on constructal law adopted to modify an actual mini-size straight channels heat sink device. Different cases were studied by dividing the channel length into three zones with the same inlet and outlet conditions.

The main objectives of this work are listed in the following bullet points:

- To describe a promising approach to reduce the high-temperature gradient levels associated with the stream direction, which presents a main downside of the micro channel-based cooling systems;
- To introduce metal inserts to develop a secondary flow that improves the heat transfer and reduces the flow maldistribution;
- To remove any local hotspots which could harm the physical structure of the semiconductor devices, consequently, electronic components' life span and performance;
- To reduce thermal management systems' power consumption without compromising cooling performance and reliability.

2. Numerical Modeling and Simulation Analysis

A computational fluid dynamics (CFD) model is established using the commercial package ANSYS Fluent 2021R1 to simulate the thermal management system's fluid flow and heat transfer. The high-performance computing facility of Maha Cluster at the University of Sharjah was used for simulation. A detailed description of the modeling procedures and case setup is discussed in the following subsections.

2.1. Physical Model and Proposed Configurations

The physical description of the presented model is shown in Figure 1. The heat source is a microelectronic silicon chip with an area of 25 mm × 25 mm, directedly bonded to the main board from the bottom side. Meanwhile, the chip is directly attached to the bottom of a cooling channel. In the current work, the chip is simulated as a constant heat source that acts at the cooling channel bottom wall. The designs used in the current work are shown in Figure 2. The channel dimensions and geometries were designated considering the constraints of 3D metal printing and pumping requirements. The figure shows two schematic graphs to identify the characteristics of the channel geometry of a traditional straight channel and one of the newly introduced cases. An isometric view with a magnification of the internal channel structure sections A-A cross-sectional view of the channel in the middle of the coolant domain is presented. As illustrated in the figure, Case 0 refers to the baseline or the standard straight channel heat sink, whereas cases 1 to 5 present the modified channels. Micro metal inserts are placed in the flow stream at different configurations, as shown in the figure, to improve the heat transfer rate in the cooling channels. As seen from the figure, in Case 1, a single row of metal inserts is introduced in the upper half of the flow path. Case 2 introduces another shorter row of metal at the corner that occupies only about one-third of the area. Case 3 introduces the same second short row as in Case 2 but in the middle. Case 4 presents a longer second row that covers two-thirds of the flow area, whereas Case 5 introduces two complete rows that cover the whole area. The proposed idea is to develop boundary layers to break any hot spots within the device. A parametric study is performed to trade off heat dissipation rates and pumping requirements. As for the coolant, water is used as the working fluid during the numerical study with thermophysical properties carefully defined as polynomial equations in temperature, as described in Table 1. The detailed heat sink and channel dimensions are listed in Table 2.

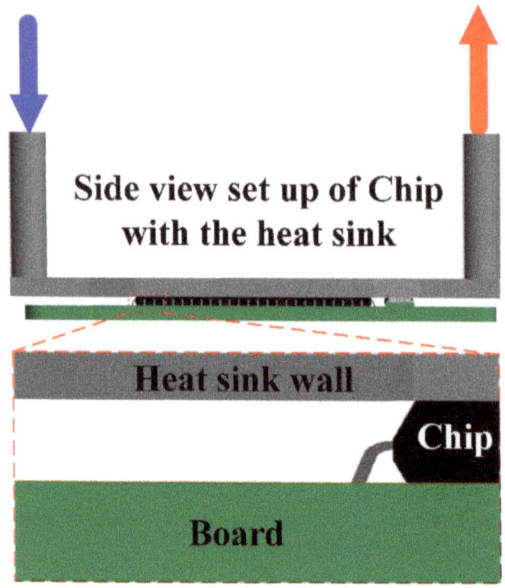

Figure 1. A schematic diagram of the proposed heat sink assembly with a chip in an integrated board.

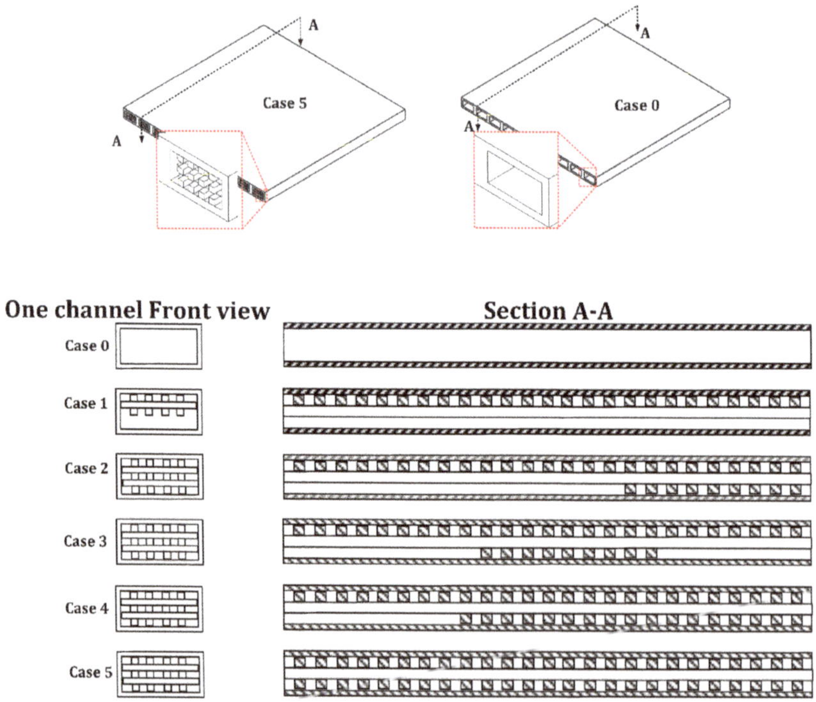

Figure 2. A schematic graph to identify the investigated channel geometry cases. A cross-sectional view of the channel in the middle of the coolant domain for Cases 0 through 5.

Table 1. Thermo-physical properties of cooling fluid as a function of the temperature [29,30].

	Property = A + B × T + C × T^2 + D × T^3 + E × T^4, (T in Absolute)			
Coefficient	**ρ (kg/m^3)**	**C_p (J/kg·K)**	**K(W/m·K)**	**μ (Pa·s)**
A	−1184.2713	5025.1	−0.785392198460943	0.41260813
B	24.66772581	−4.7536	0.00759075936901133	−0.004756694
C	−0.1032172737	0.0047	−0.00000993306796402867	0.000020683412
D	0.0001919823623	0.000006	-	−0.000000040117972
E	−0.0000001374629319		-	0.000000000029250643

Table 2. Detailed general dimensions of the cooling channel.

Term, Dimension	Value (mm)
Hest sink width	25
Hest sink Length	25
Channel height	2
Channel height	1
Walls thicknesses	0.5
Metal inserts width and height	0.2
Void width and height	0.2–0.3

2.2. Model Assumptions

In the current work, mass, momentum, and energy conservation are simultaneously solved to predict the flow characteristics. Some simplifications and assumptions are adopted to simulate the heat transfer and fluid flow characteristics of the current 3-dimensional (3D) model. Concerning the operating conditions:

- The simulations are contacted under 3D steady state steady flow conditions;
- Single-phase, incompressible, and laminar fluid flow prevails across the channel;
- The body forces and the effect of viscous heating are ignored;
- Very smooth walls (No-slip condition at walls). This assumption was based on Knudsen number (Kn) calculations. It is valid when the channel characteristic length is significantly bigger than the mean free path i.e., Kn < 0.001 [31]. In this case, as liquids are incompressible, the mean free path can be considered a constant. A good approach for water is to set the mean free path around 0.31 nm. Therefore, in microfluidic channels with characteristic length scales of more than 300 nm, we can safely assume no-slip boundary conditions;
- The gravity influence is neglected. This assumption is valid for small spaces and tight streams, as confirmed by Dang et al. [32]. As the height of the current cooling (H_{HS}) channel is only 1 mm, and the calculations are conducted under single-phase and incompressible flow circumstances, the consequence of gravity is ignored in the current simulation. Furthermore, forced convection is assumed to be dominant, with no local density differences.

2.3. Boundary Conditions

The boundary conditions used in the numerical model are depicted in Figure 3 and can be summarized as follows:

- Inlet working fluid velocity with Reynolds number (Re) spanning from 1000 to 2000;
- Inlet working fluid temperature of 293 K;
- Pressure outlet boundary;
- A heat flux of 100 W/cm^2 at the top wall;
- Symmetric left- and right-hand side walls;
- All other surfaces are considered adiabatic walls.

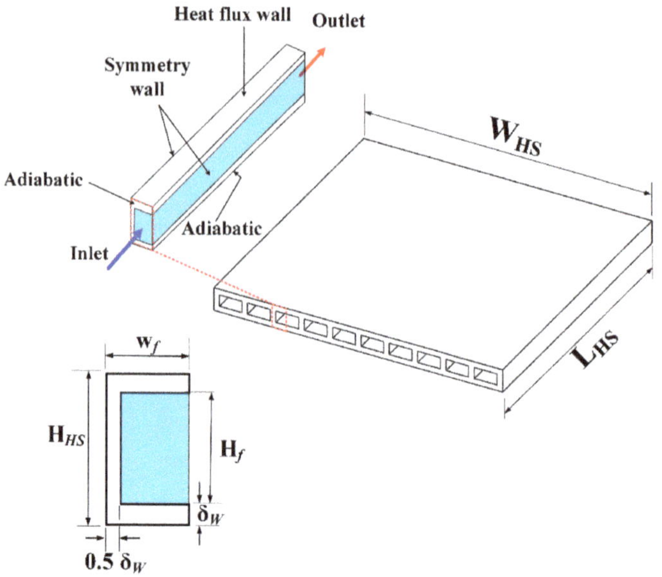

Figure 3. A schematic of the domains used in the CFD calculations and boundary conditions.

2.4. Grid Sensitivity Test

A structured mesh was generated using ANSYS Meshing. Figure 4a presents a sample of the structured mesh at different locations. A mesh-independence investigation was carried out before the model was used to study each design case. The grid sensitivity test was performed concerning the average interface temperature, temperature nonuniformity, and pressure drop. A Reynolds number of 2000, an ambient temperature of 293 K, and a heat flux of 100 W/cm^2 were selected as simulation conditions for the mesh sensitivity study. Different computational structured grid sizes were studied. All the studied grid sizes were physically controlled hexahedral configurations. The coarsest studied element size mesh was 250 μm, and the smallest mesh size was 20 μm. The coarse and extra fine mesh sizes element numbers were between 6000 and 8,000,000 elements, respectively. Figure 4b–d present the effect of changing mech size on the average wall temperature, the nonuniformity, and the pressure drop respectively. The percentage error between each consecutive mesh size was calculated to seek a < 1% error value. Therefore, a mesh of four million elements was chosen to reduce computation time while maintaining acceptable accuracy.

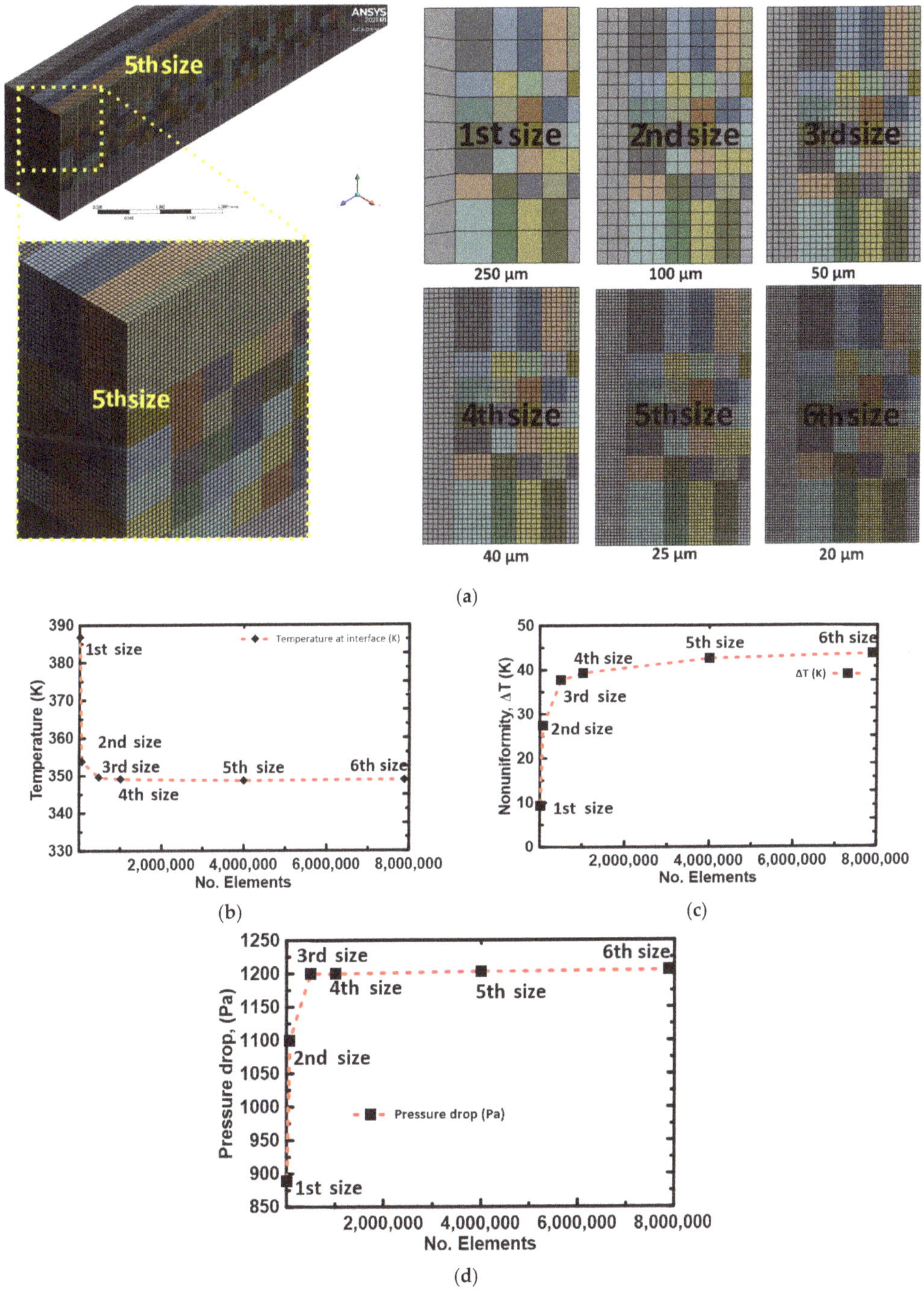

Figure 4. (**a**) Generated mesh at different locations and sensitivity test regarding; (**b**) average interface temperature; (**c**) temperature nonuniformity; and (**d**) pressure drop.

2.5. Model Validation

The CFD model is validated using previous literature investigations utilizing water as a single-phase coolant. The validation was conducted for multi-bifurcation cooling channel at different coolant flow rates, as shown in Figure 5. The numerical results are consistent with Xie et al. [33] data for the same conditions. Nusselt number and pressure drop values were compared. As can be seen from Figure 2a,b, the average variation between any two numerical values was found to be less than 5.5% for the Nusselt number and less than 2.8% for the pressure drop in all investigated cases. Also, the generated Nu values for case 0 were compared to the Sieder–Tate correlation [29] and were found to be in good agreement, as can be seen from Figure 2c. Hence, the mathematical model's dependability was confirmed and established.

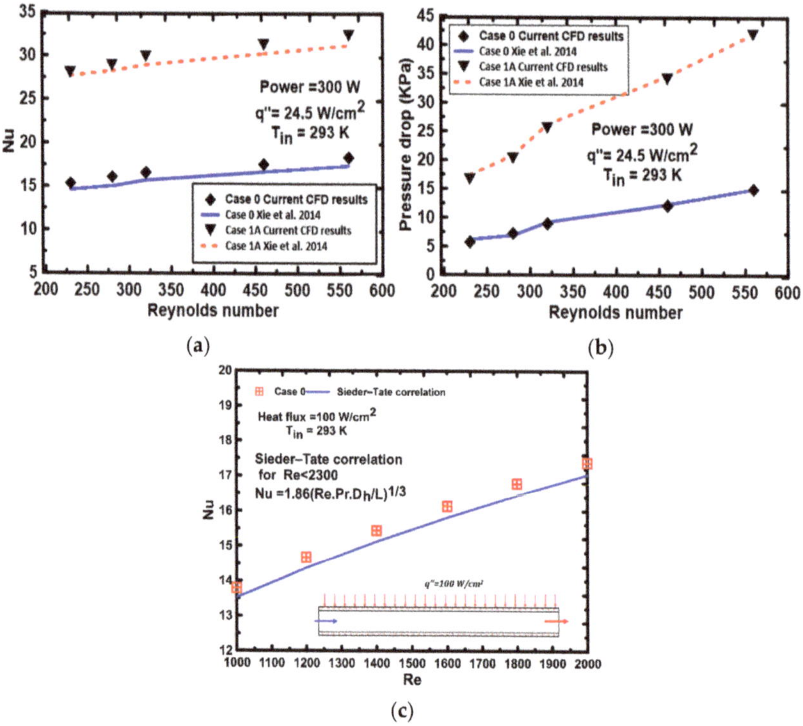

Figure 5. Comparing the current CFD data and Xie et al. [33] findings under identical conditions for (**a**) Nu and (**b**) pressure drop, respectively; and (**c**) Verifications Nu of Case 0 with Sieder–Tate correlation [29].

The error analysis was estimated using the root mean square percent deviation (RMSD) and the mean absolute percent error (MAE) [34] techniques as follows:

$$MAE(\%) = \frac{100}{n}\sum_{1}^{n}\left|\frac{X_{sim,i} - X_{exp,i}}{X_{exp,i}}\right| \quad (1)$$

$$RMSD(\%) = 100 \times \sqrt{\frac{\sum_{1}^{n}\left(\frac{X_{sim,i}-X_{exp,i}}{X_{exp,i}}\right)^2}{n}} \quad (2)$$

where $X_{sim,i}$, $X_{exp,i}$ are the predicted and experimental values. The range of errors using MAE and $RMSD$ were 0.56% and 0.7% for Nusselt number. As for the pressure drop, the error was very small (less than 0.1% for both techniques). The Nusselt number error analysis for case 0 had error of 1.85% and 1.89%, respectively.

The data obtained from the model was reduced using the equations mentioned in Appendix A.

3. Results and Discussion

Numerical studies of the heat transfer performance were carried out for all the considered cases with various metal mesh insert locations and lengths. The inlet Re was specified in the range of 1000 to 2000, and the corresponding ranges of the volumetric flow rate and inlet velocity were calculated. The convective heat transfer coefficient was evaluated in each case (Case 1 to Case 5) and compared with the reference case (Case 0). Figure 6 introduces the heat transfer coefficient and Nu change with the inlet Re for all investigated cases. Over the Re range, the heat transfer performance of Case 0 (a cooling channel without any metal insert) showed the lowest heat transfer performance.

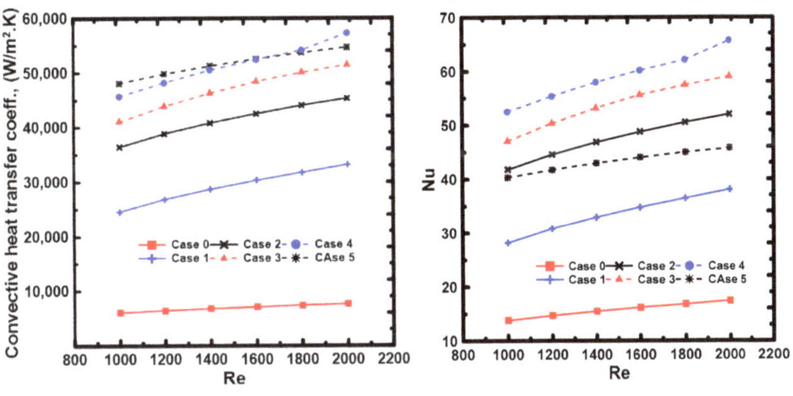

Figure 6. Comparison of (**a**) heat transfer coefficient; and (**b**) Nusselt number with the Re for all the studied cases.

As shown in Figure 6, the impact of the metal inserts on the heat transfer coefficient and Nu is significant compared to the reference case in all the proposed configurations. At the lowest inlet Re < 1000, cases 4 and 5 with a longer length of the second row of the metal insert attained a slightly higher heat transfer coefficient than that with a short length (Cases 2 and 3) and Case 1 with a single row. Although Case 5 exhibited comparable heat transfer coefficient values to those of Case 4, the Nu number values are relatively low due to the decrease in the hydraulic diameter. At the studied range, Case 4 had the highest values among all the cases. This means that Case 4 has the best thermal performance in heat transfer.

Figure 7 compares solid, fluid, and total thermal resistances between the conventional Case 0 and the enhanced Cases 1 to 5. All the thermal resistances' fluid thermal resistance decreased significantly with any increase in Re. As can be seen from the figure, the thermal resistance initially dramatically decreased with the increases of the Re, then the reduction rate diminishes for higher Re values.

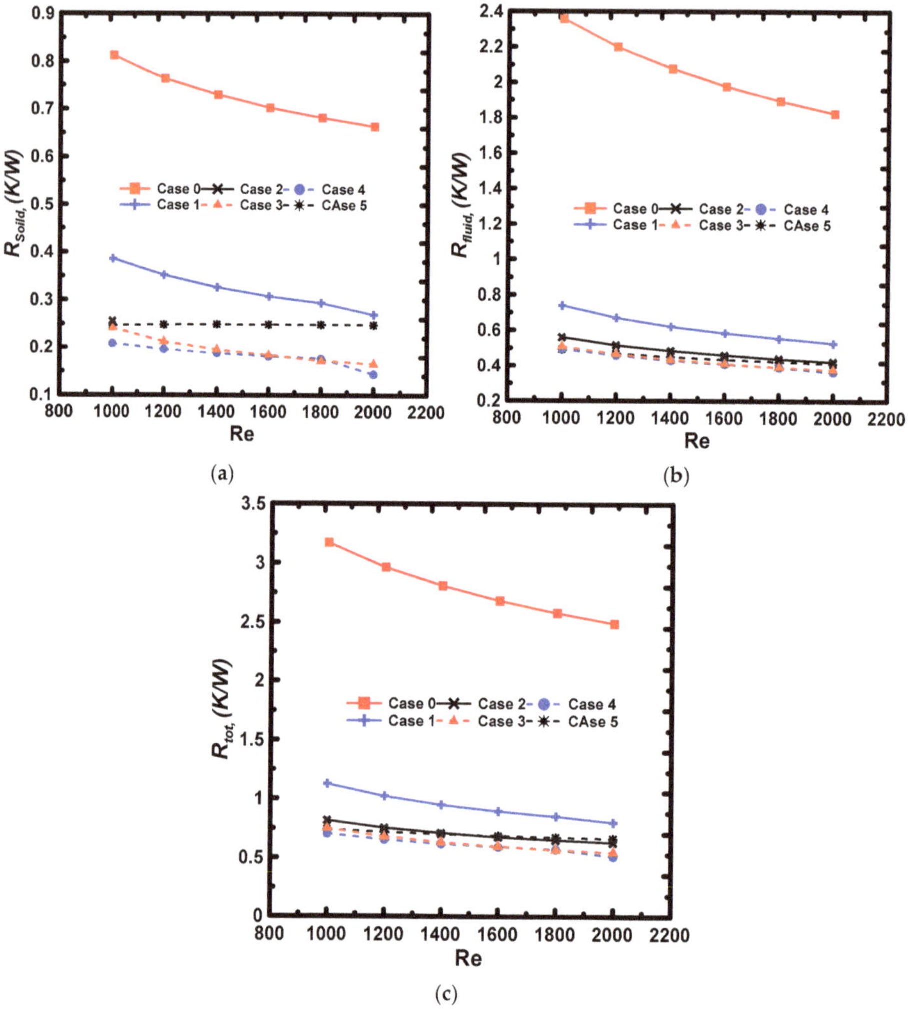

Figure 7. Comparison of thermal resistance (**a**) solid thermal resistance; (**b**) fluid thermal resistance; and (**c**) total thermal resistance.

Figure 8 shows the variations of the Nusselt number ratio (Nu/Nu_0) and the fanning friction factor ratio (f/f_0) with Re, where the denominators present the values for Case 0. From Figure 8a, increasing Re had no significant effect on the Nu/Nu_0, or the f/f_0 values within the studied range. Case 1 exhibited the lowest Nu/Nu_0 and f/f_0 values, whereas Case 4 attained the highest values of about $Nu/Nu_0 = 5$, and f/f_0 of about 15.

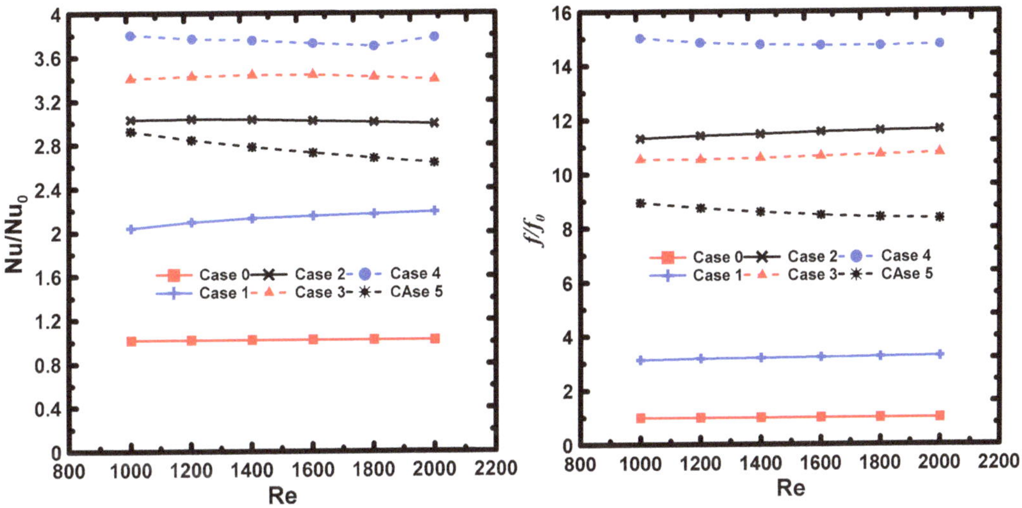

(a) (b)

Figure 8. Variations of (**a**) Nu/Nu_0; and (**b**) f/f_0 with Re.

The efficiency evaluation criterion (EEC) is used to evaluate the enhanced performance at identical Re. From Figure 9, Re had a negligible effect on the thermal performance of all the investigated cases. Again, Case 4 exhibited the best performance within the studied range. The performance factor (PF), which measures the EEC to the power of 1/3, is used to gauge the overall effectiveness of the proposed design. The value of describes how successful the thermal system is in reducing pressure drop. If the value is more than 1, it means that thermal performance is better than pressure drop, and if it is less than 1, it means the opposite.

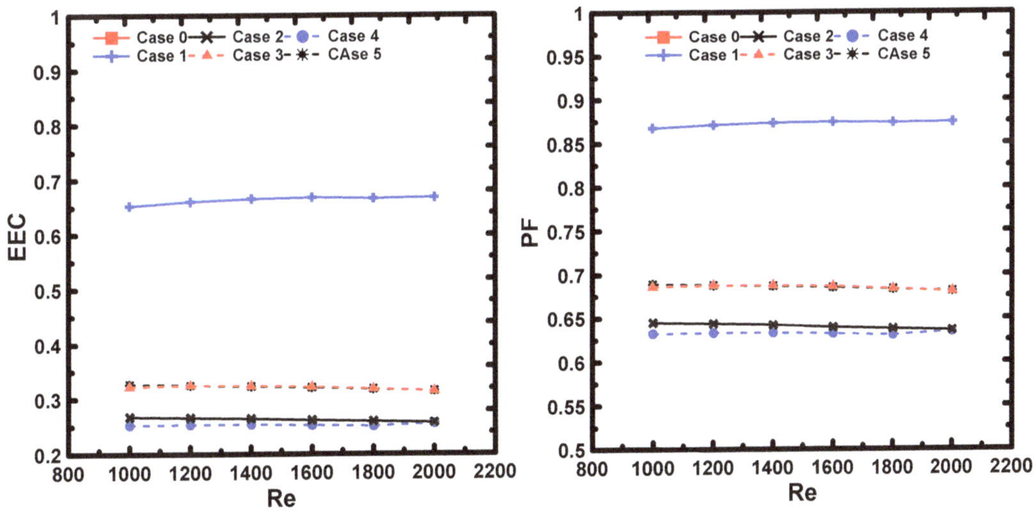

(a) (b)

Figure 9. Variations of the efficiency: (**a**) evaluation criterion EEC; and (**b**) performance factor (PF) with Re.

The nonuniformity degree within the heat source for all the proposed configurations was compared with the base case to determine the impact of the metal inserts on the wall temperature uniformity at all Re ranges and the same operating conditions. It was observed that the temperature variation was decreased for all cases with the increase in Re, thereby enhancing the temperature distribution uniformity. It is also clear from the figure that the introduction of the metal inserts considerably enhanced the wall temperature uniformity and reduced the temperature gradient with the length of the device. Cases 3 and 4 showed the lowest nonuniformity levels over the Re range. For Cases 3 and 4, the average uniformity enhancement compared to Case 0 was about 85.4% and 86.6%, respectively. For instance, when Re = 2000, nonuniformity in Case 0 was 48 K, whereas it was 5.84 K and 4.7 K in Cases 3 and 4, respectively. For Cases 1, 2, and 5, the average enhancement of wall temperature uniformity was 67.1%, 82%, and 78.3%, respectively. The metal inserts improved the heat transfer generally and deduced flow and heat maldistribution, which can be interpreted as the metal inserts regenerating the boundary conditions, which kept the heat transfer capability of the working fluid along the flow stream (Figure 10).

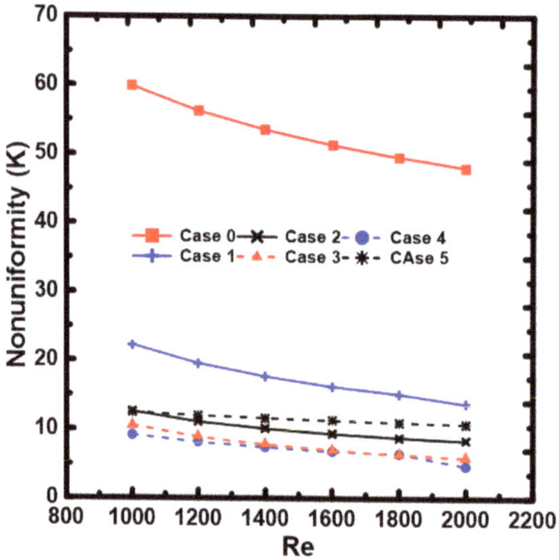

Figure 10. The effect of involving different metal inserts on temperature nonuniformity of the device when for all the considered cases.

Figure 11 shows the temperature contours of the heat source when the structured metal inserts are involved under the same heat flux of 100 W/cm^2 and Re = 2000. All the subfigures used different legends scales due to the significant difference in the obtained results. The figure shows the typical wall temperature distribution contours, which match the literature [33,35,36]. The contours show that the metal inserts considerably reduced the average wall temperature and nonuniformity. It is evident that the metal inserts' length and petition control the flow field and hotspot location. For more details, Cases 1 and 5 are comparable with Case 0 as the metal inserts were placed along the flow stream. However, in Case 5, the cooling channel is wholly packed with metal inserts, which is not the case in Case 1, where the metal inserts were placed in the top half of the channel. The difference between the two designs (Cases 1 and 5) is reflected in the pressure drop value, nonuniformity (even though the contours distribution is almost the same), and average wall temperature. For Case 2, the metal inserts had a negative effect where the hot spot occurred near the heat source center, a critical position in which any hotspots must be avoided. In the last two, Cases 3 and 4, the lowest nonuniformity levels were achieved, as depicted in the figure. However, Case 4 is preferable as the metal inserts were packed after 1/3 of the channel length from the inlet. In the regions near the inlet, the cooling fluid temperature is typically low, so boundary regeneration is not as important.

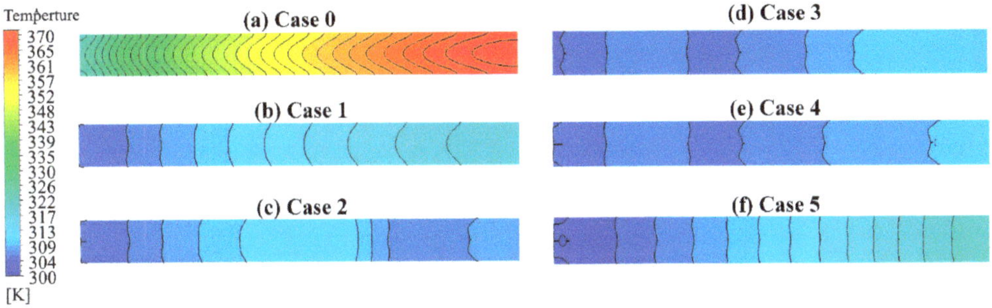

Figure 11. Temperature contour distributions over the heat source top surface for all cases, Re = 2000, heat flux = 100 W/cm^2 for the six investigated cases.

For more clarification, the streamlines of each case at Re = 2000 are shown in Figure 12. An automatically generated legend is used to show areas of high and low-velocity values. The enhancement in the average wall temperature, uniformity, and Nu mentioned earlier can be justified more with the flowing velocity streamlines. It can be noted that the flow inside the flow channel was distributed differently due to the packing of the metal inserts. Hence, higher interaction between the solid surface and the incoming flôw. The effect of distributed flow channels generated by the metal inserts is reflected in the enhancement of heat transfer and heat dissipation. As shown in Figure 12a for Case 0, the large flow channels show low resistance to the incoming flow, simplifying its stream along the channel without any distributed manner.in contrast, in Cases 1 to 5, the complex micro inserts' inner structure forces the flow to change trajectory as it flows past the flow channel, inducing vortices that enhance turbulence and heat transfer.

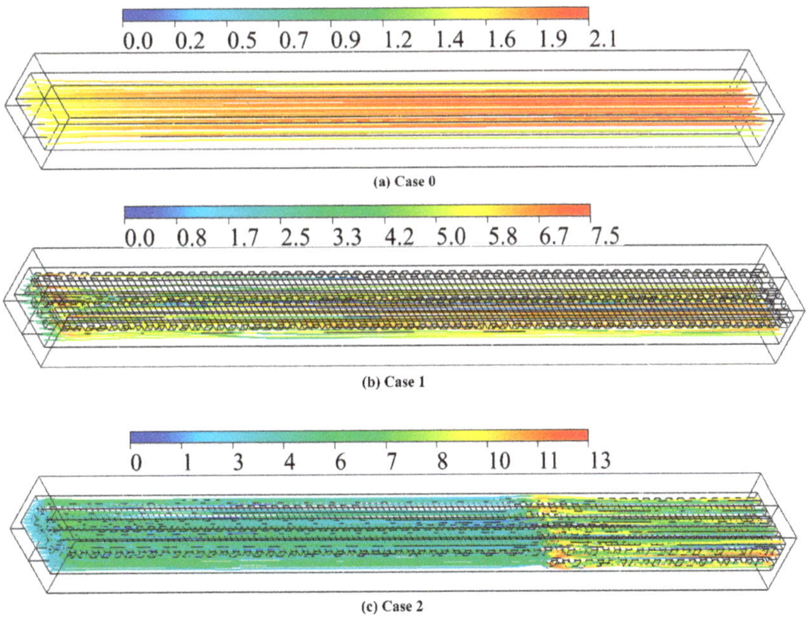

Figure 12. *Cont.*

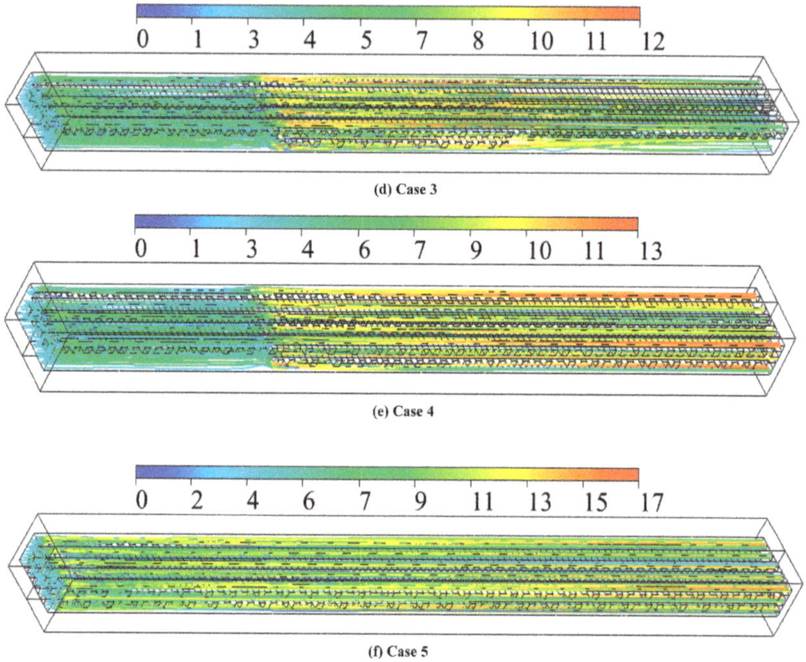

Figure 12. Velocity streamlines all the cases evaluated at Re = 2000 and heat flux = 100 W/cm^2 six investigated cases.

4. Conclusions

Thermal management of high heat flux surfaces is essential with the recent progress in the high computing devices. In this study, a new technique is proposed to improve the thermal management of high-power of these devices. The proposed technique introduces metal inserts within the flow path of a cooling channel. These inserts can be monolithically fabricated within the flow channel during the fabrication with the use of 3D metal printing technology. The intent of using the inserts is to break the flow and generate boundary layers that decrease the formation of hot spots within the flow. Five different configurations were introduced and numerically compared. The fluid flow and heat transfer characteristics of these configurations are compared. The results showed that using micro inserts in the microchannel enhances the heat transfer compared to the conventional channel without micro-metal inserts. For instance, using micro inserts on part of the bottom wall of the heated wall of the microchannel attained an average uniformity enhancement of 86.6% compared to the conventional smooth case. Further, at Re = 2000, the temperature non-uniformity decreased from 48 °C for the smooth case to 4.7 °C for the case with using metal inserts covering two-thirds of the bottom heated wall.

Author Contributions: Conceptualization, A.R.; Data curation, M.R.E.; Investigation, E.M.A.-Z.; Methodology, M.A.S.; Software, E.M.A.-Z.; Supervision, C.G.; Validation, A.R.; Writing—original draft, O.A., A.K. and M.A.S.; Writing—review & editing, M.S.S. All authors have read and agreed to the published version of the manuscript.

Funding: This research received no external funding.

Institutional Review Board Statement: Not applicable.

Informed Consent Statement: Not applicable.

Data Availability Statement: Not applicable.

Acknowledgments: The authors would like to thank Mansoura University and the Research Institute for Science and Engineering University of Sharjah.

Conflicts of Interest: The authors declare no conflict of interests.

Nomenclature

Nomenclature	
C	Specific heat, J/kg.K
D	Diameter, m
F	Friction factor
H	Height, m
K	Thermal conductivity, W/m.K
L	Length, m
Nu	Nusselt number
P	Pressure, Pa
PF	Performance factor
Pr	Prandtl number
Q	heat flux, W/m^2
R	Thermal resistance, K/W
Re	Reynolds number
T	Temperature, K
U	Axial velocity, m/s
$\vec{V}$	Velocity vector, m/s
W	Width, m
Greek symbols	
μ	Dynamic viscosity, Pa.s
P	Density, kg/m^3
Subscripts	
C	Channel wall
F	Fluid
H	Hydraulic
HS	Heat source
I	Inlet
Max	Maximum
S	Solid
S	Substrate
Tot	Total
Superscripts	
T	Total
Abbreviations	
CFD	Computational Fluid Dynamics
MAE	mean absolute error
RMSD	root mean square deviation

Appendix A. Data Reduction

To investigate the thermo-fluid performance of the presented heat exchanging technique, some dimensionless numbers should be defined, including Reynolds number, Nusselt number (Nu), and fanning friction factor (f). Those parameters are calculated using the following equations.

$$\mathrm{Re} = \frac{\rho u_i D_h}{\mu} \tag{A1}$$

$$\mathrm{Nu} = \frac{h D_h}{K} \tag{A2}$$

$$f = \frac{\Delta p}{(L/D_h)\rho u_i^2/2} \tag{A3}$$

where Δp presents the pressure drop calculated from inlet 1 [Pa], u_i is the average inlet velocity [m/s], D_h is the hydraulic diameter of the cross-sectional area [m], L is the heat source length [m], h is the convective heat transfer coefficient [W/m^2·K] which can be calculated using the equation:

$$h = \frac{qA_s}{A_c(T_w - \overline{T}_f)} \tag{A4}$$

where q represents the supplied heat flux to the substrate [W/m^2], A_s and A_c are the substrate bottom and the channel wall surface area [m^2], respectively, and T_w and $\overline{T}_f$ are the average channel wall temperature and the average fluid temperature across the channel [K], respectively.

On the other hand, the total thermal resistance (R_{tot}) [K/W], substrate thermal resistance (R_{solid}) [K/W], and fluid thermal resistance (R_f) [K/W] are defined as follows.

$$R_{tot} = \frac{T_{S,Max} - T_{in}}{qA_s} \tag{A5}$$

$$R_{Soild} = \frac{T_{S,Max} - T_w}{qA_s} \tag{A6}$$

$$R_{Fluid} = \frac{T_w - T_{in}}{qA_s} \tag{A7}$$

where $T_{s,max}$ and T_{in} is the maximum substrate temperature and the average fluid temperature at the channel inlet [K], respectively.

Finally, the efficiency evaluation criterion (EEC) is defined using the following equation to judge the thermal performance of the presented configurations.

$$EEC = \frac{Nu/Nu_0}{f/f_0} \tag{A8}$$

The performance factor (PF), which measures the EEC to the power of 1/3 is calculated as the following equation:

$$PF = \left(\frac{Nu/Nu_0}{f/f_0}\right)^{1/3} \tag{A9}$$

where Nu_0 is calculated for the Case 0 or the smooth channels [29]. On the other hand, f_o is calculated using the procedures described by Kandlikar et al. [37].

References

1. Yu, C.; Ji, Y.; Li, Y.; Liu, Z.; Chu, L.; Kuang, H.; Wang, Z. A Three-Dimensional Oscillating Heat Pipe Filled with Liquid Metal and Ammonia for High-Power and High-Heat-Flux Dissipation. *Int. J. Heat Mass Transf.* **2022**, *194*, 123096. [CrossRef]
2. Gholinia, M.; Ranjbar, A.A.; Javidan, M.; Hosseinpour, A.A. Employing a New Micro-Spray Model and (MWCNTs–SWCNTs)-H_2O Nanofluid on Si-IGBT Power Module for Energy Storage: A Numerical Simulation. *Energy Rep.* **2021**, *7*, 6844–6853. [CrossRef]
3. Jung, K.W.; Kharangate, C.R.; Lee, H.; Palko, J.; Zhou, F.; Asheghi, M.; Dede, E.M.; Goodson, K.E. Microchannel cooling strategies for high heat flux (1 kW/cm2) power electronic applications. In Proceedings of the 2017 16th IEEE Intersociety Conference on Thermal and Thermomechanical Phenomena in Electronic Systems (ITherm), Orlando, FL, USA, 30 May–2 June 2017; pp. 98–104. [CrossRef]
4. Asim, M.; Siddiqui, F.R. Hybrid Nanofluids—Next-Generation Fluids for Spray-Cooling-Based Thermal Management of High-Heat-Flux Devices. *Nanomaterials* **2022**, *12*, 507. [CrossRef] [PubMed]
5. Abdelrehim, O.; Khater, A.; Mohamad, A.A.; Radwan, A. Two-Phase Simulation of Nanofluid in a Confined Single Impinging Jet. *Case Stud. Therm. Eng.* **2019**, *14*, 100423. [CrossRef]
6. Mohammadi, M.; Taheri, A.; Passandideh-Fard, M.; Sardarabadi, M. Electronic Chipset Thermal Management Using a Nanofluid-Based Mini-Channel Heat Sink: An Experimental Study. *Int. Commun. Heat Mass Transf.* **2020**, *118*, 104836. [CrossRef]
7. Arshad, W.; Ali, H.M. Experimental Investigation of Heat Transfer and Pressure Drop in a Straight Minichannel Heat Sink Using TiO2 Nanofluid. *Int. J. Heat Mass Transf.* **2017**, *110*, 248–256. [CrossRef]
8. Righetti, G.; Zilio, C.; Doretti, L.; Longo, G.A.; Mancin, S. On the Design of Phase Change Materials Based Thermal Management Systems for Electronics Cooling. *Appl. Therm. Eng.* **2021**, *196*, 117276. [CrossRef]

9. Tuckerman, D.B.; Pease, R.F.W. High-Performance Heat Sinking for VLSI. *IEEE Electron Device Lett.* **1981**, *EDL-2*, 126–129. [CrossRef]
10. Gunnasegaran, P.; Mohammed, H.A.; Shuaib, N.H.; Saidur, R. The Effect of Geometrical Parameters on Heat Transfer Characteristics of Microchannels Heat Sink with Different Shapes. *Int. Commun. Heat Mass Transf.* **2010**, *37*, 1078–1086. [CrossRef]
11. Wang, H.; Chen, Z.; Gao, J. Influence of Geometric Parameters on Flow and Heat Transfer Performance of Micro-Channel Heat Sinks. *Appl. Therm. Eng.* **2016**, *107*, 870–879. [CrossRef]
12. Hung, T.C.; Huang, Y.X.; Yan, W.M. Thermal Performance of Porous Microchannel Heat Sink: Effects of Enlarging Channel Outlet. *Int. Commun. Heat Mass Transf.* **2013**, *48*, 86–92. [CrossRef]
13. Chai, L.; Wang, L.; Bai, X. Thermohydraulic Performance of Microchannel Heat Sinks with Triangular Ribs on Sidewalls—Part 2: Average Fluid Flow and Heat Transfer Characteristics. *Int. J. Heat Mass Transf.* **2019**, *128*, 634–648. [CrossRef]
14. Chai, L.; Xia, G.D.; Wang, H.S. Parametric Study on Thermal and Hydraulic Characteristics of Laminar Flow in Microchannel Heat Sink with Fan-Shaped Ribs on Sidewalls—Part 3: Performance Evaluation. *Int. J. Heat Mass Transf.* **2016**, *97*, 1091–1101. [CrossRef]
15. Xia, G.; Ma, D.; Zhai, Y.; Li, Y.; Liu, R.; Du, M. Experimental and Numerical Study of Fluid Flow and Heat Transfer Characteristics in Microchannel Heat Sink with Complex Structure. *Energy Convers. Manag.* **2015**, *105*, 848–857. [CrossRef]
16. Ghani, I.A.; Kamaruzaman, N.; Sidik, N.A.C. Heat Transfer Augmentation in a Microchannel Heat Sink with Sinusoidal Cavities and Rectangular Ribs. *Int. J. Heat Mass Transf.* **2017**, *108*, 1969–1981. [CrossRef]
17. Chai, L.; Xia, G.; Zhou, M.; Li, J. Numerical Simulation of Fluid Flow and Heat Transfer in a Microchannel Heat Sink with Offset Fan-Shaped Reentrant Cavities in Sidewall. *Int. Commun. Heat Mass Transf.* **2011**, *38*, 577–584. [CrossRef]
18. Li, Y.F.; Xia, G.D.; Ma, D.D.; Jia, Y.T.; Wang, J. Characteristics of Laminar Flow and Heat Transfer in Microchannel Heat Sink with Triangular Cavities and Rectangular Ribs. *Int. J. Heat Mass Transf.* **2016**, *98*, 17–28. [CrossRef]
19. Sui, Y.; Teo, C.J.; Lee, P.S. Direct Numerical Simulation of Fluid Flow and Heat Transfer in Periodic Wavy Channels with Rectangular Cross-Sections. *Int. J. Heat Mass Transf.* **2012**, *55*, 73–88. [CrossRef]
20. Ermagan, H.; Rafee, R. Numerical Investigation into the Thermo-Fluid Performance of Wavy Microchannels with Superhydrophobic Walls. *Int. J. Therm. Sci.* **2018**, *132*, 578–588. [CrossRef]
21. Lin, L.; Zhao, J.; Lu, G.; Wang, X.D.; Yan, W.M. Heat Transfer Enhancement in Microchannel Heat Sink by Wavy Channel with Changing Wavelength/Amplitude. *Int. J. Therm. Sci.* **2017**, *118*, 423–434. [CrossRef]
22. Sui, Y.; Teo, C.J.; Lee, P.S.; Chew, Y.T.; Shu, C. Fluid Flow and Heat Transfer in Wavy Microchannels. *Int. J. Heat Mass Transf.* **2010**, *53*, 2760–2772. [CrossRef]
23. Mohammed, H.A.; Gunnasegaran, P.; Shuaib, N.H. Numerical Simulation of Heat Transfer Enhancement in Wavy Microchannel Heat Sink. *Int. Commun. Heat Mass Transf.* **2011**, *38*, 63–68. [CrossRef]
24. Chiam, Z.L.; Lee, P.S.; Singh, P.K.; Mou, N. Investigation of Fluid Flow and Heat Transfer in Wavy Micro-Channels with Alternating Secondary Branches. *Int. J. Heat Mass Transf.* **2016**, *101*, 1316–1330. [CrossRef]
25. Chuan, L.; Wang, X.D.; Wang, T.H.; Yan, W.M. Fluid Flow and Heat Transfer in Microchannel Heat Sink Based on Porous Fin Design Concept. *Int. Commun. Heat Mass Transf.* **2015**, *65*, 52–57. [CrossRef]
26. Gong, L.; Li, Y.; Bai, Z.; Xu, M. Thermal Performance of Micro-Channel Heat Sink with Metallic Porous/Solid Compound Fin Design. *Appl. Therm. Eng.* **2018**, *137*, 288–295. [CrossRef]
27. Ghani, I.A.; Sidik, N.A.C.; Mamat, R.; Najafi, G.; Ken, T.L.; Asako, Y.; Japar, W.M.A.A. Heat Transfer Enhancement in Microchannel Heat Sink Using Hybrid Technique of Ribs and Secondary Channels. *Int. J. Heat Mass Transf.* **2017**, *114*, 640–655. [CrossRef]
28. Shi, X.; Li, S.; Mu, Y.; Yin, B. Geometry Parameters Optimization for a Microchannel Heat Sink with Secondary Flow Channel. *Int. Commun. Heat Mass Transf.* **2019**, *104*, 89–100. [CrossRef]
29. Incropera, F.P.; DeWitt, D.P. *Fundamentals of Heat and Mass Transfer*; Wiley: New York, NY, USA, 1996; Volume 6, ISBN 0471304603.
30. Dinçer, İ.; Zamfirescu, C. *Drying Phenomena: Theory and Applications*; John Wiley & Sons: Hoboken, NJ, USA, 2016.
31. Rapp, B.E. Chapter 9—Fluids. In *Applied Mathematics and Mechanics*; Elsevier: Oxford, UK, 1989; Volume 1897, pp. 44–45. ISBN 9784431538592.
32. Dang, T.; Teng, J.T.; Chu, J.C. A Study on the Simulation and Experiment of a Microchannel Counter-Flow Heat Exchanger. *Appl. Therm. Eng.* **2010**, *30*, 2163–2172. [CrossRef]
33. Xie, G.; Zhang, F.; Sundén, B.; Zhang, W. Constructal Design and Thermal Analysis of Microchannel Heat Sinks with Multistage Bifurcations in Single-Phase Liquid Flow. *Appl. Therm. Eng.* **2014**, *62*, 791–802. [CrossRef]
34. Kandil, A.A.; Awad, M.M.; Sultan, G.I.; Salem, M.S. Investigating the Performance Characteristics of Low Concentrated Photovoltaic Systems Utilizing a Beam Splitting Device under Variable Cutoff Wavelengths. *Renew. Energy* **2022**, *196*, 375–389. [CrossRef]
35. Li, W.; Xie, Z.; Xi, K.; Xia, S.; Ge, Y. Constructal Optimization of Rectangular Microchannel Heat Sink with Porous Medium for Entropy Generation Minimization. *Entropy* **2021**, *23*, 1528. [CrossRef] [PubMed]
36. Shen, H.; Wang, C.C.; Xie, G. A Parametric Study on Thermal Performance of Microchannel Heat Sinks with Internally Vertical Bifurcations in Laminar Liquid Flow. *Int. J. Heat Mass Transf.* **2018**, *117*, 487–497. [CrossRef]
37. Kandlikar, S.G.; Garimella, S.; Li, D.; Colin, S.; King, M.R. *Heat Transfer and Fluid Flow in Minichannels and Microchannels*; Elsevier: Oxford, UK, 2006; ISBN 9780080445274.

sustainability

MDPI

Article

Enhancement of Dye Separation Performance of Eco-Friendly Cellulose Acetate-Based Membranes

Omneya A. Koriem [1,*], Alaa Mostafa Kamel [1], Waleed Shaaban [2] and Marwa F. Elkady [1,3,*]

1 Chemical and Petrochemical Engineering Department, Egypt-Japan University of Science and Technology, New Borg El-Arab City, Alexandria 21934, Egypt
2 Mechanical Engineering Department, Mansoura University, El-Mansoura 35516, Egypt
3 Fabrication Technology Department, Advanced Technology and New Materials Research Institute (ATNMRI), City of Scientific Research and Technological Applications, New Borg El-Arab City, Alexandria 21934, Egypt
* Correspondence: omneya.koriem@ejust.edu.eg (O.A.K.); marwa.elkady@ejust.edu.eg (M.F.E.)

Abstract: Many reasons have caused a worldwide water stress problem. Thus, the recycling of wastewater streams has been extensively studied. In this work, eco-friendly mixed matrix membranes (MMMs) were fabricated, characterized, and tested for the removal of two separate dyes from simulated waste streams. The environmentally friendly nano activated carbon (NAC) was extracted from water hyacinth to be impregnated as a membrane nano-filler to enhance the neat membrane performance. The extracted NAC was further studied and characterized. Cellulose acetate (CA)-based membranes were obtained by phase inversion and electrospinning mechanisms. All four synthesized blank and MMMs were characterized via scanning electron microscope (SEM) and contact angle to study their structure and hydrophilic nature, respectively. However, the membrane with optimum performance was further characterized using Fourier transfer infrared (FTIR) and X-ray diffraction (XRD). The four prepared cast and electro-spun, blank, and mixed matrix CA-based membranes showed an acceptable performance in the removal and selectivity of methylene blue (MB) dye over Congo red (CR) dye with a removal percentage ranging from 31 to 70% depending on the membrane used. It was found that the CA/NAC hybrid nanofiber membrane possessed the highest removal efficiency for MB, where the dye concentration declined from 10 to 2.92 mg/L. In contrast, the cast blank CA membrane showed the least removal percentage among the synthesized membranes with only 30% removal. As a result, this paper suggests the use of the CA/NAC hybrid membrane as an alternative and cost-effective solution for MB dye removal.

Keywords: adsorption; membrane; adsorptive membrane; activated carbon; dye removal; hybrid membrane

Citation: Koriem, O.A.; Kamel, A.M.; Shaaban, W.; Elkady, M.F. Enhancement of Dye Separation Performance of Eco-Friendly Cellulose Acetate-Based Membranes. *Sustainability* **2022**, *14*, 14665. https://doi.org/10.3390/su142214665

Academic Editor: Andrea G. Capodaglio

Received: 14 September 2022
Accepted: 1 November 2022
Published: 8 November 2022

1. Introduction

Due to globalization and the developed human lifestyle, dyes are widely used in various industrial fields. They are extensively used not only in the textile and paper industries, but also in food, pharmaceuticals, cosmetics, and many other vital industries [1]. In aqueous solutions, dyes are classified into two main categories: anionic and cationic [2]. It is estimated that around 10% of the textile industry's dyes are lost directly into water streams [3]. Even in low concentrations, most synthetic dyes have a harmful impact on flora and fauna, humans, and the environment. They are generally characterized by their toxicity, carcinogenic effects, and high color visibility [4–6]. Furthermore, they prevent the penetration of oxygen and sunlight into water systems. Consequently, they affect the photosynthetic ability of the flora, causing harm to aquatic life [7,8]. In addition, they are hardly removed from water bodies due to their high stability and water solubility [9–11]. As a result of the current water depletion situation, the available water sources need to be protected [12]. Accordingly, discharging industrial untreated waste streams containing dye contamination is considered one of the main environmental problems that require immediate intervention. As a result, many researchers are eager to discover cost-effective and eco-friendly techniques to discard dyes and other pollutants from waterways [13–16]. Up to this point, numerous methods have been utilized in dye removal from waste streams. They can be categorized into physical, chemical, biological, or hybrid methods [17]. Those methods, including but not limited to, coagulation [18], adsorption [19], photolysis [20], electrochemical [21], biological treatment [22], membrane processes [23], and their hybrids [24–26] have all been investigated. Amongst them, adsorption is used for its efficiency, easy processing, and cheap cost [27], while membranes are used for their low required energy, low required space, reduced chemical consumption, and relatively low cost of operation [6]. However, they still have some limitations and drawbacks. Thus, the hybrid adsorptive membranes concept was investigated by incorporating adsorptive nanoparticles (NPs) into the synthesized membranes, resulting in a membrane that has the ability to possess the function of both systems [28–30].

For the process to be as effective as possible, the polymeric membrane material and the suitable incorporated adsorbent should be selected carefully. According to the literature, chitosan [31], alginate [32], and cellulose acetate [33] have been previously investigated for dye removal for their eco-friendly characteristics and removal efficiency. Amongst them, cellulose acetate (CA) has been extensively used because it is a biodegradable, low-cost, easily tailored, and available biopolymer [33,34]. Moreover, it is well known for its chemical and thermal stability [35]. Despite that, its inferior adsorption efficiency lessens its application for dye removal [36]. Therefore, impregnating adsorptive particles within CA may enhance the neat membrane adsorptive capacity [37]. The mechanism of adsorptive membranes depends on two steps: rejection and adsorption. In the rejection step, the particles with sizes larger than the pore size of the membrane are rejected by the membrane's surface, or what is called the membrane's active layer. In the next step, adsorption takes place, where small particles react or attach to the impregnated adsorbent, which in turn increases the overall removal efficiency of the membrane [38]. Previously, various adsorbents were tested for dye removal, namely zeolites, clay, limestone, lignite, and graphene oxide [39,40]. Among the most commonly used adsorbents, activated carbon (AC) is recognized as a perfect adsorbent for dye removal [41]. AC has an undoubtable dye removal performance, and it has previously proven to be a competitive adsorbent. AC is well known for its high adsorption capacity, large porous structure, high available active sites, and large surface area [42,43]. Additionally, nano activated carbon (NAC) is characterized by a much higher surface area and in return better adsorption capacity over AC [44]. Nevertheless, it has some limitations due to its high cost and non-renewable source. Thus, many studies have been conducted in order to provide an alternative low-cost adsorbent [34,45]. Agricultural wastes provide a competitive raw material for obtaining cheap, efficient, and renewable ACs [46]. Water hyacinth is an unwanted waste that can cause severe damage to aquatic life [47]. The control of water hyacinth is an effort, time, and money-consuming problem. Consequently, turning this irritating plant into an effective bio-sorbent could be more valuable [48,49].

To the best of the authors' knowledge, there is no other study that has investigated the impregnation of NAC fabricated from agricultural waste with blank cast and electro-spun CA-based membranes for dye decontamination from polluted water. Accordingly, in the current study, eco-friendly CA-based membranes were synthesized via phase inversion and electrospinning techniques. Additionally, nano activated carbon (NAC) was fabricated from Egyptian water hyacinths. The effect of impregnating the synthesized NAC bio-sorbent into the cast and the nanofibrous membranes was investigated. Methylene blue (MB) and Congo red (CR) were selected as cationic and anionic dyes, respectively. The selectivity of the fabricated, cast and electro-spun, neat CA and hybrid CA/NAC membranes towards MB and CR was studied.

2. Materials and Methods

2.1. Materials

All chemicals included in this study were used as purchased without any further modification. Cellulose acetate (CA) polymer with a molecular weight of 30,000 was supplied by Lobachemie, acetone (ACS > 99.5%) and dimethylformamide (DMF > 99.9%) were purchased from Fisher, sodium hydroxide (NaOH), methylene blue (MB) dye, and Congo red (CR) dye, and distilled water were bought from Sigma Aldrich. Additionally, activated carbon was prepared from Egyptian water hyacinth, which was collected from Itay El-Baroud Drainage, Al-Buhayrah governorate, Egypt.

2.2. Synthesis of Nano Activated Carbon (NAC)

Water hyacinth roots were used as the precursor for the preparation of dye removal bio-sorbent material [50]. To remove dust or any trapped particles, the collected raw precursor was rinsed extensively with distilled water. The washed water hyacinth was then dried in an electrical oven before being crushed into a fine powder. Then, for alkaline chemical treatment, 1 g of the obtained powder was further suspended in 100 mL of 2 M NaOH for 1 h under stirring at 60 °C. The chemically treated materials were then filtered and washed several times to get rid of any basic or acidic residuals. The treated washed materials were finally dried overnight at 70 °C, followed by a carbonization process in an Across muffle furnace for 1 h at 600 °C.

2.3. Synthesis of Neat and Composite CA-Based Dope Solutions

For the neat CA membrane, a suitable amount of CA polymeric powder was dissolved in a binary solvent mixture of acetone and DMF with a ratio of 2:1, respectively, to prepare a 15 wt% of neat CA dope solution. On the other hand, 5 wt% of the nano-filler (NAC) was sonicated firstly with the same binary solvent mixture for 1 h before the CA powder was added and stirred with it to obtain the composite CA/NAC dope solution. The previous solutions were then used to fabricate the neat and composite CA-based membranes via phase inversion and electrospinning techniques.

2.3.1. Synthesis of Cast Neat and Composite CA-Based Membranes

The membrane phase inversion method has been used for many years for the fabrication of flat sheet polymeric membranes [51–53]. The previously prepared blank CA and composite CA/NAC dope solutions were cast on a clean glass plate at room temperature. A doctor knife blade with a micrometer fixed at 0.25 mm was used to prepare the membranes with the required thickness. The cast films were left to evaporate in the air for 60 s before being immersed at (0–4 °C) in a distilled water bath. The obtained flat sheet membrane was then rinsed to get rid of any residual solvent. As a final treatment step, the membrane was then annealed at 80 °C for 10 min in another distilled water bath. The fabricated membranes were preserved in distilled water for at least 24 h for further use and characterization.

2.3.2. Synthesis of Electro-Spun Nanofibrous Neat and Composite CA-Based Membranes

Thanks to the electrospinning technique, the fabrication of nanofibrous membranes is possible. Due to the outstanding properties of higher surface area as well as higher membrane porosity [54,55], the technique was investigated. The pristine CA and hybrid CA/NAC polymeric solutions were spun with the help of (NanoNC) electro-spinning equipment under different conditions of feed pumping flow rate, applied voltage, and tip-to-collector distance. The optimum nanofibrous membrane was obtained at operating conditions of 0.7 mL/h, 19 kV, and 15 cm, respectively. The fabricated nanofiber membranes were then left to dry in a drying oven overnight for further use and characterization.

2.4. Characterization of the Extracted NAC and Fabricated CA-Based Membranes

Various techniques were involved in characterizing the synthesized electro-spun and cast CA-based membranes as well as the extracted NAC. Fourier transform infrared spectroscopy (FTIR) (Bruker Vertex 70) was used to observe and confirm the presence of the functional groups of the used materials. A Shimadzu XRD-6100 X-ray diffraction (XRD) device was used to study the crystalline nature of the fabricated membranes. The spectra were recorded from 3500 to 500 cm^{-1}.

The morphology of the membranes and the extracted NAC was investigated using a scanning electron microscope (SEM) device (JOEL JSM-6010 LV). The hydrophilic/hydrophobic nature of the membranes was measured using a contact angle system (DSA 100, KRÜSS). For more accurate results, five random locations on each membrane surface were selected to measure the contact angle.

2.5. *Performance Testing of the Synthesized Electro-Spun and Cast CA-Based Membranes*

In this study, the performance of the fabricated neat and hybrid membranes in the removal of cationic (MB) and anionic (CR) dyes was investigated. All decolorization experiments were repeated 4 times and the average measurements were recorded for more accuracy. Accordingly, as shown in Figure 1, each membrane's efficiency was evaluated using an Amicon cell setup with an active filtration area of around 480 mm^2. Each membrane was divided into circular shapes similar to the active area of the cell. Two stock solutions of MB and CR, 10 ppm each, were prepared to simulate the industrial dye-containing waste streams. At room temperature, the cell was then filled with the dye solution and connected to a syringe pump to apply a small pressure that allowed the water to flow through the membrane. The filtrate was recirculated until equilibrium was reached. A sample of the filtrate of each cycle was taken and the dye concentration was measured using a spectrophotometer device.

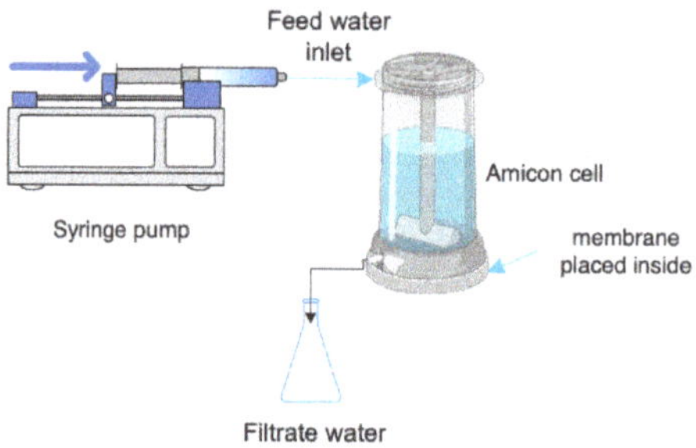

Figure 1. Amicon cell filtration setup.

The rejection factor (R) was calculated from Equation (1)

$$\%R = (1 - \frac{C_P}{C_R}) \quad (1)$$

where C_P and C_R are the dye concentrations of the filtrate and the feed solution (mg/L), respectively.

3. Results and Discussion

3.1. *Characterization of Nano Activated Carbon (NAC)*

As the functional groups of the obtained NAC have an impact on the adsorption efficiency [56], FTIR was used to recognize those functional groups. As can be seen in Figure 2a, the peak at 3401 cm^{-1} refers to the O-H functional group, which gives an indication of the bonded hydroxide in the prepared NAC. The band found at 1635 cm^{-1} could be due to C=C or C=O stretching [57]. The broad band found at 1157 cm^{-1} suggests the surface group of C-O stretching [58]. The absorption band at 616 cm^{-1} might refer to the C-H stretching or C=O bonding [59].

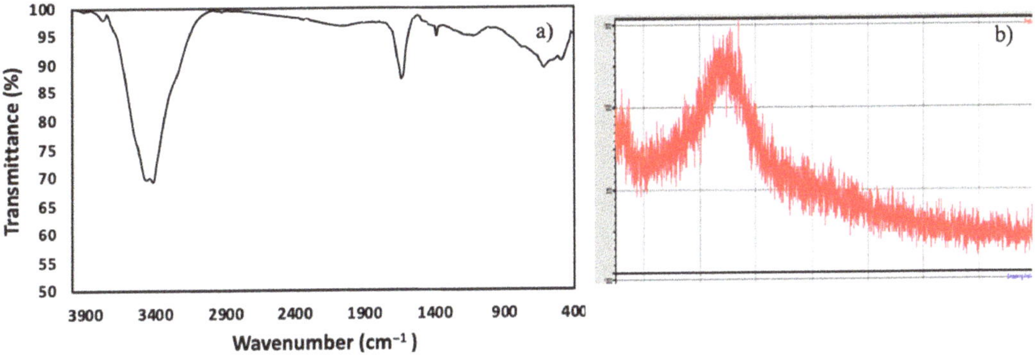

Figure 2. (**a**) FTIR spectrum (**b**) XRD pattern of the fabricated nano activated carbon NAC.

The XRD pattern was investigated in order to study the crystalline nature of the prepared NAC. As illustrated in Figure 2b, the obtained NAC has a semi-crystalline structure. In addition, the appearance of a broad peak in the range of 22° to 24° indicates the existence of carbon. On the other hand, the absence of other characteristic peaks gives an indication of the absence of any other contaminants [60].

The morphology of the prepared NAC was investigated by SEM images as presented in Figure 3a. SEM micrographs confirmed that a uniform nanosized material was obtained and the average diameter of the fabricated particles was estimated at 58 nm.

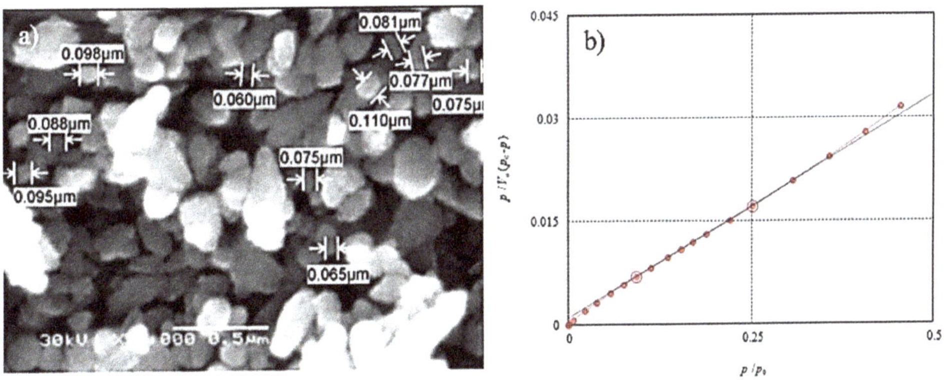

Figure 3. (**a**) SEM image (**b**) BET of the fabricated nano activated carbon (NAC).

The BET surface area of the fabricated NAC was measured to evaluate the material pore size, pore volume, and surface area. As shown in Figure 3b, N2 adsorption/desorption isotherm of the synthesized bio-sorbent was investigated. The NAC surface area was found to be 66.2 m^2/g. On the other hand, the total pore volume was measured as 0.25 cm^3/g, while its average pore size was recorded as 15.2 nm.

3.2. Characterization of Cast and Electro-Spun Neat and Composite CA-Based Membranes

As illustrated in Figure 4, the surface and cross-sectional structure of the cast neat CA and composite CA/NAC membranes was investigated via SEM images. Compared to the clear neat surface of CA, Figure 4a, the impregnated NAC clearly appears as white particles on the surface of the membrane in Figure 4b. This could give an indication of the successful dispersion of the NAC particles within the dope solution. In addition, from the cross-sectional images, it can be seen that the addition of NAC particles transformed the structure of the blank CA membrane, Figure 4c, into a more porous structure. A few longitudinal pores of the composite CA/NAC membrane are demonstrated in Figure 4d.

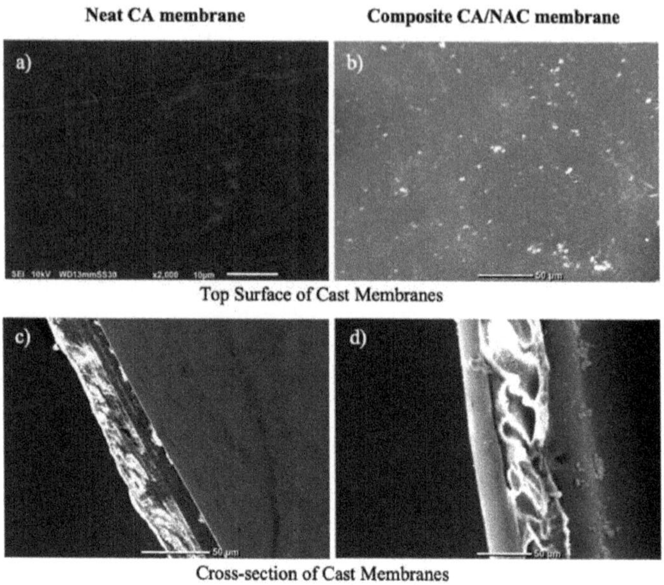

Figure 4. SEM images of cast membranes of "(**a**,**c**)" CA and "(**b**,**d**)" CA/NAC nanofiber membranes.

The optimum electro-spinning operating conditions were selected depending on the fiber shapes and diameters. The optimum conditions were used to fabricate a full matrix of CA and CA/NAC. The difference in the morphological structure between the synthesized blank and composite CA membranes can be seen in Figure 5. SEM images illustrate that both CA and CA/NAC membranes have uniform, straight, and bead-less fibers. In Figure 5a, the blank CA membrane nanofibers are shown as a uniform clear network. However, in Figure 5b, the loaded NAC can be seen as white dispersed particles and show good distribution with no agglomerations.

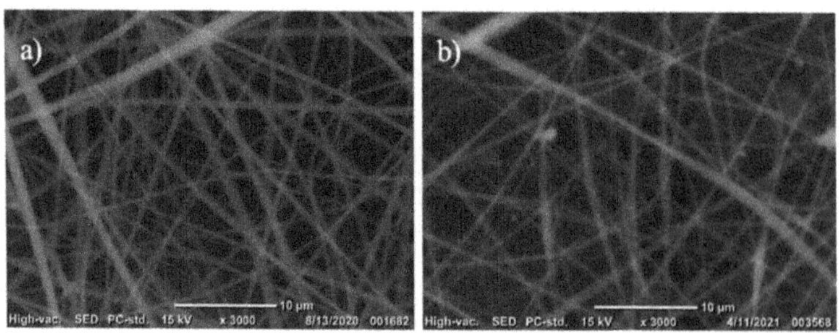

Figure 5. SEM images of "(**a**)" CA and "(**b**)"CA/NAC nanofiber membranes.

The surface wettability of the four fabricated cast and electro-spun membranes was investigated in order to study the effect of membrane surface morphology on the contact angle of each membrane. The hydrophilicity of the surface of the membrane affects the filtration flux [42]. The contact angle was expected to be less than 90° to be hydrophilic. Additionally, it was expected from the literature that the nanofibrous mat would be more hydrophilic than the cast one [61]. As displayed in Figure 6, all four membranes were found to be hydrophilic to different degrees. The contact angle of the blank cast and nanofiber membranes decreased after the addition of NAC particles. The contact angle of the blank CA cast membrane decreased from 63.5° to 58.6°, while it lowered from 60.5° for the neat CA nanofibrous membrane to become 58° for the composite CA/NAC nanofibrous membrane. This means that the addition of NAC has enhanced the membrane's hydrophilicity, which is an important and crucial parameter in water filtration.

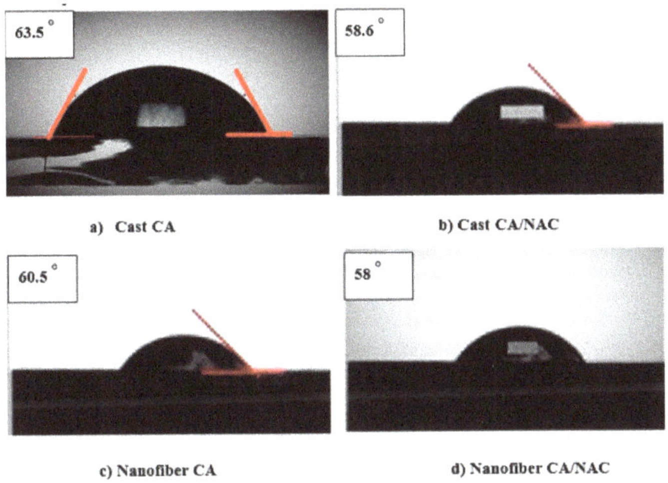

Figure 6. Contact angle of cast and electro-spun neat and composite CA-based membranes.

For nanofibrous blank CA and mixed matrix CA/NAC, FTIR spectra were further used to study the functional groups of both membranes. As demonstrated in Figure 6a, a wide band at 3395 cm^{-1} represents the presence of hydroxyl (–OH) stretching [62]. This band was shifted to 3474 cm^{-1} and its intensity decreased with the addition of NAC, as seen in Figure 7b. The carbonyl (C=O) group was observed at 1731 cm^{-1}. The adsorption peak at 1428 cm^{-1} refers to the (CH_2) deformation vibration. The characteristic (C-O-C) group can be seen at 1225 cm^{-1}. However, as represented in Figure 7b, this adsorption peak was shifted to become 1218 cm^{-1}. The stretching vibration of (C-OH) in the blank CA membrane was illustrated at 1029 cm^{-1} [63].

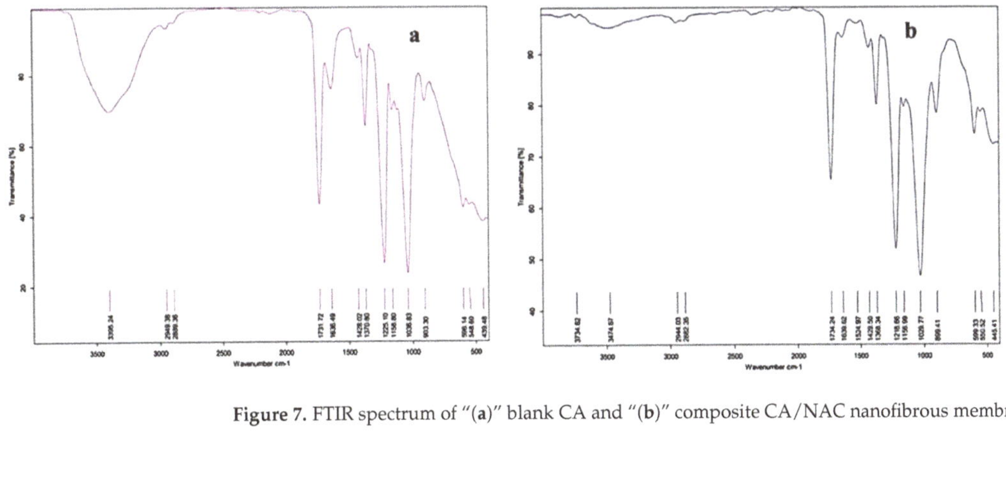

Figure 7. FTIR spectrum of "(**a**)" blank CA and "(**b**)" composite CA/NAC nanofibrous membranes.

As displayed in Figure 8, XRD was further used to give an indication of the successful impregnation of NAC. Figure 8a depicts the CA characteristic peaks at 10 and 20° [64]. Nonetheless, the intensity of the peak at 20° increased significantly.

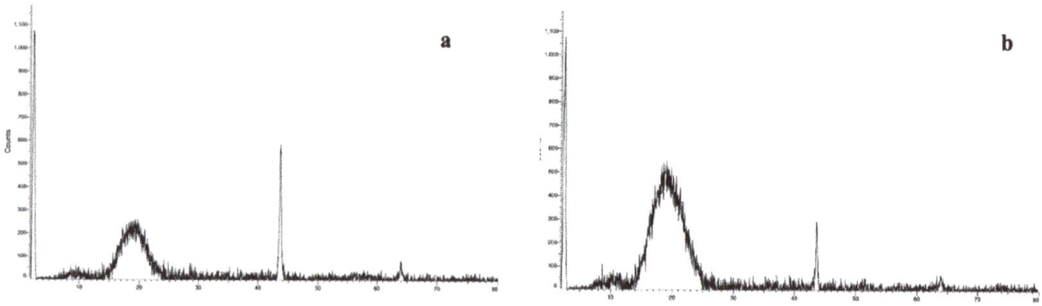

Figure 8. XRD pattern of (**a**) blank CA and (**b**) composite CA/NAC nanofibrous membranes.

This might be explained by the presence of carbon atoms and in return the impregnation of NAC particles.

3.3. Filtration Performance of Cast and Electro-Spun Neat and Composite CA-Based Membranes

The filtration performance of both blank and composite CA-based membranes for the removal of cationic MB and anionic CR was investigated using the previously mentioned Amicon cell setup. Generally, the four membranes showed a better affinity towards MB removal than CR. As illustrated in Figures 9 and 10, all the membranes had a noticeable effect on MB final concentration, while they had a negligible effect on CR final concentration.

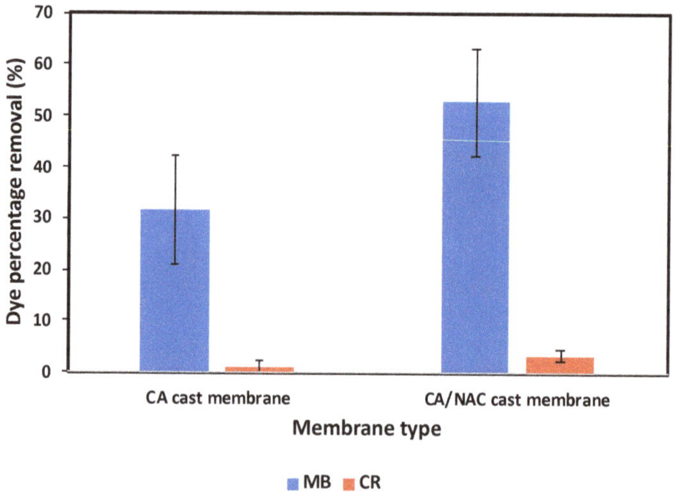

Figure 9. Performance of blank and composite cast membranes on the removal of MB and CR.

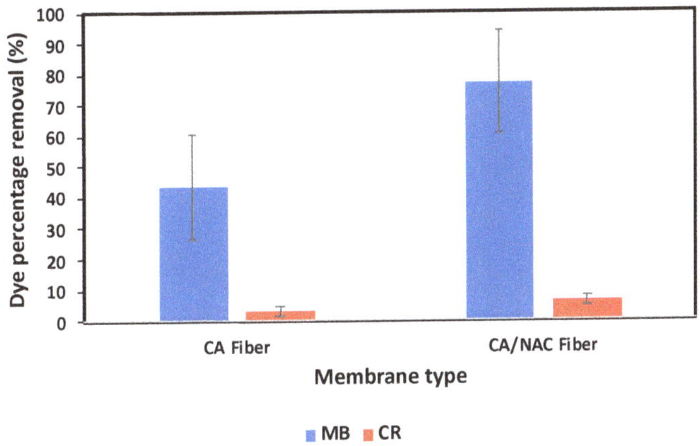

Figure 10. Performance of blank and composite nanofibrous membranes on the removal of MB and CR.

As clarified in Figure 9, the equilibrium concentration of the MB on the blank CA cast membrane was reached at 6.84 ppm with a removal percentage of 31.6%. However, the equilibrium concentration of CR on the same membrane was reached at 9.89 ppm with only 1.1% removal.

From Figure 9, it was noticed that the addition of NAC had a positive effect on the performance of the blank membrane. As illustrated, the equilibrium concentration of MB was reduced to be reached at 4.73 ppm, with around 53% removal for the cast CA/NAC composite membrane. However, this removal percentage is considered higher than that of the early reported cast CA/MWCNTs membrane. It was found that for 5 mg/L MB concentration, the recorded removal was less than 30% [65]. Unfortunately, regarding CR, the equilibrium concentration was affected slightly by the addition of NAC and was reached at 9.67 ppm to achieve a low removal percentage of only 3.3%.

On the other hand, the performances of the blank and impregnated CA-based nanofiber membranes are demonstrated in Figure 10. The fibrous membranes were also selective for MB over CR. The equilibrium concentration of MB over the blank CA fibrous membrane was found to be 5.67 ppm and after the addition of NAC, it became even lower until it reached almost 2.92 ppm. In other words, MB was removed using blank and impregnated fiber membranes by 43% and 70%, respectively.

On the contrary, CR reached the equilibrium concentration at 9.73 ppm using the blank CA fiber membrane and 9.41 ppm using the impregnated CA/NAC fiber membrane. The removal of CR using blank CA and impregnated CA/NAC fiber membranes was recorded at very small amounts of 2.7% and 5.9%, sequentially.

The presence of both the acetyl (CH_3CO^-) and the hydroxyl (OH^-) groups led to the negatively charged surface of the CA-based membranes. Thus, the selective behavior and better performance of CA-based membranes towards MB over CR could be due to the electrostatic interaction that takes place between MB molecules (positively charged) and the membrane surface (negatively charged) [30].

Regarding MB removal, Figure 11 compares the performance of the four membranes. Overall, the performance of fibers was better than that of cast membranes. At equilibrium, CA fiber removed 43.3% of methylene blue, while the blank cast membrane removed only 31%. The CA/NAC fiber presented the best behavior in removing the MB dye as it removed more than 70% of the dye. The removal of MB dye on the CA/NAC fiber at a neutral pH value was higher than that of the previously investigated CA/GO/TiO_2-NH_2 composite membrane. The literature reported that MB removal efficiency was about 60% at pH = 7 [66]. To conclude, the addition of NAC improved the performance of the blank CA membranes. This might be due to the existence of NAC resulting in an increase in the available active adsorbent sites in the fabricated adsorptive membranes. Those active sites may have captured more MB molecules within its porous structure, and this led to the enhancement of the blank CA membrane's performance [30].

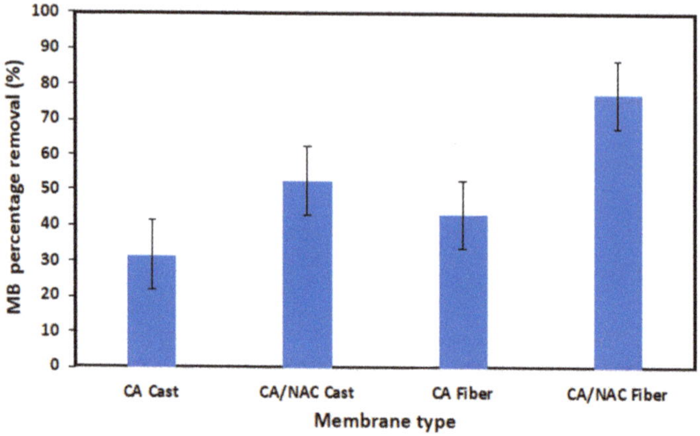

Figure 11. Comparison between the performance of the fabricated cast and nanofibrous CA-based membranes on the removal of MB dye.

4. Conclusions

The current study used water hyacinth biomass as a raw material for the extraction of nano activated carbon to be used as a filler to enhance the performance of the blank CA membrane for water treatment. Through this study, the problem of water hyacinth spreading as a biomass waste could be solved. In addition, a more useful and cheaper bio-sorbent filler can be synthesized from this waste by a simple and facile technique. In return, this will help in solving an irritating water pollution problem. The characteristics of the synthesized NAC and the four fabricated membranes were studied by SEM, FTIR, and XRD. All the used techniques have allowed the successful impregnation of NAC into CA membranes. SEM images displayed the surface characteristics of NAC as well as all the fabricated membranes. FTIR and XRD were used to investigate and confirm the differences between the neat and composite CA-based membranes. In addition, the contact angle was used to explore the membrane hydrophilicity. The results showed the hydrophilic nature of all the fabricated membranes. Moreover, it proved that the nanofibrous mat had a better hydrophilic nature than the cast one. The performance of the four membranes was evaluated using an Amicon cell setup. It was found that all the membranes had a better removal selectivity towards MB dye than CR. Furthermore, the addition of NAC enhanced the blank CA membrane dye removal efficiency. Among the four tested membranes, the composite cast and electro-spun CA/NAC recorded the best performances of 52.7 and 70%, respectively. However, the performance of the neat cast and electro-spun CA membrane achieved only 30 and 43.3%. In conclusion, the nanofibrous membrane showed a better performance than the cast membranes. Moreover, the impregnation of NAC had a positive effect on the membrane dye removal efficiency.

Author Contributions: Writing-original draft, O.A.K., A.M.K. and M.F.E.; validation, O.A.K., A.M.K., W.S. and M.F.E.; investigation, O.A.K. and A.M.K.; formal analysis, M.F.E. and W.S.; writing—review and editing, O.A.K., A.M.K., W.S. and M.F.E.; methodology, M.F.E.; supervision, M.F.E. and W.S. All authors have read and agreed to the published version of the manuscript.

Funding: This research received no external funding.

Institutional Review Board Statement: Not applicable.

Informed Consent Statement: Not applicable.

Data Availability Statement: All data are provided.

Conflicts of Interest: The authors declare no conflict of interest.

References

1. Li, Z.; Hanafy, H.; Zhang, L.; Sellaoui, L.; Netto, M.S.; Oliveira, M.L.S.; Seliem, M.K.; Dotto, G.L.; Bonilla-Petriciolet, A.; Li, Q. Adsorption of congo red and methylene blue dyes on an ashitaba waste and a walnut shell-based activated carbon from aqueous solutions: Experiments, characterization and physical interpretations. *Chem. Eng. J.* **2020**, *388*, 124263. [CrossRef]
2. Xiao, J.; Lv, W.; Xie, Z.; Tan, Y.; Song, Y.; Zheng, Q. Environmentally friendly reduced graphene oxide as a broad-spectrum adsorbent for anionic and cationic dyes: Via π-π Interactions. *J. Mater. Chem. A* **2016**, *4*, 12126–12135. [CrossRef]
3. Adesina, A.O.; Elvis, O.A.; Mohallem, N.D.S.; Olusegun, S.J. Adsorption of Methylene blue and Congo red from aqueous solution using synthesized alumina–zirconia composite. *Environ. Technol.* **2021**, *42*, 1061–1070. [CrossRef]
4. Ardila-Leal, L.D.; Poutou-Piñales, R.A.; Pedroza-Rodríguez, A.M.; Quevedo-Hidalgo, B.E. A brief history of colour, the environmental impact of synthetic dyes and removal by using laccases. *Molecules* **2021**, *26*, 3813. [CrossRef]
5. Ganapuram, B.R.; Alle, M.; Dadigala, R.; Dasari, A.; Maragoni, V.; Guttena, V. Catalytic reduction of methylene blue and Congo red dyes using green synthesized gold nanoparticles capped by *Salmalia malabarica* gum. *Int. Nano Lett.* **2015**, *5*, 215–222. [CrossRef]
6. Moradihamedani, P. Recent advances in dye removal from wastewater by membrane technology: A review. *Polym. Bull.* **2022**, *79*, 2603–2631. [CrossRef]
7. Sharma, J.; Sharma, S.; Soni, V. Classification and impact of synthetic textile dyes on Aquatic Flora: A review. *Reg. Stud. Mar. Sci.* **2021**, *45*, 101802. [CrossRef]
8. Chowdhury, S.; Pan, S.; Balasubramanian, R.; Das, P. Date Palm Based Activated Carbon for the Efficient Removal of Organic Dyes from Aqueous Environment. In *Sustainable Agriculture Reviews 34: Date Palm for Food, Medicine and the Environment*; Naushad, M., Lichtfouse, E., Eds.; Springer International Publishing: Cham, Switzerland, 2019; pp. 247–263. [CrossRef]
9. Lellis, B.; Fávaro-Polonio, C.Z.; Pamphile, J.A.; Polonio, J.C. Effects of textile dyes on health and the environment and bioremediation potential of living organisms. *Biotechnol. Res. Innov.* **2019**, *3*, 275–290. [CrossRef]
10. Raza, A.; Rehman, R.; Batool, M. Recent Review of Titania-Clay-Based Composites Emerging as Advanced Adsorbents and Photocatalysts for Degradation of Dyes over the Last Decade. *Adsorpt. Sci. Technol.* **2022**, *2022*, 3823008. [CrossRef]
11. Elessawy, N.A.; El-Sayed, E.M.; Ali, S.; Elkady, M.F.; Elnouby, M.; Hamad, H.A. One-pot green synthesis of magnetic fullerene nanocomposite for adsorption characteristics. *J. Water Process Eng.* **2020**, *34*, 101047. [CrossRef]
12. Praveen, S.; Jegan, J.; Pushpa, T.B.; Gokulan, R.; Bulgariu, L. Biochar for removal of dyes in contaminated water: An overview. *Biochar* **2022**, *4*, 10. [CrossRef]
13. Labena, A.; Abdelhamid, A.E.; Amin, A.S.; Husien, S.; Hamid, L.; Safwat, G.; Diab, A.; Gobouri, A.A.; Azab, E. Removal of Methylene Blue and Congo Red Using Adsorptive Membrane Impregnated with Dried *Ulva fasciata* and *Sargassum dentifolium*. *Plants* **2021**, *10*, 384. [CrossRef] [PubMed]
14. Yang, G.; Wu, L.; Xian, Q.; Shen, F.; Wu, J.; Zhang, Y. Removal of Congo red and methylene blue from aqueous solutions by vermicompost-derived biochars. *PLoS ONE* **2016**, *11*, e0154562. [CrossRef] [PubMed]
15. Pai, S.; Kini, M.S.; Selvaraj, R. A review on adsorptive removal of dyes from wastewater by hydroxyapatite nanocomposites. *Environ. Sci. Pollut. Res.* **2021**, *28*, 11835–11849. [CrossRef] [PubMed]
16. Elkady, M.F.; Hassan, H.S. Invention of Hollow Zirconium Tungesto-Vanadate at Nanotube Morphological Structure for Radionuclides and Heavy Metal Pollutants Decontamination from Aqueous Solutions. *Nanoscale Res. Lett.* **2015**, *10*, 474. [CrossRef]
17. Gürses, A.; Güneş, K.; Şahin, E. Chapter 5-Removal of dyes and pigments from industrial effluents. In *Advances in Green and Sustainable Chemistry*; Sharma, S.K., Ed.; Elsevier: Amsterdam, The Netherlands, 2021; pp. 135–187. [CrossRef]
18. Mcyotto, F.; Wei, Q.; Macharia, D.K.; Huang, M.; Shen, C.; Chow, C.W.K. Effect of dye structure on color removal efficiency by coagulation. *Chem. Eng. J.* **2021**, *405*, 126674. [CrossRef]
19. Khodaie, M.; Ghasemi, N.; Moradi, B.; Rahimi, M. Removal of methylene blue from wastewater by adsorption onto znclactivated corn husk carbon equilibrium studies. *J. Chem.* **2013**, *2013*, 383985. [CrossRef]
20. Mohammed, H.A.; Khaleefa, S.A.; Basheer, M.I. Photolysis of Methylene Blue Dye Using an Advanced Oxidation Process (Ultraviolet Light and Hydrogen Peroxide). *J. Eng. Sustain. Dev.* **2022**, *25*, 59–67. [CrossRef]
21. Beddai, A.A.; Badday, B.A.; Al-Yaqoobi, A.M.; Mejbel, M.K.; Hachim, Z.S.A.; Mohammed, M.K.A. Color Removal of Textile Wastewater Using Electrochemical Batch Recirculation Tubular Upflow Cell. *Int. J. Chem. Eng.* **2022**, *2022*, 4713399. [CrossRef]
22. Bhatia, D.; Sharma, N.R.; Singh, J.; Kanwar, R.S. Biological methods for textile dye removal from wastewater: A review. *Crit. Rev. Environ. Sci. Technol.* **2017**, *47*, 1836–1876. [CrossRef]
23. Kadhim, R.J.; Al-Ani, F.H.; Al-Shaeli, M.; Alsalhy, Q.F.; Figoli, A. Removal of dyes using graphene oxide (Go) mixed matrix membranes. *Membranes* **2020**, *10*, 366. [CrossRef] [PubMed]
24. Alardhi, S.M.; Albayati, T.M.; Alrubaye, J.M. A hybrid adsorption membrane process for removal of dye from synthetic and actual wastewater. *Chem. Eng. Process. Process Intensif.* **2020**, *157*, 108113. [CrossRef]
25. Hai, F.I.; Yamamoto, K.; Fukushi, K. Hybrid treatment systems for dye wastewater. *Crit. Rev. Environ. Sci. Technol.* **2007**, *37*, 315–377. [CrossRef]

26. Ledakowicz, S.; Pázdzior, K. Recent achievements in dyes removal focused on advanced oxidation processes integrated with biological methods. *Molecules* **2021**, *26*, 870. [CrossRef] [PubMed]
27. Wong, S.; Ghafar, N.A.; Ngadi, N.; Razmi, F.A.; Inuwa, I.M.; Mat, R.; Amin, N.A.S. Effective removal of anionic textile dyes using adsorbent synthesized from coffee waste. *Sci. Rep.* **2020**, *10*, 2928. [CrossRef]
28. Qalyoubi, L.; Al-Othman, A.; Al-Asheh, S. Recent progress and challenges of adsorptive membranes for the removal of pollutants from wastewater. Part II: Environmental applications. *Case Stud. Chem. Environ. Eng.* **2021**, *3*, 100102. [CrossRef]
29. Chen, Y.S.; Ooi, C.W.; Show, P.L.; Hoe, B.C.; Chai, W.S.; Chiu, C.-Y.; Wang, S.S.-S.; Chang, Y.-K. Removal of Ionic Dyes by Nanofiber Membrane Functionalized with Chitosan and Egg White Proteins: Membrane Preparation and Adsorption Efficiency. *Membranes* **2022**, *12*, 63. [CrossRef]
30. Tahazadeh, S.; Mohammadi, T.; Tofighy, M.A.; Khanlari, S.; Karimi, H.; Emrooz, H.B.M. Development of cellulose acetate/metal-organic framework derived porous carbon adsorptive membrane for dye removal applications. *J. Memb. Sci.* **2021**, *638*, 119692. [CrossRef]
31. Li, C.; Lou, T.; Yan, X.; Long, Y.Z.; Cui, G.; Wang, X. Fabrication of pure chitosan nanofibrous membranes as effective absorbent for dye removal. *Int. J. Biol. Macromol.* **2018**, *106*, 768–774. [CrossRef]
32. Ma, Y.; Qi, P.; Ju, J.; Wang, Q.; Hao, L.; Wang, R.; Sui, K.; Tan, Y. Gelatin/alginate composite nanofiber membranes for effective and even adsorption of cationic dyes. *Compos. B Eng.* **2019**, *162*, 671–677. [CrossRef]
33. Cheng, J.; Zhan, C.; Wu, J.; Cui, Z.; Si, J.; Wang, Q.; Peng, X.; Turng, L.-S. Highly Efficient Removal of Methylene Blue Dye from an Aqueous Solution Using Cellulose Acetate Nanofibrous Membranes Modified by Polydopamine. *ACS Omega* **2020**, *5*, 5389–5400. [CrossRef] [PubMed]
34. Rana, J.; Goindi, G.; Kaur, N. Potential of cellulose acetate for the removal of methylene blue dye from aqueous streams. *Int. J. Innov. Technol. Explor. Eng.* **2019**, *8*, 1379–1382. [CrossRef]
35. Rambabu, K.; Velu, S. Modified cellulose acetate ultrafiltration composite membranes for enhanced dye removal. *Int. J. Chem. Sci.* **2016**, *14*, 195–205.
36. Chen, W.; Ma, H.; Xing, B. Electrospinning of multifunctional cellulose acetate membrane and its adsorption properties for ionic dyes. In *International Journal of Biological Macromolecules*; Elsevier: Amsterdam, The Netherlands, 2020; Volume 158, pp. 1342–1351. [CrossRef]
37. Abu-Dalo, M.A.; Al-Rosan, S.A.; Albiss, B.A. Photocatalytic degradation of methylene blue using polymeric membranes based on cellulose acetate impregnated with zno nanostructures. *Polymers* **2021**, *13*, 3451. [CrossRef] [PubMed]
38. Qalyoubi, L.; Al-Othman, A.; Al-Asheh, S. Recent progress and challenges on adsorptive membranes for the removal of pollutants from wastewater. Part I: Fundamentals and classification of membranes. *Case Stud. Chem. Environ. Eng.* **2021**, *3*, 100086. [CrossRef]
39. Osagie, C.; Othmani, A.; Ghosh, S.; Malloum, A.; Esfahani, Z.K.; Ahmadi, S. Dyes adsorption from aqueous media through the nanotechnology: A review. *J. Mater. Res. Technol.* **2021**, *14*, 2195–2218. [CrossRef]
40. Homaeigohar, S.; Zillohu, A.U.; Abdelaziz, R.; Hedayati, M.K.; Elbahri, M. A novel nanohybrid nanofibrous adsorbent for water purification from dye pollutants. *Materials* **2016**, *9*, 848. [CrossRef]
41. Bhatnagar, A.; Hogland, W.; Marques, M.; Sillanpää, M. An overview of the modification methods of activated carbon for its water treatment applications. *Chem. Eng. J.* **2013**, *219*, 499–511. [CrossRef]
42. Zhao, X.; Huang, C.; Zhang, S.; Wang, C. Cellulose Acetate/Activated Carbon Composite Membrane with Effective Dye Adsorption Performance. *J. Macromol. Sci. Part B Phys.* **2019**, *58*, 909–920. [CrossRef]
43. Moosavi, S.; Lai, C.W.; Gan, S.; Zamiri, G.; Pivehzhani, O.A.; Johan, M.R. Application of efficient magnetic particles and activated carbon for dye removal from wastewater. *ACS Omega* **2020**, *5*, 20684–20697. [CrossRef]
44. Shokry, H.; Elkady, M.; Hamad, H. Nano activated carbon from industrial mine coal as adsorbents for removal of dye from simulated textile wastewater: Operational parameters and mechanism study. *J. Mater. Res. Technol.* **2019**, *8*, 4477–4488. [CrossRef]
45. Kanawade, S.M.; Gaikwad, R.W. Removal of Methylene Blue from Effluent by Using Activated Carbon and Water Hyacinth as Adsorbent. *Int. J. Chem. Eng. Appl.* **2011**, *2*, 317–319. [CrossRef]
46. Köseoğlu, E.; Akmil-Başar, C. Preparation, structural evaluation and adsorptive properties of activated carbon from agricultural waste biomass. *Adv. Powder Technol.* **2015**, *26*, 811–818. [CrossRef]
47. Malik, A. Environmental challenge vis a vis opportunity: The case of water hyacinth. *Environ. Int.* **2007**, *33*, 122–138. [CrossRef] [PubMed]
48. Mahamadi, C. Water hyacinth as a biosorbent: A review. *Afr. J. Environ. Sci. Tech.* **2012**, *5*, 1137–1145. [CrossRef]
49. Li, F.; He, X.; Srishti, A.; Song, S.; Tan, H.T.W.; Sweeney, D.J.; Ghosh, S.; Wang, C.-H. Water hyacinth for energy and environmental applications: A review. *Bioresour. Technol.* **2021**, *327*, 124809. [CrossRef] [PubMed]
50. Shokry, H.; Elkady, M.; Salama, E. Eco-friendly magnetic activated carbon nano-hybrid for facile oil spills separation. *Sci. Rep.* **2020**, *10*, 10265. [CrossRef] [PubMed]
51. Strathmann, H.; Scheible, P.; Baker, R.W. A rationale for the preparation of Loeb-Sourirajan-type cellulose acetate membranes. *J. Appl. Polym. Sci.* **1971**, *15*, 811–828. [CrossRef]
52. Sahraei, R.; Shahalizade, T.; Ghaemy, M.; Mahdavi, H. Fabrication of cellulose acetate/Fe_3O_4@GO-APTS-poly(AMPS-co-MA) mixed matrix membrane and its evaluation on anionic dyes removal. *Cellulose* **2018**, *25*, 3519–3532. [CrossRef]

53. Ghaseminezhad, S.M.; Barikani, M.; Salehirad, M. Development of graphene oxide-cellulose acetate nanocomposite reverse osmosis membrane for seawater desalination. *Compos. B Eng.* **2019**, *161*, 320–327. [CrossRef]
54. Thamer, B.M.; Aldalbahi, A.; Moydeen, M.A.; Rahaman, M.; El-Newehy, M.H. Modified electrospun polymeric nanofibers and their nanocomposites as nanoadsorbents for toxic dye removal from contaminated waters: A review. *Polymers* **2021**, *23*, 20. [CrossRef] [PubMed]
55. Subrahmanya, T.M.; Arshad, A.B.; Lin, P.T.; Widakdo, J.; Makari, H.K.; Austria, H.F.; Hu, C.C.; Lai, J.Y.; Hung, W.S. A review of recent progress in polymeric electrospun nanofiber membranes in addressing safe water global issues. *RSC Adv.* **2021**, *11*, 9638–9663. [CrossRef]
56. Elkady, M.; Shokry, H.; Hamad, H. New activated carbon from mine coal for adsorption of dye in simulated water or multiple heavy metals in real wastewater. *Materials* **2020**, *13*, 2498. [CrossRef] [PubMed]
57. Liu, Y.; Liu, X.; Dong, W.; Zhang, L.; Kong, Q.; Wang, W. Efficient Adsorption of Sulfamethazine onto Modified Activated Carbon: A Plausible Adsorption Mechanism. *Sci. Rep.* **2017**, *7*, 12437. [CrossRef] [PubMed]
58. Yu, X.; Han, Z.; Fang, S.; Chang, C.; Han, X. Optimized Preparation of High Value-Added Activated Carbon and Its Adsorption Properties for Methylene Blue. *Int. J. Chem. React. Eng.* **2019**, *17*, 20180267. [CrossRef]
59. Kebir, M.; Trari, M.; Maachi, R.; Nasrallah, N.; Amrane, A. Valorization of Inula viscosa waste extraction, modeling of isotherm, and kinetic for the tartrazine dye adsorption. *Desalination Water Treat* **2015**, *54*, 2806–2816. [CrossRef]
60. Kalyani, P.; Anitha, A. Refuse derived energy-tea derived boric acid activated carbon as an electrode material for electrochemical capacitors. *Port. Electrochim. Acta* **2013**, *31*, 165–174. [CrossRef]
61. Ding, B.; Li, C.; Hotta, Y.; Kim, J.; Kuwaki, O.; Shiratori, S. Conversion of an electrospun nanofibrous cellulose acetate mat from a super-hydrophilic to super-hydrophobic surface. *Nanotechnology* **2006**, *17*, 4332–4339. [CrossRef]
62. Namjoufar, M.; Farzi, A.; Karimi, A. Removal of Acid Brown 354 from wastewater by aminized cellulose acetate nanofibers: Experimental and theoretical study of the effect of different parameters on adsorption efficiency. *Water Sci. Technol.* **2021**, *83*, 1649–1661. [CrossRef]
63. Sudiarti, T.; Wahyuningrum, D.; Bundjali, B.; Arcana, I.M. Mechanical strength and ionic conductivity of polymer electrolyte membranes prepared from cellulose acetate-lithium perchlorate. *IOP Conf. Ser. Mater. Sci. Eng.* **2017**, *223*, 012052. [CrossRef]
64. Baniasadi, J.; Shabani, Z.; Mohammadi, T.; Sahebi, S. Enhanced performance and fouling resistance of cellulose acetate forward osmosis membrane with the spatial distribution of TiO_2 and Al_2O_3 nanoparticles. *J. Chem. Technol. Biotechnol.* **2021**, *96*, 147–162. [CrossRef]
65. Silva, M.A.; Hilliou, L.; de Amorim, M.T.P. Fabrication of pristine-multiwalled carbon nanotubes/cellulose acetate composites for removal of methylene blue. *Polym. Bull.* **2020**, *77*, 623–653. [CrossRef]
66. Aboamera, N.M.; Mohamed, A.; Salama, A.; Osman, T.A.; Khattab, A. An effective removal of organic dyes using surface functionalized cellulose acetate/graphene oxide composite nanofibers. *Cellulose* **2018**, *25*, 4155–4166. [CrossRef]

sustainability

MDPI

Article

Effect of Mushy Zone Parameter on Phase Change Behavior of Different Configurations Storage Unit: Numerical Simulation and Experimental Validation

Ahmed Saad Soliman [1], Ahmed A. Sultan [1,*] and Mohamed A. Sultan [2]

1 Mechanical Power Engineering Department, Mansoura University, El-Mansoura 35516, Egypt
2 Mechanical Engineering Department, Higher Future Institute of Engineering and Technology, El-Mansoura 35516, Egypt
* Correspondence: aasultan@mans.edu.eg

Abstract: The melting process of paraffin wax placed in storage capsules of different shapes was experimentally and numerically studied. The phase change material (PCM) was initially at 27 °C. The effect of the mushy zone parameter (A_{mush}) value on the melting process of the PCM was studied with storage capsules of different shapes (circular, vertical oval, and horizontal oval). The results of the numerical model were validated with the experimental results to obtain the optimum A_{mush} value for each shape of the latent heat storage unit. The results showed that the value of the A_{mush} has a great impact on the numerical results of the PCM melting process and changes with the shape of the storage capsule. The rate of heat transfer, convection, and fluid velocity all decrease as the A_{mush} value rises. The experimental results of the circular, vertical oval, and horizontal oval capsules match very well with the numerical model with A_{mush} values equal to 2×10^6, 1×10^5, and 1×10^6, respectively.

Keywords: natural convection; CFD; melting PCM; enthalpy-porosity method; mushy zone parameter

Citation: Soliman, A.S.; Sultan, A.A.; Sultan, M.A. Effect of Mushy Zone Parameter on Phase Change Behavior of Different Configurations Storage Unit: Numerical Simulation and Experimental Validation. *Sustainability* **2022**, *14*, 14540. https://doi.org/10.3390/su142114540

Academic Editor: Valeria Palomba

Received: 25 September 2022
Accepted: 12 October 2022
Published: 4 November 2022

1. Introduction

Energy can be stored in several forms, such as sensible, latent, and thermochemical heat storage. Phase change materials (PCMs) are used in latent heat energy storage (LHES) systems, which have more desirable characteristics than conventional heat storage systems. PCMs offer a number of beneficial characteristics, including the ability to be used as a heat source at a constant temperature, with only slight temperature fluctuations through heat recovery. They possess a low vapour pressure at operating temperature, chemical stability, non-corrosion, and a high energy density with small storage space requirements [1]. Thus, PCMs demonstrate one of the best techniques used in thermal energy storage applications [2,3]. They are used in several engineering applications, for example, the technology of electronic cooling, waste heat recovery systems [4,5], solar cell systems [6], absorption systems [7], ventilation, and air conditioning (HVAC) systems [8].

1.1. Numerical Techniques Limitations

The numerical modeling of thermal energy storage (TES) systems has recently received a great deal of interest. Although the most appropriate numerical techniques can vary greatly depending on the system, cost and time restrictions are virtually always the driving forces behind their employment. Numerical models are now increasingly necessary to accurately explain the behaviour of TES systems, enabling computational tools to assist in the solution of governing equations. As a result, a significant portion of the research on this topic shows the development of adequate numerical analysis [9]. The heat transfer mechanism through the melting process is initially controlled by conduction; then, natural convection begins, and this process of natural convection has a great impact and cannot be ignored [10,11].

Different modeling approaches for the solid–liquid phase change have been developed in recent decades. The most widely used methodology for numerically simulating the phase change phenomenon in several types of research is the enthalpy-porosity technique [12,13]. The other approaches to doing the same thing depend on varying the PCM's specific heat during the phase transition temperature [14]. The enthalpy-porosity approach can be used in all forms of the melting process, whereas the other methods are more suited to phase changes that are dominated by conduction [15]. The coupled convection–diffusion phase transition was numerically studied using the enthalpy-porosity technique [16]. This approach avoids directly following the solid–liquid contact. With appropriate momentum sink factors added to momentum equations, the solid–liquid mushy zone is viewed as a porous zone with a quantity denoted as a liquid fraction as its "porosity," which is the reason for the pressure reduction as the resulting solid material comes into existence.

1.2. Mushy Zone Effect

The thin mushy zone that divides the areas occupied by the solid and liquid PCMs contains a parameter known as the mushy zone constant that measures the resistance to the flow of liquid PCMs in this region. The literature demonstrates that the value of this constant affects simulation outcomes [17–19], and that the best value for agreement with experiments varies depending on the situation [18,20]. The cell porosity, which is calculated using the cell enthalpy, is a non-linear function of this parameter. Thus, the parameter was designed in such a way that it would be zero for entirely liquid cells and have no effect, while it would be in the same order as the other transport terms for cells going through a phase transition. The parameter's value would be high enough for totally solid cells to essentially push any velocity prediction to zero. The parameter also contains a constant known as the mushy zone parameter, A_{mush}, which captures the impact of the shape of the mushy zone [15].

The mushy zone is considered to be a semi-solid that exists at the boundary between the melted and un-melted portion of a PCM through the melting or freezing process. The amplitude of the damping is measured by the mushy zone parameter A_{mush}; the higher this value, the steeper the transition of the velocity of the material to zero as it solidifies. The predicted solution might oscillate at very high values of A_{mush} [21]. The overall phase transition process and the degree of phase interaction are both greatly influenced by the value of A_{mush}. However, the A_{mush} value may vary depending on the substance. For Rubitherm RT82, the value of A_{mush} ranges from 1×10^5 [22] to 1.6×10^6 [23] whereas for gallium it ranges from 1×10^5 to 1×10^{15} [24,25]. Several studies have found a correlation between the value of A_{mush} and the solid particle diameters within the mushy zone constant [26,27]. Thus, the choice of mushy zone constant becomes very important as the behaviour predicted using the enthalpy-porosity technique for the same material can differ significantly at different values of A_{mush}.

In previous studies, the A_{mush} value was studied for different types of phase change materials and it was observed that this value has a great impact on the melting process results of the numerical model. In the present work, the effect of the A_{mush} value on the PCM melting process was studied with different shapes of storage capsules (circular, horizontal oval, and vertical oval), which are filled with paraffin wax, which has not been previously studied. The melting process of the PCM was experimentally and numerically investigated in a circular, horizontal, and vertical oval storage units. The results that were obtained from the numerical model were validated with the experimental results to obtain the optimum value of A_{mush} for each shape of the latent heat storage unit.

The effects of A_{mush} through the storage units of circular, vertical oval, and horizontal oval capsules have not been previously studied. However, it can be used in energy-saving building applications.

2. Experimental Setup and Procedures

An image of the test rig is shown in Figure 1, while the layout is shown in Figure 2. The test rig consists of the water tank (10), which is made of galvanized sheet with a diameter of 40 cm and a height of 40 cm. The tank is used to supply the test section with hot water (11) at a constant temperature value. The tank is insulated with glass wool insulation to decrease the heat lost from hot water in the tank. The tank is equipped with a 1 kW electric heater (12) connected to the electrical main supply with a TROIDC type electric transformer (13). The electric transformer provides a voltage variation from 0 to 220 V to adjust the water temperature in the tank at a predefined temperature. The tank is equipped with a small 100-Watt water pump (8) connected to the tank bottom via an inlet connection (4) to supply the test section with the required hot water. The hot water is used to heat the capsule at different velocities using a supply valve (5) which regulates the water discharge rate to the test section (1). The valve (7) is used to bypass the excess water from the pump to the tank via a bypass connection (6), to prevent the pump from overheating.

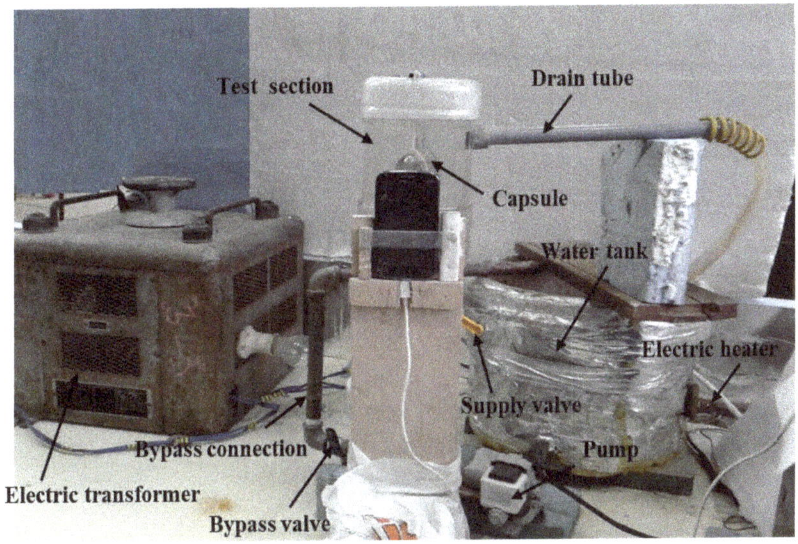

Figure 1. Photography of the experimental setup.

The test section (1) is a rectangular cross-section channel of 8 cm by 10 cm, and 15 cm in height. Hot water is brought to it via a $\frac{1}{2}$-inch diameter pipe (4) connected to the bottom of the channel through the pump (8). To drain the water into the hot water tank, a hole of 1/2 inch diameter is used and a tube (9) of the same diameter is connected to one side near the top of the channel to return the water to the hot water tank. The inlet water temperature is measured using the thermometer (3). The capsule (2) filled with PCM is placed in a horizontal position using a 1 mm diameter wire carrier.

Two Pyrex glass capsules are used, as seen in Figure 3; one has a circular cross-section with a 4 cm diameter and 6 cm length. The second capsule has an elliptical cross-section, the axes of which have dimensions of 6.2 and 2.6 cm. The two capsules are filled with the molten paraffin, then left to cool and closed with the resulting solid paraffin. The PCM thermophysical properties are represented in Table 1.

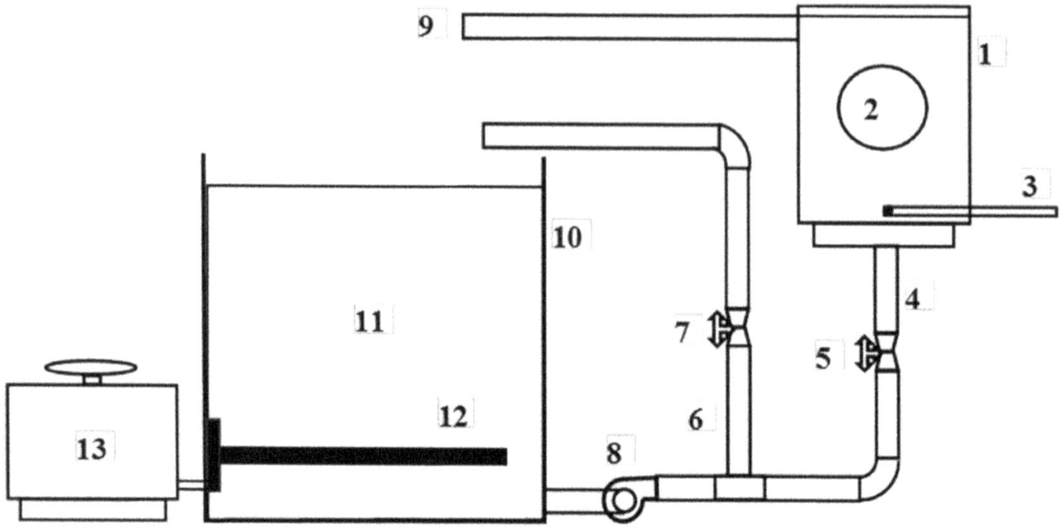

1-Test section | 2-PCM capsule | 3-Thermometer | 4-Inlet connection
5-Supply valve | 6-Bypass connection | 7-Bypass valve | 8-Water pump
9-Outlet connection | 10-Water tank | 11-Hot water | 12-Electric heater
13-Electric transformer

Figure 2. Schematic representation of the experimental setup.

Table 1. PCM thermophysical properties.

Thermophysical Properties	Paraffin Wax [11]	Thermophysical Properties	Paraffin Wax [11]
K_s (W/m.K)	0.15	CP_l (kJ/kg.K)	2.44
K_l (W/m.K)	0.15	Kinematic viscosity (m^2/s)	8.31×10^{-5}
ρ_s (kg/m^3)	890	Melting point (°C)	54.22/56.22
ρ_l (kg/m^3)	712	Heat of fusion (kJ/kg)	278.8
CP_s (kJ/kg. K)	2.384	Thermal exp. Coeff (1/k)	0.000714

To facilitate the viewing and imaging of the capsule, the test section was made from Pyrex glass. In order to measure the average temperature of the water before interring the channel, a thermometer (3) was fixed inside the channel (1). The rate of hot water passing through the test section was measured by the amount of water collected at a given time during the experiment.

Khot et al., [28] observed that the use of thermocouples inside the PCM capsule restricted the solid PCM from sinking to the bottom of the spherical capsule. Therefore, the unconstrained PCM melted faster than the constrained one. Based on the previous study, the experimental investigation was made to validate the results obtained by the Ansys Fluent software wherein there is no thermocouple embedded inside the capsules to measure the temperature distribution inside it. The image processing technique for the experimental liquid fraction photos was used to detect the value of the PCM liquid fraction.

(a) (b)

Figure 3. Photos of experimental capsules for (**a**) circular and (**b**) elliptical cross-section.

3. Experimental Measurements

First of all, we checked the system against leakage by turning on the pump and monitoring the system connections. Then, we measured the flow rate of water and adjusted the two valves (5 and 7) to reach the predetermined mean velocity of the water. After that, we regulated the electric transformer to vary the voltage supplied to the electric heater and let the system work, monitoring the reading of the thermometer until reaching the wanted temperature of hot water interring the test section. After reaching the suggested temperature, we turned the circular cross-section PCM capsule in place inside the channel, into a horizontal position. We took photos of the capsule cross-section every five minutes until the total melting of the PCM. Finally, we repeated the previous steps for the capsule of the elliptical cross-section on its largest axes, vertical and horizontal, one at a time to investigate the two radii ratios.

The cross-section area of the elliptical cross-section can be calculated as:

$$\text{Ac} = \pi \text{ a b} \tag{1}$$

where a is half the smallest axis and b is half of the largest axis of the ellipse as shown in Figure 4.

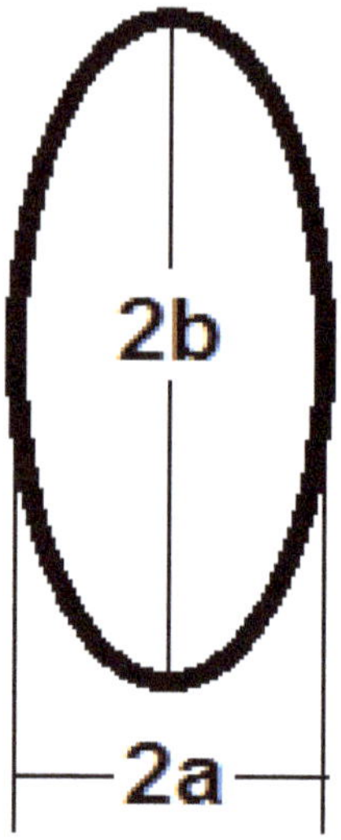

Figure 4. The main axes of the elliptical cross-section.

The surface area of the elliptical cross-section capsule can be calculated as:

$$\text{As} = \text{C l} \tag{2}$$

where l is the cylinder length and C is its perimeter.

The more accurate perimeter of the elliptical cross-section as a function of its axes using the Ramanujan second approximation [29] can be calculated as shown in the following equation:

$$C \approx \pi(a+b)\left(1+\frac{3\left(\frac{a-b}{a+b}\right)^2}{10+\sqrt{4-3\left(\frac{a-b}{a+b}\right)^2}}\right) \tag{3}$$

Two positions of the elliptical cross-section are discussed. The first position is a vertical oval with an axes ratio of Ar = 0.383, and the second position is a horizontal oval with an axes ratio of Ar = 2.6.

4. Uncertainty Analysis

Calculation and measurement errors resulting from human and instrumental errors are presented. Repeating measurements several times can help eliminate or reduce individual errors. As demonstrated in Table 2, the instrumental errors are due to the accuracy of the measuring devices. Based on the uncertainty of the primary measurements, the uncertainty of the computation results was calculated.

Table 2. Measurement device errors.

Measured Value	Measured Value Error
Temperature	+0.1 °C
Diameter	+0.5 mm
Voltage	+0.1 Volt
Resistance	+0.1 Ohm

Error propagation is estimated by using the root sum square method. Given the specific result Z, as follows:

$$Z = z\,(x1,\ x2, \ldots,\ xn) \tag{4}$$

where x_i is independently measured quantity, the uncertainty ω_i is the uncertainty of independent variables, and the uncertainty ω_z in the result, Z is assumed as follows:

$$\omega_z = \left[\left(\frac{\partial Z}{\partial x_1} .\omega_1 \right)^2 + \left(\frac{\partial Z}{\partial x_2} .\omega_2 \right)^2 + \ldots + \left(\frac{\partial Z}{\partial x_{n-1}} .\omega_{n-1} \right)^2 + \left(\frac{\partial Z}{\partial x_n} .\omega_n \right)^2 \right]^{0.5} \tag{5}$$

5. Numerical Modeling

The Boussinesq approximation, which incorporates thermal buoyancy, represents the liquid PCM density difference in the buoyancy term. As a result, the 2D transient laminar flow containing buoyancy-driven conventional governing equations can be expressed by the following equations [11]:

Continuity equation:

$$\frac{\partial \rho}{\partial t} + \frac{1}{r}\frac{\partial (r\rho V_r)}{\partial r} + \frac{1}{r}\frac{\partial (\rho V_\theta)}{\partial \theta} = 0 \tag{6}$$

Momentum equations:

Momentum r:

$$\begin{aligned} &\rho\left(\tfrac{\partial (V_r)}{\partial t} + V_r \tfrac{\partial (V_r)}{\partial r} + \tfrac{V_\theta}{r}\tfrac{\partial (V_r)}{\partial \theta} - \tfrac{V_\theta^2}{r} \right) \\ &= -\tfrac{\partial P}{\partial r} + \mu\left[\tfrac{1}{r}\tfrac{\partial}{\partial r}\left(r\tfrac{\partial (V_r)}{\partial r} \right) - \tfrac{V_r}{r^2} + \tfrac{1}{r^2}\tfrac{\partial^2 (V_r)}{\partial \theta^2} - \tfrac{2}{r^2}\tfrac{\partial (V_\theta)}{\partial \theta} \right] + \rho g \beta (\mathrm{T} - \mathrm{T_m}) + S \end{aligned} \tag{7}$$

Momentum θ:

$$\begin{aligned} &\rho\left(\tfrac{\partial (V_\theta)}{\partial t} + V_r \tfrac{\partial (V_\theta)}{\partial r} + \tfrac{V_\theta}{r}\tfrac{\partial (V_\theta)}{\partial \theta} + \tfrac{V_r V_\theta}{r} \right) \\ &= -\tfrac{1}{r}\tfrac{\partial P}{\partial \theta} + \mu\left[\tfrac{1}{r}\tfrac{\partial}{\partial r}\left(r\tfrac{\partial (V_\theta)}{\partial r} \right) - \tfrac{V_\theta}{r^2} + \tfrac{1}{r^2}\tfrac{\partial^2 (V_\theta)}{\partial \theta^2} - \tfrac{2}{r^2}\tfrac{\partial (V_r)}{\partial \theta} \right] + S \end{aligned} \tag{8}$$

The parameters s are Darcy's law damping terms added to the momentum equation as a convective phase change effect. These terms depend on both the existing constant A_{mush} and a liquid fraction (λ), which can be expressed as follows:

$$\vec{S} = \frac{(1-\lambda)^2}{(\lambda^3 + \gamma)} A_{\text{mush}}.\vec{V}$$

The minor number γ is used to avoid division by zero, and it is naturally about 10^{-3}. The mushy zone constant A_{mush} explains how steeply its velocity is condensed to zero when the material solidifies. Its value depends on the morphology of the medium. This value is frequently high and typically runs from 10^4 to 10^8. The same occurred when the local liquid fraction becomes high, and the velocity is reduced to zero. Numerous academics advise taking into account the A_{mush} value of 10^6 as the best value. This value offers a good agreement between their experimental and numerical results [30]. In this study, A_{mush} varies throughout the whole simulation, where it attains the best conformity between the experimental and numerical results.

Energy equation:

$$\frac{\partial h}{\partial t}+\frac{\partial H}{\partial t}+\nabla\cdot\left(\vec{V}h\right)=\nabla\cdot\left(\frac{k}{\rho c_p}\nabla h\right) \tag{9}$$

where H is the enthalpy of the PCM and is estimated as the summation of the sensible enthalpy, h_1, and the latent heat ΔH:

$$H = h_1 + \Delta H \tag{10}$$

$$h_1 = h_{ref} + \int_{T_{ref}}^{T} Cp\, dT \tag{11}$$

where h_{ref} is the enthalpy reference at temperature reference T_{ref} and latent heat term can be expressed in the form of the latent heat of the PCMs (L) as follows:

$$\Delta H = \lambda\, L \tag{12}$$

where ΔH can vary from zero (solid) to L (liquid). Therefore, the liquid fraction λ can be articulated as follows:

$$\lambda = \begin{cases} \frac{\Delta H}{L} = 0\ T < T_m \\ \frac{\Delta H}{L} = \frac{T - T_{solidus}}{T_{liqudus} - T_{solidus}}\ T_m < T < T_m + \Delta\, T_m \\ \frac{\Delta H}{L} = 1\ T > T_m + \Delta\, T_m \end{cases} \tag{13}$$

where T_m is the PCM melting temperature, and ΔT_m is the range between liquid and solid phase, which is expressed as the difference between liquidus temperature and solidus temperature, as demonstrated in Equation (14):

$$\Delta\, T_m = T_{liquidus} - T_{solidus} \tag{14}$$

The PCM density and thermal conductivity depend on the phase change process as follows:

$$\rho_{PCM} = \begin{cases} \rho_s\ T < T_m \\ \frac{\rho_s + \rho_l}{2}\ T_m < T < T_m + \Delta\, T_m \\ \rho_l\ T > T_m + \Delta\, T_m \end{cases} \tag{15}$$

$$k_{PCM} = \begin{cases} k_s\ T < T_m \\ \frac{k_s + k_l}{2}\ T_m < T < T_m + \Delta\, T_m \\ k_l\ T > T_m + \Delta\, T_m \end{cases} \tag{16}$$

5.1. Initial and Boundary Conditions

The previous governing equations were solved together with the following initial and boundary conditions. Initially, the system was at the ambient temperature of 27 °C, and the storage capsules were filled with solid paraffin wax, which was at a temperature lower than the melting point. The outer side of the tank was insulated; in addition, the inlet of the hot water was defined as the velocity inlet and the outlet of the hot water was defined as the pressure outlet, as shown in Figure 5.

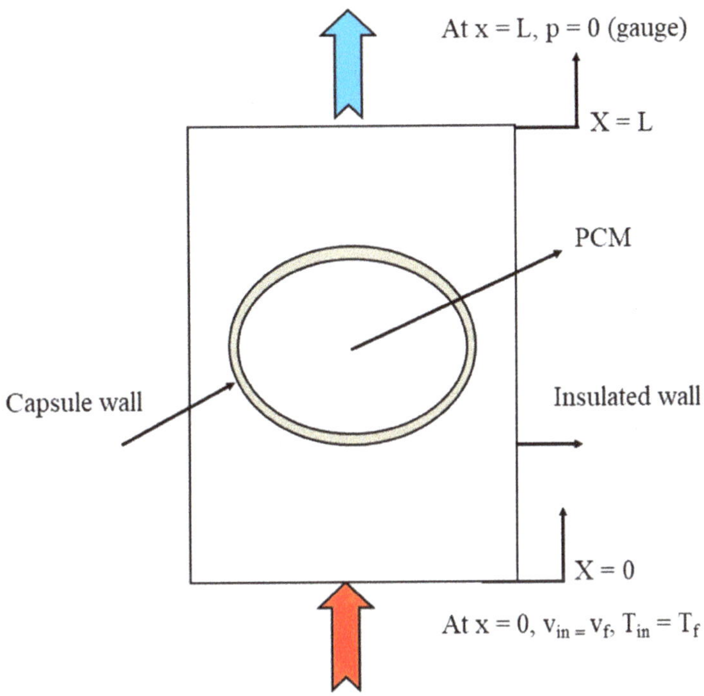

Figure 5. Test section with boundary conditions.

5.2. Numerical Procedure and Validation

The numerical simulation was established using the ANSYS 2022 R2 program. The Semi-Implicit Method for the Pressure-Linked Equations (SIMPLE) scheme was used to calculate the pressure–velocity coupling equations in the PCM. A control volume approach was used to explain the governing equations in conjunction with the boundary and initial conditions. The diffusion and convection terms are described using the central difference and QUICK techniques, respectively. The pressure–velocity coupling was handled using the SIMPLER algorithm. By solving the governing equations at each time step, the liquid fraction could be changed using Equation (13). After carefully evaluating the independence of the results for high accuracy, the size of the grid and the time steps were selected. The calculation process continued until a relative convergence criterion of 10^{-6} was satisfied by all the present case variables.

To validate the numerical model of melting in the present finite volume computational fluid dynamic (CFD) code, an initial run was established and compared to the experimental results of Soliman et al. [10,11] for a vertical cylinder that was heated by a constant heat flux, and Kamkari and Amlashi [31] for a vertical enclosure that was heated by a constant wall temperature. The computational model and the experimental study were established under the same operating conditions. Figure 6 displays the PCM temperature and liquid fractions, with the melting time for the experimental results of [11,31], respectively, with the present numerical results. As can be noted from the figure, the numerical results are in good agreement with the experimental results.

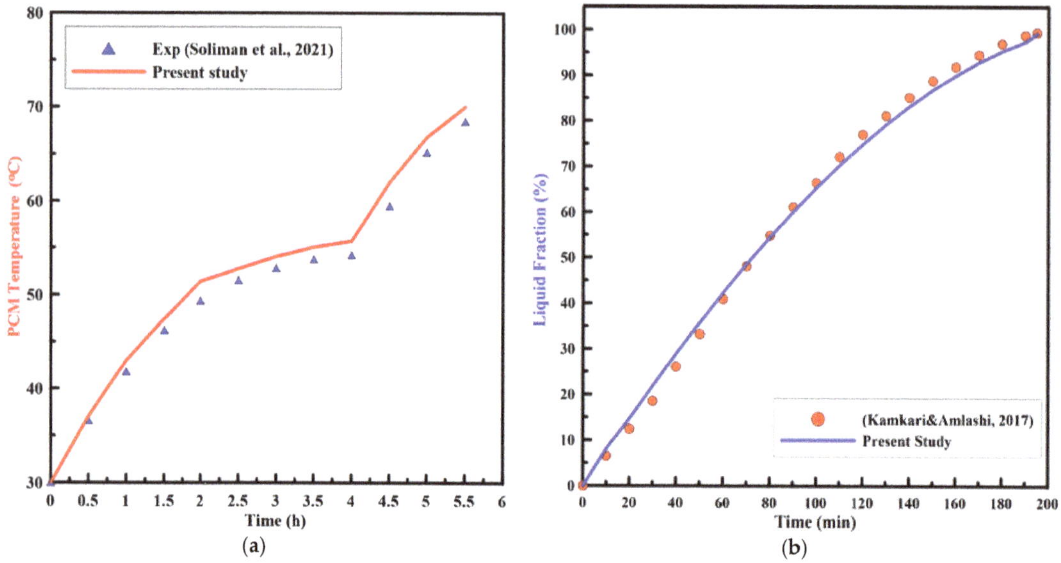

Figure 6. Experimental results of PCM (**a**) temperature [11], and (**b**) liquid fraction [31] vs. time with the present numerical study.

5.3. Mesh Dependency Study

Free triangular mesh has been adapted in the numerical model. In order to test the dependency of numerical results on the mesh element size, a simulation was run with a circular capsule of 4 cm in diameter and 6 cm in length. The capsule was initially at 300 K. At any time, t > 0, the boundary of the capsule was at 340 K. The average temperature and liquid fraction of the PCM for the circular capsule were compared for different numbers of elements size (20,062, 32,634, 45,409, 68,312, and 87,472) as shown in Figure 7a,b. It is observed from the figure that there is a slight difference between the number of elements 68,312 and 87,472. Therefore, the number of elements equal to 68,312 is selected for the present study to save time. Similarly, a grid independency test has also been carried out for other configurations. The time steps used in the analysis is 0.3 sec throughout all models.

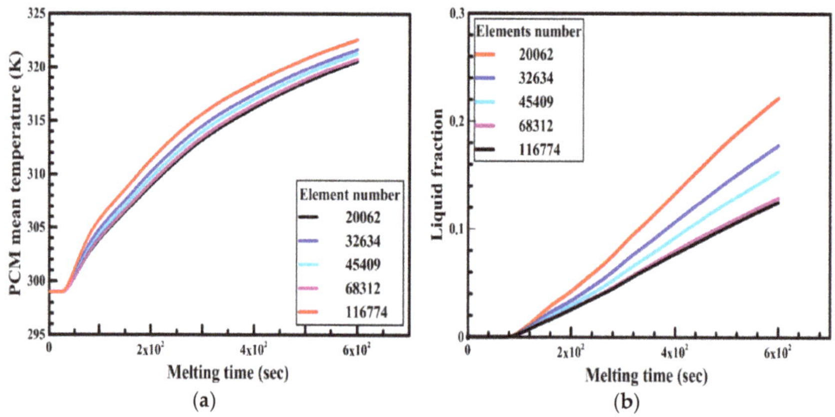

Figure 7. Effect of elements size number on the PCM (**a**) temperature and (**b**) liquid fraction.

6. Results and Discussion

In this section, firstly, a comparison between the experimental results of different shapes of storage capsules (circular, horizontal oval, and vertical oval) is established with the theoretical results of the same shape and dimensions from the numerical model. Tables 3–5 illustrate the variation of the liquid fraction contours of the PCM melting process with the photos captured during the experiments.

Table 3. Comparison between liquid fraction contours and camera photos for circular cross-section capsule.

Theoretical	Exp.	Theoretical	Exp.	Theoretical	Exp.

Time = 0 min | Time = 10 min | Time = 20 min

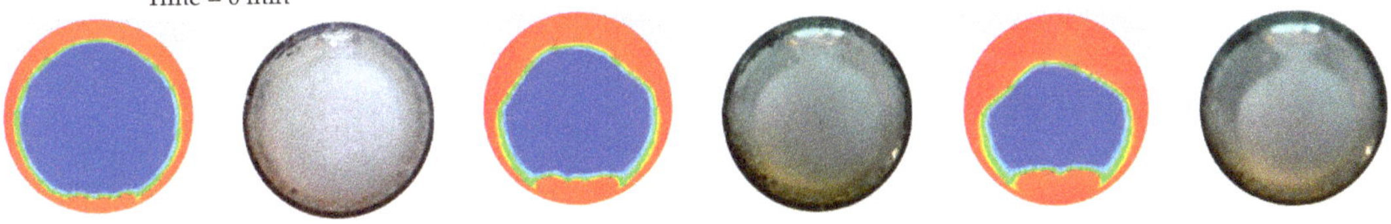

Time = 30 min | Time = 40 min | Time = 50 min

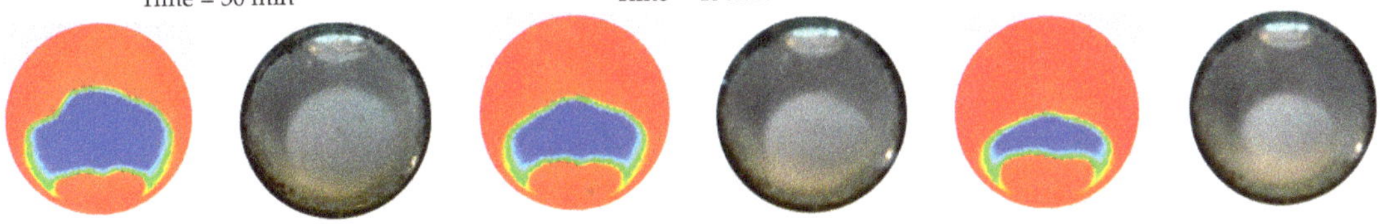

Time = 60 min | Time = 70 min | Time = 80 min

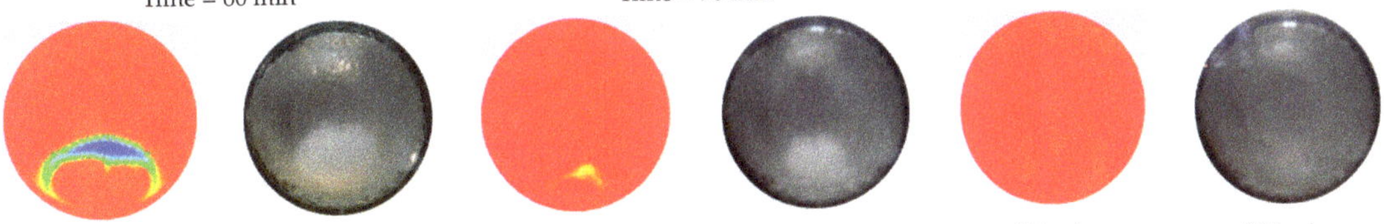

Time = 90 min | Time = 100 min | 106 min | 104 min

Table 4. Comparison between liquid fraction contours and camera photos for circular cross-section capsule of Ar = 0.383.

Theoretical	Exp.	Theoretical	Exp.	Theoretical	Exp.
				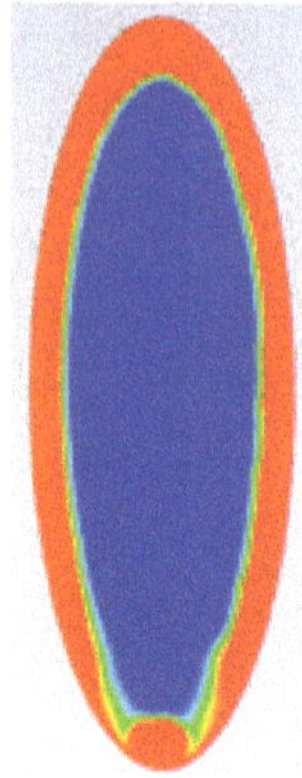	
Time = 0 min		Time = 8 min		Time = 18 min	
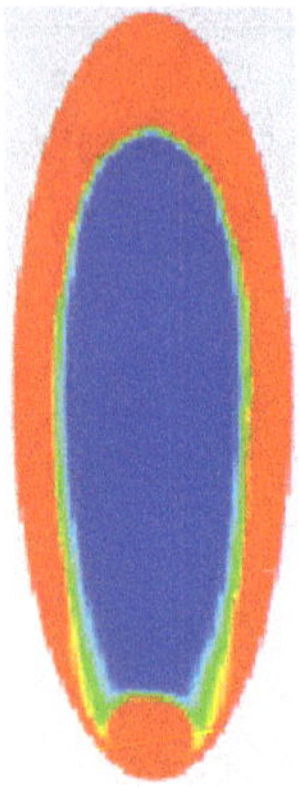		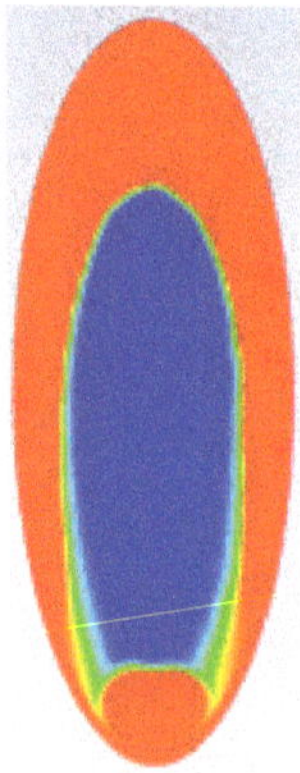		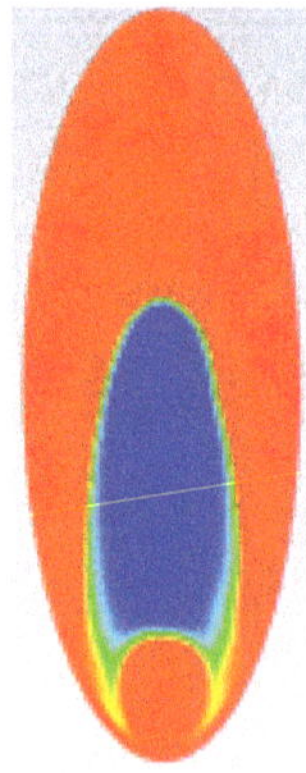	
Time = 28 min		Time = 38 min		Time = 48 min	

Time = 58 min		Time = 63 min		67 min	66 min

Table 5. Comparison between liquid fraction contours and camera photos for circular cross-section capsule of Ar = 2.6.

Theoretical	Exp.	Theoretical	Exp.
		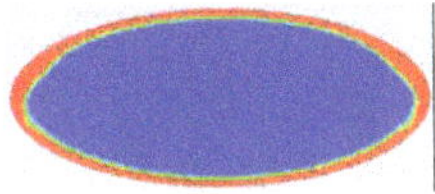	
Time = 0 min		Time= 7 min	
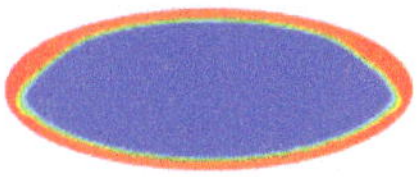		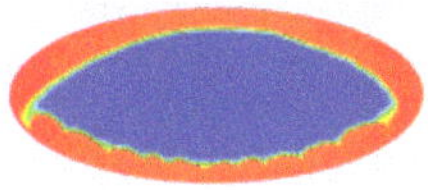	
Time = 15 min		Time = 27 min	
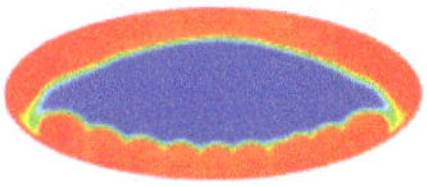		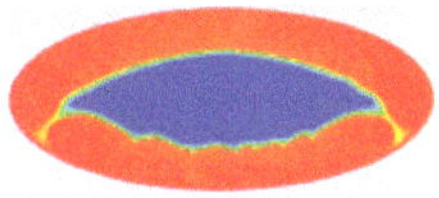	
Time = 37 min		Time = 47 min	
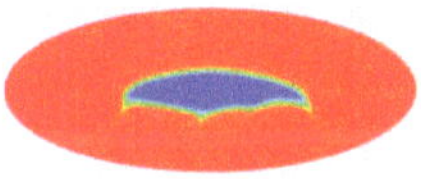		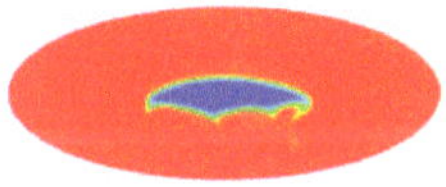	
Time = 57 min		Time = 59 min	

66 min	64 min		

Secondly, the effect of the mushy zone parameter on the PCM melting process is studied and validated with the experimental results to obtain the optimum value of A_{mush} for each shape. Figures 8–10 illustrate the PCM liquid fraction with the melting time of the experimental result and the numerical results with different values of the mushy zone parameter.

Table 3 represents the comparison between the liquid fraction contours of a circular capsule with the experimental photos. The circular cross-section capsule is 4 cm in diameter (axes ratio of 1) and 6 cm in length for heat transfer fluid (HTF) initial temperature of 340 K and a velocity of 0.003 m/s. Within the experimental photos, the white part is a solid PCM and the transparent part is a liquid PCM. Within the numerical contours, the blue color signifies the solid PCM and the red color denotes the liquid PCM. The solid PCM sinks to the bottom of the capsule due to the force of gravity for both the numerical and experimental results. The experimental total melting time is less than that of the theoretical one by 2 min.

Table 4 represents the comparison between the liquid fraction contours of the elliptical cross-section capsule with the experimental photos. The elliptical cross-section capsule is in a vertical position with an axes ratio of 0.383 for the HTF initial temperature of 340 K and a velocity of 0.003 m/s. The experimental total melting time is less than that of the theoretical one by 1 min.

Table 5 represents the comparison between the liquid fraction contours of the elliptical cross-section capsule with the experimental photos. The elliptical cross-section capsule is in a horizontal position with an axes ratio of 2.6 for the HTF initial temperature of 340 K and a velocity of 0.003 m/s. The experimental total melting time is less than that of the theoretical one by 2 min.

It is demonstrated in the figures that there is a fair agreement between the experimental and theoretical results and the shape of molten and solid paraffin. The fair agreement between experimental and theoretical results ensures the validity of the theoretical model and the accuracy of the different thermal properties of the paraffin wax.

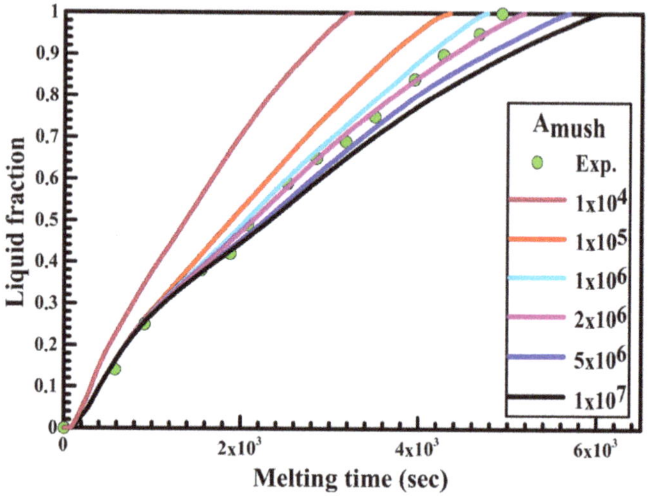

Figure 8. Experimental and numerical liquid fraction vs. time for a circular capsule.

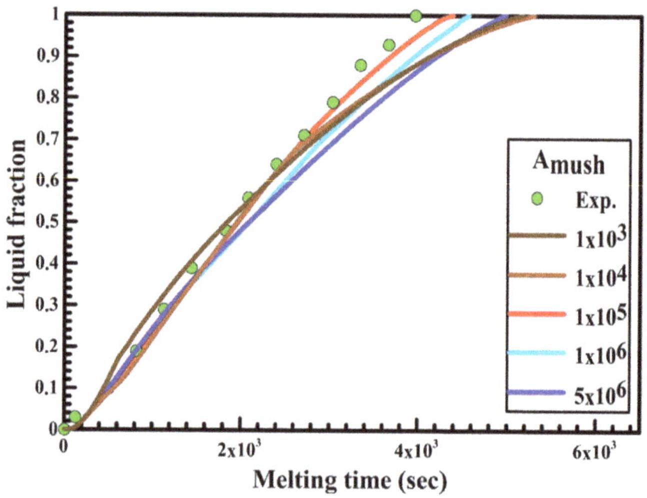

Figure 9. Experimental and numerical liquid fraction vs. time for a vertical oval capsule.

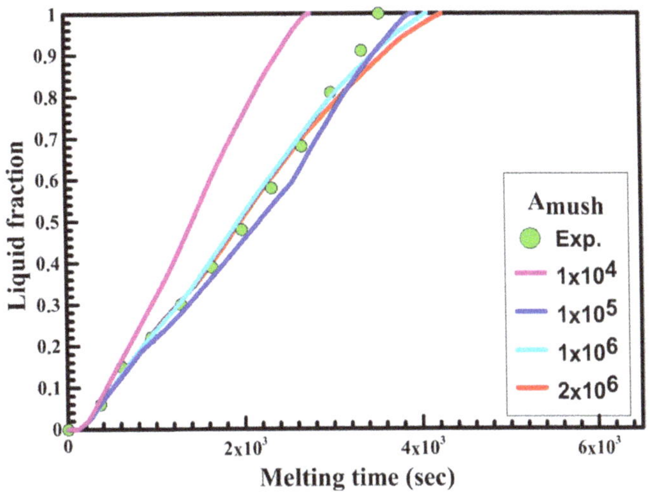

Figure 10. Experimental and numerical liquid fraction vs. time for a horizontal oval capsule.

The experimental investigation was made to validate the results obtained by the Ansys Fluent software. Khot et al. [28] observed that using the thermocouples inside the PCM throughout the melting process restricted the solid PCM from moving, which affected the melting process. Therefore, there is no thermocouple embedded inside the circular and oval capsules to measure the temperature distribution inside it. The experimental results of the liquid fraction are calculated using the image processing method, which enables the determination of the percentage of the solid "white color" and the liquid "gray color".

Figure 8 shows the relation between liquid fraction and time for both experimental and numerical results for a circular capsule at the HTF inlet temperature of 340 K and a velocity of 0.003 m/s. The numerical results are obtained with different values of A_{mush} (1×10^4, $1 \times 10^5, 2 \times 10^6, 5 \times 10^6, 1 \times 10^7$). It is observed from the figure that the melting rate of the numerical results is slow at $A_{mush} = 1 \times 10^7$, while at $A_{mush} = 1 \times 10^4$ the melting rate of the numerical results increases, and its value is greater than the experimental results.

The previous phenomena are due to the great effect of A_{mush} on the convection effect. Increasing the A_{mush} value leads to a reduction in the convection effect and the melting rate. For the circular capsule, the experimental results have a good agreement with the numerical results at the value of A_{mush} equal to 2×10^6.

Figure 9 displays the relation between liquid fraction and time for both experimental and numerical results for a vertical oval capsule at the HTF inlet temperature of 340 K and a velocity of 0.003 m/s. The numerical results are obtained with different numbers of A_{mush} ($1 \times 10^4, 1 \times 10^5, 1 \times 10^6, 5 \times 10^6$). It can be observed from the figure that the melting rate of the numerical results is slow at $A_{mush} = 5 \times 10^6$, while at $A_{mush} = 1 \times 10^6$, and 5×10^6 the melting rate of the numerical results is close to the experimental results. For the vertical oval capsule, the experimental results have a good agreement with the numerical results at the value of A_{mush} equal to 1×105.

Figure 10 illustrates the relation between liquid fraction and time for both experimental and numerical results for a horizontal oval capsule at HTF inlet temperature of 340 K and a velocity of 0.003 m/s. The numerical results are obtained with different numbers of A_{mush} ($1 \times 10^4, 1 \times 10^5, 1 \times 10^6, 2 \times 10^6$).

It can be observed from Figure 10 that the melting rate of the numerical results increases and its value is greater than the experimental results at $A_{mush} = 1 \times 10^4$, while at $A_{mush} = 1 \times 10^6$ and 2×10^6, the melting rate of the numerical results is close to the experimental results. For the horizontal oval capsule, the experimental results have a good agreement with the numerical results at the value of A_{mush} equal to 1×10^6.

From the present study, it can be observed that the mushy zone parameter (A_{mush}) has a great impact on the numerical model results of the PCM melting process and changes with the shape of the storage unit. The present study provides a range of the mushy zone parameter for the numerical modeling of the melting process in the circular, horizontal oval, and vertical oval capsules in future work to obtain more accurate results. Table 6 represents the optimum number of the mushy zone parameter for circular, vertical oval, and horizontal oval capsules.

Table 6. Optimum number of the mushy zone parameter for circular, vertical oval, and horizontal oval capsules.

Cases	Shape	Dimensions	A_{mush} Range	Outcomes
Circular capsule		Ar = 1	1×10^4 1×10^5 2×10^6 5×10^6 1×10^7	Good agreement with the experimental results at A_{mush} equal to 2×10^6.
Vertical oval capsule		Ar = 0.383	1×10^4 1×10^5 1×10^6 5×10^6	Good agreement with the numerical results at A_{mush} equal to 1×10^5.
Horizontal oval capsule		Ar = 2.6	1×10^4 1×10^5 1×10^6 2×10^6	Good agreement with the numerical results at A_{mush} equal to 1×10^6.

7. Conclusions

The melting process of a phase change material (PCM) was experimentally and numerically investigated in storage capsules of different shapes (circular, vertical oval, and horizontal oval). The numerical model was established to study the effect of the A_{mush} (mushy zone parameter) on the thermal performance of the PCM melting process. The model was validated with previous related work, and the results exhibited good agreement. The results of the numerical model were validated with the experimental results to obtain the optimum value of A_{mush} for each shape of the latent heat storage unit. According to the present experimental/numerical results of the PCM melting process inside different capsules, the following conclusions can be drawn:

1. The value of A_{mush} has a great impact on the numerical model results of the PCM melting process and changes with the shape of the storage unit;
2. The rate of heat transfer, convection, and fluid velocity all decrease as the value of A_{mush} rises;
3. The experimental results of the circular, vertical oval, and horizontal oval capsules match well with the numerical model at the value of A_{mush} equal to 2×10^6, 1×10^5, and 1×10^6, respectively.

The effect of nanoparticles on the mushy zone parameter should be taken into account in future research due to their great effect upon the enhancement of the pelting process.

Author Contributions: Credit author statement A.S.S.: idea, methodology, writing the original draft, writing the initial draft, and preparing literature survey, validation. A.A.S.: computer programming, writing the initial draft, writing, reviewing, editing, and supervision. M.A.S.: computer programming, writing the initial draft, writing, reviewing, and editing. All authors have read and agreed to the published version of the manuscript.

Funding: The publication fees of this article have been supported by Mansoura University.

Institutional Review Board Statement: Not applicable.

Acknowledgments: The author would like to gratefully acknowledge the support of Mansoura University, Egypt, on the development of this research by providing access to their facility.

Conflicts of Interest: The authors declare no conflict of interest.

References

1. Hosseini, M.J.; Ranjbar, A.A.; Sedighi, K.; Rahimi, M. A Combined Experimental and Computational Study on the Melting Behavior of a Medium Temperature Phase Change Storage Material inside Shell and Tube Heat Exchanger. *Int. Commun. Heat Mass Transf.* **2012**, *39*, 1416–1424. [CrossRef]
2. Guerraiche, D.; Bougriou, C.; Guerraiche, K.; Valenzuela, L.; Driss, Z. Experimental and Numerical Study of a Solar Collector Using Phase Change Material as Heat Storage. *J. Energy Storage* **2020**, *27*, 101133. [CrossRef]
3. Haghighi, A.; Babapoor, A.; Azizi, M.; Javanshir, Z.; Ghasemzade, H. Optimization of the Thermal Performance of PCM Nanocomposites. *J. Energy Manag. Technol.* **2019**, *4*, 14–19.
4. Soliman, A.S.; Zhu, S.; Xu, L.; Dong, J.; Cheng, P. Efficient Waste Heat Recovery System for Diesel Engines Using Nano-Enhanced Phase Change Materials. *Case Stud. Therm. Eng.* **2021**, *28*, 101309. [CrossRef]
5. Soliman, A.S.; Radwan, A.; Xu, L.; Dong, J.; Cheng, P. Energy Harvesting in Diesel Engines to Avoid Cold Start-up Using Phase Change Materials. *Case Stud. Therm. Eng.* **2022**, *31*, 101807. [CrossRef]
6. Soliman, A.S.; Xu, L.; Dong, J.; Cheng, P. A Novel Heat Sink for Cooling Photovoltaic Systems Using Convex/Concave Dimples and Multiple PCMs. *Appl. Therm. Eng.* **2022**, *215*, 119001. [CrossRef]
7. Soliman, A.S.; Zhu, S.; Xu, L.; Dong, J.; Cheng, P. Design of an H2O-LiBr Absorption System Using PCMs and Powered by Automotive Exhaust Gas. *Appl. Therm. Eng.* **2021**, *191*, 116881. [CrossRef]
8. Dhaidan, N.S.; Khodadadi, J.M. Melting and Convection of Phase Change Materials in Different Shape Containers: A Review. *Renew. Sustain. Energy Rev.* **2015**, *43*, 449–477. [CrossRef]
9. Fadl, M.; Eames, P.C. Numerical Investigation of the Influence of Mushy Zone Parameter Amush on Heat Transfer Characteristics in Vertically and Horizontally Oriented Thermal Energy Storage Systems. *Appl. Therm. Eng.* **2019**, *151*, 90–99. [CrossRef]
10. Soliman, A.S.; Zhu, S.; Xu, L.; Dong, J.; Cheng, P. Melting Enhancement of Nano-Phase Change Material in Cylindrical Enclosure Using Convex/Concave Dimples: Numerical Simulation with Experimental Validation. *J. Energy Storage* **2021**, *44*, 103470. [CrossRef]
11. Soliman, A.S.; Zhu, S.; Xu, L.; Dong, J.; Cheng, P. Numerical Simulation and Experimental Verification of Constrained Melting of Phase Change Material in Cylindrical Enclosure Subjected to a Constant Heat Flux. *J. Energy Storage* **2021**, *35*, 102312. [CrossRef]
12. Voller, V.R.; Cross, M.; Markatos, N.C. An Enthalpy Method for Convection/Diffusion Phase Change. *Int. J. Numer. Methods Eng.* **1987**, *24*, 271–284. [CrossRef]
13. Voller, V.R.; Prakash, C. A Fixed Grid Numerical Modeling Methodology for Convection-Diffusion Mushy Region Phase-Change Problems. *Int. J. Heat Mass Transf.* **1987**, *30*, 1709–1719. [CrossRef]
14. Ehms, J.H.N.; de Césaro Oliveski, R.; Rocha, L.A.O.; Biserni, C.; Garai, M. Fixed Grid Numerical Models for Solidification and Melting of Phase Change Materials (PCMs). *Appl. Sci.* **2019**, *9*, 4334.
15. Singh, V.K.; Patel, A. Effect of Mushy Zone Constant on the Melting of a Solid-Liquid PCM under Hyper-Gravity Conditions. *Int. Commun. Heat Mass Transf.* **2022**, *134*, 105993. [CrossRef]
16. Brent, A.D.; Voller, V.R.; Reid, K.J. Enthalpy-Porosity Technique for Modeling Convection-Diffusion Phase Change: Application to the Melting of a Pure Metal. *Numer. Heat Transf.* **1988**, *5720*, 297–318.
17. Kumar, M.; Krishna, D.J. Influence of Mushy Zone Constant on Thermohydraulics of a PCM. *Energy Procedia* **2017**, *109*, 314–321. [CrossRef]
18. Shmueli, H.; Ziskind, G.; Letan, R. Melting in a Vertical Cylindrical Tube: Numerical Investigation and Comparison with Experiments. *Int. J. Heat Mass Transf.* **2010**, *53*, 4082–4091. [CrossRef]
19. Arena, S.; Casti, E.; Gasia, J.; Cabeza, L.F.; Cau, G. Numerical Simulation of a Finned-Tube LHTES System: Influence of the Mushy Zone Constant on the Phase Change Behaviour. *Energy Procedia* **2017**, *126*, 517–524. [CrossRef]
20. Hosseinizadeh, S.F.; Rabienataj Darzi, A.A.; Tan, F.L.; Khodadadi, J.M. Unconstrained Melting inside a Sphere. *Int. J. Therm. Sci.* **2013**, *63*, 55–64. [CrossRef]
21. Ansys. *"Ansys Fluent 12.0 User's Guide." Fluent Documentation*; Ansys: Canonsburg, PA, USA, 2009.

22. Abdulateef, A.M.; Abdulateef, J.; Mat, S.; Sopian, K.; Elhub, B.; Mussa, M.A. Experimental and Numerical Study of Solidifying Phase-Change Material in a Triplex-Tube Heat Exchanger with Longitudinal/Triangular Fins. *Int. Commun. Heat Mass Transf.* **2018**, *90*, 73–84. [CrossRef]
23. Nithyanandam, K.; Pitchumani, R. Computational Modeling of Dynamic Response of a Latent Thermal Energy Storage System with Embedded Heat Pipes. *J. Sol. Energy Eng. Trans. ASME* **2014**, *136*, 011010. [CrossRef]
24. Yang, X.H.; Tan, S.C.; Liu, J. Numerical Investigation of the Phase Change Process of Low Melting Point Metal. *Int. J. Heat Mass Transf.* **2016**, *100*, 899–907. [CrossRef]
25. Kousksou, T.; Mahdaoui, M.; Ahmed, A.; Msaad, A.A. Melting over a Wavy Surface in a Rectangular Cavity Heated from Below. *Energy* **2014**, *64*, 212–219. [CrossRef]
26. Tabassum, T.; Hasan, M.; Begum, L. 2-D Numerical Investigation of Melting of an Impure PCM in the Arbitrary-Shaped Annuli. *Int. J. Therm. Sci.* **2017**, *114*, 296–319. [CrossRef]
27. Libeer, W.; Ramos, F.; Newton, C.; Alipanahrostami, M.; Depcik, C.; Li, X. Two-Phase Heat and Mass Transfer of Phase Change Materials in Thermal Management Systems. *Int. J. Heat Mass Transf.* **2016**, *100*, 215–223. [CrossRef]
28. Khot, S.A.; Sane, N.K.; Gawali, B.S. Experimental Investigation of Phase Change Phenomena of Paraffin Wax inside a Capsule. *Int. J. Eng. Trends Technol.* **2011**, *2*, 67–71.
29. Villarino, M.B. Ramanujan's Perimeter of an Ellipse. *arXiv* **2005**, arXiv:0506384.
30. Emam, M.; Ahmed, M. Cooling Concentrator Photovoltaic Systems Using Various Configurations of Phase-Change Material Heat Sinks. *Energy Convers. Manag.* **2018**, *158*, 298–314. [CrossRef]
31. Kamkari, B.; Amlashi, H.J. Numerical Simulation and Experimental Verification of Constrained Melting of Phase Change Material in Inclined Rectangular Enclosures. *Int. Commun. Heat Mass Transf.* **2017**, *88*, 211–219. [CrossRef]

sustainability

MDPI

Article

Simulation Assessment of Inlet Parameters and Membrane-Surface-Structure Effects on CO_2 Absorption Flux in Membrane Contactors

Amin Mojarad Garehbagh [1], Saeid Rajabzadeh [2,3,*], Mahmoud A. Shouman [4], Mohamed R. Elmarghany [4], Mohamed S. Salem [4], Nasrul Arahman [5], Toraj Mohammadi [1] and Hideto Matsuyama [3]

[1] Centre of Excellence for Membrane Science and Technology, Department of Chemical, Petroleum and Gas Engineering, Iran University of Science and Technology (IUST), Narmak, Tehran 16846-13114, Iran
[2] School of Civil and Environmental Engineering, University of Technology Sydney (UTS), City Campus, Broadway, NSW 2007, Australia
[3] Research Center for Membrane and Film Technology, Department of Chemical Science and Engineering, Kobe University, Rokkodaicho 1-1, Nada, Kobe 657-8501, Japan
[4] Mechanical Power Engineering Department, Faculty of Engineering, Mansoura University, Mansoura 35516, Egypt
[5] Department of Chemical Engineering, Universitas Syiah Kuala, Banda Aceh 23111, Indonesia
* Correspondence: rajabzadehk@people.kobe-u.ac.jp

Abstract: The management of global carbon dioxide (CO_2) emissions is considered one of the main environmental problems facing the modern world. One of the potential techniques for CO_2 capture is absorption, using membrane contactor modules. Most of the previous research that dealt with membrane contactor simulations considered the whole membrane surface as the active reaction surface. However, in this paper, a more realistic model of the membrane-contactor module is presented, taking into account the effects of the pore size and surface porosity. CO_2 absorption into the monoethanolamine (MEA) solution in hollow fiber membrane-contactor modules was numerically investigated. A computational fluid dynamics simulation was established using essential basic fluid dynamics and mass transfer equations in reactive mode. An algorithmic function was used to present the relations between the CO_2 absorption flux and the hollow fiber length, membrane surface pore size, and porosity. The simulation results were compared to previously obtained experimental results without using any fitting parameters, and a good agreement was found with an average error of 8.5%. The validated simulation was then used to predict the effects of the MEA inlet velocity and concentration, the membrane surface pore size, and porosity on the total CO_2 absorption flux. A maximum absorption flux of about 1.8 mol/m^2·s was achieved at an MEA concentration of 4 M with a pore size of 0.2 microns, a surface porosity of 1%, and an inlet velocity of 0.25 m/s. The extrapolation technique was then used to predict the values of the absorption flux at longer fiber lengths. The concentration profiles around the pores at the gas–liquid contact surface of the membrane were obtained and presented. The proposed model exhibited excellent potential to evaluate the effective reaction surface in hollow fiber membrane contactors. This model could be considered the first step to obtaining accurate predictions of the membrane contactor gas absorption performance based on its surface structure.

Keywords: CO_2 absorption; MEA; membrane contactor; CFD; porosity; pore size

Citation: Garehbagh, A.M.; Rajabzadeh, S.; Shouman, M.A.; Elmarghany, M.R.; Salem, M.S.; Arahman, N.; Mohammadi, T.; Matsuyama, H. Simulation Assessment of Inlet Parameters and Membrane-Surface-Structure Effects on CO_2 Absorption Flux in Membrane Contactors. *Sustainability* **2022**, *14*, 14527. https://doi.org/10.3390/su142114527

Academic Editors: Mohamed El-Alfy, Ahmed El Kenawy, Petra-Manuela Schuwerack and Zhongfeng Xu

Received: 26 September 2022
Accepted: 31 October 2022
Published: 4 November 2022

1. Introduction

In recent decades, the development of industrial activities has increased greenhouse gas concentration in the atmosphere, including carbon dioxide (CO_2). Increased CO_2 emissions are significant contributors to catastrophic environmental changes. Carbon dioxide gas contributes to around 55% of the observed global warming [1]. This vital issue sparked interest in CO_2 capture [2]. Hollow fiber membrane contactor (HFMC) technology is a potential candidate for greenhouse gases (especially CO_2) abatement. It has considerable advantages over conventional CO_2 recovery techniques [1,3,4]. In HFMC modules, the membrane lumen, bulk, and module shell enable direct contact between gas and liquid phases without the dispersion of one phase into the other. While acid gases flow on one side of the membrane, filling the membrane bulk structure to reach the other side, absorbent liquid flows on the other surface of the membrane. Tiny liquid–gas interfaces are formed in the openings of the pores adjacent to the absorbent liquid. It is a crucial point in the gas–liquid contacting process to avoid entering absorbent liquid into the membrane pores. This phenomenon is known as wetting in HFMC, and it sharply and strongly decreases the module performance [5].

Yan et al. [6] compared the performance of several chemical and physical absorbents and concluded that chemical absorbents have better performance because of their higher absorption rate due to their fast reaction with CO_2. Several kinds of chemicals were used to absorb CO_2 in the liquid phase, including organic and inorganic bases [7–12]. Among the reported chemical absorbents, different types were used. The amine solutions exhibited fast reaction rates, resulting in a higher CO_2 absorption flux, minor damage to the membrane structure, and less toxicity and safety issues [7,11]. Monoethanolamine (MEA) solution was widely used in HFMC processes with considerable reported advantages. According to previous investigations, MEA exhibited higher CO_2 removal performance, compared to other absorbents, such as sodium hydroxide (NaOH), triethanolamine (TEA), diethanolamine (DEA), and 2-amino-2-methyl-1-propanol (AMP) [8,10].

CO_2 capture techniques were previously investigated, taking into account various parameters, such as solvent absorption, pressure, and temperature-swing adsorption in the presence of a variety of solid sorbents, membranes, and cryogenic distillation [13]. Using MEA as a liquid absorbent showed high potential as a technique for carbon capture. On the other hand, with the development in the production of ceramic and metallic membranes, effective membranes could be fabricated with significantly effective separation, compared to the liquid absorption technique. A study was held to investigate the feasibility of using membrane-based carbon capture technology compared to the amine-based absorption technique [14]. The study compared the use of three plants: two of them use amine liquid absorption, and the third is a membrane-based plant. The latter showed high potential compared to the other two. A CO_2 avoided cost, which is an economic indicator to estimate the additional cost of processing the carbon capture facility, of 46 USD/tons was established using a membrane-based plant compared to 58 and 71 USD/tons exhibited by the other two plants. The merits of membrane-based plants may include compactness, as membranes may provide 1000 times more interfacial area per unit volume compared to absorption–adsorption processes [2]. Other advantages are modularity, ease of installation, flexibility, lower cost with lesser energy consumption, and fewer chemical requirements [15,16].

Kreulen et al. [17,18] established that a very thin layer covers the membrane contactor surface, and a gas–liquid reaction occurs at this thin film layer at the mouth of the pores. The thin film gas-absorbed layer theory is dominated by physical absorption, even for very low porosity membranes. Even in the cases of membranes with very low surface porosity (<3%) and pore size of 5 nm, when water was used as the absorbent liquid (physical absorption), the membrane surface pore size and porosity had little to no effect on the gas absorption flux. Therefore, this assumption was used in most gas–liquid membrane contacting simulations. Although the membrane surface pore size and porosity effects were very well evaluated for physical gas absorption by Kreulen et al. [18], no such evaluation was performed for chemical absorption processes. While the effects of membrane surface pore size and porosity on the performance of the membranes used in liquid filtration are very well studied in many studies [19,20], this effect is still not clear for gas absorption processes, and more evaluation is still needed. In our previous studies [21,22], we considered that all membrane surfaces were active for the reaction based on the aforementioned studies [17,18]. According to our previous studies, at high MEA concentrations (higher than 2 mol/L and especially for 4 mol/L) and low porosity membranes, the assumption of a thin film of a saturated MEA layer at the reaction interface was found to be lacking. Simulation results considerably deviated from the experimental results. Comite et al. [23] observed a very similar trend when they used a membrane with a high porosity of 60%. Although they considered the effect of amine reaction products on the CO_2 absorption flux, the effects of the membrane surface porosity and pore size were also neglected.

In this study, we systematically considered the membrane surface porosity and pore size effects on the CO_2 gas absorption flux. In previously reported papers, all the membrane surfaces were considered active areas for reaction. In the current study, we tried to fulfill the gas–liquid membrane contacting simulation without considering the assumption of covering thin film over the pores. Our simulation results, without using any fitting parameters, show that at high MEA concentration, membrane surface pore size and porosity play key roles in determining the absorption flux. Simulation results were compared with experimental results obtained in our previous study [21,22]. The agreement between the simulation and experimental results was satisfactory.

2. Computational Fluid Dynamics (CFD) Simulation in Reactive Mode

In this study, a 3D model was established using the commercial software package COMSOL Multiphysics 5.4 to assess the effects of the membrane surface pore size and porosity on the CO_2 absorption into MEA solution. Since many experimental and simulation studies [23–25] were performed on CO_2 absorption into the MEA solution, this system was selected for more reliable access to the established physical properties and previous experimental results published, including the authors' [22]. This led to more dependable simulation results with less deviation from the experimental data. Generally, an HFMC configuration usually comprises several hollow fibers, each performing independently. Therefore, investigating a single fiber performance can be sufficient to predict the performance of the whole HFMC module [23,24,26,27]. The proposed model is based on the assumption of non-wetted mode, where the CO_2 gas completely fills the membrane pores. In this model, the CO_2 was fed into the shell side of the HFMC, while the MEA solution went through the lumen. Due to the non-wetted assumption, the contact between the two fluids only occurred over the membrane pores surface within the lumen side, as shown in Figures 1 and 2. In the simulation, the CO_2 was transported from the shell to the lumen side through tiny cylinders representing the membrane pores. The effects of the membrane surface pore size and porosity on the CO_2 absorption flux at the interface of the MEA and CO_2 were numerically evaluated.

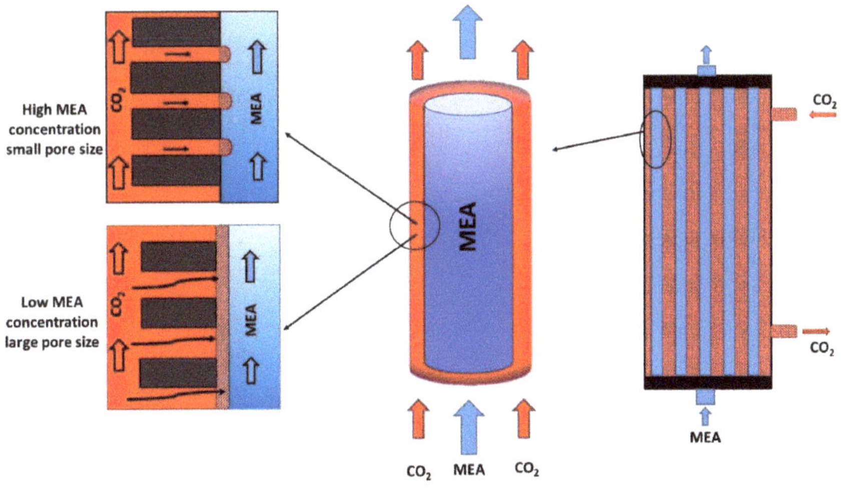

Figure 1. A schematic diagram of a membrane contactor module.

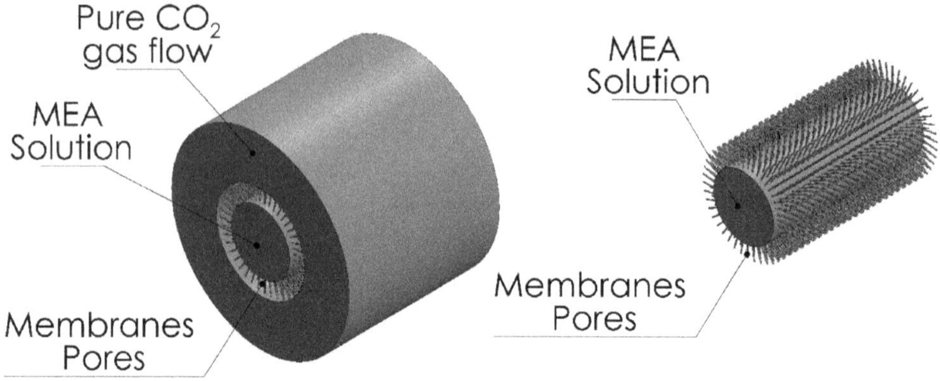

Figure 2. A schematic diagram of the feed flow in a hollow fiber membrane; The cylinders represent the membrane pores in the model extended in the membrane thickness.

2.1. *Simulation Assumptions*

The main simulation assumptions are summarized as follows:

- Feed gas on the shell side is pure CO_2 to simplify the mass transfer resistance on the gas side. Otherwise, gas phase resistance becomes important, and the whole membrane with tortuosity and pore size should be considered, which is practically impossible to run, even with a supercomputer.
- Laminar and a steady-state hydrodynamically fully developed MEA solution is considered for feed flow. Although this assumption might not be ideal for the entrance length, it is the most suitable assumption. Otherwise, it would take too much processing time and power without much change in results.
- Isothermal feed flow all over the module.
- A very smooth wall (no-slip condition at the wall).
- Gravity effects are neglected in this simulation as evident in previous studies [28–30].
- Henry's law was applied to calculate the CO_2 concentration at the interface of the gas–liquid contact at the mouth of the pores (i.e., thermodynamic equilibrium state).
- The non-wetted mode was assumed, indicating that the pores are filled with CO_2 with no absorbent entry into the pores.
- The reaction between CO_2 and the MEA is very fast (instantaneous), as the reaction parameters satisfied the instantaneous reaction regime conditions, as can be seen in [22].
- Axis symmetry condition is assumed at the hollow fiber axis.

2.2. *Governing Equations*

Based on the previous assumptions, the steady-state material balance equations can be expressed as follows [18]:

$$D_{CO_2}\left(\frac{\partial^2 C_{CO_2}}{\partial r^2}+\frac{1}{r}\frac{\partial C_{CO_2}}{\partial r}\right)-u_z\frac{\partial C_{CO_2}}{\partial z}-k\,C_{CO_2}C_{MEA}=0 \tag{1}$$

$$D_{MEA}\left(\frac{\partial^2 C_{MEA}}{\partial r^2}+\frac{1}{r}\frac{\partial C_{MEA}}{\partial r}\right)-u_z\frac{\partial C_{MEA}}{\partial z}-2\,k\,C_{CO_2}C_{MEA}=0 \tag{2}$$

where D presents the species diffusivity (m^2/s), C is the species concentration ($mole/m^3$), and k is the reaction rate coefficient [$m^3/mol\cdot s$]. The CO_2 concentration (mol/m^3) inside the absorbent at the pores surfaces mouths is calculated by Equation (3):

$$C_{CO_2}=P_{CO_2}\cdot H_{CO_2} \tag{3}$$

where P represents the CO_2 pressure (atm), and H is Henry's constant (mol/m^3·atm).

The velocity distribution for the MEA, $u(r)$, can be estimated as a flow in a circular pipe using Equation (4) [31].

$$u(r) = u_{max}\left(1 - \left(\frac{r}{R}\right)^2\right) \quad (4)$$

where r is the radius (m), R is the fiber inner radius (m), and u_{max} (m/s) is the maximum liquid velocity in the membrane lumen, which can be obtained by dividing the liquid flow rate by the area of the lumen of the HFMC.

The reaction between the gas and the MEA, which is assumed to occur at the lumen side and be instantaneous, can be expressed using Equation (5). The reaction rate constant of the reaction reported by Hikita et al. [32] was used in the simulation (5.92 m^3/mol. s at 25 °C):

$$CO_2 + 2RNH_2 \leftrightarrow RNHCOO^- + RNH^+{}_3 \quad (5)$$

$$rate = \frac{-dC_{CO_2}}{dt} = -\left(\frac{1}{2}\right)\left(\frac{-dC_{MEA}}{dt}\right) = k\, C_{CO_2} C_{MEA} \quad (6)$$

Table 1 summarizes the different physical properties that are used in this simulation, considering the effect of amine products on the physical properties of the MEA solution [22].

Table 1. Physical properties at different MEA concentrations.

MEA Concentration (M)	1	2	3	4
D_{CO_2} (m^2/s)	1.66×10^{-9}	1.397×10^{-9}	1.17×10^{-9}	9.49×10^{-10}
D_{MEA} (m^2/s)	9.45×10^{-10}	7.96×10^{-10}	6.68×10^{-10}	5.41×10^{-10}
H_{CO_2} (mol/m^3·kPa)	0.217	0.135	0.1385	0.142
k (m^3/mol·s)			5.92	

The CO_2 absorption flux (J) (mol/m^2. s) can be estimated using Equation (7) [25], assuming instantaneous reaction, thus no CO_2 in the MEA outflow.

$$J_{CO_2} = \frac{1}{2\pi RL}\left(\frac{1}{2}\left[\int_0^R 2\pi r u(r)\left(C_{MEA-inlet} - \left(C_{MEA-interface}(r)\right)_{z=L}\right)dr + 2\int_0^R 2\pi r u(r)\left(C_{co_2-interface}(r)\right)_{z=L} dr\right]\right) \quad (7)$$

For the simulation, the following boundary conditions were used:

- Zero CO_2 concentration is assumed at the hollow fiber inlet with the MEA concentration equal to the feed concentration.
- CO_2 concentration is calculated using Equation (3) with no MEA flux at the pore mouths.
- Outflow condition is also considered at the fiber outlet.
- The rest of the membrane surfaces are considered walls with no-slip conditions.

2.3. Simulation Time Reduction

Firstly, to reduce the simulation time while maintaining an acceptable level of accuracy, several cuts of hollow fiber were created with different angles, as shown in Figure 3. The chosen angles were 90°, 45°, 27°, 9°, and 2.5°. Additionally, several fiber lengths ranging from 100 microns up to 7200 microns were investigated. As described in Table 2, HFMC absorption flux with a very short length was calculated to evaluate the effect of cutting angle. Later on, calculations for longer lengths of the HFMC were performed, and the results were extrapolated for the full length of the fiber. The extrapolated results were then compared to the experimental results as will be shown in the following section of the manuscript. The main used input parameters are summarized in Table 3. The absorption flux was estimated for each case and is shown in Table 3.

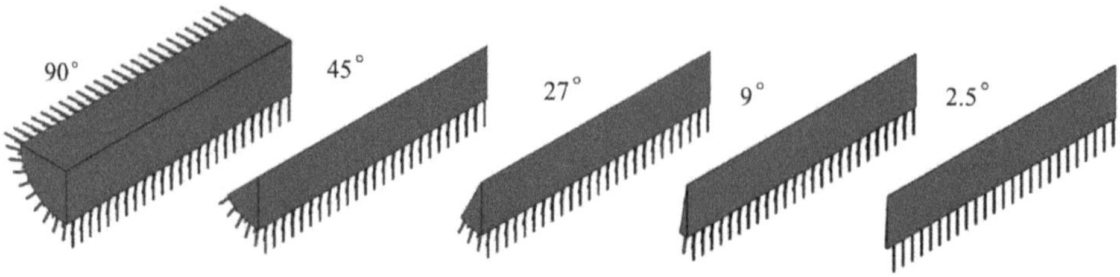

Figure 3. Different angle cuts of the hollow fibers.

Table 2. Absorption flux simulation values at different angle cuts and lengths.

Cutting Angle (°) / Fiber Length (μm)	90	45	27	9	2.5
	Absorption Flux (mol/m^2·s)				
100	0.932	0.931	0.931	0.930	0.930
300	0.392	0.392	0.390	0.390	0.389
600	-	0.260	0.259	0.258	0.258
900	-	-	0.24	0.24	0.24
1800	-	-	-	0.199	0.198
3600	-	-	-	0.174	0.174
7200	-	-	-	-	0.161

Table 3. Main input parameters for the mesh dependency test.

MEA Concentration (mol/L)	Absorbent Velocity (m/s)	Pore Size (μm)	Surface Porosity
2	0.25	0.1	1%

As shown in Table 3, the hollow fiber cutting angle seems to have a negligible effect on the CO_2 absorption flux simulated values at different lengths, with a maximum error of about 0.8%. Therefore, a 2.5° angle was used to expedite the simulation process. The results were extended over the whole fiber to decrease the simulation time. As it is clear from Table 3, the CO_2 absorption flux is extraordinarily high at the entrance of the membrane, and it sharply decreases by increasing the membrane length. This is because, at the entrance, no layer was yet formed across the membrane surface to decrease the diffusion rate, hence decreasing the flux.

2.4. Mesh Dependency Test

A mesh dependency test was performed to determine the proper mesh size for the required simulation. Four fibers with a length of 1 mm and 2.5° cut and different mesh sizes from 430,327 elements to 2,958,891 elements were investigated. The absorption flux and MEA concentration profile at the fiber outlet were estimated for each case. The mesh dependency test results are shown in Figure 4. Considering the absorption flux (Figure 4a), the error between the mesh with 1,427,212 elements and the finer mesh with 2,958,891 elements is about 4.5%. Therefore, this mesh size (1,427,212 elements) was chosen for completing the results, thus allowing for faster calculations while maintaining reasonable accuracy. The difference in the MEA profile for different mesh sizes was completely negligible for all mesh sizes as shown in Figure 4b. Although not shown here, the MEA concentration profile in this study is very similar to our previous MEA concentration profile [22].

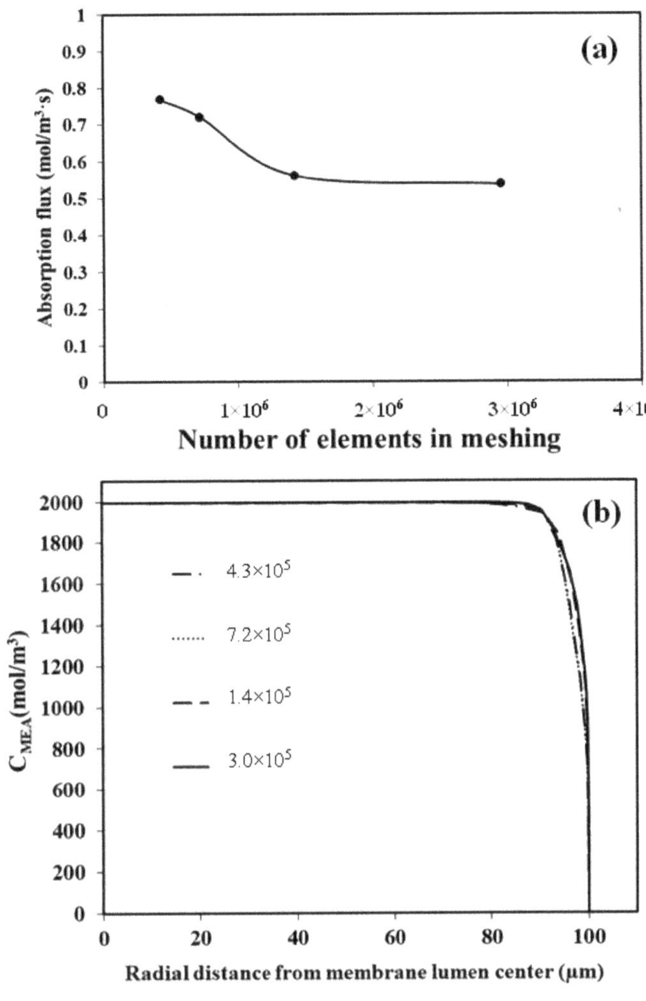

Figure 4. Effect of mesh size on (**a**) absorption flux and (**b**) MEA concentration profile.

3. Results and Discussion

3.1. Simulation Validation

Instead of actually simulating the absorption flux in the full length of the hollow fiber, which would have needed too many calculations, memory, computing power, and time, another approach was chosen to estimate the CO_2 absorption flux. The absorption flux of several hollow fibers with different short lengths was calculated. Then, the calculated CO_2 absorption flux was used to calculate the coefficients of a mathematical function that could be used to present the relation between the actual hollow fiber length in the experimental condition and the calculated absorption flux. This function was then used to extrapolate the absorption flux of the actual length, and the estimated results were then compared to experimental results in our previous studies. The simulation results of the absorption flux at different fiber lengths for different MEA concentrations are shown in Figure 5. By increasing the fiber length, the CO_2 absorption flux was noticed to sharply decrease at first, and then its change became almost flat. We think the initial extraordinarily high CO_2 absorption flux is due to the non-fully developed MEA concentration boundary layer at the membrane entrance. By increasing the length, the flow regime became fully developed, and the concentration boundary layer formed, thus decreasing the CO_2 contact with a fresh, non-saturated MEA solution. Subsequently, the CO_2 absorption flux sharply decreased. Based on the results in Figure 5, for more than 1000 μm of the hollow fiber length, the trend of the CO_2 absorption flux becomes almost linear in the logarithmic figure. The trendline for the relationship between the fiber length and the absorption fluxes was noticed to be similar to an algorithmic relation (y = A + B * ln(x)). Thus, the curve-fitting technique was used to estimate the function coefficients A and B values at different MEA concentrations. These values were calculated and are tabulated in Table 4.

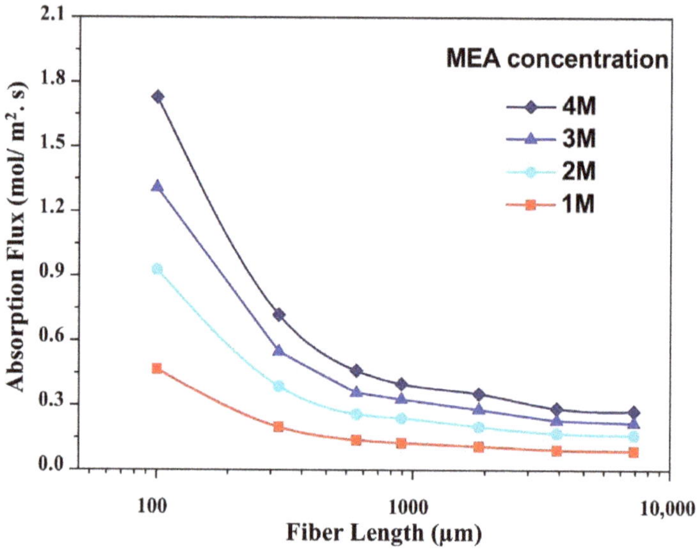

Figure 5. The estimated absorption flux at different fiber lengths for four different MEA concentrations; The main parameters for this calculation are membrane surface porosity of 1%, a pore diameter of 0.2 μm, and an MEA average velocity of 0.25 m/s.

Table 4. Algorithmic function coefficients for different MEA concentrations.

MEA Concentration (mol/L)	1	2	3	4
A	0.2634	0.5194	0.7055	0.8807
B	−0.04652	−0.09283	−0.1269	−0.1592

Kumar et al. [26] found that when a physical absorption was applied for HFMC, by increasing the absorbent inlet velocity from 0.1 to 0.5 m/s, the effect of the membrane length on the local CO_2 absorption flux decreased sharply. At an inlet velocity of 0.5 m/s, no considerable change was noticed in the local CO_2 absorption flux over the length of the HFMC. It was reported by Shirazian et al. [12] that when a fast reaction occurs between CO_2 and the chemical absorbent, the CO_2 removal efficiency does not change with the membrane length very much. By decreasing the reaction rate, the membrane length effect becomes considerable. Since MEA was used for CO_2 absorption in our study, and the reaction rate for CO_2 in MEA is instantaneous, extrapolating the CO_2 absorption flux over the hollow fiber membrane length looks reasonable. Although not shown here, as it is still under current investigation, the calculation time and memory were considerably decreased when a different software was used to make the HFMC simulation, and the results were then exported to Comsol. Then, we were able to extend our calculation to nearly 15 cm length of the fiber (15 times longer than the current calculation). The new simulation and calculation results showed similar results to the current calculation and extrapolation that was performed here with less than 8% error.

Using the estimated relation, the absorption flux was estimated under the same conditions as the previous experimental results, and the simulation results were compared to the actual experimental results of Rajabzadeh et al. [22]. The comparison results are shown in Figure 6. The simulation results show an acceptable agreement with the experimental results in all ranges of the studied MEA concentration, with a maximum error of less than 20%, which validates the proposed simulation. This agreement becomes clearer at high MEA concentration (4 mol/L). Comparing our current calculation with that of the calculation we performed in our previous work with all surfaces as active layers available for reaction shows that the current calculation considering the pore size and porosity shows much better agreement with experimental results. Thus, it can be concluded that the simulation and the curve fitting techniques used could estimate the CO_2 absorption flux of the membrane contactor with an acceptable degree of accuracy. It is worth mentioning that the effect of the amine reaction products is already included in our previous simulation. Therefore, we firmly believe that changing the CO_2 absorption flux, especially at a high MEA concentration (4 mol/L), is related to the effect of membrane surface porosity.

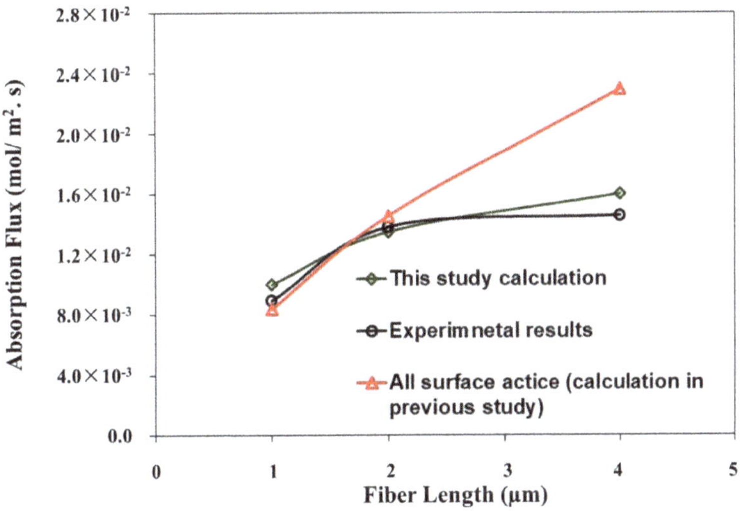

Figure 6. Comparing the simulation results in the current study with that of the experimental results and simulation without considering the effect of the membrane surface porosity [22]; fiber length of 27 cm, a pore diameter of 0.2 µm, and an MEA average velocity of 0.25 m/s.

Figure 7 shows the CO_2 concentration distribution contours over the membrane pores at two different MEA concentrations of 1 and 4 mol/L. It is noteworthy that by increasing the absorbent concentration, the CO_2 concentration around the pores decreases. It approaches the saturation concentration just at the mouth of the pores, and the chance for CO_2 to diffuse around the pores sharply decreases. This is because the reaction rate of the MEA with CO_2 considerably increases. On the other hand, at low MEA concentrations, the CO_2 finds the chance to diffuse around the pores. Therefore, there is a chance to make a saturated CO_2-absorbed layer over the pores which covers the membrane surface, thus making the membrane structure effect negligible.

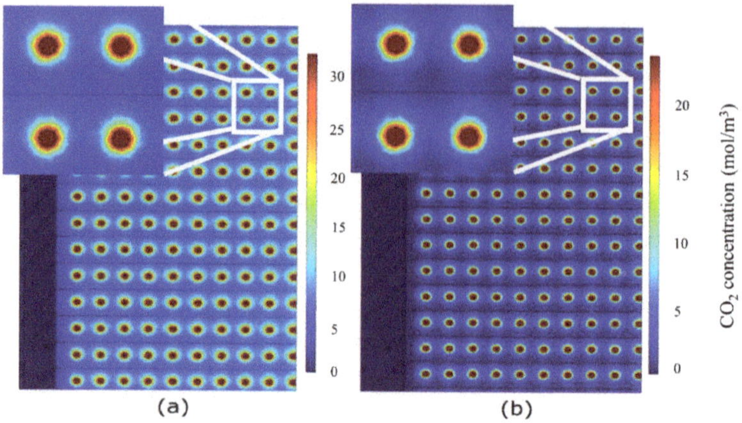

Figure 7. CO_2 concentration distribution over the membrane surface (absorbent concentration: (**a**) 1 mol/L and (**b**) 4 mol/L with pores diameter of 0.2 μm. circles in the figure show a hollow fiber pore mouth.

3.2. Effect of Average Absorbent Velocity

In order to further verify our developed simulation, we tried to compare our simulation results with that of the experimental results [22] when the average absorbent velocity is changed. The results are shown in Figure 8. As can be clearly seen, the simulation results are in good agreement with the experimental results. Hence, this confirms the ability of the simulation and extrapolating techniques to estimate the CO_2 absorption flux of the membrane contactor with an acceptable degree of accuracy.

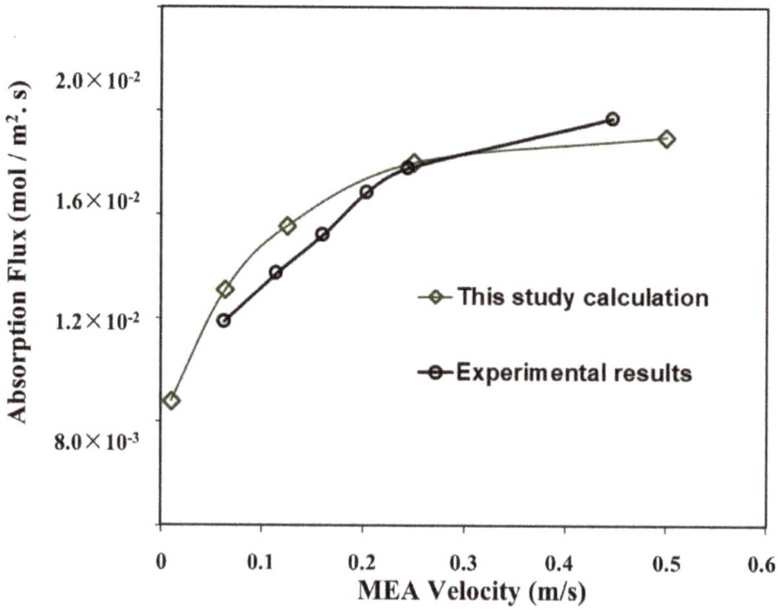

Figure 8. Simulation results versus experimental results at different absorbent average velocities; Main parameters for simulations: Fiber length of 27 cm, a membrane surface porosity of 1%, a pore diameter of 0.2 μm, and an absorbent concentration of 2 mol/L.

Figure 9 presents the CO_2 concentration distribution contours over the membrane pores at two different inlet absorbent velocities. As can be seen, by increasing the absorbent flow rate, the CO_2 concentration around the pores decreases. The CO_2 concentration approaches the saturation concentration just at the mouth of the pores, and the chance for CO_2 to diffuse around the pores decreases sharply. By increasing the absorbent velocity, the thickness of the concentration boundary layer of MEA at the membrane surface decreases, and the CO_2 reaction rate increases since the chance of contacting CO_2 with fresh MEA increases. Contrary to the high absorbent average velocity, CO_2 finds the chance to diffuse slightly more around the pores at a low absorbent velocity. This condition increases the chance of making a saturated CO_2-absorbed layer on the pores covering the membrane surface, decreasing the effect of the membrane structure. However, this effect is not as clear as changing the absorbent concentration.

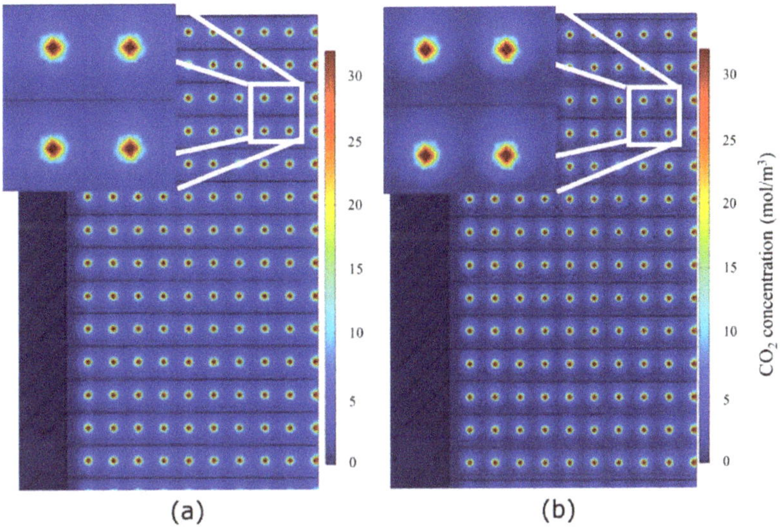

Figure 9. CO_2 concentration distribution over the membrane surface; absorbent average velocity: (**a**) 0.062 m/s and (**b**) 0.5 m/s, pore diameter of 0.1 μm, and an absorbent concentration of 2 mol/L.

3.3. *The Effect of Changing the Porosity and Pore Size*

The model was then used to study the effect of the membrane porosity and pore size on the CO_2 absorption flux at four different MEA solution concentrations. The pore size increased from 0.05 μm to 0.2 μm while keeping the number of pores the same, thus, increasing the surface porosity with the increase in pore size. The simulation results are shown in Figure 10. Increasing the surface porosity by increasing the membrane surface pore size was found to significantly enhance the CO_2 absorption flux, as shown in Figure 10. Increasing CO_2 absorption flux by increasing the surface porosity is entirely plausible since it provides a higher chance of contact between CO_2 and MEA and results in higher CO_2 absorption flux. When the chemical reaction is used for HFMC gas absorption at low surface porosity at the interface, it is the absorption-controlling step.

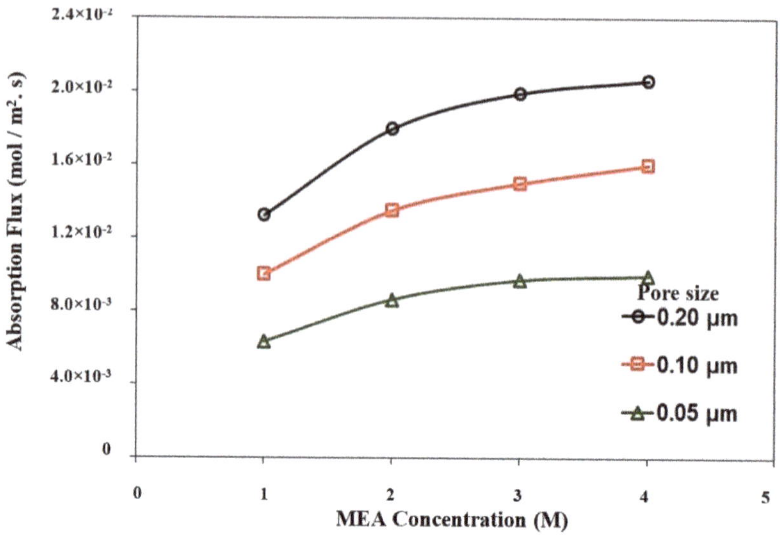

Figure 10. The effect of the membrane surface porosity on the CO_2 absorption flux. The number of pores was kept constant, and the pore size varied from 0.05 µm to 0.2 µm.

The effect of the membrane pore size on the CO_2 absorption flux at a fixed surface porosity of 1% was evaluated by changing the pore size diameter with changing the number of pores to keep the porosity fixed at 1%. Figure 11 shows the effect of the membrane surface pore size on the CO_2 absorption flux at fixed surface porosity for two different pore sizes of 0.1 µm and 0.2 µm. Additionally, Figure 12 presents the flux contours around the pores for two different pore sizes, namely 0.1 and 0.2 microns. It should be mentioned here that the number of pores in the simulation increases exponentially by decreasing the pore diameter. In this study, when the pore size is decreased to 0.05 µm, the calculation time and needed memory are increased exponentially. Thus, we could not perform the calculation for 0.05 µm pore size, as it was far beyond our workstation capabilities. Surprisingly, we can see that membranes with a smaller pore size showed a higher gas absorption flux. This trend sounds interesting because it is strongly in line with our raised hypothesis that when the pore size is large, the possibility of forming the thin film layer of MEA with dissolved CO_2 near the pore mouth increases. When this layer forms, the CO_2 cannot contact fresh MEA, and the reaction rate and CO_2 absorption flux decrease. On the other hand, when the pore size is small, the chance for CO_2 to diffuse around the pore mouth and make a thin layer decreases as soon as CO_2 exits from the pore and contacts the fresh MEA. Therefore, the reaction rate and CO_2 absorption flux increase. This phenomenon is schematically shown in Figure 1. It can be concluded that using a membrane with a small pore size at a fixed surface porosity is much more beneficial for the CO_2 absorption flux. However, fabrication costs would probably increase. A future economic study is required to estimate the optimal pore size which can achieve high performance at a reasonable cost.

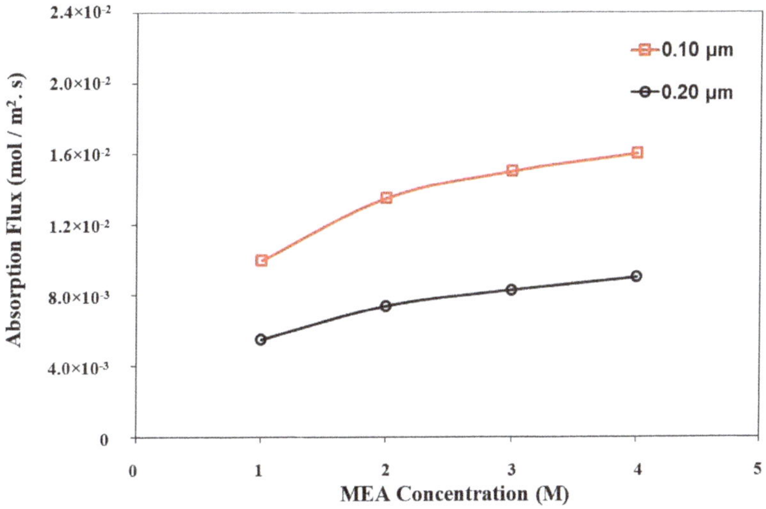

Figure 11. The effect of the membrane pore size on the CO_2 absorption flux. The surface porosity was fixed at 1%, and the pore size varied from 0.1 µm to 0.2 µm.

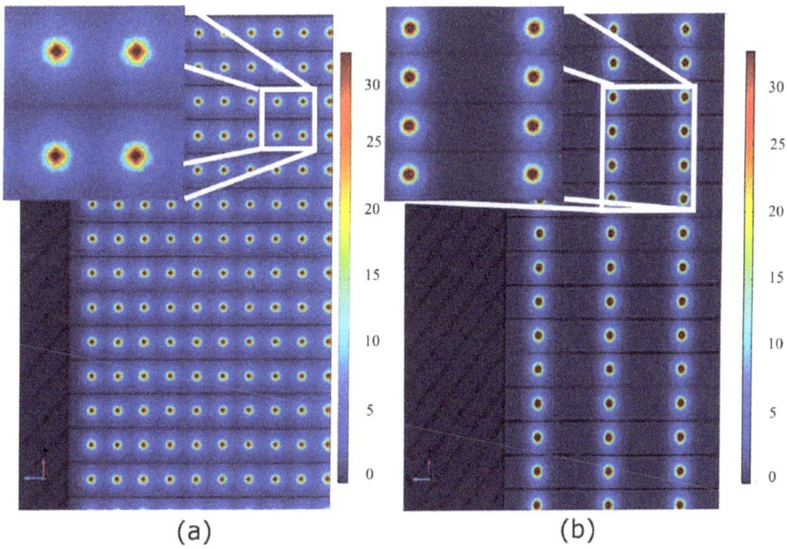

Figure 12. CO_2 concentration distribution over the membrane surface: pore diameter: (**a**) 0.1 µm and (**b**) 0.2 µm, an absorbent average velocity of 0.062 m/s, and an absorbent concentration of 2 mol/L.

4. Conclusions and Future Work

A numerical simulation was established to simulate the absorption of CO_2 into MEA within hollow fiber membrane contactor modules. The simulation was used to establish a relationship between the absorption flux and the fiber length, which was used to estimate the CO_2 absorption flux at long hollow fibers. The simulation results were validated using experimental data from our previously published paper. The simulation results were in good agreement with the experimental results with a maximum error of less than 20%. The validated simulation was then used to investigate the effects of the absorbent inlet velocity and concentration and membrane surface pore size and porosity at the liquid–gas interface on the CO_2 absorption flux and the CO_2 concentration profiles. Using high absorbent velocity, MEA concentration, porosity, and small pore size at a fixed-surface porosity was beneficial for obtaining high CO_2 absorption flux. The CO_2 concentration contours showed that by using high absorbent velocity, MEA concentration, low porosity, and small pore size at a fixed surface porosity, the CO_2 concentration is limited just to the mouth of the pores of the membrane. In this situation, the possibility of CO_2 diffusion around the mouth of the pore decreases, and the CO_2 absorption flux increases. Although it needs more detailed investigations, the established simulation showed the potential to be efficiently utilized to study the absorption phenomenon within the membrane contactors.

Future investigations regarding the model may include a full parametric investigation to study the limits of the proposed model and the effects of changing the pore size and porosity of the membrane on the model results compared to the experimental results. Using the same model for other absorption pairs is also an important study. Applying the fitting over long hollow fiber lengths is also in the works. The effect of impurities in the CO_2 stream should also be investigated, as it may block the membrane pores which may decrease the absorption flux and affect the module performance. Additionally, the effects of long-term operation on the module performance are a very vital parameter for practical applications.

Author Contributions: Conceptualization, S.R. and T.M.; methodology, M.A.S. and M.S.S.; software, A.M.G.; validation, M.R.E. and N.A.; formal analysis, S.R.; writing—original draft preparation, A.M.G.; writing—review and editing, S.R. and M.S.S.; supervision, T.M. and H.M. All authors have read and agreed to the published version of the manuscript.

Funding: This work was partially supported by Kobe University Strategic International Collaborative Research Grant (Type B Fostering Joint Research). The experimental part of this research was partially funded by Iran National Science Foundation (INSF) (Grant No. 96008182).

Institutional Review Board Statement: Not applicable.

Informed Consent Statement: Not applicable.

Data Availability Statement: The data supporting reported results are available upon request.

Acknowledgments: The authors would like to thank Mansoura University for their support in finishing and publishing this manuscript.

Conflicts of Interest: The authors declare no conflict of interest.

References

1. Gabelman, A.; Hwang, S.T. Hollow Fiber Membrane Contactors. *J. Membr. Sci.* **1999**, *159*, 61–106. [CrossRef]
2. Sreenivasulu, B.; Gayatri, D.V.; Sreedhar, I.; Raghavan, K.V. A Journey into the Process and Engineering Aspects of Carbon Capture Technologies. *Renew. Sustain. Energy Rev.* **2015**, *41*, 1324–1350. [CrossRef]
3. Li, J.L.; Chen, B.H. Review of CO_2absorption Using Chemical Solvents in Hollow Fiber Membrane Contactors. *Sep. Purif. Technol.* **2005**, *41*, 109–122. [CrossRef]
4. Mansourizadeh, A.; Ismail, A.F. Hollow Fiber Gas-Liquid Membrane Contactors for Acid Gas Capture: A Review. *J. Hazard. Mater.* **2009**, *171*, 38–53. [CrossRef] [PubMed]
5. Mosadegh-Sedghi, S.; Rodrigue, D.; Brisson, J.; Iliuta, M.C. Wetting Phenomenon in Membrane Contactors—Causes and Prevention. *J. Membr. Sci.* **2014**, *452*, 332–353. [CrossRef]
6. Yan, Y.; Zhang, Z.; Zhang, L.; Chen, Y.; Tang, Q. Dynamic Modeling of Biogas Upgrading in Hollow Fiber Membrane Contactors. *Energy Fuels* **2014**, *28*, 5745–5755. [CrossRef]
7. Wang, R.; Li, D.F.; Liang, D.T. Modeling of CO_2 Capture by Three Typical Amine Solutions in Hollow Fiber Membrane Contactors. *Chem. Eng. Process. Process Intensif.* **2004**, *43*, 849–856. [CrossRef]
8. Nakhjiri, A.T.; Heydarinasab, A.; Bakhtiari, O.; Mohammadi, T. The Effect of Membrane Pores Wettability on CO_2 Removal from CO_2/CH_4 Gaseous Mixture Using NaOH, MEA and TEA Liquid Absorbents in Hollow Fiber Membrane Contactor. *Chin. J. Chem. Eng.* **2018**, *26*, 1845–1861. [CrossRef]
9. Rosli, A.; Shoparwe, N.F.; Ahmad, A.L.; Low, S.C.; Lim, J.K. Dynamic Modelling and Experimental Validation of CO_2 Removal Using Hydrophobic Membrane Contactor with Different Types of Absorbent. *Sep. Purif. Technol.* **2019**, *219*, 230–240. [CrossRef]

10. Rongwong, W.; Jiraratananon, R.; Atchariyawut, S. Experimental Study on Membrane Wetting in Gas-Liquid Membrane Contacting Process for CO_2 Absorption by Single and Mixed Absorbents. *Sep. Purif. Technol.* **2009**, *69*, 118–125. [CrossRef]
11. Shoukat, U.; Pinto, D.D.D.; Knuutila, H.K. Study of Various Aqueous and Non-Aqueous Amine Blends for Hydrogen Sulfide Removal from Natural Gas. *Processes* **2019**, *7*, 160. [CrossRef]
12. Shirazian, S.; Nakhjiri, A.T.; Heydarinasab, A.; Ghadiri, M. Theoretical Investigations on the Effect of Absorbent Type on Carbon Dioxide Capture in Hollow-Fiber Membrane Contactors. *PLoS ONE* **2020**, *15*, e0236367. [CrossRef] [PubMed]
13. Aaron, D.; Tsouris, C. Separation of CO_2 from Flue Gas: A Review. *Sep. Sci. Technol.* **2005**, *40*, 321–348. [CrossRef]
14. Ji, G.; Zhao, M. Membrane Separation Technology in Carbon Capture. In *Recent Advances in Carbon Capture and Storage*; Yun, Y., Ed.; InTech: Rijeka, Croatia, 2017; Chapter 3. ISBN 978-953-51-3006-2.
15. Seader, J.D.; Henley, E.J.; Roper, D.K. Separation Process Principles. *Choice Rev. Online* **1999**, *36*, 36–5112. [CrossRef]
16. He, X.; Hägg, M.B. Membranes for Environmentally Friendly Energy Processes. *Membranes* **2012**, *2*, 706–726. [CrossRef] [PubMed]
17. Kreulen, H.; Smolders, C.A.; Versteeg, G.F.; van Swaaij, W.P.M. Microporous Hollow Fibre Membrane Modules as Gas-Liquid Contactors. Part 1. Physical Mass Transfer Processes. A Specific Application: Mass Transfer in Highly Viscous Liquids. *J. Membr. Sci.* **1993**, *78*, 197–216. [CrossRef]
18. Kreulen, H.; Smolders, C.A.; Versteeg, G.F.; van Swaaij, W.P.M. Microporous Hollow Fibre Membrane Modules as Gas-Liquid Contactors Part 2. Mass Transfer with Chemical Reaction. *J. Membr. Sci.* **1993**, *78*, 217–238. [CrossRef]
19. Yuliwati, E.; Ismail, A.F. Effect of Additives Concentration on the Surface Properties and Performance of PVDF Ultrafiltration Membranes for Refinery Produced Wastewater Treatment. *Desalination* **2011**, *273*, 226–234. [CrossRef]
20. Yuliwati, E.; Ismail, A.F.; Matsuura, T.; Kassim, M.A.; Abdullah, M.S. Effect of Modified Pvdf Hollow Fiber Submerged Ultrafiltration Membrane for Refinery Wastewater Treatment. *Desalination* **2011**, *283*, 214–220. [CrossRef]
21. Rajabzadeh, S.; Yoshimoto, S.; Teramoto, M.; Al-Marzouqi, M.; Ohmukai, Y.; Maruyama, T.; Matsuyama, H. Effect of Membrane Structure on Gas Absorption Performance and Long-Term Stability of Membrane Contactors. *Sep. Purif. Technol.* **2013**, *108*, 65–73. [CrossRef]
22. Rajabzadeh, S.; Yoshimoto, S.; Teramoto, M.; Al-Marzouqi, M.; Matsuyama, H. CO_2 absorption by Using PVDF Hollow Fiber Membrane Contactors with Various Membrane Structures. *Sep. Purif. Technol.* **2009**, *69*, 210–220. [CrossRef]
23. Albarracin Zaidiza, D.; Wilson, S.G.; Belaissaoui, B.; Rode, S.; Castel, C.; Roizard, D.; Favre, E. Rigorous Modelling of Adiabatic Multicomponent CO_2 Post-Combustion Capture Using Hollow Fibre Membrane Contactors. *Chem. Eng. Sci.* **2016**, *145*, 45–58. [CrossRef]
24. Kumar, P.S.; Hogendoorn, J.A.; Feron, P.H.M.; Versteeg, G.F. Approximate Solution to Predict the Enhancement Factor for the Reactive Absorption of a Gas in a Liquid Flowing through a Microporous Membrane Hollow Fiber. *J. Membr. Sci.* **2003**, *213*, 231–245. [CrossRef]
25. Albarracin Zaidiza, D.; Billaud, J.; Belaissaoui, B.; Rode, S.; Roizard, D.; Favre, E. Modeling of CO_2 Post-Combustion Capture Using Membrane Contactors, Comparison between One- and Two-Dimensional Approaches. *J. Membr. Sci.* **2014**, *455*, 64–74. [CrossRef]
26. Kumar, P.S.; Hogendoorn, J.A.; Feron, P.H.M.; Versteeg, G.F. New Absorption Liquids for the Removal of CO_2 from Dilute Gas Streams Using Membrane Contactors. *Chem. Eng. Sci.* **2002**, *57*, 1639–1651. [CrossRef]
27. Rezakazemi, M.; Niazi, Z.; Mirfendereski, M.; Shirazian, S.; Mohammadi, T.; Pak, A. CFD Simulation of Natural Gas Sweetening in a Gas–Liquid Hollow-Fiber Membrane Contactor. *Chem. Eng. J.* **2011**, *168*, 1217–1226. [CrossRef]
28. Yuan, C.; Li, L.; Li, Y.; Pan, Z.; Zhang, N.; Borhani, T.N.; Zhang, Z. Modeling of CO_2 Absorption into 4-Diethylamino-2-Butanol Solution in a Membrane Contactor under Wetting or Non-Wetting Conditions. *Carbon Capture Sci. Technol.* **2022**, 100069. [CrossRef]
29. Ismail, M.S.; Mohamed, A.M.; Poggio, D.; Walker, M.; Pourkashanian, M. Modelling Mass Transport within the Membrane of Direct Contact Membrane Distillation Modules Used for Desalination and Wastewater Treatment: Scrutinising Assumptions. *J. Water Process Eng.* **2022**, *45*, 102460. [CrossRef]
30. Zhang, X.; Koirala, R.; Date, A.; Jegatheesan, V. Modelling and Simulation of Flux Prediction and Salinity Variation in Direct Contact Membrane Distillation for Seawater Desalination and Brine Treatment. *Desalination* **2022**, *540*, 116021. [CrossRef]
31. Bird, R.B.; Stewart, W.E.; Lightfoot, E.N. *Transport Phenomena, Revised Se*; John Wiley & Sons: New York, NY, USA, 2006.
32. Hikita, H.; Asai, S.; Ishikawa, H.; Honda, M. The Kinetics of Reactions of Carbon Dioxide with Monoethanolamine, Diethanolamine and Triethanolamine by a Rapid Mixing Method. *Chem. Eng. J.* **1977**, *13*, 7–12. [CrossRef]

sustainability

MDPI

Article

Synthesis of Urchin-Shaped Gold Nanoparticles Utilizing Green Reducing and Capping Agents at Different Preparation Conditions: An In Vitro Study

Mohamed S. Salem [1], Mohamed R. Elmarghany [1,*], Noha Salem [2] and Norhan Nady [2,*]

1 Mechanical Power Engineering Department, Faculty of Engineering, Mansoura University, Mansoura 35516, Egypt
2 Polymeric Materials Research Department, City of Scientific Research and Technological Applications (SRTA-City), Borg El-Arab City, Alexandria 21934, Egypt
* Correspondence: mohamed_ragab@mans.edu.eg (M.R.E.); norhan.nady77@yahoo.com (N.N.); Tel.: +20-01007121799 (M.R.E.); +20-1090918521 (N.N.)

Abstract: Employing environmentally friendly reducing and capping materials to synthesize gold nanoparticles is an exciting research point. However, the used materials usually need a long reduction time that can take days. In this work, the instantaneous production of small-sized (less than 20 nm) gold nanoparticles is investigated using ascorbic acid, gelatin, and a mixture of the two agents at different preparation conditions (at room temperature; 20 ± 3 °C and near boiling temperature; 95 ± 3 °C). Particle size analysis, as well as transmission electron microscopy, were used to assess the produced particles' physical characteristics. The structural changes and optical characteristics of the nanoparticles were monitored using UV–visible spectroscopy. Fourier Transform Infrared spectroscopy (FTIR) was used to establish the presence of a gelatin coating over the gold nanoparticles. The morphology of the produced nanoparticles at 95 ± 3 °C was spherical with a size ranging from 8–18 nm, whereas urchin-shaped nanoparticles ranging from 24–100 nm were formed at 20 ± 3 °C reaction temperature. The presence of hydroxyl and amine groups associated with the gelatin was confirmed using FTIR. This could be a step for wider usage of green synthesized nanogold particles in several applications.

Keywords: gold nanoparticles; instant nanosynthesis; urchin-shaped nanogold; green capping agent; biocompatible nanogold

Citation: Salem, M.S.; Elmarghany, M.R.; Salem, N.; Nady, N. Synthesis of Urchin-Shaped Gold Nanoparticles Utilizing Green Reducing and Capping Agents at Different Preparation Conditions: An In Vitro Study. *Sustainability* **2022**, *14*, 16838. https://doi.org/10.3390/su142416838

Academic Editor: Farid Menaa

Received: 30 September 2022
Accepted: 13 December 2022
Published: 15 December 2022

1. Introduction

Nanotechnology research has recently emerged as a vital key to obtaining novel, desirable properties from well-known conventional materials [1]. Nanomaterials have at least one dimension within the nanometer range [2]. Although almost any material in existence could exist within the nanoscale, noble metals-based nanoparticles hold a special value. They generally have some exceptional characteristics that can be tailored for many useful applications.

Gold is considered one of the most famed noble metals to be prepared in nanoscale forms [3]. Gold nanoparticles drew increasing interest due to their excellent optical, electronic, and magnetic properties [4], as well as their biocompatibility [5] and non-toxicity [6]. Gold-based nanomaterials have been previously prepared in different nanostructures like nanoparticles [7,8], nanorods [9,10], nanowires [11,12], nanostars [13,14], nanospheres [15,16], and nanotriangles [17,18]. Nanogold-based materials were utilized for different purposes, such as catalysis [19], energy storage and conversion [20], biomedical therapeutics [20–22], drug delivery [23], photonics [24], and biosensing [25,26].

Composite nanostructures that combine gold with other materials, such as iron [27], silver [28], and graphene [28], were also investigated and prepared. Moreover, researchers managed to prepare fairly complicated composite morphologies. For example, Nghiem et al. [29] managed to prepare a composite comprising a core–shell nanostructure, combining a silica core with a gold shell.

Nanostructure attributes are greatly influenced by their physicochemical characteristics, which include particle volume, form, and composition. Furthermore, for composite nanostructures, the properties are also affected by the molecular distribution within the particle [30]. Tuning and manipulating these characteristics allows control or maximization of the desired attribute to be suitable for a specific application. In the case of catalysis, for example, catalytic activity and selectivity are strongly influenced by the surface area, composition, and shape of the nanostructure [31]. On the other hand, when the object is to enhance the optical attributes of a nanostructure for imaging functions, controlling those characteristics would impact absorption and scattering efficiency [32].

It is evident that the manipulation of these nanostructures during the synthesis process is essential to produce predictable and distinct physical and chemical properties. This is critical not only for optimizing their performances but also for understanding the structure-performance relations. This can be partly achieved through the nanostructure controllable preparation method.

Microemulsion [33], reversed micelles [34], seeding growth [35], sonochemistry [36], photochemistry [37], radiolysis [38], and chemical reduction [39] techniques have all been used for gold nanostructure preparation. Several chemical reductants have been used to reduce the gold precursor (chloroauric acid, $HAuCl_4$), such as 1-amino-2-naphthol-4-sulphonic acid [40], monosodium glutamate [41], sodium borohydride [42], amine [43], etc. Recently, natural materials have been used as green reductants, such as ascorbic acid [44], piper beetle leaf broth [45], a polyphenol extracted from leaf buds of camellia Sinensis tea [46], aspartame [47], Rhizopus oryzae fungal strain [48], and gelatin [3].

One of the earliest and simplest methods for preparing gold nanoparticles is the Turkevich technique [49], which uses sodium citrate to both reduce the gold precursor ($HAuCl_4$) and prevent further particle development and retain the formed gold in the nano range as well as prevent the nanogold from aggregation. The second role of the used sodium citrate is a capping agent that can offset the Van der Waals tractive forces through steric and electrostatic repulsions, hence reducing the aggregation phenomenon. However, the stability and dispersity of the produced gold nanoparticles are often poor [50]. In recent years, another technique was developed to adjust the reaction pH utilizing a buffer rather than a solution to present nanoparticles with high monodispersity [51]. Also, ethylenediaminetetraacetic acid (EDTA) was used to improve the uniformity of the produced nanogold with better size control [52].

Gold nanoparticles have excellent potential for the immobilization of biomolecules because they are biocompatible and bind readily to a range of biomolecules, such as amino acids, proteins, DNA, and enzymes. Nanogold membranes can be used as templates for the immobilization of enzymes [53]. Consequently, the bioconjugate material could be easily separated after reaction by mild centrifugation and easily reused. This membrane can also be used in nanoplasmonic sensors [54].

As can be seen, the search continues for producing proficient gold nanostructures using green reducing and capping agents that combine sustainability and efficiency [55]. In the literature, some authors used surfactants to obtain such small sizes that resulted in chemical contamination. Others claimed to produce nanoparticles instantaneously without quantifying the exact reaction time. On the other hand, most of the published research did not obtain uniform shapes but mostly a mixture of different shapes [56]. In this work, the instantaneous fabrication of gold nanoparticles with two different morphologies using a green, reducing agent (ascorbic acid) and a green capping agent (gelatin) along with distilled water as solvent was investigated and presented. Ascorbic acid (B_1), gelatin (B_2), as well as a mixture of both (M) were tested as reducing agents of the gold precursor. Two different morphologies were obtained under the sole influence of changing the reaction temperature. Particle size analysis, as well as transmission electron microscopy, were used to assess the produced particles' physical characteristics. The structural changes and optical characteristics of the nanoparticles were monitored using UV–visible spectroscopy. Moreover, the crystal nature of the prepared gold nanoparticles was determined by X-ray

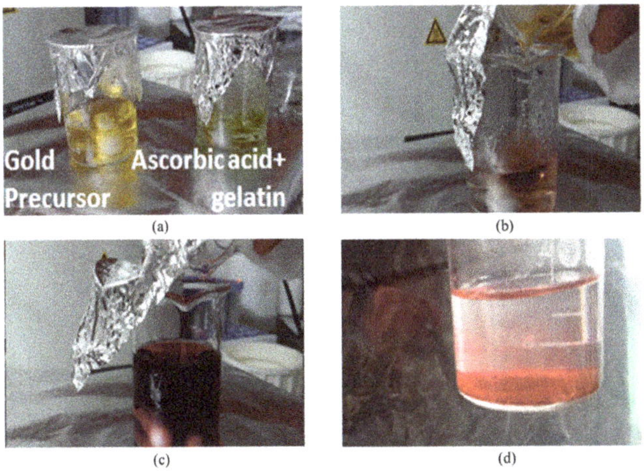

Figure 2. Preparing gold nanoparticles using a mixture of ascorbic acid and gelatin at 95 ± 3 °C (**a**) before mixing, (**b**) one second after mixing starts, (**c**) after mixing for one minute, and (**d**) gold nanoparticles precipitation after adding ethanol.

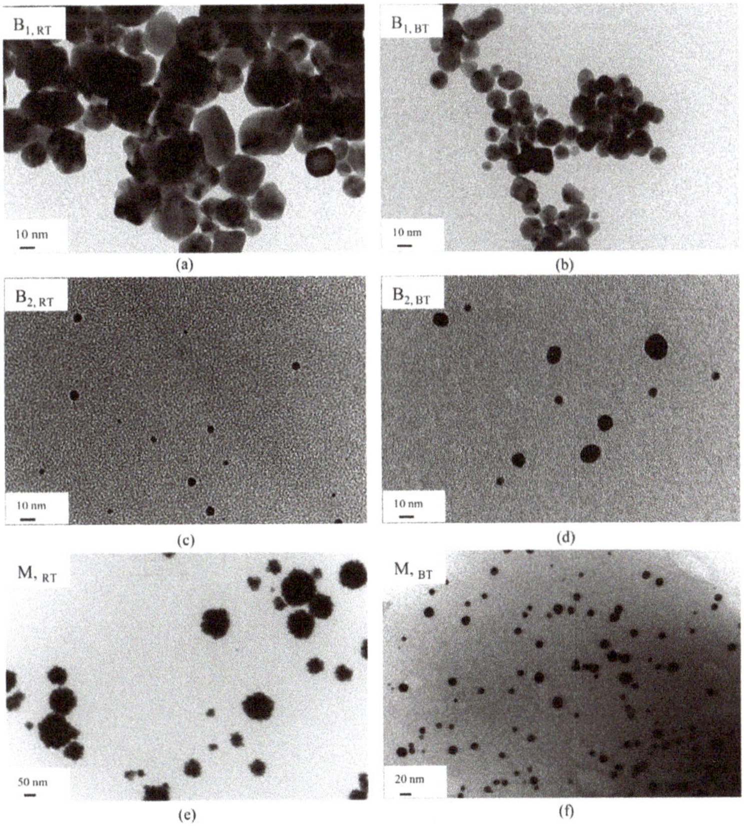

Figure 3. TEM images of the formed gold nanoparticles using only ascorbic acid at; (**a**) 20 ± 3 °C ($B_{1,RT}$), (**b**) 95 ± 3 °C ($B_{1,BT}$), only gelatin at (**c**) 20 ± 3 °C ($B_{2,RT}$), (**d**) 95 ± 3 °C ($B_{2,BT}$), and a mixture at (**e**) 20 ± 3 °C ($M_{,RT}$), and (**f**) 95 ± 3 °C ($M_{,BT}$).

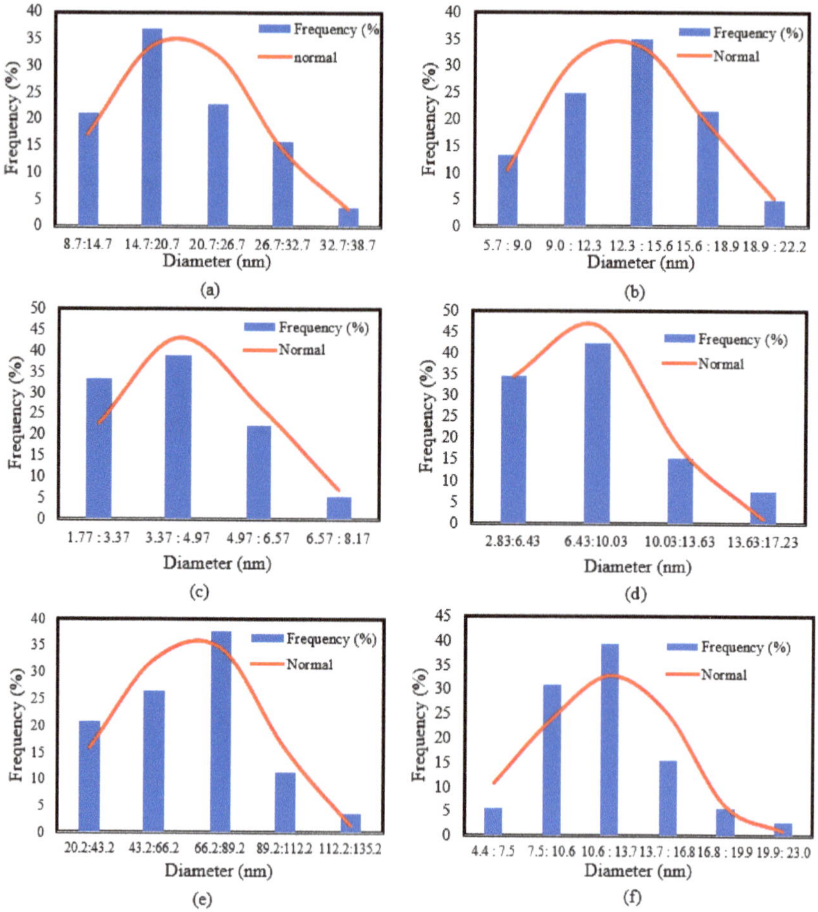

Figure 4. Histograms of the particle size distribution via TEM images of the formed gold nanoparticles using only ascorbic acid at (**a**) 20 ± 3 °C ($B_{1,RT}$), (**b**) 95 ± 3 °C ($B_{1,BT}$); only gelatin at (**c**) 20 ± 3 °C ($B_{2,RT}$), (**d**) 95 ± 3 °C ($B_{2,BT}$), and a mixture at (**e**) 20 ± 3 °C ($M_{,RT}$), and (**f**) 95 ± 3 °C ($M_{,BT}$).

Table 1. Summary of the mean value and standard deviation of the obtained gold nanoparticles at two different reaction temperatures.

Sample	Mean Value (nm)	Standard Deviation (nm)
$B_{1,RT}$	20.3	6.6
$B_{1,BT}$	14.0	3.6
$B_{2,RT}$	4.3	1.5
$B_{2,BT}$	7.2	3.2
$M_{,RT}$	65.6	25.1
$M_{,BT}$	11.8	3.6

The obtained results show that the higher temperature of the reaction resulted in the almost instantaneous creation of gold nanoparticles in a uniform shape. This is in full agreement with previous research [57–59], in which the reaction temperature decisively controlled the formation of small gold nanoparticles. Generally, the temperature accelerates the nucleation process and with the aid of the used capping agent, the gold nanoparticles are able to maintain their small size. However, the lower reaction temperature causes a limited nucleation process and provides a chance for particle growth to yield a larger particle size, which the naked eye can notice after longer reaction times (15–30 min). On the other hand, at a relatively low pH medium (pH 3.3 ± 002), the formation of gold nanoparticles is carried out through an intermediate $[AuCl_3(OH)]^-$ which undergoes a LaMer burst nucleation within about 10 s [60,61]. However, at higher pH mediums, a much longer nucleation time (around 60 s) is required, and the particles undergo reduction through $[AuCl_2(OH)]_2{}^-$and $[AuCl(OH)]_3{}^-$ followed by a much slower growth process. When coupling both effects of high temperature and pH, a uniform low-size range of gold nanoparticles could be produced [58,61].

UV–visible spectroscopy can check for morphological changes and optical properties as nanoparticles display surface plasmon resonance (SPR) varying with its form and size at different frequencies [57,61]. The reaction was executed inside the device cuvette for 30 s and the measurement of the mixed solution was done immediately for higher accuracy. Figure 5a shows a UV–visible spectrum of an aqueous solution of gold precursor in water at room temperature (20 ± 3 °C), mixed with only ascorbic acid ($B_{1,RT}$), and mixed with only gelatin ($B_{2,RT}$). The figure also presents the spectra of the gold prosecutor mixed with both agents at room temperature ($M_{,RT}$) as well as near boiling temperature ($M_{,BT}$). All the measurements took place within 2 min from the reaction beginning (i.e., 30 s for mixing and around 60 s for measuring). As noted, there are no peaks for the used gold precursor solution in water or the solution of gold precursor with the ascorbic acid. However, the presence of gelatin with the gold precursor shows a surface plasmon resonance broad band at around 570–600 nm. In the case of the gold prosecutor with the mixture, a strong absorption band around 500–600 nm is observed that corresponds to the excitation of surface plasmon vibrations in the gold nanoparticles. The presence of both ascorbic acid and gelatin shows a broader SPR beak at around 507–667 nm. It can be seen that the peak was broadened with urchin-shaped gold nanoparticles, as illustrated in TEM images in the previous section, which was prepared at RT (20 ± 3 °C). This broad peak can be attributed to the existence of transversal and longitudinal SPR [62]. The sharp increase in the UV band at a wavelength of about 500 nm may be due to the instantaneous formation of the nanogold particles during the measurement, as the reaction was processed within the measuring cuvette for maximum accuracy. The unsmooth surface causes a greater enhancement of the electromagnetic field at the ends of the urchin-shaped nanoparticle spikes supplemented by a red shift in the peak [57].

Additionally, the uneven distribution of the surface electron layers in the urchin-shaped gold nanoparticles is what causes the difference in absorption qualities between them and the spherical ones [45,63]. On the other hand, the much less broad SPR absorption band of gold nanoparticle that was prepared near boiling temperature (95 ± 3 °C) appeared at 540 nm. Figure 5b shows there was not a noticeable difference in the SPR after 12 min of mixing the gold precursor and both ascorbic acid and gelatin together at room temperature (the measuring was executed with 2-min intervals).

Furthermore, Figure 5c shows the same determined SPR band after 12 min mixing between the gold precursor mixed with both ascorbic acid and gelatin at near boiling temperature (95 ± 3 °C). Regarding using only ascorbic acid with the gold precursor near BT ($B_{1,BT}$), a small SPR band was formed compared to ($B_{1,RT}$) in Figure 5a, which can be ascribed to the fast growth of gold nanoparticles with the ascorbic acid with heating.

It is considered common knowledge that gold nanoparticles smaller than 2 nm do not demonstrate SPR [57,64]. Such small gold nanoparticles can be determined by using a particle size analyzer as shown in Figure 6. The use of either ascorbic acid or gelatin at near boiling temperature (95 ± 3 °C) can produce gold nanoparticles. To stop the growth of gold particles and/or to reduce the nanoparticle aggregation, the addition of pure ethanol with a ratio of 3:1 ethanol/reaction solution was done after 2 min of mixing at BT and after 15 min mixing at RT. The effect of ethanol is strongly illustrated by a reduction in the particle size of the formed gold nanoparticles produced using only the reducing agent (ascorbic acid) at 95 ± 3 °C from 257–421 nm to a much narrower 6–13 nm size, whereas the addition of ethanol to the solution of gold precursor and gelatin at BT resulted in shifting the particle size range from 45–74 nm to 95–200 nm.

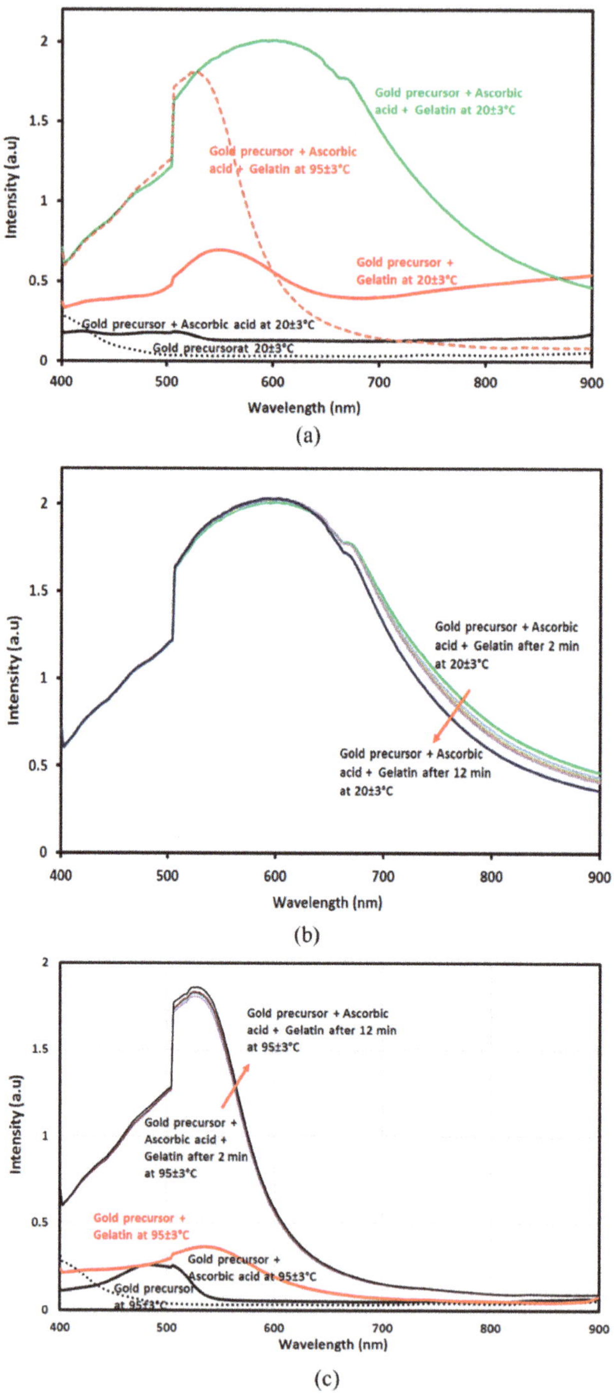

Figure 5. UV–visible spectra of the formed gold nanoparticles at two different temperatures (RT; 20 ± 3 °C and near BT; 95 ± 3 °C) after 2 min reaction time (**a**) and after 12 min reaction time ((**b**); 20 ± 3 °C and (**c**); 95 ± 3 °C).

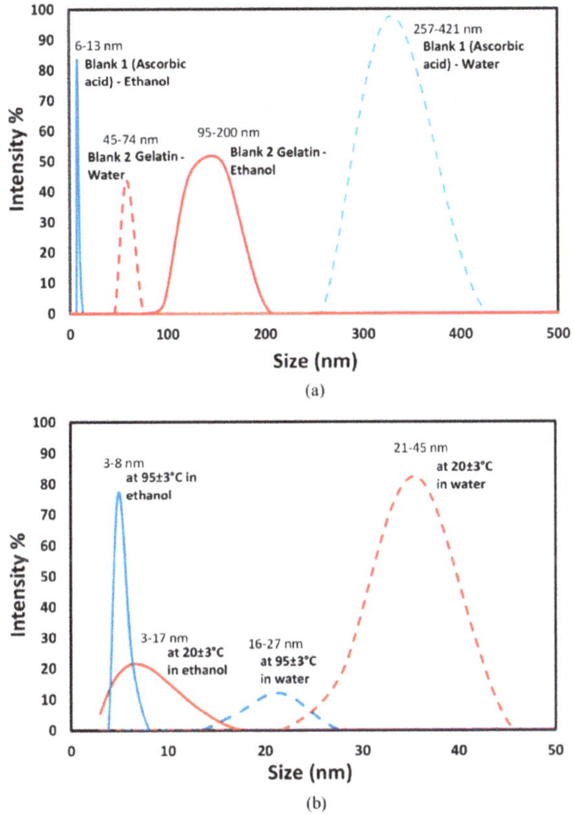

Figure 6. Particle size analyzer of the formed gold nanoparticles suspended in the reaction solution, and after adding ethanol. The nanoparticles were synthesized using (**a**) only ascorbic acid ($B_{1,BT}$) and only gelatin ($B_{2,BT}$). (**b**) Using both ascorbic acid and gelatin at both room temperature ($M_{,RT}$), and near the boiling point ($M_{,BT}$).

It seems the role of ethanol is different in the two cases; in the case of using ascorbic acid, the used ethanol may reduce the particle size growth, but in the case of using gelatin, the used ethanol reduced the role of gelatin as a capping agent and enhanced the particle aggregation. The degradation effect of the ethanolic solution on the ascorbic acid was previously proven [64] in which the higher ethanol concentration (>50%) promotes lowering the water activity environment that supports the degradation of ascorbic acid into a transitional compound; L-xylosone, and then to other carbon compounds [65]. Therefore, adding the ethanol solution will lead to stopping the reducing reaction and result in a reduction in the particle size of the produced nanogold. Meanwhile, ethanol with a high concentration (>50%) works on the precipitation of gelatin into particles [66]. Therefore, the effectiveness of the gelatin as a capping agent is weakened, and a larger gold particle size is formed. The results of $B_{1,RT}$, and $B_{2,RT}$ were not within the measuring limit, which may be attributed to the formation of an extremely small particle size (out of the measuring limits of the particle size analyzer; <3 nm). In the case of using both ascorbic acid and gelatin near BT as shown in Figure 6B, the size of produced nanoparticle shifted from 16–27 nm to 3–8 nm. In the case of using both ascorbic acid and gelatin at RT, the size of produced urchin-shaped gold nanoparticles shifted from 21–45 nm to 3–17 nm. From the obtained results, we can assume that the addition of ethanol may be a tool to control the size of the gold nanoparticles according to the targeted final gold particle size.

The existence of gold nanoparticles was verified using XRD. Figure 7 shows diffraction peaks for the used gold precursor in distilled water at 95 ± 3 °C reaction temperature and the green synthesized nanoparticles using a mixture of ascorbic acid and gelatin as a reducing agent and a capping agent, respectively at 95 ± 3 °C reaction temperature. The determined four strong Bragg diffraction peaks at 38.18°, 44.02°, 65.75°, and 78.02° can be assigned to the (111), (200), (220), and (311) planes [67,68]. It is possible to interpret these diffraction peaks as a face-centered cubic of gold and a natural crystal structure [58,59].

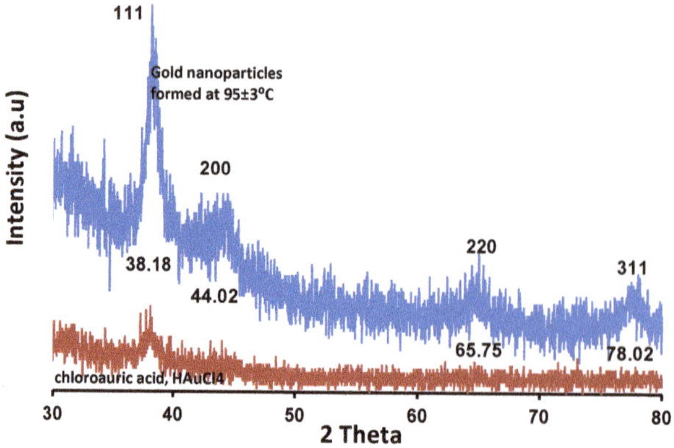

Figure 7. XRD diffraction patterns of gold precursor (hydrogen tetrachlorourate) and the formed gold nanoparticles at 95 ± 3 °C after adding ethanol, decantation, and drying in a desiccator.

The FT-IR spectrum shown in Figure 8 illustrates the existence of the following groups: the strong, broad band observed at 3414 cm^{-1} suggests the existence of the O-H group and primary amine O-H band. The band at 2953 cm^{-1} may correspond to C-H stretching vibrations of the alkanes group, and the band at 1641 cm^{-1} indicates the existence of amide I. The narrow peak at 1541 cm^{-1} can be assigned for the presence of amide II. The peak at 1437 cm^{-1} may be attributed to –N-H. The peak at 1257 cm^{-1} can be assigned to –C-O. The narrow peak at 1084 cm^{-1} can be attributed to the presence of C-N stretching vibrations of aliphatic amines. This FT-IR spectrum indicates the presence of gelatin as a capping agent for gold nanoparticles even after the addition of ethanol to the formed gold nanoparticles. FT-IR proposed that both carboxylic acid groups and amine groups share the conjugation with gold surfaces as well as the possibility of further conjugation with other molecules.

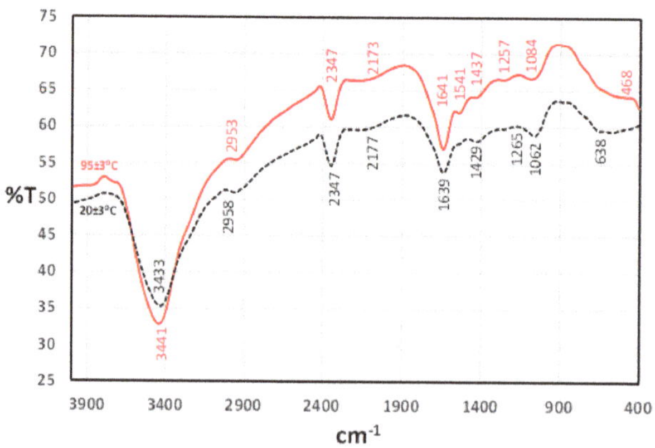

Figure 8. FT-IR spectra of gold nanoparticles formed near BT (red solid line) and at RT (black dashed line).

4. Conclusions

Spherical gold nanoparticles and urchin-shaped gold nanoparticles were successfully and instantly produced using a green-natured reducing agent (ascorbic acid), and a green-natured capping agent (gelatin). Gelatin could be used as both a reducing and capping agent as reported in previous research [67], but the formed gold nanoparticles are not urchin-shaped. In in vivo applications, urchin-shaped particles are preferred to spherical ones because they have a lower background and more effectively transmit near-infrared light through biological tissues [58]. For that, the authors strongly propose utilizing the prepared unique urchin-shaped nanogold in disease diagnostics and therapeutical applications.

Table 2 shows that using both ascorbic acid and gelatin together resulted in an instant production of two different morphologies of gold nanoparticles as the effect of the reaction temperature: at RT (20 ± 3 °C) and near BT (95 ± 3 °C). The role of gelatin is mainly as a capping agent, although it can be considered as a second reducing agent. The mixing time in a mixture of ascorbic acid and gelatin was shorter than those of either only ascorbic acid or only gelatin, especially at a room temperature reaction, due to the higher reaction rates indicated by the change in the solution reaction. Using both gelatin and ascorbic acid together resulted in a spontaneous reaction that can be used to shift the formed gold particles to extremely small sizes. The use of ethanol effectively reduced the aggregation of the particles. However, it had the opposite effect in the case of using gelatin alone. Regarding the used analysis techniques, the particle size analyzer was proved unsuitable for irregular (urchin-shaped) nanoparticles. However, it could be used as an indication of the formed, small, and smooth morphology.

From this study, it is clear that the synthesis of biocompatible gold nanoparticles using both ascorbic acid and gelatin together as reducing and capping agents, respectively, is an instant, flexible, and green method. They produced spherical and urchin-shaped gold nanoparticles, which are candidates for use in medical, pharmaceutical, food, and cosmetics applications.

Table 2. Summary of the size range and morphology of the obtained gold nanoparticles at two different reaction temperatures.

Preparation Temperature	Property	Ascorbic Acid (B_1) pH 3.6 ± 0.02			Gelatin (B_2) pH 3.1 ± 0.01			Ascorbic Acid and Gelatin (M) pH 3.3 ± 0.02		
		TEM (water)	Particle size		TEM (water)	Particle size		TEM (water)	Particle size	
			Water	Eth.		water	Eth.		water	Eth.
RT (20 ± 3 °C)	Determined Size (nm)	9–34	-	-	2–8	-	-	20–120	21–45	3–17
	Morphology	Irregular			Spherical/Urchin mix			Urchin		
	Mixing time	30 min			30 min			15 min		
BT (95 ± 3 °C)	Determined Size (nm)	5–20	257–421	6–13	3–14	45–74	95–200	5–22	16–27	3–8
	Morphology	Spherical								
	Mixing time	10 min			20 min			2 min		

Author Contributions: Conceptualization, N.N.; methodology, N.N.; software, N.N., M.S.S., N.S. and M.R.E.; validation, N.N., M.S.S. and M.R.E.; formal analysis, N.N. and N.S.; investigation, N.N. and M.R.E.; resources, N.N.; data curation, N.N, M.S.S. and N.S.; writing—original draft preparation, N.N. and M.R.E.; writing—review and editing, N.N, M.S.S. and M.R.E.; visualization, M.S.S. and N.S.; supervision, N.N.; project administration, N.N.; funding acquisition, N.N. All authors have read and agreed to the published version of the manuscript.

Funding: This research received no external funding.

Institutional Review Board Statement: Not applicable.

Informed Consent Statement: Not applicable.

Data Availability Statement: All data generated or analyzed during this study are included in this published article.

Acknowledgments: The publication fees of this article have been supported by Mansoura University.

Conflicts of Interest: The authors declare no conflict of interest.

References

1. Boisselier, E.; Astruc, D. Gold Nanoparticles in Nanomedicine: Preparations, Imaging, Diagnostics, Therapies and Toxicity. *Chem. Soc. Rev.* **2009**, *38*, 1759–1782. [CrossRef]
2. Jain, P.K.; Huang, X.; El-Sayed, I.H.; El-Sayed, M.A. Noble Metals on the Nanoscale: Optical and Photothermal Properties and Some Applications in Imaging, Sensing, Biology, and Medicine. *Acc. Chem. Res.* **2008**, *41*, 1578–1586. [CrossRef]
3. Zhou, J.; Ralston, J.; Sedev, R.; Beattie, D.A. Functionalized Gold Nanoparticles: Synthesis, Structure and Colloid Stability. *J. Colloid Interface Sci.* **2009**, *331*, 251–262. [CrossRef] [PubMed]
4. Sardar, R.; Funston, A.M.; Mulvaney, P.; Murray, R.W. Gold Nanoparticles: Past, Present, and Future. *Langmuir* **2009**, *25*, 13840–13851. [CrossRef] [PubMed]
5. Shukla, R.; Bansal, V.; Chaudhary, M.; Basu, A.; Bhonde, R.R.; Sastry, M. Biocompatibility of Gold Nanoparticles and Their Endocytotic Fate inside the Cellular Compartment: A Microscopic Overview. *Langmuir* **2005**, *21*, 10644–10654. [CrossRef] [PubMed]
6. Alkilany, A.M.; Murphy, C.J. Toxicity and Cellular Uptake of Gold Nanoparticles: What We Have Learned so Far? *J. Nanoparticle Res.* **2010**, *12*, 2313–2333. [CrossRef] [PubMed]
7. Rodríguez-Fernández, J.; Pérez-Juste, J.; García De Abajo, F.J.; Liz-Marzán, L.M. Seeded Growth of Submicron Au Colloids with Quadrupole Plasmon Resonance Modes. *Langmuir* **2006**, *22*, 7007–7010. [CrossRef]
8. Hammami, I.; Alabdallah, N.M. Gold Nanoparticles: Synthesis Properties and Applications. *J. King Saud Univ.* **2021**, *33*, 101560. [CrossRef]
9. Vigderman, L.; Khanal, B.P.; Zubarev, E.R. Functional Gold Nanorods: Synthesis, Self-Assembly, and Sensing Applications. *Adv. Mater.* **2012**, *24*, 4811–4841. [CrossRef]
10. Pellas, V.; Hu, D.; Mazouzi, Y.; Mimoun, Y.; Blanchard, J.; Guibert, C.; Salmain, M.; Boujday, S. Gold Nanorods for LSPR Biosensing: Synthesis, Coating by Silica, and Bioanalytical Applications. *Biosensors* **2020**, *10*, 146. [CrossRef]
11. Kim, F.; Sohn, K.; Wu, J.; Huang, J. Chemical Synthesis of Gold Nanowires in Acidic Solutions. *J. Am. Chem. Soc.* **2008**, *130*, 14442–14443. [CrossRef] [PubMed]
12. Jiang, X.; Qiu, X.; Fu, G.; Sun, J.; Huang, Z.; Sun, D.; Xu, L.; Zhou, J.; Tang, Y. Highly Simple and Rapid Synthesis of Ultrathin Gold Nanowires with (111)-Dominant Facets and Enhanced Electrocatalytic Properties. *J. Mater. Chem. A* **2018**, *6*, 17682–17687. [CrossRef]
13. Senthil Kumar, P.; Pastoriza-Santos, I.; Rodríguez-González, B.; Javier García De Abajo, F.; Liz-Marzán, L.M. High-Yield Synthesis and Optical Response of Gold Nanostars. *Nanotechnology* **2008**, *19*, 15606. [CrossRef]
14. Andreiuk, B.; Nicolson, F.; Clark, L.M.; Panikkanvalappil, S.R.; Kenry; Rashidian, M.; Harmsen, S.; Kircher, M.F. Design and Synthesis of Gold Nanostars-Based SERS Nanotags for Bioimaging Applications. *Nanotheranostics* **2022**, *6*, 10–30. [CrossRef] [PubMed]
15. Yoon, J.H.; Selbach, F.; Langolf, L.; Schlücker, S. Ideal Dimers of Gold Nanospheres for Precision Plasmonics: Synthesis and Characterization at the Single-Particle Level for Identification of Higher Order Modes. *Small* **2018**, *14*, 1702754. [CrossRef] [PubMed]
16. Xuan, Y.; Zhang, R.Y.; Zhao, D.H.; Zhang, X.S.; An, J.; Cheng, K.; Hou, X.L.; Song, X.L.; Zhao, Y.D.; Yang, X.Q. Ultrafast Synthesis of Gold Nanosphere Cluster Coated by Graphene Quantum Dot for Active Targeting PA/CT Imaging and near-Infrared Laser/PH-Triggered Chemo-Photothermal Synergistic Tumor Therapy. *Chem. Eng. J.* **2019**, *369*, 87–99. [CrossRef]
17. Scarabelli, L.; Coronado-Puchau, M.; Giner-Casares, J.J.; Langer, J.; Liz-Marzán, L.M. Monodisperse Gold Nanotriangles: Size Control, Large-Scale Self-Assembly, and Performance in Surface-Enhanced Raman Scattering. *ACS Nano* **2014**, *8*, 5833–5842. [CrossRef]
18. Koetz, J. The Effect of Surface Modification of Gold Nanotriangles for Surface-Enhanced Raman Scattering Performance. *Nanomaterials* **2020**, *10*, 2187. [CrossRef]
19. Carabineiro, S.A.C. Supported Gold Nanoparticles as Catalysts for the Oxidation of Alcohols and Alkanes. *Front. Chem.* **2019**, *7*, 702. [CrossRef]
20. Anjana, P.M.; Bindhu, M.R.; Rakhi, R.B. Green Synthesized Gold Nanoparticle Dispersed Porous Carbon Composites for Electrochemical Energy Storage. *Mater. Sci. Energy Technol.* **2019**, *2*, 389–395. [CrossRef]
21. Vasam, M.; Punagoti, R.A.; Mourya, R. Biomedical Applications of Gold Nanoparticles. *Nanotechnol. Life Sci.* **2021**, *15*, 41–59. [CrossRef]

22. Nawaz, A.; Ali, S.M.; Rana, N.F.; Tanweer, T.; Batool, A.; Webster, T.J.; Menaa, F.; Riaz, S.; Rehman, Z.; Batool, F.; et al. Ciprofloxacin-Loaded Gold Nanoparticles against Antimicrobial Resistance: An In Vivo Assessment. *Nanomaterials* **2021**, *11*, 3152. [CrossRef] [PubMed]
23. Riaz, S.; Fatima Rana, N.; Hussain, I.; Tanweer, T.; Nawaz, A.; Menaa, F.; Janjua, H.A.; Alam, T.; Batool, A.; Naeem, A.; et al. Effect of Flavonoid-Coated Gold Nanoparticles on Bacterial Colonization in Mice Organs. *Nanomaterials* **2020**, *10*, 1769. [CrossRef] [PubMed]
24. Budaszewski, D.; Chychłowski, M.; Budaszewska, A.; Bartosewicz, B.; Jankiewicz, B.; Woliński, T.R. Enhanced Efficiency of Electric Field Tunability in Photonic Liquid Crystal Fibers Doped with Gold Nanoparticles. *Opt. Express* **2019**, *27*, 14260–14269. [CrossRef] [PubMed]
25. Li, Y.; Schluesener, H.J.; Xu, S. Gold Nanoparticle-Based Biosensors. *Gold Bull.* **2010**, *43*, 29–41. [CrossRef]
26. Soares, A.L.; Hryniewicz, B.M.; Deller, A.E.; Volpe, J.; Marchesi, L.F.; Souto, D.E.P.; Vidotti, M. Electrodes Based on PEDOT Nanotubes Decorated with Gold Nanoparticles for Biosensing and Energy Storage. *ACS Appl. Nano Mater.* **2021**, *4*, 9945–9956. [CrossRef]
27. Teng, Y.; Shi, J.; Pong, P.W.T. Sensitive and Specific Colorimetric Detection of Cancer Cells Based on Folate-Conjugated Gold-Iron-Oxide Composite Nanoparticles. *ACS Appl. Nano Mater.* **2019**, *2*, 7421–7431. [CrossRef]
28. Homik, Z.; Kopniczky, J.; Smausz, T.; Berkesi, D.; Hopp, B. Formation of Gold/Silver Composite Nanoparticles by Pulsed Laser Ablation of Gold–Silver Layered Films in Liquid. *Appl. Phys. A Mater. Sci. Process.* **2022**, *128*, 797. [CrossRef]
29. Nghiem, T.H.L.; Le, T.N.; Do, T.H.; Vu, T.T.D.; Do, Q.H.; Tran, H.N. Preparation and Characterization of Silica-Gold Core-Shell Nanoparticles. *J. Nanoparticle Res.* **2013**, *15*, 2091. [CrossRef]
30. Xia, Y.; Xiong, Y.; Lim, B.; Skrabalak, S.E. Shape-Controlled Synthesis of Metal Nanocrystals: Simple Chemistry Meets Complex Physics? *Angew. Chem. Int. Ed.* **2009**, *48*, 60–103. [CrossRef]
31. Zecchina, A.; Groppo, E.; Bordiga, S. Selective Catalysis and Nanoscience: An Inseparable Pair. *Chem.-A Eur. J.* **2007**, *13*, 2440–2460. [CrossRef] [PubMed]
32. Nehl, C.L.; Hafner, J.H. Shape-Dependent Plasmon Resonances of Gold Nanoparticles. *J. Mater. Chem.* **2008**, *18*, 2415–2419. [CrossRef]
33. Rajapantulu, A.; Bandyopadhyaya, R. Formation of Gold Nanoparticles in Water-in-Oil Microemulsions: Experiment, Mechanism, and Simulation. *Langmuir* **2021**, *37*, 6623–6631. [CrossRef] [PubMed]
34. Revina, A.A.; Chernyshova, K.F.; Tabachkova, N.Y.; Parkhomenko, Y.N. Gold Nanoparticles in Reverse Micellar Solutions: Preparation, Optical Properties, and Dimensional Characteristics. *Russ. Chem. Bull.* **2019**, *68*, 1164–1170. [CrossRef]
35. Yahaya, M.L.; Zakaria, N.D.; Noordin, R.; Abdul Razak, K. Synthesis of Large and Stable Colloidal Gold Nanoparticles (AuNPs) by Seeding-Growth Method. *Mater. Today Proc.* **2022**, *66*, 2943–2947. [CrossRef]
36. Dheyab, M.A.; Aziz, A.A.; Jameel, M.S. Recent Advances in Inorganic Nanomaterials Synthesis Using Sonochemistry: A Comprehensive Review on Iron Oxide, Gold and Iron Oxide Coated Gold Nanoparticles. *Molecules* **2021**, *26*, 2453. [CrossRef]
37. Bianchi, P.; Petit, G.; Monbaliu, J.C.M. Scalable and Robust Photochemical Flow Process towards Small Spherical Gold Nanoparticles. *React. Chem. Eng.* **2020**, *5*, 1224–1236. [CrossRef]
38. Bondaz, L.; Fontaine, P.; Muller, F.; Pantoustier, N.; Perrin, P.; Morfin, I.; Goldmann, M.; Cousin, F. Controlled Synthesis of Gold Nanoparticles in Copolymers Nanomolds by X-ray Radiolysis. *Langmuir* **2020**, *36*, 6132–6144. [CrossRef]
39. Daruich De Souza, C.; Ribeiro Nogueira, B.; Rostelato, M.E.C.M. Review of the Methodologies Used in the Synthesis Gold Nanoparticles by Chemical Reduction. *J. Alloy. Compd.* **2019**, *798*, 714–740. [CrossRef]
40. Rawat, P.; Rajput, Y.S.; Bharti, M.K.; Sharma, R. A Method for Synthesis of Gold Nanoparticles Using 1-Amino-2-Naphthol-4-Sulphonic Acid as Reducing Agent. *Curr. Sci.* **2016**, *110*, 2297–2300. [CrossRef]
41. Johan, M.R.; Chong, L.C.; Hamizi, N.A. Preparation and Stabilization of Monodisperse Colloidal Gold by Reduction with Monosodium Glutamate and Poly (Methyl Methacrylate). *Int. J. Electrochem. Sci.* **2012**, *7*, 4567–4573.
42. Yang, S.K.; Kim, Y. Nanogold Particles Produced by $NaBH_4$ Reduction of Gold Salt in the Presence of Laponite Sol. *Bull. Korean Chem. Soc.* **2013**, *34*, 363–364. [CrossRef]
43. Newman, J.D.S.; Blanchard, G.J. Formation of Gold Nanoparticles Using Amine Reducing Agents. *Langmuir* **2006**, *22*, 5882–5887. [CrossRef] [PubMed]
44. Khan, Z.; Singh, T.; Hussain, J.I.; Hashmi, A.A. Au(III)-CTAB Reduction by Ascorbic Acid: Preparation and Characterization of Gold Nanoparticles. *Colloids Surf. B Biointerfaces* **2013**, *104*, 11–17. [CrossRef]
45. Mallikarjuna, K.; Sushma, N.J.; Deva Prasad Raju, B. Novel, Fast, Bio-Derivatized Sonochemical Synthesis of Gold Nanoparticles by Using Piper Betle Leaf Broth as a Reducing and Capping Agent. In *Springer Proceedings in Physics*; Springer: Berlin/Heidelberg, Germany, 2013; Volume 143, pp. 41–49. ISBN 9783642342158.
46. Boruah, S.K.; Boruah, P.K.; Sarma, P.; Medhi, C.; Medhi, O.K. A Study on the Electrospinning Behaviour and Nanofibre Morphology of Anionically Charged Lignin. *Adv. Mater. Lett.* **2012**, *3*, 481–486. [CrossRef]
47. Wu, S.; Yan, S.; Qi, W.; Huang, R.; Cui, J.; Su, R.; He, Z. Green Synthesis of Gold Nanoparticles Using Aspartame and Their Catalytic Activity for P-Nitrophenol Reduction. *Nanoscale Res. Lett.* **2015**, *10*, 213. [CrossRef] [PubMed]
48. Das, S.K.; Das, A.R.; Guha, A.K. Gold Nanoparticles: Microbial Synthesis and Application in Water Hygiene Management. *Langmuir* **2009**, *25*, 8192–8199. [CrossRef]

49. Turkevich, J.; Stevenson, P.C.; Hillier, J. A Study of the Nucleation and Growth Processes in the Synthesis of Colloidal Gold. *Discuss. Faraday Soc.* **1951**, *11*, 55–75. [CrossRef]
50. Verma, H.N.; Singh, P.; Chavan, R.M. Gold Nanoparticle: Synthesis and Characterization. *Vet. World* **2014**, *7*, 72–77. [CrossRef]
51. Schulz, F.; Homolka, T.; Bastús, N.G.; Puntes, V.; Weller, H.; Vossmeyer, T. Little Adjustments Significantly Improve the Turkevich Synthesis of Gold Nanoparticles. *Langmuir* **2014**, *30*, 10779–10784. [CrossRef]
52. Dozol, H.; Mériguet, G.; Ancian, B.; Cabuil, V.; Xu, H.; Wang, D.; Abou-Hassan, A. On the Synthesis of Au Nanoparticles Using EDTA as a Reducing Agent. *J. Phys. Chem. C* **2013**, *117*, 20958–20966. [CrossRef]
53. Chae, S.Y.; Son, S.; Lee, M.; Jang, M.K.; Nah, J.W. Deoxycholic Acid-Conjugated Chitosan Oligosaccharide Nanoparticles for Efficient Gene Carrier. *J. Control. Release* **2005**, *109*, 330–344. [CrossRef] [PubMed]
54. Rodrigues, M.S.; Borges, J.; Lopes, C.; Pereira, R.M.S.; Vasilevskiy, M.I.; Vaz, F. Gas Sensors Based on Localized Surface Plasmon Resonances: Synthesis of Oxide Films with Embedded Metal Nanoparticles, Theory and Simulation, and Sensitivity Enhancement Strategies. *Appl. Sci.* **2021**, *11*, 5388. [CrossRef]
55. Uzair, B.; Liaqat, A.; Iqbal, H.; Menaa, B.; Razzaq, A.; Thiripuranathar, G.; Fatima Rana, N.; Menaa, F. Green and Cost-Effective Synthesis of Metallic Nanoparticles by Algae: Safe Methods for Translational Medicine. *Bioengineering* **2020**, *7*, 129. [CrossRef]
56. Britto Hurtado, R.; Cortez-Valadez, M.; Ramírez-Rodríguez, L.P.; Larios-Rodriguez, E.; Alvarez, R.A.B.; Rocha-Rocha, O.; Delgado-Beleño, Y.; Martinez-Nuñez, C.E.; Arizpe-Chávez, H.; Hernández-Martínez, A.R.; et al. Instant Synthesis of Gold Nanoparticles at Room Temperature and SERS Applications. *Phys. Lett. A* **2016**, *380*, 2658–2663. [CrossRef]
57. Nguyen, D.T.; Kim, D.J.; So, M.G.; Kim, K.S. Experimental Measurements of Gold Nanoparticle Nucleation and Growth by Citrate Reduction of $HAuCl_4$. *Adv. Powder Technol.* **2010**, *21*, 111–118. [CrossRef]
58. Neupane, M.P.; Lee, S.J.; Park, I.S.; Lee, M.H.; Bae, T.S.; Kuboki, Y.; Uo, M.; Watari, F. Synthesis of Gelatin-Capped Gold Nanoparticles with Variable Gelatin Concentration. *J. Nanoparticle Res.* **2011**, *13*, 491–498. [CrossRef]
59. Nellore, J.; Pauline, P.C.; Amarnath, K. Biogenic Synthesis by Sphearanthus Amaranthoids; towards the Efficient Production of the Biocompatible Gold Nanoparticles. *Dig. J. Nanomater. Biostructures* **2012**, *7*, 123–133.
60. Robb, D.T.; Privman, V. Model of Nanocrystal Formation in Solution by Burst Nucleation and Diffusional Growth. *Langmuir* **2008**, *24*, 26–35. [CrossRef]
61. Thanh, N.T.K.; Maclean, N.; Mahiddine, S. Mechanisms of Nucleation and Growth of Nanoparticles in Solution. *Chem. Rev.* **2014**, *114*, 7610–7630. [CrossRef]
62. Mountrichas, G.; Pispas, S.; Kamitsos, E.I. Effect of Temperature on the Direct Synthesis of Gold Nanoparticles Mediated by Poly(Dimethylaminoethyl Methacrylate) Homopolymer. *J. Phys. Chem. C* **2014**, *118*, 22754–22759. [CrossRef]
63. Lim, S.; Gunasekaran, S.; Imm, J.Y. Gelatin-Templated Gold Nanoparticles as Novel Time-Temperature Indicator. *J. Food Sci.* **2012**, *77*, N45–N49. [CrossRef] [PubMed]
64. Mock, A.J.; Kang, J. Comment on Degradation of Ascorbic Acid in Ethanolic Solutions. *J. Agric. Food Chem.* **2013**, *61*, 2580–2582. [CrossRef] [PubMed]
65. Yuan, J.P.; Chen, F. Degradation of Ascorbic Acid in Aqueous Solution. *J. Agric. Food Chem.* **1998**, *46*, 5078–5082. [CrossRef]
66. Mohanty, B.; Bohidar, H.B. Systematic of Alcohol-Induced Simple Coacervation in Aqueous Gelatin Solutions. *Biomacromolecules* **2003**, *4*, 1080–1086. [CrossRef]
67. Suarasan, S.; Focsan, M.; Soritau, O.; Maniu, D.; Astilean, S. One-Pot, Green Synthesis of Gold Nanoparticles by Gelatin and Investigation of Their Biological Effects on Osteoblast Cells. *Colloids Surf. B Biointerfaces* **2015**, *132*, 122–131. [CrossRef]
68. Allouche, J.; Soulé, S.; Dupin, J.C.; Masse, S.; Coradin, T.; Martinez, H. Design of Gold Nanoshells via a Gelatin-Mediated Self-Assembly of Gold Nanoparticles on Silica Cores. *RSC Adv.* **2014**, *4*, 63234–63237. [CrossRef]

sustainability MDPI

Article

Numerical Investigation of the Ribs' Shape, Spacing, and Height on Heat Transfer Performance of Turbulent Flow in a Flat Plate Heat Exchanger

Ahmed Saad Soliman [1,*], Li Xu [2], Junguo Dong [2] and Ping Cheng [2]

1 Mechanical Power Engineering Department, Mansoura University, El-Mansoura 35516, Egypt
2 School of Environmental and Chemical Engineering, Shanghai University, Shanghai 200444, China
* Correspondence: asaad200@mans.edu.eg

Abstract: This paper presents a numerical study of detailed heat transfer and flow field characteristics in a flat plate heat exchanger with different types of ribs (rectangular and triangular) with different heights (H: 2.5, 5, 7.5, and 10 mm) and spacings (S: 75, 95, 128, and 195 mm). A comprehensive numerical model was established and validated with experimental data in the literature, resulting in good agreement. The effects of the height, spacing, and shape of the ribs in addition to thermal-hydraulic parameters were investigated over two values of the Reynolds number (9880–3210). The results concluded that the highest values of the thermal-hydraulic parameters are 1.62 and 1.84 for the hot and cold air sides, in the rectangular ribs, with a thickness, height, and spacing of 75 mm, and 95 mm, respectively. It is worth mentioning that the rectangular ribs have the maximum value of thermal performance at the high and low values of the Reynolds number of turbulent flows. The present design can be used in waste heat recovery systems.

Keywords: numerical simulation; heat transfer enhancement; turbulent flow; plate heat exchanger

Citation: Soliman, A.S.; Xu, L.; Dong, J.; Cheng, P. Numerical Investigation of the Ribs' Shape, Spacing, and Height on Heat Transfer Performance of Turbulent Flow in a Flat Plate Heat Exchanger. *Sustainability* **2022**, *14*, 15143. https://doi.org/10.3390/su142215143

Academic Editor: Sergio Nardini

Received: 30 September 2022
Accepted: 14 November 2022
Published: 15 November 2022

1. Introduction

Heat transfer augmentation techniques are essential in many industrial applications because they allow for compact heat exchangers with high thermal performance [1].

Several studies have shown interest in improving heat transfer in engineering applications. There are typically three methods for improving heat transfer: Active, passive, and compound techniques. In the active technique, an external source of energy is required to enhance the heat transfer such as surface vibration, magnetic field, jet impingement, etc. In the passive technique, the configuration of the system has some modifications to improve the heat transfer such as ribs and dimples, rough surfaces, nanoparticles, etc. In the compound technique, the combination of active and passive methods occurs [2].

The passive technique can be divided into three groups (using nanoparticles, geometry modifications, and a combination between nanoparticles and geometry modifications) as follows:

Firstly, nanoparticles are added to the fluid to create nanofluids. Nanofluids have recently been seen as a promising choice for thermal fluids and can be effectively used for heat transfer. The nanofluids significantly enhance the heat exchanger's thermal performance in various applications, such as CPU (central processing unit) coolers [3] and annulus and plate heat exchangers [4]. On the other hand, the nanoparticles were also added to phase-change materials (PCMs) to enhance their thermal conductivity in different applications [5,6].

Secondly, geometry modifications are made by using dimples [7] or roughening the heat transfer surfaces using grooves, ribs or wires, or helically corrugated tubes [8–10]. Artificial roughness is used in several applications such as evaporators, steam condensers, blades of gas turbines, cooling channels, heat exchangers, and solar air heaters [11,12].

Due to the secondary flow areas that form near the wall and decrease the thickness of the thermal boundary layer in a way that increases the rate of heat transfer, these modified geometries produce good mixing in the fluid flow.

Unfortunately, adopting these methods to speed up the heat transfer rate sometimes results in an increase in the pressure drop as a side effect [8,9]. The impact of the channel rib and flow features on heat transfer and friction factor characteristics has been demonstrated by several experimental and numerical studies [13,14].

Moon et al. [15] numerically investigated the configuration of 16 ribs with various cross-sectional shapes and found that boot-shaped rib turbulators have the maximum heat transfer performance with a similar pressure penalty to that of the square rib. In addition, the size of the recirculating zone is directly impacted by the slope of the front surface of the rib, making it a crucial component in determining the performance of heat transfer.

Pingan et al. [11] numerically studied the heat transfer of the airflow in a channel with various rib shapes (triangular, semi-circular, and rectangular) using a standard k-ε turbulent model with enhanced wall treatment. It was found that the maximum and minimum values of the average Nusselt number of the channel occurred with the triangular and rectangular ribs, respectively.

Finally, several researchers concentrated on utilizing both previous methods of enhancing heat transfer (using ribs and nanoparticles). For instance, Wongcharee and Eiamsa-ard [16] conducted an experimental study on the flow of a turbulent CuO-water nanofluid inside a corrugated tube. It was observed that the heat transfer for this geometry increases by 1.57 times compared to the pure fluid inside the simple tube when a 0.7% volume fraction of CuO is used.

Manca et al. [17] numerically investigated the turbulent forced convection of an Al_2O_3-water nanofluid in a ribbed channel with different rib shapes (square and rectangular). It was observed that although there is a greater pressure drop penalty, the enhancement of heat transfer grows with the volume fraction of the particles and Reynolds number. The pressure drop and heat transfer rate are improved by using the ribbed surfaces and adding nanoparticles to the base fluid.

Overall, the spanwise vortices created by the ribs break up the boundary layer and cause flow separation and reattachment in the inter-rib area, which causes turbulence. A higher rate of convective heat transfer results from the high magnitude of turbulent transport because it improves heat exchange through good mixing. However, the rib turbulators also increase the pressure drop and the consumption of energy, which has an impact on the thermal system's total thermo-hydraulic performance [18]. Therefore, finding the optimum rib design is crucial in order to maximize heat transfer while minimizing pressure drop.

As previously mentioned, the flow behavior in ribbed channels was studied to produce a significant enhancement in heat transfer rates. The flow recirculation patterns around a single rib are described by Cho et al. [19]. A large recirculation zone is created at the lee-side regions downstream of the rib, accompanied by a small vortex in the vicinity of the bottom corner of the rib. These recirculation zones encompass a sizable portion of the rib and bottom wall, which is frequently regarded as having poor heat transfer and locally low surface Nusselt numbers [20].

Consequently, the optimum design is the case that increases the performance of heat transfer with a small value of pressure drop. Therefore, to obtain the best design conditions, a trade-off between the increase in heat transfer and the decrease in the pressure drop should be performed.

It can be observed from the previous literature review that most of the studies focused separately on the effect of the rib shape, spacing, and height in a duct or a tube with one fluid. The optimum rib shape in a counter flat plate heat exchanger (hot and cold flow) for a complete heat transfer process between two turbulent fluids has not been studied before.

This research gap is the aim of this paper, and the present design can be used in waste heat recovery systems. In the present paper, the influence of the different rib shapes (rectangular and triangular) with different heights (H: 2.5, 5, 7.5, and 10 mm) and spacing (S: 75, 95, 128, and 195 mm) is investigated on the performance of the heat transfer and pressure drop. The thermal-hydraulic parameter, which combines an enhancement of the rate of heat transfer and the pressure drop due to friction, is also studied on hot and cold sides to obtain the best design of the flat plate heat exchanger.

2. Theoretical Modeling

The current work studied the different types of rib shapes, heights, and spaces to optimize the design that can improve the thermal performance of the heat exchanger. The flat plate heat exchanger consisted of five channels: Three channels for the hot air flow (at Re = 3210 and T = 260 °C) and two channels for the cold air flow (at Re = 9880 and T = 27 °C). The previous conditions were selected to match the velocity of the hot fluid (exhaust gases) and cold fluid (intake air) in waste heat recovery systems, especially for a cold start-up in diesel engines as in Ref [6,21]. The length and width of the heat exchanger were 400 mm and 180 mm, respectively, with a 30 mm height for the hot and cold channels. A counterflow was created between the hot and cold air, which can increase the heat transfer between them. The cold air passes through two channels (L × W × H: 400 × 180 × 30 mm) and the hot air passes through three channels of the same dimensions. The separation wall between the hot and cold channels was made of aluminum with a 4 mm thickness. Two types of ribs (rectangular and triangular shapes) were studied to compare the results and obtain the optimum design.

Different numbers of ribs with increasing spacings (S: 75, 95, 128, and 195 mm) and heights (H: 2.5, 5, 7.5, and 10 mm) in two shapes (rectangular and triangular) were simulated in cases A, B, C, and D. The rectangular ribs were represented in case A, and the triangular ribs with sloping in the direction of the cold airflow and the hot airflow were represented in cases B and C, respectively. The previously mentioned cases with case D (no ribs) are displayed in Figure 1, and the geometry of the triangular and rectangular ribbed flat plate heat exchangers is presented in Figure 2.

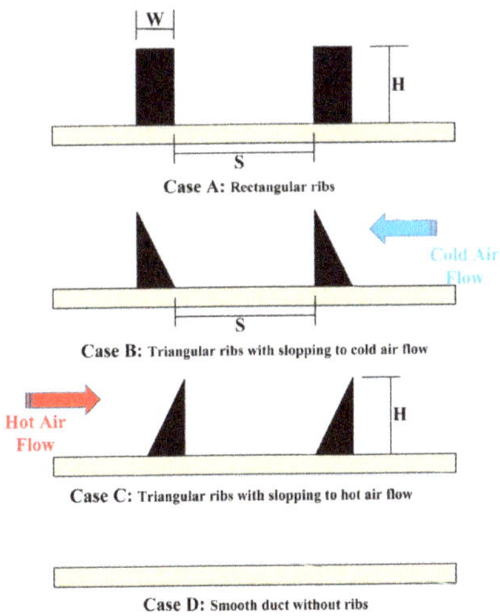

Figure 1. Rib configurations.

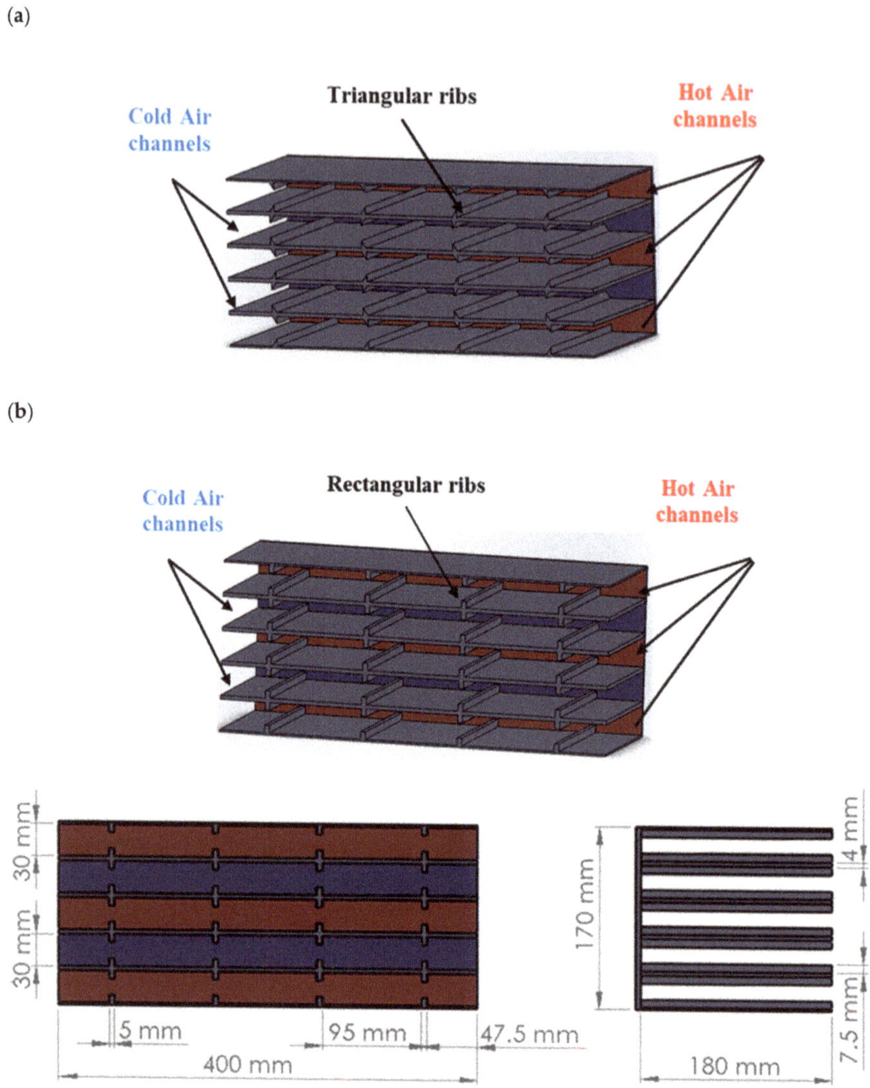

Figure 2. Heat exchanger with (**a**) triangular and (**b**) rectangular ribs.

The turbulent renormalization group model (RNG k-ε model) displays the flow regime of the turbulent flow of the cold and hot air. The governing equations that rule the heat transfer and the motion of a fluid are the continuity, momentum, and energy equations, which can be described as follows [6,21]:

Continuity:

$$\frac{\partial \rho}{\partial t} + \frac{\partial}{\partial x_i}(\rho u_i) = 0 \tag{1}$$

Momentum:

$$\frac{\partial}{\partial t}(u_i) + \frac{\partial}{\partial x_j}(u_i u_j) = -\frac{1}{\rho}\frac{\partial P}{\partial x_i} + \frac{\partial}{\partial x_j}\left(\nu S_{ij} - \overline{u'_i u'_j}\right) \tag{2}$$

where S_{ij} is the main strain rate and is estimated by the following equation:

$$S_{ij} = \frac{1}{2}\left(\frac{\partial u_i}{\partial x_j} + \frac{\partial u_j}{\partial x_i}\right) \tag{3}$$

$\overline{u'_i u'_j}$ represents the unknown turbulent or Reynolds-stress tensor and u'_i signifies the velocity fluctuation in the i-direction.

Energy:

$$\frac{\partial}{\partial t}(\rho E) + \frac{\partial}{\partial x_i}[u_i(\rho E + p)] = \frac{\partial}{\partial x_j}\left[\left(k + \frac{c_p \mu_t}{\mathrm{Pr}_t}\right)\frac{\partial T}{\partial x_j} + u_i(\tau_{ij})_{\mathrm{eff}}\right] \tag{4}$$

where E indicates the total energy, (τ_{ij}) eff denotes the stress tensor, k represents the thermal conductivity (W/m °C), and $\mathrm{Pr_t}$ is a turbulent Prandtl number.

$$(\tau_{ij})_{eff} = \mu_{eff}\left[\frac{\partial u_j}{\partial x_j} + \frac{\partial u_i}{\partial x_j}\right] - \frac{2}{3}\mu_{eff}\frac{\partial u_i}{\partial x_i}\delta_{ij} \tag{5}$$

The description of turbulence using two transport partial differential equations is given by the k-ε model (PDEs). It can be divided into three types, standard, renormalization group (RNG), and realizable k-ε models. The theory of the renormalization group is a statistical technique that is used to develop the RNG model. The transport equation of k (turbulence kinetic energy) and ε (turbulence dissipation rate) for steady flow can be defined as follows [22]:

$$\frac{\partial}{\partial t}(\rho k) + \frac{\partial}{\partial x_i}(\rho k u_i) = \frac{\partial}{\partial x_j}\left[\alpha_k \mu_{eff}\frac{\partial k}{\partial x_j}\right] + G_k + G_b - \rho_\varepsilon - Y_M + S_k \tag{6}$$

and

$$\frac{\partial}{\partial t}(\rho \varepsilon) + \frac{\partial}{\partial x_i}(\rho \varepsilon u_i) = \frac{\partial}{\partial x_j}\left[\alpha_\varepsilon \mu_{eff}\frac{\partial \varepsilon}{\partial x_j}\right] + C_{1\varepsilon}\frac{\varepsilon}{k}(G_k + C_{3\varepsilon}G_b) - C_{2\varepsilon}\rho\frac{\varepsilon^2}{k} + S_\varepsilon \tag{7}$$

where:

μ_{eff} denotes the effective viscosity at the high values of Reynolds numbers.

α_k and α_ε indicate the inverse effective Prandtl numbers ($\alpha_k = \alpha_\varepsilon = 1.393$).

G_k is the generation of turbulence kinetic energy due to the mean velocity gradients.

G_b is the generation of turbulence kinetic energy due to buoyancy.

Y_M is the effect of the fluctuating dilatation incompressible turbulence on the overall dissipation rate.

S_k and S_ε are user-defined source terms.

$$\mu_{eff} = \rho C_\mu \frac{k^2}{\varepsilon} \tag{8}$$

The present model is established by the following model constants: $C_\mu = 0.0845$, $C_{1\varepsilon} = 1.44$, and $C_{2\varepsilon} = 1.92$.

The enhancement of the heat transfer rate can be calculated in terms of the thermal-hydraulic parameter (η), which is described by the following equation [23]:

$$\eta = \frac{\frac{Nu}{Nu_s}}{\left(\frac{f}{f_s}\right)^{\left(\frac{1}{3}\right)}} \tag{9}$$

where f and fs represent the friction coefficients for the ribbed channel and the smooth channel, respectively. In addition, Nu and Nu_s are the Nusselt numbers for the ribbed channel and the smooth channel, respectively.

The thermal-hydraulic parameter combines the enhancement of the rate of heat transfer and the pressure drop due to friction to obtain the best design of the flat plate heat exchanger.

3. Numerical Simulation

Numerical simulations were carried out to analyze the process of heat transfer using commercial code ANSYS 2022 R2, thereby employing a finite volume approach to discretize the governing equations.

3.1. Boundary and Initial Conditions

The entire device was initially at an ambient temperature of 27 °C. For the inlet of both the hot and cold airflow, a uniform velocity distribution with uniform temperature was created. The boundary conditions for the inlets of the duct were estimated to be a velocity inlet. At the same time, the pressure outlets were selected for the duct's exits and the outer walls of the heat exchangers were insulated.

3.2. Solution Procedures

The turbulence model for hot and cold airflows was considered along with initial and boundary conditions. The numerical simulation of the model was developed using ANSYS 2022 R2 software. The pressure–velocity coupling equations were determined using the semi-implicit method for the pressure-linked equations (SIMPLE) scheme. The governing equations in conjunction with the boundary and initial conditions were described using a control volume approach. The central difference and QUICK schemes were used to describe the diffusion and convection terms, respectively. For velocity–pressure coupling, the SIMPLER algorithm was applied. Iterative calculations were nonstop until all problem variables satisfied a relative convergence criterion of 10^{-6}.

3.3. Model Validation

The present model was numerically solved, and the predicted results were compared with the experimentally published results of the turbulence model. Under the same operating conditions, the numerical results were validated.

The turbulent flow through the hot and cold air channels was fabricated, and the numerical model recognized in this work was validated against the numerical study by Menge et al. [24]. Figure 3 illustrates the boundary conditions of the published model [24] as follows: The inlet boundary is defined as a velocity inlet with a uniform velocity of 1 m/s. In addition, at the inlet, the relative turbulent intensity is equal to 5%. The outlet boundary is defined as a pressure outlet with an average static reference pressure of 0 Pa. The Solver was double precision, pressure based, and implicit. The pressure–velocity coupling was SIMPLE, and the turbulent kinetic energy and the turbulent dissipation rate were second-ordered upwind with a convergence criterion of 1×10^{-6}.

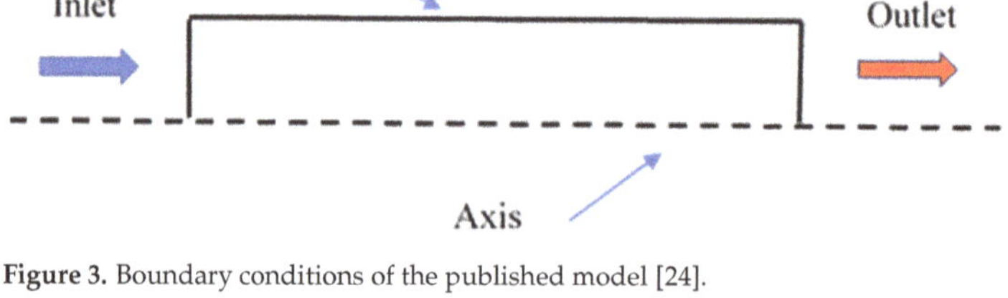

Figure 3. Boundary conditions of the published model [24].

Figure 4 indicates the relationship between the axial velocity and the pipe inlet position to the point at which the entire pipe reaches a constant value using the present numerical model and the published model. As seen in the figure, they match each other well.

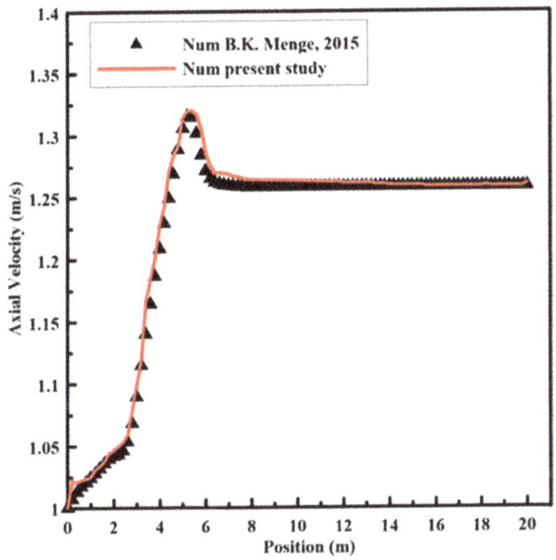

Figure 4. Numerical axial velocity of [24] and the present model versus the distance from the pipe inlet.

3.4. Mesh Generation and Mesh Independence Test

ANSYS-Design Modeler 2022 R2 software was used to create the fluid domain for a fully developed channel with a certain height (*h*). The dimensions were chosen to confirm a fully developed flow and to facilitate a comparison with Blasius. The fluid region was discretized by using the finite element method through the ANSYS-Mesh 2022 R2. Two different techniques are used to distribute the cell through the cold and hot channels. Firstly, through the channel length in the x-direction, the cells were uniformly distributed. Secondly, through the channel height in the y-direction, the cells were non-uniformly distributed. The grid was recognized so that it could capture the velocity distribution in all near-wall regions of the flow. ANSYS-Mesh 2022 R2 software was used to generate the mesh independence test with a non-uniform range of cells. The cell number started with 20 to 50 cells, which were distributed through the height of the channel with a growth rate (GR) equal to 1.1. The RNG k-ε model was employed with enhanced wall treatment, which can be simulated in the near-wall region. Using this option, the height of the first cell near the wall, which is called y^+, is acceptable when $y^+ < 4$ [22]. A numerical study was established to study the two-dimensional (2-D) flow and heat transfer in the ribbed duct at two values of Reynolds numbers when using rectangular and triangular ribs.

In FLUENT, the finite volume technique was used to solve the continuity, momentum, and energy equations that define heat transfer and fluid flow under given boundary conditions. The continuity and momentum equation residuals have been estimated to be 10^{-4} and 10^{-6} for the energy equation.

Figure 5 compares the skin coefficient values (C_f) at different cell numbers per height for both hot and cold channels, and the skin coefficient drops with the increase in the cell number. The relationship between the skin coefficients of both hot and cold channels at different values of the first cell height y^+ ($y^+ = u_\tau y/\nu$) is shown in Figure 6, and the skin coefficient declines when reducing the first cell height.

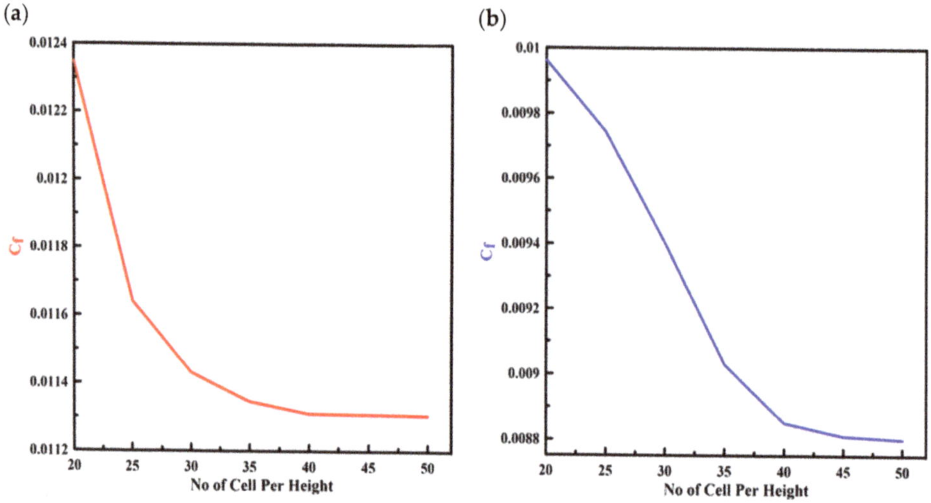

Figure 5. Skin coefficient with the number of cells per height for (**a**) hot air and (**b**) cold air.

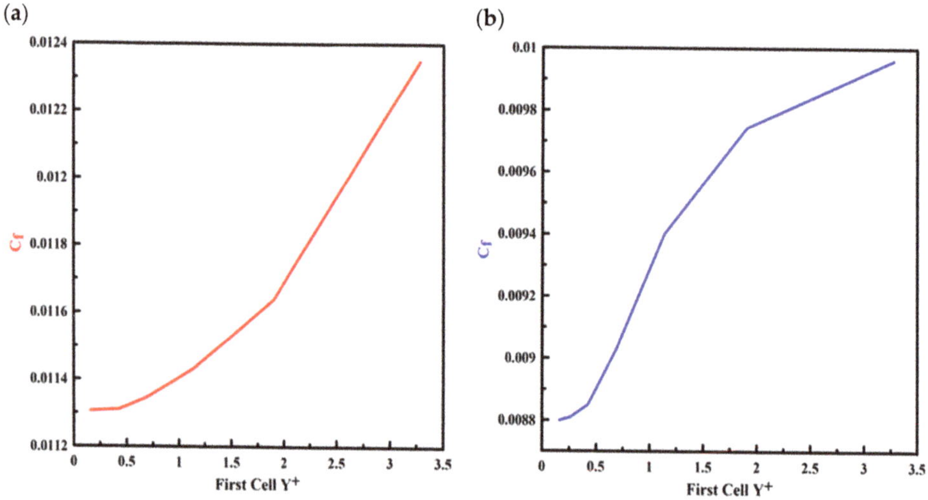

Figure 6. Skin coefficient versus the height of the first cell (y^+) for (**a**) hot air and (**b**) cold air.

It is observed from the figures that, for both hot and cold air flows, the skin coefficient value shows a slight change when the number of cells per height increases above 45 cells. Thus, the optimal cell number per channel height with a small error in the results is 45 cells at a growth rate of 1.1 and y^+ of 0.26.

4. Results and Discussion

The obtained results elucidate the effect of rectangular and triangular ribs on the thermal performance of the flat plate heat exchanger and compare them to obtain the optimum design.

4.1. Flow Structure in Ribbed Duct

The main concept of designing ribbed ducts is to achieve higher flow turbulence and hence greater heat transfer. The predicted velocity contours for the rectangular and triangular ribs are shown in Table 1. It can be observed that both hot and cold airflow accelerates and decelerates through the passage due to the contraction and expansion caused by the ribs.

Table 1. Velocity contours for the rectangular and triangular ribs.

	Rectangular Ribs	Triangular Ribs
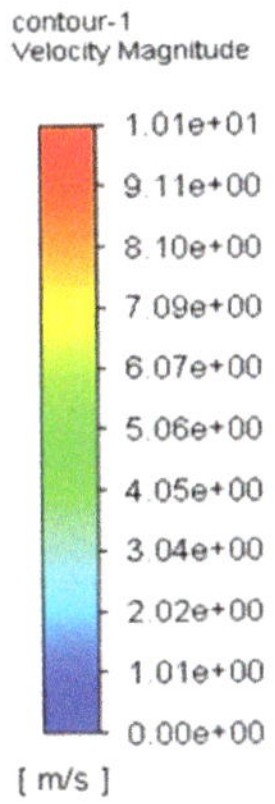	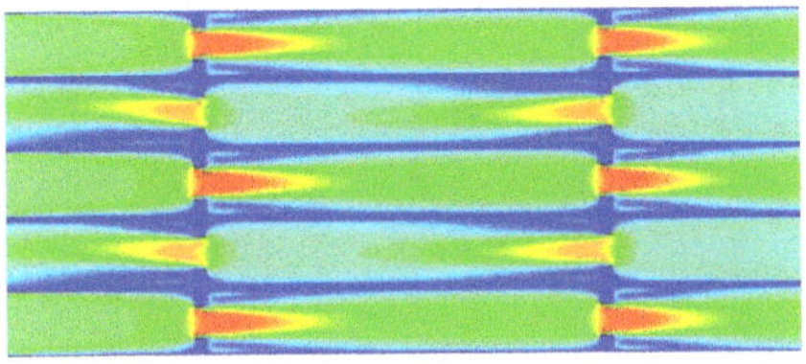	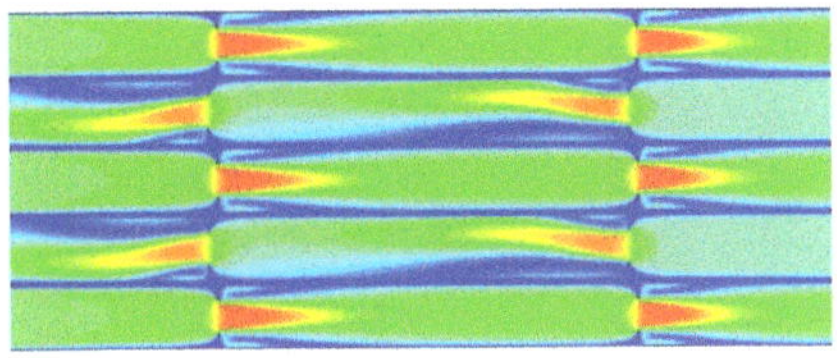
	195 mm spacing	
	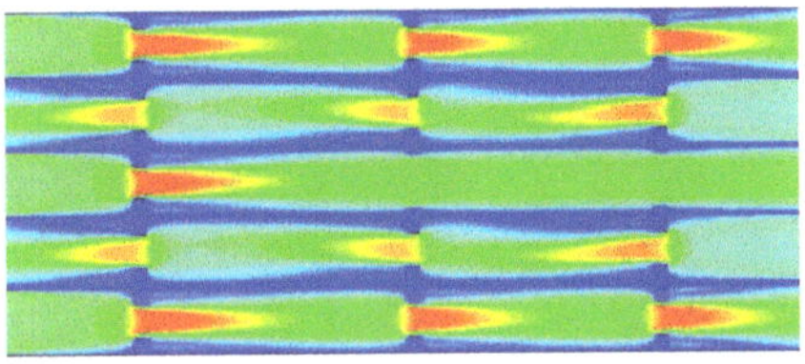	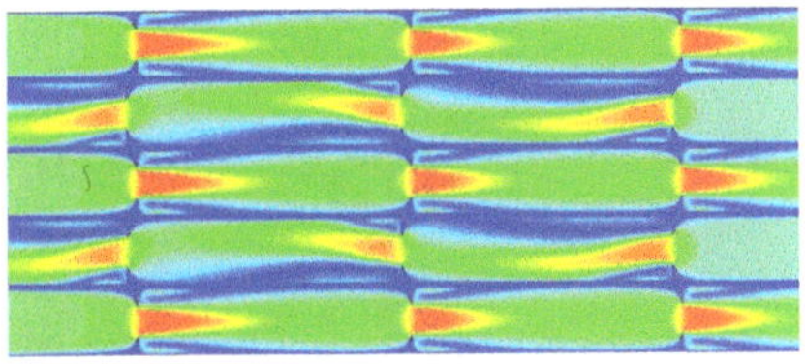
	128 mm spacing	
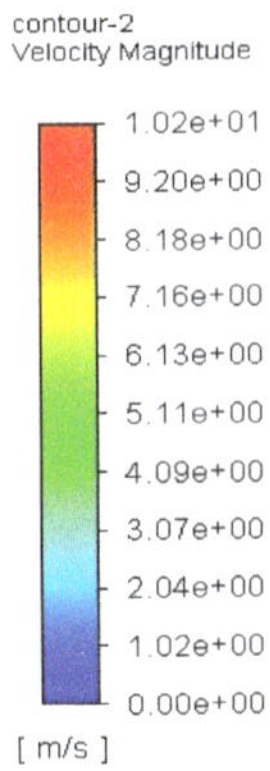	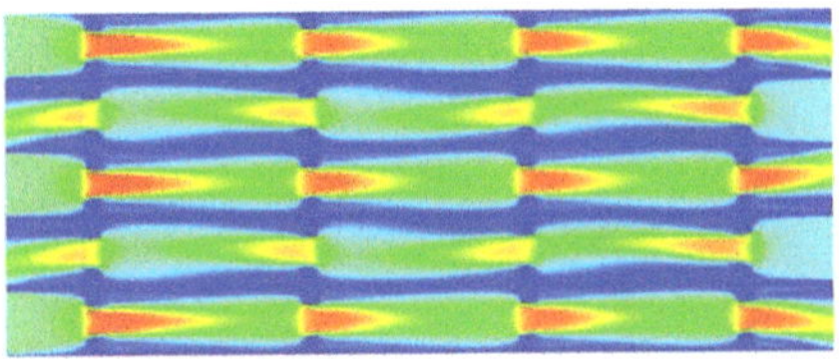	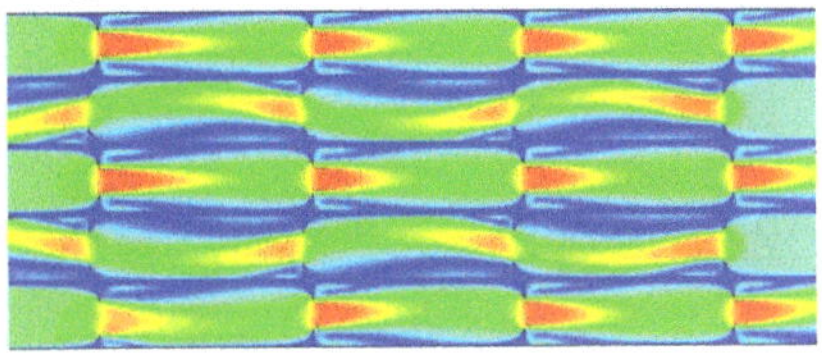
	95 mm spacing	
	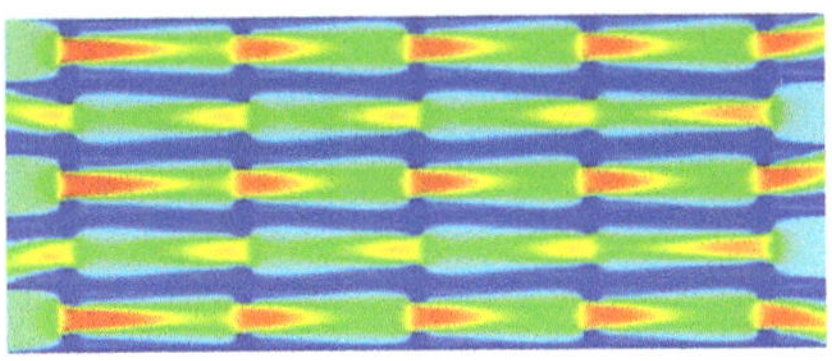	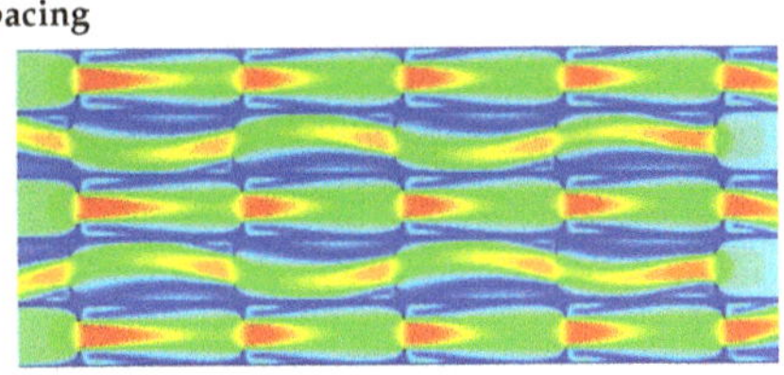
	75 mm spacing	

In the triangular ribs, the fluid motion of a highly turbulent Reynolds number in the cold airflow has a wavy style due to the slope of the ribs. In addition, the recirculation regions developed by the ribs are increased with the increase in the Reynolds number as mentioned in Ref [25]. However, through the rectangular ribs, the fluid motion through the high- and low-turbulence Reynolds numbers is stable. The flow separates and then reattaches as mentioned in Refs [26,27]; therefore, there is non-complete contact with the wall at some points on the cold air side (high Reynolds number), but on the hot air side (low Reynolds number), it is uniform in the triangular ribs.

In the rectangular ribs, when the flow is on the upper side, a large recirculation zone is created at the lee-side regions on the lower side of the rib. Moreover, when the flow is on the lower side, a large recirculation zone is created at the lee-side regions on the upper side of the rib. For both of the previous positions, there is a small vortex in the vicinity of the bottom corner of the rib. This description of the flow is in agreement with that mentioned in Ref [19]. These recirculation zones encompass a sizable portion of the rib and bottom wall, which is frequently regarded as having poor heat transfer and locally low surface Nusselt numbers [20]. On the other hand, the low Reynolds number in the hot airflow has no wavy motion through the velocity contours.

Table 2 shows the contours of the turbulence intensity for all cases of the RNG turbulence model. It can be realized from these contours that the turbulence intensity is higher downstream of the upper edge of the ribs where the major shear layer region is located due to the non-uniform velocity and motion of the fluid through the triangular ribs as mentioned before due to the separation and reattaching of the fluids. This motion makes non-complete contact with the wall at some points on the cold air side (high Reynolds number) but the hot air side (low Reynolds number) is uniform. The turbulence intensity is also non-uniform, especially in the highly turbulent Reynolds number as in the cold airflow.

Table 2. Turbulence intensity contours for the rectangular and triangular ribs.

<table>
<tr><th></th><th>Rectangular Ribs</th><th>Triangular Ribs</th></tr>
<tr><td rowspan="4">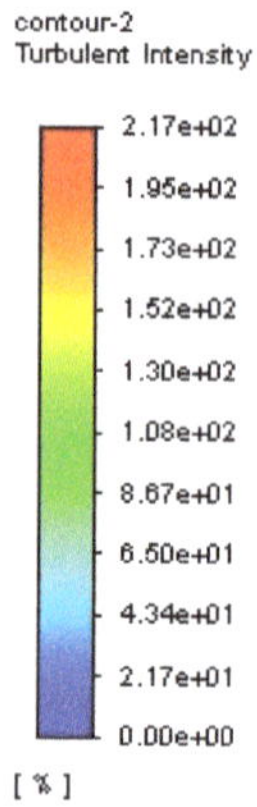
</td><td>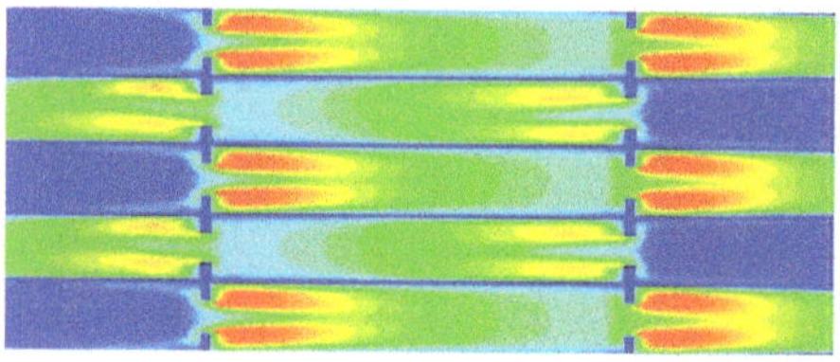</td><td>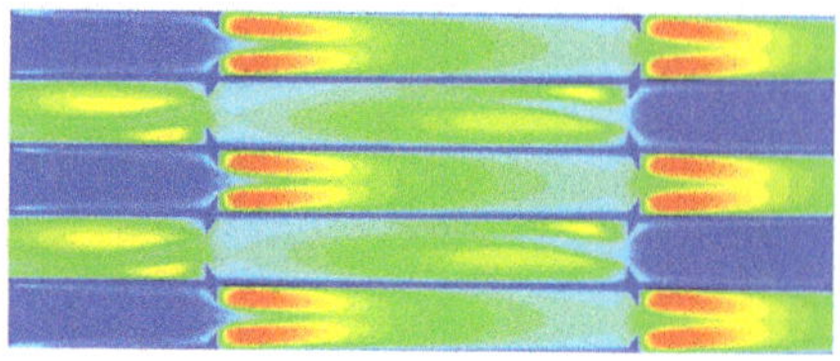</td></tr>
<tr><td colspan="2">195 mm spacing</td></tr>
<tr><td>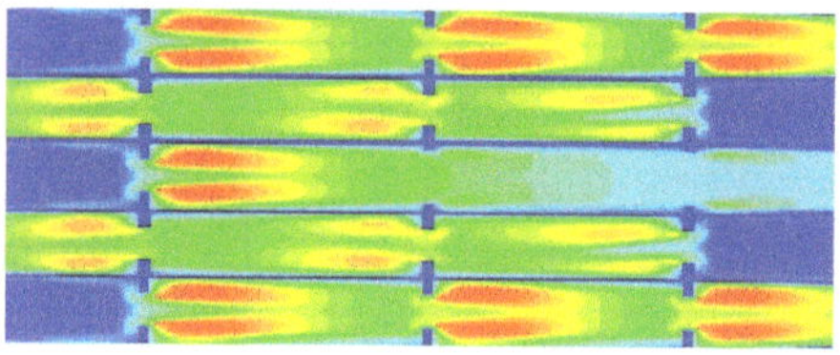</td><td>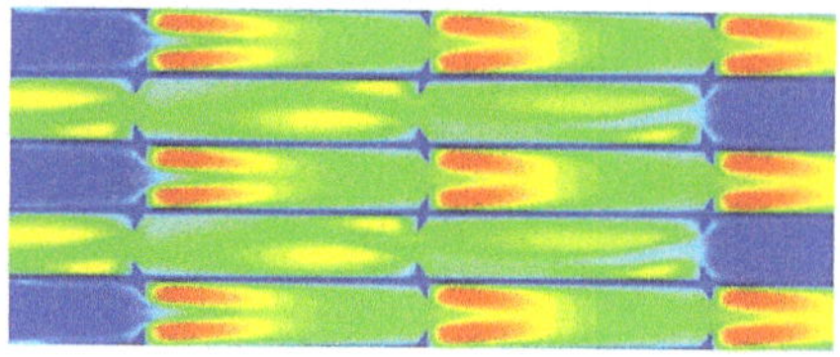</td></tr>
<tr><td colspan="2">128 mm spacing</td></tr>
<tr><td rowspan="4">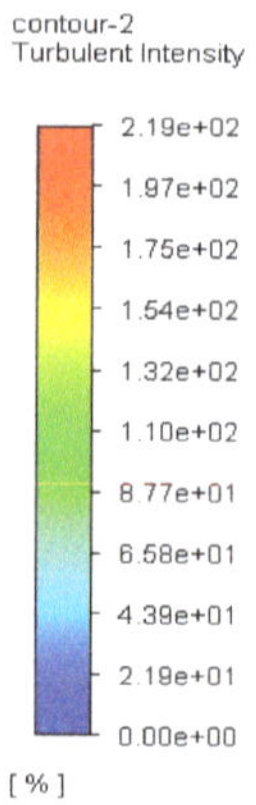
</td><td>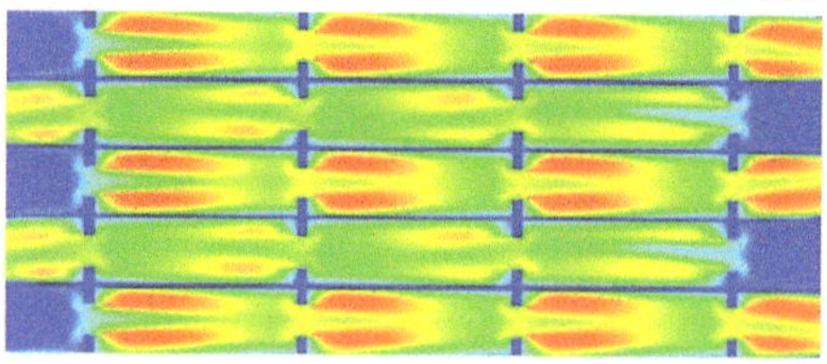</td><td>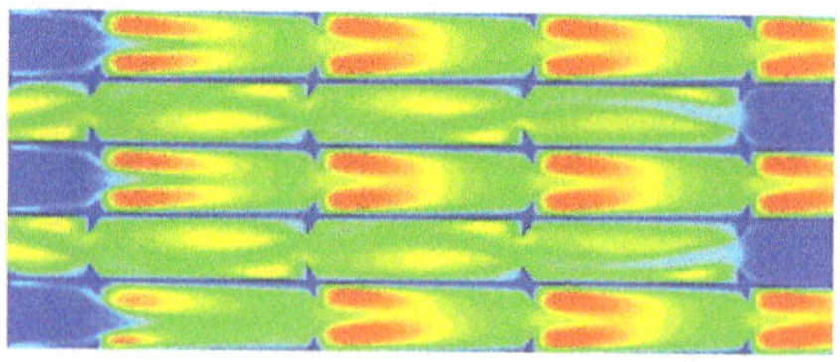</td></tr>
<tr><td colspan="2">95 mm spacing</td></tr>
<tr><td>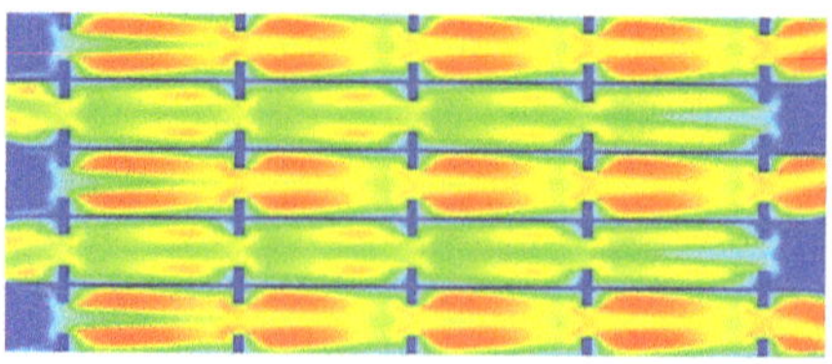</td><td>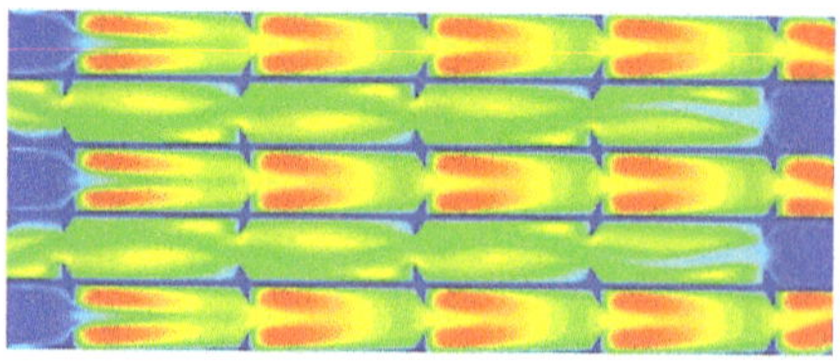</td></tr>
<tr><td colspan="2">75 mm spacing</td></tr>
</table>

As shown in the contours, the weak and robust intensity is established according to the cold air position after the ribs. The cold-air-side turbulence intensity in rectangular ribs is higher than in triangular ribs due to the aforementioned intensity fluctuations. However, on the hot air side, the slope of the triangular ribs causes the intensity after the ribs to be more concentrated than the rectangular ones.

The predicted contours of the temperature distribution for the ribbed cross-sectional passages are shown in Table 3. It was found that the temperature of both fluids near the wall was most affected by the ribs. Due to the effect on the circulation generated by the ribs, the heat transfer between the cold and hot fluid flows is increased.

Table 3. Temperature contours for the rectangular and triangular ribs.

Rectangular Ribs	Triangular Ribs

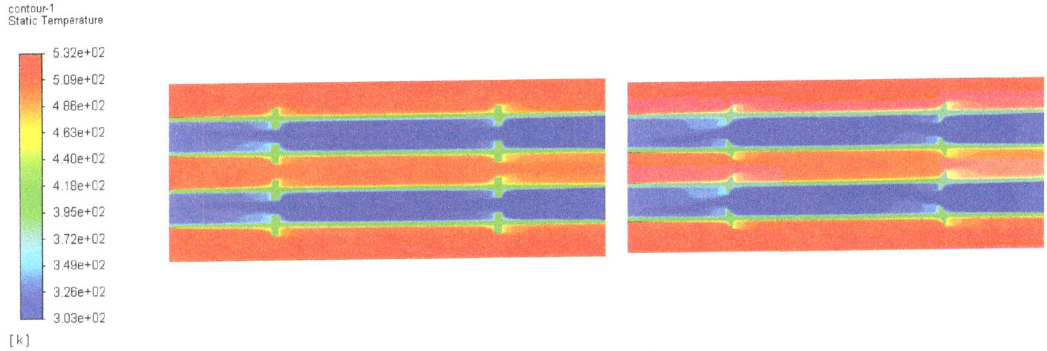

195 mm spacing

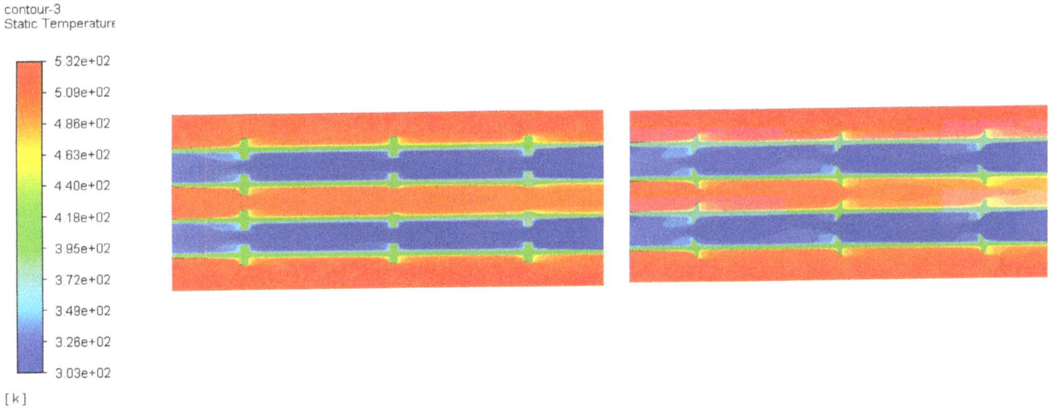

128 mm spacing

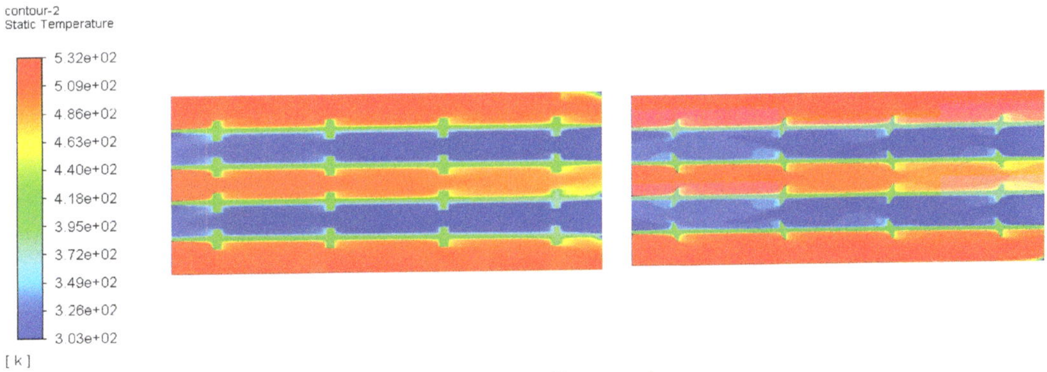

95 mm spacing

Table 3. *Cont.*

	Rectangular Ribs	Triangular Ribs

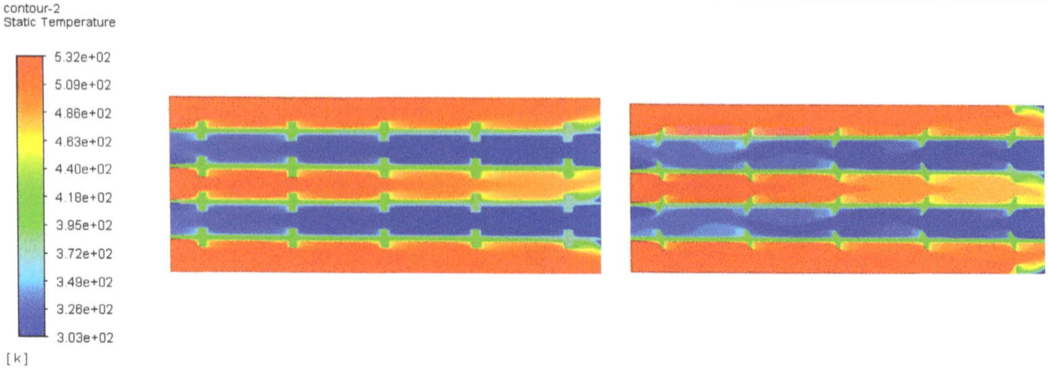

75 mm spacing

4.2. The Ribs' Height Effect

Figure 7 illustrates the variation of the convection heat transfer coefficient for the hot and cold channels with different heights, spacings, and shapes of ribs. The figure also demonstrates that the convection heat transfer coefficient increases with the increase in the rib height, and the maximum value occurs at a spacing of 75 mm for the rectangular and triangular ribs. The convection heat transfer coefficient in the triangular ribs is higher than in the rectangular ones, especially for the hot air side.

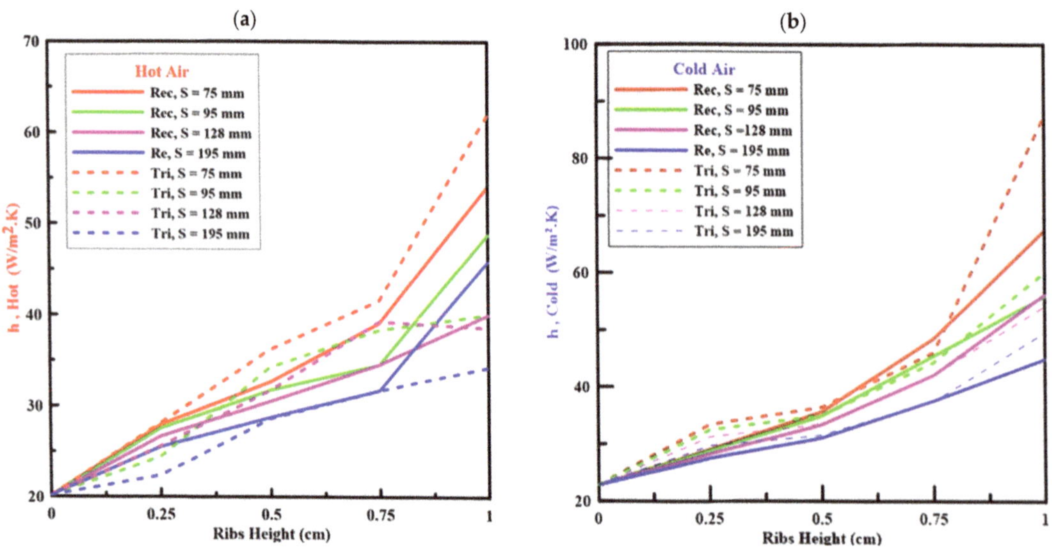

Figure 7. Convection heat transfers versus cell height for all cases of (**a**) hot and (**b**) cold air sides.

For both rectangular and triangular ribs, the convection heat transfer coefficient gradually increases until the rib height reaches 7.5 mm. At this rib height, there is a significant increase in the convection coefficient due to the higher flow turbulence as mentioned in Tables 1–3.

Figure 8 shows the pumping power variations through the channel of the hot and cold air sides with the rib height for different shapes of ribs. It can also be seen that the pumping power increases with the increase in rib height, and the maximum value occurs at 75 mm of rib spacing for the rectangular and triangular ribs.

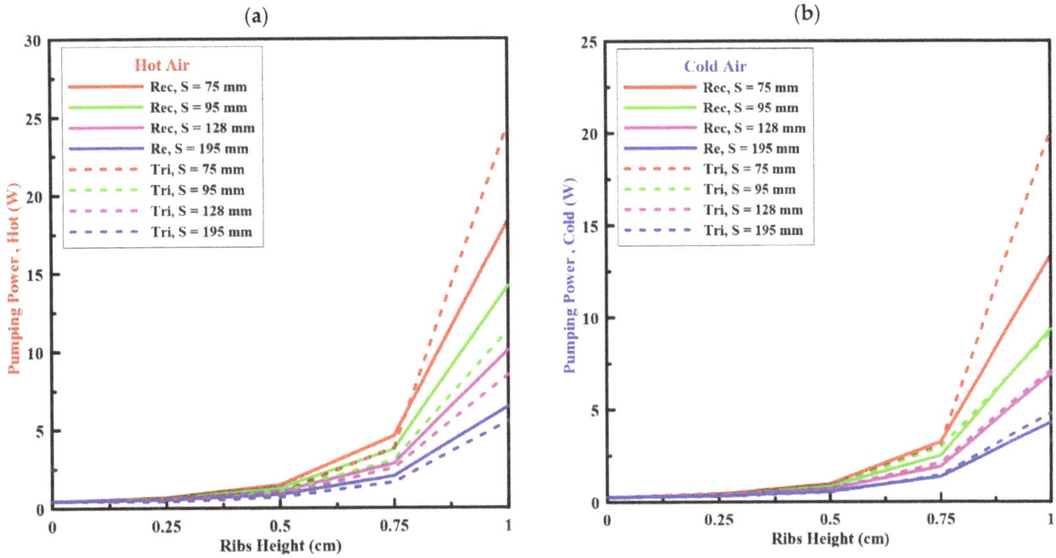

Figure 8. Pumping power required with cell height for the (**a**) hot and (**b**) cold air sides.

For the hot and cold air sides, the pumping power for rectangular and triangular ribs almost has the same value and gradually increases with the rib height until it reaches 7.5 mm. At this rib height, there is a significant increase in the pumping power due to the big increase in the friction value.

It can be observed that increasing the rib's height and number increases the turbulence intensity and velocity, which can improve the thermal performance of the flat plate heat exchanger. However, the enhancement of heat transfer by using ribs occurs with an increase in the friction value. Therefore, it is essential to calculate the thermal-hydraulic parameter, which combines the enhancement in the heat transfer and pressure drop due to friction through the channel, to obtain the best number and height of the cell.

Figure 9 shows the disparity of the overall heat transfer coefficient at different heights of ribs, as the overall heat transfer coefficient increases when increasing the height of ribs. The overall heat transfer has a rapid increase after the rib height of 7.5 mm in both shapes due to the increase in the turbulence intensity in the fluid flow because of increasing the rib height. The overall heat transfer coefficient has a small increase with the rib height at the spacing of 195 mm in the two shapes of ribs. The other ribs' spacings have slight differences between the triangular and rectangular ribs until reaching the 7.5 mm rib height. The overall heat transfer coefficient in the rectangular ribs is greater than the triangular ribs in the spacing cases of 95 and 128 mm due to the separation of the fluid in the triangular rib after this height. The spacing of 75 mm (5 ribs) creates a small distance between ribs, therefore the effect of fluid separation in triangular ribs decreases and has a higher overall heat transfer value than in rectangular ribs.

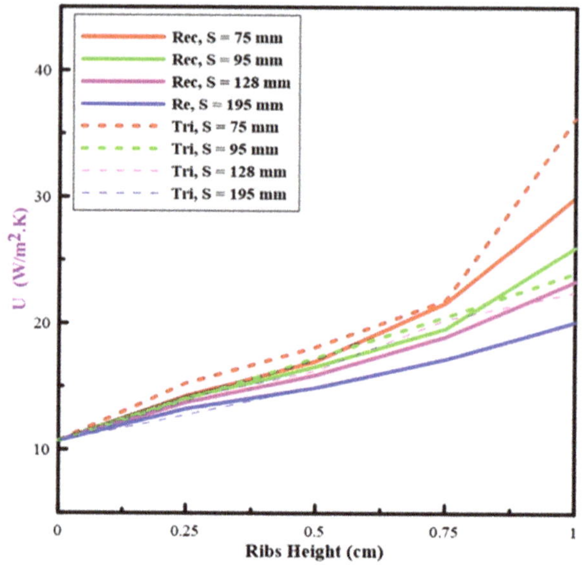

Figure 9. Overall heat transfer coefficient versus cell height for all cases.

Figure 10 presents the thermal-hydraulic parameter with the rib's height. With an increasing rib height, the thermal-hydraulic parameter increases until it reaches the highest value and then begins to decrease for both hot and cold airflow due to the increase in the value of the friction. At this point, the fiction value is higher than the improvement in the heat transfer rate.

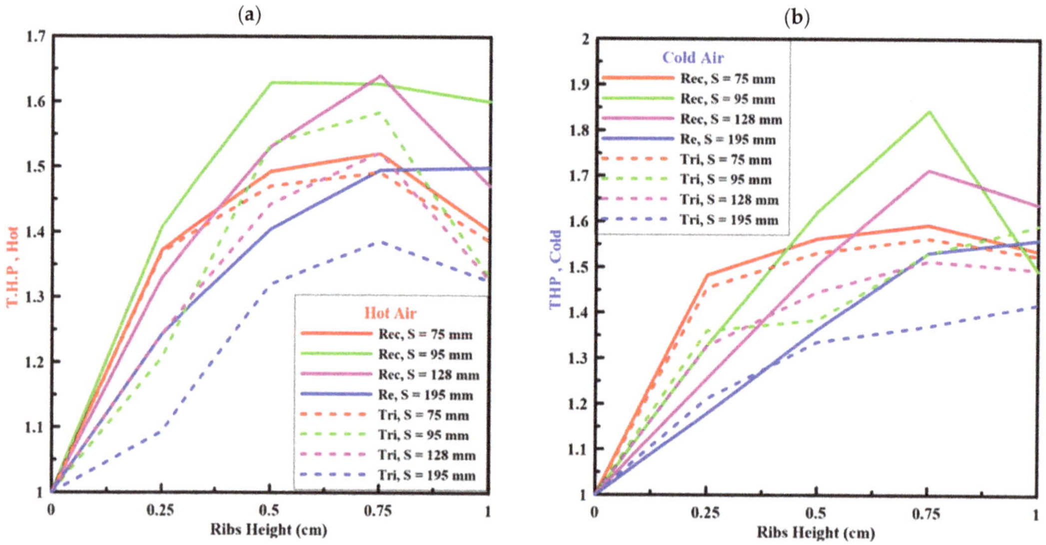

Figure 10. Thermal hydraulic parameter with the cell height for (**a**) hot and (**b**) cold air sides.

The thermal-hydraulic parameter value with rectangular ribs is higher than the value of triangular shapes. The maximum thermal-hydraulic parameters are 1.62 and 1.84 for the hot and cold airflow at a height of 75 mm and a spacing of 95 mm, respectively. According to the previous results, ribs with a height and spacing of 75 mm and 95 mm, respectively, were selected as the optimal design for the flat plate heat exchanger.

5. Conclusions

A numerical simulation for the heat transfer and fluid flow in the ribbed flat plate heat exchanger (rectangular and triangular ribs) is developed for various spacing distances and heights, and the following results are drawn from this work:

1. The ribs' shape and the value of the Reynolds number have a great effect on the fluid motion. Therefore, the motion of a high-Reynolds-number fluid as in the cold airflow has a wavy style when passing through the triangular ribs. Due to that, there is non-complete contact with the wall at some points on the cold air side.
2. The highest values of the thermal-hydraulic parameters are 1.62 and 1.84 for the hot air and cold air sides, in the rectangular ribs, with a thickness, height, and spacing of 5mm, 75 mm, and 95 mm, respectively.

It is worth mentioning that the rectangular ribs have the maximum value of thermal performance at high and low values of the Reynolds number of turbulent flows compared to the triangular ribs. The present design of a flat plate heat exchanger can be used in waste heat recovery systems.

Author Contributions: P.C.: Idea, methodology, writing—original draft, writing the initial draft, supervision. A.S.S.: Writing—original draft, idea, computer program, writing the initial draft, validation, writing—review and editing, data writing, writing, review, and editing. L.X.: Presentation of the published work, supervision. J.D.: Writing—original draft, writing—review and editing, writing the initial draft, preparing literature survey. All authors have read and agreed to the published version of the manuscript.

Funding: This research received no external funding.

Institutional Review Board Statement: Not applicable.

Acknowledgments: The author would like to gratefully acknowledge the support of Mansoura University, Egypt, in the development of this research by providing access to their facility.

Conflicts of Interest: The authors declare no conflict of interest.

References

1. Promvonge, P.; Skullong, S. Augmented Heat Transfer in Tubular Heat Exchanger Fitted with V-Baffled Tapes. *Int. J. Therm. Sci.* **2020**, *155*, 106429. [CrossRef]
2. Ajarostaghi, S.S.M.; Zaboli, M.; Javadi, H.; Badenes, B.; Urchueguia, J.F. A Review of Recent Passive Heat Transfer Enhancement Methods. *Energies* **2022**, *15*, 986. [CrossRef]
3. Sarafraz, M.M.; Hormozi, F. Comparatively Experimental Study on the Boiling Thermal Performance of Metal Oxide and Multi-Walled Carbon Nanotube Nanofluids. *Powder Technol.* **2016**, *287*, 412–430. [CrossRef]
4. Sarafraz, M.M.; Nikkhah, V.; Madani, S.A.; Jafarian, M.; Hormozi, F. Low-Frequency Vibration for Fouling Mitigation and Intensification of Thermal Performance of a Plate Heat Exchanger Working with CuO/Water Nanofluid. *Appl. Therm. Eng.* **2017**, *121*, 388–399. [CrossRef]
5. Soliman, A.S.; Zhu, S.; Xu, L.; Dong, J.; Cheng, P. Melting Enhancement of Nano-Phase Change Material in Cylindrical Enclosure Using Convex/Concave Dimples: Numerical Simulation with Experimental Validation. *J. Energy Storage* **2021**, *44*, 103470. [CrossRef]
6. Soliman, A.S.; Zhu, S.; Xu, L.; Dong, J.; Cheng, P. Efficient Waste Heat Recovery System for Diesel Engines Using Nano-Enhanced Phase Change Materials. *Case Stud. Therm. Eng.* **2021**, *28*, 101390. [CrossRef]
7. Soliman, A.S.; Xu, L.; Dong, J.; Cheng, P. A Novel Heat Sink for Cooling Photovoltaic Systems Using Convex/Concave Dimples and Multiple PCMs. *Appl. Therm. Eng.* **2022**, *215*, 119001. [CrossRef]
8. Rafee, R. Entropy Generation Calculation for Laminar Fully Developed Forced Flow and heat Transfer of Nanofluids inside Annuli. *J. Heat Mass Transf. Res.* **2014**, *1*, 25–33.
9. Duangthongsuk, W.; Wongwises, S. An Experimental Study on the Heat Transfer Performance and Pressure Drop of TiO_2-Water Nanofluids Flowing under a Turbulent Flow Regime. *Int. J. Heat Mass Transf.* **2010**, *53*, 334–344. [CrossRef]
10. Soliman, A.S.; Zhu, S.; Xu, L.; Dong, J.; Cheng, P. Design of an H_2O-LiBr Absorption System Using PCMs and Powered by Automotive Exhaust Gas. *Appl. Therm. Eng.* **2021**, *191*, 116881. [CrossRef]

11. Li, Z.; Lu, J.; Tang, G.; Liu, Q.; Wu, Y. Effects of Rib Geometries and Property Variations on Heat Transfer to Supercritical Water in Internally Ribbed Tubes. *Appl. Therm. Eng.* **2015**, *78*, 303–314. [CrossRef]
12. Durga Prasad, P.V.; Gupta, A.V.S.S.K.S.; Sreeramulu, M.; Sundar, L.S.; Singh, M.K.; Sousa, A.C.M. Experimental Study of Heat Transfer and Friction Factor of Al_2O_3 Nanofluid in U-Tube Heat Exchanger with Helical Tape Inserts. *Exp. Therm. Fluid Sci.* **2015**, *62*, 141–150. [CrossRef]
13. Furlani, L.; Armellini, A.; Casarsa, L. Rotational Effects on the Flow Field inside a Leading Edge Impingement Cooling Passage. *Exp. Therm. Fluid Sci.* **2016**, *76*, 57–66. [CrossRef]
14. Ma, C.; Chen, X.; Wang, J.; Zang, S.; Ji, Y. An Experimental Investigation of Heat Transfer Characteristics for Steam Cooling and Air Cooling in a Rectangular Channel Roughened with Parallel Ribs. *Exp. Therm. Fluid Sci.* **2015**, *64*, 142–151. [CrossRef]
15. Moon, M.A.; Park, M.J.; Kim, K.Y. Evaluation of Heat Transfer Performances of Various Rib Shapes. *Int. J. Heat Mass Transf.* **2014**, *71*, 275–284. [CrossRef]
16. Wongcharee, K.; Eiamsa-ard, S. Heat Transfer Enhancement by Using CuO/Water Nanofluid in Corrugated Tube Equipped with Twisted Tape. *Int. Commun. Heat Mass Transf.* **2012**, *39*, 251–257. [CrossRef]
17. Manca, O.; Nardini, S.; Ricci, D. A Numerical Study of Nanofluid Forced Convection in Ribbed Channels. *Appl. Therm. Eng.* **2012**, *37*, 280–292. [CrossRef]
18. Casarsa, L.; Arts, T. Experimental Investigation of the Aerothermal Performance of a High Blockage Rib-Roughened Cooling Channel. *J. Turbomach.* **2005**, *127*, 580–588. [CrossRef]
19. Cho, H.H.; Lee, S.Y.; Wu, S.J. The Combined Effects of Rib Arrangements and Discrete Ribs on Local Heat/Mass Transfer in a Square Duct. In *Turbo Expo: Power for Land, Sea, and Air*; American Society of Mechanical Engineers: New York, NY, USA, 2001.
20. Ligrani, P.M.; Mahmood, G.I. Spatially Resolved Heat Transfer and Friction Factors in a Rectangular Channel with 45-Deg Angled Crossed-Rib Turbulators. *J. Turbomach.* **2003**, *125*, 575–584. [CrossRef]
21. Soliman, A.S.; Radwan, A.; Xu, L.; Dong, J.; Cheng, P. Energy Harvesting in Diesel Engines to Avoid Cold Start-up Using Phase Change Materials. *Case Stud. Therm. Eng.* **2022**, *31*, 101807. [CrossRef]
22. ANSYS. *Ansys Fluent 12.0 User's Guide*; Fluent Documentatation; Ansys: Canonsburg, PA, USA, 2009.
23. Nanan, K.; Piriyarungrod, N.; Thianpong, C.; Wongcharee, K.; Eiamsa-ard, S. Numerical and Experimental Investigations of Heat Transfer Enhancement in Circular Tubes with Transverse Twisted-Baffles. *Heat Mass Transf.* **2016**, *52*, 2177–2192. [CrossRef]
24. Menge, B.K. Analysis of Turbulent Flow in a Pipe at Constant Reynolds Number Using Computational Fluid Dynamics. Ph.D. Thesis, Jomo Kenyatta University of Agriculture and Technology, Juja, Kenya, 2014.
25. Abdurazzaq, T.; Togun, H.; Kazi, S.; Mariah Adam, N. Numerical Simulation on Heat Transfer Enhancement in Channel by Triangular Ribs. *Int. J. Mech. Ind. Sci. Eng.* **2013**, *7*, 434–438.
26. Wang, H.T.; Lee, W.B.; Chan, J.; To, S. Numerical and Experimental Analysis of Heat Transfer in Turbulent Flow Channels with Two-Dimensional Ribs. *Appl. Therm. Eng.* **2015**, *75*, 623–634. [CrossRef]
27. Manca, O.; Nardini, S.; Ricci, D. Numerical Analysis of Water Forced Convection in Channels with Differently Shaped Transverse Ribs. *J. Appl. Math.* **2011**, *2011*, 323485. [CrossRef]

MDPI
St. Alban-Anlage 66
4052 Basel
Switzerland
Tel. +41 61 683 77 34
Fax +41 61 302 89 18
www.mdpi.com

Sustainability Editorial Office
E-mail: sustainability@mdpi.com
www.mdpi.com/journal/sustainability

www.ingramcontent.com/pod-product-compliance
Lightning Source LLC
Chambersburg PA
CBHW061505220926
51850CB00004B/179
* 9 7 8 3 0 3 6 5 6 5 8 5 9 *